To Caroline, Tatiana and Olga

Contents

Introduction

The algebraic theory of quadratic forms, i.e., the study of quadratic forms over arbitrary fields, really began with the pioneering work of Witt. In his paper [**139**], Witt considered the totality of nondegenerate symmetric bilinear forms over an arbitrary field F of characteristic different from 2. Under this assumption, the theory of symmetric bilinear forms and the theory of quadratic forms are essentially the same. His work allowed him to form a ring $W(F)$, now called the Witt ring, arising from the isometry classes of such forms. This work set the stage for further study. From the viewpoint of ring theory, Witt gave a presentation of this ring as a quotient of the integral group ring where the group consists of the nonzero square classes of the field F. Three methods of study arise: ring theoretic, field theoretic, i.e., the relationship of $W(F)$ and $W(K)$ where K is a field extension of F, and algebraic geometric. In this book, we will develop all three methods. Historically, the powerful approach using algebraic geometry has been the last to be developed. This volume attempts to show its usefulness.

The theory of quadratic forms lay dormant until the work of Cassels and then of Pfister in the 1960's when it was still under the assumption of the field being of characteristic different from 2. Pfister employed the first two methods, ring theoretic and field theoretic, as well as a nascent algebraic geometric approach. In his postdoctoral thesis [**110**] Pfister determined many properties of the Witt ring. His study bifurcated into two cases: formally real fields, i.e., fields in which -1 is not a sum of squares and nonformally real fields. In particular, the Krull dimension of the Witt ring is one in the formally real case and zero otherwise. This makes the study of the interaction of bilinear forms and orderings an imperative, hence the importance of looking at real closures of the base field resulting in extensions of Sylvester's work and Artin-Schreier theory. Pfister determined the radical, zero-divisors, and spectrum of the Witt ring. Even earlier, in [**108**], he discovered remarkable forms, now called Pfister forms. These are forms that are tensor products of binary forms that represent one. Pfister showed that scalar multiples of these were precisely the forms that become hyperbolic over their function field. In addition, the nonzero value set of a Pfister form is a group and in fact the group of similarity factors of the form. As an example, this applies to the quadratic form that is a sum of 2^n squares. Pfister also used it to show that in a nonformally real field, the least number $s(F)$ so that -1 is a sum of $s(F)$ squares is always a power of 2 (cf. [**109**]). Interest in and problems about other arithmetic field invariants have also played a role in the development of the theory.

The nondegenerate even-dimensional symmetric bilinear forms determine an ideal $I(F)$ in the Witt ring of F, called the fundamental ideal. Its powers $I^n(F) := \big(I(F)\big)^n$, each generated by appropriate Pfister forms, give an important filtration of $W(F)$. The problem then arises: What ring theoretic properties respect this filtration? From $W(F)$ one also forms the graded ring $GW(F)$ associated to $I(F)$ and asks the same question.

Using Matsumoto's presentation of $K_2(F)$ of a field (cf. [**98**]), Milnor gave an *ad hoc* definition of a graded ring $K_*(F) := \bigoplus_{n\geq 0} K_n(F)$ of a field in [**106**]. From the viewpoint of Galois cohomology, this was of great interest as there is a natural map, called the norm residue map, from $K_n(F)$ to the Galois cohomology group $H^n(\Gamma_F, \mu_m^{\otimes n})$ where Γ_F is the absolute Galois group of F and m is relatively prime to the characteristic of F. For the case $m = 2$, Milnor conjectured this map to be an epimorphism with kernel $2K_n(F)$ for all n. Voevodsky proved this conjecture in [**136**]. Milnor also related his algebraic K-ring of a field to quadratic form theory by asking if $GW(F)$ and $K_*(F)/2K_*(F)$ are isomorphic. This was solved in the affirmative by Orlov, Vishik, and Voevodsky in [**107**]. Assuming these results, one can answer some of the questions that have arisen about the filtration of $W(F)$ induced by the fundamental ideal.

In this book, we do not restrict ourselves to fields of characteristic different from 2. Historically the cases of fields of characteristic different from 2 and 2 have been studied separately. Usually the case of characteristic different from 2 is investigated first. In this book, we shall give characteristic free proofs whenever possible. This means that the study of symmetric bilinear forms and the study of quadratic forms must be done separately, then interrelated. We not only present the classical theory characteristic free but we also include many results not proven in any text as well as some previously unpublished results to bring the classical theory up to date.

We shall also take a more algebraic geometric viewpoint than has historically been done. Indeed, the final two parts of the book will be based on such a viewpoint. In our characteristic free approach, this means a firmer focus on quadratic forms which have nice geometric objects attached to them rather than on bilinear forms. We do this for a variety of reasons.

First, one can associate to a quadratic form a number of algebraic varieties: the quadric of isotropic lines in a projective space and, more generally, for an integer $i > 0$, the variety of isotropic subspaces of dimension i. More importantly, basic properties of quadratic forms can be reformulated in terms of the associated varieties: a quadratic form is isotropic if and only if the corresponding quadric has a rational point. A nondegenerate quadratic form is hyperbolic if and only if the variety of maximal totally isotropic subspaces has a rational point.

Not only are the associated varieties important but also the morphisms between them. Indeed, if φ is a quadratic form over F and L/F a finitely generated field extension, then there is a variety Y over F with function field L, and the form φ is isotropic over L if and only if there is a rational morphism from Y to the quadric of φ.

Working with correspondences rather than just rational morphisms adds further depth to our study, where we identify morphisms with their graphs. Working with these leads to the category of Chow correspondences. This provides greater flexibility because we can view correspondences as elements of Chow groups and apply the rich machinery of that theory: pull-back and push-forward homomorphisms, Chern classes of vector bundles, and Steenrod operations. For example, suppose we wish to prove that a property A of quadratic forms implies a property B. We translate the properties A and B to "geometric" properties A' and B' for the existence of certain cycles on certain varieties. Starting with cycles satisfying

A' we can then attempt to apply the operations over the cycles as above to produce cycles satisfying B'.

All the varieties listed above are projective homogeneous varieties under the action of the orthogonal group or special orthogonal group of φ, i.e., the orthogonal group acts transitively on the varieties. It is not surprising that the properties of quadratic forms are reflected in the properties of the special orthogonal groups. For example, if φ is of dimension $2n$ (with $n \geq 2$) or $2n + 1$ (with $n \geq 1$), then the special orthogonal group is a semisimple group of type D_n or B_n. The classification of semisimple groups is characteristic free. This explains why most important properties of quadratic forms hold in all characteristics.

Unfortunately, bilinear forms are not "geometric". We can associate varieties to a bilinear form, but it would be a variety of the associated quadratic form. Moreover, in characteristic 2 the automorphism group of a bilinear form is not semisimple.

In this book we sometimes give several proofs of the same results — one is classical, another is geometric. (This can be the same proof, but written in geometric language.) For example, this is done for Springer's theorem and the Separation Theorem.

The first part of the text will derive classical results under this new setting. It is self-contained, needing minimal prerequisites except for Chapter VII. In this chapter we shall assume the results of Voevodsky in [**136**] and Orlov-Vishik-Voevodsky in [**107**] for fields of characteristic not 2, and Kato in [**78**] for fields of characteristic 2 on the solution for the analog of the Milnor Conjecture in algebraic K-theory. We do give new proofs for the case $n = 2$.

Prerequisites for the second two parts of the text will be more formidable. A reasonable background in algebraic geometry will be assumed. For the convenience of the reader appendices have been included as an aid. Unfortunately, we cannot give details of [**136**] or [**107**] as it would lead us away from the methods at hand.

The first part of this book covers the "classical" theory of quadratic forms, i.e., without heavy use of algebraic geometry, bringing it up to date. As the characteristic of a field is not deemed to be different from 2, this necessitates a bifurcation of the theory into the theory of symmetric bilinear forms and the theory of quadratic forms. The introduction of these subjects is given in the first two chapters.

Chapter I investigates the foundations of the theory of symmetric bilinear forms over a field F. Two major consequences of dealing with arbitrary characteristic are that such forms may not be diagonalizable and that nondegenerate isotropic planes need not be hyperbolic. With this taken into account, standard Witt theory, to the extent possible, is developed. In particular, Witt decomposition still holds, so that the Witt ring can be constructed in the usual way as well as the classical group presentation of the Witt ring $W(F)$. This presentation is generalized to the fundamental ideal $I(F)$ of even-dimensional forms in $W(F)$ and then to the the second power $I^2(F)$ of $I(F)$, a theme returned to in Chapter VII. The Stiefel-Whitney invariants of bilinear forms are introduced along with their relationship with the invariants $\bar{e}_n : I^n(F)/I^{n+1}(F) \to K_n(F)/2K_{n+1}(F)$ for $n = 1, 2$. The theory of bilinear Pfister forms is introduced and some basic properties developed. Following [**32**], we introduce chain p-equivalence and linkage of Pfister forms as well as introducing annihilators of Pfister forms in the Witt ring.

Chapter II investigates the foundations of the theory of quadratic forms over a field F. Because of the arbitrary characteristic assumption on the field F, the definition of nondegenerate must be made more carefully, and quadratic forms are far from having orthogonal bases in general. Much of Witt theory, however, goes through as the Witt Extension Theorem holds for quadratic forms under fairly weak assumptions, hence Witt Decomposition. The Witt group $I_q(F)$ of nondegenerate even-dimensional quadratic forms is defined and shown to be a $W(F)$-module. The theory of quadratic Pfister forms is introduced and some results analogous to that of the bilinear case are introduced. Moreover, cohomological invariants of quadratic Pfister forms are introduced and some preliminary results about them and their extension to the appropriate filtrant of the Witt group of quadratic forms are discussed. In addition, the classical quadratic form invariants, discriminant and Clifford invariant, are defined.

Chapter III begins the utilization of function field techniques in the study of quadratic forms, all done without restriction of characteristic. The classical Cassels-Pfister theorem is established. Values of anisotropic quadratic and bilinear forms over a polynomial ring are investigated, special cases being the representation of one form as a subform of another and various norm principles due to Knebusch (cf. [**82**]). To investigate norm principles of similarity factors due to Scharlau (cf. [**119**]), quadratic forms over valuation rings and transfer maps are introduced.

Chapter IV introduces algebraic geometric methods, i.e., looking at the theory under the base extension of the function field of a fixed quadratic form. In particular, the notion of domination of one form by another is introduced where an anisotropic quadratic form φ is said to dominate an anisotropic quadratic form ψ (both of dimension of at least two) if $\varphi_{F(\psi)}$ is isotropic. The geometric properties of Pfister forms are developed, leading to the Arason-Pfister Hauptsatz that nonzero anisotropic quadratic (respectively, symmetric bilinear) forms in $I_q^n(F)$ (respectively, $I^n(F)$) are of dimension at least 2^n and its application to linkage of Pfister forms. Knebusch's generic tower of an anisotropic quadratic form is introduced and the $W(F)$-submodules $J_n(F)$ of $I_q(F)$ are defined by the notion of degree. These submodules turn out to be precisely the corresponding $I_q^n(F)$ (to be shown in Chapter VII). Hoffmann's Separation Theorem that if φ and ψ are two anisotropic quadratic forms over F with $\dim\varphi \leq 2^n < \dim\psi$ for some $n \geq 0$, then $\varphi_{F(\psi)}$ is anisotropic is proven as well as Fitzgerald's theorem characterizing quadratic Pfister forms. In addition, excellent forms and extensions are discussed. In particular, Arason's result that the extension of a field by the function field of a nondegenerate 3-dimensional quadratic form is excellent is proven. The chapter ends with a discussion of central simple algebras over the function field of a quadric.

Chapter V studies symmetric bilinear and quadratic forms under field extensions. The chapter begins with the study of the structure of the Witt ring of a field F based on the work of Pfister. After dispensing with the nonformally real F, we turn to the study over a formally real field utilizing the theory of pythagorean fields and the pythagorean closure of a field, leading to the Local-Global Theorem of Pfister and its consequences for structure of the Witt ring over a formally real field. The total signature map from the Witt ring to the ring of continuous functions from the order space of a field to the integers is then carefully studied, in particular, the approximation of elements in this ring of functions by quadratic forms. The behavior of quadratic and bilinear forms under quadratic extensions (both separable and inseparable) is then investigated. The special case of the torsion of the Witt ring under such extensions is studied. A detailed investigation of torsion Pfister forms is begun, leading to the theorem of Krüskemper which implies that if K/F is a quadratic field extension with $I^n(K) = 0$, then $I^n(F)$ is torsion-free.

Chapter VI studies u-invariants, their behavior under field extensions, and values that they can take. Special attention is given to the case of formally real fields.

Chapter VII establishes consequences of the result of Orlov-Vishik-Voevodsky in [**107**] which we assume in this chapter. In particular, answers and generalizations of results from the previous chapters are established. For fields of characteristic not 2, the ideals $I^n(F)$ and $J_n(F)$ are shown to be identical. The annihilators of Pfister forms in the Witt ring are shown to filter through the $I^n(F)$, i.e., the intersection of such annihilators and $I^n(F)$ are generated by Pfister forms in the intersection. A consequence is that torsion in $I^n(F)$ is generated by torsion n-fold Pfister forms, solving a conjecture of Lam. A presentation for the group structure of the $I^n(F)$'s is determined, generalizing that given for $I^2(F)$ in Chapter I. Finally, it is shown that if K/F is a finitely generated field extension of transcendence degree m, then $I^n(K)$ torsion-free implies the same for $I^{n-m}(F)$.

In Chapter VIII, we give a new elementary proof of the theorem in [**100**] that the second cohomological invariant is an isomorphism in the case that the characteristic of the field is different from 2 (the case of characteristic 2 having been done in Chapter II). This is equivalent to the degree two case of the Milnor Conjecture in [**106**] stating that the *norm residue homomorphism*

$$h^n_F : K_n(F)/2K_n(F) \to H^n(F, \mathbb{Z}/2\mathbb{Z})$$

is an isomorphism for every integer n. The Milnor Conjecture was proven in full by V. Voevodsky in [**136**]. Unfortunately, the scope of this book does not allow us to prove this beautiful result as the proof requires motivic cohomology and Steenrod operations developed by Voevodsky. In Chapter VIII, we give an "elementary" proof of the degree two case of the Milnor Conjecture that does not rely either on a specialization argument or on higher K-theory as did the original proof of this case in [**100**].

In the second part of the book, we develop the needed tools in algebraic geometry that will be applied in the third part. The main object studied in Part Two is the Chow group of algebraic cycles modulo rational equivalence on an algebraic scheme. Using algebraic cycles, we introduce the category of correspondences.

In Chapter IX (following the approach of [**117**] given by Rost), we develop the K-homology and K-cohomology theories of schemes over a field. This generalizes the Chow groups. We establish functorial properties of these theories (pull-back, push-forward, deformation and Gysin homomorphisms), introduce Euler and Chern classes of vector bundles, and prove basic results such as the Homotopy Invariance and Projective Bundle Theorems. We apply these results to Chow groups in the next chapter.

Chapter XI is devoted to the study of Steenrod operations on Chow groups modulo 2 over fields of characteristic not 2. Steenrod operations for motivic cohomology modulo a prime integer p of a scheme X were originally constructed by

Voevodsky in [**137**]. The reduced power operations (but not the Bockstein operation) restrict to the Chow groups of X. An "elementary" construction of the reduced power operations modulo p on Chow groups (requiring equivariant Chow groups) was given by Brosnan in [**20**].

In Chapter XII, we introduce the notion of a Chow motive that is due to Grothendieck. Many (co)homology theories defined on the category of smooth complete varieties, such as Chow groups and more generally the K-(co)homology groups, take values in the category of abelian groups. But the category of smooth complete varieties itself does not have the structure of an additive category as we cannot add morphisms of varieties. The category of Chow motives, however, is an additive tensor category. This additional structure gives more flexibility when working with regular and rational morphisms.

In the third part of the book we apply algebraic geometric methods to the further study of quadratic forms. In Chapter XIII, we prove preliminary facts about algebraic cycles on quadrics and their powers. We also introduce shell triangles and diagrams of cycles, the basic combinatorial objects associated to a quadratic form. The corresponding pictures of these shell triangles simplify visualization of algebraic cycles and operations over the cycles.

In Chapter XIV, we study the Izhboldin dimension of smooth projective quadrics. It is defined as the integer

$$\dim_{\mathrm{Izh}}(X) := \dim X - \mathfrak{i}_1(X) + 1,$$

where $\mathfrak{i}_1(X)$ is the first Witt index of the quadric X. The Izhboldin dimension behaves better than the classical dimension with respect to splitting properties. For example, if X and Y are anisotropic smooth projective quadrics and Y is isotropic over the function field $F(X)$, then $\dim_{\mathrm{Izh}} X \leq \dim_{\mathrm{Izh}} Y$ but $\dim X$ may be bigger than $\dim Y$.

Chapter XV is devoted to applications of the Steenrod operations. The following problems are solved:

(1) All possible values of the first Witt index of quadratic forms are determined.
(2) All possible values of dimensions of anisotropic quadratic forms in $I^n(F)$ are determined.
(3) It is shown that excellent forms have the smallest height among all quadratic forms of a given dimension.

In Chapter XVI, we study the variety of maximal isotropic subspaces of a quadratic forms. A discrete invariant $J(\varphi)$ of a quadratic form φ is introduced. We also introduce the notion of canonical dimension and compute it for projective quadrics and varieties of totally isotropic subspaces.

In the last chapter we study motives of smooth projective quadrics in the category of correspondences and motives.

This book could not have been written without the help and encouragement of many of our friends, collaborators, and students nor the many researchers whose work was essential to producing this volume. These are too numerous to be listed here individually. We do wish to mention those who gave us valuable advice in the preparation and writing of this tome and those who helped proofread the manuscript: Jón Arason, Ricardo Baeza, Alex Boisvert, Detlev Hoffmann, Bryant Mathews, Claus Schubert, Jean-Pierre Tignol, Alexander Vishik, Maksim Zhykhovich. In addition, we would like to thank the referees who made valuable suggestions to improve the manuscript and Sergei Gelfand and the American Mathematical Society for its encouragement. It also gives us great pleasure to thank the National Science Foundation (grant DMS 0652316), as well as the Collaborative Research Centre 701 of the Bielefeld University, École Polytechnique Fédérale de Lausanne, Institute for Advanced Study in Princeton (The James D. Wolfensohn Fund and The Ellentuck Fund), Institut des Hautes Études Scientifiques, Institut Universitaire de France, and Max-Planck-Institut für Mathematik in Bonn, for their generous support. Finally, we wish to thank our families who put up with us through the long process of bringing this volume to fruition.

Part 1

Classical theory of symmetric bilinear forms and quadratic forms

CHAPTER I

Bilinear Forms

1. Foundations

The study of $(n \times n)$-matrices over a field F leads to various classification problems. Of special interest is to classify alternating and symmetric matrices. If A and B are two such matrices, we say that they are congruent if $A = P^t B P$ for some invertible matrix P. For example, it is well-known that symmetric matrices are diagonalizable if the characteristic of F is different from 2. So the problem of classifying matrices up to congruence reduces to the study of a congruence class of a matrix in this case. The study of alternating and symmetric bilinear forms over an arbitrary field is the study of this problem in a coordinate-free approach. Moreover, we shall, whenever possible, give proofs independent of characteristic. In this section, we introduce the definitions and notation needed throughout the text and prove that we have a Witt Decomposition Theorem (cf. Theorem 1.27 below) for such forms. As we make no assumption on the characteristic of the underlying field, this makes the form of this theorem more delicate.

Definition 1.1. Let V be a finite dimensional vector space over a field F. A *bilinear form* on V is a map $\mathfrak{b} : V \times V \to F$ satisfying for all $v, v', w, w' \in V$ and $c \in F$,

$$\begin{aligned} \mathfrak{b}(v + v', w) &= \mathfrak{b}(v, w) + \mathfrak{b}(v', w), \\ \mathfrak{b}(v, w + w') &= \mathfrak{b}(v, w) + \mathfrak{b}(v, w'), \\ \mathfrak{b}(cv, w) &= c\mathfrak{b}(v, w) = \mathfrak{b}(v, cw). \end{aligned}$$

The bilinear form is called *symmetric* if $\mathfrak{b}(v, w) = \mathfrak{b}(w, v)$ for all $v, w \in V$ and is called *alternating* if $\mathfrak{b}(v, v) = 0$ for all $v \in V$. If $\mathfrak{b}$ is an alternating form, expanding $\mathfrak{b}(v + w, v + w)$ shows that $\mathfrak{b}$ is *skew symmetric*, i.e., that $\mathfrak{b}(v, w) = -\mathfrak{b}(w, v)$ for all $v, w \in V$. In particular, every alternating form is symmetric if $\operatorname{char} F = 2$. We call $\dim V$ the *dimension* of the bilinear form and also write it as $\dim \mathfrak{b}$. We write $\mathfrak{b}$ is a bilinear form over F if $\mathfrak{b}$ is a bilinear form on a finite dimensional vector space over F and denote the underlying space by $V_{\mathfrak{b}}$.

Let $V^* := \operatorname{Hom}_F(V, F)$ denote the dual space of V. A bilinear form $\mathfrak{b}$ on V is called *nondegenerate* if $l : V \to V^*$ defined by $v \mapsto l_v : w \mapsto \mathfrak{b}(v, w)$ is an isomorphism. An *isometry* $f : \mathfrak{b}_1 \to \mathfrak{b}_2$ between two bilinear forms $\mathfrak{b}_i$, $i = 1, 2$, is a linear isomorphism $f : V_{\mathfrak{b}_1} \to V_{\mathfrak{b}_2}$ such that $\mathfrak{b}_1(v, w) = \mathfrak{b}_2\big(f(v), f(w)\big)$ for all $v, w \in V_{\mathfrak{b}_1}$. If such an isometry exists, we write $\mathfrak{b}_1 \simeq \mathfrak{b}_2$ and say that $\mathfrak{b}_1$ and $\mathfrak{b}_2$ are *isometric*.

Let $\mathfrak{b}$ be a bilinear form on V. Let $\{v_1, \ldots, v_n\}$ be a basis for V. Then $\mathfrak{b}$ is determined by the matrix $\big(\mathfrak{b}(v_i, v_j)\big)$ and the form is nondegenerate if and only if $\big(\mathfrak{b}(v_i, v_j)\big)$ is invertible. Conversely, any matrix B in the $n \times n$ matrix ring $\mathbb{M}_n(F)$

determines a bilinear form based on V. If $\mathfrak{b}$ is symmetric (respectively, alternating), then the associated matrix is symmetric (respectively, alternating where a square matrix (a_{ij}) is called *alternating* if $a_{ij} = -a_{ji}$ and $a_{ii} = 0$ for all i, j). Let $\mathfrak{b}$ and $\mathfrak{b}'$ be two bilinear forms with matrices B and B' relative to some bases. Then $\mathfrak{b} \simeq \mathfrak{b}'$ if and only if $B' = A^t B A$ for some invertible matrix A, i.e., the matrices B' and B are *congruent*. As $\det B' = \det B \cdot (\det A)^2$ and $\det A \neq 0$, the determinant of B' coincides with the determinant of B up to squares. We define the *determinant of* a nondegenerate bilinear form $\mathfrak{b}$ by $\det \mathfrak{b} := \det B \cdot F^{\times 2}$ in $F^\times / F^{\times 2}$, where B is a matrix representation of $\mathfrak{b}$ and $F^\times$ is the multiplicative group of F (and more generally, $R^\times$ denotes the unit group of a ring R). So the det is an invariant of the isometry class of a nondegenerate bilinear form.

The set $\operatorname{Bil}(V)$ of *bilinear forms* on V is a vector space over F. The space $\operatorname{Bil}(V)$ contains the subspaces $\operatorname{Alt}(V)$ of *alternating forms* on V and $\operatorname{Sym}(V)$ of *symmetric bilinear forms* on V. The correspondence of bilinear forms and matrices given above defines a linear isomorphism $\operatorname{Bil}(V) \to \mathbb{M}_{\dim V}(F)$. If $\mathfrak{b} \in \operatorname{Bil}(V)$, then $\mathfrak{b} - \mathfrak{b}^t$ is alternating where the bilinear form $\mathfrak{b}^t$ is defined by $\mathfrak{b}^t(v, w) = \mathfrak{b}(w, v)$ for all $v, w \in V$. Since every alternating $n \times n$ matrix is of the form $B - B^t$ for some B, the linear map $\operatorname{Bil}(V) \to \operatorname{Alt}(V)$ given by $\mathfrak{b} \mapsto \mathfrak{b} - \mathfrak{b}^t$ is surjective. Therefore, we have an exact sequence of vector spaces

$$0 \to \operatorname{Sym}(V) \to \operatorname{Bil}(V) \to \operatorname{Alt}(V) \to 0. \tag{1.2}$$

Exercise 1.3. Construct natural isomorphisms

$$\operatorname{Bil}(V) \simeq (V \otimes_F V)^* \simeq V^* \otimes_F V^*, \qquad \operatorname{Sym}(V) \simeq S^2(V)^*,$$
$$\operatorname{Alt}(V) \simeq \Lambda^2(V)^* \simeq \Lambda^2(V^*)$$

and show that the exact sequence (1.2) is dual to the standard exact sequence

$$0 \to \Lambda^2(V) \to V \otimes_F V \to S^2(V) \to 0$$

where $\Lambda^2(V)$ is the exterior square of V and $S^2(V)$ is the symmetric square of V.

If $\mathfrak{b}, \mathfrak{c} \in \operatorname{Bil}(V)$, we say the two bilinear forms $\mathfrak{b}$ and $\mathfrak{c}$ are *similar* if $\mathfrak{b} \simeq a\mathfrak{c}$ for some $a \in F^\times$.

Let V be a finite dimensional vector space over F and let $\lambda = \pm 1$. Define the *hyperbolic λ-bilinear form* of V to be the form $\mathbb{H}_\lambda(V) = \mathfrak{b}_{\mathbb{H}_\lambda}$ on $V \oplus V^*$ defined by

$$\mathfrak{b}_{\mathbb{H}_\lambda}(v_1 + f_1, v_2 + f_2) := f_1(v_2) + \lambda f_2(v_1)$$

for all $v_1, v_2 \in V$ and $f_1, f_2 \in V^*$. If $\lambda = 1$, the form $\mathbb{H}_\lambda(V)$ is a symmetric bilinear form and if $\lambda = -1$, it is an alternating bilinear form. A bilinear form $\mathfrak{b}$ is called a *hyperbolic bilinear form* if $\mathfrak{b} \simeq \mathbb{H}_\lambda(W)$ for some finite dimensional F-vector space W and some $\lambda = \pm 1$. The hyperbolic form $\mathbb{H}_\lambda(F)$ is called the *hyperbolic plane* and denoted $\mathbb{H}_\lambda$. It has the matrix representation

$$\begin{pmatrix} 0 & 1 \\ \lambda & 0 \end{pmatrix}$$

in the appropriate basis. If $\mathfrak{b} \simeq \mathbb{H}_\lambda$, then $\mathfrak{b}$ has the above matrix representation in some basis $\{e, f\}$ of $V_\mathfrak{b}$. We call e, f a *hyperbolic pair*. Hyperbolic forms are nondegenerate.

Let $\mathfrak{b}$ be a bilinear form on V and $W \subset V$ a subspace. The restriction of $\mathfrak{b}$ to W is a bilinear form on W and is called a *subform* of $\mathfrak{b}$. We denote this form by $\mathfrak{b}|_W$.

1.A. Structure theorems for bilinear forms. Let $\mathfrak{b}$ be a symmetric or alternating bilinear form on V. We say $v, w \in V$ are *orthogonal* if $\mathfrak{b}(v, w) = 0$. Let $W, U \subset V$ be subspaces. Define the *orthogonal complement* of W by

$$W^\perp := \{v \in V \mid \mathfrak{b}(v, w) = 0 \text{ for all } w \in W\}.$$

This is a subspace of V. We say W is *orthogonal* to U if $W \subset U^\perp$, equivalently $U \subset W^\perp$. If $V = W \oplus U$ is a direct sum of subspaces with $W \subset U^\perp$, we write $\mathfrak{b} = \mathfrak{b}|_W \perp \mathfrak{b}|_U$ and say $\mathfrak{b}$ is the (*internal*) *orthogonal sum* of $\mathfrak{b}|_W$ and $\mathfrak{b}|_U$. The subspace $V^\perp$ is called the *radical* of $\mathfrak{b}$ and denoted by $\operatorname{rad} \mathfrak{b}$. The form $\mathfrak{b}$ is nondegenerate if and only if $\operatorname{rad} \mathfrak{b} = 0$.

If K/F is a field extension, let $V_K := K \otimes_F V$, a vector space over K. We have the standard embedding $V \to V_K$ by $v \mapsto 1 \otimes v$. Let $\mathfrak{b}_K$ denote the extension of $\mathfrak{b}$ to V_K, so $\mathfrak{b}_K(a \otimes v, c \otimes w) = ac\mathfrak{b}(v, w)$ for all $a, c \in K$ and $v, w \in V$. The form $\mathfrak{b}_K$ is of the same type as $\mathfrak{b}$. Moreover, $\operatorname{rad} \mathfrak{b}_K = (\operatorname{rad} \mathfrak{b})_K$, hence $\mathfrak{b}$ is nondegenerate if and only if $\mathfrak{b}_K$ is nondegenerate.

Let $^- : V \to \overline{V} := V/\operatorname{rad} \mathfrak{b}$ be the canonical epimorphism. Define $\overline{\mathfrak{b}}$ to be the bilinear form on $\overline{V}$ determined by $\overline{\mathfrak{b}}(\overline{v_1}, \overline{v_2}) := \mathfrak{b}(v_1, v_2)$ for all $v_1, v_2 \in V$. Then $\overline{\mathfrak{b}}$ is a nondegenerate bilinear form of the same type as $\mathfrak{b}$. Note also that if $f : \mathfrak{b}_1 \to \mathfrak{b}_2$ is an isometry of symmetric or alternative bilinear forms, then $f(\operatorname{rad} \mathfrak{b}_1) = \operatorname{rad} \mathfrak{b}_2$.

We have:

Lemma 1.4. *Let $\mathfrak{b}$ be a symmetric or alternating bilinear form on V. Let W be any subspace of V such that $V = \operatorname{rad} \mathfrak{b} \oplus W$. Then $\mathfrak{b}|_W$ is nondegenerate and*

$$\mathfrak{b} = \mathfrak{b}|_{\operatorname{rad} \mathfrak{b}} \perp \mathfrak{b}|_W = 0|_{\operatorname{rad} \mathfrak{b}} \perp \mathfrak{b}|_W$$

with $\mathfrak{b}|_W \simeq \overline{\mathfrak{b}}$, the form induced on $V/\operatorname{rad} \mathfrak{b}$. In particular, $\mathfrak{b}|_W$ is unique up to isometry.

The lemma above shows that it is sufficient to classify nondegenerate bilinear forms. In general, if $\mathfrak{b}$ is a symmetric or alternating bilinear form on V and $W \subset V$ is a subspace, then we have an exact sequence of vector spaces

$$0 \to W^\perp \to V \xrightarrow{l_W} W^*,$$

where l_W is defined by $v \mapsto l_v|_W : x \mapsto \mathfrak{b}(v, x)$. Hence $\dim W^\perp \geq \dim V - \dim W$. It is easy to determine when this is an equality.

Proposition 1.5. *Let $\mathfrak{b}$ be a symmetric or alternating bilinear form on V. Let W be any subspace of V. Then the following are equivalent:*

(1) $W \cap \operatorname{rad} \mathfrak{b} = 0$.
(2) $l_W : V \to W^*$ *is surjective.*
(3) $\dim W^\perp = \dim V - \dim W$.

PROOF. (1) holds if and only if the map $l_W^* : W \to V^*$ is injective, if and only if the map $l_W : V \to W^*$ is surjective, and if and only if (3) holds. □

Note that the conditions (1)-(3) hold if either $\mathfrak{b}$ or $\mathfrak{b}|_W$ is nondegenerate.

A key observation is:

Proposition 1.6. *Let $\mathfrak{b}$ be a symmetric or alternating bilinear form on V. Let W be a subspace such that $\mathfrak{b}|_W$ is nondegenerate. Then $\mathfrak{b} = \mathfrak{b}|_W \perp \mathfrak{b}|_{W^\perp}$. In particular, if $\mathfrak{b}$ is also nondegenerate, then so is $\mathfrak{b}|_{W^\perp}$.*

PROOF. By Proposition 1.5, $\dim W^\perp = \dim V - \dim W$, hence $V = W \oplus W^\perp$. The result follows. $\square$

Corollary 1.7. *Let $\mathfrak{b}$ be a symmetric bilinear form on V. Let $v \in V$ satisfy $\mathfrak{b}(v,v) \neq 0$. Then $\mathfrak{b} = \mathfrak{b}|_{Fv} \perp \mathfrak{b}_{(Fv)^\perp}$.*

Let $\mathfrak{b}_1$ and $\mathfrak{b}_2$ be two symmetric or alternating bilinear forms on V_1 and V_2 respectively. Then their *external orthogonal sum*, denoted by $\mathfrak{b}_1 \perp \mathfrak{b}_2$, is the form on $V_1 \oplus V_2$ given by

$$(\mathfrak{b}_1 \perp \mathfrak{b}_2)\big((v_1, v_2), (w_1, w_2)\big) := \mathfrak{b}_1(v_1, w_1) + \mathfrak{b}_2(v_2, w_2)$$

for all $v_i, w_i \in V_i$, $i = 1, 2$.

If n is a nonnegative integer and $\mathfrak{b}$ is a symmetric or alternating bilinear form over F, abusing notation, we let

$$n\mathfrak{b} := \underbrace{\mathfrak{b} \perp \cdots \perp \mathfrak{b}}_{n}.$$

In particular, if n is a nonnegative integer, we do not interpret $n\mathfrak{b}$ with n viewed in the field.

For example, $\mathbb{H}_\lambda(V) \simeq n\mathbb{H}_\lambda$ for any n-dimensional vector space V over F.

It is now easy to complete the classification of alternating forms.

Proposition 1.8. *Let $\mathfrak{b}$ be a nondegenerate alternating form on V. Then $\dim V = 2n$ for some n and $\mathfrak{b} \simeq n\mathbb{H}_{-1}$, i.e., $\mathfrak{b}$ is hyperbolic.*

PROOF. Let $0 \neq v \in V$. Then there exists $w \in V$ such that $\mathfrak{b}(v,w) = a \neq 0$. Replacing w by $a^{-1}w$, we see that v, w is a hyperbolic pair in the space $W = Fv \oplus Fw$, so $\mathfrak{b}|_W$ is a hyperbolic subform of $\mathfrak{b}$, in particular, nondegenerate. Therefore, $\mathfrak{b} = \mathfrak{b}|_W \perp \mathfrak{b}|_{W^\perp}$ by Proposition 1.6. The result follows by induction on $\dim \mathfrak{b}$. $\square$

The proof shows that every nondegenerate alternating form $\mathfrak{b}$ on V has a *symplectic basis*, i.e., a basis $\{v_1, \dots, v_{2n}\}$ for V satisfying $\mathfrak{b}(v_i, v_{n+i}) = 1$ for all $i \in [1,\ n] := \{i \in \mathbb{Z} \mid 1 \leq i \leq n\}$ and $\mathfrak{b}(v_i, v_j) = 0$ if $i \leq j$ and $j \neq n + i$.

We turn to the classification of the isometry type of symmetric bilinear forms. By Lemma 1.4, Corollary 1.7, and induction, we therefore have the following:

Corollary 1.9. *Let $\mathfrak{b}$ be a symmetric bilinear form on V. Then*

$$\mathfrak{b} = \mathfrak{b}|_{\operatorname{rad}\mathfrak{b}} \perp \mathfrak{b}|_{V_1} \perp \cdots \perp \mathfrak{b}|_{V_n} \perp \mathfrak{b}|_W$$

with V_i a 1-dimensional subspace of V and $\mathfrak{b}|_{V_i}$ nondegenerate for all $i \in [1,\ n]$ and $\mathfrak{b}|_W$ a nondegenerate alternating subform on a subspace W of V.

If $\operatorname{char} F \neq 2$, then, in the corollary, $\mathfrak{b}|_W$ is symmetric and alternating hence $W = \{0\}$. In particular, every symmetric bilinear form $\mathfrak{b}$ has an *orthogonal basis*, i.e., a basis $\{v_1, \dots, v_n\}$ for $V_\mathfrak{b}$ satisfying $\mathfrak{b}(v_i, v_j) = 0$ if $i \neq j$. The form is nondegenerate if and only if $\mathfrak{b}(v_i, v_i) \neq 0$ for all i.

If $\operatorname{char} F = 2$, by Proposition 1.8, the alternating form $\mathfrak{b}|_W$ in the corollary above has a symplectic basis and satisfies $\mathfrak{b}|_W \simeq n\mathbb{H}_1$ for some n.

Let $a \in F$. Denote the bilinear form on F given by $\mathfrak{b}(v,w) = avw$ for all $v, w \in F$ by $\langle a \rangle_b$ or simply $\langle a \rangle$. In particular, $\langle a \rangle \simeq \langle b \rangle$ if and only if $a = b = 0$ or $aF^{\times 2} = bF^{\times 2}$ in $F^\times / F^{\times 2}$. Denote

$$\langle a_1 \rangle \perp \cdots \perp \langle a_n \rangle \quad \text{by} \quad \langle a_1, \dots, a_n \rangle_b \quad \text{or simply by} \quad \langle a_1, \dots, a_n \rangle.$$

We call such a form a *diagonal form*. A symmetric bilinear form $\mathfrak{b}$ isometric to a diagonal form is called *diagonalizable*. Consequently, $\mathfrak{b}$ is diagonalizable if and only if $\mathfrak{b}$ has an orthogonal basis. Note that $\det\langle a_1, \dots, a_n\rangle = a_1 \cdots a_n F^{\times 2}$ if $a_i \in F^\times$ for all i. Corollary 1.9 says that every bilinear form $\mathfrak{b}$ on V satisfies

$$\mathfrak{b} \simeq r\langle 0\rangle \perp \langle a_1, \dots, a_n\rangle \perp \mathfrak{b}'$$

with $r = \dim(\operatorname{rad}\mathfrak{b})$ and $\mathfrak{b}'$ an alternating form and $a_i \in F^\times$ for all i. In particular, if $\operatorname{char} F \neq 2$, then every symmetric bilinear form is diagonalizable.

Example 1.10. Let $a, b \in F^\times$. Then $\langle 1, a\rangle \simeq \langle 1, b\rangle$ if and only if $aF^{\times 2} = \det\langle 1, a\rangle = \det\langle 1, b\rangle = bF^{\times 2}$.

1.B. Values and similarities of bilinear forms. We study the values that a bilinear form can take as well as the similarity factors. We begin with some notation.

Definition 1.11. Let $\mathfrak{b}$ be a bilinear form on V over F. Let

$$D(\mathfrak{b}) := \{\mathfrak{b}(v, v) \mid v \in V \text{ with } \mathfrak{b}(v, v) \neq 0\},$$

the *set on nonzero values* of $\mathfrak{b}$ and

$$G(\mathfrak{b}) := \{a \in F^\times \mid a\mathfrak{b} \simeq \mathfrak{b}\},$$

a group called the *group of similarity factors of* $\mathfrak{b}$. Also set

$$\widetilde{D}(\mathfrak{b}) := D(\mathfrak{b}) \cup \{0\}.$$

We say that elements in $\widetilde{D}(\mathfrak{b})$ are *represented by* $\mathfrak{b}$.

For example, $G(\mathbb{H}_1) = F^\times$. A symmetric bilinear form is called *round* if $G(\mathfrak{b}) = D(\mathfrak{b})$. In particular, if $\mathfrak{b}$ is round, then $D(\mathfrak{b})$ is a group.

Remark 1.12. If $\mathfrak{b}$ is a symmetric bilinear form and $a \in D(\mathfrak{b})$, then $\mathfrak{b} \simeq \langle a\rangle \perp \mathfrak{c}$ for some symmetric bilinear form $\mathfrak{c}$ by Corollary 1.7.

Lemma 1.13. *Let $\mathfrak{b}$ be a bilinear form. Then*

$$D(\mathfrak{b}) \cdot G(\mathfrak{b}) \subset D(\mathfrak{b}).$$

In particular, if $1 \in D(\mathfrak{b})$, then $G(\mathfrak{b}) \subset D(\mathfrak{b})$.

Proof. Let $a \in G(\mathfrak{b})$ and $b \in D(\mathfrak{b})$. Let $\lambda : \mathfrak{b} \to a\mathfrak{b}$ be an isometry and $v \in V_{\mathfrak{b}}$ satisfy $b = \mathfrak{b}(v, v)$. Then $\mathfrak{b}\big(\lambda(v), \lambda(v)\big) = a\mathfrak{b}(v, v) = ab$. □

Example 1.14. Let $K = F[t]/(t^2 - a)$ with $a \in F$, where $F[t]$ is the polynomial ring over F. So $K = F \oplus F\theta$ as a vector space over F where θ denotes the class of t in K. If $z = x + y\theta$ with $x, y \in F$, write $\overline{z} = x - y\theta$. Let $s : K \to F$ be the F-linear functional defined by $s(x + y\theta) = x$. Then $\mathfrak{b}$ defined by $\mathfrak{b}(z_1, z_2) = s(z_1\overline{z}_2)$ is a binary symmetric bilinear form on K. Let $N(z) = z\overline{z}$ for $z \in K$. Then $D(\mathfrak{b}) = \{N(z) \neq 0 \mid z \in K\} = \{N(z) \mid z \in K^\times\}$. If $z \in K$, then $\lambda_z : K \to K$ given by $w \to zw$ is an F-linear isomorphism if and only if $N(z) \neq 0$. Suppose that λ_z is an F-isomorphism. As

$$\mathfrak{b}(\lambda_z z_1, \lambda_z z_2) = \mathfrak{b}(zz_1, zz_2) = N(z)s(z_1\overline{z}_2) = N(z)\mathfrak{b}(z_1, z_2),$$

we have an isometry $N(z)\mathfrak{b} \simeq \mathfrak{b}$ for all $z \in K^\times$. In particular, $\mathfrak{b}$ is round. Computing $\mathfrak{b}$ on the orthogonal basis $\{1, \theta\}$ for K shows that $\mathfrak{b}$ is isometric to the bilinear form $\langle 1, -a\rangle$. If $a \in F^\times$, then $\mathfrak{b} \simeq \langle 1, -a\rangle$ is nondegenerate.

Remark 1.15. (1) Let $\mathfrak{b}$ be a binary symmetric bilinear form on V. Suppose there exists a basis $\{v, w\}$ for V satisfying $\mathfrak{b}(v,v) = 0$, $\mathfrak{b}(v,w) = 1$, and $\mathfrak{b}(w,w) = a \neq 0$. Then $\mathfrak{b}$ is nondegenerate as the matrix corresponding to $\mathfrak{b}$ in this basis, is invertible. Moreover, $\{w, -av + w\}$ is an orthogonal basis for V and, using this basis, we see that $\mathfrak{b} \simeq \langle a, -a\rangle$.

(2) Suppose that $\operatorname{char} F \neq 2$. Let $\mathfrak{b} = \langle a, -a\rangle$ with $a \in F^\times$ and $\{e, g\}$ an orthogonal basis for $V_\mathfrak{b}$ satisfying $a = \mathfrak{b}(e,e) = -\mathfrak{b}(f,f)$. Evaluating on the basis $\{e+f, \frac{1}{2a}(e-f)\}$ shows that $\mathfrak{b} \simeq \mathbb{H}_1$. In particular, $\langle a, -a\rangle \simeq \mathbb{H}_1$ for all $a \in F^\times$. Moreover, $\langle a, -a\rangle \simeq \mathbb{H}_1$ is round and universal, where a nondegenerate symmetric bilinear form $\mathfrak{b}$ is called *universal* if $D(\mathfrak{b}) = F^\times$.

(3) Suppose that $\operatorname{char} F = 2$. As $\mathbb{H}_1 = \mathbb{H}_{-1}$ is alternating while $\langle a, a\rangle$ is not, $\langle a, a\rangle \not\simeq \mathbb{H}_1$ for any $a \in F^\times$. Moreover, $\mathbb{H}_1$ is not round since $D(\mathbb{H}_1) = \emptyset$. As $D\big(\langle a, a\rangle\big) = D\big(\langle a\rangle\big) = aF^{\times 2}$, we have $G\big(\langle a, a\rangle\big) = F^{\times 2}$ by Lemma 1.13. In particular, $\langle a, a\rangle$ is round if and only if $a \in F^{\times 2}$, and $\langle a, a\rangle \simeq \langle b, b\rangle$ if and only if $aF^{\times 2} \simeq bF^{\times 2}$.

(4) Witt Cancellation holds if $\operatorname{char} F \neq 2$, i.e., if there exists an isometry of symmetric bilinear forms $\mathfrak{b} \perp \mathfrak{b}' \simeq \mathfrak{b} \perp \mathfrak{b}''$ over F with $\mathfrak{b}$ nondegenerate, then $\mathfrak{b}' \simeq \mathfrak{b}''$. (Cf. Theorem 8.4 below.) If $\operatorname{char} F = 2$, this is false in general. For example,

$$\langle 1, 1, -1\rangle \simeq \langle 1\rangle \perp \mathbb{H}_1$$

over any field. Indeed if $\mathfrak{b}$ is 3-dimensional on V and V has an orthogonal basis $\{e, f, g\}$ with $\mathfrak{b}(e,e) = 1 = \mathfrak{b}(f,f)$ and $\mathfrak{b}(g,g) = -1$, then the right hand side arises from the basis $\{e+f+g, e+g, -f-g\}$. But by (3), $\langle 1, -1\rangle \not\simeq \mathbb{H}_1$ if $\operatorname{char} F = 2$. Multiplying the equation above by any $a \in F^\times$, we also have

$$\langle a, a, -a\rangle \simeq \langle a\rangle \perp \mathbb{H}_1. \tag{1.16}$$

Proposition 1.17. *Let $\mathfrak{b}$ be a symmetric bilinear form. If $D(\mathfrak{b}) \neq \emptyset$, then $\mathfrak{b}$ is diagonalizable. In particular, a nonzero symmetric bilinear form is diagonalizable if and only if it is not alternating.*

PROOF. If $a \in D(\mathfrak{b})$, then

$$\mathfrak{b} \simeq \langle a\rangle \perp \mathfrak{b}_1 \simeq \langle a\rangle \perp \operatorname{rad}\mathfrak{b}_1 \perp \mathfrak{c}_1 \perp \mathfrak{c}_2$$

with $\mathfrak{b}_1$ a symmetric bilinear form by Corollary 1.7, and $\mathfrak{c}_1$ a nondegenerate diagonal form, and $\mathfrak{c}_2$ a nondegenerate alternating form by Corollary 1.9. By the remarks following Corollary 1.9, we have $\mathfrak{c}_2 = 0$ if $\operatorname{char} F \neq 2$ and $\mathfrak{c}_2 = m\mathbb{H}_1$ for some integer m if $\operatorname{char} F = 2$. By (1.16), we conclude that $\mathfrak{b}$ is diagonalizable in either case.

If $\mathfrak{b}$ is not alternating, then $D(\mathfrak{b}) \neq \emptyset$, hence $\mathfrak{b}$ is diagonalizable. Conversely, if $\mathfrak{b}$ is diagonalizable, it cannot be alternating as it is not the zero form. □

Corollary 1.18. *Let $\mathfrak{b}$ be a symmetric bilinear form over F. Then $\mathfrak{b} \perp \langle 1\rangle$ is diagonalizable.*

Let $\mathfrak{b}$ be a symmetric bilinear form on V. A vector $v \in V$ is called *anisotropic* if $\mathfrak{b}(v,v) \neq 0$ and *isotropic* if $v \neq 0$ and $\mathfrak{b}(v,v) = 0$. We call $\mathfrak{b}$ *anisotropic* if there are no isotropic vectors in V and *isotropic* otherwise.

Corollary 1.19. *Every anisotropic bilinear form is diagonalizable.*

Note that an anisotropic symmetric bilinear form is nondegenerate as its radical is trivial.

Example 1.20. Let F be a *quadratically closed* field, i.e., every element in F is a square. Then, up to isometry, 0 and $\langle 1 \rangle$ are the only anisotropic forms over F. In particular, this applies if F is algebraically closed.

An anisotropic form may not be anisotropic under base extension. However, we do have:

Lemma 1.21. *Let $\mathfrak{b}$ be an anisotropic bilinear form over F. If K/F is purely transcendental, then $\mathfrak{b}_K$ is anisotropic.*

PROOF. First suppose that $K = F(t)$, the field of rational functions over F in the variable t. Suppose that $\mathfrak{b}_{F(t)}$ is isotropic. Then there exist a vector $0 \neq v \in V_{\mathfrak{b}_{F(t)}}$ such that $\mathfrak{b}_{F(t)}(v, v) = 0$. Multiplying by an appropriate nonzero polynomial, we may assume that $v \in F[t] \otimes_F V$. Write $v = v_0 + t \otimes v_1 + \cdots + t^n \otimes v_n$ with $v_1, \ldots, v_n \in V$ and $v_n \neq 0$. As the t^{2n} coefficient $\mathfrak{b}(v_n, v_n)$ of $\mathfrak{b}(v, v)$ must vanish, v_n is an isotropic vector of $\mathfrak{b}$, a contradiction.

If K/F is finitely generated, then the result follows by induction on the transcendence degree of K over F. In the general case, if $\mathfrak{b}_K$ is isotropic there exists a finitely generated purely transcendental extension K_0 of F in K with $\mathfrak{b}_{K_0}$ isotropic, a contradiction. □

1.C. Metabolic bilinear forms. Let $\mathfrak{b}$ be a symmetric bilinear form on V. A subspace $W \subset V$ is called a *totally isotropic* subspace of $\mathfrak{b}$ if $\mathfrak{b}|_W = 0$, i.e., if $W \subset W^{\perp}$. If $\mathfrak{b}$ is isotropic, then it has a nonzero totally isotropic subspace. Suppose that $\mathfrak{b}$ is nondegenerate and W is a totally isotropic subspace. Then $\dim W + \dim W^{\perp} = \dim V$ by Proposition 1.5, hence $\dim W \leq \frac{1}{2} \dim V$. We say that W is a *lagrangian* for b if we have an equality $\dim W = \frac{1}{2} \dim V$, equivalently $W^{\perp} = W$. A nondegenerate symmetric bilinear form is called *metabolic* if it has a lagrangian. In particular, every metabolic form is even-dimensional. Clearly, an orthogonal sum of metabolic forms is metabolic.

Example 1.22. (1) Symmetric hyperbolic forms are metabolic.

(2) The form $\mathfrak{b} \perp (-\mathfrak{b})$ is metabolic if $\mathfrak{b}$ is any nondegenerate symmetric bilinear form.

(3) A 2-dimensional metabolic space is nothing but a nondegenerate isotropic plane. A metabolic plane is therefore either isomorphic to $\langle a, -a \rangle$ for some $a \in F^{\times}$ or to the hyperbolic plane $\mathbb{H}_1$ by Remark 1.15. In particular, the determinant of a metabolic plane is $-F^{\times 2}$. If $\operatorname{char} F \neq 2$, then $\langle a, -a \rangle \simeq \mathbb{H}_1$ by Remark 1.15, so in this case, every metabolic plane is hyperbolic.

Lemma 1.23. *Let $\mathfrak{b}$ be an isotropic nondegenerate symmetric bilinear form over V. Then every isotropic vector belongs to a 2-dimensional metabolic subform.*

PROOF. Suppose that $\mathfrak{b}(v, v) = 0$ with $v \neq 0$. As $\mathfrak{b}$ is nondegenerate, there exists a $u \in V$ such that $\mathfrak{b}(u, v) \neq 0$. Then $\mathfrak{b}|_{Fv \oplus Fu}$ is metabolic. □

Corollary 1.24. *Every metabolic form is an orthogonal sum of metabolic planes. In particular, if $\mathfrak{b}$ is a metabolic form over F, then* $\det \mathfrak{b} = (-1)^{\frac{\dim \mathfrak{b}}{2}} F^{\times 2}$.

PROOF. We induct on the dimension of a metabolic form $\mathfrak{b}$. Let $W \subset V = V_{\mathfrak{b}}$ be a lagrangian. By Lemma 1.23, a nonzero vector $v \in W$ belongs to a metabolic plane $P \subset V$. It follows from Proposition 1.6 that $\mathfrak{b} = \mathfrak{b}|_P \perp \mathfrak{b}|_{P^{\perp}}$ and by dimension count that $W \cap P^{\perp}$ is a lagrangian of $\mathfrak{b}|_{P^{\perp}}$ as W cannot lie in $P^{\perp}$. By the induction hypothesis, $\mathfrak{b}|_{P^\perp}$ is an orthogonal sum of metabolic planes. The second statement follows from Example 1.22(3). □

Corollary 1.25. *If* $\operatorname{char} F \neq 2$, *the classes of metabolic and hyperbolic forms coincide. In particular, every isotropic nondegenerate symmetric bilinear form is universal.*

Proof. This follows from Remark 1.15(2) and Lemma 1.23. □

Lemma 1.26. *Let* $\mathfrak{b}$ *and* $\mathfrak{b}'$ *be two symmetric bilinear forms. If* $\mathfrak{b} \perp \mathfrak{b}'$ *and* $\mathfrak{b}'$ *are both metabolic, then so is* $\mathfrak{b}$.

Proof. By Corollary 1.24, we may assume that $\mathfrak{b}'$ is 2-dimensional. Let W be a lagrangian for $\mathfrak{b} \perp \mathfrak{b}'$. Let $p : W \to V_{\mathfrak{b}'}$ be the projection and $W_0 = \operatorname{Ker}(p) = W \cap V_{\mathfrak{b}}$. Suppose that p is not surjective. Then $\dim W_0 \geq \dim W - 1$, hence W_0 is a lagrangian of $\mathfrak{b}$ and $\mathfrak{b}$ is metabolic.

So we may assume that p is surjective. Then $\dim W_0 = \dim W - 2$. As $\mathfrak{b}'$ is metabolic, it is isotropic. Choose an isotropic vector $v' \in V_{\mathfrak{b}'}$ and a vector $w \in W$ such that $p(w) = v'$, i.e., $w = v + v'$ for some $v \in V_{\mathfrak{b}}$. In particular, $\mathfrak{b}(v, v) = (\mathfrak{b} \perp \mathfrak{b}')(w, w) - \mathfrak{b}'(v', v') = 0$. Since $W_0 \subset V_{\mathfrak{b}}$, we have v' is orthogonal to W_0, hence v is also orthogonal to W_0. If we show that $v' \notin W$, then $v \notin W_0$ and $W_0 \oplus Fv$ is a lagrangian of $\mathfrak{b}$ and $\mathfrak{b}$ is metabolic.

So suppose $v' \in W$. There exists $v'' \in V_{\mathfrak{b}'}$ such that $\mathfrak{b}'(v', v'') \neq 0$ as $\mathfrak{b}'$ is nondegenerate. Since p is surjective, there exists $w'' \in W$ with $w'' = u'' + v''$ for some $u'' \in V_{\mathfrak{b}}$. As W is totally isotropic,

$$0 = (\mathfrak{b} \perp \mathfrak{b}')(v', w'') = (\mathfrak{b} \perp \mathfrak{b}')(v', u'' + v'') = \mathfrak{b}'(v', v''),$$

a contradiction. □

1.D. Witt Theory. We have the following form of the classical Witt Decomposition Theorem (cf. [**139**]) for symmetric bilinear forms over a field of arbitrary characteristic.

Theorem 1.27 (Bilinear Witt Decomposition Theorem). *Let* $\mathfrak{b}$ *be a nondegenerate symmetric bilinear form on* V. *Then there exist subspaces* V_1 *and* V_2 *of* V *such that* $\mathfrak{b} = \mathfrak{b}|_{V_1} \perp \mathfrak{b}|_{V_2}$ *with* $\mathfrak{b}|_{V_1}$ *anisotropic and* $\mathfrak{b}|_{V_2}$ *metabolic. Moreover,* $\mathfrak{b}|_{V_1}$ *is unique up to isometry.*

Proof. We prove existence of the decomposition by induction on $\dim \mathfrak{b}$. If $\mathfrak{b}$ is isotropic, there is a metabolic plane $P \subset V$ by Lemma 1.23. As $\mathfrak{b} = \mathfrak{b}|_P \perp \mathfrak{b}|_{P^\perp}$, the proof of existence follows by applying the induction hypothesis to $\mathfrak{b}|_{P^\perp}$.

To prove uniqueness, assume that $\mathfrak{b}_1 \perp \mathfrak{b}_2 \simeq \mathfrak{b}_1' \perp \mathfrak{b}_2'$ with $\mathfrak{b}_1$ and $\mathfrak{b}_1'$ both anisotropic and $\mathfrak{b}_2$ and $\mathfrak{b}_2'$ both metabolic. We show that $\mathfrak{b}_1 \simeq \mathfrak{b}_1'$. The form

$$\mathfrak{b}_1 \perp (-\mathfrak{b}_1') \perp \mathfrak{b}_2 \simeq \mathfrak{b}_1' \perp (-\mathfrak{b}_1') \perp \mathfrak{b}_2'$$

is metabolic, hence $\mathfrak{b}_1 \perp (-\mathfrak{b}_1')$ is metabolic by Lemma 1.26. Let W be a lagrangian of $\mathfrak{b}_1 \perp (-\mathfrak{b}_1')$. Since $\mathfrak{b}_1$ is anisotropic, the intersection $W \cap V_{\mathfrak{b}_1}$ is trivial. Therefore, the projection $W \to V_{\mathfrak{b}_1'}$ is injective and $\dim W \leq \dim \mathfrak{b}_1'$. Similarly, $\dim W \leq \dim \mathfrak{b}_1$. Consequently, $\dim \mathfrak{b}_1 = \dim W = \dim \mathfrak{b}_1'$ and the projections $p : W \to V_{\mathfrak{b}_1}$ and $p' : W \to V_{\mathfrak{b}_1'}$ are isomorphisms. Let $w = v + v' \in W$, where $v \in V_{\mathfrak{b}_1}$ and $v' \in V_{\mathfrak{b}_1'}$. As

$$0 = \big(\mathfrak{b}_1 \perp (-\mathfrak{b}_1')\big)(w, w) = \mathfrak{b}_1(v, v) - \mathfrak{b}_1'(v', v'),$$

the isomorphism $p' \circ p^{-1} : V_{\mathfrak{b}_1} \to V_{\mathfrak{b}_1'}$ is an isometry between $\mathfrak{b}_1$ and $\mathfrak{b}_1'$. □

Let $\mathfrak{b} = \mathfrak{b}|_{V_1} \perp \mathfrak{b}|_{V_2}$ be the decomposition of the nondegenerate symmetric bilinear form $\mathfrak{b}$ on V in the theorem. The anisotropic form $\mathfrak{b}|_{V_1}$, unique up to isometry, will be denoted by $\mathfrak{b}_{an}$ and called the *anisotropic part* of $\mathfrak{b}$. Note that the metabolic form $\mathfrak{b}|_{V_2}$ in Theorem 1.27 is not unique in general by Remark 1.15(4). However, its dimension is unique and even. Define the *Witt index* of $\mathfrak{b}$ to be $\mathfrak{i}(\mathfrak{b}) := (\dim V_2)/2$.

Remark 1.15(4) also showed that the Witt Cancellation Theorem does not hold for nondegenerate symmetric bilinear forms in characteristic 2. The obstruction is the metabolic forms. We have, however, the following:

Corollary 1.28 (Witt Cancellation). *Let $\mathfrak{b}$, $\mathfrak{b}_1$, $\mathfrak{b}_2$ be nondegenerate symmetric bilinear forms satisfying $\mathfrak{b}_1 \perp \mathfrak{b} \simeq \mathfrak{b}_2 \perp \mathfrak{b}$. If $\mathfrak{b}_1$ and $\mathfrak{b}_2$ are anisotropic, then $\mathfrak{b}_1 \simeq \mathfrak{b}_2$.*

PROOF. We have $\mathfrak{b}_1 \perp \mathfrak{b} \perp (-\mathfrak{b}) \simeq \mathfrak{b}_2 \perp \mathfrak{b} \perp (-\mathfrak{b})$ with $\mathfrak{b} \perp (-\mathfrak{b})$ metabolic. By Theorem 1.27, $\mathfrak{b}_1 \simeq \mathfrak{b}_2$. □

2. The Witt and Witt-Grothendieck rings of symmetric bilinear forms

In this section, we construct the Witt ring. The orthogonal sum induces an additive structure on the isometry classes of symmetric bilinear forms. Defining the tensor product of symmetric bilinear forms (corresponding to the classical Kronecker product of matrices) turns this set of isometry classes into a semi-ring. The Witt Decomposition Theorem leads to a nice description of the Grothendieck ring in terms of isometry classes of anisotropic symmetric bilinear forms. The Witt ring $W(F)$ is the quotient of this ring by the ideal generated by the hyperbolic plane.

Let $\mathfrak{b}_1$ and $\mathfrak{b}_2$ be symmetric bilinear forms over F. The *tensor product* of $\mathfrak{b}_1$ and $\mathfrak{b}_2$ is defined to be the symmetric bilinear form $\mathfrak{b} := \mathfrak{b}_1 \otimes \mathfrak{b}_2$ with underlying space $V_{\mathfrak{b}_1} \otimes_F V_{\mathfrak{b}_2}$ with the form $\mathfrak{b}$ defined by

$$\mathfrak{b}(v_1 \otimes v_2, w_1 \otimes w_2) = \mathfrak{b}_1(v_1, w_1) \cdot \mathfrak{b}_2(v_2, w_2)$$

for all $v_1, w_1 \in V_{\mathfrak{b}_1}$ and $v_2, w_2 \in V_{\mathfrak{b}_2}$. For example, if $a \in F$, then $\langle a \rangle \otimes \mathfrak{b}_1 \simeq a\mathfrak{b}_1$.

Lemma 2.1. *Let $\mathfrak{b}_1$ and $\mathfrak{b}_2$ be two nondegenerate bilinear forms over F. Then*

(1) *$\mathfrak{b}_1 \perp \mathfrak{b}_2$ is nondegenerate.*
(2) *$\mathfrak{b}_1 \otimes \mathfrak{b}_2$ is nondegenerate.*
(3) *$\mathbb{H}_1(V) \otimes \mathfrak{b}_1$ is hyperbolic for all finite dimensional vector spaces V.*

PROOF. (1), (2): Let $V_i = V_{\mathfrak{b}_i}$ for $i = 1, 2$. The $\mathfrak{b}_i$ induce isomorphisms $l_i : V_i \to V_i^*$ for $i = 1, 2$, hence $\mathfrak{b}_1 \perp \mathfrak{b}_2$ and $\mathfrak{b}_1 \otimes \mathfrak{b}_2$ induce isomorphisms $l_1 \oplus l_2 : V_1 \oplus V_2 \to (V_1 \oplus V_2)^*$ and $l_1 \otimes l_2 : V_1 \otimes_F V_2 \to (V_1 \otimes_F V_2)^*$ respectively.

(3): Let $\{e, f\}$ be a hyperbolic pair for $\mathbb{H}_1$. Then the linear map $(F \oplus F^*) \otimes_F V_1 \to V_1 \oplus V_1^*$ induced by $e \otimes v \mapsto v$ and $f \otimes v \mapsto l_v : w \mapsto \mathfrak{b}(w, v)$ is an isomorphism and induces the isometry $\mathbb{H}_1 \otimes \mathfrak{b} \to \mathbb{H}_1(V)$. □

It follows that the isometry classes of nondegenerate symmetric bilinear forms over F is a semi-ring under orthogonal sum and tensor product. The Grothendieck ring of this semi-ring is called the *Witt-Grothendieck* ring of F and denoted by $\widehat{W}(F)$. (Cf. Scharlau [**121**] or Lang [**90**] for the definition and construction of the Grothendieck group and ring.) In particular, every element in $\widehat{W}(F)$ is a difference of two isometry classes of nondegenerate symmetric bilinear forms over F. If $\mathfrak{b}$ is a nondegenerate symmetric bilinear form over F, we shall also write $\mathfrak{b}$ for the class in $\widehat{W}(F)$. Thus, if $\alpha \in \widehat{W}(F)$, there exist nondegenerate symmetric bilinear forms $\mathfrak{b}_1$ and $\mathfrak{b}_2$ over F such that $\alpha = \mathfrak{b}_1 - \mathfrak{b}_2$ in $\widehat{W}(F)$. By definition, we have

$$\mathfrak{b}_1 - \mathfrak{b}_2 = \mathfrak{b}_1' - \mathfrak{b}_2' \text{ in } \widehat{W}(F)$$

if and only if there exists a nondegenerate symmetric bilinear form $\mathfrak{b}''$ over F such that

$$\mathfrak{b}_1 \perp \mathfrak{b}_2' \perp \mathfrak{b}'' \simeq \mathfrak{b}_1' \perp \mathfrak{b}_2 \perp \mathfrak{b}''. \tag{2.2}$$

As any hyperbolic form $\mathbb{H}_1(V)$ is isometric to $(\dim V)\mathbb{H}_1$ over F, the ideal consisting of the hyperbolic forms over F in $\widehat{W}(F)$ is the principal ideal $\mathbb{H}_1$ by Lemma 2.1(3). The quotient $W(F) := \widehat{W}(F)/(\mathbb{H}_1)$ is called the *Witt ring* of nondegenerate symmetric bilinear forms over F. Elements in $W(F)$ are called *Witt classes.* Abusing notation, we shall also write $\mathfrak{b} \in W(F)$ for the Witt class of $\mathfrak{b}$ and often call it just the *class* of $\mathfrak{b}$. The operations in $W(F)$ (and $\widehat{W}(F)$) shall be denoted by $+$ and $\cdot$.

By (1.16), we have

$$\langle a, -a\rangle = 0 \text{ in } W(F)$$

for all $a \in F^\times$ and in all characteristics. In particular, $\langle -1\rangle = -\langle 1\rangle = -1$ in $W(F)$, hence the additive inverse of the Witt class of any nondegenerate symmetric bilinear form $\mathfrak{b}$ in $W(F)$ is represented by the form $-\mathfrak{b}$. It follows that if $\alpha \in W(F)$ then there exists a nondegenerate bilinear form $\mathfrak{b}$ such that $\alpha = \mathfrak{b}$ in $W(F)$.

Exercise 2.3 (cf. Scharlau [**121**], p. 22). Let $\mathfrak{b}$ be a metabolic symmetric bilinear form and V a lagrangian of $\mathfrak{b}$. Show that

$$\mathfrak{b} \perp (-\mathfrak{b}) \simeq \mathbb{H}(V) \perp (-\mathfrak{b}).$$

In particular, $\mathfrak{b} = \mathbb{H}(V)$ in $\widehat{W}(F)$. Use this to give another proof that $\mathfrak{c} + (-\mathfrak{c}) = 0$ in $W(F)$ for every nondegenerate form $\mathfrak{c}$.

The Witt Cancellation Theorem 1.28 allows us to conclude the following.

Proposition 2.4. *Let $\mathfrak{b}_1$ and $\mathfrak{b}_2$ be anisotropic symmetric bilinear forms. Then the following are equivalent:*

(1) $\mathfrak{b}_1 \simeq \mathfrak{b}_2$.
(2) $\mathfrak{b}_1 = \mathfrak{b}_2$ *in* $\widehat{W}(F)$.
(3) $\mathfrak{b}_1 = \mathfrak{b}_2$ *in* $W(F)$.

PROOF. The implications (1) $\Rightarrow$ (2) $\Rightarrow$ (3) are easy.

(3) $\Rightarrow$ (1): By definition of the Witt ring, $\mathfrak{b}_1 + n\mathbb{H} = \mathfrak{b}_2 + m\mathbb{H}$ in $\widehat{W}(F)$ for some $n, m \geq 0$. It follows from the definition of the Grothendieck-Witt ring that

$$\mathfrak{b}_1 \perp n\mathbb{H} \perp \mathfrak{b} \simeq \mathfrak{b}_2 \perp m\mathbb{H} \perp \mathfrak{b}$$

for some nondegenerate form $\mathfrak{b}$. Thus $\mathfrak{b}_1 \perp n\mathbb{H} \perp \mathfrak{b} \perp (-\mathfrak{b}) \simeq \mathfrak{b}_2 \perp m\mathbb{H} \perp \mathfrak{b} \perp (-\mathfrak{b})$ and $\mathfrak{b}_1 \simeq \mathfrak{b}_2$ by Corollary 1.28. $\square$

We also have:

Corollary 2.5. $\mathfrak{b} = 0$ *in* $W(F)$ *if and only if* $\mathfrak{b}$ *is metabolic.*

It follows from Proposition 2.4 that every Witt class in $W(F)$ contains (up to isometry) a unique anisotropic form. As every anisotropic bilinear form is diagonalizable by Corollary 1.19, we have a ring epimorphism

$$\mathbb{Z}[F^\times/F^{\times 2}] \to W(F) \quad \text{given by} \quad \sum_i n_i(a_iF^{\times 2}) \mapsto \sum_i n_i\langle a_i\rangle. \tag{2.6}$$

Proposition 2.7. *A homomorphism of fields $F \to K$ induces ring homomorphisms*

$$r_{K/F} : \widehat{W}(F) \to \widehat{W}(K) \quad \text{and} \quad r_{K/F} : W(F) \to W(K).$$

If K/F is purely transcendental, then these maps are injective.

PROOF. Let $\mathfrak{b}$ be symmetric bilinear form over F. Define $r_{K/F}(\mathfrak{b})$ on $K \otimes_F V_\mathfrak{b}$ by

$$r_{K/F}(\mathfrak{b})(x \otimes v, y \otimes w) = xy\mathfrak{b}(v, w)$$

for all $x, y \in K$ and for all $v, w \in V_\mathfrak{b}$. This construction is compatible with orthogonal sums and tensor products of symmetric bilinear forms.

As $r_{K/F}(\mathfrak{b})$ is hyperbolic if $\mathfrak{b}$ is, it follows that $\mathfrak{b} \mapsto r_{K/F}(\mathfrak{b})$ induces the desired maps. These are ring homomorphisms.

The last statement follows by Lemma 1.21. □

If K/F is a field extension, the ring homomorphisms $r_{K/F}$ defined above are called *restriction* homomorphisms. Of course, the maps $r_{K/F}$ are the unique homomorphisms such that $r_{K/F}(\mathfrak{b}) = \mathfrak{b}_K$.

3. Chain equivalence

Two nondegenerate diagonal symmetric bilinear forms $\mathfrak{a} = \langle a_1, a_2, \dots, a_n\rangle$ and $\mathfrak{b} = \langle b_1, b_2, \dots, b_n\rangle$, are called *simply chain equivalent* if either $n = 1$ and $a_1F^{\times 2} = b_1F^{\times 2}$ or $n \geq 2$ and $\langle a_i, a_j\rangle \simeq \langle b_i, b_j\rangle$ for some indices $i \neq j$ and $a_k = b_k$ for every $k \neq i, j$. Two nondegenerate diagonal forms $\mathfrak{a}$ and $\mathfrak{b}$ are called *chain equivalent* (we write $\mathfrak{a} \approx \mathfrak{b}$) if there is a chain of forms $\mathfrak{b}_1 = \mathfrak{a}, \mathfrak{b}_2, \dots, \mathfrak{b}_m = \mathfrak{b}$ such that $\mathfrak{b}_i$ and $\mathfrak{b}_{i+1}$ are simply chain equivalent for all $i \in [1, m-1]$. Clearly, $\mathfrak{a} \approx \mathfrak{b}$ implies $\mathfrak{a} \simeq \mathfrak{b}$.

Note that as the symmetric group S_n is generated by transpositions, we have

$$\langle a_1, a_2, \dots, a_n\rangle \approx \langle a_{\sigma(1)}, a_{\sigma(2)}, \dots, a_{\sigma(n)}\rangle$$

for every $\sigma \in S_n$.

Lemma 3.1. *Every nondegenerate diagonal form is chain equivalent to an orthogonal sum of an anisotropic diagonal form and metabolic binary diagonal forms $\langle a, -a\rangle$, $a \in F^\times$.*

PROOF. By induction, it is sufficient to prove that any isotropic diagonal form $\mathfrak{b}$ is chain equivalent to $\langle a, -a\rangle \perp \mathfrak{b}'$ for some diagonal form $\mathfrak{b}'$ and $a \in F^\times$. Let $\{v_1, \dots, v_n\}$ be the orthogonal basis of $\mathfrak{b}$ and set $\mathfrak{b}(v_i, v_i) = a_i$. Choose an isotropic vector v with the smallest number k of nonzero coordinates. Changing the order of the v_i if necessary, we may assume that $v = \sum_{i=1}^k c_iv_i$ for nonzero $c_i \in F$ and $k \geq 2$. We prove the statement by induction on k. If $k = 2$, the restriction of $\mathfrak{b}$ to the plane $Fv_1 \oplus Fv_2$ is metabolic and therefore is isomorphic to $\langle a, -a\rangle$ for some $a \in F^\times$ by Example 1.22(3), hence $\mathfrak{b} \approx \langle a, -a\rangle \perp \langle a_3, \dots, a_n\rangle$.

If $k > 2$, the vector $v_1' = c_1v_1 + c_2v_2$ is anisotropic. Complete v_1' to an orthogonal basis $\{v_1', v_2'\}$ of $Fv_1 \oplus Fv_2$ by Corollary 1.7 and set $a_i' = \mathfrak{b}(v_i', v_i')$, $i = 1, 2$. Then

$\langle a_1, a_2\rangle \simeq \langle a'_1, a'_2\rangle$ and $\mathfrak{b} \approx \langle a'_1, a'_2, a_3, \dots, a_n\rangle$. The vector v has $k-1$ nonzero coordinates in the orthogonal basis $\{v'_1, v'_2, v_3, \dots, v_n\}$. Applying the induction hypothesis to the diagonal form $\langle a'_1, a'_2, a_3, \dots, a_n\rangle$ completes the proof. □

Lemma 3.2 (Witt Chain Equivalence). *Two anisotropic diagonal forms are chain equivalent if and only if they are isometric.*

PROOF. Let $\{v_1, \dots, v_n\}$ and $\{u_1, \dots, u_n\}$ be two orthogonal bases of the bilinear form $\mathfrak{b}$ with $\mathfrak{b}(v_i, v_i) = a_i$ and $\mathfrak{b}(u_i, u_i) = b_i$. We must show that $\langle a_1, \dots, a_n\rangle \approx \langle b_1, \dots, b_n\rangle$. We do this by double induction on n and the number k of nonzero coefficients of u_1 in the basis $\{v_i\}$. Changing the order of the v_i if necessary, we may assume that $u_1 = \sum_{i=1}^{k} c_i v_i$ for some nonzero $c_i \in F$.

If $k = 1$, i.e., $u_1 = c_1 v_1$, the two $(n-1)$-dimensional subspaces generated by the v_i's and u_i's respectively, with $i \geq 2$, coincide. By the induction hypothesis and Witt Cancellation (Corollary 1.28), $\langle a_2, \dots, a_n\rangle \approx \langle b_2, \dots, b_n\rangle$, hence $\langle a_1, a_2, \dots, a_n\rangle \approx \langle a_1, b_2, \dots, b_n\rangle \approx \langle b_1, b_2, \dots, b_n\rangle$.

If $k \geq 2$, set $v'_1 = c_1 v_1 + c_2 v_2$. As $\mathfrak{b}$ is anisotropic, $a'_1 = \mathfrak{b}(v'_1, v'_1)$ is nonzero. Choose an orthogonal basis $\{v'_1, v'_2\}$ of $Fv_1 \oplus Fv_2$ and set $a'_2 = \mathfrak{b}(v'_2, v'_2)$. We have $\langle a_1, a_2\rangle \simeq \langle a'_1, a'_2\rangle$. The vector u_1 has $k-1$ nonzero coordinates in the basis $\{v'_1, v'_2, v_3, \dots, v_n\}$. By the induction hypothesis

$$\langle a_1, a_2, a_3, \dots, a_n\rangle \approx \langle a'_1, a'_2, a_3, \dots, a_n\rangle \approx \langle b_1, b_2, b_3, \dots, b_n\rangle. \qquad \square$$

Exercise 3.3. Prove that a diagonalizable metabolic form $\mathfrak{b}$ is isometric to $\langle 1, -1\rangle \otimes \mathfrak{b}'$ for some diagonalizable bilinear form $\mathfrak{b}'$.

4. Structure of the Witt ring

In this section, we give a presentation of the Witt-Grothendieck and Witt rings. The classes of even-dimensional anisotropic symmetric bilinear forms generate an ideal $I(F)$ in the Witt ring. We also derive a presentation for it and its square, $I(F)^2$.

4.A. The presentation of $\widehat{W}(F)$ and $W(F)$. We turn to determining presentations of $\widehat{W}(F)$ and $W(F)$. The generators will be the isometry classes of nondegenerate 1-dimensional symmetric bilinear forms. The defining relations arise from the following:

Lemma 4.1. *Let $a, b \in F^\times$ and $z \in D\big(\langle a, b\rangle\big)$. Then $\langle a, b\rangle \simeq \langle z, abz\rangle$. In particular, if $a + b \neq 0$, then*

$$\langle a, b\rangle \simeq \langle a+b, ab(a+b)\rangle. \tag{4.2}$$

PROOF. By Corollary 1.7, we have $\langle a, b\rangle \simeq \langle z, d\rangle$ for some $d \in F^\times$. Comparing determinants, we must have $abF^{\times 2} = dzF^{\times 2}$ so $dF^{\times 2} = abzF^{\times 2}$. □

The isometry (4.2) is often called the *Witt relation.*

Define an abelian group $W'(F)$ by generators and relations. Generators are isometry classes of nondegenerate 1-dimensional symmetric bilinear forms. For any $a \in F^\times$ we write $[a]$ for the generator — the isometry class of the form $\langle a\rangle$. Note that $[ax^2] = [a]$ for every $a, x \in F^\times$. The relations are

$$[a] + [b] = [a+b] + [ab(a+b)] \tag{4.3}$$

for all $a, b \in F^\times$ such that $a + b \neq 0$.

Lemma 4.4. *If $\langle a,b\rangle \simeq \langle c,d\rangle$, then $[a]+[b]=[c]+[d]$ in $W'(F)$.*

PROOF. As $\langle a,b\rangle \simeq \langle c,d\rangle$, we have $abF^{\times 2} = \det\langle a,b\rangle = \det\langle c,d\rangle = cdF^{\times 2}$ and $d = abcz^2$ for some $z \in F^{\times}$. Since $c \in D\big(\langle a,b\rangle\big)$, there exist $x,y \in F$ satisfying $c = ax^2+by^2$. If $x=0$ or $y=0$, the statement is obvious, so we may assume that $x,y \in F^{\times}$. It follows from (4.3) that

$$[a]+[b] = [ax^2]+[by^2] = [c]+[ax^2by^2c] = [c]+[d]. \qquad \square$$

Lemma 4.5. *We have $[a]+[-a]=[b]+[-b]$ in $W'(F)$ for all $a,b \in F^{\times}$.*

PROOF. We may assume that $a+b \neq 0$. From (4.3), we have

$$[-a]+[a+b] = [b]+[-ab(a+b)], \quad [-b]+[a+b] = [a]+[-ab(a+b)].$$

The result follows. $\square$

If $\operatorname{char} F \neq 2$, the forms $\langle a,-a\rangle$ and $\langle b,-b\rangle$ are isometric by Remark 1.15(2). Therefore, in this case Lemma 4.5 follows from Lemma 4.4.

Lemma 4.6. *If $\langle a_1,\dots,a_n\rangle \approx \langle b_1,\dots,b_n\rangle$, then $[a_1]+\cdots+[a_n] = [b_1]+\cdots+[b_n]$ in $W'(F)$.*

PROOF. We may assume that the forms are simply chain equivalent. In this case the statement follows from Lemma 4.4. $\square$

Theorem 4.7. *The Grothendieck-Witt group $\widehat{W}(F)$ is generated by the isometry classes of 1-dimensional symmetric bilinear forms that are subject to the defining relations $\langle a\rangle + \langle b\rangle = \langle a+b\rangle + \langle ab(a+b)\rangle$ for all $a,b \in F^{\times}$ such that $a+b \neq 0$.*

PROOF. It suffices to prove that the homomorphism $W'(F) \to \widehat{W}(F)$ taking $[a]$ to $\langle a\rangle$ is an isomorphism. As $\mathfrak{b} \perp \langle 1\rangle$ is diagonalizable for any nondegenerate symmetric bilinear form $\mathfrak{b}$ by Corollary 1.18, the map is surjective. An element in the kernel is given by the difference of two diagonal forms $\mathfrak{b} = \langle a_1,\dots,a_n\rangle$ and $\mathfrak{b}' = \langle a_1',\dots,a_n'\rangle$ such that $\mathfrak{b} = \mathfrak{b}'$ in $\widehat{W}(F)$. By the definition of $\widehat{W}(F)$ and Corollary 1.18, there is a diagonal form $\mathfrak{b}''$ such that $\mathfrak{b} \perp \mathfrak{b}'' \simeq \mathfrak{b}' \perp \mathfrak{b}''$. Replacing $\mathfrak{b}$ and $\mathfrak{b}'$ by $\mathfrak{b} \perp \mathfrak{b}''$ and $\mathfrak{b}' \perp \mathfrak{b}''$ respectively, we may assume that $\mathfrak{b} \simeq \mathfrak{b}'$. It follows from Lemma 3.1 that $\mathfrak{b} \approx \mathfrak{b}_1 \perp \mathfrak{b}_2$ and $\mathfrak{b}' \approx \mathfrak{b}_1' \perp \mathfrak{b}_2'$, where $\mathfrak{b}_1, \mathfrak{b}_1'$ are anisotropic diagonal forms and $\mathfrak{b}_2, \mathfrak{b}_2'$ are orthogonal sums of metabolic planes $\langle a,-a\rangle$ for various $a \in F^{\times}$. It follows from the Corollary 1.28 that $\mathfrak{b}_1 \simeq \mathfrak{b}_1'$, and therefore, $\mathfrak{b}_1 \approx \mathfrak{b}_1'$ by Lemma 3.2. Note that the dimension of $\mathfrak{b}_2$ and $\mathfrak{b}_2'$ are equal. By Lemmas 4.5 and 4.6, we conclude that $[a_1]+\cdots+[a_n] = [a_1']+\cdots+[a_n']$ in $W'(F)$. $\square$

Since the Witt class in $W(F)$ of the hyperbolic plane $\mathbb{H}_1$ is equal to $\langle 1,-1\rangle$ by Remark 1.15(4), Theorem 4.7 yields:

Theorem 4.8. *The Witt group $W(F)$ is generated by the isometry classes of 1-dimensional symmetric bilinear forms that are subject to the following defining relations:*

(1) $\langle 1\rangle + \langle -1\rangle = 0$.
(2) $\langle a\rangle + \langle b\rangle = \langle a+b\rangle + \langle ab(a+b)\rangle$ *for all $a,b \in F^{\times}$ such that $a+b \neq 0$.*

If $\operatorname{char} F \neq 2$, the above is the well-known presentation of the Witt-Grothendieck and Witt groups first demonstrated by Witt in [**139**].

4.B. The presentation of $I(F)$. The Witt-Grothendieck and Witt rings have a natural filtration that we now describe. Define the *dimension map*

$$\dim : \widehat{W}(F) \to \mathbb{Z} \quad \text{given by} \quad \dim x = \dim \mathfrak{b}_1 - \dim \mathfrak{b}_2 \ \text{ if } \ x = \mathfrak{b}_1 - \mathfrak{b}_2.$$

This is a well-defined map by (2.2).

We let $\widehat{I}(F)$ denote the kernel of this map. As

$$\langle a\rangle - \langle b\rangle = \big(\langle 1\rangle - \langle b\rangle\big) - \big(\langle 1\rangle - \langle a\rangle\big) \ \text{ in } \ \widehat{W}(F)$$

for all $a, b \in F^\times$, the elements $\langle 1\rangle - \langle a\rangle$ with $a \in F^\times$ generate $\widehat{I}(F)$ as an abelian group.

It follows that $\widehat{W}(F)$ is generated by the elements $\langle 1\rangle$ and $\langle 1\rangle - \langle x\rangle$ with $x \in F^\times$. Let $I(F)$ denote the image of $\widehat{I}(F)$ in $W(F)$. If $a \in F^\times$, write $\langle\langle a\rangle\rangle_b$ or simply $\langle\langle a\rangle\rangle$ for the binary symmetric bilinear form $\langle 1, -a\rangle_b$. As $\widehat{I}(F) \cap (\mathbb{H}_1) = 0$, we have $I(F) \simeq \widehat{I}(F)/\big(\widehat{I}(F) \cap (\mathbb{H}_1)\big) \simeq \widehat{I}(F)$. Then the map $\widehat{W}(F) \to W(F)$ induces an isomorphism

$$\widehat{I}(F) \to I(F) \quad \text{given by} \quad \langle 1\rangle - \langle x\rangle \mapsto \langle\langle x\rangle\rangle.$$

In particular, $I(F)$ is the ideal in $W(F)$ consisting of the Witt classes of even-dimensional forms. It is called the *fundamental ideal* of $W(F)$ and is generated by the classes $\langle\langle a\rangle\rangle$ with $a \in F^\times$. Note that if $F \to K$ is a homomorphism of fields, then $r_{K/F}\big(\widehat{I}(F)\big) \subset \widehat{I}(K)$ and $r_{K/F}\big(I(F)\big) \subset I(K)$.

The relations in Theorem 4.8 can be rewritten as

$$\langle\langle a\rangle\rangle + \langle\langle b\rangle\rangle = \langle\langle a+b\rangle\rangle + \langle\langle ab(a+b)\rangle\rangle$$

for $a, b \in F^\times$ with $a + b \neq 0$. We conclude:

Corollary 4.9. *The group $I(F)$ is generated by the isometry classes of dimension 2 symmetric bilinear forms $\langle\langle a\rangle\rangle$ with $a \in F^\times$ subject to the defining relations:*

(1) $\langle\langle 1\rangle\rangle = 0$.
(2) $\langle\langle a\rangle\rangle + \langle\langle b\rangle\rangle = \langle\langle a+b\rangle\rangle + \langle\langle ab(a+b)\rangle\rangle$ *for all* $a, b \in F^\times$ *such that* $a + b \neq 0$.

Let $\widehat{I}^n(F) := \widehat{I}(F)^n$ be the nth power of $\widehat{I}(F)$. Then $\widehat{I}^n(F)$ maps isomorphically onto $I^n(F) := I(F)^n$, the nth power of $I(F)$ in $W(F)$. It defines the filtration

$$W(F) \supset I(F) \supset I^2(F) \supset \cdots \supset I^n(F) \supset \cdots$$

in which we shall be interested.

For convenience, we let $\widehat{I}^0(F) = \widehat{W}(F)$ and $I^0(F) = W(F)$.

We denote the tensor product $\langle\langle a_1\rangle\rangle \otimes \langle\langle a_2\rangle\rangle \otimes \cdots \otimes \langle\langle a_n\rangle\rangle$ by

$$\langle\langle a_1, a_2, \dots, a_n\rangle\rangle_b \quad \text{or simply by} \quad \langle\langle a_1, a_2, \dots, a_n\rangle\rangle$$

and call a form isometric to such a tensor product a *bilinear n-fold Pfister form*. (We call any form isometric to $\langle 1\rangle$ a *0-fold* Pfister form.) For $n \geq 1$, the isometry classes of bilinear n-fold Pfister forms generate $I^n(F)$ as an abelian group.

We shall be interested in relations between isometry classes of Pfister forms in $W(F)$. We begin with a study of 1- and 2-fold Pfister forms.

Example 4.10. We have $\langle\langle a\rangle\rangle + \langle\langle b\rangle\rangle = \langle\langle ab\rangle\rangle + \langle\langle a, b\rangle\rangle$ in $W(F)$. In particular, $\langle\langle a\rangle\rangle + \langle\langle b\rangle\rangle \equiv \langle\langle ab\rangle\rangle \mod I^2(F)$.

As the hyperbolic plane is 2-dimensional, the dimension invariant induces a map

$$e_0 : W(F) \to \mathbb{Z}/2\mathbb{Z} \quad \text{defined by} \quad e_0(\mathfrak{b}) = \dim \mathfrak{b} \bmod 2.$$

Clearly, this is a homomorphism with kernel the fundamental ideal $I(F)$, so it induces an isomorphism

$$\bar{e}_0 : W(F)/I(F) \to \mathbb{Z}/2\mathbb{Z}. \tag{4.11}$$

By Corollary 1.24, we have a map

$$e_1 : I(F) \to F^\times/F^{\times 2} \quad \text{defined by} \quad e_1(\mathfrak{b}) = (-1)^{\frac{\dim \mathfrak{b}}{2}} \det \mathfrak{b}$$

called the *signed determinant.*

The map e_1 is a homomorphism as $\det(\mathfrak{b} \perp \mathfrak{b}') = \det \mathfrak{b} \cdot \det \mathfrak{b}'$ and surjective as $e_1(\langle\langle a\rangle\rangle) = aF^{\times 2}$. Clearly, $e_1\big(\langle\langle a, b\rangle\rangle\big) = F^{\times 2}$, so e_1 induces an epimorphism

$$\bar{e}_1 : I(F)/I^2(F) \to F^\times/F^{\times 2}. \tag{4.12}$$

We have

Proposition 4.13. *The kernel of e_1 is $I^2(F)$ and the map $\bar{e}_1 : I(F)/I^2(F) \to F^\times/F^{\times 2}$ is an isomorphism.*

PROOF. Let $f_1 : F^\times/F^{\times 2} \to I(F)/I^2(F)$ be given by $aF^{\times 2} \mapsto \langle\langle a\rangle\rangle + I^2(F)$. This is a homomorphism by Example 4.10 inverse to $\bar{e}_1$, since $I(F)$ is generated by $\langle\langle a\rangle\rangle$, $a \in F^\times$. □

For fields of characteristic different than 2 this is Pfister's characterization of $I^2(F)$ (cf. [**110**, Satz 13, Kor.]).

4.C. The presentation of $I^2(F)$. We turn to $I^2(F)$.

Lemma 4.14. *Let $a, b \in F^\times$. Then $\langle\langle a, b\rangle\rangle = 0$ in $W(F)$ if and only if either $a \in F^{\times 2}$ or $b \in D\big(\langle\langle a\rangle\rangle\big)$. In particular, $\langle\langle a, 1-a\rangle\rangle = 0$ in $W(F)$ for any $a \neq 1$ in $F^\times$.*

PROOF. Suppose that $\langle\langle a\rangle\rangle$ is anisotropic. Then $\langle\langle a, b\rangle\rangle = 0$ in $W(F)$ if and only if $b\langle\langle a\rangle\rangle \simeq \langle\langle a\rangle\rangle$ by Proposition 2.4 if and only if $b \in G\big(\langle\langle a\rangle\rangle\big) = D\big(\langle\langle a\rangle\rangle\big)$ by Example 1.14. □

Isometries of bilinear 2-fold Pfister forms are easily established using isometries of binary forms. For example, we have

Lemma 4.15. *Let $a, b \in F^\times$ and $x, y \in F$. Let $z = ax^2 + by^2 \neq 0$. Then:*

1. *$\langle\langle a, b\rangle\rangle \simeq \langle\langle a, b(y^2 - ax^2)\rangle\rangle$ if $y^2 - ax^2 \neq 0$.*
2. *$\langle\langle a, b\rangle\rangle \simeq \langle\langle z, -ab\rangle\rangle$.*
3. *$\langle\langle a, b\rangle\rangle \simeq \langle\langle z, abz\rangle\rangle$.*
4. *If z is a square in F, then $\langle\langle a, b\rangle\rangle$ is metabolic. In particular, if* $\operatorname{char} F \neq 2$, *then $\langle\langle a, b\rangle\rangle \simeq 2\mathbb{H}_1$.*

PROOF. (1): Let $w = y^2 - ax^2$. We have

$$\langle\langle a, b\rangle\rangle \simeq \langle 1, -a, -b, ab\rangle \simeq \langle 1, -a, -by^2, abx^2\rangle \simeq \langle 1, -a, -bw, abw\rangle \simeq \langle\langle a, bw\rangle\rangle.$$

(2): We have

$$\langle\langle a, b\rangle\rangle \simeq \langle 1, -a, -b, ab\rangle \simeq \langle 1, -ax^2, -by^2, ab\rangle \simeq \langle 1, -z, -zab, ab\rangle \simeq \langle\langle z, -ab\rangle\rangle.$$

(3) follows from (1)and (2) while (4) follows from (2) and Remark 1.15(2). □

Explicit examples of such isometries are:

Example 4.16. Let $a, b \in F^\times$, then:

(1) $\langle\langle a, 1\rangle\rangle$ is metabolic.
(2) $\langle\langle a, -a\rangle\rangle$ is metabolic.
(3) $\langle\langle a, a\rangle\rangle \simeq \langle\langle a, -1\rangle\rangle$.
(4) $\langle\langle a, b\rangle\rangle + \langle\langle a, -b\rangle\rangle = \langle\langle a, -1\rangle\rangle$ in $W(F)$.

We turn to a presentation of $I^2(F)$ first done for fields of characteristic not 2 in [**30**] and rediscovered by Suslin (cf. [**129**]). It is different from that for $I(F)$ as we need a new generating relation. Indeed, the analogue of the Witt relation will be a consequence of our new relation and a metabolic relation.

Let $\underline{I}_2(F)$ be the abelian group generated by all the isometry classes $[\mathfrak{b}]$ of bilinear 2-fold Pfister forms $\mathfrak{b}$ subject to the generating relations:

(1) $\big[\langle\langle 1, 1\rangle\rangle\big] = 0$.
(2) $\big[\langle\langle ab, c\rangle\rangle\big] + \big[\langle\langle a, b\rangle\rangle\big] = \big[\langle\langle a, bc\rangle\rangle\big] + \big[\langle\langle b, c\rangle\rangle\big]$ for all $a, b, c \in F^\times$.

We call the second relation the *cocycle relation.*

Remark 4.17. The cocycle relation holds in $I^2(F)$: Let $a, b, c \in F^\times$. Then

$$\begin{aligned}\langle\langle ab, c\rangle\rangle + \langle\langle a, b\rangle\rangle &= \langle 1, -ab, -c, abc\rangle + \langle 1, -a, -b, ab\rangle \\ &= \langle 1, 1, -c, abc, -a, -b\rangle \\ &= \langle 1, -a, -bc, abc\rangle + \langle 1, -b, -c, bc\rangle \\ &= \langle\langle a, bc\rangle\rangle + \langle\langle b, c\rangle\rangle\end{aligned}$$

in $I^2(F)$.

We begin by showing that the analogue of the Witt relation is a consequence of the other two relations.

Lemma 4.18. *The relations*

(1) $\big[\langle\langle a, 1\rangle\rangle\big] = 0$,
(2) $\big[\langle\langle a, c\rangle\rangle\big] + \big[\langle\langle b, c\rangle\rangle\big] = \big[\langle\langle (a+b), c\rangle\rangle\big] + \big[\langle\langle (a+b)ab, c\rangle\rangle\big]$

hold in $\underline{I}_2(F)$ *for all* $a, b, c \in F^\times$ *if* $a + b \neq 0$.

PROOF. Applying the cocycle relation to $a, a, 1$ shows that

$$\big[\langle\langle 1, 1\rangle\rangle\big] + \big[\langle\langle a, a\rangle\rangle\big] = \big[\langle\langle a, a\rangle\rangle\big] + \big[\langle\langle a, 1\rangle\rangle\big].$$

The first relation now follows. Applying Lemma 4.15 and the cocycle relation to a, c, c shows that

$$\begin{aligned}\big[\langle\langle -a, c\rangle\rangle\big] + \big[\langle\langle a, c\rangle\rangle\big] &= \big[\langle\langle ac, c\rangle\rangle\big] + \big[\langle\langle a, c\rangle\rangle\big] \\ &= \big[\langle\langle -a, c\rangle\rangle\big] + \big[\langle\langle a, c\rangle\rangle\big] = \big[\langle\langle -1, c\rangle\rangle\big]\end{aligned} \tag{4.19}$$

for all $c \in F^\times$.

Applying the cocycle relation to $a(a+b), a, c$ yields

$$\big[\langle\langle a+b, c\rangle\rangle\big] + \big[\langle\langle a(a+b), a\rangle\rangle\big] = \big[\langle\langle a(a+b), ac\rangle\rangle\big] + \big[\langle\langle a, c\rangle\rangle\big] \tag{4.20}$$

and to $a(a+b), b, c$ yields

$$\big[\langle\langle ab(a+b), c\rangle\rangle\big] + \big[\langle\langle a(a+b), b\rangle\rangle\big] = \big[\langle\langle a(a+b), bc\rangle\rangle\big] + \big[\langle\langle b, c\rangle\rangle\big]. \tag{4.21}$$

Adding the equations (4.20) and (4.21) and then using the isometries

$$\langle\langle a(a+b), a\rangle\rangle \simeq \langle\langle a(a+b), -b\rangle\rangle \quad \text{and} \quad \langle\langle a(a+b), ac\rangle\rangle \simeq \langle\langle a(a+b), -bc\rangle\rangle$$

derived from Lemma 4.15, followed by using equation (4.19) yields

$$\begin{aligned}
&\big[\langle\langle a,c\rangle\rangle\big] + \big[\langle\langle b,c\rangle\rangle\big] - \big[\langle\langle (a+b),c\rangle\rangle\big] - \big[\langle\langle a+b)ab,c\rangle\rangle\big] \\
&= \big[\langle\langle a(a+b),a\rangle\rangle\big] + \big[\langle\langle a(a+b),b\rangle\rangle\big] - \big[\langle\langle a(a+b),ac\rangle\rangle\big] - \big[\langle\langle a(a+b),bc\rangle\rangle\big] \\
&= \big[\langle\langle a(a+b),-b\rangle\rangle\big] + \big[\langle\langle a(a+b),b\rangle\rangle\big] - \big[\langle\langle a(a+b),-bc\rangle\rangle\big] - \big[\langle\langle a(a+b),bc\rangle\rangle\big] \\
&= \big[\langle\langle a(a+b),-1\rangle\rangle\big] - \big[\langle\langle a(a+b),-1\rangle\rangle\big] = 0. \qquad \square
\end{aligned}$$

Theorem 4.22. *The ideal $I^2(F)$ is generated as an abelian group by the isometry classes $\langle\langle a,b\rangle\rangle$ of bilinear 2-fold Pfister forms for all $a,b \in F^\times$ subject to the generating relations:*

(1) $\langle\langle 1,1\rangle\rangle = 0$.

(2) $\langle\langle ab,c\rangle\rangle + \langle\langle a,b\rangle\rangle = \langle\langle a,bc\rangle\rangle + \langle\langle b,c\rangle\rangle$ *for all $a,b,c \in F^\times$.*

PROOF. Clearly, we have well-defined homomorphisms

$$g : \underline{I}_2(F) \to I^2(F) \quad \text{induced by} \quad [\mathfrak{b}] \mapsto \mathfrak{b}$$

and

$$j : \underline{I}_2(F) \to I(F) \quad \text{induced by} \quad \big[\langle\langle a,b\rangle\rangle\big] \mapsto \langle\langle a\rangle\rangle + \langle\langle b\rangle\rangle - \langle\langle ab\rangle\rangle,$$

the latter being the composition with the inclusion $I^2(F) \subset I(F)$ using Example 4.10.

We show that the map $g : \underline{I}_2(F) \to I^2(F)$ is an isomorphism. Define

$$\gamma : F^\times/F^{\times 2} \times F^\times/F^{\times 2} \to \underline{I}_2(F) \quad \text{by} \quad (aF^{\times 2}, bF^{\times 2}) \mapsto \big[\langle\langle a,b\rangle\rangle\big].$$

This is clearly well-defined. For convenience, write (a) for $aF^{\times 2}$. Using (2), we see that

$$\begin{aligned}
&\gamma\big((b),(c)\big) - \gamma\big((ab),(c)\big) + \gamma\big((a),(bc)\big) - \gamma\big((a),(b)\big) \\
&\qquad = \big[\langle\langle b,c\rangle\rangle\big] - \big[\langle\langle ab,c\rangle\rangle\big] + \big[\langle\langle a,bc\rangle\rangle\big] - \big[\langle\langle a,b\rangle\rangle\big] = 0,
\end{aligned}$$

so γ is a 2-cocycle. By Lemma 4.18, we have $\big[\langle\langle 1,a\rangle\rangle\big] = 0$ in $\underline{I}_2(F)$, so γ is a normalized 2-cocycle. The map γ defines an extension $N = (F^\times/F^{\times 2}) \times \underline{I}_2(F)$ of $\underline{I}_2(F)$ by $F^\times/F^{\times 2}$ with

$$\big((a),\alpha\big) + \big((b),\beta\big) = \big((ab), \alpha + \beta + [\langle\langle a,b\rangle\rangle]\big).$$

As γ is symmetric, N is abelian. Let

$$h : N \to I(F) \quad \text{be defined by} \quad \big((a),\alpha\big) \mapsto \langle\langle a\rangle\rangle + j(\alpha).$$

We see that the map h is a homomorphism:

$$\begin{aligned}
h\big(((a),\alpha) + ((b),\beta)\big) &= h\big((ab), \alpha + \beta + [\langle\langle a,b\rangle\rangle]\big) \\
&= \langle\langle ab\rangle\rangle + j(\alpha) + j(\beta) + j\big([\langle\langle a,b\rangle\rangle]\big) \\
&= \langle\langle a\rangle\rangle + \langle\langle b\rangle\rangle + j(\alpha) + j(\beta) \\
&= h\big((a),\alpha\big) + h\big((b),\beta\big).
\end{aligned}$$

Thus we have a commutative diagram

$$\begin{array}{ccccccccc}
0 & \longrightarrow & \underline{I}_2(F) & \longrightarrow & N & \longrightarrow & F^\times/F^{\times 2} & \longrightarrow & 0 \\
& & {\scriptstyle g}\big\downarrow & & {\scriptstyle h}\big\downarrow & & {\scriptstyle f_1}\big\downarrow & & \\
0 & \longrightarrow & I^2(F) & \longrightarrow & I(F) & \longrightarrow & I(F)/I^2(F) & \longrightarrow & 0
\end{array}$$

where f_1 is the isomorphism inverse of $\bar{e}_1$ in Proposition 4.13.

Let

$$k : I(F) \to N \quad \text{be induced by} \quad \langle\langle a\rangle\rangle \mapsto \big((a), 0\big).$$

Using Lemma 4.15 and Corollary 4.9, we see that k is well-defined as

$$\begin{aligned}
\big((a),0\big) + \big((b),0\big) &= \big((ab), [\langle\langle a,b\rangle\rangle]\big) = \big((ab), [\langle\langle a+b, ab(a+b)\rangle\rangle]\big) \\
&= \big((a+b),0\big) + \big((ab(a+b),0\big)
\end{aligned}$$

if $a+b \neq 0$. As

$$\begin{aligned}
&k\big(\langle\langle a,b\rangle\rangle\big) = k\big(\langle\langle a\rangle\rangle + \langle\langle b\rangle\rangle - \langle\langle ab\rangle\rangle\big) = \big((a),0\big) + \big((b),0\big) - \big((ab),0\big) \\
&= \big((ab),[\langle\langle a,b\rangle\rangle]\big) - \big((ab),0\big) = \big((ab),0\big) + \big(1,[\langle\langle, a,b\rangle\rangle]\big) - \big((ab),0\big) = \big(1,[\langle\langle, a,b\rangle\rangle]\big),
\end{aligned}$$

we have

$$(k \circ h)\big((c), [\langle\langle a,b\rangle\rangle]\big) = k\big(\langle\langle c\rangle\rangle + \langle\langle a,b\rangle\rangle\big) = \big((c), [\langle\langle a,b\rangle\rangle]\big)\big).$$

Hence $k \circ h$ is the identity on N. As $(h \circ k)\big(\langle\langle a\rangle\rangle\big) = \langle\langle a\rangle\rangle$, the composition $h \circ k$ is the identity on $I(F)$. Thus h is an isomorphism, hence so is g. $\square$

5. The Stiefel-Whitney map

In this section, we investigate Stiefel-Whitney maps. In the case of fields of characteristic different from 2, this was first defined by Milnor. We shall use facts about Milnor K-theory (cf. §100). We write

$$k_*(F) := \coprod_{n\geq 0} k_n(F)$$

for the graded ring $K_*(F)/2K_*(F)$. Abusing notation, if $\{a_1, \dots, a_n\}$ is a symbol in $K_n(F)$, we shall also write it for its coset $\{a_1, \dots, a_n\} + 2K_n(F)$.

The associated graded ring

$$GW_*(F) = \coprod_{n\geq 0} I^n(F)/I^{n+1}(F)$$

of $W(F)$ with respect to the fundamental ideal $I(F)$ is called the *graded Witt ring of symmetric bilinear forms.* Note that since $2 \cdot I^n(F) = \langle 1,1\rangle \cdot I^n(F) \subset I^{n+1}(F)$ we have $2 \cdot GW_*(F) = 0$.

By Example 4.10, the map $F^\times \to I(F)/I^2(F)$ defined by $a \mapsto \langle\langle a\rangle\rangle + I^2(F)$ is a homomorphism. By the definition of the Milnor ring and Lemma 4.14, this map gives rise to a graded ring homomorphism

$$f_* : k_*(F) \to GW_*(F) \tag{5.1}$$

taking the symbol $\{a_1, a_2, \dots, a_n\}$ to $\langle\langle a_1, a_2, \dots, a_n\rangle\rangle + I^{n+1}(F)$. Since the graded ring $GW_*(F)$ is generated by the degree one component $I(F)/I^2(F)$, the map f_* is surjective.

Note that the map $f_0 : k_0(F) \to W(F)/I(F)$ is the inverse of the map $\bar{e}_0$ and the map $f_1 : k_1(F) \to I(F)/I^2(F)$ is the inverse of the map $\bar{e}_1$ (cf. Proposition 4.13).

Lemma 5.2. *Let $\langle\langle a,b\rangle\rangle$ and $\langle\langle c,d\rangle\rangle$ be isometric bilinear 2-fold Pfister forms. Then $\{a,b\}=\{c,d\}$ in $k_2(F)$.*

PROOF. If the form $\langle\langle a,b\rangle\rangle$ is metabolic, then $b\in D(\langle\langle a\rangle\rangle)$ or $a\in F^{\times 2}$ by Lemma 4.14. In particular, if $\langle\langle a,b\rangle\rangle$ is metabolic, then $\{a,b\}=0$ in $k_2(F)$. Therefore, we may assume that $\langle\langle a,b\rangle\rangle$ is anisotropic. Using Witt Cancellation 1.28, we see that $c=ax^2+by^2-abz^2$ for some $x,y,z\in F$. If $c\notin aF^{\times 2}$, let $w=y^2-az^2\neq 0$. Then $\langle\langle a,b\rangle\rangle\simeq\langle\langle a,bw\rangle\rangle\simeq\langle\langle c,-abw\rangle\rangle$ by Lemma 4.15 and $\{a,b\}=\{a,bw\}=\{c,-abw\}$ in $k_2(F)$ by Lemma 100.3. Hence we may assume that $a=c$. By Witt Cancellation, $\langle -b,ab\rangle\simeq\langle -d,ad\rangle$, so $bd\in D(\langle\langle a\rangle\rangle)$, i.e., $bd=x^2-ay^2$ in F for some $x,y\in F$. Thus $\{a,b\}=\{a,d\}$ by Lemma 100.3. □

Proposition 5.3. *The homomorphism*

$$e_2: I^2(F)\to k_2(F)\quad \text{given by}\quad \langle\langle a,b\rangle\rangle\mapsto\{a,b\}$$

is a well-defined surjection with $\mathrm{Ker}(e_2)=I^3(F)$. *Moreover, e_2 induces an isomorphism*

$$\bar{e}_2: I^2(F)/I^3(F)\to k_2(F).$$

PROOF. By Lemma 5.2 and the presentation of $I^2(F)$ in Theorem 4.22, the map is well-defined. Since

$$\langle\langle a,b,c\rangle\rangle=\langle\langle a,c\rangle\rangle+\langle\langle b,c\rangle\rangle-\langle\langle ab,c\rangle\rangle,$$

we have $I^3(F)\subset\mathrm{Ker}(e_2)$. As $\bar{e}_2$ and f_2 are inverses of each other, the result follows. □

Let $\mathfrak{F}(F)$ be the free abelian group on the set of isometry classes of nondegenerate 1-dimensional symmetric bilinear forms. Then we have a group homomorphism

$$w:\mathfrak{F}(F)\to k_*(F)[[t]]^\times\quad\text{given by}\quad\langle a\rangle\mapsto 1+\{a\}t.$$

If $a,b\in F^\times$ satisfy $a+b\neq 0$, then by Lemma 100.3, we have

$$\begin{aligned}w(\langle a\rangle+\langle b\rangle)&=(1+\{a\}t)(1+\{b\}t)\\&=1+(\{a\}+\{b\})t+\{a,b\}t^2\\&=1+\{ab\}t+\{a,b\}t^2\\&=1+\{ab(a+b)^2\}t+\{a+b,ab(a+b)\}t^2\\&=w(\langle a+b\rangle+\langle ab(a+b)\rangle).\end{aligned}$$

In particular, w factors through the relation $\langle a\rangle+\langle b\rangle=\langle a+b\rangle+\langle ab(a+b)\rangle$ for all $a,b\in F^\times$ satisfying $a+b\neq 0$, hence induces a group homomorphism

$$w:\widehat{W}(F)\to k_*(F)[[t]]^\times \tag{5.4}$$

by Theorem 4.7 called the *total Stiefel-Whitney map.* If $\mathfrak{b}$ is a nondegenerate symmetric bilinear form and α is its class in $\widehat{W}(F)$ define the *total Stiefel-Whitney class* $w(\mathfrak{b})$ of $\mathfrak{b}$ to be $w(\alpha)$.

Example 5.5. If $\mathfrak{b}$ is a metabolic plane, then $\mathfrak{b}=\langle a\rangle+\langle -a\rangle$ in $\widehat{W}(F)$ for some $a\in F^\times$. (Note the hyperbolic plane equals $\langle 1\rangle+\langle -1\rangle$ in $\widehat{W}(F)$ by Example 1.15(4), so $w(\mathfrak{b})=1+\{-1\}t$ as $\{a,-a\}=1$ in $k_2(F)$ for any $a\in F^\times$.)

Lemma 5.6. *Let* $\alpha = (\langle 1\rangle - \langle a_1\rangle)\cdots(\langle 1\rangle - \langle a_n\rangle)$ *in* $\widehat{W}(F)$. *Let* $m = 2^{n-1}$. *Then*

$$w(\alpha) = \big(1 + \{a_1, \dots, a_n, \underbrace{-1, \dots, -1}_{m-n}\}t^m\big)^{-1}.$$

PROOF. As

$$\alpha = \sum_{\varepsilon} s_\varepsilon \langle a_1^{\varepsilon_1} \cdots a_n^{\varepsilon_n}\rangle,$$

where the sum runs over all $\varepsilon = (\varepsilon_1, \dots, \varepsilon_n) \in \{0,1\}^n$ and $s_\varepsilon = (-1)^{\sum_i \varepsilon_i}$, we have

$$w(\alpha) = \prod_{\varepsilon} \big(1 + \sum_i \varepsilon_i \{a_i\} t\big)^{s_\varepsilon}.$$

Let

$$h = h(t_1, \dots, t_n) = \prod_{\varepsilon} \big(1 + \varepsilon_1 t_1 t + \cdots + \varepsilon_n t_n t\big)^{-s_\varepsilon}$$

in $\big((\mathbb{Z}/2\mathbb{Z})[[t]]\big)[[t_1, \dots, t_n]]$, the ring of power series over $\mathbb{Z}/2\mathbb{Z}$ in variables

$$t, t_1, \dots, t_n.$$

Substituting zero for any t_i in h, yields one, so

$$h = 1 + t_1 \cdots t_n g(t_1, \dots, t_n)t^n \quad \text{for some} \quad g \in \big((\mathbb{Z}/2\mathbb{Z})[[t]]\big)[[t_1, \dots, t_n]].$$

As $\{a, a\} = \{a, -1\}$, we have

$$\begin{aligned} w(\alpha)^{-1} &= 1 + \{a_1, \dots, a_n\} g(\{a_1\}, \dots, \{a_n\})t^n \\ &= 1 + \{a_1, \dots, a_n\} g(\{-1\}, \dots, \{-1\})t^n. \end{aligned}$$

We have with s a variable,

$$1 + g(s, \dots, s)t^n = h(s, \dots, s) = \prod_{\varepsilon} \big(1 + \sum_{i \geq 1} \varepsilon_i st\big)^{-s_\varepsilon} = (1 + st)^m = 1 + s^m t^m$$

as $\sum \varepsilon_i = 1$ in $\mathbb{Z}/2\mathbb{Z}$ exactly m times, so $g(s, \dots, s) = (st)^{m-n}$ and the result follows. □

Let $w_0(\alpha) = 1$ and

$$w(\alpha) = 1 + \sum_{i \geq 1} w_i(\alpha) t^i$$

for $\alpha \in \widehat{W}(F)$. The map $w_i : \widehat{W}(F) \to k_i(F)$ is called the ith *Stiefel-Whitney class*. Let $\alpha, \beta \in \widehat{W}(F)$. As $w(\alpha + \beta) = w(\alpha)w(\beta)$, for every $n \geq 0$, we have the *Whitney Sum Formula*

$$w_n(\alpha + \beta) = \sum_{i+j=n} w_i(\alpha) w_j(\beta). \tag{5.7}$$

Remark 5.8. Let K/F be a field extension and $\alpha \in \widehat{W}(F)$. Then

$$\operatorname{res}_{K/F}\big(w_i(\alpha)\big) = w_i(\alpha_K) \quad \text{in} \quad k_i(F) \quad \text{for all} \quad i.$$

Corollary 5.9. *Let* $m = 2^{n-1}$. *Then* $w_j\big(\widehat{I}^n(F)\big) = 0$ *for* $j \in [1,\ m-1]$ *and* $w_m : \widehat{I}^n(F) \to k_m(F)$ *is a group homomorphism mapping* $(\langle 1\rangle - \langle a_1\rangle)\cdots(\langle 1\rangle - \langle a_n\rangle)$ *to* $\{a_1, \dots, a_n, \underbrace{-1, \dots, -1}_{m-n}\}$.

PROOF. Let $\alpha = (\langle 1\rangle - \langle a_1\rangle)\cdots(\langle 1\rangle - \langle a_n\rangle)$. By Lemma 5.6, we have $w_i(\alpha) = 0$ for $i \in [1,\ m-1]$. The result follows from the Whitney formula (5.7). □

Let $j : \widehat{I}(F) \to I(F)$ be the isomorphism sending $\langle 1\rangle - \langle a\rangle \mapsto \langle\langle a\rangle\rangle$. Let $\tilde{w}_m$ be the composition

$$I^n(F) \xrightarrow{j^{-1}} \widehat{I}^n(F) \xrightarrow{w_m|_{\widehat{I}^n(F)}} k_m(F).$$

Corollary 5.9 shows that $\tilde{w}_i = e_i$ for $i = 1, 2$. The map $\tilde{w}_m : I^n(F) \to k_m(F)$ is a group homomorphism with $I^{n+1}(F) \subset \operatorname{Ker}(\tilde{w}_m)$ so it induces a homomorphism

$$\bar{w}_m : I^n(F)/I^{n+1}(F) \to k_m(F).$$

We have $\bar{w}_i = \bar{e}_i$ for $i = 1, 2$. The composition $\bar{w}_m \circ f_n$ is multiplication by $\{\underbrace{-1, \dots, -1}_{m-n}\}$. In particular, $\bar{w}_1$ and $\bar{w}_2$ are isomorphisms, i.e.,

$$I^2(F) = \operatorname{Ker}(\tilde{w}_1) \quad \text{and} \quad I^3(F) = \operatorname{Ker}(\tilde{w}_2) \tag{5.10}$$

and

$$\widehat{I}^2(F) = \operatorname{Ker}(w_1|_{\widehat{I}(F)}) \quad \text{and} \quad \widehat{I}^3(F) = \operatorname{Ker}(w_2|_{\widehat{I}^2(F)}). \tag{5.11}$$

This gives another proof for Propositions 4.13 and 5.3.

Remark 5.12. Let $\operatorname{char} F \neq 2$ and $h_F^2 : k_2(F) \to H^2(F)$ be the norm-residue homomorphism defined in §101. If $\mathfrak{b}$ is a nondegenerate symmetric bilinear form, then $h_2 \circ w_2(\mathfrak{b})$ is the classical *Hasse-Witt invariant* of $\mathfrak{b}$. (Cf. [**89**], Definition V.3.17, [**121**], Definition 2.12.7.) More generally, the Stiefel-Whitney classes defined above are compatible with Stiefel-Whitney classes defined by Delzant w_i' in [**27**], i.e., $h_i \circ w_i = w_i'$ for all $i \geq 0$.

Example 5.13. Suppose that K is a real-closed field. (Cf. §95.) Then $k_i(K) = \mathbb{Z}/2\mathbb{Z}$ for all $i \geq 0$ and $\widehat{W}(K) = \mathbb{Z} \oplus \mathbb{Z}\xi$ with $\xi = \langle -1\rangle$ and $\xi^2 = 1$. The Stiefel-Whitney map $w : \widehat{W}(F) \to k_*(K)[[t]]^\times$ is then the map $n + m\xi \mapsto (1+t)^m$. In particular, if $\mathfrak{b}$ is a nondegenerate form, then $w(\mathfrak{b})$ determines the signature of $\mathfrak{b}$. Hence if $\mathfrak{b}$ and $\mathfrak{c}$ are two nondegenerate symmetric bilinear forms over K, we have $\mathfrak{b} \simeq \mathfrak{c}$ if and only if $\dim \mathfrak{b} = \dim \mathfrak{c}$ and $w(\mathfrak{b}) = w(\mathfrak{c})$.

It should be noted that if $\mathfrak{b} = \langle\langle a_1, \dots, a_n\rangle\rangle$ that $w(\mathfrak{b})$ is not equal to $w(\alpha) = \tilde{w}([\mathfrak{b}])$ where $\alpha = (\langle 1\rangle - \langle a_1\rangle)\cdots(\langle 1\rangle - \langle a_n\rangle)$ in $\widehat{W}(F)$ as the following exercise shows.

Exercise 5.14. Let $m = 2^{n-1}$. If $\mathfrak{b}$ is the bilinear n-fold Pfister form $\langle\langle a_1, \dots, a_n\rangle\rangle$, then

$$w(\mathfrak{b}) = 1 + \big(\{\underbrace{-1, \dots, -1}_{m}\} + \{a_1, \dots, a_n, \underbrace{-1, \dots, -1}_{m-n}\}\big)t^m.$$

The following fundamental theorem was proved by Orlov-Vishik-Voevodsky [**107**] in the case that $\operatorname{char} F \neq 2$ and by Kato [**78**] in the case that $\operatorname{char} F = 2$.

Fact 5.15. *The map $f_* : k_*(F) \to GW_*(F)$ is a ring isomorphism.*

For $i = 0, 1, 2$, we have proven that f_i is an isomorphism in (4.11), Proposition 4.13 , and Proposition 5.3, respectively.

6. Bilinear Pfister forms

The isometry classes of tensor products of nondegenerate binary symmetric bilinear forms representing one are quite interesting. These forms, called Pfister forms, whose properties over fields of characteristic different from 2 were discovered by Pfister in [**108**] and were named after him in [**32**], generate a filtration of the Witt ring by the powers of its fundamental ideal $I(F)$. Properties of these forms in the case of characteristic 2 were first studied by Baeza in [**15**]. In this section, we derive the main elementary properties of these forms.

By Example 1.14, a bilinear 1-fold Pfister form $\mathfrak{b} = \langle\langle a \rangle\rangle$, $a \in F^\times$, is round, i.e., $D(\langle\langle a \rangle\rangle) = G(\langle\langle a \rangle\rangle)$. Because of this, the next proposition shows that there are many round forms and, in particular, bilinear Pfister forms are round.

Proposition 6.1. *Let $\mathfrak{b}$ be a round bilinear form and let $a \in F^\times$. Then:*

(1) *The form $\langle\langle a \rangle\rangle \otimes \mathfrak{b}$ is also round.*
(2) *If $\langle\langle a \rangle\rangle \otimes \mathfrak{b}$ is isotropic, then either $\mathfrak{b}$ is isotropic or $a \in D(\mathfrak{b})$.*

PROOF. Set $\mathfrak{c} = \langle\langle a \rangle\rangle \otimes \mathfrak{b}$.

(1): Since $1 \in D(\mathfrak{b})$, it suffices to prove that $D(\mathfrak{c}) \subset G(\mathfrak{c})$. Let c be a nonzero value of $\mathfrak{c}$. Write $c = x - ay$ for some $x, y \in \widetilde{D}(\mathfrak{b})$. If $y = 0$, we have $c = x \in D(\mathfrak{b}) = G(\mathfrak{b}) \subset G(\mathfrak{c})$. Similarly, $y \in G(\mathfrak{c})$ if $x = 0$, hence $c = -ay \in G(\mathfrak{c})$ as $-a \in G(\langle\langle a \rangle\rangle) \subset G(\mathfrak{c})$.

Now suppose that x and y are nonzero. Since $\mathfrak{b}$ is round, $x, y \in G(\mathfrak{b})$ and, therefore,

$$\mathfrak{c} = \mathfrak{b} \perp (-a\mathfrak{b}) \simeq \mathfrak{b} \perp (-ayx^{-1})\mathfrak{b} = \langle\langle ayx^{-1} \rangle\rangle \otimes \mathfrak{b}.$$

By Example 1.14, we know that $1 - ayx^{-1} \in G(\langle\langle ayx^{-1} \rangle\rangle) \subset G(\mathfrak{c})$. Since $x \in G(\mathfrak{b}) \subset G(\mathfrak{c})$, we have $c = (1 - ayx^{-1})x \in G(\mathfrak{c})$.

(2): Suppose that $\mathfrak{b}$ is anisotropic. Since $\mathfrak{c} = \mathfrak{b} \perp (-a\mathfrak{b})$ is isotropic, there exist $x, y \in D(\mathfrak{b})$ with $x - ay = 0$. Therefore $a = xy^{-1} \in D(\mathfrak{b})$ as $D(\mathfrak{b})$ is closed under multiplication. □

Corollary 6.2. *Bilinear Pfister forms are round.*

PROOF. 0-fold Pfister forms are round. □

Corollary 6.3. *A bilinear Pfister form is either anisotropic or metabolic.*

PROOF. Suppose that $\mathfrak{c}$ is an isotropic bilinear Pfister form. We show that $\mathfrak{c}$ is metabolic by induction on the dimension of the $\mathfrak{c}$. We may assume that $\mathfrak{c} = \langle\langle a \rangle\rangle \otimes \mathfrak{b}$ for a Pfister form $\mathfrak{b}$. If $\mathfrak{b}$ is metabolic, then so is $\mathfrak{c}$. By the induction hypothesis, we may assume that $\mathfrak{b}$ is anisotropic. By Proposition 6.1 and Corollary 6.2, $a \in D(\mathfrak{b}) = G(\mathfrak{b})$. Therefore $a\mathfrak{b} \simeq \mathfrak{b}$ hence the form $\mathfrak{c} \simeq \mathfrak{b} \perp (-a\mathfrak{b}) \simeq \mathfrak{b} \perp (-\mathfrak{b})$ is metabolic. □

Remark 6.4. Note that the only metabolic 1-fold Pfister form is $\langle\langle 1 \rangle\rangle$. If $\operatorname{char} F \neq 2$, there is only one metabolic bilinear n-fold Pfister form for all $n \geq 1$, viz., the hyperbolic one. It is universal by Corollary 1.25. If $\operatorname{char} F = 2$, then there may exist many metabolic n-fold Pfister forms for $n > 1$.

Example 6.5. If $\operatorname{char} F = 2$, a bilinear Pfister form $\langle\langle a_1, \dots, a_n \rangle\rangle$ is anisotropic if and only if $a_1, \dots, a_n$ are 2-independent. Indeed, $[F^2(a_1, \dots, a_n) : F^2] < 2^n$ if and only if $\langle\langle a_1, \dots, a_n \rangle\rangle$ is isotropic.

Corollary 6.6. *Let* $\operatorname{char} F \neq 2$ *and* $z \in F^\times$. *Then* $2^n\langle\langle z\rangle\rangle = 0$ *in* $W(F)$ *if and only if* $z \in D(2^n\langle 1\rangle)$.

PROOF. If $z \in D(2^n\langle 1\rangle)$, then the Pfister form $2^n\langle\langle z\rangle\rangle$ is isotropic hence metabolic by Corollary 6.3.

Conversely, suppose that $2^n\langle\langle z\rangle\rangle$ is metabolic. Then $2^n\langle 1\rangle = 2^n\langle z\rangle$ in $W(F)$. If $2^n\langle 1\rangle$ is isotropic, it is universal as $\operatorname{char} F \neq 2$, so $z \in D(2^n\langle 1\rangle)$. If $2^n\langle 1\rangle$ is anisotropic, then $2^n\langle 1\rangle \simeq 2^n\langle z\rangle$ by Proposition 2.4, so $z \in G(2^n\langle 1\rangle) = D(2^n\langle 1\rangle)$ by Corollary 6.2. $\square$

As additional corollaries, we have the following two theorems of Pfister (cf. [**109**]). The first generalizes the well-known 2-, 4-, and 8-square theorems arising from quadratic extensions, quaternion algebras, and Cayley algebras.

Corollary 6.7. $D(2^n\langle 1\rangle)$ *is a group for every nonnegative integer* n.

The *level* of a field F is defined to be

$$s(F) := \min\{n \mid \text{the element } \; -1 \; \text{ is a sum of } n \text{ squares}\}$$

or infinity if no such integer exists.

Corollary 6.8. *The level* $s(F)$ *of a field* F, *if finite, is a power of two.*

PROOF. Suppose that $s(F)$ is finite. Then $2^n \leq s(F) < 2^{n+1}$ for some n. By Proposition 6.1(2), with $\mathfrak{b} = 2^n\langle 1\rangle$ and $a = -1$, we have $-1 \in D(\mathfrak{b})$. Hence $s(F) = 2^n$. $\square$

In [**109**], Pfister also showed that there exist fields of level 2^n for all $n \geq 0$. (Cf. Lemma 31.3 below.)

6.A. Chain p-equivalence of bilinear Pfister forms. Since the isometry classes of 2-fold Pfister forms are easy to deal with, we use them to study n-fold Pfister forms. We follow the development in [**32**] which we extend to all characteristics. The case of characteristic 2 was also independently done by Arason and Baeza in [**6**].

Definition 6.9. Let $a_1, \dots, a_n, b_1, \dots, b_n \in F^\times$ with $n \geq 1$. We say that the forms $\langle\langle a_1, \dots, a_n\rangle\rangle$ and $\langle\langle b_1, \dots, b_n\rangle\rangle$ are *simply p-equivalent* if $n = 1$ and $a_1F^{\times 2} = b_1F^{\times 2}$ or $n \geq 2$ and there exist $i, j \in [1,\ n]$ such that

$$\langle\langle a_i, a_j\rangle\rangle \simeq \langle\langle b_i, b_j\rangle\rangle \quad \text{with} \quad i \neq j \quad \text{and} \quad a_l = b_l \quad \text{for all} \quad l \neq i, j.$$

We say bilinear n-fold Pfister forms $\mathfrak{b}$ and $\mathfrak{c}$ are *chain p-equivalent* if there exist bilinear n-fold Pfister forms $\mathfrak{b}_0, \dots, \mathfrak{b}_m$ for some m such that $\mathfrak{b} = \mathfrak{b}_0$, $\mathfrak{c} = \mathfrak{b}_m$ and $\mathfrak{b}_i$ is simply p-equivalent to $\mathfrak{b}_{i+1}$ for each $i \in [0,\ m-1]$.

Chain p-equivalence is clearly an equivalence relation on the set of anisotropic bilinear forms of the type $\langle\langle a_1, \dots, a_n\rangle\rangle$ with $a_1, \dots, a_n \in F^\times$ and is denoted by $\approx$. As transpositions generate the symmetric group, we have $\langle\langle a_1, \dots, a_n\rangle\rangle \approx \langle\langle a_{\sigma(1)}, \dots, a_{\sigma(n)}\rangle\rangle$ for every permutation σ of $\{1, \dots, n\}$. We shall show the following result:

Theorem 6.10. *Let* $\langle\langle a_1, \dots, a_n\rangle\rangle$ *and* $\langle\langle b_1, \dots, b_n\rangle\rangle$ *be anisotropic. Then*

$$\langle\langle a_1, \dots, a_n\rangle\rangle \simeq \langle\langle b_1, \dots, b_n\rangle\rangle$$

if and only if

$$\langle\langle a_1, \dots, a_n\rangle\rangle \approx \langle\langle b_1, \dots, b_n\rangle\rangle.$$

Of course, we need only show isometric anisotropic bilinear Pfister forms are chain p-equivalent. We shall do this in a number of steps. If $\mathfrak{b}$ is an n-fold Pfister form, then we can write $\mathfrak{b} \simeq \mathfrak{b}' \perp \langle 1 \rangle$. If $\mathfrak{b}$ is anisotropic, then $\mathfrak{b}'$ is unique up to isometry and we call it the *pure subform* of $\mathfrak{b}$.

Lemma 6.11. *Suppose that $\mathfrak{b} = \langle\langle a_1, \dots, a_n \rangle\rangle$ is anisotropic. Let $-b \in D(\mathfrak{b}')$. Then there exist $b_2, \dots, b_n \in F^\times$ such that $\mathfrak{b} \approx \langle\langle b, b_2, \dots, b_n \rangle\rangle$.*

PROOF. We induct on n, the case $n = 1$ being trivial. Let $\mathfrak{c} = \langle\langle a_1, \dots, a_{n-1} \rangle\rangle$ so $\mathfrak{b}' \simeq \mathfrak{c}' \perp -a_n\mathfrak{c}$ by Witt Cancellation 1.28. Write

$$-b = -x + a_n y \quad \text{with} \ -x \in \widetilde{D}(\mathfrak{c}'), \quad -y \in \widetilde{D}(\mathfrak{c}).$$

If $y = 0$, then $x \neq 0$ and we finish by induction, so we may assume that $0 \neq y = y_1 + z^2$ with $-y_1 \in \widetilde{D}(\mathfrak{c}')$ and $z \in F$. If $y_1 \neq 0$, then $\mathfrak{c} \approx \langle\langle y_1, \dots, y_{n-1} \rangle\rangle$ for some $y_i \in F^\times$ and, using Lemma 4.15, we get

$$(6.12) \quad \mathfrak{c} \approx \langle\langle y_1, \dots, y_{n-1}, a_n \rangle\rangle \approx \langle\langle y_1, \dots, y_{n-1}, -a_n y \rangle\rangle \approx \langle\langle a_1, \dots, a_{n-1}, -a_n y \rangle\rangle.$$

This is also true if $y_1 = 0$. If $x = 0$, we are done. If not $\mathfrak{c} \approx \langle\langle x, x_2, \dots, x_{n-1} \rangle\rangle$ for some $x_i \in F^\times$ and

$$\begin{aligned} \mathfrak{b} &\approx \langle\langle x, x_2, \dots, x_{n-1}, -a_n y \rangle\rangle \approx \langle\langle a_n x y, x_2, \dots, x_{n-1}, -a_n y + x \rangle\rangle \\ &\approx \langle\langle a_n x y, x_2, \dots, x_{n-1}, b \rangle\rangle \end{aligned}$$

by Lemma 4.15(2) as needed. □

The argument to establish equation (6.12) yields:

Corollary 6.13. *Let $\mathfrak{b} = \langle\langle x_1, \dots, x_n \rangle\rangle$ and $y \in D(\mathfrak{b})$. Let $z \in F^\times$. If $\mathfrak{b} \otimes \langle\langle z \rangle\rangle$ is anisotropic, then $\langle\langle x_1, \dots, x_n, z \rangle\rangle \approx \langle\langle x_1, \dots, x_n, yz \rangle\rangle$.*

We also have the following generalization of Lemma 4.14:

Corollary 6.14. *Let $\mathfrak{b}$ be an anisotropic bilinear Pfister form over F and let $a \in F^\times$. Then $\langle\langle a \rangle\rangle \cdot \mathfrak{b} = 0$ in $W(F)$ if and only if either $a \in F^{\times 2}$ or $\mathfrak{b} \simeq \langle\langle b \rangle\rangle \otimes \mathfrak{c}$ for some $b \in D(\langle\langle a \rangle\rangle)$ and bilinear Pfister form $\mathfrak{c}$. In the latter case, $\langle\langle a, b \rangle\rangle$ is metabolic.*

PROOF. Clearly $\langle\langle a, b \rangle\rangle = 0$ in $W(F)$ if $b \in D(\langle\langle a \rangle\rangle)$. Conversely, suppose that $\langle\langle a \rangle\rangle \otimes \mathfrak{b} = 0$. Hence $a \in G(\mathfrak{b}) = D(\mathfrak{b})$ by Corollary 6.2. Write $a = x^2 - b$ for some $x \in F$ and $-b \in \widetilde{D}(\mathfrak{b}')$. If $b = 0$, then $a \in F^{\times 2}$. Otherwise, $b \in D(\langle\langle a \rangle\rangle)$ and $\mathfrak{b} \simeq \langle\langle b \rangle\rangle \otimes \mathfrak{c}$ for some bilinear Pfister form $\mathfrak{c}$ by Lemma 6.11. □

The following generalization of Lemma 6.11 is very useful in computation and is the key to proving further relations among Pfister forms.

Proposition 6.15. *Let $\mathfrak{b} = \langle\langle a_1, \dots, a_m \rangle\rangle$ and $\mathfrak{c} = \langle\langle b_1, \dots, b_n \rangle\rangle$ be such that $\mathfrak{b} \otimes \mathfrak{c}$ is anisotropic. Let $-c \in D(\mathfrak{b} \otimes \mathfrak{c}')$, then*

$$\langle\langle a_1, \dots, a_m, b_1, \dots, b_n \rangle\rangle \approx \langle\langle a_1, \dots, a_m, c_1, c_2, \dots, c_{n-1}, c \rangle\rangle$$

for some $c_1, \dots, c_{n-1} \in F^\times$.

PROOF. We induct on n. If $n = 1$, then $-c = y b_1$ for some $-y \in D(\mathfrak{b})$ and this case follows by Corollary 6.13, so we assume that $n > 1$. Let $\mathfrak{d} = \langle\langle b_1, \dots, b_{n-1} \rangle\rangle$.

Then $\mathfrak{c}' \simeq b_n\mathfrak{d} \perp \mathfrak{d}'$ so $\mathfrak{b}\mathfrak{c}' \simeq b_n\mathfrak{b} \otimes \mathfrak{d} \perp \mathfrak{b} \otimes \mathfrak{d}'$. Write $0 \neq -c = b_n y - z$ with $-y \in \widetilde{D}(\mathfrak{b} \otimes \mathfrak{c})$ and $-z \in \widetilde{D}(\mathfrak{b} \otimes \mathfrak{c}')$. If $z = 0$, then $x \neq 0$ and

$$\langle\langle a_1, \ldots, a_m, b_1, \ldots, b_n\rangle\rangle \approx \langle\langle a_1, \ldots, a_m, b_1, \ldots, b_{n-1}, -yb_n\rangle\rangle$$

by Corollary 6.13 and we are done. So we may assume that $z \neq 0$. By induction

$$\langle\langle a_1, \ldots, a_m, b_1, \ldots, b_{n-1}\rangle\rangle \approx \langle\langle a_1, \ldots, a_m, c_1, c_2, \ldots, c_{n-2}, z\rangle\rangle$$

for some $c_1, \ldots, c_{n-2} \in F^\times$. If $y = 0$, tensoring this by $\langle 1, -b_n\rangle$ completes the proof, so we may assume that $y \neq 0$. Then

$$\begin{aligned}
&\langle\langle a_1, \ldots, a_m, b_1, \ldots, b_n\rangle\rangle \approx \langle\langle a_1, \ldots, a_m, b_1, \ldots, b_{n-1}, -yb_n\rangle\rangle \\
&\approx \langle\langle a_1, \ldots, a_m, c_1, \ldots, c_{n-2}, z, -yb_n\rangle\rangle \approx \langle\langle a_1, \ldots, a_m, c_1, \ldots, c_{n-2}, z - yb_n, zyb_n\rangle\rangle \\
&\approx \langle\langle a_1, \ldots, a_m, c_1, \ldots, c_{n-2}, c, zyb_n\rangle\rangle
\end{aligned}$$

by Lemma 4.15(2). This completes the proof. □

Corollary 6.16 (Common Slot Property). *Let*

$$\langle\langle a_1, \ldots, a_{n-1}, x\rangle\rangle \text{ and } \langle\langle b_1, \ldots, b_{n-1}, y\rangle\rangle$$

be isometric anisotropic bilinear forms. Then there exists a $z \in F^\times$ satisfying

$$\langle\langle a_1, \ldots, a_{n-1}, z\rangle\rangle = \langle\langle a_1, \ldots, a_{n-1}, x\rangle\rangle \text{ and } \langle\langle b_1, \ldots, b_{n-1}, z\rangle\rangle = \langle\langle b_1, \ldots, b_{n-1}, y\rangle\rangle.$$

PROOF. Let $\mathfrak{b} = \langle\langle a_1, \ldots, a_{n-1}\rangle\rangle$ and $\mathfrak{c} = \langle\langle b_1, \ldots, b_{n-1}\rangle\rangle$. As $x\mathfrak{b} - y\mathfrak{c} = \mathfrak{b}' - \mathfrak{c}'$ in $W(F)$, the form $x\mathfrak{b} \perp -y\mathfrak{c}$ is isotropic. Hence there exists a $z \in D(x\mathfrak{b}) \cap D(y\mathfrak{c})$. The result follows by Proposition 6.15. □

A nondegenerate symmetric bilinear form $\mathfrak{b}$ is called a *general bilinear n-fold Pfister form* if $\mathfrak{b} \simeq a\mathfrak{c}$ for some $a \in F^\times$ and bilinear n-fold Pfister form $\mathfrak{c}$. As Pfister forms are round, a general Pfister form is a Pfister form if and only if it represents one.

Corollary 6.17. *Let $\mathfrak{c}$ and $\mathfrak{b}$ be general anisotropic bilinear Pfister forms. If $\mathfrak{c}$ is a subform of $\mathfrak{b}$, then $\mathfrak{b} \simeq \mathfrak{c} \otimes \mathfrak{d}$ for some bilinear Pfister form $\mathfrak{d}$.*

PROOF. If $\mathfrak{c} = c\mathfrak{c}_1$ for some Pfister form $\mathfrak{c}_1$ and $c \in F^\times$, then $\mathfrak{c}_1$ is a subform of $c\mathfrak{b}$. In particular, $c\mathfrak{b}$ represents one, so it is a Pfister form. Replacing $\mathfrak{b}$ by $c\mathfrak{b}$ and $\mathfrak{c}$ by $c\mathfrak{c}$, we may assume both are Pfister forms.

Let $\mathfrak{c} \simeq \langle\langle a_1, \ldots, a_n\rangle\rangle$ with $a_i \in F^\times$. By Witt Cancellation 1.28, we have $\mathfrak{c}'$ is isometric to a subform of $\mathfrak{b}'$, hence $\mathfrak{b} \simeq \langle\langle a_1\rangle\rangle \otimes \mathfrak{d}_1$ for some Pfister form $\mathfrak{d}_1$ by Lemma 6.11. By induction, there exists a Pfister form $\mathfrak{d}_k$ satisfying $\mathfrak{b} \simeq \langle\langle a_1, \ldots, a_k\rangle\rangle \otimes \mathfrak{d}_k$. By Witt Cancellation 1.28, we have $\langle\langle a_1, \ldots, a_k\rangle\rangle \otimes \langle\langle a_{k+1}, \ldots, a_n\rangle\rangle'$ is a subform of $\langle\langle a_1, \ldots, a_k\rangle\rangle \otimes \mathfrak{d}_k'$ so $-a_{k+1} \in D\big(\langle\langle a_1, \ldots, a_k\rangle\rangle \otimes \mathfrak{d}_k'\big)$. By Proposition 6.15, we complete the induction step. □

Let $\mathfrak{b}$ and $\mathfrak{c}$ be general Pfister forms. We say that $\mathfrak{c}$ *divides* $\mathfrak{b}$ if $\mathfrak{b} \simeq \mathfrak{c} \otimes \mathfrak{d}$ for some Pfister form $\mathfrak{d}$. The corollary says that $\mathfrak{c}$ divides $\mathfrak{b}$ if and only if it is isometric to a subform of $\mathfrak{b}$.

We now prove Theorem 6.10.

PROOF. Let $\mathfrak{a} = \langle\langle a_1, \ldots, a_n\rangle\rangle$ and $\mathfrak{b} = \langle\langle b_1, \ldots, b_n\rangle\rangle$ be isometric over F. Clearly we may assume that $n > 1$. By Lemma 6.11, we have $\mathfrak{a} \approx \langle\langle b_1, a_2', \ldots, a_n'\rangle\rangle$

for some $a'_i \in F^\times$. Suppose that we have shown $\mathfrak{a} \approx \langle\langle b_1, \dots, b_m, a'_{m+1}, \dots, a'_n \rangle\rangle$ for some m. By Witt Cancellation 1.28,

$$\langle\langle b_1, \dots, b_m \rangle\rangle \otimes \langle\langle b_{m+1}, \dots, b_n \rangle\rangle' \simeq \langle\langle b_1, \dots, b_m \rangle\rangle \otimes \langle\langle a'_{m+1}, \dots, a'_n \rangle\rangle',$$

so $-b_{m+1} \in D\big(\langle\langle b_1, \dots, b_m \rangle\rangle \otimes \langle\langle a'_{m+1}, \dots, a'_n \rangle\rangle'\big)$. By Proposition 6.15, we have

$$\mathfrak{a} \approx \langle\langle b_1, \dots, b_{m+1}, a''_{m+2}, \dots, a''_n \rangle\rangle$$

for some $a''_i \in F^\times$. This completes the induction step. □

We need the following theorem of Arason and Pfister (cf. [**10**]):

Theorem 6.18 (Hauptsatz). *Let $0 \neq \mathfrak{b}$ be an anisotropic form lying in $I^n(F)$. Then* $\dim \mathfrak{b} \geq 2^n$.

We shall prove this theorem in Theorem 23.7 below. Using it we show:

Corollary 6.19. *Let $\mathfrak{b}$ and $\mathfrak{c}$ be two anisotropic general bilinear n-fold Pfister forms. If $\mathfrak{b} \equiv \mathfrak{c} \bmod I^{n+1}(F)$, then $\mathfrak{b} \simeq a\mathfrak{c}$ for some $a \in F^\times$. In addition, if $D(\mathfrak{b}) \cap D(\mathfrak{c}) \neq \emptyset$, then $\mathfrak{b} \simeq \mathfrak{c}$.*

Proof. Choose $a \in F^\times$ such that $\mathfrak{b} \perp -a\mathfrak{c}$ is isotropic. By the Hauptsatz, this form must be metabolic. By Proposition 2.4, we have $\mathfrak{b} \simeq a\mathfrak{c}$.

Suppose that $x \in D(\mathfrak{b}) \cap D(\mathfrak{c})$. Then $\mathfrak{b} \perp -\mathfrak{c}$ is isotropic and one can take $a = 1$. □

Theorem 6.20. *Let $a_1, \dots, a_n, b_1, \dots, b_n \in F^\times$. The following are equivalent:*

(1) $\langle\langle a_1, \dots, a_n \rangle\rangle = \langle\langle b_1, \dots, b_n \rangle\rangle$ in $W(F)$.
(2) $\langle\langle a_1, \dots, a_n \rangle\rangle \equiv \langle\langle b_1, \dots, b_n \rangle\rangle \bmod I^{n+1}(F)$.
(3) $\{a_1, \dots, a_n\} = \{b_1, \dots, b_n\}$ in $K_n(F)/2K_n(F)$.

Proof. Let $\mathfrak{b} = \langle\langle a_1, \dots, a_n \rangle\rangle$ and $\mathfrak{c} = \langle\langle b_1, \dots, b_n \rangle\rangle$. As metabolic Pfister forms are trivial in $W(F)$ and any bilinear n-fold Pfister form lying in $I^{n+1}(F)$ must be metabolic by the Hauptsatz 6.18, we may assume that $\mathfrak{b}$ and $\mathfrak{c}$ are both anisotropic.

(2) ⇒ (1) follows from Corollary 6.19.

(1) ⇒ (3): By Theorem 6.10, we have $\langle\langle a_1, \dots, a_n \rangle\rangle \approx \langle\langle b_1, \dots, b_n \rangle\rangle$, so it suffices to show that (3) holds if

$$\langle\langle a_i, a_j \rangle\rangle \simeq \langle\langle b_i, b_j \rangle\rangle \text{ with } i \neq j \quad \text{and} \quad a_l = b_l \text{ for all } l \neq i, j.$$

As $\{a_i, a_j\} = \{b_i, b_j\}$ by Proposition 5.3, statement (3) follows.

(3) ⇒ (2) follows from (5.1). □

Arason proved and used the Common Slot Property 6.16 in [**4**] to give an independent proof in the case of characteristic different from 2 of Theorem 6.20 which was first proven in [**32**].

6.B. Linkage of bilinear Pfister forms. We derive some further properties of bilinear Pfister forms that we shall need later.

Proposition 6.21. *Let $\mathfrak{b}_1$ and $\mathfrak{b}_2$ be two anisotropic general bilinear Pfister forms. Let $\mathfrak{c}$ be a general r-fold Pfister form with $r \geq 0$ that is isometric to a subform of $\mathfrak{b}_1$ and to a subform of $\mathfrak{b}_2$. If $\mathfrak{i}(\mathfrak{b}_1 \perp -\mathfrak{b}_2) > 2^r$, then there exists a k-fold Pfister form $\mathfrak{d}$ with $k > 0$ such that $\mathfrak{c} \otimes \mathfrak{d}$ is isometric to a subform of $\mathfrak{b}_1$ and to a subform of $\mathfrak{b}_2$. Furthermore, $\mathfrak{i}(\mathfrak{b}_1 \perp -\mathfrak{b}_2) = 2^{r+k}$.*

PROOF. By Corollary 6.17, there exist Pfister forms $\mathfrak{d}_1$ and $\mathfrak{d}_2$ such that $\mathfrak{b}_1 \simeq \mathfrak{c} \otimes \mathfrak{d}_1$ and $\mathfrak{b}_2 \simeq \mathfrak{c} \otimes \mathfrak{d}_2$. Let $\mathfrak{b} = \mathfrak{b}_1 \perp -\mathfrak{b}_2$. As $\mathfrak{b}$ is isotropic, $\mathfrak{b}_1$ and $\mathfrak{b}_2$ have a common nonzero value. Dividing the $\mathfrak{b}_i$ by this nonzero common value, we may assume that the $\mathfrak{b}_i$ are Pfister forms. We have

$$\mathfrak{b} \simeq \mathfrak{c} \otimes (\mathfrak{d}_1' \perp -\mathfrak{d}_2') \perp (\mathfrak{c} \perp -\mathfrak{c}).$$

The form $\mathfrak{c} \perp -\mathfrak{c}$ is metabolic by Example 1.22(2) and $\mathfrak{i}(\mathfrak{b}) > \dim \mathfrak{c}$. Therefore, the form $\mathfrak{c} \otimes (\mathfrak{d}_1' \perp -\mathfrak{d}_2')$ is isotropic, hence there is $a \in D(\mathfrak{c} \otimes \mathfrak{d}_1') \cap D(\mathfrak{c} \otimes \mathfrak{d}_2')$. By Proposition 6.15, we have $\mathfrak{b}_1 \simeq \mathfrak{c} \otimes \langle\langle -a \rangle\rangle \otimes \mathfrak{e}_1$ and $\mathfrak{b}_2 \simeq \mathfrak{c} \otimes \langle\langle -a \rangle\rangle \otimes \mathfrak{e}_2$ for some bilinear Pfister forms $\mathfrak{e}_1$ and $\mathfrak{e}_2$. As

$$\mathfrak{b} \simeq \mathfrak{c} \otimes (\mathfrak{e}_1' \perp -\mathfrak{e}_2') \perp \big(\mathfrak{c} \otimes \langle\langle -a \rangle\rangle \perp -\mathfrak{c} \otimes \langle\langle -a \rangle\rangle\big),$$

either $\mathfrak{i}(\mathfrak{b}) = 2^{r+1}$ or we may repeat the argument. The result follows. □

If a general bilinear r-fold Pfister form $\mathfrak{c}$ is isometric to a common subform of two general Pfister forms $\mathfrak{b}_1$ and $\mathfrak{b}_2$, we call it a *linkage* of $\mathfrak{b}_1$ and $\mathfrak{b}_2$ and say that $\mathfrak{b}_1$ and $\mathfrak{b}_2$ are *r-linked.* The integer $m = \max\{r \mid \mathfrak{b}_1 \text{ and } \mathfrak{b}_2 \text{ are } r\text{-linked}\}$ is called the *linkage number* of $\mathfrak{b}_1$ and $\mathfrak{b}_2$. The proposition says that $\mathfrak{i}(\mathfrak{b}_1 \perp -\mathfrak{b}_2) = 2^m$. If $\mathfrak{b}_1$ and $\mathfrak{b}_2$ are n-fold Pfister forms and $r = n-1$, we say that $\mathfrak{b}_1$ and $\mathfrak{b}_2$ are *linked.* By Corollary 6.17 the linkage of any pair of bilinear Pfister forms is a divisor of each. The theory of linkage was first developed in [**32**].

If $\mathfrak{b}$ is a nondegenerate symmetric bilinear form over F, then the *annihilator* of $\mathfrak{b}$ in $W(F)$,

$$\operatorname{ann}_{W(F)}(\mathfrak{b}) := \{\mathfrak{c} \in W(F) \mid \mathfrak{b} \cdot \mathfrak{c} = 0\}$$

is an ideal in $W(F)$. When $\mathfrak{b}$ is a Pfister form this ideal has a nice structure that we now establish. First note that if $\mathfrak{b}$ is an anisotropic Pfister form and $x \in D(\mathfrak{b})$, then, as $\mathfrak{b}$ is round by Corollary 6.2, we have $\langle\langle x \rangle\rangle \otimes \mathfrak{b} \simeq \mathfrak{b} \perp -x\mathfrak{b} \simeq \mathfrak{b} \perp -\mathfrak{b}$ is metabolic. It follows that $\langle\langle x \rangle\rangle \in \operatorname{ann}_{W(F)}(\mathfrak{b})$. We shall show that these binary forms generate $\operatorname{ann}_{W(F)}(\mathfrak{b})$. This will follow from the next result, also known as the Pfister-Witt Theorem.

Proposition 6.22. *Let $\mathfrak{b}$ be an anisotropic bilinear Pfister form and $\mathfrak{c}$ a nondegenerate symmetric bilinear form. Then there exists a symmetric bilinear form $\mathfrak{d}$ satisfying all of the following:*

(1) $\mathfrak{b} \cdot \mathfrak{c} = \mathfrak{b} \cdot \mathfrak{d}$ *in* $W(F)$.
(2) $\mathfrak{b} \otimes \mathfrak{d}$ *is anisotropic. Moreover,* $\dim \mathfrak{d} \leq \dim \mathfrak{c}$ *and* $\dim \mathfrak{d} \equiv \dim \mathfrak{c} \mod 2$.
(3) $\mathfrak{c} - \mathfrak{d}$ *lies in the subgroup of* $W(F)$ *generated by* $\langle\langle x \rangle\rangle$ *with* $x \in D(\mathfrak{b})$.

PROOF. We prove this by induction on $\dim \mathfrak{c}$. By the Witt Decomposition Theorem 1.27, we may assume that $\mathfrak{c}$ is anisotropic. Hence $\mathfrak{c}$ is diagonalizable by Corollary 1.19, say $\mathfrak{c} = \langle x_1, \dots, x_n \rangle$ with $x_i \in F^\times$. If $\mathfrak{b} \otimes \mathfrak{c}$ is anisotropic, the result is trivial, so assume it is isotropic. Therefore, there exist $a_1, \dots, a_n \in \widetilde{D}(\mathfrak{b})$ not all zero such that $a_1x_1 + \cdots + a_nx_n = 0$. Let $b_i = a_i$ if $a_i \neq 0$ and $b_i = 1$ otherwise. In particular, $b_i \in G(\mathfrak{b})$ for all i. Let $\mathfrak{e} = \langle b_1x_1, \dots, b_nx_n \rangle$. Then $\mathfrak{c} - \mathfrak{e} = x_1\langle\langle b_1 \rangle\rangle + \cdots + x_n\langle\langle b_n \rangle\rangle$ with each $b_i \in D(\mathfrak{b})$ as $\mathfrak{b}$ is round by Corollary 6.2. Since $\mathfrak{e}$ is isotropic, we have $\mathfrak{b} \cdot \mathfrak{c} = \mathfrak{b} \cdot (\mathfrak{e})_{an}$ in $W(F)$. As $\dim(\mathfrak{e})_{an} < \dim \mathfrak{c}$, by the induction hypothesis there exists $\mathfrak{d}$ such that $\mathfrak{b} \otimes \mathfrak{d}$ is anisotropic and $\mathfrak{e} - \mathfrak{d}$ and therefore $\mathfrak{c} - \mathfrak{d}$ lies in the subgroup of $W(F)$ generated by $\langle\langle x \rangle\rangle$ with $x \in D(\mathfrak{b})$. As $\mathfrak{b} \otimes \mathfrak{d}$ is anisotropic, it follows by (1) that $\dim \mathfrak{d} \leq \dim \mathfrak{c}$. It follows from (3) that the dimension of $\mathfrak{c} - \mathfrak{d}$ is even. □

Corollary 6.23. *Let $\mathfrak{b}$ be an anisotropic bilinear Pfister form. Then* $\operatorname{ann}_{W(F)}(\mathfrak{b})$ *is generated by* $\langle\langle x\rangle\rangle$ *with* $x \in D(\mathfrak{b})$.

If $\mathfrak{b}$ is 2-dimensional, we obtain stronger results first established in [**34**].

Lemma 6.24. *Let $\mathfrak{b}$ be a binary anisotropic bilinear form over F and $\mathfrak{c}$ an anisotropic bilinear form over F such that $\mathfrak{b} \otimes \mathfrak{c}$ is isotropic. Then $\mathfrak{c} \simeq \mathfrak{d} \perp \mathfrak{e}$ for some symmetric binary bilinear form $\mathfrak{d}$ annihilated by $\mathfrak{b}$ and symmetric bilinear form $\mathfrak{e}$ over F.*

PROOF. Let $\{e, f\}$ be a basis for $V_{\mathfrak{b}}$. By assumption there exist vectors $v, w \in V_{\mathfrak{c}}$ such that $e \otimes v + f \otimes w$ is an isotropic vector for $\mathfrak{b} \otimes \mathfrak{c}$. Choose a 2-dimensional subspace $W \subset V_{\mathfrak{c}}$ containing v and w. Since $\mathfrak{c}$ is anisotropic, so is $\mathfrak{c}|_W$. In particular, $\mathfrak{c}|_W$ is nondegenerate, hence $\mathfrak{c} = \mathfrak{c}|_W \perp \mathfrak{c}|_{W^\perp}$ by Proposition 1.6. As $\mathfrak{b} \otimes (\mathfrak{c}|_W)$ is an isotropic general 2-fold Pfister form, it is metabolic by Corollary 6.3. □

Proposition 6.25. *Let $\mathfrak{b}$ be a binary anisotropic bilinear form over F and $\mathfrak{c}$ an anisotropic form over F. Then there exist symmetric bilinear forms $\mathfrak{c}_1$ and $\mathfrak{c}_2$ over F such that $\mathfrak{c} \simeq \mathfrak{c}_1 \perp \mathfrak{c}_2$ with $\mathfrak{b} \otimes \mathfrak{c}_2$ anisotropic and $\mathfrak{c}_1 \simeq \mathfrak{d}_1 \perp \cdots \perp \mathfrak{d}_n$ where each $\mathfrak{d}_i$ is a binary bilinear form annihilated by $\mathfrak{b}$. In particular, if $\det \mathfrak{d}_i = -d_i F^{\times 2}$, then $d_i \in D(\mathfrak{b})$ for each i.*

PROOF. The first statement of the proposition follows from the lemma and the second from its proof. □

Corollary 6.26. *Let $\mathfrak{b}$ be a binary anisotropic bilinear form over F and $\mathfrak{c}$ an anisotropic form over F annihilated by $\mathfrak{b}$. Then $\mathfrak{c} \simeq \mathfrak{d}_1 \perp \cdots \perp \mathfrak{d}_n$ for some symmetric binary forms $\mathfrak{d}_i$ annihilated by $\mathfrak{b}$ for $i \in [1,\ n]$.*

CHAPTER II

Quadratic Forms

7. Foundations

In this section, we introduce the basic properties of quadratic forms over an arbitrary field F. Their study arose from the investigation of homogeneous polynomials of degree two. If the characteristic of F is different from 2, then this study and that of symmetric bilinear forms are essentially the same as the diagonal of a symmetric bilinear form is a quadratic form and each determines the other by the polar identity. However, they are different when the characteristic of F is 2. In general, quadratic forms unlike bilinear forms have a rich geometric flavor. When studying symmetric bilinear forms, we saw that one could easily reduce to the study of nondegenerate forms. For quadratic forms, the situation is more complex. The polar form of a quadratic form no longer determines the quadratic form when the underlying field is of characteristic 2. However, the radical of the polar form is invariant under field extension. This leads to two types of quadratic forms. One is the study of totally singular quadratic forms, i.e., those whose polar bilinear form is zero. Such quadratic forms need not be trivial in the case of characteristic 2. The other extreme is when the radical of the polar form is as small as possible (which means of dimension zero or one), this gives rise to nondegenerate quadratic forms. As in the study of bilinear forms, certain properties are not invariant under base extension. The most important of these is anisotropy. Analogous to the bilinear case, an anisotropic quadratic form is one having no nontrivial zero, i.e., no isotropic vectors. Every vector that is isotropic for the quadratic form is isotropic for its polar form. If the characteristic is 2, the converse is false as every vector is an isotropic vector of the polar form. As in the previous chapter, we shall base this study on a coordinate free approach and strive to give uniform proofs in a characteristic free fashion.

Definition 7.1. Let V be a finite dimensional vector space over F. A *quadratic form* on V is a map $\varphi : V \to F$ satisfying:

(1) $\varphi(av) = a^2\varphi(v)$ for all $v \in V$ and $a \in F$.

(2) (Polar Identity) $\mathfrak{b}_\varphi : V \times V \to F$ defined by
$$\mathfrak{b}_\varphi(v, w) = \varphi(v + w) - \varphi(v) - \varphi(w)$$
is a bilinear form.

The bilinear form $\mathfrak{b}_\varphi$ is called the *polar form of* of φ. We call $\dim V$ the *dimension* of the quadratic form and also write it as $\dim \varphi$. We write φ is a quadratic form over F if φ is a quadratic form on a finite dimensional vector space over F and denote the underlying space by V_φ.

Note that the polar form of a quadratic form is automatically symmetric and even alternating if $\operatorname{char} F = 2$. If $\mathfrak{b} : V \times V \to F$ is a bilinear form (not necessarily symmetric), let $\varphi_{\mathfrak{b}} : V \to F$ be defined by $\varphi_{\mathfrak{b}}(v) = \mathfrak{b}(v,v)$ for all $v \in V$. Then $\varphi_{\mathfrak{b}}$ is a quadratic form and its polar form $\mathfrak{b}_{\varphi_{\mathfrak{b}}}$ is $\mathfrak{b} + \mathfrak{b}^t$. We call $\varphi_{\mathfrak{b}}$ the *associated quadratic form* of $\mathfrak{b}$.

In particular, if $\mathfrak{b}$ is symmetric, the composition $\mathfrak{b} \mapsto \varphi_{\mathfrak{b}} \mapsto \mathfrak{b}_{\varphi_{\mathfrak{b}}}$ is multiplication by 2 as is the composition $\varphi \mapsto \mathfrak{b}_{\varphi} \mapsto \varphi_{\mathfrak{b}_{\varphi}}$.

Let φ and ψ be two quadratic forms. An *isometry* $f : \varphi \to \psi$ is a linear map $f : V_\varphi \to V_\psi$ such that $\varphi(v) = \psi\big(f(v)\big)$ for all $v \in V_\varphi$. If such an isometry exists, we write $\varphi \simeq \psi$ and say that φ and ψ are *isometric.*

Example 7.2. If φ is a quadratic form over F and $v \in V$ satisfies $\varphi(v) \neq 0$, then the (*hyperplane*) *reflection*

$$\tau_v : \varphi \to \varphi \quad \text{given by} \quad w \mapsto w - \mathfrak{b}_\varphi(v,w)\varphi(v)^{-1}v$$

is an isometry.

Let V be a finite dimensional vector space over F. Define the *hyperbolic form* of V to be the form $\mathbb{H}(V) = \varphi_{\mathbb{H}}$ on $V \oplus V^*$ defined by

$$\varphi_{\mathbb{H}}(v,f) := f(v)$$

for all $v \in V$ and $f \in V^*$. Note that the polar form of $\varphi_{\mathbb{H}}$ is $\mathfrak{b}_{\varphi_{\mathbb{H}}} = \mathbb{H}_1(V)$. If φ is a quadratic form isometric to $\mathbb{H}(W)$ for some vector space W, we call φ a *hyperbolic form.* The form $\mathbb{H}(F)$ is called the *hyperbolic plane* and we denote it simply by $\mathbb{H}$. If $\varphi \simeq \mathbb{H}$, two vectors $e, f \in V_\varphi$ satisfying $\varphi(e) = \varphi(f) = 0$ and $\mathfrak{b}_\varphi(e,f) = 1$ are called a *hyperbolic pair.*

Let φ be a quadratic form on V and let $\{v_1, \dots, v_n\}$ be a basis for V. Let $a_{ii} = \varphi(v_i)$ for all i and

$$a_{ij} = \begin{cases} \mathfrak{b}_\varphi(v_i, v_j) & \text{for all } i < j, \\ 0 & \text{for all } i > j. \end{cases}$$

As

$$\varphi(\sum_{i=1}^n x_i v_i) = \sum_{i,j} a_{ij} x_i x_j,$$

the homogeneous polynomial on the right hand side as well as the matrix (a_{ij}) determined by φ completely determines φ.

Notation 7.3. (1) Let $a \in F$. The quadratic form on F given by $\varphi(v) = av^2$ for all $v \in F$ will be denoted by $\langle a \rangle_q$ or simply $\langle a \rangle$.

(2) Let $a, b \in F$. The 2-dimensional quadratic form on F^2 given by $\varphi(x,y) = ax^2 + xy + by^2$ will be denoted by $[a,b]$. The corresponding matrix for φ in the standard basis is

$$A = \begin{pmatrix} a & 1 \\ 0 & b \end{pmatrix},$$

while the corresponding matrix for $\mathfrak{b}_\varphi$ is

$$\begin{pmatrix} 2a & 1 \\ 1 & 2b \end{pmatrix} = A + A^t.$$

Remark 7.4. Let φ be a quadratic form on V over F. Then the associated polar form $\mathfrak{b}_\varphi$ is not the zero form if and only if there are two vectors v, w in V satisfying $b(v,w) = 1$. In particular, if φ is a nonzero binary form, then $\varphi \simeq [a,b]$ for some $a, b \in F$.

Example 7.5. Let $\{e, f\}$ be a hyperbolic pair for $\mathbb{H}$. Using the basis $\{e, ae + f\}$, we have $\mathbb{H} \simeq [0,0] \simeq [0,a]$ for any $a \in F$.

Example 7.6. Let $\operatorname{char} F = 2$ and $\wp : F \to F$ be the Artin-Schreier map $\wp(x) = x^2 + x$. Let $a \in F$. Then the quadratic form $[1, a]$ is isotropic if and only if $a \in \wp(F)$.

Let V be a finite dimension vector space over F. The set Quad(V) of quadratic forms on V is a vector space over F. We have linear maps

$$\mathrm{Bil}(V) \to \mathrm{Quad}(V) \quad \text{given by} \quad \mathfrak{b} \mapsto \varphi_{\mathfrak{b}}$$

and

$$\mathrm{Quad}(V) \to \mathrm{Sym}(V) \quad \text{given by} \quad \varphi \mapsto \mathfrak{b}_\varphi.$$

Restricting the first map to $\mathrm{Sym}(V)$ and composing shows the compositions

$$\mathrm{Sym}(V) \to \mathrm{Quad}(V) \to \mathrm{Sym}(V) \quad \text{and} \quad \mathrm{Quad}(V) \to \mathrm{Sym}(V) \to \mathrm{Quad}(V)$$

are multiplication by 2. In particular, if $\operatorname{char} F \neq 2$ the map $\mathrm{Quad}(V) \to \mathrm{Sym}(V)$ given by $\varphi \mapsto \frac{1}{2}\mathfrak{b}_\varphi$ is an isomorphism inverse to the map $\mathrm{Sym}(V) \to \mathrm{Quad}(V)$ by $\mathfrak{b} \mapsto \varphi_{\mathfrak{b}}$. For this reason, we shall usually identify quadratic forms and symmetric bilinear forms over a field of characteristic different from 2.

The correspondence between quadratic forms on a vector space V of dimension n and matrices defines a linear isomorphism $\mathrm{Quad}(V) \to \mathbb{T}_n(F)$, where $\mathbb{T}_n(F)$ is the vector space of $n \times n$ upper-triangular matrices. Therefore by the surjectivity of the linear epimorphism $\mathbb{M}_n(F) \to \mathbb{T}_n(F)$ given by $(a_{ij}) \mapsto (b_{ij})$ with $b_{ij} = a_{ij} + a_{ji}$ for all $i < j$, and $b_{ii} = a_{ii}$ for all i, and $b_{ij} = 0$ for all $j < i$ implies that the linear map $\mathrm{Bil}(V) \to \mathrm{Quad}(V)$ given by $\mathfrak{b} \mapsto \varphi_{\mathfrak{b}}$ is also surjective. We, therefore, have an exact sequence

$$0 \to \mathrm{Alt}(V) \to \mathrm{Bil}(V) \to \mathrm{Quad}(V) \to 0.$$

Exercise 7.7. The natural exact sequence

$$0 \to \Lambda^2(V^*) \to V^* \otimes_F V^* \to S^2(V^*) \to 0$$

can be identified with the sequence above via the isomorphism

$$S^2(V^*) \to \mathrm{Quad}(V) \quad \text{given by} \quad f \cdot g \mapsto \varphi_{f \cdot g} : v \mapsto f(v)g(v).$$

If $\varphi, \psi \in$ Quad(V), we say φ is *similar* to ψ if there exists an $a \in F^\times$ such that $\varphi \simeq a\psi$.

Let φ be a quadratic form on V. A vector $v \in V$ is called *anisotropic* if $\varphi(v) \neq 0$ and *isotropic* if $v \neq 0$ and $\varphi(v) = 0$. We call φ *anisotropic* if there are no isotropic vectors in V and *isotropic* if there are.

If $W \subset V$ is a subspace, the restriction of φ on W is the quadratic form whose polar form is given by $\mathfrak{b}_{\varphi|_W} = \mathfrak{b}_\varphi|_W$. It is denoted by $\varphi|_W$ and called a *subform* of φ. Define $W^\perp$ to be the orthogonal complement of W relative to the polar form of φ. The space W is called a *totally isotropic subspace* if $\varphi|_W = 0$. If this is the case, then $\mathfrak{b}_\varphi|_W = 0$.

Example 7.8. If F is algebraically closed, then any homogeneous polynomial in more than one variable has a nontrivial zero. In particular, up to isometry, the only anisotropic quadratic forms over F are 0 and $\langle 1 \rangle$.

Remark 7.9. Let φ be a quadratic form on V over F. If $\varphi = \varphi_{\mathfrak{b}}$ for some symmetric bilinear form $\mathfrak{b}$, then φ is isotropic if and only if $\mathfrak{b}$ is. In addition, if $\operatorname{char} F \neq 2$, then φ is isotropic if and only if its polar form $\mathfrak{b}_\varphi$ is. However, if $\operatorname{char} F = 2$, then every $0 \neq v \in V$ is an isotropic vector for $\mathfrak{b}_\varphi$.

Let ψ be a subform of a quadratic form φ. The restriction of φ on $(V_\psi)^\perp$ (with respect to the polar form $\mathfrak{b}_\varphi$) is denoted by $\psi^\perp$ and is called the *complementary form* of ψ in φ. If $V_\varphi = W \oplus U$ is a direct sum of vector spaces with $W \subset U^\perp$, we write $\varphi = \varphi|_W \perp \varphi|_U$ and call it an *internal orthogonal sum.* So $\varphi(w+u) = \varphi(w) + \varphi(u)$ for all $w \in W$ and $u \in U$. Note that $\varphi|_U$ is a subform of $(\varphi|_W)^\perp$.

Remark 7.10. Let φ be a quadratic form with $\operatorname{rad} \mathfrak{b}_\varphi = 0$. If ψ is a subform of φ, then by Proposition 1.5, we have $\dim \psi^\perp = \dim \varphi - \dim \psi$ and therefore $\psi^{\perp\perp} = \psi$.

Let φ be a quadratic form on V. We say that φ is *totally singular* if its polar form $\mathfrak{b}_\varphi$ is zero. If $\operatorname{char} F \neq 2$, then φ is totally singular if and only if φ is the zero quadratic form. If $\operatorname{char} F = 2$ this may not be true. Define the *quadratic radical* of φ by

$$\operatorname{rad} \varphi := \{v \in \operatorname{rad} \mathfrak{b}_\varphi \mid \varphi(v) = 0\}.$$

This is a subspace of $\operatorname{rad} \mathfrak{b}_\varphi$. We say that φ is *regular* if $\operatorname{rad} \varphi = 0$. If $\operatorname{char} F \neq 2$, then $\operatorname{rad} \varphi = \operatorname{rad} \mathfrak{b}_\varphi$. In particular, φ is regular if and only if its polar form is nondegenerate. If $\operatorname{char} F = 2$, this may not be true.

Example 7.11. Every anisotropic quadratic form is regular.

Clearly, if $f : \varphi \to \psi$ is an isometry of quadratic forms, then $f(\operatorname{rad} \mathfrak{b}_\varphi) = \operatorname{rad} \mathfrak{b}_\psi$ and $f(\operatorname{rad} \varphi) = \operatorname{rad} \psi$.

Let φ be a quadratic form on V and $\overline{} : V \to \overline{V} = V/\operatorname{rad} \varphi$ the canonical epimorphism. Let $\overline{\varphi}$ denote the quadratic form on $\overline{V}$ given by $\overline{\varphi}(\overline{v}) := \varphi(v)$ for all $v \in V$. In particular, the restriction of $\overline{\varphi}$ to $\operatorname{rad} \mathfrak{b}_\varphi / \operatorname{rad} \varphi$ determines an anisotropic quadratic form. We have:

Lemma 7.12. *Let φ be a quadratic form on V and W any subspace of V satisfying $V = \operatorname{rad} \varphi \oplus W$. Then*

$$\varphi = \varphi|_{\operatorname{rad} \varphi} \perp \varphi|_W = 0|_{\operatorname{rad} \varphi} \perp \varphi|_W$$

with $\varphi|_W \simeq \overline{\varphi}$ the induced quadratic form on $V/\operatorname{rad} \varphi$. In particular, $\varphi|_W$ is unique up to isometry.

If φ is a quadratic form, the form $\varphi|_W$, unique up to isometry will be called its *regular part.* The subform $\varphi|_W$ in the lemma is regular but $\mathfrak{b}_{\varphi|_W}$ may be degenerate if $\operatorname{char} F = 2$. To obtain a further orthogonal decomposition of a quadratic form, we need to look at the regular part. The key is the following:

Proposition 7.13. *Let φ be a regular quadratic form on V. Suppose that V contains an isotropic vector v. Then there exists a 2-dimensional subspace W of V containing v such that $\varphi|_W \simeq \mathbb{H}$.*

PROOF. As $\operatorname{rad} \varphi = 0$, we have $v \notin \operatorname{rad} \mathfrak{b}_\varphi$. Thus there exists a vector $w \in V$ such that $a = \mathfrak{b}_\varphi(v, w) \neq 0$. Replacing v by $a^{-1}v$, we may assume that $a = 1$. Let $W = Fv \oplus Fw$. Then $v, w - \varphi(w)v$ is a hyperbolic pair. □

We say that any isotropic regular quadratic form *splits off* a hyperbolic plane.

If K/F is a field extension, let φ_K be the quadratic form on V_K defined by $\varphi_K(x \otimes v) := x^2 \varphi(v)$ for all $x \in K$ and $v \in V$ with polar form $\mathfrak{b}_{\varphi_K} = (\mathfrak{b}_\varphi)_K$. Although $(\operatorname{rad} \mathfrak{b}_\varphi)_K = \operatorname{rad}(\mathfrak{b}_\varphi)_K$, we only have $(\operatorname{rad} \varphi)_K \subset \operatorname{rad} \varphi_K$ with inequality possible.

Remark 7.14. If K/F is a field extension and φ a quadratic form over F, then φ is regular if φ_K is.

The following is a useful observation. The proof analogous to that for Lemma 1.21 shows:

Lemma 7.15. *Let φ be an anisotropic quadratic form over F. If K/F is purely transcendental, then φ_K is anisotropic.*

7.A. Nondegenerate quadratic forms. To define nondegeneracy, we use the following lemma.

Lemma 7.16. *Let φ be a quadratic form over F. Then the following are equivalent:*

(1) *φ_K is regular for every field extension K/F.*
(2) *φ_K is regular over an algebraically closed field K containing F.*
(3) *φ is regular and* $\dim \operatorname{rad} \mathfrak{b}_\varphi \leq 1$.

PROOF. (1) $\Rightarrow$ (2) is trivial.

(2) $\Rightarrow$ (3): As $(\operatorname{rad}\varphi)_K \subset \operatorname{rad}\varphi_K = 0$, we have $\operatorname{rad}\varphi = 0$. To show the second statement, we may assume that F is algebraically closed. As $\varphi|_{\operatorname{rad}\mathfrak{b}_\varphi} = \overline{\varphi}|_{\operatorname{rad}\mathfrak{b}_\varphi/\operatorname{rad}\varphi}$ is anisotropic and over an algebraically closed field, any quadratic form of dimension greater than one is isotropic, $\dim \operatorname{rad}\mathfrak{b}_\varphi \leq 1$.

(3) $\Rightarrow$ (1): Suppose that $\operatorname{rad}\varphi_K \neq 0$. As $\operatorname{rad}\varphi_K \subset \operatorname{rad}\mathfrak{b}_{\varphi_K}$ and $\operatorname{rad}\mathfrak{b}_{\varphi_K} = (\operatorname{rad}\mathfrak{b}_\varphi)_K$ is of dimension at most one, we have $\operatorname{rad}\varphi_K = (\operatorname{rad}\mathfrak{b}_\varphi)_K$. Let $0 \neq v \in \operatorname{rad}\mathfrak{b}_\varphi$. Then $v \in \operatorname{rad}\varphi_K$, hence $\varphi(v) = 0$ contradicting $\operatorname{rad}\varphi = 0$. □

Definition 7.17. A quadratic form φ over F is called *nondegenerate* if the equivalent conditions of the lemma are satisfied.

Remark 7.18. If K/F is a field extension, then φ is nondegenerate if and only if φ_K is nondegenerate by Lemma 7.16.

This definition of a nondegenerate quadratic form agrees with the one given in [**86**]. It is different than that found in some other texts. The geometric characterization of this definition of nondegeneracy explains our definition. In fact, if φ is a nonzero quadratic form on V of dimension at least two, then the following are equivalent:

(1) The quadratic form φ is nondegenerate.
(2) The projective quadric X_φ associated to φ is smooth. (Cf. Proposition 22.1.)
(3) The even Clifford algebra $C_0(\varphi)$ (cf. §11 below) of φ is separable (i.e., is a product of finite dimensional simple algebras each central over a separable field extension of F). (Cf. Proposition 11.6.)
(4) The group scheme $\mathbf{SO}(\varphi)$ of all isometries of φ identical on $\operatorname{rad}\varphi$ is reductive (semi-simple if $\dim\varphi \geq 3$ and simple if $\dim\varphi \geq 5$). (Cf. [**86**, Chapter VI].)

Proposition 7.19. (1) *The form $\langle a\rangle$ is nondegenerate if and only if $a \in F^\times$.*

(2) *The form $[a,b]$ is nondegenerate if and only if $1 - 4ab \neq 0$. In particular, this binary quadratic form as well as its polar form is always nondegenerate if* $\operatorname{char} F = 2$.

(3) *Hyperbolic forms are nondegenerate.*

(4) *Every binary isotropic nondegenerate quadratic form is isomorphic to $\mathbb{H}$.*

PROOF. (1) and (3) are clear.

(2): This follows by computing the determinant of the matrix representing the polar form corresponding to $[a, b]$. (Cf. Notation 7.3.)

(4) follows by Proposition 7.13. □

Let φ_i be a quadratic form on V_i for $i = 1, 2$. Then their *external orthogonal sum*, denoted by $\varphi_1 \perp \varphi_2$, is the form on $V_1 \oplus V_2$ given by

$$(\varphi_1 \perp \varphi_2)\big((v_1, v_2)\big) := \varphi_1(v_1) + \varphi_2(v_2)$$

for all $v_i \in V_i$, $i = 1, 2$. Note that $\mathfrak{b}_{\varphi_1 \perp \varphi_2} = \mathfrak{b}_{\varphi_1} \perp \mathfrak{b}_{\varphi_2}$.

Remark 7.20. Let $\operatorname{char} F \neq 2$. Let φ and ψ be quadratic forms over F.

(1) The form φ is nondegenerate if and only if φ is regular.

(2) If φ and ψ are both nondegenerate, then $\varphi \perp \psi$ is nondegenerate as $\mathfrak{b}_{\varphi \perp \psi} = \mathfrak{b}_\varphi \perp \mathfrak{b}_\psi$.

Remark 7.21. Let $\operatorname{char} F = 2$. Let φ and ψ be quadratic forms over F.

(1) If $\dim \varphi$ is even, then φ is nondegenerate if and only if its polar form $\mathfrak{b}_\varphi$ is nondegenerate.

(2) If $\dim \varphi$ is odd, then φ is nondegenerate if and only if $\dim \operatorname{rad} \mathfrak{b}_\varphi = 1$ and $\varphi|_{\operatorname{rad} \mathfrak{b}_\varphi}$ is nonzero.

(3) If φ and ψ are nondegenerate quadratic forms over F at least one of which is of even dimension, then $\varphi \perp \psi$ is nondegenerate.

The important analogue of Proposition 1.6 is immediate (using Lemma 7.16 for the last statement):

Proposition 7.22. *Let φ be a quadratic form on V. Let W be a vector subspace such that $\mathfrak{b}_{\varphi|_W}$ is a nondegenerate bilinear form. Then $\varphi|_W$ is nondegenerate and $\varphi = \varphi|_W \perp \varphi|_{W^\perp}$. In particular, $(\varphi|_W)^\perp = \varphi|_{W^\perp}$. Further, if φ is also nondegenerate, then so is $\varphi|_{W^\perp}$.*

Example 7.23. Suppose that $\operatorname{char} F = 2$ and $a, b, c \in F$. Let $\varphi = [c, a] \perp [c, b]$ and $\{e, f, e', f'\}$ be a basis for V_φ satisfying $\varphi(e) = c = \varphi(e')$, $\varphi(f) = a$, $\varphi(f') = b$, and $\mathfrak{b}_\varphi(e, f) = 1 = \mathfrak{b}(e', f')$. Then in the basis $\{e, f + f', e + e', f'\}$, we have

$$[c, a] \perp [c, b] \simeq [c, a + b] \perp \mathbb{H}$$

by Example 7.5.

If n is a nonnegative integer and φ is a quadratic form over F, we let

$$n\varphi := \underbrace{\varphi \perp \cdots \perp \varphi}_{n}.$$

In particular, if n is an integer, we do not interpret $n\varphi$ with n viewed in the field. For example, if V is an n-dimensional vector space, $\mathbb{H}(V) \simeq n\mathbb{H}$.

We denote $\langle a_1 \rangle_q \perp \cdots \perp \langle a_n \rangle_q$ by

$$\langle a_1, \dots, a_n \rangle_q \quad \text{or simply} \quad \langle a_1, \dots, a_n \rangle.$$

So $\varphi \simeq \langle a_1, \dots, a_n \rangle$ for some $a_i \in F$ if and only if V_φ has an orthogonal basis. If V_φ has an orthogonal basis, we say φ is *diagonalizable*.

Remark 7.24. Suppose that $\operatorname{char} F = 2$ and φ is a quadratic form over F. Then φ is diagonalizable if and only if φ is totally singular, i.e., its polar form $\mathfrak{b}_\varphi = 0$. If this is the case, then every basis for V_φ is orthogonal. In particular, there are no diagonalizable nondegenerate quadratic forms of dimension greater than one.

Exercise 7.25. A quadratic form φ is diagonalizable if and only if $\varphi = \varphi_{\mathfrak{b}}$ for some symmetric bilinear form $\mathfrak{b}$.

Example 7.26. Suppose that $\operatorname{char} F \neq 2$. If $a \in F^\times$, then $\langle a, -a\rangle \simeq \mathbb{H}$.

Example 7.27. (Cf. Example 1.10.) Let $\operatorname{char} F = 2$ and $\varphi = \langle 1, a\rangle$ with $a \neq 0$. If $\{e, f\}$ is the basis for V_φ with $\varphi(e) = 1$ and $\varphi(f) = a$, then computing on the orthogonal basis $\{e, xe + yf\}$ with $x, y \in F$, $y \neq 0$ shows $\varphi \simeq \langle 1, x^2 + ay^2\rangle$. Consequently, $\langle 1, a\rangle \simeq \langle 1, b\rangle$ if and only if $b = x^2 + ay^2$ with $y \neq 0$.

7.B. Structure theorems for quadratic forms. We wish to decompose a quadratic form over a field F into an orthogonal sum of nice subforms. We begin with nondegenerate quadratic forms with large totally isotopic subspaces. Unlike the case of symmetric bilinear forms in characteristic 2, 2-dimensional nondegenerate isotropic quadratic forms are hyperbolic.

Proposition 7.28. *Let φ be an $2n$-dimensional nondegenerate quadratic form on V. Suppose that V contains a totally isotropic subspace W of dimension n. Then $\varphi \simeq n\mathbb{H}$. Conversely, every hyperbolic form of dimension $2n$ contains a totally isotropic subspace of dimension n.*

PROOF. Let $0 \neq v \in W$. Then by Proposition 7.13 there exists a 2-dimensional subspace V_1 of V containing v with $\varphi|_{V_1}$ a nondegenerate subform isomorphic to $\mathbb{H}$. By Proposition 7.22, this subform splits off as an orthogonal summand. Since $\varphi|_{V_1}$ is nondegenerate, $W \cap V_1$ is 1-dimensional, so $\dim W \cap V_1^\perp = n - 1$. The first statement follows by induction applied to the totally isotropic subspace $W \cap V_1^\perp$ of $V_1^\perp$. The converse is easy. □

We turn to splitting off anisotopic subforms of regular quadratic forms. It is convenient to write these decompositions separately for fields of characteristic 2 and not 2.

Proposition 7.29. *Let $\operatorname{char} F \neq 2$ and φ a quadratic form on V over F. Then there exists an orthogonal basis for V. In particular, there exist 1-dimensional subspaces $V_i \subset V$, $1 \leq i \leq n$ for some n and an orthogonal decomposition*

$$\varphi = \varphi|_{\operatorname{rad} \mathfrak{b}_\varphi} \perp \varphi|_{V_1} \perp \cdots \perp \varphi|_{V_n}$$

with $\varphi|_{V_1} \simeq \langle a_i\rangle$, $a_i \in F^\times$ for all $i \in [1,\, n]$. In particular,

$$\varphi \simeq r\langle 0\rangle \perp \langle a_1, \dots, a_n\rangle$$

with $r = \dim \operatorname{rad} \mathfrak{b}_\varphi$.

PROOF. We may assume that $\varphi \neq 0$. Hence there exists an anisotropic vector $0 \neq v \in V$. As $\mathfrak{b}_{\varphi|_{Fv}}$ is nondegenerate, $\varphi|_{Fv}$ splits off as an orthogonal summand of φ by Proposition 7.22. The result follows easily by induction. □

Corollary 7.30. *Suppose that $\operatorname{char} F \neq 2$. Then every quadratic form over F is diagonalizable.*

Proposition 7.31. *Let* char $F = 2$ *and* φ *a quadratic form on* V *over* F. *Then there exists* 2*-dimensional subspaces* $V_i \subset V$, $1 \le i \le n$ *for some* n, *a subspace* $W \subset \operatorname{rad} \mathfrak{b}_\varphi$, *and an orthogonal decomposition*

$$\varphi = \varphi|_{\operatorname{rad}(\varphi)} \perp \varphi|_W \perp \varphi|_{V_1} \perp \cdots \perp \varphi|_{V_n}$$

with $\varphi|_{V_i} \simeq [a_i, b_i]$ *nondegenerate,* $a_i, b_i \in F$ *for all* $i \in [1,\ n]$. *Moreover,* $\varphi|_W$ *is anisotropic, diagonalizable, and is unique up to isometry. In particular,*

$$\varphi \simeq r\langle 0\rangle \perp \langle c_1, \dots, c_s\rangle \perp [a_1, b_1] \perp \cdots \perp [a_n, b_n]$$

with $r = \dim \operatorname{rad} \varphi$ *and* $s = \dim W$, $c_i \in F^\times$, $1 \le i \le s$.

PROOF. Let $W \subset V$ be a subspace such that $\operatorname{rad} \mathfrak{b}_\varphi = \operatorname{rad} \varphi \oplus W$ and $V' \subset V$ a subspace such that $V = \operatorname{rad} \mathfrak{b}_\varphi \oplus V'$. Then $\varphi = \varphi|_{\operatorname{rad}(\varphi)} \perp \varphi|_W \perp \varphi|_{V'}$. The form $\varphi|_W$ is diagonalizable as $\mathfrak{b}_{\varphi|_W} = 0$ and anisotropic as $W \cap \operatorname{rad} \varphi = 0$. By Lemma 7.12, the form $\varphi|_W = (\varphi|_{\operatorname{rad} \mathfrak{b}_\varphi})|_W$ is unique up to isometry. So to finish we need only show that $\varphi|_{V'}$ is an orthogonal sum of nondegenerate binary subforms of the desired isometry type. We may assume that $V' \neq \{0\}$. Let $0 \neq v \in V'$. Then there exists $0 \neq v' \in V'$ such that $c = \mathfrak{b}_\varphi(v, v') \neq 0$. Replacing v' by $c^{-1}v'$, we may assume that $\mathfrak{b}_\varphi(v, v') = 1$. In particular, $\varphi|_{Fv \oplus Fv'} \simeq [\varphi(v), \varphi(v')]$. As $[\varphi(v), \varphi(v')]$ and its polar form are nondegenerate by Proposition 7.19, the subform $\varphi|_{Fv \oplus Fv'}$ is an orthogonal direct summand of φ by Proposition 7.22. The decomposition follows by Lemma 7.12 and induction. $\square$

Corollary 7.32. *Let* char $F = 2$ *and let* φ *be a nondegenerate quadratic form over* F.

(1) *If* $\dim \varphi = 2n$, *then*

$$\varphi \simeq [a_1, b_1] \perp \cdots \perp [a_n, b_n]$$

for some $a_i, b_i \in F$, $i \in [1,\ n]$.

(2) *If* $\dim \varphi = 2n + 1$, *then*

$$\varphi \simeq \langle c\rangle \perp [a_1, b_1] \perp \cdots \perp [a_n, b_n]$$

for some $a_i, b_i \in F$, $i \in [1,\ n]$, *and* $c \in F^\times$ *unique up to* $F^{\times 2}$.

Example 7.33. Suppose that F is quadratically closed of characteristic 2. Then every anisotropic form is isometric to 0, $\langle 1\rangle$ or $[1, a]$ with $a \in F \setminus \wp(F)$ where $\wp : F \to F$ is the Artin-Schreier map.

Exercise 7.34. Every nondegenerate quadratic form over a separably closed field F is isometric to $n\mathbb{H}$ or $\langle a\rangle \perp n\mathbb{H}$ for some $n \ge 0$ and $a \in F^\times$.

8. Witt's Theorems

As with the bilinear case, the classical Witt theorems are more delicate to ascertain over fields of arbitrary characteristic. We shall give characteristic free proofs of these. The basic Witt theorem is the Witt Extension Theorem (cf. Theorem 8.3 below). Witt's original theorem (cf. [**139**]) has been generalized in various ways. We use one similar to that given by Kneser (cf. [**80**, Th. 1.2.2]). We construct the quadratic Witt group of even-dimensional anisotropic quadratic forms and use the Witt theorems to study this group.

To obtain further decompositions of a quadratic form, we need generalizations of the classical Witt theorems for bilinear forms over fields of characteristic different from 2.

Let φ be a quadratic form on V. Let v and v' in V satisfy $\varphi(v) = \varphi(v')$. If the vector $\bar{v} = v - v'$ is anisotropic, then the reflection (cf. Example 7.2) $\tau_{\bar{v}} : \varphi \to \varphi$ satisfies

$$\tau_{\bar{v}}(v) = v'. \tag{8.1}$$

What if $\bar{v}$ is isotropic?

Lemma 8.2. *Let φ be a quadratic form on V with polar form $\mathfrak{b}$. Let v and v' lie in V and $\bar{v} = v - v'$. Suppose that $\varphi(v) = \varphi(v')$ and $\varphi(\bar{v}) = 0$. If $w \in V$ is anisotropic and satisfies that both $\mathfrak{b}(w, v)$ and $\mathfrak{b}(w, v')$ are nonzero, then the vector $w' = v - \tau_w(v')$ is anisotropic and $(\tau_w \circ \tau_{w'})(v) = v'$.*

PROOF. As $w' = \bar{v} + \mathfrak{b}(v', w)\varphi(w)^{-1}w$, we have

$$\begin{aligned}\varphi(w') &= \varphi(\bar{v}) + \mathfrak{b}\big(\bar{v}, \mathfrak{b}(v', w)\varphi(w)^{-1}w\big) + \mathfrak{b}(v', w)^2\varphi(w)^{-1} \\ &= \mathfrak{b}(v, w)\mathfrak{b}(v', w)\varphi(w)^{-1} \neq 0.\end{aligned}$$

It follows from (8.1) that $\tau_{w'}(v) = \tau_w(v')$, hence the result. □

Theorem 8.3 (Witt Extension Theorem). *Let φ and φ' be isometric quadratic forms on V and V' respectively. Let $W \subset V$ and $W' \subset V'$ be subspaces such that $W \cap \operatorname{rad} \mathfrak{b}_\varphi = 0$ and $W' \cap \operatorname{rad} \mathfrak{b}_{\varphi'} = 0$. Suppose that there is an isometry $\alpha : \varphi|_W \to \varphi'|_{W'}$. Then there exists an isometry $\tilde{\alpha} : \varphi \to \varphi'$ such that $\tilde{\alpha}(W) = W'$ and $\tilde{\alpha}|_W = \alpha$.*

PROOF. It is sufficient to treat the case $V = V'$ and $\varphi = \varphi'$. Let $\mathfrak{b}$ denote the polar form of φ. We proceed by induction on $n = \dim W$, the case $n = 0$ being obvious. Suppose that $n > 0$. In particular, φ is not identically zero. Let $u \in V$ satisfy $\varphi(u) \neq 0$. As $\dim W \cap (Fu)^\perp \geq n - 1$, there exists a subspace $W_0 \subset W$ of codimension one with $W_0 \subset (Fu)^\perp$. Applying the induction hypothesis to $\beta = \alpha|_{W_0} : \varphi|_{W_0} \to \varphi|_{\alpha(W_0)}$, there exists an isometry $\tilde{\beta} : \varphi \to \varphi$ satisfying $\tilde{\beta}(W_0) = \alpha(W_0)$ and $\tilde{\beta}|_{W_0} = \beta$. Replacing W' by $\tilde{\beta}^{-1}(W')$, we may assume that $W_0 \subset W'$ and $\alpha|_{W_0}$ is the identity.

Let v be any vector in $W \setminus W_0$ and set $v' = \alpha(v) \in W'$. It suffices to find an isometry γ of φ such that $\gamma(v) = v'$ and $\gamma|_{W_0} = \mathrm{Id}$, the identity on W_0. Let $\bar{v} = v - v'$ as above and $S = W_0^\perp$. Note that for every $w \in W_0$, we have $\alpha(w) = w$, hence

$$\mathfrak{b}(\bar{v}, w) = \mathfrak{b}(v, w) - \mathfrak{b}\big(\alpha(v), \alpha(w)\big) = 0,$$

i.e., $\bar{v} \in S$.

Suppose that $\varphi(\bar{v}) \neq 0$. Then $\tau_{\bar{v}}(v) = v'$ using (8.1). Moreover, $\tau_{\bar{v}}(w) = w$ for every $w \in W_0$ as $\bar{v}$ is orthogonal to W_0. Then $\gamma = \tau_{\bar{v}}$ works. So we may assume that $\varphi(\bar{v}) = 0$. We have

$$0 = \varphi(\bar{v}) = \varphi(v) - \mathfrak{b}(v, v') + \varphi(v') = \mathfrak{b}(v, v) - \mathfrak{b}(v, v') = \mathfrak{b}(v, \bar{v}),$$

i.e., $\bar{v}$ is orthogonal to v. Similarly, $\bar{v}$ is orthogonal to v'.

By Proposition 1.5, the map $l_W : V \to W^*$ is surjective. In particular, there exists $u \in V$ such that $\mathfrak{b}(u, W_0) = 0$ and $\mathfrak{b}(u, v) = 1$. In other words, v is not orthogonal to S, i.e., the intersection $H = (Fv)^\perp \cap S$ is a subspace of codimension one in S. Similarly, $H' = (Fv')^\perp \cap S$ is also a subspace of codimension one in S. Note that $\bar{v} \in H \cap H'$.

Suppose that there exists an anisotropic vector $w \in S$ such that $w \notin H$ and $w \notin H'$. By Lemma 8.2, we have $(\tau_w \circ \tau_{w'})(v) = v'$ where

$$w' = v - \tau_w(v') = \bar{v} + \mathfrak{b}(v', w)\varphi(w)^{-1}w \in S.$$

As $w, w' \in S$, the map $\tau_w \circ \tau_{w'}$ is the identity on W_0. Setting $\gamma = \tau_w \circ \tau_{w'}$ produces the desired extension. Consequently, we may assume that $\varphi(w) = 0$ for every $w \in S \setminus (H \cup H')$.

Case 1: $|F| > 2$.

Let $w_1 \in H \cap H'$ and $w_2 \in S \setminus (H \cup H')$. Then $aw_1 + w_2 \in S \setminus (H \cup H')$ for any $a \in F$, so by assumption

$$0 = \varphi(aw_1 + w_2) = a^2\varphi(w_1) + a\mathfrak{b}(w_1, w_2) + \varphi(w_2).$$

Since $|F| > 2$, we must have $\varphi(w_1) = \mathfrak{b}(w_1, w_2) = \varphi(w_2) = 0$. So $\varphi(H \cap H') = 0$, $\varphi\big(S \setminus (H \cup H')\big) = 0$, and $H \cap H'$ is orthogonal to $S \setminus (H \cup H')$, (i.e., $\mathfrak{b}(x, y) = 0$ for all $x \in H \cap H'$ and $y \in S \setminus (H \cup H')$).

Let $w \in H$ and $w' \in S \setminus (H \cup H')$. As $|F| > 2$, we see that $w + aw' \in S \setminus (H \cup H')$ for some $a \in F$. Hence the set $S \setminus (H \cup H')$ generates S. Consequently, $H \cap H'$ is orthogonal to S. In particular, $\mathfrak{b}(\bar{v}, S) = 0$. Thus $H = H'$. It follows that $\varphi(H) = 0$ and $\varphi(S \setminus H) = 0$, hence $\varphi(S) = 0$, a contradiction. This finishes the proof in this case.

Case 2: $F = \mathbb{F}_2$, the field of two elements.

As $H \cup H' \neq S$, there exists a $w \in S$ such that $\mathfrak{b}(w, v) \neq 0$ and $\mathfrak{b}(w, v') \neq 0$. As $F = \mathbb{F}_2$, this means that $\mathfrak{b}(w, v) = 1 = \mathfrak{b}(w, v')$. Moreover, by our assumptions, $\varphi(\bar{v}) = 0$ and $\varphi(w) = 0$. Consider the linear map

$$\gamma : V \to V \quad \text{given by} \quad \gamma(x) = x + \mathfrak{b}(\bar{v}, x)w + \mathfrak{b}(w, x)\bar{v}.$$

Note that $\mathfrak{b}(w, \bar{v}) = \mathfrak{b}(w, v) + \mathfrak{b}(w, v') = 1 + 1 = 0$. A simple calculation shows that $\gamma^2 = \mathrm{Id}$ and $\varphi\big(\gamma(x)\big) = \varphi(x)$ for any $x \in V$, i.e., γ is an isometry. Moreover, $\gamma(v) = v + \bar{v} = v'$. Finally, $\gamma|_{W_0} = \mathrm{Id}$ since w and $\bar{v}$ are orthogonal to W_0. □

Theorem 8.4 (Witt Cancellation Theorem). *Let φ, φ' be quadratic forms on V and V' respectively, and ψ, ψ' quadratic forms on W and W' respectively, with* $\operatorname{rad} \mathfrak{b}_\psi = 0 = \operatorname{rad} \mathfrak{b}_{\psi'}$. *If*

$$\varphi \perp \psi \simeq \varphi' \perp \psi' \quad \text{and} \quad \psi \simeq \psi',$$

then $\varphi \simeq \varphi'$.

PROOF. Let $f : \psi \to \psi'$ be an isometry. By the Witt Extension Theorem, f extends to an isometry $\tilde{f} : \varphi \perp \psi \to \varphi' \perp \psi'$. As $\tilde{f}$ takes $V = W^\perp$ to $V' = (W')^\perp$, the result follows. □

Witt Cancellation together with our previous computations allow us to derive the decomposition that we want.

Theorem 8.5 (Witt Decomposition Theorem). *Let φ be a quadratic form on V. Then there exist subspaces V_1 and V_2 of V such that $\varphi = \varphi|_{\operatorname{rad}\varphi} \perp \varphi|_{V_1} \perp \varphi|_{V_2}$ with $\varphi|_{V_1}$ anisotropic and $\varphi|_{V_2}$ hyperbolic. Moreover, $\varphi|_{V_1}$ and $\varphi|_{V_2}$ are unique up to isometry.*

PROOF. We know that $\varphi = \varphi|_{\mathrm{rad}\,\varphi} \perp \varphi|_{V'}$ with $\varphi_{V'}$ on V' unique up to isometry. Therefore, we can assume that φ is regular. Suppose that $\varphi_{V'}$ is isotropic. By Proposition 7.13, we can split off a subform as an orthogonal summand isometric to the hyperbolic plane. The desired decomposition follows by induction. As every hyperbolic form is nondegenerate, the Witt Cancellation Theorem shows the uniqueness of $\varphi|_{V_1}$ up to isometry hence $\varphi|_{V_2}$ is unique by dimension count. □

8.A. Witt equivalence. Using the Witt Decomposition Theorem 8.5, we can define an equivalence of quadratic forms over a field F.

Definition 8.6. Let φ be a quadratic form on V and $\varphi = \varphi|_{\mathrm{rad}\,\varphi} \perp \varphi|_{V_1} \perp \varphi|_{V_2}$ be the decomposition in the theorem. The anisotropic form $\varphi|_{V_1}$, unique up to isometry, will be denoted φ_{an} and called the *anisotropic part* of φ. As φ_{V_2} is hyperbolic, $\dim V_2 = 2n$ for some unique nonnegative number n. The integer n is called the *Witt index* of φ and denoted by $\mathfrak{i}_0(\varphi)$. We say that two quadratic forms φ and ψ are *Witt-equivalent* and write $\varphi \sim \psi$ if $\dim \mathrm{rad}\,\varphi = \dim \mathrm{rad}\,\psi$ and $\varphi_{an} \simeq \psi_{an}$. Equivalently, $\varphi \sim \psi$ if and only if $\varphi \perp n\mathbb{H} \simeq \psi \perp m\mathbb{H}$ for some n and m.

Note that if $\varphi \sim \psi$, then $\varphi_K \sim \psi_K$ for any field extension K/F.

Witt cancellation does not hold in general for nondegenerate quadratic forms in characteristic 2. We show in the next result, Proposition 8.8, that

$$[a, b] \perp \langle a \rangle \simeq \mathbb{H} \perp \langle a \rangle \tag{8.7}$$

if $\mathrm{char}\,F = 2$ for all $a, b \in F$ with $a \neq 0$. But $[a, b] \simeq \mathbb{H}$ if and only if $[a, b]$ is isotropic by Proposition 7.19(4). Although Witt cancellation does not hold in general in characteristic 2, we can use the following:

Proposition 8.8. *Let ρ be a nondegenerate quadratic form of even dimension over a field F of characteristic 2. Then $\rho \perp \langle a \rangle \sim \langle a \rangle$ for some $a \in F^\times$ if and only if $\rho \sim [a, b]$ for some $b \in F$.*

PROOF. The case $\mathrm{char}\,F \neq 2$ is easy, so we can assume that $\mathrm{char}\,F = 2$. Let $\varphi = [a, b] \perp \langle a \rangle$ with $a, b \in F$ and $a \neq 0$. Clearly, φ is isotropic and it is nondegenerate as $\varphi|_{\mathrm{rad}\,\mathfrak{b}_\varphi} = \langle a \rangle$. It follows by Proposition 7.13 that $[a, b] \perp \langle a \rangle \simeq \mathbb{H} \perp \langle a \rangle \sim \langle a \rangle$. Since $\rho \sim [a, b]$, we have $\rho \perp \langle a \rangle \sim \langle a \rangle$.

Conversely, suppose that $\rho \perp \langle a \rangle \sim \langle a \rangle$ for some $a \in F^\times$. We prove the statement by induction on $n = \dim \rho$. If $n = 0$, we can take $b = 0$. So assume that $n > 0$. We may also assume that ρ is anisotropic. By assumption, the form $\rho \perp \langle a \rangle$ is isotropic. Therefore, $a = \rho(v)$ for some $v \in V_\rho$, and we can find a decomposition $\rho = \rho' \perp [a, d]$ for some nondegenerate form ρ' of dimension $n - 2$ and $b \in F$. As $[a, d] \perp \langle a \rangle \simeq \mathbb{H} \perp \langle a \rangle$ by the first part of the proof, we have

$$\langle a \rangle \sim \rho \perp \langle a \rangle = \rho' \perp [a, d] \perp \langle a \rangle \sim \rho' \perp \langle a \rangle.$$

By the induction hypothesis, $\rho' \simeq [a, c]$ for some $c \in F$. Therefore, by Example 7.23,

$$\rho = \rho' \perp [a, d] \sim [a, c] \perp [a, d] \simeq [a, c + d] \perp \mathbb{H} \sim [a, c + d].$$

□

Remark 8.9. Let φ and ψ be quadratic forms over F.

(1) If φ is nondegenerate and anisotropic over F and K/F a purely transcendental extension, then φ_K remains anisotropic by Lemma 7.15. In particular, for any nondegenerate φ, we have $\mathfrak{i}_0(\varphi) = \mathfrak{i}_0(\varphi_K)$.

(2) Let $a \in F^\times$. Then $\varphi \simeq a\psi$ if and only if $\varphi_{an} \simeq a\psi_{an}$ as any form similar to a hyperbolic form is hyperbolic.

(3) If $\operatorname{char} F = 2$, the quadratic form φ_{an} may be degenerate. This is not possible if $\operatorname{char} F \neq 2$.

(4) If $\operatorname{char} F \neq 2$, then every symmetric bilinear form corresponds to a quadratic form, hence the Witt theorems hold for symmetric bilinear forms in characteristic different from 2.

8.B. Totally isotropic subspaces. Given a regular quadratic form φ on V, we show that every totally isotopic subspace of V lies in a maximal isotropic subspace of V of dimension equal to the Witt index of φ.

Lemma 8.10. *Let φ be a regular quadratic form on V with $W \subset V$ a totally isotropic subspace of dimension m. If ψ is the quadratic form on $W^\perp/W$ induced by the restriction of φ on $W^\perp$, then $\varphi \simeq \psi \perp m\mathbb{H}$.*

PROOF. As $W \cap \operatorname{rad} \mathfrak{b}_\varphi \subset \operatorname{rad} \varphi$, the intersection $W \cap \operatorname{rad} \mathfrak{b}_\varphi$ is trivial. Thus the map $V \to W^*$ by $v \mapsto l_v|_W : w \mapsto \mathfrak{b}_\varphi(v, w)$ is surjective by Proposition 1.5 and $\dim W^\perp = \dim V - \dim W$. Let $W' \subset V$ be a subspace mapping isomorphically onto W^*. Clearly, $W \cap W' = \{0\}$. Let $U = W \oplus W'$.

We show the form $\varphi|_U$ is hyperbolic. The subspace $W \oplus W'$ is nondegenerate with respect to $\mathfrak{b}_\varphi$. Indeed, let $0 \neq v = w + w' \in W \oplus W'$. If $w' \neq 0$, there exists a $w_0 \in W$ such that $\mathfrak{b}_\varphi(w', w_0) \neq 0$, hence $\mathfrak{b}_\varphi(v, w_0) \neq 0$. If $w' = 0$, there exists $w_0' \in W'$ such that $\mathfrak{b}_\varphi(w, w_0') \neq 0$, hence $\mathfrak{b}_\varphi(v, w_0') \neq 0$. Thus by Proposition 7.28, the form $\varphi|_U$ is isometric to $m\mathbb{H}$ where $m = \dim W$.

By Proposition 7.22, we have $\varphi = \varphi|_{U^\perp} \perp \varphi|_U \simeq \varphi|_{U^\perp} \perp m\mathbb{H}$. As W and $U^\perp$ are subspaces of $W^\perp$ and $U \cap W^\perp = W$, we have $W^\perp = W \oplus U^\perp$. Thus $W^\perp/W \simeq U^\perp$ and the result follows. □

Proposition 8.11. *Let φ be a regular quadratic form on V. Then every totally isotropic subspace of V is contained in a totally isotropic subspace of dimension $\mathfrak{i}_0(\varphi)$.*

PROOF. Let $W \subset V$ be a totally isotropic subspace of V. We may assume that it is a maximal totally isotropic subspace. In the notation in the proof of Lemma 8.10, we have $\varphi = \varphi|_{U^\perp} \perp \varphi|_U$ with $\varphi|_U \simeq m\mathbb{H}$ where $m = \dim W$. The form $\varphi|_{U^\perp}$ is anisotropic by the maximality of W, hence must be φ_{an} by the Witt Decomposition Theorem 8.5. In particular, $\dim W = \mathfrak{i}_0(\varphi)$. □

Corollary 8.12. *Let φ be a regular quadratic form on V. Then every totally isotropic subspace W of V has dimension at most $\mathfrak{i}_0(\varphi)$ with equality if and only if W is a maximal totally isotropic subspace of V.*

Let ρ be a nondegenerate quadratic form and φ a subform of ρ. If $\mathfrak{b}_\varphi$ is nondegenerate, then $\rho = \varphi \perp \varphi^\perp$, hence $\rho \perp (-\varphi) \sim \varphi^\perp$. However, in general, $\rho \neq \varphi \perp \varphi^\perp$. We do always have:

Lemma 8.13. *Let ρ be a nondegenerate quadratic form of even dimension and φ a regular subform of ρ. Then $\rho \perp (-\varphi) \sim \varphi^\perp$.*

PROOF. Let W be the subspace defined by $W = \{(v, v) \mid v \in V_\varphi\}$ of $V_\rho \oplus V_\varphi$. Clearly W is totally isotropic with respect to the form $\rho \perp (-\varphi)$ on $V_\rho \oplus V_\varphi$. By the proof of Lemma 8.10, we have $\dim W^\perp/W = \dim V_\rho \oplus V_\varphi - 2\dim W =$

$\dim V_\rho - \dim V_\varphi$. By Remark 7.10, we also have $\dim V_\varphi^\perp = \dim V_\rho - \dim V_\varphi$. It follows that the linear map $W^\perp/W \to V_\varphi^\perp$ defined by $(v, v') \mapsto v - v'$ is an isometry. On the other hand, by Lemma 8.10, the form on $W^\perp/W$ is Witt-equivalent to $\rho \perp (-\varphi)$. □

Let V and W be vector spaces over F. Let $\mathfrak{b}$ be a symmetric bilinear form on W and φ be a quadratic form on V. The *tensor product* of $\mathfrak{b}$ and φ is the quadratic form $\mathfrak{b} \otimes \varphi$ on $W \otimes_F V$ defined by

$$(\mathfrak{b} \otimes \varphi)(w \otimes v) = \mathfrak{b}(w, w) \cdot \varphi(v) \tag{8.14}$$

for all $w \in W$ and $v \in V$ with the polar form of $\mathfrak{b} \otimes \varphi$ equal to $\mathfrak{b} \otimes \mathfrak{b}_\varphi$. For example, if $a \in F$, then $\langle a \rangle_b \otimes \varphi \simeq a\varphi$.

Example 8.15. If $\mathfrak{b}$ is a symmetric bilinear form, then $\varphi_\mathfrak{b} \simeq \mathfrak{b} \otimes \langle 1 \rangle_q$.

Lemma 8.16. *Let $\mathfrak{b}$ be a nondegenerate symmetric bilinear form over F and φ a nondegenerate quadratic form over F. In addition, assume that* $\dim \varphi$ *is even if characteristic of F is* 2. *Then:*

(1) *The quadratic form $\mathfrak{b} \otimes \varphi$ is nondegenerate.*
(2) *If either φ or $\mathfrak{b}$ is hyperbolic, then $\mathfrak{b} \otimes \varphi$ is hyperbolic.*

Proof. (1): The bilinear form $\mathfrak{b}_\varphi$ is nondegenerate by Remark 7.20 and by Remark 7.21 if characteristic of F is not 2 or 2, respectively. By Lemma 2.1, the form $\mathfrak{b} \otimes \mathfrak{b}_\varphi$ is nondegenerate, hence so is $\mathfrak{b} \otimes \varphi$.

(2): Using Proposition 7.28, we see that $V_{\mathfrak{b} \otimes \varphi}$ contains a totally isotropic space of dimension $\frac{1}{2} \dim(\mathfrak{b} \otimes \varphi)$. □

As the orthogonal sum of even-dimensional nondegenerate quadratic forms over F is nondegenerate, the isometry classes of even-dimensional nondegenerate quadratic forms over F form a monoid under orthogonal sum. The quotient of the Grothendieck group of this monoid by the subgroup generated by the image of the hyperbolic plane is called the *quadratic Witt group* and will be denoted by $I_q(F)$. The tensor product of a bilinear with a quadratic form induces a $W(F)$-module structure on $I_q(F)$ by Lemma 8.16.

Remark 8.17. Let φ and ψ be two nondegenerate even-dimensional quadratic forms over F. By the Witt Decomposition Theorem 8.5,

$$\varphi \simeq \psi \quad \text{if and only if} \quad \varphi = \psi \text{ in } I_q(F) \text{ and } \dim \varphi = \dim \psi.$$

Remark 8.18. Let $F \to K$ be a homomorphism of fields. Analogous to Proposition 2.7, this map induces the *restriction homomorphism*

$$r_{K/F} : I_q(F) \to I_q(K).$$

It is a group homomorphism. If K/F is purely transcendental, the restriction map is injective by Lemma 7.15.

Suppose that $\operatorname{char} F \neq 2$. Then we have an isomorphism $I(F) \to I_q(F)$ given by $\mathfrak{b} \mapsto \varphi_\mathfrak{b}$. We will use the correspondence $\mathfrak{b} \mapsto \varphi_\mathfrak{b}$ to identify bilinear forms in $W(F)$ with quadratic forms. In particular, we shall view the class of a quadratic form in the Witt ring of bilinear forms when $\operatorname{char} F \neq 2$.

9. Quadratic Pfister forms I

As in the bilinear case, there is a special class of forms built from tensor products of forms. If the characteristic of F is different from 2, these forms can be identified with the bilinear Pfister forms. If the characteristic is 2, these forms arise as the tensor product of a bilinear Pfister form and a binary quadratic form of the type $[1, a]$ and were first introduced by Baeza (cf. [**15**]). In general, the quadratic 1-fold Pfister forms are just the norm forms of a quadratic étale F-algebra and the 2-fold quadratic Pfister forms are just the reduced norm forms of quaternion algebras. These forms as their bilinear analogue satisfy the property of being round. In this section, we begin their study.

9.A. Values and similarities of quadratic forms. Analogously to the symmetric bilinear case, we study the values that a quadratic form can take as well as the similarity factors. We begin with some notation.

Let φ be a quadratic form on V over F. Let

$$D(\varphi) := \{\varphi(v) \mid v \in V,\ \varphi(v) \neq 0\},$$

the *set on nonzero values* of φ and

$$G(\varphi) := \{a \in F^\times \mid a\varphi \simeq \varphi\},$$

a group called the *group of similarity factors of* $\mathfrak{b}$. If $D(\varphi) = F^\times$, we say that φ is *universal.* Also set

$$\widetilde{D}(\varphi) := D(\varphi) \cup \{0\}.$$

We say that elements in $\widetilde{D}(\varphi)$ are *represented by* φ. For example, $G(\mathbb{H}) = F^\times$ (as for bilinear hyperbolic planes) and $D(\mathbb{H}) = F^\times$. In particular, if φ is an regular isotropic quadratic form over F, then φ is universal by Proposition 7.13.

The analogous proof of Lemma 1.13 shows:

Lemma 9.1. *Let φ be a quadratic form. Then*

$$D(\varphi) \cdot G(\varphi) \subset D(\varphi).$$

In particular, if $1 \in D(\varphi)$, then $G(\varphi) \subset D(\varphi)$.

The relationship between values and similarities of a symmetric bilinear form and its associated quadratic form is given by the following:

Lemma 9.2. *Let $\mathfrak{b}$ a symmetric bilinear form on F and $\varphi = \varphi_{\mathfrak{b}}$. Then:*

(1) $D(\varphi) = D(\mathfrak{b})$.
(2) $G(\mathfrak{b}) \subset G(\varphi)$.

PROOF. (1): By definition, $\varphi(v) = \mathfrak{b}(v, v)$ for all $v \in V$.

(2): Let $a \in G(\mathfrak{b})$ and $\lambda : \mathfrak{b} \to a\mathfrak{b}$ an isometry. Then $\varphi(\lambda(v)) = \mathfrak{b}(\lambda(v), \lambda(v)) = a\mathfrak{b}(v, v) = a\varphi(v)$ for all $v \in V$. □

A quadratic form is called *round* if $G(\varphi) = D(\varphi)$. In particular, if φ is round, then $D(\varphi)$ is a group. For example, any hyperbolic form is round.

A basic example of round forms arises from quadratic F-algebras (cf. §98.B):

Example 9.3. Let K be a quadratic F-algebra. Then there exists an involution on K denoted by $x \mapsto \bar{x}$ and a quadratic norm form $\varphi = \mathrm{N}$ given by $x \mapsto x\bar{x}$ (cf. §98.B). We have $\varphi(xy) = \varphi(x)\varphi(y)$ for all $x, y \in K$. If $x \in K$ with $\varphi(x) \neq 0$, then $x \in K^\times$. Hence the map $K \to K$ given by multiplication by x is an F-isomorphism and $\varphi(x) \in G(\varphi)$. Thus $D(\varphi) \subset G(\varphi)$. As $1 \in D(\varphi)$, we have $G(\varphi) \subset D(\varphi)$. In particular, φ is round.

Let K be a quadratic étale F-algebra. So $K = F_a$ for some $a \in F$ (cf. Examples 98.2 and 98.3). Denote the norm form N of F_a in Example 9.3 by $\langle\langle a]]$ and call it a *quadratic* 1-*fold Pfister form.* In particular, it is round. Explicitly, we have:

Example 9.4. For F_a a quadratic étale F algebra, we have:

(1) (Cf. Example 98.2.) If $\operatorname{char} F \neq 2$, then $F_a = F[t]/(t^2 - a)$ with $a \in F^\times$ and the quadratic form $\langle\langle a]] = \langle 1, -a\rangle_q \simeq \langle\langle a\rangle\rangle_b \otimes \langle 1\rangle_q$ is the norm form of F_a.

(2) (Cf. Example 98.3.) If $\operatorname{char} F = 2$, then $F_a = F[t]/(t^2 + t + a)$ with $a \in F$ and the quadratic form $\langle\langle a]] = [1, a]$ is the norm form of F_a. In particular, $\langle\langle a]] \simeq \langle\langle x^2 + x + a]]$ for any $x \in F$.

9.B. Quadratic Pfister forms and round forms. Let $n \geq 1$. A quadratic form isometric to a quadratic form of the type

$$\langle\langle a_1, \dots, a_n]] := \langle\langle a_1, \dots, a_{n-1}\rangle\rangle_b \otimes \langle\langle a_n]]$$

for some $a_1, \dots, a_{n-1} \in F^\times$ and $a_n \in F$ (with $a_n \neq 0$ if $\operatorname{char} F \neq 2$) is called a *quadratic n-fold Pfister form.* It is convenient to call the form isometric to $\langle 1\rangle_q$ a 0-*fold* Pfister form. Every quadratic n-fold Pfister form is nondegenerate by Lemma 8.16. We let

$$\begin{aligned}
P_n(F) &:= \{\varphi \mid \varphi \text{ a quadratic } n\text{-fold Pfister form}\},\\
P(F) &:= \bigcup P_n(F),\\
GP_n(F) &:= \{a\varphi \mid a \in F^\times,\ \varphi \text{ a quadratic } n\text{-fold Pfister form}\},\\
GP(F) &:= \bigcup GP_n(F).
\end{aligned}$$

Forms in $GP_n(F)$ are called *general quadratic n-fold Pfister forms.*

If $\operatorname{char} F \neq 2$, the form $\langle\langle a_1, \dots, a_n]]$ is the associated quadratic form of the bilinear Pfister form $\langle\langle a_1, \dots, a_n\rangle\rangle_b$ by Example 9.4 (1). We shall also use the notation $\langle\langle a_1, \dots, a_n\rangle\rangle$ for the quadratic Pfister form $\langle\langle a_1, \dots, a_n]]$ in this case.

The class of an n-fold Pfister form ($n > 0$) belongs to

$$I_q^n(F) := I^{n-1}(F) \cdot I_q(F).$$

As $[a, b] = a[1, ab]$ for all $a, b \in F$ with $a \neq 0$, every nondegenerate binary quadratic form is a general 1-fold Pfister form. In particular, $GP_1(F)$ generates $I_q(F)$. It follows that $GP_n(F)$ generates $I_q^n(F)$ as an abelian group. In fact, as

$$a\langle\langle b, c]] = \langle\langle ab, c]] - \langle\langle a, c]] \tag{9.5}$$

for all $a, b \in F^\times$ and $c \in F$ (with $c \neq 0$ if $\operatorname{char} F \neq 2$), $P_n(F)$ generates $I_q^n(F)$ as an abelian group for $n > 1$.

Note that in the case $\operatorname{char} F \neq 2$, under the identification of $I(F)$ with $I_q(F)$, the group $I^n(F)$ corresponds to $I_q^n(F)$ and a bilinear Pfister form $\langle\langle a_1, \dots, a_n\rangle\rangle_b$ corresponds to the quadratic Pfister form $\langle\langle a_1, \dots, a_n\rangle\rangle$.

Using the material in §98.E, we have the following example.

Example 9.6. Let A be a quaternion F-algebra.

(1) (Cf. Example 98.10.) Suppose that $\operatorname{char} F \neq 2$. If $A = \left(\frac{a,b}{F}\right)$, then the reduced quadratic norm form is equal to the quadratic form $\langle 1, -a, -b, ab\rangle = \langle\langle a, b\rangle\rangle$.

(2) (Cf. Example 98.11.) Suppose that char $F = 2$. If $A = \left[\frac{a,b}{F}\right]$, then the reduced quadratic norm form is equal to the quadratic form $[1, ab] \perp [a, b]$. This form is hyperbolic if $a = 0$ and is isomorphic to $\langle 1, a\rangle_b \otimes [1, ab] = \langle\langle a, ab]]$ otherwise.

Example 9.7. Let L/F be a separable quadratic field extension and $Q = (L/F, b)$, i.e., $Q = L \oplus Lj$ a quaternion F-algebra with $j^2 = b \in F^\times$ (cf. §98.E). For any $q = l + l'j \in Q$, we have $\mathrm{Nrd}_Q(q) = \mathrm{N}_L(l) - b\,\mathrm{N}_L(l')$. Therefore, $\mathrm{Nrd}_Q \simeq \langle\langle b\rangle\rangle \otimes \mathrm{N}_L$.

Proposition 9.8. *Let φ be a round quadratic form and $a \in F^\times$. Then*

(1) *The form $\langle\langle a\rangle\rangle \otimes \varphi$ is also round.*
(2) *If φ is regular, then the following are equivalent:*
   (i) *$\langle\langle a\rangle\rangle \otimes \varphi$ is isotropic.*
   (ii) *$\langle\langle a\rangle\rangle \otimes \varphi$ is hyperbolic.*
   (iii) *$a \in D(\varphi)$.*

PROOF. Set $\psi = \langle\langle a\rangle\rangle \otimes \varphi$.

(1): Since $1 \in D(\varphi)$, it suffices to prove that $D(\psi) \subset G(\psi)$. Let c be a nonzero value of ψ. Write $c = x - ay$ for some $x, y \in \widetilde{D}(\varphi)$. If $y = 0$, we have $c = x \in D(\varphi) = G(\varphi) \subset G(\psi)$. Similarly, $y \in G(\psi)$ if $x = 0$, hence $c = -ay \in G(\psi)$ as $-a \in G(\langle\langle a\rangle\rangle) \subset G(\psi)$.

Now suppose that x and y are nonzero. Since φ is round, $x, y \in G(\varphi)$ and, therefore,

$$\psi = \varphi \perp (-a\varphi) \simeq \varphi \perp (-ayx^{-1})\varphi = \langle\langle ayx^{-1}\rangle\rangle \otimes \varphi.$$

By Example 1.14, we know that $1 - ayx^{-1} \in G(\langle\langle ayx^{-1}\rangle\rangle) \subset G(\psi)$. Since $x \in G(\varphi) \subset G(\psi)$, we have $c = (1 - ayx^{-1})x \in G(\psi)$.

(2): (i) $\Rightarrow$ (iii): If φ is isotropic, then φ is universal by Proposition 7.13. So suppose that φ is anisotropic. Since $\psi = \varphi \perp (-a\varphi)$ is isotropic, there exist $x, y \in D(\varphi)$ such that $x - ay = 0$. Therefore $a = xy^{-1} \in D(\varphi)$ as $D(\varphi)$ is closed under multiplication.

(iii) $\Rightarrow$ (ii): As φ is round, $a \in D(\varphi) = G(\varphi)$ and $\langle\langle a\rangle\rangle \otimes \varphi$ is hyperbolic.

(ii) $\Rightarrow$ (i) is trivial. □

Corollary 9.9. *Quadratic Pfister forms are round.*

Corollary 9.10. *A quadratic Pfister form is either anisotropic or hyperbolic.*

PROOF. Suppose that ψ is an isotropic quadratic n-fold Pfister form. If $n = 1$, the result follows by Proposition 7.19(4). So assume that $n > 1$. Then $\psi \simeq \langle\langle a\rangle\rangle \otimes \varphi$ for a quadratic Pfister form φ and the result follows by Proposition 9.8. □

Let char $F = 2$. We need another characterization of hyperbolic Pfister forms in this case. Let $\wp : F \to F$ be the Artin-Schreier map defined by $\wp(x) = x^2 + x$. (Cf. §98.B.) For a quadratic 1-fold Pfister form we have $\langle\langle d]]$ is hyperbolic if and only if $d \in \mathrm{im}(\wp)$ by Example 98.3. More generally, we have:

Lemma 9.11. *Let $\mathfrak{b}$ be an anisotropic bilinear Pfister form and $d \in F$. Then $\mathfrak{b} \otimes \langle\langle d]]$ is hyperbolic if and only if $d \in \mathrm{im}(\wp) + \widetilde{D}(\mathfrak{b}')$.*

PROOF. Suppose that $\mathfrak{b} \otimes \langle\langle d]]$ is hyperbolic and therefore isotropic. Let $\{e, f\}$ be the standard basis of $\langle\langle d]]$. Let $v \otimes e + w \otimes f$ be an isotropic vector of $\mathfrak{b} \otimes \langle\langle d]]$ where $v, w \in V_{\mathfrak{b}}$. We have $a + b + cd = 0$ where $a = \mathfrak{b}(v, v)$, $b = \mathfrak{b}(v, w)$, and $c = \mathfrak{b}(w, w)$.

As $\mathfrak{b}$ is anisotropic, we have $w \neq 0$, i.e., $c \neq 0$. Suppose first that $v = sw$ for some $s \in F$. Then $0 = a + b + cd = c(s^2 + s + d)$, hence $d = s^2 + s \in \operatorname{im}(\wp)$.

Now suppose that v and w generate a 2-dimensional subspace W of $V_{\mathfrak{b}}$. The determinant of $\mathfrak{b}|_W$ is equal to $xF^{\times 2}$ where $x = b^2 + bc + c^2 d$. Hence $\mathfrak{b}|_W \simeq c\langle\langle x\rangle\rangle$ by Example 1.10. As $c \in D(\mathfrak{b}) = G(\mathfrak{b})$ by Corollary 6.2, the form $\langle\langle x\rangle\rangle$ is isometric to a subform of $\mathfrak{b}$. By the Bilinear Witt Cancellation (Corollary 1.28), we have $\langle x\rangle$ is a subform of $\mathfrak{b}'$, i.e., $x \in D(\mathfrak{b}')$. Hence $(b/c)^2 + (b/c) + d = x/c^2 \in D(\mathfrak{b}')$ and therefore $d \in \operatorname{im}(\wp) + D(\mathfrak{b}')$.

Conversely, let $d = x + y$ where $x \in \operatorname{im}(\wp)$ and $y \in \widetilde{D}(\mathfrak{b}')$. If $y = 0$, then $\langle\langle d]]$ is hyperbolic, hence so is $\mathfrak{b} \otimes \langle\langle d]]$. So suppose that $y \neq 0$. By Lemma 6.11 there is a bilinear Pfister form $\mathfrak{c}$ such that $\mathfrak{b} \simeq \mathfrak{c} \otimes \langle\langle y\rangle\rangle$. Therefore, $\mathfrak{b} \otimes \langle\langle d]] \simeq \mathfrak{c} \otimes \langle\langle y, d\rangle\rangle$ is hyperbolic as $\langle\langle y, d]] \simeq \langle\langle y, y]]$ by Example 98.3 which is hyperbolic. □

9.C. Annihilators. If φ is a nondegenerate quadratic form over F, then the annihilator of φ in $W(F)$

$$\operatorname{ann}_{W(F)}(\varphi) := \{\mathfrak{c} \in W(F) \mid \mathfrak{c} \cdot \varphi = 0\}$$

is an ideal. When φ is a Pfister form this ideal has the structure that we had when φ was a bilinear anisotropic Pfister form. Indeed, the same proof yielding Proposition 6.22 and Corollary 6.23 shows:

Theorem 9.12. *Let φ be anisotropic quadratic Pfister form. Then $\operatorname{ann}_{W(F)}(\varphi)$ is generated by binary symmetric bilinear forms $\langle\langle x\rangle\rangle_{\mathfrak{b}}$ with $x \in D(\varphi)$.*

As in the bilinear case, if φ is 2-dimensional, we obtain stronger results. Indeed, the same proofs for the corresponding results show:

Lemma 9.13 (cf. Lemma 6.24). *Let φ be a binary anisotropic quadratic form over F and $\mathfrak{c}$ an anisotropic bilinear form over F such that $\mathfrak{c} \otimes \varphi$ is isotropic. Then $\mathfrak{c} \simeq \mathfrak{d} \perp \mathfrak{e}$ for some binary bilinear form $\mathfrak{d}$ annihilated by φ and bilinear form $\mathfrak{e}$ over F.*

Proposition 9.14 (cf. Proposition 6.25). *Let φ be a binary anisotropic quadratic form over F and $\mathfrak{c}$ an anisotropic bilinear form over F. Then there exist bilinear forms $\mathfrak{c}_1$ and $\mathfrak{c}_2$ over F such that $\mathfrak{c} \simeq \mathfrak{c}_1 \perp \mathfrak{c}_2$ with $\mathfrak{c}_2 \otimes \varphi$ anisotropic and $\mathfrak{c}_1 \simeq \mathfrak{d}_1 \perp \cdots \perp \mathfrak{d}_n$ where each $\mathfrak{d}_i$ is a binary bilinear form annihilated by φ. In particular, $-\det \mathfrak{d}_i \in D(\varphi)$ for each i.*

Corollary 9.15 (cf. Corollary 6.26). *Let φ be a binary anisotropic quadratic form over F and $\mathfrak{c}$ an anisotropic bilinear form over F annihilated by $\mathfrak{b}$. Then $\mathfrak{c} \simeq \mathfrak{d}_1 \perp \cdots \perp \mathfrak{d}_n$ for some binary bilinear forms $\mathfrak{d}_i$ annihilated by $\mathfrak{b}$ for $i \in [1,\ n]$.*

10. Totally singular forms

Totally singular forms in characteristic different from 2 are zero forms, but in characteristic 2 they become interesting. In this section, we look at totally singular forms in characteristic 2. In particular, throughout most of this section, $\operatorname{char} F = 2$.

Let $\operatorname{char} F = 2$. Let φ be a quadratic form over F. Then φ is a totally singular form if and only if it is diagonalizable. Moreover, if this is the case, then every basis of V_φ is orthogonal by Remark 7.24 and $\widetilde{D}(\varphi)$ is a vector space over the field F^2.

We investigate the F-subspace $\big(\widetilde{D}(\varphi)\big)^{1/2}$ of $F^{1/2}$. Define an F-linear map

$$f : V_\varphi \to \big(\widetilde{D}(\varphi)\big)^{1/2} \quad \text{given by} \quad f(v) = \sqrt{\varphi(v)}.$$

Then f is surjective and $\operatorname{Ker}(f) = \operatorname{rad}\varphi$. Let $\widetilde{\varphi}$ be the quadratic form on $\big(\widetilde{D}(\varphi)\big)^{1/2}$ over F defined by $\widetilde{\varphi}(\sqrt{a}) = a$. Clearly $\widetilde{\varphi}$ is anisotropic. Consequently, if $\bar{\varphi}$ is the quadratic form induced on $V_\varphi/\operatorname{rad}\varphi$ by φ, then f induces an isometry between $\bar{\varphi}$ and $\widetilde{\varphi}$. Moreover, $\widetilde{\varphi} \simeq \varphi_{an}$. Therefore, if $\operatorname{char} F = 2$, the correspondence $\varphi \mapsto \widetilde{D}(\varphi)$ gives rise to a bijection

Isometry classes of totally singular anisotropic quadratic forms	$\xrightarrow{\sim}$	Finite dimensional F^2-subspaces of F

Moreover, for any totally singular quadratic form φ, we have

$$\dim \varphi_{an} = \dim \widetilde{D}(\varphi)$$

and if φ and ψ are two totally singular quadratic forms, then

$$\varphi \simeq \psi \quad \text{if and only if} \quad D(\varphi) = D(\psi) \text{ and } \dim\varphi = \dim\psi.$$

We also have $\widetilde{D}(\varphi \perp \psi) = \widetilde{D}(\varphi) + \widetilde{D}(\psi)$.

Example 10.1. If F is a separably closed field of characteristic 2, the anisotropic quadratic forms are diagonalizable, hence totally singular.

Note that if $\mathfrak{b}$ is an alternating bilinear form and ψ is a totally singular quadratic form, then $\mathfrak{b}\otimes\psi = 0$. It follows that the tensor product of totally singular quadratic forms $\varphi \otimes \psi := \mathfrak{c} \otimes \psi$ is well-defined where $\mathfrak{c}$ is a bilinear form with $\varphi = \varphi_{\mathfrak{c}}$. The space $\widetilde{D}(\varphi \otimes \psi)$ is spanned by $D(\varphi)\cdot D(\psi)$ over F^2.

Proposition 10.2. *Let* $\operatorname{char} F = 2$. *If φ is a totally singular quadratic form, then*

$$G(\varphi) = \{a \in F^\times \mid aD(\varphi) \subset D(\varphi)\}.$$

Proof. The inclusion "$\subset$" follows from Lemma 9.1. Conversely, let $a \in F^\times$ satisfy $aD(\varphi) \subset D(\varphi)$. Then the F-linear map $g : \big(\widetilde{D}(\varphi)\big)^{1/2} \to \big(\widetilde{D}(\varphi)\big)^{1/2}$ defined by $g(b) = \sqrt{a}\, b$ is an isometry between $\widetilde{\varphi}$ and $a\widetilde{\varphi}$. Therefore $a \in G(\widetilde{\varphi}) = G(\varphi)$. □

It follows from Proposition 10.2 that $\widetilde{G}(\varphi) := G(\varphi) \cup \{0\}$ is a subfield of F containing F^2 and $\widetilde{D}(\varphi)$ is a vector space over $\widetilde{G}(\varphi)$.

It is also convenient to introduce a variant of the notion of Pfister forms in all characteristics. A quadratic form φ is called a *quasi-Pfister form* if there exists a bilinear Pfister form $\mathfrak{b}$ with $\varphi \simeq \varphi_{\mathfrak{b}}$, i.e., $\varphi \simeq \langle\langle a_1, \dots, a_n \rangle\rangle_b \otimes \langle 1 \rangle_q$. for some $a_1, \dots, a_n \in F^\times$. Denote

$$\langle\langle a_1, \dots, a_n \rangle\rangle_b \otimes \langle 1 \rangle_q \quad \text{by} \quad \langle\langle a_1, \dots, a_n \rangle\rangle_q.$$

If $\operatorname{char} F \neq 2$, then the classes of quadratic Pfister and quasi-Pfister forms coincide. If $\operatorname{char} F = 2$ every quasi-Pfister form is totally singular. Quasi-Pfister forms have some properties similar to those for quadratic Pfister forms.

Corollary 10.3. *Quasi-Pfister forms are round.*

Proof. Let $\mathfrak{b}$ be a bilinear Pfister form. As $\langle 1 \rangle_q$ is a round quadratic form, the quadratic form $\mathfrak{b} \otimes \langle 1 \rangle_q$ is round by Proposition 9.8. □

Remark 10.4. Let $\operatorname{char} F = 2$. Let $\rho = \langle\langle a_1, \dots, a_n \rangle\rangle_q$ be an anisotropic quasi-Pfister form. Then $\widetilde{D}(\rho)$ is equal to the field $F^2(a_1, \dots, a_n)$ of degree 2^n over F^2. Conversely, every field K such that $F^2 \subset K \subset F$ with $[K : F^2] = 2^n$ is generated by n elements and therefore $K = \widetilde{D}(\rho)$ for an anisotropic n-fold quasi-Pfister form ρ. Thus we get a bijection

Isometry classes of anisotropic n-fold quasi-Pfister forms	$\simeq$	Fields K with $F^2 \subset K \subset F$ and $[K : F^2] = 2^n$

Let φ be an anisotropic totally singular quadratic form. Then $K = \widetilde{G}(\varphi)$ is a field with $K \cdot \widetilde{D}(\varphi) \subset \widetilde{D}(\varphi)$. We have $[K : F^2] < \infty$ and $\widetilde{D}(\varphi)$ is a vector space over K. Let $\{b_1, \dots, b_m\}$ be a basis of $\widetilde{D}(\varphi)$ over K and set $\psi = \langle b_1, \dots, b_m \rangle_q$. Choose an anisotropic n-fold quasi-Pfister form ρ such that $\widetilde{D}(\rho) = \widetilde{G}(\varphi)$. As $\widetilde{D}(\varphi)$ is the vector space spanned by $K \cdot D(\psi)$ over F^2, we have $\varphi \simeq \rho \otimes \psi$. In fact, ρ is the largest quasi-Pfister divisor of φ, i.e., quasi-Pfister form of maximal dimension such that $\varphi \simeq \rho \otimes \psi$.

11. The Clifford algebra

To each quadratic form φ, one associates a $\mathbb{Z}/2\mathbb{Z}$-graded algebra by factoring the tensor algebra on V_φ by the relation $\varphi(v) = v^2$. This algebra, called the Clifford algebra generalizes the exterior algebra. In this section, we study the basic properties of Clifford algebras.

Let φ be a quadratic form on V over F. Define the *Clifford algebra* of φ to be the factor algebra $C(\varphi)$ of the tensor algebra

$$T(V) := \coprod_{n \geq 0} V^{\otimes n}$$

modulo the ideal I generated by $(v \otimes v) - \varphi(v)$ for all $v \in V$. We shall view vectors in V as elements of $C(\varphi)$ via the natural F-linear map $V \to C(\varphi)$. Note that $v^2 = \varphi(v)$ in $C(\varphi)$ for every $v \in V$. The Clifford algebra of φ has a natural $\mathbb{Z}/2\mathbb{Z}$-grading

$$C(\varphi) = C_0(\varphi) \oplus C_1(\varphi)$$

as I is homogeneous if degree is viewed modulo two. The subalgebra $C_0(\varphi)$ is called the *even Clifford algebra of* φ. We have $\dim C(\varphi) = 2^{\dim \varphi}$ and $\dim C_0(\varphi) = 2^{\dim \varphi - 1}$. If K/F is a field extension, $C(\varphi_K) = C(\varphi)_K$ and $C_0(\varphi_K) = C_0(\varphi)_K$.

Lemma 11.1. *Let φ be a quadratic form on V over F with polar form $\mathfrak{b}$. Let $v, w \in V$. Then $\mathfrak{b}(v, w) = vw + wv$ in $C(\varphi)$. In particular, v and w are orthogonal if and only if $vw = -wv$ in $C(\varphi)$.*

PROOF. This follows from the polar identity. □

Example 11.2. (1) The Clifford algebra of the zero quadratic form on V coincides with the exterior algebra $\Lambda(V)$.

(2) $C_0(\langle a \rangle) = F$.

(3) If $\operatorname{char} F \neq 2$, then the Clifford algebra of the quadratic form $\langle a, b \rangle$ is $C(\langle a, b \rangle) = \left(\frac{a,b}{F}\right)$ and $C_0(\langle a, b \rangle) = F_{-ab}$. In particular, $C_0(\langle\langle b \rangle\rangle) = F_b$.

(4) If $\operatorname{char} F = 2$, then $C\big([a,b]\big) = \left[\frac{a,b}{F}\right]$ and $C_0\big([a,b]\big) = F_{ab}$. In particular, $C_0\big(\langle\langle b]]\big) = F_b$.

By construction, the Clifford algebra satisfies the following universal property: For any F-algebra A and any F-linear map $f : V \to A$ satisfying $f(v)^2 = \varphi(v)$ for all $v \in V$, there exists a unique F-algebra homomorphism $\tilde{f} : C(\varphi) \to A$ satisfying $\tilde{f}(v) = f(v)$ for all $v \in V$.

Example 11.3. Let $C(\varphi)^{op}$ denote the Clifford algebra of φ with the opposite multiplication. The canonical linear map $V \to C(\varphi)^{op}$ extends to an involution $^{-} : C(\varphi) \to C(\varphi)$ given by the algebra isomorphism $C(\varphi) \to C(\varphi)^{op}$. Note that if $x = v_1 v_2 \cdots v_n$, then $\bar{x} = v_n \cdots v_2 v_1$.

Proposition 11.4. *Let φ be a quadratic form on V over F and let $a \in F^\times$. Then*

(1) *$C_0(a\varphi) \simeq C_0(\varphi)$, i.e., the even Clifford algebras of similar quadratic forms are isomorphic.*

(2) *Let $\varphi = \langle a \rangle \perp \psi$. Then $C_0(\varphi) \simeq C(-a\psi)$.*

PROOF. (1): Set $K = F[t]/(t^2 - a) = F \oplus F\bar{t}$ with $\bar{t}$ the image of t in K. Since $(v \otimes \bar{t})^2 = \varphi(v) \otimes \bar{t}^2 = a\varphi(v) \otimes 1$ in $C(\varphi)_K = C(\varphi) \otimes_F K$, there is an F-algebra homomorphism $\alpha : C(a\varphi) \to C(\varphi)_K$ taking $v \in V$ to $v \otimes \bar{t}$ by the universal property of the Clifford algebra $a\varphi$. Since

$$(v \otimes \bar{t})(v' \otimes \bar{t}) = vv' \otimes \bar{t}^2 = avv' \otimes 1 \in C(\varphi) \subset C(\varphi)_K,$$

the map α restricts to an F-algebra homomorphism $C_0(a\varphi) \to C_0(\varphi)$. As this map is clearly a surjective map of algebras of the same dimension, it is an isomorphism.

(2): Let $V = Fv \oplus W$ with $\varphi(v) = a$ and $W \subset (Fv)^\perp$. Since

$$(vw)^2 = -v^2 w^2 = -\varphi(v)\psi(w) = -a\psi(w)$$

for every $w \in W$, the map $W \to C_0(\varphi)$ defined by $w \mapsto vw$ extends to an F-algebra isomorphism $C(-a\psi) \xrightarrow{\sim} C_0(\varphi)$ by the universal property of Clifford algebras. □

Let φ be a quadratic form on V over F. Applying the universal property of Clifford algebras to the natural linear map $V \to V/\operatorname{rad}\varphi \to C(\bar{\varphi})$, where $\bar{\varphi}$ is the induced quadratic form on $V/\operatorname{rad}\varphi$, we get a surjective F-algebra homomorphism $C(\varphi) \to C(\bar{\varphi})$ with kernel $(\operatorname{rad}\varphi)C(\varphi)$. Consequently, we get canonical isomorphisms

$$C(\bar{\varphi}) \simeq C(\varphi)/(\operatorname{rad}\varphi)C(\varphi),$$
$$C_0(\bar{\varphi}) \simeq C_0(\varphi)/(\operatorname{rad}\varphi)C_1(\varphi).$$

Example 11.5. Let $\varphi = \mathbb{H}(W)$ be the hyperbolic form of the vector space $V = W \oplus W^*$ with W a nonzero vector space. Then

$$C(\varphi) \simeq \operatorname{End}_F\big(\Lambda(W)\big),$$

where the exterior algebra $\Lambda(W)$ of V is considered as a vector space (cf. [**86**, Prop. 8.3]). Moreover,

$$C_0(\varphi) = \operatorname{End}_F\big(\Lambda_0(W)\big) \times \operatorname{End}_F\big(\Lambda_1(W)\big),$$

where

$$\Lambda_0(W) := \bigoplus_{i \geq 0} \Lambda^{2i}(W) \quad \text{and} \quad \Lambda_1(W) := \bigoplus_{i \geq 0} \Lambda^{2i+1}(W).$$

In particular, $C(\varphi)$ is a split central simple F-algebra and the center of $C_0(\varphi)$ is the split quadratic étale F-algebra $F \times F$. Note also that the natural F-linear map $V \to C(\varphi)$ is injective.

Proposition 11.6. *Let φ be a quadratic form over F.*

(1) *If $\dim \varphi \geq 2$ is even, then the following conditions are equivalent:*
   - (a) *φ is nondegenerate.*
   - (b) *$C(\varphi)$ is central simple.*
   - (c) *$C_0(\varphi)$ is separable with center $Z(\varphi)$, a quadratic étale algebra.*

(2) *If $\dim \varphi \geq 3$ is odd, then the following conditions are equivalent:*
   - (a) *φ is nondegenerate.*
   - (b) *$C_0(\varphi)$ is central simple.*

PROOF. We may assume that F is algebraically closed. Suppose first that φ is nondegenerate and even-dimensional. Then φ is hyperbolic; and, by Example 11.5, the algebra $C(\varphi)$ is a central simple F-algebra and $C_0(\varphi)$ is a separable F-algebra whose center is the split quadratic étale F-algebra $F \times F$.

Conversely, suppose that the even Clifford algebra $C_0(\varphi)$ is separable or $C(\varphi)$ is central simple. Let $v \in \operatorname{rad} \varphi$. The ideals $I = vC_1(\varphi)$ in $C_0(\varphi)$ and $J = vC(\varphi)$ in $C(\varphi)$ satisfy $I^2 = 0 = J^2$. Consequently, $I = 0$ or $J = 0$ as $C_0(\varphi)$ is semi-simple or $C(\varphi)$ is central simple and therefore $v = 0$ thus $\operatorname{rad} \varphi = 0$. Thus φ is nondegenerate.

Now suppose that $\dim \varphi$ is odd. Write $\varphi = \langle a \rangle \perp \psi$ for some $a \in F$ and an even-dimensional form ψ. Let $v \in V_\varphi$ be a nonzero vector satisfying $\varphi(v) = a$ with v orthogonal to V_ψ. If φ is nondegenerate, then $a \neq 0$ and ψ is nondegenerate. It follows from Proposition 11.4(2) and the first part of the proof that the algebra $C_0(\varphi) \simeq C(-a\psi)$ is central simple.

Conversely, suppose that the algebra $C_0(\varphi)$ is central simple. As $\dim \varphi \geq 3$, the subspace $I := vC_1(\varphi)$ of $C_0(\varphi)$ is nonzero. If $a = 0$, then I is a nontrivial ideal of $C_0(\varphi)$, a contradiction to the simplicity of $C_0(\varphi)$. Thus $a \neq 0$ and by Proposition 11.4(2), $C_0(\varphi) \simeq C(-a\psi)$. Hence by the first part of the proof, the form ψ is nondegenerate. Therefore, φ is also nondegenerate. □

Lemma 11.7. *Let φ be a nondegenerate quadratic form of positive even dimension. Then $yx = \bar{x}y$ for every $x \in Z(\varphi)$ and $y \in C_1(\varphi)$.*

PROOF. Let $v \in V_\varphi$ be an anisotropic vector hence a unit in $C(\varphi)$. Since conjugation by v on $C(\varphi)$ stabilizes $C_0(\varphi)$, it stabilizes the center of $C_0(\varphi)$, i.e., $vZ(\varphi)v^{-1} = Z(\varphi)$. As $C(\varphi)$ is a central algebra, conjugation by v induces a nontrivial automorphism on $Z(\varphi)$ given by $x \mapsto \bar{x}$ otherwise, since $C_1(\varphi) = C_0(\varphi)v$, the full algebra $C(\varphi)$ would commute with $Z(\varphi)$. Thus $vx = \bar{x}v$ for all $x \in Z(\varphi)$. Let $y \in C_1(\varphi)$. Writing y in the form $y = zv$ for some $z \in C_0(\varphi)$, we have $yx = zvx = z\bar{x}v = \bar{x}zv = \bar{x}y$ for every $x \in Z(\varphi)$. □

Corollary 11.8. *Let φ be a nondegenerate quadratic form of positive even dimension. If a is a norm for the quadratic étale algebra $Z(\varphi)$ then $C(a\varphi) \simeq C(\varphi)$.*

PROOF. Let $x \in Z(\varphi)$ satisfy $\mathrm{N}(x) = a$. By Lemma 11.7, we have $(vx)^2 = \mathrm{N}(x)v^2 = a\varphi(v)$ in $C(\varphi)$ for every $v \in V$. By the universal property of the Clifford algebra $a\varphi$, there is an algebra homomorphism $\alpha : C(a\varphi) \to C(\varphi)$ mapping v to vx. Since both algebras are simple of the same dimension, α is an isomorphism. □

12. Binary quadratic forms and quadratic algebras

In §98.B and §98.E we review the theory of quadratic and quaternion algebras. In this section, we study the relationship between these algebras and quadratic forms.

If A is a quadratic F-algebra, we let Tr_A and N_A denote the trace form and the quadratic norm form of A, respectively. (cf. §98.B). Note that N_A is a binary form representing 1.

Conversely, if φ is a binary quadratic form over F, then the even Clifford algebra $C_0(\varphi)$ is a quadratic F-algebra. We have defined two maps

Quadratic F-algebras	$\longrightarrow$ $\longleftarrow$	Binary quadratic forms representing 1

Proposition 12.1. *The two maps above induce mutually inverse bijections of the set of isomorphism classes of quadratic F-algebras and the set of isometry classes of binary quadratic forms representing one. Under these bijections, we have:*

(1) *Quadratic étale algebras correspond to nondegenerate binary forms.*
(2) *Quadratic fields correspond to anisotropic binary forms.*
(3) *Semisimple algebras correspond to regular binary quadratic forms.*

PROOF. Let A be a quadratic F-algebra. We need to show that $A \simeq C_0(\mathrm{N}_A)$. We have $C_1(\mathrm{N}_A) = A$. Therefore, the map $\alpha : A \to C_0(\mathrm{N}_A)$ defined by $x \mapsto 1 \cdot x$ (where dot denotes the product in the Clifford algebra) is an F-linear isomorphism. We shall show that α is an algebra isomorphism, i.e., $(1 \cdot x) \cdot (1 \cdot y) = 1 \cdot xy$ for all $x, y \in A$. The equality holds if $x \in F$ or $y \in F$. Since A is 2-dimensional over F, it suffices to check the equality when $x = y$ and it does not lie in F. We have $1 \cdot x + x \cdot 1 = \mathrm{N}_A(x+1) - \mathrm{N}_A(x) - \mathrm{N}_A(1) = \mathrm{Tr}_A(x)$, so

$$(1 \cdot x) \cdot (1 \cdot x) = (1 \cdot x) \cdot (\mathrm{Tr}_A(x) - x \cdot 1) = 1 \cdot \mathrm{Tr}_A(x)x - 1 \cdot \mathrm{N}_A(x) \cdot 1 = 1 \cdot x^2$$

as needed.

Conversely, let φ be a binary quadratic form on V representing 1. We shall show that the norm form for the quadratic F-algebra $C_0(\varphi)$ is isometric to φ. Let $v_0 \in V$ be a vector satisfying $\varphi(v_0) = 1$. Let $f : V \to C_0(\varphi)$ be the F-linear isomorphism defined by $f(v) = v \cdot v_0$. The quadratic equation (98.1) for $v \cdot v_0 \in C_0(\varphi)$ in §98.B becomes

$$(v \cdot v_0)^2 = v \cdot (\mathfrak{b}(v, v_0) - v \cdot v_0) \cdot v_0 = \mathfrak{b}(v, v_0)(v \cdot v_0) - \varphi(v),$$

so $\mathrm{N}_{C_0(\varphi)}(v \cdot v_0) = \varphi(v)$; hence

$$\mathrm{N}_{C_0(\varphi)}\big(f(v)\big) = \mathrm{N}_{C_0(\varphi)}(v \cdot v_0) = \varphi(v),$$

i.e., f is an isometry of φ with the norm form of $C_0(\varphi)$ as needed.

In order to prove that quadratic étale algebras correspond to nondegenerate binary forms it is sufficient to assume that F is algebraically closed. Then a quadratic étale algebra A is isomorphic to $F \times F$ and therefore $\mathrm{N}_A \simeq \mathbb{H}$. Conversely, by Example 11.5, $C_0(\mathbb{H}) \simeq F \times F$.

If a quadratic F-algebra A is a field, then obviously the norm form N_A is anisotropic. Conversely, if N_A is anisotropic, then for every nonzero $a \in A$ we have $a\bar{a} = \mathrm{N}_A(a) \neq 0$, therefore a is invertible, i.e., A is a field.

Statement (3) follows from statements (1) and (2), since a quadratic F-algebra is semisimple if and only if it is either a field or $F \times F$, and a binary quadratic form is regular if and only if it is anisotropic or hyperbolic. $\square$

Corollary 12.2. (1) *Let A and B be quadratic F-algebras. Then A and B are isomorphic if and only if the norm forms N_A and N_B are isometric.*

(2) *Let φ and ψ be nonzero binary quadratic forms. Then φ and ψ are similar if and only if the even Clifford algebras $C_0(\varphi)$ and $C_0(\psi)$ are isomorphic.*

Corollary 12.3. *Let φ be an anisotropic binary quadratic form and let K/F be a quadratic field extension. Then φ_K is isotropic if and only if $K \simeq C_0(\varphi)$.*

PROOF. By Proposition 12.1, the form φ_K is isotropic if and only if the 2-dimensional even Clifford K-algebra $C_0(\varphi_K) = C_0(\varphi) \otimes K$ is not a field. The latter is equivalent to $K \simeq C_0(\varphi)$. $\square$

We now consider the relationship between quaternion and Clifford algebras.

Proposition 12.4. *Let Q be a quaternion F-algebra and φ the reduced norm quadratic form of Q. Then $C(\varphi) \simeq \mathbb{M}_2(Q)$.*

PROOF. For every $x \in Q$, let m_x be the matrix $\left(\begin{smallmatrix} 0 & x \\ \bar{x} & 0 \end{smallmatrix}\right)$ in $\mathbb{M}_2(Q)$. Since $m_x^2 = x\bar{x} = \operatorname{Nrd}(x) = \varphi(x)$, the F-linear map $Q \to \mathbb{M}_2(Q)$ defined by $x \mapsto m_x$ extends to an F-algebra homomorphism $\alpha : C(\varphi) \to \mathbb{M}_2(Q)$ by the universal property of Clifford algebras. As $C(\varphi)$ is a central simple algebra of dimension $16 = \dim \mathbb{M}_2(Q)$, the map α is an isomorphism. $\square$

Corollary 12.5. *Two quaternion algebras are isomorphic if and only if their reduced norm quadratic forms are isometric. In particular, a quaternion algebra is split if and only if its reduced norm quadratic form is hyperbolic.*

Exercise 12.6. Let Q be a quaternion F-algebra and let φ' be the restriction of the reduced norm quadratic form to the space Q' of pure quaternions. Prove that φ' is nondegenerate and $C_0(\varphi')$ is isomorphic to Q. Conversely, prove that any 3-dimensional nondegenerate quadratic form ψ is similar to the restriction of the reduced norm quadratic form of $C_0(\psi)$ to the space of pure quaternions. Therefore, there is a natural bijection between the set of isomorphism classes of quaternion algebras over F and the set of similarity classes of 3-dimensional nondegenerate quadratic forms over F.

13. The discriminant

A major objective is to define sufficiently many invariants of quadratic forms. The first and simplest such invariant is the dimension. In this section, using quadratic étale algebras, we introduce a second invariant, the discriminant, of a nondegenerate quadratic form.

Let φ be a nondegenerate quadratic form over F of positive even dimension. The center $Z(\varphi)$ of $C_0(\varphi)$ is a quadratic étale F-algebra. The class of $Z(\varphi)$ in $\text{Ét}_2(F)$, the group of isomorphisms classes of quadratic étale F-algebras under the operation $\cdot$ induced by $\star$ (cf. §98.B), is called the *discriminant* of φ and will be denoted by $\operatorname{disc}(\varphi)$. Define the discriminant of the zero form to be trivial.

Example 13.1. By Example 11.2, we have $\operatorname{disc}(\langle a, b\rangle) = F_{-ab}$ if $\operatorname{char} F \neq 2$ and $\operatorname{disc}([a, b]) = F_{ab}$ if $\operatorname{char} F = 2$. It follows from Example 11.5 that the discriminant of a hyperbolic form is trivial.

The discriminant is a complete invariant for the similarity class of a nondegenerate binary quadratic form, i.e.,

Proposition 13.2. *Two nondegenerate binary quadratic forms are similar if and only if their discriminants are equal.*

PROOF. Let $\operatorname{disc}(\varphi) = \operatorname{disc}(\psi)$, i.e., $C_0(\varphi) \simeq C_0(\psi)$. Write $\varphi = a\varphi'$ and $\psi = b\psi'$, where φ' and ψ' both represent 1. By Proposition 12.1, the forms φ' and ψ' are the norm forms for $C_0(\varphi') = C_0(\varphi)$ and $C_0(\psi') = C_0(\psi)$, respectively. Since these algebras are isomorphic, we have $\varphi' \simeq \psi'$. ☐

Corollary 13.3. *A nondegenerate binary quadratic form φ is hyperbolic if and only if* $\operatorname{disc}(\varphi)$ *is trivial.*

Lemma 13.4. *Let φ and ψ be nondegenerate quadratic forms of even dimension over F. Then* $\operatorname{disc}(\varphi \perp \psi) = \operatorname{disc}(\varphi) \cdot \operatorname{disc}(\psi)$.

PROOF. The even Clifford algebra $C_0(\varphi \perp \psi)$ coincides with

$$\big(C_0(\varphi) \otimes_F C_0(\psi)\big) \oplus \big(C_1(\varphi) \otimes_F C_1(\psi)\big)$$

and contains $Z(\varphi) \otimes_F Z(\psi)$. By Lemma 11.7, we have $yx = \bar{x}y$ for every $x \in Z(\varphi)$ and $y \in C_1(\varphi)$. Similarly, $wz = \bar{z}w$ for every $z \in Z(\psi)$ and $w \in C_1(\psi)$. Therefore, the center of $C_0(\varphi \perp \psi)$ coincides with the subalgebra $Z(\varphi) \star Z(\psi)$ of all stable elements of $Z(\varphi) \otimes_F Z(\psi)$ under the automorphism $x \otimes y \mapsto \bar{x} \otimes \bar{y}$. ☐

Example 13.5. (1) Let $\operatorname{char} F \neq 2$. Then

$$\operatorname{disc}(\langle a_1, a_2, \dots, a_{2n}\rangle) = F_c$$

where $c = (-1)^n a_1 a_2 \dots a_{2n}$. For this reason, the discriminant is often called the *signed determinant* when the characteristic of F is different from 2.

(2) Let $\operatorname{char} F = 2$. Then

$$\operatorname{disc}([a_1, b_1] \perp \cdots \perp [a_n, b_n]) = F_c$$

where $c = a_1 b_1 + \cdots + a_n b_n$. The discriminant in the characteristic 2 case is often called the *Arf invariant.*

Proposition 13.6. *Let ρ be a nondegenerate quadratic form over F. If* $\operatorname{disc}(\rho) = 1$ *and $\rho \perp \langle a\rangle \sim \langle a\rangle$ for some $a \in F^\times$, then $\rho \sim 0$.*

PROOF. We may assume that the characteristic of F is 2. By Proposition 8.8, we have $\rho \sim [a, b]$ for some $b \in F$. Therefore, $\operatorname{disc}([a, b])$ is trivial and $[a, b] \sim 0$. ☐

It follows from Lemma 13.4 and Example 11.5 that the map

$$e_1 : I_q(F) \to \text{Ét}_2(F)$$

taking a form φ to $\operatorname{disc}(\varphi)$ is a well-defined group homomorphism.

The analogue of Proposition 4.13 is true, viz.,

Theorem 13.7. *The homomorphism e_1 is surjective with kernel $I_q^2(F)$.*

PROOF. The surjectivity follows from Example 13.1. Since similar forms have isomorphic even Clifford algebras, for any $\varphi \in I_q(F)$ and $a \in F^\times$, we have

$$e_1\big(\langle\langle a\rangle\rangle \cdot \varphi\big) = e_1(\varphi) \cdot e_1(-a\varphi) = 0.$$

Therefore, $e_1\big(I_q^2(F)\big) = 0$.

Let $\varphi \in I_q(F)$ be a form with trivial discriminant. We show by induction on $\dim \varphi$ that $\varphi \in I_q^2(F)$. The case $\dim \varphi = 2$ follows from Corollary 13.3. Suppose that $\dim \varphi \geq 4$. Write $\varphi = \rho \perp \psi$ with ρ a binary form. Let $a \in F^\times$ be chosen so that the form $\varphi' = a\rho \perp \psi$ is isotropic. Then the class of φ' in $I_q(F)$ is represented by a form of dimension less than $\dim \varphi$. As $\operatorname{disc}(\varphi') = \operatorname{disc}(\varphi)$ is trivial, $\varphi' \in I_q^2(F)$ by induction. Since $\rho \equiv a\rho \mod I_q^2(F)$, φ also lies in $I_q^2(F)$. □

Remark 13.8. One can also define a discriminant like invariant for all nondegenerate quadratic forms. Let φ be a nondegenerate quadratic form. Define the *determinant* $\det \varphi$ of φ to be $\det \mathfrak{b}_\varphi$ in $F^\times/F^{\times 2}$ if the bilinear form $\mathfrak{b}_\varphi$ is nondegenerate. If $\operatorname{char} F = 2$ and $\dim \varphi$ is odd (the only remaining case), define $\det \varphi$ to be $aF^{\times 2}$ in $F^\times/F^{\times 2}$ where $a \in F^\times$ satisfies $\varphi|_{\operatorname{rad} \mathfrak{b}_\varphi} \simeq \langle a\rangle$.

Remark 13.9. Let φ be a nondegenerate even-dimensional quadratic form with trivial discriminant over F, i.e., $\varphi \in I_q^2(F)$. Then $Z(\varphi) \simeq F \times F$, in particular, $C(\varphi)$ is not a division algebra, i.e., $C(\varphi) \simeq M_2\big(C^+(\varphi)\big)$ for a central simple F-algebra $C^+(\varphi)$ uniquely determined up to isomorphism. Moreover, $C_0(\varphi) \simeq C^+(\varphi) \times C^+(\varphi)$.

14. The Clifford invariant

A more delicate invariant of a nondegenerate even-dimensional quadratic form arises from its associated Clifford algebra.

Let φ be a nondegenerate even-dimensional quadratic form over F. The Clifford algebra $C(\varphi)$ is then a central simple F-algebra. Denote by $\operatorname{clif}(\varphi)$ the class of $C(\varphi)$ in the Brauer group $\operatorname{Br}(F)$. It follows from Example 11.3 that $\operatorname{clif}(\varphi) \in \operatorname{Br}_2(F)$. We call $\operatorname{clif}(\varphi)$ the *Clifford invariant of* φ.

Example 14.1. Let φ be the reduced norm form of a quaternion algebra Q. It follows from Proposition 12.4 that $\operatorname{clif}(\varphi) = [Q]$.

Lemma 14.2. *Let φ and ψ be two nondegenerate even-dimensional quadratic forms over F. If* $\operatorname{disc}(\varphi)$ *is trivial, then* $\operatorname{clif}(\varphi \perp \psi) = \operatorname{clif}(\varphi) \cdot \operatorname{clif}(\psi)$.

PROOF. Let $e \in Z(\varphi)$ be a nontrivial idempotent and set $s = e - \bar{e} = 1 - 2e$. We have $\bar{s} = -s$, $s^2 = 1$, and $vs = \bar{s}v = -sv$ for every $v \in V_\varphi$ by Lemma 11.7. Therefore, in the Clifford algebra of $\varphi \perp \psi$, we have $(v \otimes 1 + s \otimes w)^2 = \varphi(v) + \psi(w)$ for all $v \in V_\varphi$ and $w \in V_\psi$. It follows from the universal property of the Clifford algebra that the F-linear map $V_\varphi \oplus V_\psi \to C(\varphi) \otimes_F C(\psi)$ defined by $v \oplus w \mapsto v \otimes 1 + s \otimes w$ extends to an F-algebra homomorphism $C(\varphi \perp \psi) \to C(\varphi) \otimes_F C(\psi)$. This map is an isomorphism as the Clifford algebra of an even-dimensional form is central simple. □

Theorem 14.3. *The map*

$$e_2 : I_q^2(F) \to \operatorname{Br}_2(F)$$

taking a form φ to $\operatorname{clif}(\varphi)$ *is a well-defined group homomorphism. Moreover,* $I_q^3(F) \subset \operatorname{Ker}(e_2)$.

PROOF. It follows from Lemma 14.2 that e_2 is well-defined. Next let $\varphi \in I_q^2(F)$ and $a \in F^\times$. Since $\operatorname{disc}(\varphi)$ is trivial, it follows from Corollary 11.8 that $C(a\varphi) \simeq C(\varphi)$. Therefore, $e_2(\langle\langle a\rangle\rangle \otimes \varphi) = e_2(\varphi) - e_2(a\varphi) = 0$. □

We shall, in fact, prove that $I_q^3(F) = \operatorname{Ker}(e_2)$ in §16 for fields of characteristic 2 and Chapter VIII for fields of characteristic not 2.

15. Chain p-equivalence of quadratic Pfister forms

We saw that anisotropic bilinear Pfister forms $\langle\langle a_1, \dots, a_n\rangle\rangle$ and $\langle\langle b_1, \dots, b_n\rangle\rangle$ were p-chain equivalent if and only if they were isometric. This equivalence relation was based on isometries of 2-fold Pfister forms. In this section, we prove the analogous result for quadratic Pfister forms. This was first proven in the case of characteristic 2 by Aravire and Baeza in [**12**]. To begin we therefore need to establish isometries of quadratic 2-fold Pfister forms in characteristic 2. This is given by the following:

Lemma 15.1. *Let F be a field of characteristic 2. Then in $I_q(F)$, we have*

(1) $\langle\langle a, b + b']] = \langle\langle a, b]] + \langle\langle a, b']]$.

(2) $\langle\langle aa', b]] \equiv \langle\langle a, b]] + \langle\langle a', b]] \mod I_q^3(F)$.

(3) $\langle\langle a + x^2, b]] = \langle\langle a, \dfrac{ab}{a + x^2}]]$.

(4) $\langle\langle a + a', b]] \equiv \langle\langle a, \dfrac{ab}{a + a'}]] + \langle\langle a', \dfrac{a'b}{a + a'}]] \mod I_q^3(F)$.

PROOF. (1): This follows by Example 7.23.

(2): Follows from the equality $\langle\langle a\rangle\rangle + \langle\langle a'\rangle\rangle = \langle\langle aa'\rangle\rangle + \langle\langle a, a'\rangle\rangle$ in the Witt ring of bilinear forms by Example 4.10.

(3): Let $c = b/(a + x^2)$ and

$$A = \begin{bmatrix} a, c \\ F \end{bmatrix} \quad \text{and} \quad B = \begin{bmatrix} a + x^2, c \\ F \end{bmatrix}.$$

By Corollary 12.5, it suffices to prove that $A \simeq B$. Let $\{1, i, j, ij\}$ be the standard basis of A, i.e., $i^2 = a$, $j^2 = b$ and $ij + ji = 1$. Considering the new basis $\{1, i + x, j, (i + x)j\}$ with $(i + x)^2 = a + x^2$ shows that $A \simeq B$.

(4): We have by (1)–(3):

$$\begin{aligned}
\langle\langle a + a', b]] &\equiv \langle\langle \frac{a}{a'} + 1, b]] + \langle\langle a', b]] = \langle\langle \frac{a}{a'}, \frac{ab}{a + a'}]] + \langle\langle a', b]] \\
&\equiv \langle\langle a, \frac{ab}{a + a'}]] + \langle\langle a', \frac{ab}{a + a'}]] + \langle\langle a', b]] \\
&= \langle\langle a, \frac{ab}{a + a'}]] + \langle\langle a', \frac{a'b}{a + a'}]] \mod I_q^3(F).
\end{aligned}$$

□

The definition of chain p-equivalence for quadratic Pfister forms is slightly more involved than that for bilinear Pfister forms.

Definition 15.2. Let $a_1, \dots, a_{n-1}, b_1, \dots, b_{n-1} \in F^\times$ and $a_n, b_n \in F$ with $n \geq 1$. We assume that a_n and b_n are nonzero if $\operatorname{char} F \neq 2$. Let $\varphi = \langle\langle a_1, \dots, a_{n-1}, a_n]]$ and $\psi = \langle\langle b_1, \dots, b_{n-1}, b_n]]$. We say that the quadratic Pfister forms φ and ψ are *simply p-equivalent* if either $n = 1$ and $\langle\langle a_1]] \simeq \langle\langle b_1]]$ or $n \geq 2$ and there exist i and j with $1 \leq i < j \leq n$ satisfying

$$\langle\langle a_i, a_j\rangle\rangle \simeq \langle\langle b_i, b_j\rangle\rangle \quad \text{with } j < n \quad \text{and} \quad a_l = b_l \text{ for all } l \neq i, j \quad \text{or}$$
$$\langle\langle a_i, a_j]] \simeq \langle\langle b_i, b_j]] \quad \text{with } j = n \quad \text{and} \quad a_l = b_l \text{ for all } l \neq i, j.$$

We say that φ and ψ are *chain p-equivalent* if there exist quadratic n-fold Pfister forms $\varphi_0, \dots, \varphi_m$ for some m such that $\varphi = \varphi_0$, $\psi = \varphi_m$ and φ_i is simply p-equivalent to φ_{i+1} for each $i \in [0,\ m-1]$.

Theorem 15.3. *Let*

$$\varphi = \langle\langle a_1, \dots, a_{n-1}, a_n]] \quad \textit{and} \quad \psi = \langle\langle b_1, \dots, b_{n-1}, b_n]]$$

be anisotropic quadratic n-fold Pfister forms. Then $\varphi \approx \psi$ if and only if $\varphi \simeq \psi$.

We shall prove this result in a series of steps. Suppose that $\varphi \simeq \psi$. The case $\operatorname{char} F \neq 2$ was considered in Theorem 6.10, so we may also assume that $\operatorname{char} F = 2$. As before the map $\wp : F \to F$ is defined by $\wp(x) = x^2 + x$ when $\operatorname{char} F = 2$.

Lemma 15.4. *Let $\operatorname{char} F = 2$. If $\mathfrak{b} = \langle\langle a_1, \dots, a_n\rangle\rangle$ is an anisotropic bilinear Pfister form and $d_1, d_2 \in F$, then $\mathfrak{b} \otimes \langle\langle d_1]] \simeq \mathfrak{b} \otimes \langle\langle d_2]]$ if and only if $\mathfrak{b} \otimes \langle\langle d_1]] \approx \mathfrak{b} \otimes \langle\langle d_2]]$.*

PROOF. Assume that $\mathfrak{b} \otimes \langle\langle d_1]] \simeq \mathfrak{b} \otimes \langle\langle d_2]]$. It follows from Example 7.23 that the form $\mathfrak{b} \otimes \langle\langle d_1 + d_2]]$ is Witt-equivalent to $\mathfrak{b} \otimes \langle\langle d_1]] \perp \mathfrak{b} \otimes \langle\langle d_2]]$ and hence is hyperbolic. By Lemma 9.11, we have $d_1 + d_2 = x + y$ where $x \in \operatorname{im}(\wp)$ and $y \in \widetilde{D}(\mathfrak{b}')$. If $y = 0$, then $\langle\langle d_1]] \simeq \langle\langle d_2]]$ and we are done. So suppose that $y \neq 0$. By Lemma 6.11, there is a bilinear Pfister form $\mathfrak{c}$ such that $\mathfrak{b} \approx \mathfrak{c} \otimes \langle\langle y\rangle\rangle$. As $\langle\langle y, d_1]] \simeq \langle\langle y, d_2]]$, we have

$$\mathfrak{b} \otimes \langle\langle d_1]] \approx \mathfrak{c} \otimes \langle\langle y, d_1]] \approx \mathfrak{c} \otimes \langle\langle y, d_2]] \approx \mathfrak{b} \otimes \langle\langle d_2]]. \qquad \square$$

Lemma 15.5. *Let $\operatorname{char} F = 2$. If $\rho = \langle\langle b, b_2, \dots, b_n, d]]$ is a quadratic Pfister form, then for every $a \in F^\times$ and $z \in D(\rho)$, we have $\langle\langle a\rangle\rangle \otimes \rho \approx \langle\langle az\rangle\rangle \otimes \rho$.*

PROOF. We induct on $\dim \rho$. Let $\eta = \langle\langle b_2, \dots b_n, d]]$. We have $z = x + by$ with $x, y \in \widetilde{D}(\eta)$. If $y = 0$, then $x = z \neq 0$ and by the induction hypothesis $\langle\langle a\rangle\rangle \otimes \eta \approx \langle\langle az\rangle\rangle \otimes \eta$, hence

$$\langle\langle a\rangle\rangle \otimes \rho = \langle\langle a, b\rangle\rangle \otimes \eta \approx \langle\langle az, b\rangle\rangle \otimes \eta \approx \langle\langle az\rangle\rangle \otimes \rho.$$

If $x = 0$, then $z = by$ and by the induction hypothesis $\langle\langle a\rangle\rangle \otimes \eta \approx \langle\langle ay\rangle\rangle \otimes \eta$, hence

$$\langle\langle a\rangle\rangle \otimes \rho = \langle\langle a, b\rangle\rangle \otimes \eta \approx \langle\langle ay, b\rangle\rangle \otimes \eta \approx \langle\langle az, b\rangle\rangle \otimes \eta \approx \langle\langle az\rangle\rangle \otimes \rho.$$

Now suppose that both x and y are nonzero. As η is round, $xy \in D(\eta)$. By the induction hypothesis and Lemma 4.15,

$$\begin{aligned}\langle\langle a\rangle\rangle \otimes \rho = \langle\langle a, b\rangle\rangle \otimes \eta &\approx \langle\langle a, ab\rangle\rangle \otimes \eta \approx \langle\langle ax, aby\rangle\rangle \otimes \eta \\ &\approx \langle\langle az, bxy\rangle\rangle \otimes \eta \approx \langle\langle az, b\rangle\rangle \otimes \eta = \langle\langle az\rangle\rangle \otimes \rho. \qquad \square\end{aligned}$$

Lemma 15.6. *Let $\operatorname{char} F = 2$. Let $\mathfrak{b} = \langle\langle a_1, \dots, a_m\rangle\rangle$ be a bilinear Pfister form, $\rho = \langle\langle b_1, \dots, b_n, d]]$ a quadratic Pfister form with $n \geq 0$ (if $n = 0$, then $\rho = \langle\langle d]]$), and $c \in F^\times$. Suppose there exists an $x \in D(\mathfrak{b})$ with $c+x \neq 0$ satisfying $\mathfrak{b} \otimes \langle\langle c+x\rangle\rangle \otimes \rho$ is anisotropic. Then*

$$\mathfrak{b} \otimes \langle\langle c + x\rangle\rangle \otimes \rho \approx \mathfrak{b} \otimes \langle\langle c\rangle\rangle \otimes \psi$$

for some quadratic Pfister form ψ.

PROOF. We proceed by induction on the dimension of $\mathfrak{b}$. Suppose $\mathfrak{b} = \langle 1\rangle$. Then $x = y^2$ for some $y \in F$. It follows from Lemma 15.1 that $\langle\langle c + y^2, d]] \simeq \langle\langle c, cd/(c + y^2)]]$, hence $\langle\langle c + y^2\rangle\rangle \otimes \rho \approx \langle\langle c\rangle\rangle \otimes \langle\langle b_1, \dots, b_n, cd/(c + y^2)]]$.

So we may assume that $\dim \mathfrak{b} > 1$. Let $\mathfrak{c} = \langle\langle a_2, \dots, a_m\rangle\rangle$ and $a = a_1$. We have $x = y + az$ where $y, z \in \widetilde{D}(\mathfrak{c})$. If $c = az$, then $c + x = y$ belongs to $D(\mathfrak{b})$, so the form $\mathfrak{b} \otimes \langle\langle c + x\rangle\rangle$ would be metabolic contradicting hypothesis.

Let $d := c + az$. We have $d \neq 0$. By the induction hypothesis,

$$\mathfrak{c} \otimes \langle\langle d + y\rangle\rangle \otimes \rho \approx \mathfrak{c} \otimes \langle\langle d\rangle\rangle \otimes \mu \quad \text{and} \quad \mathfrak{c} \otimes \langle\langle ac + a^2z\rangle\rangle \otimes \mu \approx \mathfrak{c} \otimes \langle\langle ac\rangle\rangle \otimes \psi$$

for some quadratic Pfister forms μ and ψ. Hence by Lemma 4.15,

$$\begin{aligned} \mathfrak{b}\otimes\langle\langle c + x\rangle\rangle \otimes \rho &= \mathfrak{b} \otimes \langle\langle d + y\rangle\rangle \otimes \rho = \mathfrak{c} \otimes \langle\langle a, d + y\rangle\rangle \otimes \rho \\ &\approx \mathfrak{c} \otimes \langle\langle a, d\rangle\rangle \otimes \mu = \mathfrak{c} \otimes \langle\langle a, c + az\rangle\rangle \otimes \mu \approx \mathfrak{c} \otimes \langle\langle a, ac + a^2z\rangle\rangle \otimes \mu \\ &\approx \mathfrak{c} \otimes \langle\langle a, ac\rangle\rangle \otimes \psi \approx \mathfrak{c} \otimes \langle\langle a, c\rangle\rangle \otimes \psi = \mathfrak{b} \otimes \langle\langle c\rangle\rangle \otimes \psi. \end{aligned}$$ □

If $\mathfrak{b}$ is a bilinear Pfister form over a field F, then $\mathfrak{b} \simeq \mathfrak{b}' \perp \langle 1\rangle$ with the pure subform $\mathfrak{b}'$ unique up to isometry. For a quadratic Pfister form over a field of characteristic 2, the analogue of this is not true. So, in this case, we have to modify our notion of a pure subform of a quadratic Pfister form. So suppose that $\operatorname{char} F = 2$. Let $\varphi = \mathfrak{b} \otimes \langle\langle d]]$ be a quadratic Pfister form with $\mathfrak{b} = \mathfrak{b}' \perp \langle 1\rangle$, a bilinear Pfister form. We have $\varphi = \langle\langle d]] \perp \varphi^\circ$ with $\varphi^\circ = \mathfrak{b}' \otimes \langle\langle d]]$. The form φ° depends on the presentation of $\mathfrak{b}$. Let $\varphi' := \langle 1\rangle \perp \mathfrak{b}' \otimes \langle\langle d]]$. This form coincides with the complementary form $\langle 1\rangle^\perp$ in φ. The form φ' is uniquely determined by φ up to isometry. Indeed, by Witt Extension Theorem 8.3, for any two vectors $v, w \in V_\varphi$ with $\varphi(v) = \varphi(w) = 1$ there is an auto-isometry α of φ such that $\alpha(v) = w$. Therefore, the orthogonal complements of Fv and Fw are isometric. We call the form φ' the *pure subform of* φ.

Proposition 15.7. *Let $\rho = \langle\langle b_1, \dots, b_n, d]]$ be a quadratic Pfister form, $n \geq 1$, and $\mathfrak{b} = \langle\langle a_1, \dots, a_m\rangle\rangle$ a bilinear Pfister form. Set $\varphi = \mathfrak{b} \otimes \rho$. Suppose that φ is anisotropic. Let $c \in D(\mathfrak{b} \otimes \rho') \setminus D(\mathfrak{b})$. Then $\varphi \approx \mathfrak{b} \otimes \langle\langle c\rangle\rangle \otimes \psi$ for some quadratic Pfister form ψ.*

PROOF. Let $\rho = \langle\langle b\rangle\rangle \otimes \eta$ with $b = b_1$ and $\eta = \langle\langle b_2, \dots, b_n, d]]$ (with $\eta = \langle\langle d]]$ if $n = 1$). We proceed by induction on $\dim \rho$. Note that if $n = 1$, we have $\eta' = \langle 1\rangle$ and $D(\mathfrak{b} \otimes \eta') = D(\mathfrak{b})$. As

$$\mathfrak{b} \otimes \rho' = (\mathfrak{b} \otimes \eta') \perp (b\mathfrak{b} \otimes \eta),$$

we have $c = x + by$ with $x \in \widetilde{D}(\mathfrak{b} \otimes \eta')$, and $y \in \widetilde{D}(\mathfrak{b} \otimes \eta)$.

If $y = 0$, then $c = x \in D(\mathfrak{b} \otimes \eta') \setminus D(\mathfrak{b})$. In particular, $n > 1$. By the induction hypothesis, $\mathfrak{b} \otimes \eta \approx \mathfrak{b} \otimes \langle\langle c\rangle\rangle \otimes \mu$ for some quadratic Pfister form μ. Hence

$$\varphi = \mathfrak{b} \otimes \rho = \mathfrak{b} \otimes \langle\langle b\rangle\rangle \otimes \eta \approx \mathfrak{b} \otimes \langle\langle c\rangle\rangle \otimes \langle\langle b\rangle\rangle \otimes \mu.$$

Now suppose that $y \neq 0$. By Lemma 15.5,

$$\varphi = \mathfrak{b} \otimes \rho = \mathfrak{b} \otimes \langle\langle b\rangle\rangle \otimes \eta \approx \mathfrak{b} \otimes \langle\langle by\rangle\rangle \otimes \eta. \tag{15.8}$$

Assume that $x \notin \widetilde{D}(\mathfrak{b})$. In particular, $n > 1$. By the induction hypothesis, $\mathfrak{b} \otimes \eta \approx \mathfrak{b} \otimes \langle\langle x\rangle\rangle \otimes \mu$ for some quadratic Pfister form μ. Therefore, by Lemma 4.15,

$$\varphi \approx \mathfrak{b} \otimes \langle\langle by\rangle\rangle \otimes \eta \approx \mathfrak{b} \otimes \langle\langle x, by\rangle\rangle \otimes \mu \approx \mathfrak{b} \otimes \langle\langle c, bxy\rangle\rangle \otimes \mu.$$

Finally, we assume that $x \in \widetilde{D}(\mathfrak{b})$. By Lemma 15.6 and (15.8),

$$\varphi \approx \mathfrak{b} \otimes \langle\langle by \rangle\rangle \otimes \eta = \mathfrak{b} \otimes \langle\langle c + x \rangle\rangle \otimes \eta \approx \mathfrak{b} \otimes \langle\langle c \rangle\rangle \otimes \psi$$

for a quadratic Pfister form ψ. □

PROOF OF THEOREM 15.3. Let φ and ψ be isometric anisotropic quadratic n-fold Pfister forms over F as in the statement of Theorem 15.3. We must show that $\varphi \approx \psi$. We may assume that $\operatorname{char} F = 2$.

Claim: For every $r \in [0,\ n-1]$, there exist a bilinear r-fold Pfister form $\mathfrak{b}$ and quadratic $(n-r)$-fold Pfister forms ρ and μ such that $\varphi \approx \mathfrak{b} \otimes \rho$ and $\psi \approx \mathfrak{b} \otimes \mu$.

We prove the claim by induction on r. The case $r = 0$ is obvious. Suppose we have such $\mathfrak{b}, \rho$ and μ for some $r < n - 1$. Write $\rho = \langle\langle c \rangle\rangle \otimes \psi$ for some $c \in F^\times$ and quadratic Pfister form ψ, so $\varphi \approx \mathfrak{b} \otimes \langle\langle c \rangle\rangle \otimes \psi$. Note that as φ is anisotropic, we have $c \in D(\mathfrak{b} \otimes \rho') \setminus D(\mathfrak{b})$.

The form $\mathfrak{b} \otimes \langle 1 \rangle$ is isometric to subforms of φ and ψ. As $\mathfrak{b}_\varphi$ and $\mathfrak{b}_\psi$ are nondegenerate, by the Witt Extension Theorem 8.3, an isometry between these subforms extends to an isometry between φ and ψ. This isometry induces an isometry of orthogonal complements $\mathfrak{b} \otimes \rho'$ and $\mathfrak{b} \otimes \mu'$. Therefore, we have $c \in D(\mathfrak{b} \otimes \rho') \setminus D(\mathfrak{b}) = D(\mathfrak{b} \otimes \mu') \setminus D(\mathfrak{b})$. It follows from Proposition 15.7 that $\psi \approx \mathfrak{b} \otimes \langle\langle c \rangle\rangle \otimes \sigma$ for some quadratic Pfister form σ. Thus $\psi \approx \mathfrak{b} \otimes \langle\langle c \rangle\rangle \otimes \sigma$ and the claim is established.

Applying the claim in the case $r = n - 1$, we find a bilinear $(n-1)$-fold Pfister form $\mathfrak{b}$ and elements $d_1, d_2 \in F$ such that $\varphi \approx \mathfrak{b} \otimes \langle\langle d_1]]$ and $\psi \approx \mathfrak{b} \otimes \langle\langle d_2]]$. By Lemma 15.4, we have $\mathfrak{b} \otimes \langle\langle d_1]] \approx \mathfrak{b} \otimes \langle\langle d_2]]$, hence $\varphi \approx \psi$. □

16. Cohomological invariants

A major problem in the theory of quadratic forms was to determine the relationship between quadratic forms and Galois cohomology. In this section, using the cohomology groups defined in §101, we introduce the problem.

Let $H^n(F)$ be the groups defined in §101. In particular,

$$H^n(F) \simeq \begin{cases} \text{Ét}_2(F), & \text{if } n = 1, \\ \text{Br}_2(F), & \text{if } n = 2. \end{cases}$$

Let φ be a quadratic n-fold Pfister form. Suppose that $\varphi \simeq \langle\langle a_1, \dots, a_n]]$. Define the *cohomological invariant* of φ to be the class $e_n(\varphi)$ in $H^n(F)$ given by

$$e_n(\varphi) = \{a_1, a_2, \dots, a_{n-1}\} \cdot [F_{a_n}],$$

where $[F_c]$ is the class of the étale quadratic extension F_c/F in $\text{Ét}_2(F) \simeq H^1(F)$.

The cohomological invariant e_n is well-defined on quadratic n-fold Pfister forms:

Proposition 16.1. *Let φ and ψ be quadratic n-fold Pfister forms. If $\varphi \simeq \psi$, then $e_n(\varphi) = e_n(\psi)$ in $H^n(F)$.*

PROOF. This follows from Theorems 6.20 and 15.3. □

As in the bilinear case, if we use the Hauptsatz 23.7 below, we even have that if

$$\varphi \equiv \psi \bmod I_q^{n+1}(F), \qquad \text{then} \qquad e_n(\varphi) = e_n(\psi)$$

in $H^n(F)$. (Cf. Corollary 23.9 below.) In fact, we shall also show by elementary means in Proposition 24.6 below that if φ_1, φ_2 and φ_3 are general quadratic n-fold

Pfister forms such that $\varphi_1 + \varphi_2 + \varphi_3 \in I_q^{n+1}(F)$, then $e_n(\varphi_1) + e_n(\varphi_2) + e_n(\varphi_3) = 0 \in H^n(F)$.

We call the extension of e_n to a group homomorphism $e_n : I_q^n(F) \to H^n(F)$ the *nth cohomological invariant* of $I_q^n(F)$.

Fact 16.2. *The nth cohomological invariant e_n exists for all fields F and for all $n \geq 1$. Moreover,* $\operatorname{Ker}(e_n) = I_q^{n+1}(F)$. *Furthermore, there is a unique isomorphism*

$$\bar{e}_n : I_q^n(F)/I_q^{n+1}(F) \to H^n(F)$$

satisfying $\bar{e}_n\big(\varphi + I_q^{n+1}(F)\big) = e_n(\varphi)$ for every quadratic n-fold Pfister quadratic form φ.

Special cases of Fact 16.2 can be proven by elementary methods. Indeed, we have already shown that the invariant e_1 is well-defined on all of $I_q(F)$ and coincides with the discriminant in Theorem 13.7 and e_2 is well-defined on all of $I_q^2(F)$ and coincides with the Clifford invariant by Theorem 14.3. Then by Theorems 13.7 and 14.3 the maps $\bar{e}_1$ and $\bar{e}_2$ are well-defined. For fields of characteristic different from 2, e_3 was shown to be well-defined, hence $\bar{e}_3$ by Arason in [**4**], and $\bar{e}_3$ an isomorphism in [**103**], and independently by Rost [**115**]; and Jacob and Rost showed that e_4 was well-defined in [**65**].

Suppose that $\operatorname{char} F \neq 2$. Then the identification of bilinear and quadratic forms leads to the composition

$$h_F^n : K_n(F)/2K_n(F) \xrightarrow{f_n} I^n(F)/I^{n+1}(F) = I_q^n(F)/I_q^{n+1}(F) \xrightarrow{\bar{e}_n} H^n(F),$$

where h_F^n is the norm residue homomorphism of degree n defined in §101.

Milnor conjectured that h_F^n was an isomorphism for all n in [**106**]. Voevodsky proved the Milnor Conjecture in [**136**]. As stated in Fact 5.15 the map f_n is an isomorphism for all n. In particular, e_n is well-defined and $\bar{e}_n$ is an isomorphism for all n.

If $\operatorname{char} F = 2$, Kato proved Fact 16.2 in [**78**].

We have proven that $\bar{e}_1$ is an isomorphism in Theorem 13.7. We shall prove that h_F^2 is an isomorphism in Chapter VIII below if the characteristic of F is different from 2. It follows that $\bar{e}_2$ is an isomorphism. We now turn to the proof that $\bar{e}_2$ is an isomorphism if $\operatorname{char} F = 2$. The following statement was first proven by Sah in [**118**].

Theorem 16.3. *Let* $\operatorname{char} F = 2$. *Then* $\bar{e}_2 : I_q^2(F)/I_q^3(F) \to \operatorname{Br}_2(F)$ *is an isomorphism.*

Proof. The classes of quaternion algebras generate the group $\operatorname{Br}_2(F)$ by [**1**, Ch. VII, Th. 30]. It follows that $\bar{e}_2$ is surjective. So we need only show that $\bar{e}_2$ is injective.

Let $\alpha \in I_q^2(F)$ satisfy $e_2(\alpha) = 1$. Write α in the form $\sum_{i=1}^n d_i\langle\langle a_i, b_i]]$. By assumption, the product of all $\begin{bmatrix} a_i, c_i \\ F \end{bmatrix}$ with $c_i = b_i/a_i$ is trivial in $\operatorname{Br}(F)$.

We prove by induction on n that $\alpha \in I_q^3(F)$. If $n = 1$, we have $\alpha = \langle\langle a_1, b_1]]$ and $e_2(\alpha) = \begin{bmatrix} a_1, c_1 \\ F \end{bmatrix} = 1$. Therefore, the reduced norm form α of the split quaternion algebra $\begin{bmatrix} a_1, c_1 \\ F \end{bmatrix}$ is hyperbolic by Corollary 12.5, hence $\alpha = 0$.

In the general case, let $L = F(a_1^{1/2}, \dots, a_{n-1}^{1/2})$. The field L splits $\begin{bmatrix} a_i, c_i \\ F \end{bmatrix}$ for all $i \in [1,\ n-1]$ and hence splits $\begin{bmatrix} a_n, c_n \\ F \end{bmatrix}$. By Lemma 98.16, we have $\begin{bmatrix} a_n, c_n \\ F \end{bmatrix} = \begin{bmatrix} c, d \\ F \end{bmatrix}$, where c is the square of an element of L, i.e., c is the sum of elements of the form $g^2 m$ where $g \in F$ and m is a monomial in the a_i, $i \in [1,\ n-1]$. It follows from Corollary 12.5 that $\langle\langle a_n, b_n]] = \langle\langle c, cd]]$. By Lemma 15.1, we have $\langle\langle c, cd]]$ is congruent modulo $I_q^3(F)$ to the sum of 2-fold Pfister forms $\langle\langle a_i, f_i]]$ with $i \in [1,\ n-1]$, $f_i \in F$. Therefore, we may assume that $\alpha = \sum_{i=1}^{n-1} \langle\langle a_i, b_i']]$ for some b_i'. By the induction hypothesis, $\alpha \in I_q^3(F)$. □

CHAPTER III

Forms over Rational Function Fields

17. The Cassels-Pfister Theorem

Given a quadratic form φ over a field over F, it is natural to consider values of the form over the rational function field $F(t)$. The Cassels-Pfister Theorem shows that whenever φ represents a polynomial over $F(t)$, then it already does so when viewed as a quadratic form over the polynomial ring $F[t]$. This results in specialization theorems. As an n-dimensional quadratic form ψ can be viewed as a polynomial in $F[T] := F[t_1, \dots, t_n]$, one can also ask when is $\psi(T)$ a value of $\varphi_{F(T)}$? If both the forms are anisotropic, we shall also show in this section the fundamental result that this is true if and only if ψ is a subform of φ.

Let φ be an anisotropic quadratic form on V over F and $\mathfrak{b}$ its polar form. Let v and u be two distinct vectors in V and set $w = v - u$. Let τ_w be the reflection with respect to w as defined in Example 7.2. Then $\varphi\big(\tau_w(v)\big) = \varphi(v)$ as τ_w is an isometry and

$$\tau_w(v) = u + \frac{\varphi(u) - \varphi(v)}{\varphi(w)} w \tag{17.1}$$

as $\mathfrak{b}_\varphi(v, w) = -\mathfrak{b}_\varphi(v, -w) = -\varphi(u) + \varphi(v) + \varphi(w)$ by definition.

Notation 17.2. If $T = (t_1, \dots, t_n)$ is a tuple of independent variables, let

$$F[T] := F[t_1, \dots, t_n] \quad \text{and} \quad F(T) := F(t_1, \dots, t_n).$$

If V is a finite dimensional vector space over F, let

$$V[T] := F[T] \otimes_F V \quad \text{and} \quad V(T) := V_{F(T)} := F(T) \otimes_F V.$$

Note that $V(T)$ is also the localization of $V[T]$ at $F[T] \setminus \{0\}$. In particular, if $v \in V(T)$, there exist $w \in V[T]$ and a nonzero $f \in F[T]$ satisfying $v = w/f$. For a single variable t, we let $V[t] := F[t] \otimes_F V$ and $V(t) := V_{F(t)} := F(t) \otimes_F V$.

The following general form of the Classical Cassels-Pfister Theorem is true.

Theorem 17.3 (Cassels-Pfister Theorem). *Let φ be a quadratic form on V and let $h \in F[t] \cap D(\varphi_{F(t)})$. Then there is a $w \in V[t]$ such that $\varphi(w) = h$.*

PROOF. Suppose first that φ is anisotropic. Let $v \in V(t)$ satisfy $\varphi(v) = h$. There is a nonzero polynomial $f \in F[t]$ such that $fv \in V[t]$. Choose v and f so that $\deg(f)$ is the smallest possible. It suffice to show that f is constant. Suppose $\deg(f) > 0$.

Using the analog of the Division Algorithm, we can divide the polynomial vector fv by f to get $fv = fu + r$, where $u, r \in V[t]$ and $\deg(r) < \deg(f)$. If $r = 0$, then

$v = u \in V[t]$ and f is constant; so we may assume that $r \neq 0$. In particular, $\varphi(r) \neq 0$ as φ is anisotropic. Set $w = v - u = r/f$ and consider the vector

$$\tau_w(v) = u + \frac{\varphi(u) - h}{\varphi(r)/f}\, r \tag{17.4}$$

as in (17.1). We have $\varphi\big(\tau_w(v)\big) = h$. We show that $f' := \varphi(r)/f$ is a polynomial. As

$$f^2 h = \varphi(fv) = \varphi(fu + r) = f^2\varphi(u) + f\mathfrak{b}_\varphi(u, r) + \varphi(r),$$

we see that $\varphi(r)$ is divisible by f. Equation (17.4) implies that $f'\tau_w(v) \in V[t]$ and the definition of r yields

$$\deg(f') = \deg\varphi(r) - \deg(f) < 2\deg(f) - \deg(f) = \deg(f),$$

a contradiction to the minimality of $\deg(f)$.

Now suppose that φ is isotropic. By Lemma 7.12, we may assume that $\operatorname{rad}\varphi = 0$. In particular, a hyperbolic plane splits off as an orthogonal direct summand of φ by Lemma 7.13. Let e, e' be a hyperbolic pair for this hyperbolic plane. Then

$$\varphi(he + e') = \mathfrak{b}_\varphi(he, e') = h\mathfrak{b}_\varphi(e, e') = h. \qquad \square$$

The theorem above was first proved by Cassels in [**22**] for the form $\langle 1, \dots, 1\rangle$ over a field of characteristic not 2. This was generalized by Pfister to nondegenerate forms over such fields in [**108**] and used to prove the results through Corollary 17.13 below in that case.

Corollary 17.5. *Let $\mathfrak{b}$ be a symmetric bilinear form on V and let $h \in F[t] \cap D(\varphi_{F(t)})$. Then there is a $v \in V[t]$ such that $\mathfrak{b}(v, v) = h$.*

PROOF. Let φ be $\varphi_\mathfrak{b}$, i.e., $\varphi(v) = \mathfrak{b}(v, v)$ for all $v \in V$. The result follows from the Cassels-Pfister Theorem for φ. $\square$

Corollary 17.6. *Let $f \in F[t]$ be a sum of n squares in $F(t)$. Then f is a sum of n squares in $F[t]$.*

Corollary 17.7 (Substitution Principle). *Let φ be a quadratic form over F and $h \in D(\varphi_{F(T)})$ with $T = (t_1, \dots, t_n)$. Suppose that $h(x)$ is defined for $x \in F^n$ and $h(x) \neq 0$, then $h(x) \in D(\varphi)$.*

PROOF. As $h(x)$ is defined, we can write $h = f/g$ with $f, g \in F[T]$ and $g(x) \neq 0$. Replacing h by $g^2 h$, we may assume that $h \in F[T]$. Let $T' = (t_1, \dots, t_{n-1})$ and $x = (x_1, \dots, x_n)$. By the theorem, there exists $v(T', t_n) \in V(T')[t_n]$ satisfying $\varphi\big(v(T', t_n)\big) = h(T', t_n)$. Evaluating t_n at x_n shows that $h(T', x_n) = \varphi\big(v(T', x_n)\big) \in D(\varphi_{F(T')})$. The conclusion follows by induction on n. $\square$

As above, we also deduce:

Corollary 17.8 (Bilinear Substitution Principle). *Let $\mathfrak{b}$ be a symmetric bilinear form over F and $h \in D(\mathfrak{b}_{F(T)})$ with $T = (t_1, \dots, t_n)$. Suppose that $h(x)$ is defined for $x \in F^n$ and $h(x) \neq 0$, then $h(x) \in D(\mathfrak{b})$.*

We shall need the following slightly more general version of the Cassels-Pfister Theorem.

Proposition 17.9. *Let φ be an anisotropic quadratic form on V. Suppose that $s \in V$ and $v \in V(t)$ satisfy $\varphi(v) \in F[t]$ and $\mathfrak{b}_\varphi(s, v) \in F[t]$. Then there is $w \in V[t]$ such that $\varphi(w) = \varphi(v)$ and $\mathfrak{b}_\varphi(s, w) = \mathfrak{b}_\varphi(s, v)$.*

PROOF. It suffices to show the value $\mathfrak{b}_\varphi(s,v)$ does not change when v is modified in the course of the proof of Theorem 17.3. Choose $v_0 \in V[t]$ satisfying $\mathfrak{b}_\varphi(s,v_0) = \mathfrak{b}_\varphi(s,v)$.

Let $f \in F[t]$ be a nonzero polynomial such that $fv \in V[t]$. As the remainder r on dividing fv and $fv - fv_0$ by f is the same and $fv - fv_0 \in (F(t)s)^\perp$, we have $r \in (F(t)s)^\perp$. Therefore, $\mathfrak{b}_\varphi(s, \tau_r(v)) = \mathfrak{b}_\varphi(s,v)$. □

Lemma 17.10. *Let φ be an anisotropic quadratic form and ρ a nondegenerate binary anisotropic quadratic form satisfying $\rho(t_1,t_2) + d \in D(\varphi_{F(t_1,t_2)})$ for some $d \in F$. Then $\varphi \simeq \rho \perp \mu$ for some form μ and $d \in \widetilde{D}(\mu)$.*

PROOF. Let $\rho(t_1,t_2) = at_1^2 + bt_1t_2 + ct_2^2$. As $\rho(t_1,t_2) + dt_3^2$ is a value of φ over $F(t_1,t_2,t_3)$, there is a $u \in V = V_\varphi$ such that $\varphi(u) = a$ by the Substitution Principle 17.7. Applying the Cassels-Pfister Theorem 17.3 to the form $\varphi_{F(t_2)}$, we find a $v \in V_{F(t_2)}[t_1]$ such that $\varphi(v) = at_1^2 + bt_1t_2 + ct_2^2 + d$. Since φ is anisotropic, we have $\deg_{t_1} v \le 1$, i.e., $v(t_1) = v_0 + v_1t_1$ for some $v_0, v_1 \in V_{F(t_2)}$. Expanding we get

$$\varphi(v_0) = a, \quad \mathfrak{b}(v_0,v_1) = bt_2, \quad \varphi(v_1) = ct_2^2 + d,$$

where $\mathfrak{b} = \mathfrak{b}_\varphi$. Clearly, $v_0 \notin \operatorname{rad}(\mathfrak{b}_{F(t_2)})$.

We claim that $u \notin \operatorname{rad}(\mathfrak{b})$. We may assume that $u \neq v_0$ and therefore

$$0 \neq \varphi(u - v_0) = \varphi(u) + \varphi(v_0) - \mathfrak{b}(u,v_0) = \mathfrak{b}(u, u - v_0)$$

as $\varphi_{F(t_2)}$ is anisotropic by Lemma 7.15, hence the claim follows.

By the Witt Extension Theorem 8.3, there is an isometry γ of $\varphi_{F(t_2)}$ satisfying $\gamma(v_0) = u$. Replacing v_0 and v_1 by $u = \gamma(v_0)$ and $\gamma(v_1)$ respectively, we may assume that $v_0 \in V$.

Applying Proposition 17.9 to the vectors v_0 and v_1, we find $w \in V[t_2]$ satisfying $\varphi(w) = ct_2^2 + d$ and $\mathfrak{b}(v_0,w) = bt_2$. In a similar fashion, we have $w = w_0 + w_1t_2$ with $w_0, w_1 \in V$. Expanding, we have

$$\varphi(v_0) = a, \quad \mathfrak{b}(v_0,w_1) = b, \quad \varphi(w_1) = c,$$

$$\varphi(w_0) = d, \quad \mathfrak{b}(v_0,w_0) = 0, \quad b(w_0,w_1) = 0.$$

It follows that if W is the subspace generated by v_0 and w_1, then $\varphi|_W \simeq \rho$ and $d \in \widetilde{D}(\mu)$ where $\mu = \varphi|_{W^\perp}$. □

Corollary 17.11. *Let φ and ψ be two anisotropic quadratic forms over F with* $\dim\psi = n$. *Let $T = (t_1,\dots,t_n)$. Suppose that $\psi(T) \in D(\varphi_{F(T)})$. If $\psi = \rho \perp \sigma$ with ρ a nondegenerate binary form and $T' = (t_3,\dots,t_n)$, then $\varphi \simeq \rho \perp \mu$ for some form μ and $\mu(T') \in \widetilde{D}_{F(T')}(\varphi_{F(T')})$.*

Theorem 17.12 (Representation Theorem). *Let φ and ψ be two anisotropic quadratic forms over F with* $\dim\psi = n$. *Let $T = (t_1,\dots,t_n)$. Then the following are equivalent:*

(1) *$D(\psi_K) \subset D(\varphi_K)$ for every field extension K/F.*
(2) *$\psi(T) \in D(\varphi_{F(T)})$.*
(3) *ψ is isometric to a subform of φ.*

In particular, if any of the above conditions hold, then $\dim\psi \le \dim\varphi$.

PROOF. (1) $\Rightarrow$ (2) and (3) $\Rightarrow$ (1) are trivial.

(2) $\Rightarrow$ (3): Applying the structure results, Propositions 7.31 and 7.29, we can write $\psi = \psi_1 \perp \psi_2$, where ψ_1 is an orthogonal sum of nondegenerate binary forms and ψ_2 is diagonalizable. Repeated application of Corollary 17.11 allows us to reduce to the case $\psi = \psi_2$, i.e., $\psi = \langle a_1, \dots, a_n \rangle$ is diagonalizable.

We induct on n. The case $n = 1$ follows from the Substitution Principle 17.7. Suppose that $n = 2$. Then we have $a_1t^2 + a_2 \in D(\varphi_{F(t)})$. By the Cassels-Pfister Theorem, there is a $v \in V[t]$ where $V = V_\varphi$ satisfying $\varphi(v) = a_1t^2 + a_2$. As φ is anisotropic, we have $v = v_1 + v_2t$ for $v_1, v_2 \in V$ and therefore $\varphi(v_1) = a_1$, $\varphi(v_2) = a_2$, and $\mathfrak{b}(v_1, v_2) = 0$. The restriction of φ on the subspace spanned by v_1 and v_2 is isometric to ψ.

In the general case, set $T = (t_1, t_2, \dots, t_n)$, $T' = (t_2, \dots, t_n)$, $b = a_2t^2 + \dots + a_nt_n^2$. As $a_1t^2 + b$ is a value of φ over $F(T')(t)$, by the case considered above, there are vectors $v_1, v_2 \in V_{F(T')}$ satisfying

$$\varphi(v_1) = a_1, \quad \varphi(v_2) = b \quad \text{and} \quad \mathfrak{b}(v_1, v_2) = 0.$$

It follows from the Substitution Principle 17.7 that there is $w \in V$ such that $\varphi(w) = a_1$.

We claim that there is an isometry γ of φ over $F(T')$ such that $\varphi(v_1) = w$. We may assume that $w \neq v_1$ as $\varphi_{F(T')}$ is anisotropic by Lemma 7.15. We have

$$0 \neq \varphi(w - v_1) = \varphi(w) + \varphi(v_1) - \mathfrak{b}(w, v_1) = \mathfrak{b}(w, w - v_1) = \mathfrak{b}(v_1 - w, v_1),$$

therefore w and v_1 do not belong to $\operatorname{rad}\mathfrak{b}$. The claim follows by the Witt Extension Theorem 8.3.

Replacing v_1 and v_2 by $\gamma(v_1) = w$ and $\gamma(v_2)$ respectively, we may assume that $v_1 \in V$. Set $W = (Fv_1)^\perp$. Note that $v_2 \in W_{F(T')}$, hence b is a value of $\varphi|_W$ over $F(T')$. By the induction hypothesis applied to the forms $\psi' = \langle a_2, \dots, a_n \rangle$ and $\varphi|_W$, there is a subspace $V' \subset W$ such that $\varphi|_{V'} \simeq \langle a_2, \dots, a_n \rangle$. Note that v_1 is orthogonal to V' and $v_1 \notin V'$ as ψ is anisotropic. Therefore, the restriction of φ on the subspace $Fv_1 \oplus V'$ is isometric to ψ. □

A field F is called *formally real* if -1 is not a sum of squares. In particular, $\operatorname{char} F = 0$ if this is the case (cf. §95).

Corollary 17.13. *Suppose that F is formally real and $T = (t_0, \dots, t_n)$. Then $t_0^2 + t_1^2 + \dots + t_n^2$ is not a sum of n squares in $F(T)$.*

PROOF. If this is false, then $t_0^2 + t_1^2 + \dots + t_n^2 \in D\big(n\langle 1 \rangle\big)$. As $(n+1)\langle 1 \rangle$ is anisotropic, this contradicts the Representation Theorem. □

The ideas above also allow us to develop a test for simultaneous zeros for quadratic forms. This was proven independently by Amer in [**2**] and Brumer in [**21**] for fields of characteristic not 2 and by Leep for arbitrary fields in [**92**].

Theorem 17.14. *Let φ and ψ be two quadratic forms on a vector space V over F. Then the form $\varphi_{F(t)} + t\psi_{F(t)}$ on $V(t)$ over $F(t)$ is isotropic if and only if φ and ψ have a common isotropic vector in V.*

PROOF. Clearly, a common isotropic vector for φ and ψ is also an isotropic vector for $\rho := \varphi_{F(t)} + t\psi_{F(t)}$.

Conversely, let ρ be isotropic. There exists a nonzero $v \in V[t]$ such that $\rho(v) = 0$. Choose such a v of the smallest degree. We claim that $\deg v = 0$, i.e.,

$v \in V$. If we show this, the equality $\varphi(v) + t\psi(v) = 0$ implies that v is a common isotropic vector for φ and ψ.

Suppose $n := \deg v > 0$. Write $v = w + t^n u$ with $u \in V$ and $w \in V[t]$ of degree less than n. Note that by assumption $\rho(u) \neq 0$. Consider the vector

$$v' = \rho(u) \cdot \tau_u(v) = \rho(u)v - \mathfrak{b}_\rho(v,u)u \in V[t].$$

As $\rho(v) = 0$, we have $\rho(v') = 0$. It follows from the equality

$$\rho(w)v - \mathfrak{b}_\rho(v,w)w = \rho(v - t^n u)v - \mathfrak{b}_\rho(v, v - t^n u)(v - t^n u) = t^{2n}\big(\rho(u)v - \mathfrak{b}_\rho(v,u)u\big)$$

that

$$v' = \frac{\rho(w)v - \mathfrak{b}_\rho(v,w)w}{t^{2n}}.$$

Note that $\deg \rho(w) \leq 2n - 1$ and $\deg \mathfrak{b}_\rho(v,w) \leq 2n$. Therefore, $\deg v' < n$, contradicting the minimality of n. □

18. Values of forms

Let φ be an anisotropic quadratic form over F. Let $p \in F[T] = F[t_1, \dots, t_n]$ be irreducible and $F(p)$ the quotient field of $F[T]/(p)$. In this section, we determine what it means for $\varphi_{F(p)}$ to be isotropic. We base our presentation on ideas of Knebusch in [**82**]. This result has consequences for finite extensions K/F. In particular, the classical Springer's Theorem that forms remain anisotropic under odd degree extensions follows as well as a norm principle about values of φ_K.

Order the group $\mathbb{Z}^n$ lexicographically, i.e., $(i_1, \dots, i_n) < (j_1, \dots, j_n)$ if for the first integer k satisfying $i_k \neq j_k$ with $1 \leq k \leq n$, we have $i_k < j_k$. If $\boldsymbol{i} = (i_1, \dots, i_n)$ in $\mathbb{Z}^n$ and $a \in F^\times$, write $aT^{\boldsymbol{i}}$ for $at_1^{i_1} \cdots t_n^{i_n}$ and call $\boldsymbol{i}$ the *degree* of $aT^{\boldsymbol{i}}$. Let $f = aT^{\boldsymbol{i}} +$ monomials of lower degree in $F[T]$ with $a \in F^\times$. The term $aT^{\boldsymbol{i}}$ is called the *leading term* of f. We define the *degree* $\deg f$ of f to be $\boldsymbol{i}$, the degree of the leading term, and the *leading coefficient* f^* of f to be a, the coefficient of the leading term. Let T_f denote $T^{\boldsymbol{i}}$ if $\boldsymbol{i}$ is the degree of the leading term of f. Then $f = f^*T_f + f'$ with $\deg f' < \deg T_f$. For convenience, we view $\deg 0 < \deg f$ for every nonzero $f \in F[T]$. Note that $\deg(fg) = \deg f + \deg g$ and $(fg)^* = f^*g^*$. If $h \in F(T)^\times$ and $h = f/g$ with $f, g \in F[T]$, let $h^* = f^*/g^*$.

Let V be a finite dimensional vector space over F. For every nonzero $v \in V[T]$ define the *degree* $\deg v$, the *leading vector* v^*, and the *leading term* v^*T_v in a similar fashion. Let $\deg 0 < \deg v$ for any nonzero $v \in V[T]$. So if $v \in V[T]$ is nonzero, we have $v = v^*T_v + v'$ with $\deg v' < \deg T_v$.

Lemma 18.1. *Let φ be a quadratic form on V over F and $g \in F[T]$. Suppose that $g \in D(\varphi_{F(T)})$. Then $g^* \in D(\varphi)$. If, in addition, φ is anisotropic, then* $\deg g \in 2\mathbb{Z}^n$.

PROOF. Since φ on V and the induced quadratic form $\bar{\varphi}$ on $V/\operatorname{rad}\varphi$ have the same values, we may assume that $\operatorname{rad}(\varphi) = 0$. In particular, if φ is isotropic, it is universal, so we may assume that φ anisotropic.

Let $g = \varphi(v)$ with $v \in V(T)$. Write $v = w/f$ with $w \in V[T]$ and nonzero $f \in F[T]$. Then $f^2 g = \varphi(w)$. As $(f^2g)^* = (f^*)^2 g^*$, we may assume that $v \in F[T]$. Let $v = v^*T_v + v'$ with $\deg v' < \deg v$. Then

$$\begin{aligned} g &= \varphi(v^*T_v) + \mathfrak{b}_\varphi(v^*T_v, v') + \varphi(v') = \varphi(v^*)T_v^2 + \mathfrak{b}_\varphi(v^*, v')T_v + \varphi(v') \\ &= \varphi(v^*)T_v^2 + \text{ terms of lower degree.} \end{aligned}$$

As φ is anisotropic, we must have $\varphi(v^*) \neq 0$, hence $g^* = \varphi(v^*) \in D(\varphi)$. As the leading term of g is $\varphi(v^*)T_v^2$, the second statement also follows. □

Let $v \in V[T]$. Suppose that $f \in F[T]$ satisfies $\deg_{t_1} f > 0$. Let $T' = (t_2, \dots, t_n)$. Viewing $v \in V(T')[t_1]$, the analog of the usual division algorithm produces an equation

$$v = fw' + r' \quad \text{with} \quad w', r' \in V_{F(T')}[t_1] \quad \text{and} \quad \deg_{t_1} r' < \deg_{t_1} f.$$

Clearing denominators in $F[T']$, we get

$$hv = fw + r$$

$$\text{(18.2)} \qquad \text{with} \quad w, r \in V[T], \ 0 \neq h \in F[T'] \quad \text{and} \quad \deg_{t_1} r < \deg_{t_1} f,$$
$$\text{so } \deg h < \deg f, \ \deg r < \deg f.$$

If φ is a quadratic form over F let $\langle D(\varphi) \rangle$ denote the subgroup in $F^\times$ generated by $D(\varphi)$.

Theorem 18.3 (Quadratic Value Theorem). *Let φ be an arbitrary anisotropic quadratic form on V and $f \in F[T]$ a nonzero polynomial. Then the following conditions are equivalent:*

(1) $f^* f \in \langle D(\varphi_{F(T)}) \rangle$.

(2) *There exists an $a \in F^\times$ such that $af \in \langle D(\varphi_{F(T)}) \rangle$.*

(3) *$\varphi_{F(p)}$ is isotropic for each irreducible divisor p occurring to an odd power in the factorization of f.*

PROOF. (1) ⇒ (2) is trivial.

(2) ⇒ (3): Let $af \in \langle D(\varphi_{F(T)}) \rangle$, i.e., there are

$$0 \neq h \in F[T] \quad \text{and} \quad v_1, \dots, v_m \in V[T]$$

such that $ah^2 f = \prod \varphi(v_i)$. Let p be an irreducible divisor of f to an odd power. Write $v_i = p^{k_i} v_i'$ so that v_i' is not divisible by p. Dividing out both sides by p^{2k}, with $k = \sum k_i$, we see that the product $\prod \varphi(v_i')$ is divisible by p. Hence the residue of one of the $\varphi(v_i')$ is trivial in the residue field $F(p)$ while the residue of v_i' is not trivial. Therefore, $f_{F(p)}$ is isotropic.

(3) ⇒ (1): We proceed by induction on n and $\deg f$. The statement is obvious if $f = f^*$. In the general case, we may assume that f is irreducible. Therefore, by assumption $\varphi_{F(f)}$ is isotropic. In particular, we see that there exists a vector $v \in V_\varphi[T]$ such that $f \mid \varphi(v)$ and $f \nmid v$. If $\deg_{t_1} f = 0$, let $T' = (t_2, \dots, t_n)$ and let L denote the quotient field of $F[T']/(f)$. Then $F(f) = L(t_1)$ so φ_L is isotropic by Lemma 7.15 and we are done by induction on n. Therefore, we may assume that $\deg_{t_1} f > 0$. By (18.2), there exist $0 \neq h \in F[T]$ and $w, r \in V[T]$ such that $hv = fw + r$ with $\deg h < \deg f$ and $\deg r < \deg f$. As

$$\varphi(hv) = \varphi(fw + r) = f^2 \varphi(w) + f \mathfrak{b}_\varphi(w, r) + \varphi(r),$$

we have $f \mid \varphi(r)$. If $r = 0$, then $f \mid hv$. But f is irreducible and $f \nmid v$, so $f \mid h$. This is impossible as $\deg h < \deg f$. Thus $r \neq 0$. Let $\varphi(r) = fg$ for some $g \in F[T]$. As φ is anisotropic $g \neq 0$. So we have $fg \in D(\varphi_{F(T)})$, hence also $(fg)^* = f^* g^* \in D(\varphi)$ by Lemma 18.1.

Let p be an irreducible divisor occurring to an odd power in the factorization of g. As $\deg \varphi(r) < 2 \deg f$, we have $\deg g < \deg f$, hence p occurs with the same multiplicity in the factorization of fg. By (2) ⇒ (3), applied to the polynomial

fg, the form $\varphi_{F(p)}$ is isotropic. Hence the induction hypothesis implies that $g^*g \in \langle D(\varphi_{F(T)})\rangle$. Consequently, $f^*f = f^{*2} \cdot (f^*g^*)^{-1} \cdot g^*g \cdot fg \cdot g^{-2} \in \big\langle D(\varphi_{F(T)})\big\rangle$. $\square$

Theorem 18.4 (Bilinear Value Theorem). *Let $\mathfrak{b}$ be an anisotropic symmetric bilinear form on V and $f \in F[T]$ a nonzero polynomial. Then the following conditions are equivalent:*

1. $f^*f \in \big\langle D(\mathfrak{b}_{F(T)})\big\rangle$.
2. *There exists an $a \in F^\times$ such that $af \in \big\langle D(\mathfrak{b}_{F(T)})\big\rangle$.*
3. *$\mathfrak{b}_{F(p)}$ is isotropic for each irreducible divisor p occurring to an odd power in the factorization of f.*

PROOF. Let $\varphi = \varphi_{\mathfrak{b}}$. As $D(\mathfrak{b}_K) = D(\varphi_K)$ for every field extension K/F by Lemma 9.2 and $\mathfrak{b}_K$ is isotropic if and only if φ_K is isotropic, the result follows by the Quadratic Value Theorem 18.3. $\square$

A basic result in Artin-Schreier theory is that an ordering on a formally real field extends to an ordering on a finite algebraic extension of odd degree, equivalently if the bilinear form $n\langle 1\rangle$ is anisotropic over F for any integer n, it remains so over any finite extension of odd degree. Witt conjectured in [**139**] that any anisotropic symmetric bilinear form remains anisotropic under an odd degree extension (if $\operatorname{char} F \neq 2$). This was first shown to be true by Springer in [**126**]. This is in fact true without a characteristic assumption for both quadratic and symmetric bilinear forms.

Corollary 18.5 (Springer's Theorem). *Let K/F be a finite extension of odd degree. Suppose that φ (respectively, $\mathfrak{b}$) is an anisotropic quadratic form (respectively, symmetric bilinear form) over F. Then φ_K (respectively, $\mathfrak{b}_K$) is anisotropic.*

PROOF. By induction on $[K : F]$, we may assume that $K = F(\theta)$ is a primitive extension. Let p be the minimal polynomial of θ over F. Suppose that φ_K is isotropic. Then $ap \in \big\langle D(\varphi_{F(t)})\big\rangle$ for some $a \in F^\times$ by the Quadratic Value Theorem 18.3. It follows that p has even degree by Lemma 18.1, a contradiction. If $\mathfrak{b}$ is a symmetric bilinear form over F, applying the above to the quadratic form $\varphi_{\mathfrak{b}}$ shows the theorem also holds in the bilinear case. $\square$

We shall give another proof of Springer's Theorem in Corollary 71.3 below.

Corollary 18.6. *If K/F is an extension of odd degree, then $r_{K/F} : W(F) \to W(K)$ and $r_{K/F} : I_q(F) \to I_q(K)$ are injective.*

Corollary 18.7. *Let φ and ψ be two quadratic forms on a vector space V over F having no common isotropic vector in V. Then for any field extension K/F of odd degree the forms φ_K and ψ_K have no common isotropic vector in V_K.*

PROOF. This follows from Springer's Theorem and Theorem 17.14. $\square$

Exercise 18.8. Let $\operatorname{char} F \neq 2$ and K/F be a finite purely inseparable field extension. Then $r_{K/F} : W(F) \to W(K)$ is an isomorphism.

Corollary 18.9. *Let $K = F(\theta)$ be an algebraic extension of F and p the (monic) minimal polynomial of θ over F. Let φ be a regular quadratic form over F. Suppose that there exists a $c \in F$ such that $p(c) \notin \big\langle D(\varphi)\big\rangle$. Then φ_K is anisotropic.*

PROOF. As $\operatorname{rad}\varphi = 0$, if φ were isotropic it would be universal. Thus φ is anisotropic. In particular, p is not linear, hence $p(c) \neq 0$. Suppose that φ_K is isotropic. By the Quadratic Value Theorem 18.3, we have $p \in \langle D(\varphi_{F(t)})\rangle$. By the Substitution Principle 17.7, we have $p(c) \in \langle D(\varphi)\rangle$ for all $c \in F$, a contradiction. □

As another consequence, we obtain the following theorem first proved by Knebusch in [**81**].

Theorem 18.10 (Value Norm Principle). *Let φ be a quadratic form over F and K/F a finite field extension. Then $\mathrm{N}_{K/F}(D(\varphi_K)) \subset \langle D(\varphi)\rangle$.*

PROOF. Let $V = V_\varphi$. Since the form φ on V and the induced form $\bar{\varphi}$ on $V/\operatorname{rad}(\varphi)$ have the same values, we may assume that $\operatorname{rad}(\varphi) = 0$. If φ is isotropic, then φ splits off a hyperbolic plane. In particular, φ is universal and the statement is obvious. Thus we may assume that φ is anisotropic. Moreover, we may assume that $\dim\varphi \geq 2$ and $1 \in D(\varphi)$.

Case 1: φ_K is isotropic.

Let $x \in D(\varphi_K)$. Suppose that $K = F(x)$. Let $p \in F[t]$ denote the (monic) minimal polynomial of x so $K = F(p)$. It follows from the Quadratic Value Theorem 18.3 that $p \in \langle D(\varphi_{F(t)})\rangle$ and $\deg p$ is even. In particular, $\mathrm{N}_{K/F}(x) = p(0)$ and by the Substitution Principle 17.7,

$$\mathrm{N}_{K/F}(x) = p(0) \in \langle D(\varphi)\rangle.$$

If $F(x) \subsetneq K$, let $m = [K : F(x)]$. If m is even, then $\mathrm{N}_{K/F}(x) \in F^{\times 2} \subset \langle D(\varphi)\rangle$. If m is odd, then $\varphi_{F(x)}$ is isotropic by Springer's Theorem 18.5. Applying the above argument to the field extension $F(x)/F$ yields

$$\mathrm{N}_{K/F}(x) = \mathrm{N}_{F(x)/F}(x)^m \in \langle D(\varphi)\rangle$$

as needed.

Case 2: φ_K is anisotropic.

Let $x \in D(\varphi_K)$. Choose vectors $v, v_0 \in V_K$ such that $\varphi_K(v) = x$ and $\varphi_K(v_0) = 1$. Let $V' \subset V_K$ be a 2-dimensional subspace (over K) containing v and v_0. The restriction φ' of φ_K to V' is a binary anisotropic quadratic form over K representing x and 1. It follows from Proposition 12.1 that the even Clifford algebra $L = C_0(\varphi')$ is a quadratic field extension of K and $x = \mathrm{N}_{L/K}(y)$ for some $y \in L^\times$. Moreover, since $C_0(\varphi'_L) = C_0(\varphi') \otimes_K L = L \otimes_K L$ is not a field, by the same proposition, φ' and therefore φ is isotropic over L. Applying Case 1 to the field extension L/F yields

$$\mathrm{N}_{K/F}(x) = \mathrm{N}_{K/F}(\mathrm{N}_{L/K}(y)) = \mathrm{N}_{L/F}(y) \in \langle D(\varphi)\rangle. \qquad \square$$

Theorem 18.11 (Bilinear Value Norm Principle). *Let $\mathfrak{b}$ be a symmetric bilinear form over F and let K/F be a finite field extension. Then $\mathrm{N}_{K/F}(D(\mathfrak{b}_K)) \subset \langle D(\mathfrak{b})\rangle$.*

PROOF. As $D(\mathfrak{b}_E) = D(\varphi_{\mathfrak{b}_E})$ for any field extension E/F, this follows from the quadratic version of the theorem. □

19. Forms over a discrete valuation ring

We wish to look at similarity factors of bilinear and quadratic forms. To do so we need a few facts about such forms over a discrete valuation ring (DVR) which we now establish. These results are based on the work of Springer in [**127**] in the case of fields of characteristic different from 2.

Throughout this section, R will be a DVR with quotient field K, residue field $\bar{K}$, and prime element π. If V is a free R-module of finite rank, then the definition of a (symmetric) bilinear form and quadratic form on V is analogous to the field case. In particular, we can associate to every quadratic form its *polar form* $\mathfrak{b}_\varphi : (v,w) \mapsto \varphi(v+w) - \varphi(v) - \varphi(w)$. Orthogonal complements are defined in the usual way. Orthogonal sums of bilinear (respectively, quadratic) forms are defined as in the field case. We use analogous notation as in the field case when clear. If $F \to R$ is a ring homomorphism and φ is a quadratic form over F, we let $\varphi_R = R \otimes_F \varphi$.

A bilinear form $\mathfrak{b}$ on V is *nondegenerate* if $l : V \to \operatorname{Hom}_R(V,R)$ defined by $v \mapsto l_v : w \to b(v,w)$ is an isomorphism. As in the field case, we have the crucial

Proposition 19.1. *Let R be a DVR. Let V be a free R-module of finite rank and W a submodule of V. If φ is a quadratic form on V with $\mathfrak{b}_\varphi|_W$ nondegenerate, then $\varphi = \varphi|_W \perp \varphi|_{W^\perp}$.*

PROOF. As $\mathfrak{b}_\varphi|_W$ is nondegenerate, $W \cap W^\perp = \{0\}$ and if $v \in V$, there exists $w' \in W$ such that the linear map $W \to F$ by $w \mapsto \mathfrak{b}_\varphi(v,w)$ is given by $\mathfrak{b}_\varphi(v,w) = \mathfrak{b}_\varphi(w',w)$ for all $w \in W$. Consequently, $v = w + (v - w') \in W \oplus W^\perp$ and the result follows. □

Hyperbolic quadratic forms and planes are also defined in an analogous way. We let $\mathbb{H}$ denote the quadratic hyperbolic plane.

If R is a DVR and V a vector space over the quotient field K of R. A vector $v \in V$ is called *primitive* if it is not divisible by a prime element π, i.e., the image $\bar{v}$ of v in $\bar{K} \otimes_R V$ is not zero.

Arguing as in Proposition 7.13, we have

Lemma 19.2. *Let R be a DVR. Let φ be a quadratic form on V whose polar form is nondegenerate. Suppose that V contains an isotropic vector v. Then there exists a submodule W of V containing v such that $\varphi|_W \simeq \mathbb{H}$.*

PROOF. Dividing v by π^n for an appropriate choice of n, we may assume that v is primitive. It follows easily that V/Rv is torsion-free, hence free. In particular, $V \to V/Rv$ splits, therefore, Rv is a direct summand of V. Let $f : V \to R$ be an R-linear map satisfying $f(v) = 1$. As $l : V \to \operatorname{Hom}_R(V,R)$ is an isomorphism, there exists an element $w \in V$ such that $f = l_w$, hence $\mathfrak{b}_\varphi(v,w) = 1$. Let $W = Rv \oplus Rw$. Then v, $w - \varphi(w)v$ is a hyperbolic pair. □

By induction, we conclude:

Corollary 19.3. *Suppose that R is a DVR. Let φ be a quadratic form on V over R whose polar form is nondegenerate. Then $\varphi = \varphi|_{V_1} \perp \varphi|_{V_2}$ with V_1, V_2 submodules of V satisfying $\varphi|_{V_1}$ is anisotropic and $\varphi|_{V_2} \simeq m\mathbb{H}$ for some $m \geq 0$.*

Associated to a quadratic form φ on V over R are two forms: φ_K on $K \otimes_R V$ over K and $\bar{\varphi} = \varphi_{\bar{K}}$ on $\bar{K} \otimes_R V$ over $\bar{K}$.

Lemma 19.4. *Suppose that R is a complete DVR. Let φ be an anisotropic quadratic form over R whose associated bilinear form $\mathfrak{b}_\varphi$ is nondegenerate. Then $\bar{\varphi}$ is also anisotropic.*

PROOF. Let $\{v_1, \dots, v_n\}$ be a basis for V_φ and $t_1, \dots, t_n$ the respective coordinates. If $w \in V_\varphi$, then $\frac{\partial \varphi}{\partial t_i}(w) = b_\varphi(v_i, w)$. In particular, if $\bar{w} \neq \bar{0}$, there exists an i such that $\bar{b}_\varphi(\bar{v}_i, \bar{w}) \neq 0$. It follows by Hensel's lemma that φ would be isotropic if $\bar{\varphi}$ is. □

Lemma 19.5. *Let φ and ψ be two quadratic forms over a DVR R such that $\bar{\varphi}$ and $\bar{\psi}$ are anisotropic over $\bar{K}$. Then $\varphi_K \perp \pi\psi_K$ is anisotropic over K.*

PROOF. Suppose that $\varphi(u) + \pi\psi(v) = 0$ for some $u \in V_\varphi$ and $v \in V_\psi$ with at least one of u and v primitive. Reducing modulo π, we have $\bar{\varphi}(\bar{u}) = 0$. Since $\bar{\varphi}$ is anisotropic, $u = \pi w$ for some w. Therefore, $\pi\varphi(w) + \psi(v) = 0$ and reducing modulo π we get $\bar{\psi}(\bar{v}) = 0$. Since $\bar{\psi}$ is also anisotropic, v is divisible by π, a contradiction. □

Corollary 19.6. *Let φ and ψ be anisotropic forms over F. Then $\varphi_{F(t)} \perp t\psi_{F(t)}$ is anisotropic.*

PROOF. In the lemma, let $R = F[t]_{(t)}$, a DVR, $\pi = t$ a prime. As $\overline{\varphi_R} = \varphi$ and $\overline{\psi_R} = \psi$, the result follows from the lemma. □

Proposition 19.7. *Let φ be a quadratic form over a complete DVR R such that the associated bilinear form $\mathfrak{b}_\varphi$ is nondegenerate. Suppose that $\varphi_K \simeq \pi\varphi_K$. Then $\bar{\varphi}$ is hyperbolic.*

PROOF. Write $\varphi = \psi \perp n\mathbb{H}$ with ψ anisotropic. By Lemma 19.4, we have $\bar{\psi}$ is anisotropic. The form

$$\varphi_K \perp (-\pi\varphi_K) \simeq \psi_K \perp (-\pi\psi_K) \perp 2n\mathbb{H}$$

is hyperbolic and $\psi_K \perp (-\pi\psi_K)$ is anisotropic over K by Lemma 19.5. We must have $\psi = 0$ by uniqueness of the Witt decomposition over K, hence $\varphi = n\mathbb{H}$ is hyperbolic. It follows that $\bar{\varphi}$ is hyperbolic. □

Proposition 19.8. *Let φ be a nondegenerate quadratic form over F of even dimension. Let $f \in F[T]$ and $p \in F[T]$ an irreducible polynomial factor of f of odd multiplicity. If $\varphi_{F(T)} \simeq f\varphi_{F(T)}$, then $\varphi_{F(p)}$ is hyperbolic.*

PROOF. Let R denote the completion of the DVR $F[T]_{(p)}$ and let K be its quotient field. The residue field of R coincides with $F(p)$. Modifying f by a square, we may assume that $f = up$ for some $u \in R^\times$. As $\varphi_{F(T)} \simeq f\varphi_{F(T)}$, we have $\varphi_{F(T)} \simeq up\varphi_{F(T)}$. Applying Proposition 19.7 to the form φ_R and $\pi = up$ yields $\overline{(\varphi_R)} = \varphi_{F(p)}$ is hyperbolic. □

We shall also need the following:

Proposition 19.9. *Let R be a DVR with quotient field K. Let φ and ψ be two quadratic forms on V and W over R, respectively, such that their respective residue forms $\bar{\varphi}$ and $\bar{\psi}$ are anisotropic. If $\varphi_K \simeq \psi_K$, then $\varphi \simeq \psi$ (over R).*

PROOF. Let $f : V_K \to W_K$ be an isometry between φ_K and ψ_K. It suffices to prove that $f(V) \subset W$ and $f^{-1}(W) \subset V$. Suppose that there exists a $v \in V$ such that $f(v)$ is not in W. Then $f(v) = w/\pi^k$ for some primitive $w \in W$ and $k > 0$. Since f is an isometry, we have $\psi(w) = \pi^{2k}\varphi(v)$, i.e., $\psi(w)$ is divisible by π, hence $\bar{w}$ is an isotropic vector of $\bar{\psi}$, a contradiction. Analogously, $f^{-1}(W) \subset V$. □

19.A. Residue homomorphisms. If R is a DVR, then for each $x \in K^\times$ we can write $x = u\pi^n$ for some $u \in R^\times$ and $n \in \mathbb{Z}$.

Lemma 19.10. *Let R be a DVR with quotient field K and residue field $\bar{K}$. Let π be a prime element in R. There exist group homomorphisms*

$$\partial : W(K) \to W(\bar{K}) \quad \textit{and} \quad \partial_\pi : W(K) \to W(\bar{K})$$

satisfying

$$\partial(\langle u\pi^n\rangle) = \begin{cases} \langle \bar{u}\rangle & n \text{ is even} \\ 0 & n \text{ is odd} \end{cases} \quad \textit{and} \quad \partial_\pi(\langle u\pi^n\rangle) = \begin{cases} 0 & n \text{ is even,} \\ \langle \bar{u}\rangle & n \text{ is odd,} \end{cases}$$

for $u \in R^\times$ and $n \in \mathbb{Z}$.

PROOF. It suffices to prove the existence of ∂ as we can take $\partial_\pi = \partial \circ \lambda_\pi$ where λ_π is the group homomorphism $\lambda_\pi : W(K) \to W(K)$ given by $\mathfrak{b} \to \pi\mathfrak{b}$.

By Theorem 4.8, it suffices to check that the generating relations of the Witt ring are respected. As $\langle\bar{1}\rangle + \langle\overline{-1}\rangle = 0$ in $W(\bar{K})$, it suffices to show if $a, b \in R$ with $a + b \neq 0$, then

$$\partial(\langle a\rangle) + \partial(\langle b\rangle) = \partial(\langle a+b\rangle) + \partial(\langle ab(a+b)\rangle) \tag{19.11}$$

in $W(\bar{K})$.

Let

$$a = a_0\pi^n, \quad b = b_0\pi^m, \quad a + b = \pi^l c_0 \quad \text{with } a_0, b_0, c_0 \in R^\times,$$

and $m, n, l \in \mathbb{Z}$ satisfying $\min\{m, n\} \leq l$. We may assume that $n \leq m$.

Suppose that $n < m$. Then

$$a + b = \pi^n a_0(1 + \pi^{m-n}\frac{b_0}{a_0}) \quad \text{and} \quad ab(a+b) = \pi^{2n+m} b_0 a_0^2 (1 + \frac{b_0}{a_0}\pi^{m-n}).$$

In particular, $\partial(\langle a\rangle) = \partial(\langle a+b\rangle)$ and $\partial(\langle b\rangle) = \partial(\langle ab(a+b)\rangle)$ as needed.

Suppose that $n = m$. If $n = l$, then $a_0 + b_0 \in R^\times$ and the result follows by the Witt relation in $W(\bar{K})$.

So suppose that $n < l$. Then $\bar{a}_0 = -\bar{b}_0$ so the left hand side of (19.11) is zero. If l is odd, then $\partial(\langle a+b\rangle) = 0 = \partial(\langle ab(a+b)\rangle)$ as needed. So we may assume that l is even. Then $\langle a+b\rangle \simeq \langle c_0\rangle$ and $\langle ab(a+b)\rangle \simeq \langle a_0 b_0 c_0\rangle$ over K. Hence the right hand side of (19.11) is $\langle \bar{c}_0\rangle + \langle \bar{a}_0\bar{b}_0\bar{c}_0\rangle = \langle \bar{c}_0\rangle + \langle -\bar{c}_0\rangle = 0$ in $W(\bar{K})$ also. □

The map $\partial : W(K) \to W(\bar{K})$ in the lemma is not dependent on the choice on the prime element π. It is called the *first residue homomorphism* with respect to R. The map $\partial_\pi : W(K) \to W(\bar{K})$ does depend on π. It is called the *second residue homomorphism* with respect to R and π.

Remark 19.12. Let R be a DVR with quotient field K and residue field $\bar{K}$. Let π be a prime element in R. If $\mathfrak{b}$ is a nondegenerate diagonalizable bilinear form over K, we can write $\mathfrak{b}$ as

$$\mathfrak{b} \simeq \langle u_1, \ldots, u_n\rangle \perp \pi\langle v_1, \ldots, v_m\rangle$$

for some $u_i, v_j \in R^\times$. Then $\partial(\mathfrak{b}) = \langle \bar{u}_1, \dots, \bar{u}_n \rangle$ in $W(\bar{K})$ and $\partial_\pi(\mathfrak{b}) = \langle \bar{v}_1, \dots, \bar{v}_m \rangle$ in $W(\bar{K})$.

Example 19.13. Let R be a DVR with quotient field K and residue field $\bar{K}$. Let π be a prime element in R. Let $\mathfrak{b} = \langle\langle a_1, \dots, a_n \rangle\rangle$ be an anisotropic n-fold Pfister form over K. Then we may assume that $a_i = \pi^{j_i} u_i$ with $j_i = 0$ or 1 and $u_i \in R^\times$ for all i. By Corollary 6.13, we may assume that $a_i \in R^\times$ for all $i > 1$. As $\mathfrak{b} = -a_1\langle\langle a_2, \dots, a_n \rangle\rangle \perp \langle\langle a_2, \dots, a_n \rangle\rangle$, if $a_1 \in R^\times$, then $\partial(\mathfrak{b}) = \langle\langle \bar{a}_1, \dots, \bar{a}_n \rangle\rangle$ and $\partial_\pi(\mathfrak{b}) = 0$, and if $a_1 = \pi u_1$, then $\partial(\mathfrak{b}) = \langle\langle \bar{a}_2, \dots, \bar{a}_n \rangle\rangle$ and $\partial_\pi(\mathfrak{b}) = -\bar{u}_1\langle\langle \bar{a}_2, \dots, \bar{a}_n \rangle\rangle$.

As n-fold Pfister forms generate $I^n(F)$, we have, by the example, the following:

Lemma 19.14. *Let R be a DVR with quotient field K and residue field $\bar{K}$. Let π be a prime element in R. Then for every $n \geq 1$:*

(1) $\partial\big(I^n(K)\big) \subset I^{n-1}(\bar{K})$.
(2) $\partial_\pi\big(I^n(K)\big) \subset I^{n-1}(\bar{K})$.

Exercise 19.15. Suppose that R is a complete DVR with quotient field K and residue field $\bar{K}$. If $\operatorname{char} \bar{K} \neq 2$, then the residue homomorphisms induce split exact sequences of groups:

$$0 \to W(\bar{K}) \to W(K) \to W(\bar{K}) \to 0$$

and

$$0 \to I^n(\bar{K}) \to I^n(K) \to I^{n-1}(\bar{K}) \to 0.$$

20. Similarities of forms

Let φ be an anisotropic quadratic form over F. Let $p \in F[T] := F[t_1, \dots, t_n]$ be irreducible and $F(p)$ the quotient field of $F[T]/(p)$. In this section, we determine what it means for $\varphi_{F(p)}$ to be hyperbolic. We also establish the analogous result for anisotropic bilinear forms over F. We saw that a form becoming isotropic over $F(p)$ was related to the values it represented over the polynomial ring $F[T]$. We shall see that hyperbolicity is related to the similarity factors of the form over $F[T]$. We shall also deduce norm principles for similarity factors of a form over F first established by Scharlau in [**119**] and Knebusch in [**82**]. To establish these results, we introduce the transfer of forms from a finite extension of F to F, an idea introduced by Scharlau (cf. [**119**]) for quadratic forms over fields of characteristic different from 2 and Baeza (cf. [**15**]) in arbitrary characteristic.

20.A. Transfer of bilinear and quadratic forms. Let K/F be a finite field extension and $s : K \to F$ an F-linear functional. If $\mathfrak{b}$ is a symmetric bilinear form on V over K define the *transfer $s_*(\mathfrak{b})$ of $\mathfrak{b}$ induced by s* to be the symmetric bilinear form on V over F given by

$$s_*(\mathfrak{b})(v, w) = s\big(\mathfrak{b}(v, w)\big) \quad \text{for all} \quad v, w \in V.$$

If φ is a quadratic form on V over K, define the *transfer $s_*(\varphi)$ of φ induced by s* to be the quadratic form on V over F given by $s_*(\varphi)(v) = s\big(\varphi(v)\big)$ for all $v \in V$ with polar form $s_*(\mathfrak{b}_\varphi)$.

Note that $\dim s_*(\mathfrak{b}) = [K : F] \dim \mathfrak{b}$.

Lemma 20.1. *Let K/F be a finite field extension and $s : K \to F$ an F-linear functional. The transfer s_* factors through orthogonal sums and preserves isometries.*

Proof. Let $v, w \in V_{\mathfrak{b}}$. If $\mathfrak{b}(v,w) = 0$, then $s_*(\mathfrak{b})(v,w) = s\big(\mathfrak{b}(v,w)\big) = 0$. Thus $s_*(\mathfrak{b} \perp \mathfrak{c}) = s_*(\mathfrak{b}) \perp s_*(\mathfrak{c})$. If $\sigma : \mathfrak{b} \to \mathfrak{b}'$ is an isometry, then

$$s_*(\mathfrak{b}')\big(\sigma(v), \sigma(w)\big) = s\big(\mathfrak{b}'(\sigma(v), \sigma(w))\big) = s\big(\mathfrak{b}(v,w)\big) = s_*(\mathfrak{b})(v,w),$$

so $\sigma : s_*(\mathfrak{b}) \to s_*(\mathfrak{b}')$ is also an isometry. □

Proposition 20.2 (Frobenius Reciprocity). *Suppose that K/F is a finite extension of fields and $s : K \to F$ an F-linear functional. Let $\mathfrak{b}$ and $\mathfrak{c}$ be symmetric bilinear forms over F and K, respectively, and φ and ψ quadratic forms over F and K respectively. Then there exist canonical isometries:*

$$s_*(\mathfrak{b}_K \otimes_K \mathfrak{c}) \simeq \mathfrak{b} \otimes_F s_*(\mathfrak{c}), \tag{20.3a}$$

$$s_*(\mathfrak{b}_K \otimes_K \psi) \simeq \mathfrak{b} \otimes_F s_*(\psi), \tag{20.3b}$$

$$s_*(\mathfrak{c} \otimes_K \varphi_K) \simeq s_*(\mathfrak{c}) \otimes_F \varphi. \tag{20.3c}$$

In particular,

$$s_*(\mathfrak{b}_K) \simeq \mathfrak{b} \otimes_F s_*\big(\langle 1 \rangle_b\big).$$

Proof. (a): The canonical F-linear map $V_{\mathfrak{b}_K} \otimes_K V_{\mathfrak{c}} \to V_{\mathfrak{b}} \otimes_F V_{\mathfrak{c}}$ given by $(a \otimes v) \otimes w \mapsto v \otimes aw$ is an isometry. Indeed,

$$\begin{aligned} s(\mathfrak{b}_K \otimes \mathfrak{c})\big((a \otimes v) \otimes w, (a' \otimes v') \otimes w'\big) &= s\big(aa'\mathfrak{b}(v,v')\mathfrak{c}(w,w')\big) \\ &= \mathfrak{b}(v,v')s\big(\mathfrak{c}(aw, a'w')\big) \\ &= (\mathfrak{b} \otimes s\mathfrak{c})(v \otimes aw, v' \otimes a'w'). \end{aligned}$$

The last statement follows from the first by setting $\mathfrak{c} = \langle 1 \rangle$.

(b) and (c) are proved in a similar fashion. □

Lemma 20.4. *Let K/F be a finite field extension and $s : K \to F$ a nonzero F-linear functional.*

(1) *If $\mathfrak{b}$ is a nondegenerate symmetric bilinear form on V over K, then $s_*(\mathfrak{b})$ is nondegenerate on V over F.*
(2) *If φ is an even-dimensional nondegenerate quadratic form on V over K, then $s_*(\varphi)$ is nondegenerate on V over F.*

Proof. Suppose that $0 \neq v \in V$. As $\mathfrak{b}$ is nondegenerate, there exists a $w \in V$ such that $1 = \mathfrak{b}(v,w)$. As s is not zero, there exists a $c \in K$ such that $0 \neq s(c) = s_*(\mathfrak{b})(v, cw)$. This shows (1). Statement (2) follows from (1) and Remark 7.21(1). □

Corollary 20.5. *Let K/F be a finite extension of fields and $s : K \to F$ a nonzero F-linear functional.*

(1) *If $\mathfrak{c}$ is a bilinear hyperbolic form over K, then $s_*(\mathfrak{c})$ is a hyperbolic form over F.*
(2) *If φ is a quadratic hyperbolic form over K, then $s_*(\varphi)$ is a hyperbolic form over F.*

PROOF. (1): As s_* respects orthogonality, we may assume that $\mathfrak{c} = \mathbb{H}_1$. By Frobenius Reciprocity,

$$s_*(\mathbb{H}_1) \simeq s_*\big((\mathbb{H}_1)_K\big) \simeq (\mathbb{H}_1)_F \otimes s_*\big(\langle 1\rangle\big).$$

As $s_*\big(\langle 1\rangle\big)$ is nondegenerate by Lemma 20.4, we have $s_*(\mathbb{H}_1)$ is hyperbolic by Lemma 2.1.

(2): This follows in the same way as (1) using Lemma 8.16. □

Let K/F be a finite field extension and $s : K \to F$ a nonzero F-linear functional. By Lemmas 20.4 and 20.5, the functional s induces group homomorphisms

$$s_* : \widehat{W}(K) \to \widehat{W}(F), \quad s_* : W(K) \to W(F), \quad \text{and} \quad s_* : I_q(K) \to I_q(F)$$

called *transfer maps*. Let $\mathfrak{b}$ and $\mathfrak{c}$ be nondegenerate symmetric bilinear forms over F and K, respectively, and φ and ψ nondegenerate quadratic forms over F and K, respectively. By Frobenius Reciprocity, we have

$$s_*\big(r_{K/F}(\mathfrak{b}) \cdot \mathfrak{c}\big) = \mathfrak{b} \cdot s_*(\mathfrak{c})$$

in $\widehat{W}(F)$ and $W(F)$, i.e., $s_* : \widehat{W}(K) \to \widehat{W}(F)$ is a $\widehat{W}(F)$-module homomorphism and $s_* : W(K) \to W(F)$ is a $W(F)$-module homomorphism where we view $W(K)$ as a $W(F)$-module via $r_{K/F}$. Furthermore,

$$s_*\big(r_{K/F}(\mathfrak{b}) \cdot \psi\big) = \mathfrak{b} \cdot s_*(\psi) \quad \text{and} \quad s_*\big(\mathfrak{c} \cdot r_{K/F}(\varphi)\big) = s_*(\mathfrak{c}) \cdot \varphi$$

in $I_q(F)$. Note that $s_*\big(I(K)\big) \subset I(F)$.

Corollary 20.6. *Let K/F be a finite field extension and $s : K \to F$ a nonzero F-linear functional. Then the compositions*

$$s_* r_{K/F} : \widehat{W}(F) \to \widehat{W}(F), \; s_* r_{K/F} : W(F) \to W(F), \; \text{and } s_* r_{K/F} : I_q(F) \to I_q(F)$$

are given by multiplication by $s_*\big(\langle 1\rangle_b\big)$, i.e., $\mathfrak{b} \mapsto \mathfrak{b} \cdot s_*\big(\langle 1\rangle_b\big)$ for a nondegenerate symmetric bilinear form $\mathfrak{b}$ and $\varphi \mapsto s_*\big(\langle 1\rangle_b\big) \cdot \varphi$ for a nondegenerate quadratic form φ.

Corollary 20.7. *Let K/F be a field extension and $s : K \to F$ a nonzero F-linear functional. Then $\operatorname{im}(s_*)$ is an ideal in $\widehat{W}(F)$ (respectively, $W(F)$) and is independent of s.*

PROOF. By Frobenius Reciprocity, $\operatorname{im}(s_*)$ is an ideal. Suppose that $s_1 : K \to F$ is another nonzero F-linear functional. Let $K \to \operatorname{Hom}_F(K, F)$ be the F-isomorphism given by $a \mapsto \big(x \mapsto s(ax)\big)$. Hence there exists a unique $a \in K^\times$ such that $s_1(x) = s(ax)$ for all $x \in K$. Consequently, $(s_1)_*(\mathfrak{b}) = s_*(a\mathfrak{b})$ for all nondegenerate symmetric bilinear forms $\mathfrak{b}$ over K. □

Let $K = F(x)/F$ be an extension of degree n and $a = N_{K/F}(x) \in F^\times$ the norm of x. Let

(20.8)
$$s : K \to F \text{ be the } F\text{-linear functional defined by}$$
$$s(1) = 1 \text{ and } s(x^i) = 0 \text{ for all } i \in [1,\, n-1].$$

Then $s(x^n) = (-1)^{n+1}a$.

Lemma 20.9. *The transfer induced by the F-linear functional s in (20.8) satisfies*

$$s_*\big(\langle 1\rangle_b\big) = \begin{cases} \langle 1\rangle_b & \text{if } n \text{ is odd,} \\ \langle 1, -a\rangle_b & \text{if } n \text{ is even.} \end{cases}$$

PROOF. Let $\mathfrak{b} = s_*(\langle 1\rangle)$. Let $V \subset K$ be the F-subspace spanned by x^i with $i \in [1,\ n]$, a nondegenerate subspace. Then $V^\perp = F$, consequently $K = F \oplus V$.

First suppose that $n = 2m + 1$ is odd. The subspace of W spanned by x^i, $i \in [1,\ m]$, is a lagrangian of $\mathfrak{b}|_V$, hence $\mathfrak{b}|_V$ is metabolic and $\mathfrak{b} = \mathfrak{b}|_{V^\perp} = \langle 1\rangle$ in $W(F)$.

Next suppose that $n = 2m$ is even. We have

$$\mathfrak{b}(x^i, x^j) = \begin{cases} 0 & \text{if } i+j < n, \\ -a & \text{if } i+j = n. \end{cases}$$

It follows that $\det \mathfrak{b} = (-1)^m aF^{\times 2}$ and the subspace $W' \subset W$ spanned by all x^i with $i \neq m$ and $1 \leq i \leq n$ is nondegenerate. In particular, $K = W' \oplus (W')^\perp$ by Proposition 1.6. By dimension count $\dim(W')^\perp = 2$. As the subspace of W' spanned by x^i, $i \in [1,\ m-1]$, is a lagrangian of $\mathfrak{b}|_{W'}$, we have $\mathfrak{b}|_{W'}$ is metabolic. Computing determinants, yields $\mathfrak{b}|_{(W')^\perp} \simeq \langle 1, -a\rangle$, hence in $W(F)$, we have $\mathfrak{b} = \mathfrak{b}|_{(W')^\perp} = \langle 1, -a\rangle$. □

Corollary 20.10. *Suppose that $K = F(x)$ is a finite extension of even degree over F. Then $\operatorname{Ker}(r_{K/F}) \subset \operatorname{ann}_{W(F)}\big(\langle\langle N_{K/F}(x)\rangle\rangle\big)$.*

PROOF. Let s be the F-linear functional in (20.8). By Corollary 20.6 and Lemma 20.9, we have

$$\operatorname{Ker}\big(r_{K/F} : W(F) \to W(K)\big) \subset \operatorname{ann}_{W(F)}\big(s_*(\langle 1\rangle)\big) = \operatorname{ann}_{W(F)}\big(\langle\langle N_{K/F}(x)\rangle\rangle\big). \quad \square$$

Corollary 20.11. *Let K/F be a finite field extension of odd degree. Then the map $r_{K/F} : W(F) \to W(K)$ is injective.*

PROOF. If $K = F(x)$ and s is as in (20.8), then by Corollary 20.6 and Lemma 20.9, we have

$$\operatorname{Ker}\big(r_{K/F} : W(F) \to W(K)\big) \subset \operatorname{ann}_{W(F)}\big(s_*(\langle 1\rangle)\big) = \operatorname{ann}_{W(F)}\big(\langle 1\rangle\big) = 0.$$

The general case follows by induction of the odd integer $[K : F]$. □

Note that this corollary provides a more elementary proof of Corollary 18.6.

Lemma 20.12. *The transfer induced by the F-linear functional s in (20.8) satisfies*

$$s_*\big(\langle x\rangle_b\big) = \begin{cases} \langle a\rangle_b & \text{if } n \text{ is odd,} \\ 0 & \text{if } n \text{ is even.} \end{cases}$$

PROOF. Let $\mathfrak{b} = s_*(\langle x\rangle)$. First suppose that $n = 2m + 1$ is odd. Then

$$\mathfrak{b}(x^i, x^j) = \begin{cases} 0 & \text{if } i+j < n-1, \\ a & \text{if } i+j = n-1. \end{cases}$$

It follows that $\det \mathfrak{b} = (-1)^m aF^{\times 2}$ and the subspace $W \subset K$ spanned by all x^i with $i \neq m$ and $1 \leq i \leq n$ is nondegenerate. In particular, $K = W \oplus W^\perp$ by Proposition 1.6 and $W^\perp$ is 1-dimensional by dimension count. Computing determinants, we see that $\mathfrak{b}|_{W^\perp} \simeq \langle a\rangle$. As the subspace of W spanned by x^i, $i \in [0,\ m-1]$, is a lagrangian of $\mathfrak{b}|_W$, the form $\mathfrak{b}|_W$ is metabolic. Consequently, $\mathfrak{b} = \mathfrak{b}|_{W^\perp} = \langle a\rangle$ in $W(F)$.

Next, suppose that $n = 2m$ is even. The subspace of K spanned by x^i, $i \in [0,\ m-1]$, is a lagrangian of $\mathfrak{b}$ so $\mathfrak{b}$ is metabolic and $\mathfrak{b} = 0$ in $W(F)$. □

Corollary 20.13. *Let s_* be the transfer induced by the F-linear functional s in (20.8). Then $s_*\big(\langle\langle x\rangle\rangle\big) = \langle\langle a\rangle\rangle$ in $W(F)$.*

20.B. Similarity theorems. As a consequence, we get the norm principle first established by Scharlau in [**119**].

Theorem 20.14 (Similarity Norm Principle). *Let K/F be a finite field extension and φ a nondegenerate even-dimensional quadratic form over F. Then*

$$N_{K/F}\big(G(\varphi_K)\big) \subset G(\varphi).$$

PROOF. Let $x \in G(\varphi_K)$. Suppose first that $K = F(x)$. Let s be as in (20.8). As $\langle\langle x\rangle\rangle \cdot \varphi_K = 0$ in $I_q(K)$, applying the transfer $s_* : I_q(K) \to I_q(F)$ yields

$$0 = s_*\big(\langle\langle x\rangle\rangle \cdot \varphi_K\big) = s_*\big(\langle\langle x\rangle\rangle\big) \cdot \varphi = \langle\langle N_{K/F}(x)\rangle\rangle \cdot \varphi$$

in $I_q(F)$ by Frobenius Reciprocity 20.2 and Corollary 20.13. Hence $N_{K/F}(x) \in G(\varphi)$ by Remark 8.17.

In the general case, set $k = [K : F(x)]$. If k is even, we have

$$N_{K/F}(x) = N_{F(x)/F}(x)^k \in G(\varphi)$$

since $F^{\times 2} \subset G(\varphi)$. If k is odd, the homomorphism $I_q\big(F(x)\big) \to I_q(K)$ is injective by Remark 18.6, hence $\langle\langle x\rangle\rangle \cdot \varphi_{F(x)} = 0$. By the first part of the proof, $N_{F(x)/F}(x) \in G(\varphi)$. Therefore, $N_{K/F}(x) \in N_{F(x)/F}(x)F^{\times 2} \subset G(\varphi)$. □

We turn to similarities of forms over polynomial rings. As with values, Knebusch proved analogous results for similarities in [**82**].

Lemma 20.15. *Let φ be a nondegenerate quadratic form of even dimension and $p \in F[t]$ a monic irreducible polynomial (in one variable). If $\varphi_{F(p)}$ is hyperbolic, then $p \in G(\varphi_{F(t)})$.*

PROOF. Let x be the image of t in $K = F(p) = F[t]/(p)$. We have p is the norm of $t - x$ in the extension $K(t)/F(t)$. Since $\varphi_{K(t)}$ is hyperbolic, $t - x \in G(\varphi_{K(t)})$. Applying the Norm Principle 20.14 to the form $\varphi_{F(t)}$ and the field extension $K(t)/F(t)$ yields $p \in G(\varphi_{F(t)})$. □

Theorem 20.16 (Quadratic Similarity Theorem). *Let φ be a nondegenerate quadratic form of even dimension and $f \in F[T] = F[t_1, \dots, t_n]$ a nonzero polynomial. Then the following conditions are equivalent:*

(1) $f^*f \in G(\varphi_{F(T)})$.
(2) *There exists an $a \in F^\times$ such that $af \in G(\varphi_{F(T)})$.*
(3) *For any irreducible divisor p of f to an odd power, the form $\varphi_{F(p)}$ is hyperbolic.*

PROOF. (1) ⇒ (2) is trivial.

(2) ⇒ (3) follows from Proposition 19.8.

(3) ⇒ (1): We proceed by induction on the number n of variables. We may assume that f is irreducible and $\deg_{t_1} f > 0$. In particular, f is an irreducible polynomial in t_1 over the field $E = F(T') = F(t_2, \dots, t_n)$. Let $g \in F[T']$ be the leading term of f. In particular, $g^* = f^*$. As the polynomial $f' = fg^{-1}$ in $E[t_1]$ is monic irreducible and $E(f') = F(f)$, the form $\varphi_{E(f')}$ is hyperbolic. Applying Lemma 20.15 to φ_E and the polynomial f', we have $fg = f' \cdot g^2 \in G(\varphi_{F(T)})$.

Let $p \in F[T']$ be an irreducible divisor of g to an odd power. Since p does not divide f, by the first part of the proof applied to the polynomial fg, the form $\varphi_{F(p)(t_1)}$ is hyperbolic. Since the homomorphism $I_q\big(F(p)\big) \to I_q\big(F(p)(t_1)\big)$

is injective by Remark 8.18, we have $\varphi_{F(p)}$ is hyperbolic. Applying the induction hypothesis to g yields $g^*g \in G(\varphi_{F(T')})$. Therefore, $f^*f = g^*f = g^*g \cdot fg \cdot g^{-2} \cdot \in G(\varphi_{F(T)})$. □

Theorem 20.17 (Bilinear Similarity Norm Principle). *Let K/F be a finite field extension and $\mathfrak{b}$ an anisotropic symmetric bilinear form over F of positive dimension. Then*

$$N_{K/F}\big(G((\mathfrak{b}_K)_{an})\big) \subset G(\mathfrak{b}).$$

PROOF. Let $x \in G\big((\mathfrak{b}_K)_{an}\big)$. Suppose first that $K = F(x)$. Let s be as in (20.8). Let $\mathfrak{b}_K = (\mathfrak{b}_K)_{an} \perp \mathfrak{c}$ with $\mathfrak{c}$ a metabolic form over K. Then $x\mathfrak{c}$ is metabolic, so

$$\mathfrak{b}_K = (\mathfrak{b}_K)_{an} = x(\mathfrak{b}_K)_{an} = x\big((\mathfrak{b}_K)_{an} + \mathfrak{c}\big) = x\mathfrak{b}_K$$

in $W(K)$. Consequently, $\langle\langle x\rangle\rangle \cdot \mathfrak{b}_K = 0$ in $I(K)$. Applying the transfer $s_* : W(K) \to W(F)$ yields

$$0 = s_*\big(\langle\langle x\rangle\rangle \cdot \mathfrak{b}_K\big) = s_*\big(\langle\langle x\rangle\rangle\big) \cdot \mathfrak{b} = \langle\langle N_{K/F}(x)\rangle\rangle \cdot \mathfrak{b}$$

by Frobenius Reciprocity 20.2 and Corollary 20.13. Hence $N_{K/F}(x)\mathfrak{b} = \mathfrak{b}$ in $W(F)$ with both sides anisotropic. It follows from Proposition 2.4 that $N_{K/F}(x) \in G(\mathfrak{b})$.

In the general case, set $k = [K : F(x)]$. If k is even, we have

$$N_{K/F}(x) = N_{F(x)/F}(x)^k \in G(\mathfrak{b})$$

since $F^{\times 2} \subset G(\mathfrak{b})$. If k is odd, the homomorphism $W\big(F(x)\big) \to W(K)$ is injective by Corollary 18.6, hence $\langle\langle x\rangle\rangle \cdot \big(\mathfrak{b}_{F(x)}\big)_{an} = 0$ in $W\big(F(x)\big)$. Thus $x \in G\big((\mathfrak{b}_{F(x)})_{an}\big)$ by Proposition 2.4. By the first part of the proof, $N_{F(x)/F}(x) \in G(\mathfrak{b})$. Consequently, $N_{K/F}(x) \in N_{F(x)/F}(x)F^{\times 2} \subset G(\mathfrak{b})$. □

Lemma 20.18. *Let $\mathfrak{b}$ be a nondegenerate anisotropic symmetric bilinear form and $p \in F[t]$ a monic irreducible polynomial (in one variable). If $\mathfrak{b}_{F(p)}$ is metabolic, then $p \in G(\mathfrak{b}_{F(t)})$.*

PROOF. Let x be the image of t in $K = F(p) = F[t]/(p)$. We have p is the norm of $t - x$ in the extension $K(t)/F(t)$. Since $\mathfrak{b}_{K(t)}$ is metabolic, $(\mathfrak{b}_{K(t)})_{an} = 0$. Thus $x - t \in G\big((\mathfrak{b}_{K(t)})_{an}\big)$. Applying the Norm Principle 20.17 to the anisotropic form $\mathfrak{b}_{F(t)}$ and the field extension $K(t)/F(t)$ yields $p \in G(\mathfrak{b}_{F(t)})$. □

Theorem 20.19 (Bilinear Similarity Theorem). *Let $\mathfrak{b}$ be an anisotropic bilinear form of even dimension and $f \in F[T] = F[t_1, \ldots, t_n]$ a nonzero polynomial. Then the following conditions are equivalent:*

(1) $f^*f \in G(\mathfrak{b}_{F(T)})$.
(2) *There exists an $a \in F^\times$ such that $af \in G(\mathfrak{b}_{F(T)})$.*
(3) *For any irreducible divisor p of f to an odd power, the form $\mathfrak{b}_{F(p)}$ is metabolic.*

PROOF. Let $\varphi = \varphi_{\mathfrak{b}}$ be of dimension m.

(1) $\Rightarrow$ (2) is trivial.

(2) $\Rightarrow$ (3): Let p be an irreducible factor of f to an odd degree. As $F(T)$ is the quotient field of the localization $F[T]_{(p)}$ and $F[T]_{(p)}$ is a DVR, we have a group

homomorphism $\partial : W\big(F(T)\big) \to W\big(F(p)\big)$ given by Lemma 19.10. Since p is a divisor to an odd power of f,

$$\mathfrak{b}_{F(p)} = \partial(\mathfrak{b}_{F(T)}) = \partial(af\mathfrak{b}_{F(T)}) = 0$$

in $W\big(F(p)\big)$. Thus $\mathfrak{b}_{F(p)}$ is metabolic.

(3) $\Rightarrow$ (1): The proof is analogous to the proof of (3) $\Rightarrow$ (1) in the Quadratic Similarity Theorem 20.16 with Lemma 20.18 replacing Lemma 20.15 and hyperbolicity replaced by metabolicity. □

Corollary 20.20. *Let φ be a quadratic form (respectively, $\mathfrak{b}$ an anisotropic bilinear form) on V over F and $f \in F[T]$ with $T = (t_1, \dots, t_n)$. Suppose that $f \in G(\varphi_{F(T)})$ (respectively, $f \in G(\mathfrak{b}_{F(T)})$). If $f(a)$ is defined and nonzero with $a \in F^n$, then $f(a) \in G(\varphi)$.*

PROOF. We may assume that φ is anisotropic as $G(\varphi) = G(\varphi_{an})$. (Cf. Remark 8.9.) By induction, we may assume that f is a polynomial in one variable t. Let $R = F[t]_{(t-a)}$, a DVR. As $f(a) \neq 0$, we have $f \in R^\times$. Over $F(t)$, we have $\varphi_{F(t)} \simeq f\varphi_{F(t)}$, hence $\varphi_R \simeq f\varphi_R$ by Proposition 19.9. Since F is the residue class field of R, upon taking the residue forms, we see that $\varphi = f(a)\varphi$ as needed.

As in the quadratic case, we reduce to f being a polynomial in one variable. We then have $\mathfrak{b}_{F(t)} \simeq f\mathfrak{b}_{F(t)}$. Taking ∂ of this equation relative to the DVR $R = F[t]_{(t-a)}$ yields $\bar{\mathfrak{b}} = \bar{f}\bar{\mathfrak{b}} = f(a)\bar{\mathfrak{b}}$ in $W(F)$ as $f \in R^\times$. The result follows by Proposition 2.4. □

Corollary 20.21. *Let φ be an quadratic form (respectively, $\mathfrak{b}$ an anisotropic bilinear form) on V over F and $g \in F[T]$. Suppose that $g \in G(\varphi_{F(T)})$ (respectively, $g \in G(\mathfrak{b}_{F(T)})$). Then $g^* \in G(\varphi)$ (respectively, $g^* \in G(\mathfrak{b})$).*

PROOF. We may assume that φ is anisotropic as $G(\varphi) = G(\varphi_{an})$. (Cf. Remark 8.9.) By induction on the number of variables, we may assume that $g \in F[t]$. By Lemma 18.1 and Lemma 9.1, we must have $\deg g = 2r$ is even. Let $h(t) = t^{2r}g(1/t) \in G(\varphi_{F(t)})$. Then $g^* = h(0) \in G(\varphi)$ by Corollary 20.20. An analogous proof shows the result for symmetric bilinear forms (using also Lemma 9.2 to see that $\deg g$ is even). □

21. An exact sequence for $W\big(F(t)\big)$

Let $\mathbb{A}^1_F$ be the 1-dimensional affine line over F. Let $x \in \mathbb{A}^1_F$ be a closed point and $F(x)$ be the residue field of x. Then there exists a unique monic irreducible polynomial $f_x \in F[t]$ of degree $d = \deg x$ such that $F(x) = F[t]/(f_x)$. By Lemma 19.10, we have the first and second residue homomorphisms with respect to the DVR $\mathcal{O}_{\mathbb{A}^1_F, x}$ and prime element f_x:

$$W\big(F(t)\big) \xrightarrow{\partial} W\big(F(x)\big) \quad \text{and} \quad W\big(F(t)\big) \xrightarrow{\partial_{f_x}} W\big(F(x)\big).$$

Denote ∂_{f_x} by ∂_x. If $g \in F[t]$, then $\partial_x\big(\langle g\rangle\big) = 0$ unless $f_x \mid g$ in $F[t]$. It follows if $\mathfrak{b}$ is a nondegenerate bilinear form over $F(t)$ that $\partial_x(\mathfrak{b}) = 0$ for almost all $x \in \mathbb{A}^1_F$.

We have:

Theorem 21.1. *The sequence*

$$0 \to W(F) \xrightarrow{r_{F(t)/F}} W\big(F(t)\big) \xrightarrow{\boldsymbol{\partial}} \coprod_{x\in\mathbb{A}^1_F} W\big(F(x)\big) \to 0$$

is split exact where $\boldsymbol{\partial} = (\partial_x)$.

PROOF. As anisotropic bilinear forms remain anisotropic under a purely transcendental extension, $r_{F(t)/F}$ is monic. It is split by the first residue homomorphism with respect to any rational point in $\mathbb{A}^1_F$.

Let $F[t]_d := \{g \mid g \in F[t],\ \deg g \le d\}$ and $L_d \subset W\big(F(t)\big)$ the subring generated by $\langle g\rangle$ with $g \in F[t]_d$. Then $L_0 \subset L_1 \subset L_2 \subset \cdots$ and $W\big(F(t)\big) = \bigcup_d L_d$. Note that $\operatorname{im}(r_{F(t)/F}) = L_0$. Let S_d be the multiplicative monoid in $F[t]$ generated by $F[t]_d \setminus \{0\}$. As a group L_d is generated by 1-dimensional forms of the type

$$\langle f_1 \cdots f_m g\rangle \tag{21.2}$$

with distinct monic irreducible polynomials $f_1, \dots, f_m \in F[t]$ of degree d and $g \in S_{d-1}$.

Claim: The additive group L_d/L_{d-1} is generated by $\langle fg\rangle + L_{d-1}$ with $f \in F[t]$ monic irreducible of degree d and $g \in S_{d-1}$. Moreover, if $h \in F[t]_{d-1}$ satisfies $g \equiv h \mod (f)$, then $\langle fg\rangle \simeq \langle fh\rangle \mod L_{d-1}$.

We first must show that a generator of the form in (21.2) is a sum of the desired forms $\mod L_{d-1}$. By induction on m, we need only do the case $m = 2$. Let f_1, f_2 be distinct irreducible monic polynomials of degree d and $g \in S_{d-1}$. Let $h = f_1 - f_2$, so $\deg h < d$. We have

$$\langle f_1\rangle = \langle h\rangle + \langle f_2\rangle - \langle f_1 f_2 h\rangle$$

in $W\big(F(t)\big)$ by the Witt relation (4.2). Multiplying this equation by $\langle f_2 g\rangle$ and deleting squares, yields

$$\langle f_1 f_2 g\rangle = \langle f_2 g h\rangle + \langle g\rangle - \langle f_1 g h\rangle \equiv \langle f_2 g h\rangle - \langle f_1 g h\rangle \mod L_{d-1}$$

as needed.

Now suppose that $g = g_1 g_2$ with $g_1, g_2 \in F[t]_{d-1}$. As $f \nmid g$ by the Division Algorithm, there exist polynomials $q, h \in F[t]$ with $h \neq 0$ and $\deg h < d$ satisfying $g = fq + h$. It follows that $\deg q < d$. By the Witt relation (4.2), we have

$$\langle g\rangle = \langle fq\rangle + \langle h\rangle - \langle fqhg\rangle$$

in L_d, hence multiplying by $\langle f\rangle$, we have

$$\langle fg\rangle = \langle q\rangle + \langle fh\rangle - \langle qhg\rangle \equiv \langle fh\rangle \mod L_{d-1}.$$

The Claim now follows by induction on the number of factors for a general $g \in S_{d-1}$.

Let $x \in \mathbb{A}^1_F$ be of degree d and $f = f_x$. Define

$$\alpha_x : W\big(F(x)\big) \to L_d/L_{d-1} \quad \text{by} \quad \langle g + (f)\rangle \mapsto \langle g\rangle + L_{d-1} \ \text{ for } \ g \in F[t]_{d-1}.$$

We show this map is well-defined. If $h \in F[t]_{d-1}$ satisfies $gh^2 \equiv l \mod (f)$, with $l \in F[t]_{d-1}$, then $\langle fg\rangle = \langle fgh^2\rangle \equiv \langle fl\rangle \mod L_{d-1}$ by the Claim, so the map is well-defined on 1-dimensional forms. If $g_1, g_2 \in F[t]_{d-1}$ satisfy $g_1 + g_2 \neq 0$ and $h \equiv (g_1 + g_2)g_1 g_2 \mod (f)$, then

$$\langle fg_1\rangle + \langle fg_2\rangle = \langle f(g_1+g_2)\rangle + \langle fg_1g_2(g_1+g_2)\rangle \equiv \langle f(g_1+g_2)\rangle + \langle fh\rangle \mod L_{d-1}$$

by the Claim. As $\langle f\rangle + \langle -f\rangle = 0$ in $W\big(F(t)\big)$, it follows that α_x is well-defined by Theorem 4.8.

Let $x' \in \mathbb{A}^1_F$ with $\deg x' = d$. Then the composition

$$W\big(F(x)\big) \xrightarrow{\alpha_x} L_d/L_{d-1} \xrightarrow{\partial_{x'}} W\big(F(x')\big)$$

is the identity if $x = x'$, otherwise it is the zero map. It follows that the map

$$\coprod_{\deg x = d} W\big(F(x)\big) \xrightarrow{(\alpha_x)} L_d/L_{d-1}$$

is split by $(\partial_x)_{\deg x = d}$. As (α_x) is surjective by the Claim, it is an isomorphism with inverse $(\partial_x)_{\deg x = d}$. By induction on d, we check that

$$(\partial_x)_{\deg x \le d} : L_d/L_0 \longrightarrow \coprod_{\deg x \le d} W\big(F(x)\big)$$

is an isomorphism. As $L_0 = W(F)$, passing to the limit yields the result. □

It follows from Lemma 19.14 or Example 19.13 that $\partial_x\big(I^n(F(t))\big) \subset I^{n-1}\big(F(x)\big)$ for every $x \in \mathbb{A}^1_F$.

Corollary 21.3. *The sequence*

$$0 \to I^n(F) \xrightarrow{r_{F(t)/F}} I^n\big(F(t)\big) \xrightarrow{\boldsymbol{\partial}} \coprod_{x \in \mathbb{A}^1_F} I^{n-1}\big(F(x)\big) \to 0$$

is split exact for each $n \ge 1$.

PROOF. We show by induction on $d = \deg x$ that $I^{n-1}\big(F(x)\big)$ lies in $\mathrm{im}(\boldsymbol{\partial})$. Let $g_2, \dots, g_n \in F[t]$ be of degree $< d$. We need to prove that $\mathfrak{b} = \langle\langle \bar{g}_2, \dots, \bar{g}_n \rangle\rangle$ lies in $\mathrm{im}(\boldsymbol{\partial})$ where $\bar{g}_i$ is the image of g_i in $F(x)$. By Example 19.13, we have $\partial_x(\mathfrak{c}) = \mathfrak{b}$ where $\mathfrak{c} = \langle\langle -f_x, g_2, \dots, g_n \rangle\rangle$. Moreover, $\mathfrak{c} - \mathfrak{b} \in \coprod_{\deg x < d} I^{n-1}\big(F(x)\big)$ and therefore $\mathfrak{c} - \mathfrak{b} \in \mathrm{im}(\boldsymbol{\partial})$ by induction.

To finish, it suffices to show exactness at $I^n\big(F(t)\big)$. Let $\mathfrak{b} \in \mathrm{Ker}(\boldsymbol{\partial})$. By Theorem 21.1, there exists $\mathfrak{c} \in W(F)$ satisfying $r_{F(t)/F}(\mathfrak{c}) = \mathfrak{b}$. We show $\mathfrak{c} \in I^n(F)$. Let $x \in \mathbb{A}^1_F$ be a fixed rational point and $f = t - t(x)$. Define $\rho : W\big(F(t)\big) \to W(F)$ by $\rho(\mathfrak{d}) = \partial_x\big(\langle\langle -f \rangle\rangle \cdot \mathfrak{d}\big)$. By Lemma 19.14, we have $\rho(I^n\big(F(t)\big) \subset I^n(F)$ as $F(x) = F$. By Example 19.13, the composition $\rho \circ r_{F(t)/F}$ is the identity. It follows that $\mathfrak{c} = \rho(\mathfrak{b}) \in I^n(F)$ as needed. □

The above was proven by Milnor in [**106**] for fields of characteristic not 2. We wish to modify the sequence in Theorem 21.1 to the projective line $\mathbb{P}^1_F$. If $x \in \mathbb{A}^1_F$ is of degree n, let $s_x : F(x) \to F$ be the F-linear functional

$$s_x\big(t^{n-1}(x)\big) = 1 \quad \text{and} \quad s_x\big(t^i(x)\big) = 0 \text{ for } i < n-1.$$

The infinite point ∞ corresponds to the $1/t$-adic valuation. It has residue field F. The corresponding second residue homomorphism $\partial_\infty : W\big(F(t)\big) \to W(F)$ is taken with respect to the prime $1/t$. So if $0 \ne h \in F[t]$ is of degree n and has leading coefficient a, we have $\partial_\infty\big(\langle h \rangle\big) = \langle a \rangle$ if n is odd, and $\partial_\infty\big(\langle h \rangle\big) = 0$ otherwise. Define $(s_\infty)_*$ to be $-\mathrm{Id} : W(F) \to W(F)$. The following theorem was first proved by Scharlau in [**120**] for fields of characteristic not 2.

Theorem 21.4. *The sequence*

$$0 \to W(F) \xrightarrow{r_{F(t)/F}} W\big(F(t)\big) \xrightarrow{\boldsymbol{\partial}} \coprod_{x \in \mathbb{P}^1_F} W\big(F(x)\big) \xrightarrow{\mathbf{s}_*} W(F) \to 0$$

is exact with $\boldsymbol{\partial} = (\partial_x)$ and $\mathbf{s}_* = \big((s_x)_*\big)$.

PROOF. The map $(s_\infty)_*$ is $-\mathrm{Id}$. Hence by Theorem 21.1, it suffices to show $\mathbf{s}_* \circ \boldsymbol{\partial}$ is the zero map.

As 1-dimensional bilinear forms generate $W\big(F(t)\big)$, it suffices to check the result on 1-dimensional forms. Let $\langle af_1 \cdots f_n\rangle$ be a 1-dimensional form with $f_i \in F[t]$ monic of degree d_i and $a \in F^\times$ for $i \in [1,\ n]$. Let $x_i \in \mathbb{A}^1_F$ satisfy $f_i = f_{x_i}$ and $s_i = s_{x_i}$ for $1 \le i \le n$. We must show that

$$\sum_{X \in \mathbb{A}^1_F} (s_x)_* \circ \partial_x\big(\langle af_1 \cdots f_n\rangle\big) = -(s_\infty)_* \circ \partial_\infty\big(\langle af_1 \cdots f_n\rangle\big)$$

in $W(F)$. Multiplying through by $\langle a\rangle$, we may also assume that $a = 1$.

Set $A = F[t]/(f_1 \cdots f_n)$ and $d = \dim A$. Then $d = \sum d_i$. Let $^- : F[t] \to A$ be the canonical epimorphism and set $q_i = (f_1 \cdots f_n)/f_i$. We have an F-vector space homomorphism

$$\alpha : \coprod_{i=1}^{n} F(x_i) \to A \quad \text{given by} \quad \big(h_1(x_i), \dots, h_n(x_i)\big) \mapsto \sum \bar{h}_i \bar{q}_i \quad \text{for all} \quad h \in F[t].$$

We show that α is an isomorphism. As both spaces have the same dimension, it suffices to show α is monic. As the q_i are relatively prime in $F[t]$, we have an equation $\sum_{i=1}^n g_i q_i = 1$ with all $g_i \in F[t]$. Then the map

$$A \to \coprod F(x_i) \text{ given by } \bar{h} \to \big(h(x_1)g(x_1), \dots, h(x_n)g_n(x_n)\big)$$

splits α, hence α is monic as needed. Set $A_i = \alpha\big(F(x_i)\big)$ for $1 \le i \le n$.

Let $s : A \to F$ be the F-linear functional defined by $s(\bar{t}^{d-1}) = 1$ and $s(\bar{t}^i) = 0$ for $0 \le i < d-1$. Define $\mathfrak{b}$ to be the bilinear form on A over F given by $\mathfrak{b}(\bar{f}, \bar{h}) = s(\bar{f}\bar{h})$ for $f, h \in F[t]$. If $i \ne j$, we have

$$\mathfrak{b}\big(\alpha(f(x_i)), \alpha(h(x_j))\big) = \mathfrak{b}(\bar{f}\bar{q}_i, \bar{h}\bar{q}_j) = s(\bar{f}\bar{h}\bar{q}_i\bar{q}_j) = s(0) = 0$$

for all $f, h \in F[t]$. Consequently, $\mathfrak{b}|_{A_i}$ is orthogonal to $\mathfrak{b}|_{A_j}$ if $i \ne j$.

Claim: $\mathfrak{b}|_{A_i} \simeq (s_i)_*\big(\partial_{f_i}(\langle f_1 \cdots f_n\rangle)\big)$ for $i \in [1,\ n]$.

Let $g, h \in F[t]$. Write

$$q_i g h = c_0 + \cdots + c_{d_i - 1} t^{d_i - 1} + f_i p$$

for some $c_i \in F$ and $p \in F[t]$.

By definition, we have

$$(s_i)_*\big(\partial_{f_i}(\langle f_1 \cdots f_n\rangle)\big)\big(g(x_i), h(x_i)\big) = s_i\big(q_i(x_i)g(x_i)h(x_i)\big) = c_{d_i-1}.$$

As $\deg q_i = d - d_i$, we have $\deg q_i t^{d_i - 1} = d - 1$. Thus

$$\mathfrak{b}|_{A_i}\big(\alpha(g(x_i)), \alpha(h(x_i))\big) = \mathfrak{b}(\bar{g}\bar{q}_i, \bar{h}\bar{q}_i) = s(\bar{q}_i^2 \bar{g}\bar{h}) = c_{d_i-1},$$

and the claim is established.

As $\partial_f(f_1 \cdots f_n) = 0$ for all irreducible monic polynomials $f \ne f_i$, $i \in [1,\ n]$, in $F[t]$, we have, by the Claim,

$$\mathfrak{b} = \sum_{i=1}^{n} (s_i)_*\big(\partial_{x_i}(\langle f_1 \cdots f_n\rangle)\big) = \sum_{x \in \mathbb{A}^1_F} (s_x)_*\big(\partial_x(\langle f_1 \cdots f_n\rangle)\big)$$

in $W(F)$.

Suppose that $d = 2e$ is even. The form $\mathfrak{b}$ is then metabolic as it has a totally isotropic subspace of dimension e spanned by $1, \bar{t}, \dots, \bar{t}^{e-1}$. We also have $(s_\infty)_* \circ \partial_\infty(\mathfrak{b}) = 0$ in this case.

Suppose that $d = 2e + 1$. Then $\mathfrak{b}$ has a totally isotropic subspace spanned by $1, \bar{t}, \dots, \bar{t}^{e-1}$ so $\mathfrak{b} \simeq \langle a \rangle \perp \mathfrak{c}$ with $\mathfrak{c}$ metabolic by the Witt Decomposition Theorem 1.27. Computing $\det \mathfrak{b}$ on the basis $\{1, \bar{t}, \dots, t^{\bar{d}-1}\}$, we see that $\langle a \rangle = \langle 1 \rangle$. As $(s_\infty)_* \circ \partial_\infty(\mathfrak{b}) = -\langle 1 \rangle$, the result follows. □

Corollary 21.5. *Let K be a finite simple extension of F and $s : K \to F$ a nontrivial F-linear functional. Then $s_*\big(I^n(K)\big) \subset I^n(F)$ for all $n \geq 0$. Moreover, the induced map $I^n(K)/I^{n+1}(K) \to I^n(F)/I^{n+1}(F)$ is independent of the nontrivial F-linear functional s for all $n \geq 0$.*

PROOF. Let x lie in $\mathbb{A}^1_F$ with $K = F(x)$. Let $\mathfrak{b} \in I^n(K)$. By Lemma 21.3, there exists $\mathfrak{c} \in I^{n+1}\big(F(t)\big)$ such that $\partial_y(\mathfrak{c}) = 0$ for all $y \in \mathbb{A}^1_F$ unless $y = x$ in which case $\partial_x(\mathfrak{c}) = \mathfrak{b}$. It follows by Theorem 21.4 that

$$0 = \sum_{y \in \mathbb{P}^1_F} (s_y)_* \circ \partial_y(\mathfrak{c}) = (s_x)_*(\mathfrak{b}) - \partial_\infty(\mathfrak{c}).$$

By Lemma 19.14, we have $\partial_\infty(\mathfrak{c}) \in I^n(F)$, so $(s_x)_*(\mathfrak{b}) \in I^n(F)$. Suppose that $s : K \to F$ is another nontrivial F-linear functional. As in the proof of Corollary 20.7, there exists a $c \in K^\times$ such that $(s)_*(\mathfrak{c}) = (s_x)_*(c\mathfrak{c})$ for all symmetric bilinear forms $\mathfrak{c}$. In particular, $(s)_*(\mathfrak{b}) = (s_x)_*(c\mathfrak{b})$ lies in $I^n(F)$. As $\langle\langle c \rangle\rangle \cdot \mathfrak{b} \in I^{n+1}(K)$, we also have

$$s_*(\mathfrak{b}) - (s_x)_*(\mathfrak{b}) = (s_x)_*\big(\langle\langle c \rangle\rangle \cdot \mathfrak{b}\big)$$

lies in $I^{n+1}(F)$. The result follows. □

The transfer induced by distinct nontrivial F-linear functionals $K \to F$ are not, in general, equal on $I^n(F)$.

Exercise 21.6. Show that Corollary 21.5 holds for arbitrary finite extensions K/F.

Corollary 21.7. *The sequence*

$$0 \to I^n(F) \xrightarrow{r_{F(t)/F}} I^n\big(F(t)\big) \xrightarrow{\partial} \coprod_{x \in \mathbb{P}^1_F} I^{n-1}\big(F(x)\big) \xrightarrow{\mathbf{s}_*} I^{n-1}(F) \to 0$$

is exact.

CHAPTER IV

Function Fields of Quadrics

22. Quadrics

A quadratic form φ over F defines a projective quadric X_φ over F. The quadric X_φ is smooth if and only if φ is nondegenerate (cf. Proposition 22.1). The quadric X_φ encodes information about isotropy properties of φ, namely the form φ is isotropic over a field extension E/F if and only if X_φ has a point over E. In the third part of the book, we will use algebraic-geometric methods to study isotropy properties of φ.

If $\mathfrak{b}$ is a symmetric bilinear form, the quadric $X_{\varphi_\mathfrak{b}}$ reflects isotropy properties of $\mathfrak{b}$ (and of $\varphi_\mathfrak{b}$ as well). If the characteristic of F is 2, only totally singular quadratic forms arise from symmetric bilinear forms. In particular, quadrics arising from bilinear forms are not smooth. Therefore, algebraic-geometric methods have wider application in the theory of quadratic forms than in the theory of bilinear forms.

In the previous sections, we looked at quadratic forms over field extensions determined by irreducible polynomials. In particular, we were interested in when a quadratic form becomes isotropic over such a field. Viewing a quadratic form as a homogeneous polynomial of degree two, results from these sections apply.

Let φ and ψ be two anisotropic quadratic forms. In this section, we begin our study of when φ becomes isotropic or hyperbolic over the function field $F(X_\psi)$ of the integral quadric X_ψ. It is natural at this point to introduce the geometric language that we shall use, i.e., to associate to a quadratic form a projective quadric.

Let φ be a quadratic form on V. Viewing $\varphi \in S^2(V^*)$, we define the *projective quadric associated to* φ to be the closed subscheme

$$X_\varphi = \operatorname{Proj}\big(S^\bullet(V^*)/(\varphi)\big)$$

of the projective space $\mathbb{P}(V) = \operatorname{Proj} S^\bullet(V^*)$ where $S^\bullet(V^*)$ is the symmetric algebra of the dual space V^* of V. The scheme X_φ is equidimensional of dimension $\dim V - 2$ if $\varphi \neq 0$ and $\dim V \geq 2$. We define the *Witt index* of X_φ by $i_0(X_\varphi) := i_0(\varphi)$. By construction, for any field extension L/F, the set of L-points $X_\varphi(L)$ coincides with the set of isotropic lines in V_L. Therefore, $X_\varphi(L) = \emptyset$ if and only if φ_L is anisotropic.

For any field extension K/F we have $X_{\varphi_K} = (X_\varphi)_K$.

Let ψ be a subform of φ. The inclusion of vector spaces $V_\psi \subset V$ gives rise to a surjective graded ring homomorphism

$$S^\bullet(V^*)/(\varphi) \to S^\bullet({V_\psi}^*)/(\psi)$$

which in turn leads to a closed embedding $X_\psi \hookrightarrow X_\varphi$. We shall always identify X_ψ with a closed subscheme of X_φ.

Proposition 22.1. *Let φ be a nonzero quadratic form of dimension at least 2. Then the quadric X_φ is smooth if and only if φ is nondegenerate.*

PROOF. By Lemma 7.16, we may assume that F is algebraically closed. We claim that $\mathbb{P}(\operatorname{rad}\varphi)$ is the singular locus of X_φ. Let $0 \neq u \in V$ be an isotropic vector. Then the isotropic line $U = Fu \subset V$ can be viewed as a rational point of X_φ. As $\varphi(u + \varepsilon v) = 0$ if and only if u is orthogonal to v (where $\varepsilon^2 = 0$), the tangent space $T_{X,U}$ is the subspace $\operatorname{Hom}(U, U^\perp/U)$ of the tangent space $T_{\mathbb{P}(V),U} = \operatorname{Hom}(U, V/U)$ (cf. Example 104.20). In particular, the point U is regular on X if and only if $\dim T_{X,U} = \dim X = \dim V - 2$ if and only if $U^\perp \neq V$, i.e., U is not contained in $\operatorname{rad}\varphi$. Thus X_φ is smooth if and only if $\operatorname{rad}\varphi = 0$, i.e., φ is nondegenerate. □

We say that the quadratic form φ on V is *irreducible* if φ is irreducible in the ring $S^\bullet(V^*)$. If φ is nonzero and not irreducible, then $\varphi = l \cdot l'$ for some nonzero linear forms $l, l' \in V^*$. Then $\operatorname{rad}\varphi = \operatorname{Ker}(l) \cap \operatorname{Ker}(l')$ has codimension at most 2 in V. Therefore, the induced form $\bar{\varphi}$ on $V/\operatorname{rad}\varphi$ is either 1-dimensional or a hyperbolic plane. It follows that a regular quadratic form φ is irreducible if and only if $\dim\varphi \geq 3$ or $\dim\varphi = 2$ and φ is anisotropic.

If φ is irreducible, X_φ is an integral scheme. The function field $F(X_\varphi)$ is called the *function field of* φ and will be denoted by $F(\varphi)$. By definition, $F(\varphi)$ is the subfield of degree 0 elements in the quotient field of the domain $S^\bullet(V^*)/(\varphi)$. Note that the quotient field of $S^\bullet(V^*)/(\varphi)$ is a purely transcendental extension of $F(\varphi)$ of degree 1. Clearly, φ is isotropic over the quotient field of $S^\bullet(V^*)/(\varphi)$ and therefore is isotropic over $F(\varphi)$.

Example 22.2. Let σ be an anisotropic binary quadratic form. As σ is isotropic over $F(\sigma)$, it follows from Corollary 12.3 that $F(\sigma) \simeq C_0(\sigma)$.

If K/F is a field extension such that φ_K is still irreducible, we simply write $K(\varphi)$ for $K(\varphi_K)$.

Example 22.3. Let φ and ψ be irreducible quadratic forms. Then

$$F(X_\varphi \times X_\psi) \simeq F(\varphi)(\psi_{F(\varphi)}) \simeq F(\psi)(\varphi_{F(\psi)}).$$

Let φ and ψ be two irreducible regular quadratic forms. We shall be interested in when $\varphi_{F(\psi)}$ is hyperbolic or isotropic. A consequence of the Quadratic Similarity Theorem 20.16 is:

Proposition 22.4. *Let φ be a nondegenerate quadratic form of even dimension and ψ be an irreducible quadratic form of dimension n over F. Suppose that $T = (t_1, \dots, t_n)$ and $b \in D(\psi)$. Then $\varphi_{F(\psi)}$ is hyperbolic if and only if*

$$b \cdot \psi(T)\varphi_{F(T)} \simeq \varphi_{F(T)}.$$

PROOF. By the Quadratic Similarity Theorem 20.16, we have $\varphi_{F(\psi)}$ is hyperbolic if and only if $\psi^*\psi(T)\varphi_{F(T)} \simeq \varphi_{F(T)}$. Let $b \in D(\psi)$. Choosing a basis for V with first vector v satisfying $\psi(v) = b$, we have $\psi^* = b$. □

Theorem 22.5 (Subform Theorem). *Let φ be a nonzero anisotropic quadratic form and ψ be an irreducible anisotropic quadratic form such that the form $\varphi_{F(\psi)}$ is hyperbolic. Let $a \in D(\varphi)$ and $b \in D(\psi)$. Then $ab\psi$ is isometric to a subform of φ and, therefore, $\dim\psi \leq \dim\varphi$.*

PROOF. We view ψ as an irreducible polynomial in $F[T]$. The form φ is nondegenerate of even dimension by Remark 7.18, so by Corollary 22.4, we have $b\psi(T) \in G(\varphi_{F(T)})$. Since $a \in D(\varphi)$, we have $ab\psi(T) \in D(\varphi_{F(T)})$. By the Representation Theorem 17.12, we conclude that $ab\psi$ is a subform of φ. □

By the proof of the theorem and Corollary 20.20, we have

Corollary 22.6. *Let φ be an anisotropic quadratic form and ψ an irreducible anisotropic quadratic form. If $\varphi_{F(\psi)}$ is hyperbolic, then $D(\psi)D(\psi) \subset G(\varphi)$. In particular, if $1 \in D(\psi)$, then $D(\psi) \subset G(\varphi)$.*

Remark 22.7. The natural analogues of the Representation Theorem 17.12 and the Subform Theorem 22.5 are not true for bilinear forms in characteristic 2. Let $\mathfrak{b} = \langle 1, b\rangle$ and $\mathfrak{c} = \langle 1, c\rangle$ be anisotropic symmetric bilinear forms with b and $c = x^2 + by^2$ nonzero and $bF^{\times 2} \neq cF^{\times 2}$ in a field F of characteristic 2. Thus $\mathfrak{b} \not\simeq \mathfrak{c}$. However, $\varphi_{\mathfrak{b}} \simeq \varphi_{\mathfrak{c}}$ by Example 7.27. So $\varphi_{\mathfrak{c}}(t_1, t_2) \in D(\varphi_{\mathfrak{b}\,F(t_1,t_2)})$ and $\mathfrak{c}_{F(\varphi_{\mathfrak{b}})}$ is isotropic, hence metabolic, but $a\mathfrak{c}$ is not a subform of $\mathfrak{b}$ for any $a \neq 0$.

We do have, however, the following:

Corollary 22.8. *Let $\mathfrak{b}$ and $\mathfrak{c}$ be anisotropic bilinear forms with $\dim \mathfrak{c} \geq 2$ and $\mathfrak{b}$ nonzero. Let ψ be the associated quadratic form of $\mathfrak{c}$. If $\mathfrak{b}_{F(\psi)}$ is metabolic, then $\dim \mathfrak{c} \leq \dim \mathfrak{b}$.*

PROOF. Let $\varphi = \varphi_{\mathfrak{b}}$. By the Bilinear Similarity Theorem 20.19 and Lemma 9.2, we have $a\psi(T) \in G(\mathfrak{b}_{F(T)}) \subset G(\varphi_{F(T)})$ for some $a \in F^\times$ where $T = (t_1, \dots, t_{\dim \psi})$. It follows that $b\psi(T) \in D(\varphi_{F(T)})$ for some $b \in F^\times$. Consequently,

$$\dim \mathfrak{b} = \dim \varphi \geq \dim \psi = \dim \mathfrak{c}$$

by the Representation Theorem 17.12. □

We turn to the case in which a quadratic form becomes isotropic over the function field of another form or itself.

Proposition 22.9. *Let φ be an irreducible regular quadratic form. Then the field extension $F(\varphi)/F$ is purely transcendental if and only if φ is isotropic.*

PROOF. Suppose that the field extension $F(\varphi)/F$ is purely transcendental. As $\varphi_{F(\varphi)}$ is isotropic, φ is isotropic by Lemma 7.15.

Now suppose that φ is isotropic. Then $\varphi = \mathbb{H} \perp \varphi'$ for some φ' by Proposition 7.13. Let $V = V_\varphi$, $V' = V_{\varphi'}$ and let $h, h' \in V$ be a hyperbolic pair of $\mathbb{H}$. Let $\psi = \varphi|_{Fh' \oplus V'}$ with $h' \in (V')^\perp$. It suffices to show that $X_\varphi \setminus X_\psi$ is isomorphic to an affine space. Every isotropic line in $X_\varphi \setminus X_\psi$ has the form $F(h + ah' + v')$ for unique $a \in F$ and $v' \in V'$ satisfying

$$0 = \varphi(h + ah' + v') = a + \varphi(v'),$$

i.e., $a = -\varphi(v')$. Therefore, the morphism $X_\varphi \setminus X_\psi \to \mathbb{A}(V')$ taking $F(h + ah' + v')$ to v' is an isomorphism with the inverse given by $v' \mapsto F\big(h - \varphi(v')h' + v'\big)$. □

Remark 22.10. Let $\operatorname{char} F = 2$ and let φ be an irreducible totally singular form. Then the field extension $F(\varphi)/F$ is not purely transcendental even if φ is isotropic.

Proposition 22.11. *Let φ be an anisotropic quadratic form over F and K/F a quadratic field extension. Then φ_K is isotropic if and only if there is a binary subform σ of φ such that $F(\sigma) \simeq K$.*

PROOF. Let σ be a binary subform of φ with $F(\sigma) \simeq K$. Since σ is isotropic over $F(\sigma)$ we have φ isotropic over $F(\sigma) \simeq K$.

Conversely, suppose that $\varphi_K(v) = 0$ for some nonzero $v \in (V_\varphi)_K$. Since K is quadratic over F, there is a 2-dimensional subspace $U \subset V_\varphi$ with $v \in U_K$.

Therefore, the form $\sigma = \varphi|_U$ is isotropic over K. As σ is also isotropic over $F(\sigma)$, it follows from Corollary 12.3 and Example 22.2 that $F(\sigma) \simeq C_0(\sigma) \simeq K$. □

Corollary 22.12. *Let φ be an anisotropic quadratic form and σ a nondegenerate anisotropic binary quadratic form. Then $\varphi \simeq \mathfrak{b} \otimes \sigma \perp \psi$ with $\mathfrak{b}$ a nondegenerate symmetric bilinear form and $\psi_{F(\sigma)}$ anisotropic.*

PROOF. Suppose that $\varphi_{F(\sigma)}$ is isotropic. By Proposition 22.11, there is a binary subform σ' of φ with $F(\sigma') = F(\sigma)$. By Corollary 12.2 and Example 22.2, we have σ' is similar to σ. Consequently, there exists an $a \in F^\times$ such that $\varphi \simeq a\sigma \perp \psi$ for some quadratic form ψ. The result follows by induction on $\dim \varphi$. □

Recall that a field extension K/F is called *separable* if there exists and intermediate field E in K/F with E/F purely transcendental and K/E algebraic and separable. We show that regular quadratic forms remain regular after extending to a separable field extension.

Lemma 22.13. *Let φ be a regular quadratic form over F and K/F a separable (possibly infinite) field extension. Then φ_K is regular.*

PROOF. We proceed in several steps.

Case 1: $[K : F] = 2$.

Let $v \in (V_\varphi)_K$ be an isotropic vector. Then $v \in U_K$ for a 2-dimensional subspace $U \subset V_\varphi$ such that $\varphi|_U$ is similar to the norm form N of K/F (cf. Proposition 12.1). As N is nondegenerate, $v \notin \operatorname{rad} \mathfrak{b}_{\varphi_K}$, therefore, $\operatorname{rad} \varphi_K = 0$.

Case 2: K/F is of odd degree or purely transcendental.

We have $\varphi \simeq \varphi_{an} \perp n\mathbb{H}$. The anisotropic part φ_{an} stays anisotropic over K by Springer's Theorem 18.5 or Lemma 7.15, respectively; therefore φ_K is regular.

Case 3: $[K : F]$ is finite.

We may assume that K/F is Galois by Remark 7.14. Then K/F is a tower of odd degree and quadratic extensions.

Case 4: The general case.

In general, K/F is a tower of a purely transcendental and a finite separable extension. □

We turn to the function field of an irreducible quadratic form.

Lemma 22.14. *Let φ be an irreducible quadratic form over F. Then there exists a purely transcendental extension E of F with $[F(\varphi) : E] = 2$. Moreover, if φ is not totally singular, the field E can be chosen with $F(\varphi)/E$ separable. In particular, $F(\varphi)/F$ is separable.*

PROOF. Let $U \subset V_\varphi$ be an anisotropic line. The rational projection $f : X_\varphi \dashrightarrow \mathbb{P} = \mathbb{P}(V/U)$ taking a line U' to $(U + U')/U$ is a double cover, so $F(\varphi)/E$ is a quadratic field extension where E is the purely transcendental extension $F(\mathbb{P})$ of F.

Let τ be the reflection of φ with respect to a nonzero vector in U. Clearly, $f(\tau U') = f(U')$ for every line U' in X_φ. Therefore, τ induces an automorphism of every fiber of f. In particular, τ induces an automorphism of the generic fiber, hence an automorphism ε of the field $F(\varphi)$ over E.

If φ is not totally singular, we can choose U not in rad $\mathfrak{b}_\varphi$. Then the isometry τ and the automorphism ε are nontrivial. Consequently, the field extension $F(\varphi)/E$ is separable. $\square$

22.A. Domination relation. Let φ and ψ be anisotropic quadratic forms of dimension at least 2 over F. We say φ *dominates* ψ and write $\varphi \succ \psi$ if $\varphi_{F(\psi)}$ is isotropic and write $\varphi \prec\succ \psi$ if $\varphi \succ \psi$ and $\psi \succ \varphi$. For example, if ψ is a subform of φ, then $\varphi \succ \psi$.

We have $\varphi \succ \psi$ if and only if there exists a rational map $X_\psi \dashrightarrow X_\varphi$.

We show that the relation $\succ$ is transitive.

Lemma 22.15. *Let φ and ψ be anisotropic quadratic forms over F. If $\psi \succ \mu$, then there exist a purely transcendental field extension E/F and a binary subform σ of ψ_E over E such that $E(\sigma) = F(\mu)$.*

Proof. By Lemma 22.14, there exists a purely transcendental field extension E/F such that $F(\mu)$ is a quadratic extension of E. As ψ is isotropic over $F(\mu)$, it follows from Proposition 22.11 applied to the form ψ_E and the quadratic extension $F(\mu)/E$ that ψ_E contains a binary subform σ over E satisfying $E(\sigma) = F(\mu)$. $\square$

Proposition 22.16. *Let φ, ψ, and μ be anisotropic quadratic forms over F. If $\varphi \succ \psi \succ \mu$, then $\varphi \succ \mu$.*

Proof. Consider first the case when μ is a subform of ψ.

We may assume that μ is of codimension one in ψ. Let $T = (t_1, \dots, t_n)$ be the coordinates in V_ψ so that V_μ is given by $t_1 = 0$. By assumption, there is $v \in V_\varphi[T]$ such that $\varphi(v)$ is divisible by $\psi(T)$ but v is not divisible by $\psi(T)$. Since ψ is anisotropic, we have $\deg_{t_i} \psi = 2$ for every i. Applying the division algorithm by dividing v by ψ with respect to the variable t_2, we may assume that $\deg_{t_2} v \leq 1$. Moreover, dividing out a power of t_1 if necessary, we may assume that v is not divisible by t_1. Therefore, the vector $w := v|_{t_1=0} \in V_\varphi[T']$ with $T' = (t_2, \dots, t_n)$ is not zero. As $\deg_{t_2} w \leq 1$ and $\deg_{t_2} \mu = 2$, the vector w is not divisible by $\mu(T')$. On the other hand, $\varphi(w)$ is divisible by $\psi(T)|_{t_1=0} = \mu(T')$, i.e., φ is isotropic over $F(\mu)$.

Now consider the general case. By Lemma 22.15, there exist a purely transcendental field extension E/F and a binary subform σ of ψ_E over E such that $E(\sigma) = F(\mu)$. By the first part of the proof applied to the forms $\varphi_E \succ \psi_E \succ \sigma$, we have φ_E is isotropic over $E(\sigma) = F(\mu)$, i.e., $\varphi \succ \mu$. $\square$

Corollary 22.17. *Let φ, ψ, and μ be anisotropic quadratic forms over F. If $\varphi \prec\succ \psi$, then $\mu_{F(\varphi)}$ is isotropic if and only if $\mu_{F(\psi)}$ is isotropic.*

Proposition 22.18. *Let ψ and μ be anisotropic quadratic forms over F satisfying $\psi \succ \mu$. Let φ be a quadratic form such that $\varphi_{F(\psi)}$ is hyperbolic. Then $\varphi_{F(\mu)}$ is hyperbolic.*

Proof. Consider first the case when μ is a subform of ψ. Choose variables T' of μ and variables $T = (T', T'')$ of ψ so that $\mu(T') = \psi(T', 0)$. As $\varphi_{F(\psi)}$ is hyperbolic, by the Quadratic Similarity Theorem 20.16, we have $\varphi_{F(T)} \simeq a\psi(T)\varphi_{F(T)}$ over $F(T)$ for some $a \in F^\times$. Specializing the variables $T'' = 0$, we see by Corollary 20.20 that $\varphi_{F(T')} \simeq a\mu(T')\varphi_{F(T')}$ over $F(T')$, and again it follows from the Quadratic Similarity Theorem 20.16 that $\varphi_{F(\mu)}$ is hyperbolic.

Now consider the general case. By Lemma 22.15, there exist a purely transcendental field extension E/F and a binary subform σ of ψ_E over E such that $E(\sigma) = F(\mu)$. As $\varphi_{E(\psi)}$ is hyperbolic, by the first part of the proof applied to the forms $\psi_E \succ \sigma$, we have $\varphi_{E(\sigma)} = \varphi_{F(\mu)}$ is hyperbolic. □

23. Quadratic Pfister forms II

The introduction of function fields of quadrics allows us to determine the main characterization of general quadratic Pfister forms first proven by Pfister in [**108**] for fields of characteristic different from 2. They are precisely those forms that become hyperbolic over their function fields. In particular, Pfister forms can be characterized as universally round forms.

If φ is an anisotropic general quadratic Pfister form, then $\varphi_{F(\varphi)}$ is isotropic, hence hyperbolic by Corollary 9.10. We wish to show the converse of this property. We begin by looking at subforms of Pfister forms.

Lemma 23.1. *Let φ be an anisotropic quadratic form over F and ρ a subform of φ. Suppose that $D(\varphi_K)$ and $D(\rho_K)$ are groups for all field extensions K/F. Let $a = -\varphi(v)$ for some $v \in V_\rho^\perp \setminus V_\rho$. Then the form $\langle\langle a\rangle\rangle \otimes \rho$ is isometric to a subform of φ.*

PROOF. Let $T = (t_1, \dots, t_n)$ and $T' = (t_{n+1}, \dots, t_{2n})$ be $2n$ independent variables where $n = \dim \rho$. We have

$$\rho(T) - a\rho(T') = \rho(T')\Big[\frac{\rho(T)}{\rho(T')} - a\Big].$$

As $D(\rho_{F(T,T')})$ is a group, we have $\dfrac{\rho(T)}{\rho(T')} \in D(\rho_{F(T,T')})$, hence $\dfrac{\rho(T)}{\rho(T')} - a \in D(\varphi_{F(T,T')})$. As $\rho(T') \in D(\varphi_{F(T,T')})$, we have

$$\rho(T) - a\rho(T') \in D(\varphi_{F(T,T')})D(\varphi_{F(T,T')}) = D(\varphi_{F(T,T')}).$$

By the Representation Theorem 17.12, we conclude that $\langle\langle a\rangle\rangle \otimes \rho$ is a subform of φ. □

Theorem 23.2. *Let φ be a nondegenerate (respectively, totally singular) anisotropic quadratic form over F of dimension $n \geq 1$. Let $T = (t_1, \dots, t_n)$ and $T' = (t_{n+1}, \dots, t_{2n})$ be $2n$ independent variables. Then the following are equivalent:*

1. *$n = 2^k$ for some $k \geq 1$ and $\varphi \in P_k(F)$ (respectively, φ is a quadratic quasi-Pfister form).*
2. *$G(\varphi_K) = D(\varphi_K)$ for all field extensions K/F.*
3. *$D(\varphi_K)$ is a group for all field extensions K/F.*
4. *Over the rational function field $F(T, T')$, we have*
$$\varphi(T)\varphi(T') \in D(\varphi_{F(T,T')}).$$
5. *$\varphi(T) \in G(\varphi_{F(T)})$.*

PROOF. (2) ⇒ (3) ⇒ (4) are trivial.

(5) ⇐ (1) ⇒ (2): As quadratic Pfister forms are round by Corollary 9.9 and quasi-Pfister forms are round by Corollary 10.3, the implications follow.

(5) ⇒ (4): We have $\varphi(T) \in G(\varphi_{F(T)}) \subset G(\varphi_{F(T,T')})$ and $\varphi(T') \in D(\varphi_{T,T'})$. It follows by Lemma 9.1 that $\varphi(T)\varphi(T') \in D(\varphi_{F(T,T')})$.

(4) $\Rightarrow$ (3): If K/F is a field extension, then $\varphi(T)\varphi(T') \in D(\varphi_{K(T,T')})$. By the Substitution Principle 17.7, it follows that $D(\varphi_K)$ is a group.

(3) $\Rightarrow$ (1): As $1 \in D(\varphi)$, it is sufficient to show that φ is a general quadratic (quasi-) Pfister form. We may assume that $\dim \varphi \geq 2$. If φ is nondegenerate, φ contains a nondegenerate binary subform, i.e., a 1-fold general quadratic Pfister form. Let ρ be the largest quadratic general Pfister subform of φ if φ is nondegenerate and the largest quasi-Pfister form if φ is totally singular. Suppose that $\rho \neq \varphi$. If φ is nondegenerate, then $V_\rho^\perp \neq 0$ and $V_\rho^\perp \cap V_\rho = \operatorname{rad} \mathfrak{b}_\rho = 0$ and if φ is totally singular, then $V_\rho^\perp = V_\varphi$ and $V_\rho \neq V_\varphi$. In either case, there exists a $v \in V_\rho^\perp \setminus V_\rho$. Set $a = -\varphi(v)$. By Lemma 23.1, we have $\langle\langle a \rangle\rangle \otimes \rho$ is isometric to a subform of φ, a contradiction. □

Remark 23.3. Let φ be a nondegenerate isotropic quadratic form over F. As hyperbolic quadratic forms are universal and round, if φ is hyperbolic, then $\varphi(T) \in G(\varphi_{F(T)})$. Conversely, suppose $\varphi(T) \in G(\varphi_{F(T)})$. As

$$(\varphi_{F(T)})_{an} \perp \mathfrak{i}_0(\varphi)\mathbb{H} \simeq \varphi_{F(T)} \simeq \varphi(T)\varphi_{F(T)} \simeq \varphi(T)(\varphi_{F(T)})_{an} \perp \mathfrak{i}_0(\varphi)\varphi(T)\mathbb{H},$$

we have $\varphi(T) \in G\big((\varphi_{F(T)})_{an}\big)$ by Witt Cancellation 8.4. If φ was not hyperbolic, then the Subform Theorem 22.5 would imply $\dim \varphi_{F(T)} \leq \dim(\varphi_{F(T)})_{an}$, a contradiction. Consequently, $\varphi(T) \in G(\varphi_{F(T)})$ if and only if φ is hyperbolic.

Corollary 23.4. *Let φ be a nondegenerate anisotropic quadratic form of dimension at least two over F. Then the following are equivalent:*

(1) $\dim \varphi$ *is even and* $\mathfrak{i}_1(\varphi) = \dim \varphi/2$.
(2) $\varphi_{F(\varphi)}$ *is hyperbolic.*
(3) $\varphi \in GP_n(F)$ *for some* $n \geq 1$.

Proof. Statements (1) and (2) are both equivalent to $\varphi_{F(\varphi)}$ contains a totally isotropic subspace of dimension $\frac{1}{2}\dim \varphi$. Let $a \in D(\varphi)$. Replacing φ by $\langle a \rangle \varphi$ we may assume that φ represents one. By Theorem 22.4, condition (2) in the corollary is equivalent to condition (5) of Theorem 23.2, hence conditions (2) and (3) above are equivalent. □

Corollary 23.5. *Let φ and ψ be quadratic forms over F with $\varphi \in P_n(F)$ anisotropic. Suppose that there exists an F-isomorphism $F(\varphi) \simeq F(\psi)$. Then there exists an $a \in F^\times$ such that $\psi \simeq a\varphi$ over F, i.e., φ and ψ are similar over F.*

Proof. As $\varphi_{F(\varphi)}$ is hyperbolic, so is $\varphi_{F(\psi)}$. In particular, $a\psi$ is a subform of φ for some $a \in F^\times$ by the Subform Theorem 22.5. Since $F(\varphi) \simeq F(\psi)$, we have $\dim \varphi = \dim \psi$ and the result follows. □

In general, the corollary does not generalize to non-Pfister forms. Let $F = \mathbb{Q}(t_1, t_2, t_3)$. The quadratic forms $\varphi = \langle\langle t_1, t_2 \rangle\rangle \perp \langle -t_3 \rangle$ and $\varphi = \langle\langle t_1, t_3 \rangle\rangle \perp \langle -t_2 \rangle$ have isomorphic function fields but are not similar. (Cf. [**89**, Th. XII.2.15].)

Let $r : F \to K$ be a homomorphism of fields. Denote the kernel of $r_{K/F} : W(F) \to W(K)$ by $W(K/F)$ and the kernel of $r_{K/F} : I_q(F) \to I_q(K)$ by $I_q(K/F)$. If φ is a nondegenerate even-dimensional quadratic form over F, we denote by $W(F)\varphi$ the cyclic $W(F)$-module in $I_q(F)$ generated by φ.

Corollary 23.6. *Let φ be an anisotropic quadratic n-fold Pfister form with $n \geq 1$ and ψ an anisotropic quadratic form of even dimension over F. Then there is an* *isometry $\psi \simeq \mathfrak{b} \otimes \varphi$ over F for some symmetric bilinear form $\mathfrak{b}$ over F if and only if $\psi_{F(\varphi)}$ is hyperbolic. In particular, $I_q(F(\varphi)/F) = W(F)\varphi$.*

PROOF. If $\mathfrak{b}$ is a bilinear form, then $(\mathfrak{b} \otimes \varphi)_{F(\varphi)} = \mathfrak{b}_{F(\varphi)} \otimes \varphi_{F(\varphi)}$ is hyperbolic by Lemma 8.16 as $\varphi_{F(\varphi)}$ is hyperbolic by Corollary 9.10. Conversely, suppose that $\psi_{F(\varphi)}$ is hyperbolic. We induct on $\dim \psi$. Assume that $\dim \psi > 0$. By the Subform Theorem 22.5 and Proposition 7.22, we have $\psi \simeq a\varphi \perp \gamma$ for some $a \in F^\times$ and quadratic form γ. The form γ also satisfies $\gamma_{F(\varphi)}$ is hyperbolic, so the result follows by induction. □

23.A. The Hauptsatz. We next prove a fundamental fact about forms in $I^n(F)$ and $I_q^n(F)$ due to Arason and Pfister known as the Hauptsatz and proven in [**10**].

Theorem 23.7 (Hauptsatz). (1) *Let $0 \neq \varphi$ be an anisotropic quadratic form lying in $I_q^n(F)$. Then $\dim \varphi \geq 2^n$.*

(2) *Let $0 \neq \mathfrak{b}$ be an anisotropic bilinear form lying in $I^n(F)$. Then $\dim(\mathfrak{b}) \geq 2^n$.*

PROOF. (1): As $I_q^n(F)$ is additively generated by general quadratic n-fold Pfister forms, we can write $\varphi = \sum_{1=i}^r a_i\rho_i$ in $W(F)$ for some anisotropic $\rho_i \in P_n(F)$ and $a_i \in F^\times$. We prove the result by induction on r. If $r = 1$, the result is trivial as ρ_1 is anisotropic, so we may assume that $r > 1$. As $(\rho_r)_{F(\rho_r)}$ is hyperbolic by Corollary 9.10, applying the restriction map $r_{F(\rho_r)/F} : W(F) \to W\big(F(\rho_r)\big)$ to φ yields $\varphi_{F(\rho_r)} = \sum_{i=1}^{r-1} a_i(\rho_i)_{F(\rho_r)}$ in $I_q^n\big(F(\rho)\big)$. If $\varphi_{F(\rho_r)}$ is hyperbolic, then $2^n = \dim \rho \leq \dim \varphi$ by the Subform Theorem 22.5. If this does not occur, then by induction $2^n \leq \dim(\varphi_{F(\rho_r)})_{an} \leq \dim \varphi$ and the result follows.

(2): As $I^n(F)$ is additively generated by bilinear n-fold Pfister forms, we can write $\mathfrak{b} = \sum_{1=i}^r \varepsilon_i\mathfrak{c}_i$ in $W(F)$ for some $\mathfrak{c}_i$ anisotropic bilinear n-fold Pfister forms and $\varepsilon_i \in \{\pm 1\}$. Let $\varphi = \varphi_{\mathfrak{c}_r}$ be the quadratic form associated to $\mathfrak{c}_r$. Then $\varphi_{F(\varphi)}$ is isotropic, hence $(\mathfrak{c}_r)_{F(\varphi)}$ is isotropic, hence metabolic by Corollary 6.3. If $\mathfrak{b}_{F(\varphi)}$ is not metabolic, then $2^n \leq \dim(\mathfrak{b}_{F(\varphi)})_{an} \leq \dim \mathfrak{b}$ by induction on r. If $\mathfrak{b}_{F(\varphi)}$ is metabolic, then $2^n = \dim \mathfrak{c} \leq \dim \mathfrak{b}$ by Corollary 22.8. □

An immediate consequence of the Hauptsatz is a solution to a problem of Milnor, viz.,

Corollary 23.8. *$\bigcap_{i=1}^\infty I^n(F) = 0$ and $\bigcap_{i=1}^\infty I_q^n(F) = 0$.*

The proof of the Hauptsatz for bilinear forms completes the proof of Corollary 6.19 and Theorem 6.20. We have an analogous result for quadratic Pfister forms.

Corollary 23.9. *Let $\varphi, \psi \in GP_n(F)$. If $\varphi \equiv \psi \bmod I_q^{n+1}(F)$, then $\varphi \simeq a\psi$ for some $a \in F^\times$, i.e., φ and ψ are similar over F. If, in addition, $D(\varphi) \cap D(\psi) \neq \emptyset$, then $\varphi \simeq \psi$.*

PROOF. By the Hauptsatz 23.7, we may assume both φ and ψ are anisotropic. As $\langle\langle a\rangle\rangle \otimes \psi \in GP_{n+1}(F)$, we have $a\psi \equiv \psi \bmod I_q^{n+1}(F)$ for any $a \in F^\times$. Choose $a \in F^\times$ such that $\varphi \perp -a\psi$ in $I_q^{n+1}(F)$ is isotropic. By the Hauptsatz 23.7, the form $\varphi \perp -a\psi$ is hyperbolic, hence $\varphi = a\psi$ in $I_q(F)$. As both forms are anisotropic, it follows by dimension count that $\varphi \simeq a\psi$ by Remark 8.17. If $D(\varphi) \cap D(\psi) \neq \emptyset$, then we can take $a = 1$. □

If φ is a nonzero subform of dimension at least two of an anisotropic quadratic form ρ, then $\rho_{F(\varphi)}$ is isotropic. As φ must also be anisotropic, we have $\rho \succ \varphi$. For general Pfister forms, we can say more. Let ρ be an anisotropic general quadratic Pfister form. Then $\rho_{F(\rho)}$ is hyperbolic, so it contains a totally isotropic subspace of dimension $(\dim \rho)/2$. Suppose that φ is a subform of ρ satisfying $\dim \varphi > (\dim \rho)/2$. Then $\varphi_{F(\rho)}$ is isotropic, hence $\varphi \succ \rho$ also. This motivates the following:

Definition 23.10. An anisotropic quadratic form φ is called a *Pfister neighbor* if there is a general quadratic Pfister form ρ such that φ is isometric to a subform of ρ and $\dim \varphi > (\dim \rho)/2$.

For example, nondegenerate anisotropic forms of dimension at most 3 are Pfister neighbors.

Remark 23.11. Let φ be a Pfister neighbor isometric to a subform of a general quadratic Pfister form ρ with $\dim \varphi > (\dim \rho)/2$. By the above, $\varphi \prec\succ \rho$. Let σ be another general quadratic Pfister form such that φ is isometric to a subform of σ and $\dim \varphi > (\dim \sigma)/2$. As $\rho \prec\succ \varphi \prec\succ \sigma$ and $D(\rho) \cap D(\sigma) \neq \emptyset$, we have $\sigma \simeq \rho$ by the Subform Theorem 22.5. Thus the general Pfister form ρ is uniquely determined by φ up to isomorphism. We call ρ the *associated general Pfister form* of φ. If φ represents one, then ρ is a Pfister form.

24. Linkage of quadratic forms

In this section, we look at the quadratic analogue of linkage of bilinear Pfister forms. The Hauptsatz shows that anisotropic forms in $I_q^n(F)$ have dimension at least 2^n. We shall be interested in those dimensions that are realizable by anisotropic forms in $I_q^n(F)$. In this section, we determine the possible dimension of anisotropic forms that are the sum of two general quadratic Pfister forms as well as the meaning of when the sum of three general n-fold Pfister forms is congruent to zero mod $I_q^n(F)$. We shall return to and expand these results in §35 and §82.

Proposition 24.1. *Let $\varphi \in GP(F)$.*

(1) *Let $\rho \in GP_n(F)$ be a subform of φ with $n \geq 1$. Then there is a bilinear Pfister form $\mathfrak{b}$ such that $\varphi \simeq \mathfrak{b} \otimes \rho$.*
(2) *Let $\mathfrak{b}$ be a general bilinear Pfister form such that $\varphi_{\mathfrak{b}}$ is a subform of φ. Then there is $\rho \in P(F)$ such that $\varphi \simeq \mathfrak{b} \otimes \rho$.*

PROOF. We may assume that φ is anisotropic of dimension ≥ 2.

(1): Let $\mathfrak{b}$ be a bilinear Pfister form of the largest dimension such that $\mathfrak{b} \otimes \rho$ is isometric to a subform ψ of φ. As $\mathfrak{b} \otimes \rho$ in nondegenerate, $V_{\psi}^{\perp} \cap V_{\psi} = 0$. We claim that $\psi = \varphi$. Suppose not. Then $V_{\psi}^{\perp} \neq 0$, hence $V_{\psi}^{\perp} \setminus V_{\psi} \neq \emptyset$. Choose $a = -\psi(v)$ with $v \in V_{\psi}^{\perp} \setminus V_{\psi}$. Lemma 23.1 implies that $\langle\langle a \rangle\rangle \otimes \rho$ is isometric to a subform of φ, contradicting the maximality of $\mathfrak{b}$.

(2): We may assume that $\operatorname{char} F = 2$ and $\mathfrak{b}$ is a Pfister form, so $1 \in D(\varphi_{\mathfrak{b}}) \subset D(\varphi)$. Let W be a subspace of V_{φ} such that $\varphi|_W \simeq \varphi_{\mathfrak{b}}$. Choose a vector $w \in W$ such that $\varphi_{\mathfrak{b}}(w) = 1$ and write the quasi-Pfister form $\varphi_{\mathfrak{b}} = \langle 1 \rangle \perp \varphi_{\mathfrak{b}}'$ where $V_{\varphi_{\mathfrak{b}}'}$ is any complementary subspace of Fw in $V_{\varphi_{\mathfrak{b}}}$. Let $v \in V_{\varphi}$ satisfy v is orthogonal to $V_{\varphi_{\mathfrak{b}}'}$, but $\mathfrak{b}(v, w) \neq 0$. Then the restriction of φ on $W \oplus Fv$ is isometric to $\psi := \varphi_{\mathfrak{b}}' \perp [1, a]$ for some $a \in F^{\times}$. Note that ψ is isometric to subforms of both of the general Pfister forms φ and $\mu := \mathfrak{b} \otimes \langle\langle a]]$. In particular, ψ and μ are anisotropic. As $\dim\psi > \frac{1}{2}\dim\mu$, the form ψ is a Pfister neighbor of μ. Hence $\psi \prec\succ \mu$ by Remark 23.11. Since $\varphi_{F(\psi)}$ is hyperbolic by Proposition 22.18 so is $\varphi_{F(\mu)}$. It follows from the Subform Theorem 22.5 that μ is isomorphic to a subform of φ as $1 \in D(\mu) \cap D(\varphi)$. By the first statement of the proposition, there is a bilinear Pfister form $\mathfrak{c}$ such that $\varphi \simeq \mathfrak{c} \otimes \mu = \mathfrak{c} \otimes \mathfrak{b} \otimes \langle\langle a]]$. Hence $\varphi \simeq \mathfrak{b} \otimes \rho$ where $\rho = \mathfrak{c} \otimes \langle\langle a]]$. □

Let ρ be a general quadratic Pfister form. We say a general quadratic Pfister form ψ (respectively, a general bilinear Pfister form $\mathfrak{b}$) is a *divisor* ρ if $\rho \simeq \mathfrak{c} \otimes \psi$ for some bilinear Pfister form $\mathfrak{c}$ (respectively, $\rho \simeq \mathfrak{b} \otimes \mu$ for some quadratic Pfister form μ). By Proposition 24.1, any general quadratic Pfister subform of ρ is a divisor of ρ and any general bilinear Pfister form $\mathfrak{b}$ of ρ whose associated quadratic form is a subform of ρ is a divisor ρ.

Theorem 24.2. *Let $\varphi_1, \varphi_2 \in GP(F)$ be anisotropic. Let $\rho \in GP(F)$ be a form of largest dimension such that ρ is isometric to subforms of φ_1 and φ_2. Then*

$$\mathfrak{i}_0(\varphi_1 \perp -\varphi_2) = \dim \rho.$$

PROOF. Note that $\mathfrak{i}_0 := \mathfrak{i}_0(\varphi_1 \perp -\varphi_2) \geq d := \dim \rho$. We may assume that $\mathfrak{i}_0 > 1$. We claim that φ_1 and φ_2 have isometric nondegenerate binary subforms. To prove the claim let W be a 2-dimensional totally isotropic subspace of $V_{\varphi_1} \oplus V_{-\varphi_2}$. As φ_1 and φ_2 are anisotropic, the projections U_1 and U_2 of W to V_{φ_1} and $V_{-\varphi_2} = V_{\varphi_2}$, respectively, are 2-dimensional. Moreover, the binary forms $\psi_1 := \varphi_1|_{U_1}$ and $\psi_2 = \varphi_2|_{U_2}$ are isometric. We may assume that ψ_1 and ψ_2 are degenerate (and therefore, $\operatorname{char}(F) = 2$). Hence ψ_1 and ψ_2 are isometric to $\varphi_{\mathfrak{b}}$, where $\mathfrak{b}$ is a 1-fold general bilinear Pfister form. By Proposition 24.1(2), we have $\varphi_1 \simeq \mathfrak{b} \otimes \rho_1$ and $\varphi_2 \simeq \mathfrak{b} \otimes \rho_2$ for some $\rho_i \in P(F)$. Write $\rho_i = \mathfrak{c}_i \otimes \nu_i$ for bilinear Pfister forms $\mathfrak{c}_i$ and 1-fold quadratic Pfister forms ν_i. Consider quaternion algebras Q_1 and Q_2 whose reduced norm forms are similar to $\mathfrak{b} \otimes \nu_1$ and $\mathfrak{b} \otimes \nu_2$, respectively. The algebras Q_1 and Q_2 are split by a quadratic field extension that splits $\mathfrak{b}$. By Theorem 98.19, the algebras Q_1 and Q_2 have subfields isomorphic to a separable quadratic extension L/F. By Example 9.7, the reduced norm forms of Q_1 and Q_2 are divisible by the nondegenerate norm form of L/F. Hence the forms $\mathfrak{b} \otimes \nu_1$ and $\mathfrak{b} \otimes \nu_2$ and therefore φ_1 and φ_2 have isometric nondegenerate binary subforms. The claim is proven.

By the claim, ρ is a general r-fold Pfister form with $r \geq 1$. Write $\varphi_1 = \rho \perp \psi_1$ and $\varphi_2 = \rho \perp \psi_2$ for some forms ψ_1 and ψ_2. We have $\varphi_1 \perp (-\varphi_2) \simeq \psi_1 \perp (-\psi_2) \perp d\mathbb{H}$. Assume that $\mathfrak{i}_0 > d$. Then the form $\psi_1 \perp (-\psi_2)$ is isotropic, i.e., ψ_1 and ψ_2 have a common value, say $a \in F^\times$. By Lemma 23.1, the form $\langle\langle -a\rangle\rangle \otimes \rho$ is isometric to subforms of φ_1 and φ_2, a contradiction. □

Corollary 24.3. *Let $\varphi_1, \varphi_2 \in GP_n(F)$ be anisotropic forms. Then the possible values of $\mathfrak{i}_0(\varphi_1 \perp -\varphi_2)$ are $0, 1, 2, 4, \ldots, 2^n$.*

Let $\varphi_1 \in GP_m(F)$ and $\varphi_2 \in GP_n(F)$ be anisotropic forms satisfying $\mathfrak{i}(\varphi_1 \perp -\varphi_2) = 2^r > 0$. Let ρ be a general quadratic r-fold Pfister form isometric to a subform of φ_1 and to a subform of φ_2. We call ρ the *linkage* of φ_1 and φ_2 and say that φ_1 and φ_2 are *r-linked.* By Proposition 24.1, the linkage ρ is a divisor of φ_1 and φ_2. If $m = n$ and $r \geq n - 1$, we say that φ_1 and φ_2 are *linked.*

Remark 24.4. Let φ_1 and φ_2 be general quadratic Pfister forms. Suppose that φ_1 and φ_2 have isometric r-fold quasi-Pfister subforms. Then $\mathfrak{i}_0(\varphi_1 \perp -\varphi_2) \geq 2^r$

and by Theorem 24.2, the forms φ_1 and φ_2 have isometric general quadratic r-fold Pfister subforms.

For three n-fold Pfister forms, we have:

Proposition 24.5. *Let $\varphi_1, \varphi_2, \varphi_3 \in P_n(F)$. If $\varphi_1 + \varphi_2 + \varphi_3 \in I_q^{n+1}(F)$ then there exist a quadratic $(n-1)$-fold Pfister form ρ and $a_1, a_2, a_3 \in F^\times$ such that $a_1a_2a_3 = 1$ and $\varphi_i \simeq \langle\langle a_i\rangle\rangle \otimes \rho$ for $i = 1, 2, 3$. In particular, ρ is a common divisor of φ_i for $i = 1, 2, 3$.*

PROOF. We may assume that all φ_i are anisotropic Pfister forms by Corollary 9.10. In addition, we have $(\varphi_3)_{F(\varphi_3)}$ is hyperbolic. By Proposition 23.9, the form $(\varphi_1 \perp -\varphi_2)_{F(\varphi_3)}$ is also hyperbolic. As φ_3 is anisotropic, $\varphi_1 \perp -\varphi_2$ cannot be hyperbolic by the Hauptsatz 23.7. Consequently,

$$(\varphi_1 \perp -\varphi_2)_{an} \simeq a\varphi_3 \perp \tau$$

over F for some $a \in F^\times$ and a quadratic form τ by the Subform Theorem 22.5 and Proposition 7.22. As $\dim \tau < 2^n$ and $\tau \in I_q^{n+1}(F)$, the form τ is hyperbolic by Hauptsatz 23.7 and therefore $\varphi_1 - \varphi_2 = a\varphi_3$ in $I_q(F)$. It follows that $\mathfrak{i}_0(\varphi_1 \perp -\varphi_2) = 2^{n-1}$, hence φ_1 and φ_2 are linked by Theorem 24.2.

Let ρ be a linkage of φ_1 and φ_2. By Proposition 24.1, we have $\varphi_1 \simeq \langle\langle a_1\rangle\rangle \otimes \rho$ and $\varphi_2 \simeq \langle\langle a_2\rangle\rangle \otimes \rho$ for some $a_1, a_2 \in F^\times$. Then φ_3 is similar to $(\varphi_1 \perp -\varphi_2)_{an} \simeq -a_1\langle\langle a_1a_2\rangle\rangle \otimes \rho$, i.e., $\varphi_3 \simeq \langle\langle a_1a_2\rangle\rangle \otimes \rho$. □

Corollary 24.6. *Let $\varphi_1, \varphi_2, \varphi_3 \in P_n(F)$. Suppose that*

$$\varphi_1 + \varphi_2 + \varphi_3 \equiv 0 \bmod I_q^{n+1}(F). \tag{24.7}$$

Then

$$e_n(\varphi_1) + e_n(\varphi_2) + e_n(\varphi_3) = 0 \text{ in } H^n(F).$$

PROOF. By Proposition 24.5, we have $\varphi_i \simeq \langle\langle a_i\rangle\rangle \otimes \rho$ for some $\rho \in P_{n-1}(F)$ and $a_i \in F^\times$ for $i = 1, 2, 3$ satisfying $a_1a_2a_3 = 1$. It follows from Proposition 16.1 that

$$\begin{aligned} e_n(\varphi_1) + e_n(\varphi_2) + e_n(\varphi_3) &= e_n\big(\langle\langle a_1\rangle\rangle \otimes \rho\big) + e_n\big(\langle\langle a_2\rangle\rangle \otimes \rho\big) + e_n\big(\langle\langle a_3\rangle\rangle \otimes \rho\big) \\ &= \{a_1a_2a_3\}e_{n-1}(\rho) = 0. \end{aligned}$$ □

25. The submodule $J_n(F)$

By Corollary 23.4, a general quadratic Pfister form has the following "intrinsic" characterization: a nondegenerate anisotropic quadratic form φ of positive even dimension is a general quadratic Pfister form if and only if the form $\varphi_{F(\varphi)}$ is hyperbolic. We shall use this to characterize elements of $I_q^n(F)$. Let φ be a form that is nonzero in $I_q(F)$. There exists a field extension K/F such that $(\varphi_K)_{an}$ is a general quadratic n-fold Pfister form for some $n \geq 1$. The smallest possible such n is called the degree $\deg(\varphi)$ of φ. We shall see in Theorem 40.10 that $\varphi \in I_q^n(F)$ if and only if $\deg \varphi \geq n$. In this section, we shall begin the study of the degree of forms. The ideas in this section are due to Knebusch (cf. [**83**] and [**84**]).

We begin by constructing a tower of field extensions of F with $(\varphi_K)_{an}$ a general quadratic n-fold Pfister form where K is the penultimate field K in the tower.

Let φ be a nondegenerate quadratic form over F. We construct a tower of fields $F_0 \subset F_1 \subset \cdots \subset F_h$ and quadratic forms φ_k over F_k for all $k \in [0,\ h]$ as follows: We start with $F_0 := F$, $\varphi_0 := \varphi_{an}$, and inductively set $F_k := F_{k-1}(\varphi_{k-1})$,

$\varphi_k := (\varphi_{F_k})_{an}$ for $k > 0$. We stop at F_h such that $\dim \varphi_h \leq 1$. The form φ_k is called the kth *(anisotropic) kernel form* of φ. The tower of the fields F_k is called the *generic splitting tower of* φ. The integer h is called the *height* of φ and denoted by $\mathfrak{h}(\varphi)$. We have $\mathfrak{h}(\varphi) = 0$ if and only if $\dim \varphi_{an} \leq 1$.

Let $h = \mathfrak{h}(\varphi)$. For any $k \in [0, h]$, the *k-th absolute higher Witt index* $\mathfrak{j}_k(\varphi)$ of φ is defined as the integer $\mathfrak{i}_0(\varphi_{F_k})$. Clearly, one has

$$0 \leq \mathfrak{j}_0(\varphi) < \mathfrak{j}_1(\varphi) < \cdots < \mathfrak{j}_h(\varphi) = [(\dim \varphi)/2].$$

The set of integers $\{\mathfrak{j}_0(\varphi), \ldots, \mathfrak{j}_h(\varphi)\}$ is called the *splitting pattern* of φ.

Proposition 25.1. *Let φ be a nondegenerate quadratic form with $h = \mathfrak{h}(\varphi)$. The splitting pattern $\{\mathfrak{j}_0(\varphi), \ldots, \mathfrak{j}_h(\varphi)\}$ of φ coincides with the set of Witt indices $\mathfrak{i}_0(\varphi_K)$ over all field extensions K/F.*

PROOF. Let K/F be a field extension. Define a tower of fields $K_0 \subset K_1 \subset \cdots \subset K_h$ by $K_0 = K$ and $K_k = K_{k-1}(\varphi_{k-1})$ for $k > 0$. Clearly, $F_k \subset K_k$ for all k. Let $k \geq 0$ be the smallest integer such that φ_k is anisotropic over K_k. It suffices to show that $\mathfrak{i}_0(\varphi_K) = \mathfrak{j}_k(\varphi)$.

By definition of φ_k and $\mathfrak{j}_k$, we have $\varphi_{F_k} = \varphi_k \perp \mathfrak{j}_k(\varphi)\mathbb{H}$. Therefore, $\varphi_{K_k} = (\varphi_q)_{K_k} \perp \mathfrak{j}_k(\varphi)\mathbb{H}$. As φ_k is anisotropic over K_k, we have $\mathfrak{i}_0(\varphi_{K_k}) = \mathfrak{j}_k(\varphi)$.

We claim that the extension K_k/K is purely transcendental. This is clear if $k = 0$. Otherwise $K_k = K_{k-1}(\varphi_{k-1})$ is purely transcendental by Proposition 22.9, since φ_{k-1} is isotropic over K_{k-1} by the choice of k and is nondegenerate. It follows from the claim and Remark 8.9 that $\mathfrak{i}_0(\varphi_K) = \mathfrak{i}_0(\varphi_{K_k}) = \mathfrak{j}_k(\varphi)$. □

Corollary 25.2. *Let φ be a nondegenerate quadratic form over F and let K/F be a purely transcendental extension. Then the splitting patterns of φ and φ_K are the same.*

PROOF. This follows from Lemma 7.15. □

We define the *relative higher Witt index* $\mathfrak{i}_k(\varphi)$, $k \in [1, \mathfrak{h}(\varphi)]$, of a nondegenerate quadratic form φ to be the difference

$$\mathfrak{i}_k(\varphi) = \mathfrak{j}_k(\varphi) - \mathfrak{j}_{k-1}(\varphi).$$

Clearly, $\mathfrak{i}_k(\varphi) > 0$ and $\mathfrak{i}_k(\varphi) = \mathfrak{i}_r(\varphi_s)$ for any $r > 0$ and $s \geq 0$ such that $r + s = k$.

Corollary 25.3. *Let φ be a nondegenerate anisotropic quadratic form over F of dimension at least two. Then*

$$\mathfrak{i}_1(\varphi) = \mathfrak{j}_1(\varphi) = \min\{\mathfrak{i}_0(\varphi_K) \mid K/F \text{ a field extension with } \varphi_K \text{ isotropic}\}.$$

Let φ be a nondegenerate nonhyperbolic quadratic form of even dimension over F with $h = \mathfrak{h}(\varphi)$. Let $F_0 \subset F_1 \subset \cdots \subset F_h$ be the generic splitting tower of φ. The form $\varphi_{h-1} = (\varphi_{F_{h-1}})_{an}$ is hyperbolic over its function field, hence a general n-fold Pfister form for some integer $n \geq 1$ with $\mathfrak{i}_h(\varphi) = 2^{n-1}$ by Corollary 23.4. The form φ_{h-1} is called the *leading form* of φ and n is called the *degree* of φ and is denoted by $\deg \varphi$. The field F_{h-1} is called the *leading field of* φ. For convenience, we set $\deg \varphi = \infty$ if φ is hyperbolic.

Remark 25.4. Let φ be a nondegenerate quadratic form of even dimension with the generic splitting tower $F_0 \subset F_1 \subset \cdots \subset F_h$. If $\varphi_k = (\varphi_{F_k})_{an}$ with $k \in [0,\ h-1]$, then $\deg \varphi_k = \deg \varphi$.

Notation 25.5. Let φ be a nondegenerate quadratic form over F and $X = X_\varphi$. Let $k \in [0,\ \mathfrak{h}(\varphi)]$. We shall let $X_k := X_{\varphi_k}$ and also write $\mathfrak{j}_k(X)$ (respectively, $\mathfrak{i}_k(X)$) for $\mathfrak{j}_k(\varphi)$ (respectively, $\mathfrak{i}_k(\varphi)$).

It is a natural problem to classify nondegenerate quadratic forms over a field F of a given height. This is still an open problem even for forms of height two. By Corollary 23.4, we do know

Proposition 25.6. *Let φ be an even-dimensional nondegenerate anisotropic quadratic form. Then $\mathfrak{h}(\varphi) = 1$ if and only if $\varphi \in GP(F)$.*

Proposition 25.7. *Let φ be a nondegenerate quadratic form of even dimension over F and K/F a field extension with $(\varphi_K)_{an}$ an m-fold general Pfister for some $m \geq 1$. Then $m \geq \deg \varphi$. In particular, $\deg \varphi$ is the smallest integer $n \geq 1$ such that $(\varphi_K)_{an}$ is a general n-fold Pfister form over an extension K/F.*

PROOF. It follows from Proposition 25.1 that

$$(\dim \varphi - 2^m)/2 = \mathfrak{i}_0(\varphi_K) \leq \mathfrak{j}_{\mathfrak{h}(\varphi)-1}(\varphi) = (\dim \varphi - 2^{\deg \varphi})/2,$$

hence the inequality. □

Corollary 25.8. *Let φ be a nondegenerate quadratic form of even dimension over F. Then $\deg \varphi_E \geq \deg \varphi$ for any field extension E/F.*

For every $n \geq 1$ set

$$J_n(F) := \{\varphi \in I_q(F) \mid \deg \varphi \geq n\} \subset I_q(F).$$

Clearly, $J_1(F) = I_q(F)$.

Lemma 25.9. *Let $\rho \in GP_n(F)$ be anisotropic with $n \geq 1$. Let $\varphi \in J_{n+1}(F)$. Then $\deg(\rho \perp \varphi) \leq n$.*

PROOF. We may assume that φ is not hyperbolic. Let $\psi = \rho \perp \varphi$. Let $F_0, F_1, \dots, F_h$ be the generic splitting tower of φ and let $\varphi_i = (\varphi_{F_i})_{an}$. We show that ρ_{F_h} is anisotropic. Suppose not. Choose j maximal such that ρ_{F_j} is anisotropic. Then $\rho_{F_{j+1}}$ is hyperbolic so $\dim \varphi_j \leq \dim \rho$ by the Subform Theorem 22.5. Hence

$$2^n = \dim \rho \geq \dim \varphi_j \geq \deg 2^{\deg \varphi_j} = 2^{\deg \varphi} \geq 2^{n+1}$$

which is impossible. Thus $\rho_{F_\mathfrak{h}}$ is anisotropic.

As φ is hyperbolic over F_h, we have $\psi_{F_h} \sim \rho_{F_h}$. Consequently,

$$\deg \psi \leq \deg \psi_{F_h} = \deg \rho_{F_h} = n,$$

hence $\deg \psi \leq n$ as claimed. □

Corollary 25.10. *Let φ and ψ be any even-dimensional nondegenerate quadratic forms. Then $\deg(\varphi \perp \psi) \geq \min(\deg \varphi, \deg \psi)$.*

PROOF. If either φ or ψ is hyperbolic, this is trivial, so assume that both forms are not hyperbolic. We may also assume that $\varphi \perp \psi$ is not hyperbolic. Let K/F be a field extension such that $(\varphi \perp \psi)_K \sim \rho$ for some $\rho \in GP_n(K)$ where $n = \deg(\varphi \perp \psi)$. Then $\varphi_K \sim \rho \perp (-\psi_K)$. Suppose that $\deg \psi > n$. Then $\deg \psi_K > n$ and applying the lemma to the form $\rho \perp (-\psi_K)$ implies $\deg \varphi_K \leq n$. Hence $\deg \varphi \leq n = \deg(\varphi \perp \psi)$. □

Proposition 25.11. *$J_n(F)$ is a $W(F)$-submodule of $I_q(F)$ for every $n \geq 1$.*

PROOF. Corollary 25.10 shows that $J_n(F)$ is a subgroup of $I_q(F)$. Since $\deg \varphi = \deg(a\varphi)$ for all $a \in F^\times$, it follows that $J_n(F)$ is also closed under multiplication by elements of $W(F)$. □

Corollary 25.12. $I_q^n(F) \subset J_n(F)$.

PROOF. As general quadratic n-fold Pfister forms clearly lie in $J_n(F)$, the result follows from Proposition 25.11. □

Proposition 25.13. $I_q^2(F) = J_2(F)$.

PROOF. Let $\varphi \in J_2(F)$ and $\varphi_k = \varphi_{F_k}$ with F_k, $k \in [0, h]$, the generic splitting tower. As $\deg \varphi \geq 2$, the field F_k is the function field of a smooth quadric of dimension at least 2 over F_{k-1}, hence the field F_{k-1} is algebraically closed in F_k. Since the form $\varphi_h = 0$ has trivial discriminant, by descending induction on k, it follows that $\varphi = \varphi_0$ has trivial discriminant. By Theorem 13.7 we conclude that $\varphi \in I_q^2(F)$. □

Proposition 25.14. $J_3(F) = \{\varphi \mid \dim \varphi$ *is even,* $\operatorname{disc}(\varphi) = 1, \operatorname{clif}(\varphi) = 1\}$.

PROOF. Let φ be an anisotropic form of even dimension and trivial discriminant. Then $\varphi \in I_q^2(F) = J_2(F)$ by Theorem 13.7 and Proposition 25.13. Suppose φ also has trivial Clifford invariant. We must show that $\deg \varphi \geq 3$. Let K be the leading field of φ and ρ its leading form. Then $\rho \in GP_n(F)$ with $n \geq 2$. Suppose that $n = 2$. As $e_2(\rho) = 0$ in $H^2(K)$, we have ρ is hyperbolic by Corollary 12.5, a contradiction. Therefore, $\varphi \in J_3(F)$.

Let $\varphi \in J_3(F)$. Then $\varphi \in I_q^2(F)$ by Proposition 25.13. In particular, $\operatorname{disc}(\varphi) = 1$ and $\varphi = \sum_{i=1}^r \rho_i$ with $\rho_i \in GP_2(F)$, $1 \leq i \leq r$. We show that $\operatorname{clif}(\varphi) = 1$ by induction on r. Let $\rho_r = b\langle\langle a, d]]$ and $K = F_d$. Then $\varphi_K \in J_3(K)$ and satisfies $\varphi_K = \sum_{i=1}^{r-1} (\rho_i)_K$ as $(\rho_r)_K$ is hyperbolic. By induction, $\operatorname{clif}(\varphi_K) = 1$. Thus $\operatorname{clif}(\varphi)$ lies in kernel of $\operatorname{Br}(F) \to \operatorname{Br}(K)$. Therefore, the index of $\operatorname{clif}(\varphi)$ is at most two. Consequently, $\operatorname{clif}(\varphi)$ is represented by a quaternion algebra, hence there exists a 2-fold quadratic Pfister form σ satisfying $\operatorname{clif}(\varphi) = \operatorname{clif}(\sigma)$. Thus $\operatorname{clif}(\varphi + \sigma) = 1$, so $\varphi + \sigma$ lies in $J_3(F)$ by the first part of the proof. It follows that σ lies in $J_3(F)$. Therefore, $\sigma = 0$ and $\operatorname{clif}(\varphi) = 1$. □

As $\bar{e}_2$ is an isomorphism (cf. Theorem 16.3 if $\operatorname{char} F = 2$ and Chapter VIII if $\operatorname{char} F \neq 2$), we have $I^3(F) = J_3(F)$. We shall show that $I^n(F) = J_n(F)$ for all n in Theorem 40.10. In this section, we show the following due to Arason and Knebusch (cf. [**9**]).

Proposition 25.15. $I^m(F)J_n(F) \subset J_{n+m}(F)$.

PROOF. Clearly, it suffices to do the case that $m = 1$. Since 1-fold bilinear Pfister forms additively generate $I(F)$, it also suffices to show that if $\varphi \in J_n(F)$ and $a \in F^\times$, then $\langle\langle a\rangle\rangle \otimes \varphi \in J_{n+1}(F)$. Let ψ be the anisotropic part of $\langle\langle a\rangle\rangle \otimes \varphi$. We may assume that $\psi \neq 0$.

First suppose that $\psi \in GP(F)$. We prove that $\deg \psi > n$ by induction on the height h of φ. If $h = 1$, then $\varphi \in GP(F)$ and the result is clear. So assume that $h > 1$. Suppose that $\psi_{F(\varphi)}$ remains anisotropic. Applying the induction hypothesis to the form $\varphi_{F(\varphi)}$, we have

$$\deg \psi = \deg \psi_{F(\varphi)} > n.$$

If $\psi_{F(\varphi)}$ is isotropic, it is hyperbolic and therefore $\dim\psi \geq \dim\varphi$ by the Subform Theorem 22.5. As $h > 1$, we have

$$2^{\deg\psi} = \dim\psi \geq \dim\varphi > 2^{\deg\varphi} \geq 2^n,$$

hence $\deg\psi > n$.

Now consider the general case. Let K/F be a field extension such that ψ_K is Witt-equivalent to a general Pfister form and $\deg\psi_K = \deg\psi$. By the first part of the proof

$$\deg\psi = \deg\psi_K > n. \qquad \square$$

26. The Separation Theorem

There are anisotropic quadratic forms φ and ψ with $\dim\varphi < \dim\psi$ and $\varphi_{F(\psi)}$ isotropic. For example, this is the case when φ and ψ are Pfister neighbors of the same Pfister form. In this section, we show that if two anisotropic quadratic forms φ and ψ are separated by a power of two, more precisely, if $\dim\varphi \leq 2^n < \dim\psi$ for some $n \geq 0$, then $\varphi_{F(\psi)}$ remains anisotropic.

We shall need the following observation.

Remark 26.1. Let ψ be a quadratic form. Then V_ψ contains a (maximal) totally isotropic subspace of dimension $\mathfrak{i}_0'(\psi) := \mathfrak{i}_0(\psi) + \dim(\operatorname{rad}\psi)$. Define the invariant s of a form by $s(\psi) := \dim\psi - 2\mathfrak{i}_0'(\psi) = \dim\psi_{an} - \dim(\operatorname{rad}\psi)$. If two quadratic forms ψ and μ are Witt-equivalent, then $s(\psi) = s(\mu)$.

A field extension L/F is called *unirational* if there is a field extension L'/L with L'/F purely transcendental. A tower of unirational field extensions is unirational. If L/F is unirational, then every anisotropic quadratic form over F remains anisotropic over L by Lemma 7.15.

Lemma 26.2. *Let φ be an anisotropic quadratic form over F satisfying* $\dim\varphi \leq 2^n$ *for some $n \geq 0$. Then there exists a field extension K/F and an $(n+1)$-fold anisotropic quadratic Pfister form ρ over K such that*

(1) *φ_K is isometric to a subform of ρ.*
(2) *The field extension $K(\rho)/F$ is unirational.*

PROOF. Let $K_0 = F(t_1, \dots, t_{n+1})$ and let $\rho = \langle\langle t_1, \dots, t_{n+1}]]$. Then ρ is anisotropic. Indeed, by Corollary 19.6 and induction, it suffices to show $\langle\langle t]]$ is anisotropic over $F(t)$. If this is false, there is an equation $f^2 + fg + tg^2 = 0$ with $f, g \in F[t]$. Looking at the highest term of t in this equation gives either $a^2t^{2n} = 0$ or $b^2t^{2n+1} = 0$ where a, b are the leading coefficients of f, g, respectively. Neither is possible.

Consider the class $\mathcal{F}$ of field extensions E/K_0 satisfying:

(1′) ρ is anisotropic over E.
(2′) The field extension $E(\rho)/F$ is unirational.

We show that $K_0 \in \mathcal{F}$. By the above ρ is anisotropic. Let $L = K_0\big(\langle\langle 1, t_1]]\big)$. Then L/F is purely transcendental. As ρ_L is isotropic, $L(\rho)/L$ is also purely transcendental and hence so is $L(\rho)/F$. Since $K_0(\rho) \subset L(\rho)$, the field extension $K_0(\rho)/F$ is unirational.

For every field $E \in \mathcal{F}$, the form φ_E is anisotropic by (2′). As ρ_E is nondegenerate, the form $\rho_E \perp (-\varphi_E)$ is regular. We set

$$m(E) = \mathfrak{i}_0\big(\rho_E \perp (-\varphi_E)\big) = \mathfrak{i}_0'\big(\rho_E \perp (-\varphi_E)\big)$$

and let m be the maximum of the $m(E)$ over all $E \in \mathcal{F}$.

Claim 1: We have $m(E) \leq \dim \varphi$ and if $m(E) = \dim \varphi$, then φ_E is isometric to a subform of ρ_E.

Let W be a totally isotropic subspace in $V_{\rho_E} \perp V_{-\varphi_E}$ of dimension $m(E)$. Since ρ_E and φ_E are anisotropic, the projections of W to V_{ρ_E} and $V_{-\varphi_E} = V_{\varphi_E}$ are injective. This gives the inequality. Suppose that $m(E) = \dim \varphi$. Then the projection $p : W \to V_{\varphi_E}$ is an isomorphism and the composition

$$V_{\varphi_E} \xrightarrow{p^{-1}} W \to V_{\rho_E}$$

identifies φ_E with a subform of ρ_E.

Claim 2: $m = \dim \varphi$.

Assume that $m < \dim \varphi$. We derive a contradiction. Let $K \in \mathcal{F}$ be a field satisfying $m = m(K)$ and set $\tau = \big(\rho_K \perp (-\varphi_K)\big)_{an}$. As the form $\rho_K \perp (-\varphi_K)$ is regular, we have $\tau \sim \rho_K \perp (-\varphi_K)$ and

$$\dim \rho + \dim \varphi = \dim \tau + 2m. \tag{26.3}$$

Let W be a totally isotropic subspace in $V_{\rho_K} \perp V_{-\varphi_K}$ of dimension m. Let σ denote the restriction of ρ_K on $V_{\rho_K} \cap W^{\perp}$. Thus σ is a subform of ρ_K of dimension $\geq 2^{n+1} - m > 2^n$. In particular, σ is a Pfister neighbor of ρ_K. By Lemma 8.10, the natural map $V_{\rho_K} \cap W^{\perp} \to W^{\perp}/W$ identifies σ with a subform of τ.

We show that condition (2′) holds for $K(\tau)$. Since σ is a Pfister neighbor of ρ_K, the form σ and therefore τ is isotropic over $K(\rho)$. By Lemma 22.14, the extension $K(\rho)/K$ is separable, hence $\tau_{K(\rho)}$ is regular by Lemma 22.13. Therefore, by Lemma 22.9 the extension $K(\rho)(\tau)/K(\rho)$ is purely transcendental. It follows that $K(\rho)(\tau) = K(\tau)(\rho)$ is unirational over F, hence condition (2′) is satisfied.

As τ is isotropic over $K(\tau)$, we have $m\big(K(\tau)\big) > m$, hence $K(\tau) \notin \mathcal{F}$. Therefore, condition (1′) does not hold for $K(\tau)$, i.e., ρ_K is isotropic and therefore hyperbolic over $K(\tau)$. As $\emptyset \neq D(\sigma) \subset D(\rho_K) \cap D(\tau)$, the form τ is isometric to a subform of ρ_K by the Subform Theorem 22.5. Let $\tau^{\perp}$ be the complementary form of τ in ρ_K. It follows from (26.3) that

$$\dim \tau^{\perp} = \dim \rho - \dim \tau = 2m - \dim \varphi < \dim \varphi.$$

As $\rho_K \perp (-\tau) \sim \tau^{\perp}$ by Lemma 8.13,

$$\tau \perp (-\tau) \sim \rho_K \perp (-\varphi_K) \perp (-\tau) \sim \tau^{\perp} \perp (-\varphi_K). \tag{26.4}$$

We now use the invariant s defined in Remark 26.1. Since the space of $\tau \perp (-\tau)$ contains a totally isotropic subspace of dimension $\dim \tau$, it follows from (26.4) and Remark 26.1 that

$$s\big(\tau^{\perp} \perp (-\varphi_K)\big) = s\big(\tau \perp (-\tau)\big) = 0,$$

i.e., the form $\tau^{\perp} \perp (-\varphi_K)$ contains a totally isotropic subspace of half the dimension of the form. Since $\dim \varphi > \dim \tau^{\perp}$, this subspace intersects V_{φ_K} nontrivially, consequently φ_K is isotropic contradicting condition (2′). This establishes the claim.

It follows from the claims that φ_K is isometric to a subform of ρ_K. □

Theorem 26.5 (Separation Theorem). *Let φ and ψ be two anisotropic quadratic forms over F. Suppose that $\dim \varphi \leq 2^n < \dim \psi$ for some $n \geq 0$. Then $\varphi_{F(\psi)}$ is anisotropic.*

PROOF. Let ρ be an $(n+1)$-fold Pfister form over a field extension K/F as in Lemma 26.2 with φ_K a subform of ρ. By the lemma $\psi_{K(\rho)}$ is anisotropic. Suppose that $\varphi_{K(\psi)}$ is isotropic. Then $\rho_{K(\psi)}$ is isotropic, hence hyperbolic. By the Subform Theorem 22.5, there exists an $a \in F$ such that $a\psi_K$ is a subform of ρ. As $\dim \psi > \frac{1}{2} \dim \rho$, the form $a\psi_K$ is a neighbor of ρ, hence $a\psi_{K(\rho)}$ and therefore $\psi_{K(\rho)}$ is isotropic. This is a contradiction. □

Corollary 26.6. *Let φ and ψ be two anisotropic quadratic forms over F with $\dim \psi \geq 2$. If $\dim \psi \geq 2 \dim \varphi - 1$, then $\varphi_{F(\psi)}$ is anisotropic.*

The Separation Theorem was first proved by Hoffmann in [**53**] for fields of characteristic different from 2 and by Hoffmann and Laghribi in [**58**] for fields of characteristic 2. We shall give another, more geometric proof, in 79.8 below.

27. A further characterization of quadratic Pfister forms

In this section, we give a further characterization of quadratic Pfister forms due to Fitzgerald (cf. [**44**]) used to answer a question of Knebusch in [**84**] for fields of characteristic different from 2. Knebusch and Scharlau showed in [**85**] that Fitzgerald's theorem implied that if a nondegenerate anisotropic quadratic form ρ becomes hyperbolic over the function field of an irreducible anisotropic form φ satisfying $\dim \varphi > \frac{1}{3} \dim \rho$, then ρ is a general quadratic Pfister form. Hoffmann and Laghribi showed that this characterization was also true for fields of characteristic 2 in [**57**].

For a nondegenerate nonhyperbolic quadratic form ρ of even dimension, we set $N(\rho) = \dim \rho - 2^{\deg \rho}$. Since the splitting patterns of ρ and $\rho_{F(t)}$ are the same, by Corollary 25.2, we have $N(\rho_{F(t)}) = N(\rho)$.

Theorem 27.1. *Let ρ be a nonhyperbolic quadratic form and φ a subform of ρ of dimension at least 2. Suppose that:*

(1) *φ and its complementary form in ρ are anisotropic.*
(2) *$\rho_{F(\varphi)}$ is hyperbolic.*
(3) *$2 \dim \varphi > N(\rho)$.*

Then ρ is an anisotropic general Pfister form.

PROOF. Note that ρ is a nondegenerate form of even dimension by Remark 7.18 as $\rho_{F(\varphi)}$ is hyperbolic.

Claim 1: For any field extension K/F with φ_K anisotropic and ρ_K not hyperbolic, φ_K is isometric to a subform of $(\rho_K)_{an}$.

By Lemma 8.13, the form $\rho \perp (-\varphi)$ is Witt-equivalent to $\psi := \varphi^{\perp}$. In particular, $\dim \rho = \dim \varphi + \dim \psi$. Set $\rho' = (\rho_K)_{an}$. It follows from (3) that

$$\dim\bigl(\rho' \perp (-\varphi_K)\bigr) \geq 2^{\deg \rho} + \dim \varphi > \dim \rho - \dim \varphi = \dim \psi.$$

As $\rho' \perp (-\varphi_K) \sim \psi_K$ it follows that the form $\rho' \perp (-\varphi_K)$ is isotropic, therefore $D(\rho') \cap D(\varphi_K) \neq \emptyset$. Since $\rho'_{K(\varphi)}$ is hyperbolic, the form φ_K is isometric to a subform of ρ' by the Subform Theorem 22.5 as needed.

Claim 2: ρ is anisotropic.

Applying Claim 1 to $K = F$ implies that φ is isometric to a subform of $\rho' = \rho_{an}$. Let ψ' be the complementary form of φ in ρ'. By Lemma 8.13,

$$\psi' \sim \rho' \perp (-\varphi) \sim \rho \perp (-\varphi) \sim \psi.$$

As both forms ψ and ψ' are anisotropic, we have $\psi' \simeq \psi$. Hence

$$\dim \rho = \dim \varphi + \dim \psi = \dim \varphi + \dim \psi' = \dim \rho' = \dim \rho_{an}.$$

Therefore, ρ is anisotropic.

We now investigate the form $\varphi_{F(\rho)}$. Suppose it is isotropic. Then $\varphi \prec\succ \rho$, hence $\rho_{F(\rho)}$ is hyperbolic by Proposition 22.18. It follows that ρ is a general Pfister form by Corollary 23.4 and we are done. Thus we may assume that $\varphi_{F(\rho)}$ is anisotropic. Normalizing we may also assume that $1 \in D(\varphi)$. We shall prove that ρ is a Pfister form by induction on $\dim \rho$. Suppose that ρ is not a Pfister form. In particular, $\rho_1 := (\rho_{F(\rho)})_{an}$ is nonzero and $\dim \rho_1 \geq 2$. We shall finish the proof by obtaining a contradiction. Let $\varphi_1 = \varphi_{F(\rho)}$.

Note that $\deg \rho_1 = \deg \rho$ and $\dim \rho_1 < \dim \rho$, hence $N(\rho_1) < N(\rho)$.

Claim 3: ρ_1 is a Pfister form.

Applying Claim 1 to the field $K = F(\rho)$, we see that φ_1 is isometric to a subform of ρ_1. We have

$$2 \dim \varphi_1 = 2 \dim \varphi > N(\rho) > N(\rho_1).$$

By the induction hypothesis applied to the form ρ_1 and its subform, φ_1, we conclude that the form ρ_1 is a Pfister form proving the claim. In particular, $\dim \rho_1 = 2^{\deg \rho_1} = 2^{\deg \rho}$.

Claim 4: $D(\rho) = G(\rho)$.

Since $G(\rho) \subset D(\rho)$, it suffices to show if $x \in D(\rho)$, then $x \in G(\rho)$. Suppose that $x \notin G(\rho)$. Hence the anisotropic part β of the isotropic form $\langle\langle x \rangle\rangle \otimes \rho$ is nonzero. It follows from Proposition 25.15 that $\deg \beta \geq 1 + \deg \rho$.

Suppose that $\beta_{F(\rho)}$ is hyperbolic. As $\rho - \beta = -x\rho$ in $I_q(F)$, the form $\rho \perp (-\beta)$ is isotropic, hence $D(\rho) \cap D(\beta) \neq \emptyset$. It follows from this that ρ is isometric to a subform of β by the Subform Theorem 22.5. Let $\beta \simeq \rho \perp \mu \sim \rho \perp (-x\rho)$ for some form μ. By Witt cancellation, $\mu \sim -x\rho$. But $\dim \beta < 2 \dim \rho$, hence $\dim \mu < \dim \rho$. As ρ is anisotropic, this is a contradiction. It follows that the form $\beta_1 = (\beta_{F(\rho)})_{an}$ is not zero and hence $\dim \beta_1 \geq 2^{\deg \beta} \geq 2^{1+\deg \rho}$.

Since ρ is hyperbolic over $F(\varphi)$, it follows from the Subform Theorem 22.5 that φ is isometric to a subform of $x\rho$. Applying Claim 1 to the form $x\rho_{F(\rho)}$, we conclude that φ_1 is a subform of $x\rho_1$. As φ_1 is also a subform of ρ_1, the form $\langle\langle x \rangle\rangle \otimes \rho_1$ contains $\varphi_1 \perp (-\varphi_1)$ and therefore a totally isotropic subspace of dimension $\dim \varphi_1 = \dim \varphi$. Therefore, $\dim\big(\langle\langle x \rangle\rangle \otimes \rho_1\big)_{an} \leq 2 \dim \rho_1 - 2 \dim \varphi$. Consequently,

$$2^{1+\deg \rho} \leq \dim \beta_1 = \dim\big(\langle\langle x \rangle\rangle \otimes \rho_1\big)_{an} \leq 2 \dim \rho_1 - 2 \dim \varphi < 2^{1+\deg \rho},$$

a contradiction. This proves the claim.

Let $F(T) = F(t_1, \dots, t_n)$ with $n = \dim \rho$. We have $\deg \rho_{F(T)} = \deg \rho$ and $N(\rho_{F(T)}) = N(\rho)$. Working over $F(T)$ instead of F, we have the forms $\varphi_{F(T)}$ and $\rho_{F(T)}$ satisfy the conditions of the theorem. By Claim 4, we conclude that $G(\rho_{F(T)}) = D(\rho_{F(T)})$. It follows from Theorem 23.2 that ρ is a Pfister form, a contradiction. □

Corollary 27.2. *Let ρ be a nonzero anisotropic quadratic form and φ an irreducible anisotropic quadratic form satisfying $\dim \varphi > \frac{1}{3} \dim \rho$. If $\rho_{F(\varphi)}$ is hyperbolic, then $\rho \in GP(F)$.*

PROOF. As $\rho_{F(\varphi)}$ is hyperbolic, the form ρ is nondegenerate. It follows by the Subform Theorem 22.5 that $a\varphi$ is a subform of ρ for some $a \in F^\times$. As ρ is anisotropic, the complementary form of $a\varphi$ in ρ is anisotropic.

Let K be the leading field of ρ and τ its leading form. We show that φ_K is anisotropic. If $\varphi_{F(\rho)}$ is isotropic, then $\varphi \prec\succ \rho$. In particular, $\rho_{F(\rho)}$ is hyperbolic by Proposition 22.18, hence $K = F$ and φ is anisotropic by hypothesis. Thus we may assume that $\varphi_{F(\rho)}$ is anisotropic. The assertion now follows by induction on $\mathfrak{h}(\rho)$. As $\tau_{K(\varphi)} \sim \rho_{K(\varphi)}$ is hyperbolic, $\dim\varphi = \dim\varphi_K \leq \dim\tau = 2^{\deg\rho}$ by the Subform Theorem 22.5. Hence $N(\rho) = \dim\rho - 2^{\deg\rho} \leq \dim\rho - \dim\varphi < 2\dim\varphi$. The result follows by Theorem 27.1. □

The restriction $\dim\varphi > \frac{1}{3}\dim\rho$ above is best possible (cf. [**85**]).

A further application of Theorem 27.1 is given by:

Theorem 27.3. *Let φ and ψ be nondegenerate quadratic forms over F of the same odd dimension. If $\mathfrak{i}_0(\varphi_K) = \mathfrak{i}_0(\psi_K)$ for any field extension K/F, then φ and ψ are similar.*

PROOF. We may assume that φ and ψ are anisotropic and have the same determinants (cf. Remark 13.8). Let $n = \dim\varphi$. We shall show that $\varphi \simeq \psi$ by induction on n. The statement is obvious if $n = 1$, so assume that $n > 1$.

We construct a nondegenerate form ρ of dimension $2n$ and trivial discriminant containing φ such that $\varphi^\perp \simeq -\psi$ as follows: If $\operatorname{char} F \neq 2$, let $\rho = \varphi \perp (-\psi)$. If $\operatorname{char} F = 2$, write $\varphi \simeq \langle a\rangle \perp \varphi'$ and $\psi \simeq \langle a\rangle \perp \psi'$ for some $a \in F^\times$ and nondegenerate forms φ' and ψ'. Set $\rho = [a, c] \perp \varphi' \perp \psi'$, where c is chosen so that $\operatorname{disc}\rho$ is trivial.

By induction applied to the anisotropic parts of $\varphi_{F(\varphi)}$ and $\psi_{F(\varphi)}$, we have $\varphi_{F(\varphi)} \simeq \psi_{F(\varphi)}$. It follows from Witt Cancellation and Proposition 13.6 (in the case $\operatorname{char} F = 2$) that $\rho_{F(\varphi)}$ is hyperbolic. If ρ itself is not hyperbolic, then by Theorem 27.1, the form ρ is an anisotropic general Pfister form of dimension $2n$. In particular, n is a power of 2, a contradiction.

Thus ρ is hyperbolic. By Lemma 8.13, we have $-\varphi \sim \rho \perp (-\varphi) \sim \varphi^\perp \simeq -\psi$. As φ and ψ have the same dimension we conclude that $\varphi \simeq \psi$. □

28. Excellent quadratic forms

In general, if φ is a nondegenerate quadratic form and K/F a field extension then the anisotropic part of φ_K will not necessarily be isometric to a form defined over F and extended to K. Those forms over a field F whose anisotropic part over any extension field of F is defined over F are called excellent forms. These forms were first introduced by Knebusch in [**84**]. We study them in this section.

Let K/F be a field extension and ψ a quadratic form over K. We say that ψ *is defined over* F if there is a quadratic form η over F such that $\psi \simeq \eta_K$.

Theorem 28.1. *Let φ be an anisotropic nondegenerate quadratic form of dimension ≥ 2. Then φ is a Pfister neighbor if and only if the quadratic form $(\varphi_{F(\varphi)})_{an}$ is defined over F.*

PROOF. Let φ be a Pfister neighbor and let ρ be the associated general Pfister form so φ is a subform of ρ. As $\varphi_{F(\varphi)}$ is isotropic, the general Pfister form $\rho_{F(\varphi)}$ is hyperbolic by Corollary 9.10. By Lemma 8.13, the form $\varphi_{F(\varphi)}$ is Witt-equivalent to

$-(\varphi^\perp)_{F(\varphi)}$. Since $\dim \varphi^\perp < (\dim \rho)/2$, it follows by Corollary 26.6 that $(\varphi^\perp)_{F(\rho)}$ is anisotropic. By Corollary 22.17, the form $(\varphi^\perp)_{F(\varphi)}$ is also anisotropic as $\varphi \prec\succ \rho$ by Remark 23.11. Consequently, $(\varphi_{F(\varphi)})_{an} \simeq (-\varphi^\perp)_{F(\varphi)}$ is defined over F.

Suppose now that $(\varphi_{F(\varphi)})_{an} \simeq \psi_{F(\varphi)}$ for some (anisotropic) form ψ over F. Note that $\dim \psi < \dim \varphi$.

Claim: There exists a form ρ satisfying:

(1) φ is a subform of ρ.
(2) The complementary form $\varphi^\perp$ is isometric to $-\psi$.
(3) $\rho_{F(\varphi)}$ is hyperbolic.

Moreover, if $\dim \varphi \geq 3$, then ρ can be chosen in $I_q^2(F)$.

Suppose that $\dim \varphi$ is even or $\operatorname{char} F \neq 2$. Then $\rho = \varphi \perp (-\psi)$ satisfies (1), (2), and (3). As F is algebraically closed in $F(\varphi)$, if $\dim \varphi \geq 3$, we have $\operatorname{disc} \varphi = \operatorname{disc} \psi$, hence $\rho \in I_q^2(F)$.

So we may assume that $\operatorname{char} F = 2$ and $\dim \varphi$ is odd. Write $\varphi = \varphi' \perp \langle a \rangle$ and $\psi = \psi' \perp \langle b \rangle$ for nondegenerate forms φ', ψ' and $a, b \in F^\times$. Note that $\langle a \rangle$ (respectively, $\langle b \rangle$) is the restriction of φ (respectively, ψ) on $\operatorname{rad} \mathfrak{b}_\varphi$ (respectively, $\operatorname{rad} \mathfrak{b}_\psi$) by Proposition 7.31. By definition of ψ we have $\langle a \rangle_{F(\varphi)} \simeq \langle b \rangle_{F(\varphi)}$. Since $F(\varphi)/F$ is a separable field extension by Lemma 22.14, we have $\langle a \rangle \simeq \langle b \rangle$. Therefore, we may assume that $b = a$.

Choose $c \in F$ such that $\operatorname{disc}(\varphi' \perp \psi') = \operatorname{disc}[a, c]$ and set $\rho = \varphi' \perp \psi' \perp [a, c]$ so that $\rho \in I_q^2(F)$. Clearly, φ is a subform of ρ and $\varphi^\perp$ is isometric to ψ. By Lemma 8.13, $\rho \perp \varphi \sim \psi$. Since φ and ψ are Witt-equivalent over $F(\varphi)$, we have $\rho_{F(\varphi)} \perp \varphi_{F(\varphi)} \sim \varphi_{F(\varphi)}$. Cancelling the nondegenerate form $\varphi'_{F(\varphi)}$ yields

$$\rho_{F(\varphi)} \perp \langle a \rangle_{F(\varphi)} \sim \langle a \rangle_{F(\varphi)}.$$

As $\rho \in I_q^2(F)$ by Proposition 13.6, we have $\rho_{F(\varphi)} \sim 0$ establishing the claim.

As $\dim \rho = \dim \varphi + \dim \psi < 2 \dim \varphi$ and φ is anisotropic, it follows that ρ is not hyperbolic. Moreover, φ and its complement $\varphi^\perp \simeq -\psi$ are anisotropic. Consequently, ρ is a general Pfister form by Theorem 27.1, hence φ is a Pfister neighbor. □

Exercise 28.2. Let φ be a nondegenerate quadratic form of odd dimension. Then $\mathfrak{h}(\varphi) = 1$ if and only if φ is a Pfister neighbor of dimension $2^n - 1$ for some $n \geq 1$.

Theorem 28.3. *Let φ be a nondegenerate quadratic form. Then the following two conditions are equivalent:*

(1) *For any field extension K/F, the form $(\varphi_K)_{an}$ is defined over F.*
(2) *There are anisotropic Pfister neighbors $\varphi_0 = \varphi_{an}, \varphi_1, \dots, \varphi_{r-1}$ with associated general Pfister forms $\rho_0, \rho_1, \dots, \rho_{r-1}$, respectively, satisfying $\varphi_i \simeq (\rho_i \perp \varphi_{i+1})_{an}$ for all $i \in [0, r-1]$ with a form φ_r of dimension at most one.*

PROOF. (2) ⇒ (1): Let K/F be a field extension. If all general Pfister forms ρ_i are hyperbolic over K, the isomorphisms in (2) show that all the φ_i are also hyperbolic. In particular, $(\varphi_K)_{an}$ is the zero form and hence is defined over F.

Let s be the smallest integer such that $(\rho_s)_K$ is not hyperbolic. Then the forms $\varphi = \varphi_0, \varphi_1, \dots, \varphi_s$ are Witt-equivalent and $(\varphi_s)_K$ is a Pfister neighbor of the anisotropic general Pfister form $(\rho_s)_K$. In particular, $(\varphi_s)_K$ is anisotropic and therefore $(\varphi_K)_{an} = (\varphi_s)_K$ is defined over F.

(1) $\Rightarrow$ (2): We prove the statement by induction on $\dim\varphi$. We may assume that $\dim\varphi_{an}\geq 2$. By Theorem 28.1, the form φ_{an} is a Pfister neighbor. Let ρ be the associated general Pfister form of φ_{an}. Consider the negative of the complementary form $\psi=-(\varphi_{an})^{\perp}$ of φ_{an} in ρ. It follows from Lemma 8.13 that $\varphi_{an}\simeq(\rho\perp\psi)_{an}$.

We claim that the form ψ satisfies (1). Let K/F be a field extension. If ρ is hyperbolic over K, then φ_K and ψ_K are Witt-equivalent. Therefore, $(\psi_K)_{an}\simeq(\varphi_K)_{an}$ is defined over F. If ρ_K is anisotropic, then so is ψ_K, therefore $(\psi_K)_{an}=\psi_K$ is defined over F. By the induction hypothesis applied to ψ, there are anisotropic Pfister neighbors $\varphi_1=\psi,\varphi_2,\dots,\varphi_{r-1}$ with the associated general Pfister forms $\rho_1,\rho_2,\dots,\rho_{r-1}$, respectively, such that $\varphi_i\simeq(\rho_i\perp\varphi_{i+1})_{an}$ for all $i\in[1,\ r-1]$, where φ_r is a form of dimension at most 1. To finish the proof let $\varphi_0=(\varphi)_{an}$ and $\rho_0=\rho$. □

A quadratic form φ satisfying equivalent conditions of Theorem 28.3 is called *excellent*. By Lemma 8.13, the form φ_{i+1} in Theorem 28.3(2) is isometric to the negative of the complement of φ_i in ρ_i. In particular, the sequences of forms φ_i and ρ_i are uniquely determined by φ up to isometry. Note that all forms φ_i are also excellent; this allows inductive proofs while working with excellent forms.

Example 28.4. If $\operatorname{char}F\neq 2$, then the form $n\langle 1\rangle$ is excellent for every $n>0$.

Proposition 28.5. *Let φ be an excellent quadratic form. Then in the notation of Theorem 28.3 we have the following:*

(1) *The integer r coincides with the height of φ.*
(2) *If $F_0=F,F_1,\dots,F_r$ is the generic splitting tower of φ, then $(\varphi_{F_i})_{an}\simeq(\varphi_i)_{F_i}$ for all $i\in[0,\ r]$.*

PROOF. The last statement is obvious if $i=0$. As ρ_0 is hyperbolic over $F_1=F(\varphi_{an})=F(\varphi_0)$, the forms φ_{F_1} and $(\varphi_1)_{F_1}$ are Witt-equivalent. Since $\dim\varphi_1<(\dim\rho_0)/2$, the form φ_1 is anisotropic over $F(\rho_0)$ by Corollary 26.6. As $\varphi_0\prec\succ\rho_0$, the form φ_1 is also anisotropic over $F_1=F(\varphi_0)$ by Corollary 22.17. Therefore, $(\varphi_{F_1})_{an}\simeq(\varphi_1)_{F_1}$. This proves the last statement for $i=1$. Both statements of the proposition follow now by induction on r. □

The notion of an excellent form was introduced by Knebusch.

29. Excellent field extensions

A field extension E/F is called *excellent* if the anisotropic part φ_E of any quadratic form φ over F is defined over F, i.e., there is a quadratic form ψ over F satisfying $(\varphi_E)_{an}\simeq\psi_E$. The concept of an excellent field extension was introduced in [**39**] and further information about this concept can be found there.

Example 29.1. Suppose that every anisotropic form over F remains anisotropic over E. Then for every quadratic form φ over F the form $(\varphi_{an})_E$ is anisotropic and therefore is isometric to the anisotropic part of φ_E. It follows that E/F is an excellent field extension. In particular, it follows from Lemma 7.15 and Springer's Theorem 18.5 that purely transcendental field extensions and odd degree field extensions are excellent.

Example 29.2. Let E/F be a separable quadratic field extension. Then $E=F(\sigma)$, where σ is the (nondegenerate) binary norm form of E/F. It follows from Corollary 22.12 that E/F is an excellent field extension.

Example 29.3. Let E/F be a field extension such that every quadratic form over E is defined over F. Then E/F is obviously an excellent extension.

Exercise 29.4. Let E be either an algebraic closure or a separable closure of a field F. Prove that every quadratic form over E is defined over F. In particular, E/F is an excellent extension.

Let ρ be an irreducible nondegenerate quadratic form over F. If $\dim \rho = 2$, the extension $F(\rho)/F$ is separable quadratic and therefore is excellent by Example 29.2. We extend this result to nondegenerate forms of dimension 3.

Notation 29.5. Until the end of this section, let K/F be a separable quadratic field extension and let $a \in F^{\times}$. Consider the 3-dimensional quadratic form $\rho = \mathrm{N}_{K/F} \perp \langle -a \rangle$ on the space $U := K \oplus F$. Let X be the projective quadric of ρ. It is a smooth *conic curve* in $\mathbb{P}(U)$. In the projective coordinates $[s : t]$ on $K \oplus F$, the conic X is given by the equation $\mathrm{N}_{K/F}(s) = at^2$. We write E for the field $F(\rho) = F(X)$.

The intersection of X with $\mathbb{P}(K)$ is $\operatorname{Spec} F(x)$ for a point $x \in X$ of degree 2 with $F(x) \simeq K$. In fact, $\operatorname{Spec} F(x)$ is the quadric of the form $\mathrm{N}_{K/F} = \rho|_K$. Over K the norm form $\mathrm{N}_{K/F}(s)$ factors into a product $s \cdot s'$ of linear forms. Therefore, there are two rational points y and y' of the curve X_K mapping to x under the natural morphism $X_K \to X$, so that the divisor $\operatorname{div}(s/t)$ equals $y - y'$ and $\operatorname{div}(s'/t)$ equals $y' - y$. Moreover, we have

$$\mathrm{N}_{KE/E}(s/t) = \mathrm{N}_{K/F}(s)/t^2 = at^2/t^2 = a. \tag{29.6}$$

For any $n \geq 0$, let L_n be the F-subspace

$$\{f \in E^{\times} \mid \operatorname{div}(f) + nx \geq 0\} \cup \{0\}$$

of E. We have

$$F = L_0 \subset L_1 \subset L_2 \subset \cdots \subset E$$

and $L_n \cdot L_m \subset L_{n+m}$ for all $n, m \geq 0$. In particular, the union L of all L_n is a subring of E. In fact, E is the quotient field of L.

In addition, $\mathcal{O}_{X,x} \cdot L_n \subset L_n$ and $\mathfrak{m}_{X,x} \cdot L_n \subset L_{n-1}$ for every $n \geq 1$ where $\mathcal{O}_{X,x}$ is the local ring of x and $\mathfrak{m}_{X,x}$ its maximal ideal. In particular, we have the structure of a K-vector space on L_n/L_{n-1} for every $n \geq 1$.

Set $\overline{L}_n = L_n/L_{n-1}$ for $n \geq 1$ and $\overline{L}_0 = K$. The graded group $\overline{L}_*$ has the structure of a K-algebra.

The following lemma is an easy case of the Riemann-Roch Theorem.

Lemma 29.7. *In the notation above, we have* $\dim_K(\overline{L}_n) = 1$ *for all* $n \geq 0$. *Moreover,* $\overline{L}_*$ *is a polynomial ring over* K *in one variable.*

PROOF. Let $f, g \in L_n \setminus L_{n-1}$ for $n \geq 1$. Since $f = (f/g)g$ and $f/g \in (O_{X,x})^{\times}$, the images of f and g in $\overline{L}_n$ are linearly dependent over K. Hence $\dim_K(\overline{L}_n) \leq 1$. On the other hand, for a nonzero linear form l on K, we have $\operatorname{div}(l/t) = z - x$ for some $z \neq x$. Hence $(l/t)^n \in L_n \setminus L_{n-1}$ and therefore $\dim_K(\overline{L}_n) \geq 1$. Moreover, $\overline{L}_* = K[l/t]$. □

Proposition 29.8. *Let* $\varphi : V \to F$ *be an anisotropic quadratic form and suppose that for some* $n \geq 1$ *there exists*

$$v \in (V \otimes L_n) \setminus (V \otimes L_{n-1})$$

such that $\varphi(v) = 0$. *Then there exists a subspace* $W \subset V$ *of dimension* 2 *satisfying:*

(1) $\varphi|_W$ *is similar to* $\mathrm{N}_{K/F}$.
(2) *There exists a nonzero* $\tilde{v} \in V \otimes L_{n-1}$ *such that* $\tilde{\varphi}(\tilde{v}) = 0$ *where* $\tilde{\varphi}$ *is the quadratic form* $a(\varphi|_W) \perp \varphi|_{W^\perp}$ *on* V.

PROOF. Let $\bar{v}$ denote the image of v under the canonical map $V \otimes L_n \to V \otimes \overline{L}_n$. We have $\bar{v} \neq 0$ as $v \notin V \otimes L_{n-1}$. Since $\overline{L}_n$ is 2-dimensional over F by Lemma 29.7, there is a subspace $W \subset V$ of dimension 2 with $\bar{v} \in W \otimes \overline{L}_n$.

As $\bar{v}$ is an isotropic vector in $W \otimes \overline{L}_*$ and $\overline{L}_*$ is a polynomial algebra over K, we have $W \otimes K$ is isotropic. It follows from Corollary 22.12 that the restriction $\varphi|_W$ is isometric to $c\,\mathrm{N}_{K/F}$ for some $c \in F^\times$ and, in particular, nondegenerate.

By Proposition 7.22, we can write $v = w + w'$ with $w \in W \otimes L_n$ and $w' \in W^\perp \otimes L_n$. By construction of W, we have $\bar{w}' = 0$ in $V \otimes \overline{L}_n$, i.e., $w' \in V \otimes L_{n-1}$, therefore $\varphi(w') \in L_{2n-2}$. Since $0 = \varphi(v) = \varphi(w) + \varphi(w')$, we must have $\varphi(w) \in L_{2n-2}$.

We may therefore assume that $W = K$ and $\varphi|_K = c\,\mathrm{N}_{K/F}$. Thus we have $w \in K \otimes L_n \subset K \otimes E = K(X)$. Considering w as a function on X_K we have $\mathrm{div}_\infty(w) = my + m'y'$ for some $m, m' \leq n$ where div_∞ is the divisor of poles. As $w \notin W \otimes L_{n-1}$, we must have one of the numbers m and m', say m, equal to n.

Let σ be the generator of the Galois group of K/F. We have $\sigma(y) = y'$, hence $\mathrm{div}_\infty(\sigma w) = my' + m'y$ and

$$\mathrm{div}_\infty\big(\varphi(w)\big) = \mathrm{div}_\infty\big(\mathrm{N}_{K/F}(w)\big) = \mathrm{div}_\infty(w) + \mathrm{div}_\infty(\sigma w) = (m + m')(y + y').$$

As $\varphi(w) \in L_{2n-2}$, we have $m + m' \leq 2n - 2$, i.e., $m' \leq n - 2$.

Note also that

$$\mathrm{div}_\infty(ws/t) = \mathrm{div}_\infty(w) + y - y' = (m-1)y + (m'+1)y'.$$

As both $m - 1$ and $m' + 1$ are at most $n - 1$, we have $ws/t \in K \otimes L_{n-1}$.

Now let $\tilde{\varphi}$ be the quadratic form $a(\varphi|_W) \perp \varphi|_{W\perp}$ on $V = W \oplus W^\perp$ and set $\tilde{v} = a^{-1}ws/t + w' \in V \otimes L_{n-1}$. By (29.6) we have

$$\begin{aligned}\tilde{\varphi}(\tilde{v}) &= a\varphi(a^{-1}ws/t) + \varphi(w') \\ &= a^{-1}\mathrm{N}_{K(X)/F(X)}(s/t)\varphi(w) + \varphi(w') = \varphi(w) + \varphi(w') = 0.\end{aligned}$$

□

Corollary 29.9. *Let* φ *be a quadratic form over* F *such that* φ_E *is isotropic. Then there exists an isotropic quadratic form* ψ *over* F *such that* $\psi_E \simeq \varphi_E$.

PROOF. Let $v \in V \otimes E$ be an isotropic vector of φ_E. Scaling v we may assume that $v \in V \otimes L$. Choose the smallest n such that $v \in V \otimes L_n$. We induct on n. If $n = 0$, i.e., $v \in V$, the form φ is isotropic and we can take $\psi = \varphi$.

Suppose that $n \geq 1$. By Proposition 29.8, there exist a 2-dimensional subspace $W \subset V$ such that $\varphi|_W$ is similar to $N_{K/F}$ and an isotropic vector $\tilde{v} \in V \otimes L_{n-1}$ for the quadratic form $\tilde{\varphi} = a(\varphi|_W) \perp (\varphi|_{W^\perp})$ on V. As a is the norm in the quadratic extension KE/E, the forms $N_{K/F}$ and $aN_{K/F}$ are isometric over E, hence $\tilde{\varphi}_E \simeq \varphi_E$. By the induction hypothesis applied to the form $\tilde{\varphi}$, there is an isotropic quadratic form ψ over F such that $\psi_E \simeq \tilde{\varphi}_E \simeq \varphi_E$. □

Theorem 29.10. *Let* ρ *be a nondegenerate* 3*-dimensional quadratic form over* F. *Then the field extension* $F(\rho)/F$ *is excellent.*

PROOF. We may assume ρ is the form in Notation 29.5 as every nondegenerate 3-dimensional quadratic form over F is similar to such a form. Let $E = F(\rho)$ and let φ be a quadratic form over F. We show by induction on $\dim \varphi_{an}$ that $(\varphi_E)_{an}$ is defined over F. If φ_{an} is anisotropic over E we are done as $(\varphi_E)_{an} \simeq (\varphi_{an})_E$.

Suppose that φ_{an} is isotropic over E. By Corollary 29.9 applied to φ_{an}, there exists an isotropic quadratic form ψ over F such that $\psi_E \simeq (\varphi_{an})_E$. As $\dim \psi_{an} < \dim \varphi_{an}$, by the induction hypothesis, there is a quadratic form μ over F such that $(\psi_E)_{an} \simeq \mu_E$. Since $\mu_E \sim \psi_E \sim \varphi_E$, we have $(\varphi_E)_{an} \simeq \mu_E$. □

Corollary 29.11. *Let $\varphi \in GP_2(F)$. Then $F(\varphi)/F$ is excellent.*

PROOF. Let ψ be a Pfister neighbor of φ of dimension three. Let $K = F(\varphi)$ and $L = F(\psi)$. By Remark 23.11 and Proposition 22.9, the field extensions KL/K and KL/L are purely transcendental. Let ν be a quadratic form over F. By Theorem 29.10, there exists a quadratic form σ over F such that $(\nu_L)_{an} \simeq \sigma_L$. Hence

$$\big((\nu_K)_{an}\big)_{KL} \simeq (\nu_{KL})_{an} \simeq \big((\nu_L)_{an}\big)_{KL} \simeq \sigma_{KL}.$$

It follows that $(\nu_K)_{an} \simeq \sigma_K$. □

The corollary above was first proven by Arason (cf. [**39**, Appendix II]). We have followed here Rost's approach in [**116**]. This result does not generalize to higher fold Pfister forms. It is known, in general, for every $n > 2$, there exists a field F and a $\varphi \in GP_n(F)$ with $F(\varphi)/F$ not an excellent extension (cf. [**61**]).

30. Central simple algebras over function fields of quadratic forms

Let D be a finite dimensional division algebra over a field F. Let $D[t]$ denote the $F[t]$-algebra $D \otimes_F F[t]$ and $D(t)$ denote the $F(t)$-algebra $D \otimes_F F(t)$. As $D(t)$ has no zero divisors and is finite dimensional over $F(t)$, it is a division algebra. The main result in this section is Theorem 30.5. This was originally proved in [**102**] for fields of characteristic different from 2 using Swan's calculation of Grothendieck group of a smooth projective quadric. We generalize the elementary proof for this case given by Tignol in [**131**].

A subring $A \subset D(t)$ is called an *order* over $F[t]$ if it is a finitely generated $F[t]$-submodule of $D(t)$.

Lemma 30.1. *Let D be a finite dimensional division F-algebra. Then every order $A \subset D(t)$ over $F[t]$ is conjugate to a subring of $D[t]$.*

PROOF. As A is finitely generated as an $F[t]$-module, there is a nonzero $f \in F[t]$ such that $Af \subset D[t]$. The subset DAf of $D[t]$ is a left ideal. The ring $D[t]$ admits both the left and the right Euclidean algorithm relative to degree. In follows that all one-sided ideals in $D[t]$ are principal. In particular, $DAf = D[t]x$ for some $x \in D[t]$. As A is a ring, for every $y \in A$, we have

$$xy \in D[t]xy = DAfy \subset DAf = D[t]x,$$

hence $xyx^{-1} \in D[t]$. Thus $xAx^{-1} \subset D[t]$. □

Lemma 30.2. *Suppose that R is a commutative ring and S a (not necessarily commutative) R-algebra. Let $X \subset S$ be an R-submodule generated by n elements. Suppose that every $x \in X$ satisfies the equation $x^2 + ax + b = 0$ for some $a, b \in R$. Then the R-subalgebra of S generated by X can be generated as an R-module by 2^n elements.*

PROOF. Let $x_1, \dots, x_n$ be generators for the R-module X. Writing quadratic equations for every pair of generators x_i, x_j and $x_i + x_j$, we see that $x_i x_j + x_j x_i + a x_i + b x_j + c = 0$ for some $a, b, c \in R$. Therefore, the R-subalgebra of S generated by X is generated as an R-module by the monomials $x_{i_1} x_{i_2} \dots x_{i_k}$ with $i_1 < i_2 < \dots < i_k$. □

Let φ be a quadratic form on V over F and $v_0 \in V$ a vector such that $\varphi(v_0) = 1$. For every $v \in V$, the element $-vv_0$ in the even Clifford algebra $C_0(\varphi)$ satisfies the quadratic equation

$$(-vv_0)^2 + b_\varphi(v_0, v)(-vv_0) + \varphi(v) = 0. \tag{30.3}$$

Choose a subspace $U \subset V$ such that $V = Fv_0 \oplus U$. Let J be the ideal of the tensor algebra $T(U)$ generated by the elements $v \otimes v + b_\varphi(v_0, v)v + \varphi(v)$ for all $v \in U$.

Lemma 30.4. *With U as above, the F-algebra homomorphism $\alpha : T(U)/J \to C_0(\varphi)$ defined by $\alpha(v + J) = -vv_0$ is an isomorphism.*

PROOF. By Lemma 30.2, we have $\dim T(U)/J \le 2^{\dim U} = \dim C_0(\varphi)$. As α is surjective, it must be an isomorphism. □

Theorem 30.5. *Let D be a finite dimensional division F-algebra and φ an irreducible quadratic form over F. Then $D_{F(\varphi)}$ is not a division algebra if and only if there is an F-algebra homomorphism $C_0(\varphi) \to D$.*

PROOF. Scaling φ, we may assume that there is $v_0 \in V$ satisfying $\varphi(v_0) = 1$ where $V = V_\varphi$. We use the decomposition $V = Fv_0 \oplus U$ as above and set

$$l(v) = b_\varphi(v_0, v) \quad \text{for every} \quad v \in U.$$

Claim: Suppose that $D_{F(\varphi)}$ is not a division algebra. Then there is an F-linear map $f : U \to D$ satisfying the equality of quadratic maps

$$f^2 + lf + \varphi = 0. \tag{30.6}$$

(We view the left hand side as the quadratic map $v \mapsto f(v)^2 + l(v)f(v) + \varphi(v)$ on U).

If we establish the claim, then the map f extends to an F-algebra homomorphism $T(U)/J \to D$ and, by Lemma 30.4, we get an F-algebra homomorphism $C_0(\varphi) \to D$ as needed.

We prove the claim by induction on $\dim U$. Suppose that $\dim U = 1$, i.e., $U = Fv$ for some v. By Example 22.2, we have $F(\varphi) \simeq C_0(\varphi) = F \oplus Fx$ with x satisfying the quadratic equation $x^2 + ax + b = 0$ with $a = l(v)$ and $b = \varphi(v)$ by equation (30.3). Since $D_{F(\varphi)}$ is not a division algebra, there exists a nonzero element $d' + dx \in D_{F(\varphi)}$ with $d, d' \in D$ such that $(d' + dx)^2 = 0$ or equivalently $d'^2 = bd^2$ and $dd' + d'd = ad^2$. Since D is a division algebra, we have $d \ne 0$. Then the element $d'd^{-1}$ in D satisfies

$$(d'd^{-1})^2 - a(d'd^{-1}) + b = 0.$$

Therefore, the assignment $v \mapsto -d'd^{-1}$ gives rise to the desired map $f : U \to D$.

Now consider the general case $\dim U \ge 2$. Choose a decomposition

$$U = Fv_1 \oplus Fv_2 \oplus W$$

for some nonzero $v_1, v_2 \in U$ and a subspace $W \subset U$ and set $V' = Fv_0 \oplus Fv_1 \oplus W$, $U' = Fv_1 \oplus W$ so that $V' = Fv_0 \oplus U'$. Consider the quadratic form φ' on the vector space $V'_{F(t)}$ over the function field $F(t)$ defined by

$$\varphi'(av_0 + bv_1 + w) = \varphi(av_0 + bv_1 + btv_2 + w).$$

We show that the function fields $F(\varphi)$ and $F(t)(\varphi')$ are isomorphic over F. Indeed, consider the injective F-linear map $\theta : V^* \to {V'}^*_{F(t)}$ taking a linear functional z to the functional z' defined by $z'(av_0 + bv_1 + w) = z(av_0 + bv_1 + btv_2 + w)$. The map θ identifies the ring $S^\bullet(V^*)$ with a graded subring of $S^\bullet({V'}^*_{F(t)})$ so that φ is identified with φ'. Let x_1 and x_2 be the coordinate functions of v_1 and v_2 in V, respectively, and x'_1 the coordinate function of v_1 in V'. We have $x_1 = x'_1$ and $x_2 = tx'_1$ in $S^1({V'}^*_{F(t)})$. Therefore, the localization of the ring $S^\bullet(V^*)$ with respect to the multiplicative system $F[x_1, x_2] \setminus \{0\}$ coincides with the localization of $S^\bullet({V'}^*_{F(t)})$ with respect to $F(t)[x'_1] \setminus \{0\}$. Note that $F[x_1, x_2] \cap (\varphi) = 0$ and $F(t)[x'_1] \cap (\varphi') = 0$. It follows that the localizations $S^\bullet(V^*)_{(\varphi)}$ and $S^\bullet({V'}^*_{F(t)})_{(\varphi')}$ are equal. As the function fields $F(\varphi)$ and $F(t)(\varphi')$ coincide with the degree 0 components of the quotient fields of their respective localizations, the assertion follows.

Let $l'(v) = b'_\varphi(v_0, v)$, so

$$l'(av_0 + bv_1 + w) = l(av_0 + bv_1 + btv_2 + w).$$

Applying the induction hypothesis to the quadratic form φ' over $F(t)$ and the $F(t)$-algebra $D_{F(t)}$ produces an $F(t)$-linear map $f' : U'_{F(t)} \to D_{F(t)}$ satisfying

$$f'^2 + l'f' + \varphi' = 0. \tag{30.7}$$

Consider the $F[t]$-submodule $X = f'(U'_{F[t]})$ in $D_{F[t]}$. By Lemma 30.2, the $F[t]$-subalgebra generated by X is a finitely generated $F[t]$-module. It follows from Lemma 30.1 that, after applying an inner automorphism of $D_{F(t)}$, we have $f'(v) \in D_{F[t]}$ for all v. Considering the highest degree terms of f' (with respect to t) and taking into account the fact that D is a division algebra, we see that $\deg f' \le 1$, i.e., $f' = g + ht$ for two linear maps $g, h : U' \to D$. Comparison of degree 2 terms of (30.7) yields

$$h(v)^2 + bl(v_2)h(v) + b^2\varphi(v_2) = 0$$

for all $v = bv_1 + w$. In particular, h is zero on W, therefore $h(v) = bh(v_1)$. Thus (30.7) reads

$$\big(g(v) + bth(v_1)\big)^2 + l(bv_1 + btv_2)\big(g(v) + bth(v_1)\big) + \varphi(v + btv_2) = 0 \tag{30.8}$$

for every $v = bv_1 + w$. Let $f : U \to D$ be the F-linear map defined by the formula

$$f(bv_1 + cv_2 + w) = g(bv_1 + w) + ch(v_1).$$

Substituting c/b for t in (30.8), we see that (30.6) holds on all vectors $bv_1 + cv_2 + w$ with $b \ne 0$ and therefore holds as an equality of quadratic maps. The claim is proven.

We now prove the converse. Suppose that there is an F-algebra homomorphism $s : C_0(\varphi) \to D$. Consider the two F-linear maps $p, q : V \to D$ given by $p(v) = s(vv_0)$ and $q(v) = s\big(vv_0 - l(v)\big)$. We have

$$p(v)q(v) = s\big((vv_0)^2 - l(v)vv_0\big) = s\big(\varphi(v)\big) = \varphi(v)$$

by equation (30.3). It follows that p and q are injective maps if φ is anisotropic. The maps p and q stay injective over any field extension. Let L/F be a field extension such that φ_L is isotropic (e.g., $L = F(\varphi)$). Then if $v' \in V_L$ is a nonzero isotropic vector, we have $p(v')q(v') = \varphi(v') = 0$, but $p(v') \neq 0$ and $q(v') \neq 0$. It follows that D_L is not a division algebra.

It remains to consider the case when φ is isotropic. We first show that every isotropic vector $v \in V$ belongs to $\operatorname{rad} b_\varphi$. Suppose this is not true. Then there is a $u \in V$ satisfying $b_\varphi(v, u) \neq 0$. Let H be the 2-dimensional subspace generated by v and u. The restriction of φ on H is a hyperbolic plane. Let $w \in V$ be a nonzero vector orthogonal to H and let $a = \varphi(w)$. Then

$$\mathbb{M}_2(F) = C(-aH) = C_0(Fw \perp H) \subset C_0(\varphi)$$

by Proposition 11.4. The image of the matrix algebra $\mathbb{M}_2(F)$ under s is isomorphic to $\mathbb{M}_2(F)$ and therefore contains zero divisors, a contradiction proving the assertion.

Let V' be a subspace of V satisfying $V = \operatorname{rad}\varphi \oplus V'$. As every isotropic vector belongs to $\operatorname{rad} b_\varphi$, the restriction φ' of φ on V' is anisotropic. The natural map $C_0(\varphi) \to C_0(\varphi')$ induces an isomorphism $C_0(\varphi)/J \xrightarrow{\sim} C_0(\varphi')$, where $J = (\operatorname{rad}\varphi)C_1(\varphi)$. Since $x^2 = 0$ for every $x \in \operatorname{rad}\varphi$, we have $s(J) = 0$. Therefore, s induces an F-algebra homomorphism $s' : C_0(\varphi') \to D$. By the anisotropic case, D is not a division algebra over $F(\varphi')$. Since $F(\varphi)$ is a field extension of $F(\varphi')$, the algebra $D_{F(\varphi)}$ is also not a division algebra. □

Corollary 30.9. *Let D be a division F-algebra of dimension less than 2^{2n} and φ a nondegenerate quadratic form of dimension at least $2n+1$ over F. Then $D_{F(\varphi)}$ is also a division algebra.*

PROOF. Let ψ be a nondegenerate subform of φ of dimension $2n+1$. As $F(\psi)(\varphi)/F(\psi)$ is a purely transcendental extension by Proposition 22.9, we may replace φ by ψ and assume that $\dim\varphi = 2n+1$. By Proposition 11.6, the algebra $C_0(\varphi)$ is simple of dimension 2^{2n}. If $D_{F(\varphi)}$ is not a division algebra, then there is an F-algebra homomorphism $C_0(\varphi) \to D$ by Theorem 30.5. This homomorphism must be injective as $C_0(\varphi)$ is simple. But this is impossible by dimension count. □

Corollary 30.10. *Let D be a division F-algebra and let φ be a nondegenerate quadratic form over F satisfying:*

(1) *If $\dim\varphi$ is odd or $\varphi \in I_q(F) \setminus I_q^2(F)$, then $C_0(\varphi)$ is not a division algebra.*
(2) *If $\varphi \in I_q^2(F)$, then $C^+(\varphi)$ is not a division algebra over F* (cf. Remark 13.9).

Then $D_{F(\varphi)}$ is a division algebra.

PROOF. If $D_{F(\varphi)}$ is not a division algebra, there is an F-algebra homomorphism $f : C_0(\varphi) \to D$ by Theorem 30.5. If $\varphi \in I_q^2(F)$ we have $C_0(\varphi) \simeq C^+(\varphi) \times C^+(\varphi)$ by Remark 13.9. Thus in every case the image of f lies in a nondivision subalgebra of D. Therefore, D is not a division algebra, a contradiction. □

Corollary 30.11. *Let D be a division F-algebra and let $\varphi \in I_q^3(F)$ be a nonzero quadratic form. Then $D_{F(\varphi)}$ is a division algebra.*

PROOF. By Theorem 14.3, the Clifford algebra $C(\varphi)$ is split. In particular, $C^+(\varphi)$ is not a division algebra. The statement follows now from Corollary 30.10. □

CHAPTER V

Bilinear and Quadratic Forms and Algebraic Extensions

31. Structure of the Witt ring

In this section, we investigate the structure of the Witt ring of nondegenerate symmetric bilinear forms. For fields F whose level $s(F)$ is finite, i.e., nonformally real fields, the ring structure is quite simple. The Witt ring of such a field has a unique prime ideal, viz., the fundamental ideal and $W(F)$ (as an abelian group) has exponent $2s(F)$. As $s(F) = 2^n$ for some nonnegative integer this means that the Witt ring is 2-primary torsion. The case of formally real fields F, i.e., fields of infinite level, is more involved. Orderings on such a field give rise to prime ideals in $W(F)$. The torsion in $W(F)$ is still 2-primary, but this is not as easy to show. Therefore, we do the two cases separately.

31.A. Nonformally real fields. We consider the case of nonformally real fields first.

A field F is called *quadratically closed* if $F = F^2$. For example, algebraically closed fields are quadratically closed. A field of characteristic 2 is quadratically closed if and only if it is perfect. Over a quadratically closed field, the structure of the Witt ring is very simple. Indeed, we have

Lemma 31.1. *For a field F the following are equivalent:*

(1) *F is quadratically closed.*
(2) *$W(F) = \mathbb{Z}/2\mathbb{Z}$.*
(3) *$I(F) = 0$.*

PROOF. As $W(F)/I(F) = \mathbb{Z}/2\mathbb{Z}$, we have $W(F) \simeq \mathbb{Z}/2\mathbb{Z}$ if and only if $I(F) = 0$ if and only if $\langle 1, -a\rangle = 0$ in $W(F)$ for all $a \in F^\times$ if and only if $a \in F^{\times 2}$ for all $a \in F^\times$. □

Example 31.2. Let F be a finite field with $\operatorname{char} F = p$ and $|F| = q$. If $p = 2$, then $F = F^2$ and F is quadratically closed. So suppose that $p > 2$. Then $F^{\times 2} \simeq F^\times/\{\pm 1\}$ so $|F^2| = \frac{1}{2}(q+1)$. Let $F^\times/F^{\times 2} = \{F^{\times 2}, aF^{\times 2}\}$. Since the finite sets F^2 and $\{a - y^2 \mid y \in F\}$ both have $\frac{1}{2}(q+1)$ elements, they intersect nontrivially. It follows that every element in F is a sum of two squares. We have $-1 \in F^{\times 2}$ if and only if $q \equiv 1 \mod 4$.

If $q \equiv 3 \mod 4$, then $-1 \notin F^{\times 2}$ and the level $s(F)$ of F is 2. We may assume that $a = -1$. Then $\langle 1,1,1\rangle = \langle 1,-1,-1\rangle = \langle -1\rangle$ in $W(F)$ so $W(F)$ equals $\{0, \langle 1\rangle, \langle -1\rangle, \langle 1,1\rangle\}$ and is isomorphic to the ring $\mathbb{Z}/4\mathbb{Z}$.

If $q \equiv 1 \mod 4$, then -1 is a square and $W(F)$ equals $\{0, \langle 1\rangle, \langle a\rangle, \langle 1,a\rangle\}$ and is isomorphic to the group ring $\mathbb{Z}/2\mathbb{Z}[F^\times/F^{\times 2}]$.

It follows from Example 31.2 that $s(F) = 1$ or 2 for any field F of positive characteristic. In general, if F is not formally real, $s(F) = 2^n$ by Corollary 6.8. There exist fields of level 2^m for each $m \geq 1$.

Lemma 31.3. *Let $2^m \leq n < 2^{m+1}$. Suppose that F satisfies $s(F) > 2^m$, e.g., F is formally real, and $\varphi = (n+1)\langle 1 \rangle_q$. Then $s\big(F(\varphi)\big) = 2^m$.*

PROOF. As $s(F) > 1$, the characteristic of F is not 2. Since $\varphi_{F(\varphi)}$ is isotropic, it follows that $s\big(F(\varphi)\big) \leq 2^m$ by Corollary 6.8. If φ was isotropic over F, then $s(F) = s\big(F(\varphi)\big) \leq 2^m$ as $F(\varphi)/F$ is purely transcendental by Proposition 22.9. This contradicts the hypothesis. So φ is anisotropic. If $s\big(F(\varphi)\big) < 2^m$, then $\big(2^m\langle 1 \rangle\big)_{F(\varphi)}$ is a Pfister form as $\operatorname{char} F \neq 2$, hence is hyperbolic. It follows that $2^m = \dim 2^m\langle 1 \rangle \geq \dim \varphi > 2^m$ by the Subform Theorem 22.5, a contradiction. □

The ring structure of $W(F)$ due to Pfister (cf. [**110**]) is given by the following:

Proposition 31.4. *Let F be nonformally real with $s(F) = 2^n$. Then:*

(1) $\operatorname{Spec}\big(W(F)\big) = \{I(F)\}$.
(2) *$W(F)$ is a local ring of Krull dimension zero with maximal ideal $I(F)$.*
(3) $\operatorname{nil}\big(W(F)\big) = \operatorname{rad}\big(W(F)\big) = \operatorname{zd}(F) = I(F)$.
(4) $W(F)^\times = \{\mathfrak{b} \mid \dim \mathfrak{b}$ *is odd*$\}$.
(5) *$W(F)$ is connected, i.e., 0 and 1 are the only idempotents in $W(F)$.*
(6) *$W(F)$ is a 2-primary torsion group of exponent $2s(F)$.*
(7) *$W(F)$ is artinian if and only if it is noetherian if and only if $|F^\times/F^{\times 2}|$ is finite if and only if $W(F)$ is a finite ring.*

PROOF. Let $s = s(F)$. The integer $2s$ is the smallest integer such that the bilinear Pfister form $2s\langle 1 \rangle_b$ is metabolic, hence zero in the Witt ring. Therefore, $2^{n+1}\langle a \rangle = 0$ in $W(F)$ for every $a \in F^\times$. It follows that $W(F)$ is 2-primary torsion of exponent 2^{n+1}, i.e., (6) holds. As

$$\langle\langle a \rangle\rangle^{n+2} = \langle\langle a, \ldots, a \rangle\rangle = \langle\langle a, -1, \ldots, -1 \rangle\rangle = 2^{n+1}\langle\langle a \rangle\rangle = 0$$

in $W(F)$ for every $a \in F^\times$ by Example 4.16, we have $I(F)$ lies in every prime ideal. Since $W(F)/I(F) \simeq \mathbb{Z}/2\mathbb{Z}$, the fundamental ideal $I(F)$ is maximal, hence is the only prime ideal. This proves (1). As $I(F)$ is the only prime ideal (2) – (5) follows easily.

Finally, we show (7). Suppose that $W(F)$ is noetherian. Then $I(F)$ is a finitely generated $W(F)$-module so $I(F)/I^2(F)$ is a finitely generated $W(F)/I(F)$-module. As $F^\times/F^{\times 2} \simeq I(F)/I^2(F)$ by Proposition 4.13 and $\mathbb{Z}/2\mathbb{Z} \simeq W(F)/I(F)$, we have $F^\times/F^{\times 2}$ is finite. Conversely, suppose that $F^\times/F^{\times 2}$ is finite. By (2.6), we have a ring epimorphism $\mathbb{Z}[F^\times/F^{\times 2}] \to W(F)$. As the group ring $\mathbb{Z}[F^\times/F^{\times 2}]$ is noetherian, $W(F)$ is noetherian. As $2sW(F) = 0$ and $W(F)$ is generated by the classes of 1-dimensional forms, we see that $|W(F)| \leq |F^\times/F^{\times 2}|^{2s}$. Statement (7) now follows easily. □

We turn to formally real fields, i.e., those fields of infinite level. In particular, formally real fields are of characteristic 0, so the theories of symmetric bilinear forms and quadratic forms merge. The structure of the Witt ring of a formally real field is more complicated as well as more interesting. We shall use the basic algebraic and topological structure of formally real fields which can be found in Appendices §95 and §96. Recall that a formally field F is called *euclidean* if every element in $F^\times$ is a square or minus a square. So F is euclidean if and only if F is formally real and $F^\times/F^{\times 2} = \{F^{\times 2}, -F^{\times 2}\}$. In particular, every real-closed field (cf. §95) is euclidean. Sylvester's Law of Inertia for real-closed fields generalizes to euclidean fields.

Proposition 31.5 (Sylvester's Law of Inertia). *Let F be a field. Then the following are equivalent:*

(1) *F is euclidean.*
(2) *F is formally real and if $\mathfrak{b}$ is a nondegenerate symmetric bilinear form, there exists unique nonnegative integers m, n such that $\mathfrak{b} \simeq m\langle 1\rangle \perp n\langle -1\rangle$.*
(3) *$W(F) \simeq \mathbb{Z}$ as rings.*
(4) *F^2 is an ordering of F.*

PROOF. (1) $\Rightarrow$ (2): As F is formally real, $\operatorname{char} F = 0$ so every bilinear form is diagonalizable. Since $F^\times/F^{\times 2} = \{F^{\times 2}, -F^{\times 2}\}$, every nondegenerate bilinear form is isometric to $m\langle 1\rangle \perp n\langle -1\rangle$ for some nonnegative integers n and m. The integers n and m are unique by Witt Cancellation 1.28.

(2) $\Rightarrow$ (3): By (2) every anisotropic quadratic form is isometric to $r\langle 1\rangle$ for some unique integer r. As F is formally real every such form is anisotropic.

(3) $\Rightarrow$ (4): Let $\operatorname{sgn} : W(F) \to \mathbb{Z}$ be the isomorphism. Then $\operatorname{sgn}\langle 1\rangle = 1$ so $\langle 1\rangle$ has infinite order, hence F is formally real. Let $a \in F$. Then $\operatorname{sgn}\langle a\rangle = n$ for some integer n. Thus $\langle a\rangle = n\langle 1\rangle$ in $W(F)$. In particular, n is odd. Taking determinants, we must have $aF^{\times 2} = \pm F^{\times 2}$. It follows that $F^\times/F^{\times 2} = \{F^{\times 2}, -F^{\times 2}\}$. As F is formally real, $F^2 + F^2 \subset F^2$, hence F^2 is an ordering.

(4) $\Rightarrow$ (1): As F has an ordering, it is formally real. As F^2 is an ordering, $F = F^2 \cup (-F^2)$ with $-1 \notin F^2$, so F is euclidean. □

31.B. Pythagorean fields. We turn to a class of fields important in the study of formally real fields that generalize quadratically closed fields.

Let F be a euclidean field. If $\mathfrak{b}$ is a nondegenerate symmetric bilinear form, then $\mathfrak{b} \simeq m\langle 1\rangle \perp n\langle -1\rangle$ for unique nonnegative integers n and m. The integer $m - n$ is called the *signature* of $\mathfrak{b}$ and denoted $\operatorname{sgn}\mathfrak{b}$. This induces an isomorphism $\operatorname{sgn} : W(F) \to \mathbb{Z}$ taking the Witt class of $\mathfrak{b}$ to $\operatorname{sgn}\mathfrak{b}$ called the *signature map.*

Let F be a field. Set

$$D(\infty\langle 1\rangle) := \bigcup_n D(n\langle 1\rangle) = \{x \mid x \text{ is a nonzero sum of squares in } F\},$$

$$\widetilde{D}(\infty\langle 1\rangle) := D(\infty\langle 1\rangle) \cup \{0\}.$$

A field F of characteristic different from 2 is called a *pythagorean* field if every sum of squares of elements in F is itself a square, i.e., $\widetilde{D}(\infty\langle 1\rangle) = F^2$. A field F of characteristic 2 is called *pythagorean* if F is quadratically closed, i.e., perfect.

Example 31.6. Let F be a field.

(1) Every euclidean field is pythagorean.

(2) Let F be a field of characteristic different from 2 and $K = F((t))$, a *Laurent series field* over F. Then K is the quotient field of $F[[t]]$, a complete discrete valuation ring. If F is formally real, then so is K as $n\langle 1\rangle$ is anisotropic over K for all n by Lemma 19.4. Suppose that F is formally real and pythagorean. If $x_i \in K^\times$ for $i = 1, 2$, then there exist integers m_i such that $x_i = t^{m_i}(a_i + ty_i)$ with $a_i \in F^\times$ and $y_i \in F[[t]]$ for $i = 1, 2$. Suppose that $m_1 \le m_2$, then $x_1^2 + x_2^2 = t^{2m_1}(c + tz)$

with $z \in F[[t]]$ and $c = a_1^2$ if $m_1 < m_2$ and $c = a_1^2 + a_2^2$ if $m_1 = m_2$, hence c is a square in F in either case. As K is formally real, $c \neq 0$ in either case. Hence $c + tz$ is a square in K by Hensel's Lemma. It follows that K is also pythagorean. In particular, the finitely iterated Laurent series field $F_n = F((t_1)) \cdots ((t_n))$ as well as the infinitely iterated Laurent series field $F_\infty = \operatorname{colim} F_n = F((t_1)) \cdots ((t_n)) \cdots$ are formally real and pythagorean if F is.

(3) If F is not formally real and $\operatorname{char} F \neq 2$, then $s = s(F)$ is finite so the symmetric bilinear form $(s+1)\langle 1 \rangle$ is isotropic, hence universal by Corollary 1.25. Therefore, $F = \widetilde{D}\big(\infty\langle 1 \rangle\big)$. It follows that if F is not formally real, then F is pythagorean if and only if it is quadratically closed. In particular, if F is not formally real, then the field $F((t))$ is not pythagorean as t is not a square.

Exercise 31.7. Let F be a formally real pythagorean field and let $\mathfrak{b}$ be a bilinear form over F. Prove that the set $D(\mathfrak{b})$ is closed under addition.

Proposition 31.8. *Let F be a field. Then the following are equivalent:*

(1) *F is pythagorean.*
(2) *$I(F)$ is torsion-free.*
(3) *There are no anisotropic torsion binary bilinear forms over F.*

PROOF. (1) $\Rightarrow$ (2): If $s(F)$ is finite, then F is quadratically closed so $W(F) = \{0, \langle 1 \rangle\}$ and $I(F) = 0$. Therefore, we may assume that F is formally real. We show in this case that $W(F)$ is torsion-free. Let $\mathfrak{b}$ be an anisotropic bilinear form over F that is torsion in $W(F)$, say $m\mathfrak{b} = 0$ in $W(F)$ for some positive integer m. As $\mathfrak{b}$ is diagonalizable by Corollary 1.19, suppose that $\mathfrak{b} \simeq \langle a_1, \dots, a_n \rangle$ with $a_i \in F^\times$. The form $m\mathfrak{b}_i$ is isotropic so there exists a nontrivial equation $\sum_j \sum_i a_i x_{ij}^2 = 0$ in F. As F is pythagorean, there exist $x_i \in F$ satisfying $x_i^2 = \sum_j x_{ij}^2$. Since F is formally real, not all the x_i can be zero. Thus $(x_1, \dots, x_n)$ is an isotropic vector for $\mathfrak{b}$, a contradiction.

(2) $\Rightarrow$ (3) is trivial.

(3) $\Rightarrow$ (1): Let $0 \neq z \in D\big(2\langle 1 \rangle\big)$. Then $2\langle\langle z \rangle\rangle = 0$ in $W(F)$ by Corollary 6.6. By assumption, $\langle\langle z \rangle\rangle = 0$ in $W(F)$, hence $z \in F^{\times 2}$. □

Corollary 31.9. *A field F is formally real and pythagorean if and only if $W(F)$ is torsion-free.*

PROOF. Suppose that $W(F)$ is torsion-free. Then $I(F)$ is torsion-free so F is pythagorean. As $\langle 1 \rangle$ is not torsion, $s(F)$ is infinite, hence F is formally real.

Conversely, suppose that F is formally real and pythagorean. Then the proof of (1) $\Rightarrow$ (2) in Proposition 31.8 shows that $W(F)$ is torsion-free. □

Lemma 31.10. *The intersection of pythagorean fields is pythagorean.*

PROOF. Let $F = \bigcap_I F_i$ with each F_i pythagorean. If $z = x^2 + y^2$ with $x, y \in F$, then for each $i \in I$ there exist $z_i \in F_i$ with $z_i^2 = z$. In particular, $z_i = \pm z_j$ for all $i, j \in I$. Thus $z_j \in \bigcap_I F_i = F$ for every $j \in I$ and $z = z_j^2$. □

Exercise 31.11. Let K/F be a finite extension. Show that if K is pythagorean so is F. (Hint: If $\operatorname{char} F \neq 2$ and $a = 1 + x^2 \in F \setminus F^2$, let $z = a + \sqrt{a} \in K$. Show $z \in F(\sqrt{a})^2$, but $N_{F(\sqrt{a})/F}(z) \notin F^2$.)

Let F be a field and K/F an algebraic extension. We call K a *pythagorean closure* of F if K is pythagorean, and if $F \subset E \subsetneq K$ is an intermediate field, then E is not pythagorean. If $\widetilde{F}$ is an algebraic closure of F, then the intersection of all pythagorean fields between F and $\widetilde{F}$ is pythagorean by the lemma. Clearly, this is a pythagorean closure of F. In particular, a pythagorean closure is unique (after fixing an algebraic closure). We shall denote the pythagorean closure of F by F_{py}. If F is not a formally real field, then F_{py} is just the *quadratic closure* of F, i.e., a quadratically closed field K algebraic over F such that if $F \subset E \subsetneq K$, then E is not quadratically closed. For example, the quadratic closure of the field of rational numbers $\mathbb{Q}$ is the field of complex constructible numbers.

Exercise 31.12. Let E be a pythagorean closure of a field F. Prove that E/F is an excellent extension. (Hint: In the formally real case use Exercise 31.7 to show that for any quadratic form φ over F the form $(\varphi_E)_{an}$ over E takes values in F.)

We next show how to construct the pythagorean closure of a field.

Definition 31.13. Let F be a field with $\widetilde{F}$ an algebraic closure. If K/F is a finite extension in $\widetilde{F}$, then we say K/F is *admissible* if there exists a tower

$$\begin{aligned} &F = F_0 \subset F_1 \subset \cdots \subset F_n = K \quad \text{where} \\ &F_i = F_{i-1}(\sqrt{z_{i-1}}) \quad \text{with} \quad z_{i-1} \in D(2\langle 1\rangle_{F_{i-1}}) \end{aligned} \tag{31.14}$$

from F to K.

Remark 31.15. If F is a formally real field and K an admissible extension of F, then K is formally real by Theorem 95.3 in §95.

Lemma 31.16. *Suppose that* $\operatorname{char} F \neq 2$. *Let L be the union of all admissible extensions over F. Then $L = F_{py}$. If F is formally real so is F_{py}.*

PROOF. Let $\widetilde{F}$ be a fixed algebraic closure of F. If E and K are admissible extensions of F, then the compositum of EK of E and K is also an admissible extension. It follows that L is a field. If $z \in L$ satisfies $z = x^2 + y^2$, $x, y \in L$, then there exist admissible extensions E and K of F with $x \in E$ and $y \in K$. Then $EK(\sqrt{z})$ is an admissible extension of F, hence $\sqrt{z} \in EK(\sqrt{z}) \subset L$. Therefore, L is pythagorean. Let M be pythagorean with $F \subset M \subset \widetilde{F}$. We show $L \subset M$. Let K/F be admissible. Let (31.14) be a tower from F to K. By induction, we may assume that $F_i \subset M$. Therefore, $z_i \in M^2$, hence $F_{i+1} \subset M$. Consequently, $K \subset M$. It follows that $L \subset M$ so $L = F_{py}$. If F is formally real, then so is L by Remark 31.15. □

If F is an arbitrary field, then the quadratic closure of F can also be constructed by taking the union of all square root towers

$$F = F_0 \subset F_1 \subset \cdots \subset F_n = K \quad \text{where} \quad F_i = F_{i-1}(\sqrt{z_{i-1}}) \quad \text{with} \quad z_{i-1} \in F_{i-1}^{\times}$$

over F.

Recall that if K/F is a field extension, then $W(K/F)$ is the kernel of the restriction map $r_{K/F} : W(F) \to W(K)$.

Lemma 31.17. *Let* $z \in D(2\langle 1\rangle) \setminus F^{\times 2}$. *If* $K = F(\sqrt{z})$, *then*

$$W(K/F) \subset \operatorname{ann}_{W(F)}\big(2\langle 1\rangle\big).$$

PROOF. It follows from the hypothesis that $\langle\langle z\rangle\rangle$ is anisotropic, hence K/F is a quadratic extension. As z is a sum of squares and not a square, $\operatorname{char} F \neq 2$. Therefore, by Corollary 23.6, we have $W(K/F) = \langle\langle z\rangle\rangle W(F)$. By Corollary 6.6, we have $2\langle\langle z\rangle\rangle = 0$ in $W(F)$ and the result follows. □

31.C. Formally real fields. Let

$$W_t(F) := \{\mathfrak{b} \in W(F) \mid \text{there exists a positive integer } n \text{ such that } n\mathfrak{b} = 0\},$$

the additive torsion in $W(F)$. It is an ideal in $W(F)$. We have

Theorem 31.18. *Let F be a formally real field.*

(1) *$W_t(F)$ is 2-primary, i.e., all torsion elements of $W(F)$ have exponent a power of 2.*
(2) *$W_t(F) = W(F_{py}/F)$.*

PROOF. As $W(F_{py})$ is torsion-free by Corollary 31.9, the torsion subgroup $W_t(F)$ lies in $W(F_{py}/F)$, so it suffices to show $W(F_{py}/F)$ is a 2-primary torsion group. Let K be an admissible extension of F as in (31.14). Since F_{py} is the union of admissible extensions by Lemma 31.16, it suffices to show $W(K/F)$ is 2-primary torsion. By Lemma 31.17 and induction, it follows that $W(K/F) \subset \operatorname{ann}_{W(F)}(2^n\langle 1\rangle)$ as needed. □

Lemma 31.19. *Let F be a formally real field and $\mathfrak{b} \in W(F)$ satisfy $2^n\mathfrak{b} \neq 0$ in $W(F)$ for any $n \geq 0$. Let K/F be an algebraic extension that is maximal with respect to $\mathfrak{b}_K$ not having order a power of 2 in $W(K)$. Then K is euclidean. In particular,* $\operatorname{sgn} \mathfrak{b}_K \neq 0$.

PROOF. Suppose K is not euclidean. As $2^n\langle 1\rangle \neq 0$, the field K is formally real. Since K is not euclidean, there exists an $x \in K^\times$ such that $x \notin (K^\times)^2 \cup -(K^\times)^2$. In particular, both $K(\sqrt{x})/K$ and $K(\sqrt{-x})/K$ are quadratic extensions. By choice of K, there exists a positive integer n such that $\mathfrak{c} := 2^n\mathfrak{b}_K$ satisfies $\mathfrak{c}_{K(\sqrt{x})}$ and $\mathfrak{c}_{K(\sqrt{-x})}$ are metabolic, hence hyperbolic as $\operatorname{char} F \neq 2$. By Corollary 23.6, there exist forms $\mathfrak{c}_1$ and $\mathfrak{c}_2$ over K satisfying $\mathfrak{c} \simeq \langle\langle x\rangle\rangle \otimes \mathfrak{c}_1 \simeq \langle\langle -x\rangle\rangle \otimes \mathfrak{c}_2$. As $-x\langle\langle x\rangle\rangle \simeq \langle\langle x\rangle\rangle$ and $x\langle\langle -x\rangle\rangle \simeq \langle\langle -x\rangle\rangle$, we conclude that $x\mathfrak{c} \simeq \mathfrak{c} \simeq -x\mathfrak{c}$ and hence that $2\mathfrak{c} \simeq \mathfrak{c} \perp \mathfrak{c} \simeq x\mathfrak{c} \perp -x\mathfrak{c}$. Thus $2\mathfrak{c} = 0$ in $W(K)$. This means that $\mathfrak{b}_K$ is torsion of order 2^{n+1}, a contradiction. □

Proposition 31.20. *The following are equivalent:*

(1) *F can be ordered, i.e., $\mathfrak{X}(F)$, the space of orderings of F, is not empty.*
(2) *F is formally real.*
(3) *$W_t(F) \neq W(F)$.*
(4) *$W(F)$ is not a 2-primary torsion group.*
(5) *There exists an ideal $\mathfrak{A} \subset W(F)$ such that $W(F)/\mathfrak{A} \simeq \mathbb{Z}$.*
(6) *There exists a prime ideal $\mathfrak{P}$ in $W(F)$ such that* $\operatorname{char}(W(F)/\mathfrak{P}) \neq 2$.

Moreover, if F is formally real, then for any prime ideal $\mathfrak{P}$ in $W(F)$ which satisfies $\operatorname{char}\big(W(F)/\mathfrak{P}\big) \neq 2$, *the set*

$$P_{\mathfrak{P}} := \{x \in F^\times \mid \langle\langle x\rangle\rangle \in \mathfrak{P}\} \cup \{0\}$$

is an ordering of F.

PROOF. (1) $\Rightarrow$ (2) is clear.

(2) $\Rightarrow$ (3): By assumption, $-1 \notin D_F\big(n\langle 1\rangle\big)$ for any $n > 0$, so $\langle 1\rangle \notin W_t(F)$.

(3) $\Rightarrow$ (4) is trivial.

(4) $\Rightarrow$ (5): By assumption there exists $\mathfrak{b} \in W(F)$ not having order a power of 2. By Lemma 31.19, there exists K/F with K euclidean. In particular, $r_{K/F}$ is surjective. Therefore, $\mathfrak{A} = W(K/F)$ works by Lemma 31.19 and Sylvester's Law of Inertia 31.5.

(5) $\Rightarrow$ (6) is trivial.

(6) $\Rightarrow$ (1): By Proposition 31.4, the field F is formally real. We show that (6) implies the last statement. This will also prove (1). Let $\mathfrak{P}$ in $W(F)$ be a prime ideal satisfying $\operatorname{char}(W(F)/\mathfrak{P}) \neq 2$.
We must show

(i) $P_{\mathfrak{P}} \cup (-P_{\mathfrak{P}}) = F$.
(ii) $P_{\mathfrak{P}} + P_{\mathfrak{P}} \subset P_{\mathfrak{P}}$.
(iii) $P_{\mathfrak{P}} \cdot P_{\mathfrak{P}} \subset P_{\mathfrak{P}}$.
(iv) $P_{\mathfrak{P}} \cap (-P_{\mathfrak{P}}) = \{0\}$.
(v) $-1 \notin P_{\mathfrak{P}}$.

Suppose that $x \neq 0$ and both $\pm x \in P_{\mathfrak{P}}$. Then $\langle\langle -1\rangle\rangle = \langle\langle -x\rangle\rangle + \langle\langle x\rangle\rangle$ lies in $\mathfrak{P}$ so $2\langle 1\rangle + \mathfrak{P} = 0$ in $W(F)/\mathfrak{P}$, a contradiction. This shows (iv) and (v) hold. As $\langle\langle x, -x\rangle\rangle = 0$ in $W(F)$, either $\langle\langle x\rangle\rangle$ or $\langle\langle -x\rangle\rangle$ lies in $\mathfrak{P}$, so (i) holds. Next let $x, y \in P_{\mathfrak{P}}$. Then $\langle\langle xy\rangle\rangle = \langle\langle x\rangle\rangle + x\langle\langle y\rangle\rangle$ lies in $\mathfrak{P}$ so $xy \in \mathfrak{P}$ which is (iii). Finally, we show that (ii) holds, i.e., $x + y \in P_{\mathfrak{P}}$. We may assume neither x nor y is zero. This implies that $z := x + y \neq 0$, otherwise we have the equation $\langle\langle -1\rangle\rangle = \langle 1, x, -x, 1\rangle = \langle 1, -x, -y, 1\rangle = \langle\langle x\rangle\rangle + \langle\langle y\rangle\rangle$ in $W(F)$ which implies that $\langle\langle -1\rangle\rangle$ lies in $\mathfrak{P}$ contradicting (v). Since $\langle -x, -y\rangle \simeq -z\langle\langle -xy\rangle\rangle$ by Corollary 6.6, we have

$$\begin{aligned}2\langle -z\rangle = 2\langle -x, -y, zxy\rangle &= \langle -x, -y, zxy, -z, -zxy, zxy\rangle \\ &= \langle\langle x\rangle\rangle + \langle\langle y\rangle\rangle - 2\langle 1\rangle - z\langle\langle xy\rangle\rangle\end{aligned}$$

in $W(F)$. As $x, y \in P_{\mathfrak{P}}$ and $xy \in P_{\mathfrak{P}}$ by (iii), it follows that $2\langle\langle z\rangle\rangle \in \mathfrak{P}$ as needed. □

The proposition gives another proof of the Artin-Schreier Theorem that every formally real field can be ordered.

In [**94**], Lewis gave an elegant proof for the structure of the Witt ring, part of which we present in the following exercise.

Exercise 31.21. Let F be an arbitrary field. Show all of the following:

(1) (Leung) Define a polynomial $p_n(t)$ in $F[t]$ as follows:

$$p_n(t) = \begin{cases} t(t^2 - 2^2)(t^2 - 4^2)\cdots(t^2 - n^2) & \text{if } n \text{ is even,} \\ (t^2 - 1^2)(t^2 - 3^2)(t^2 - 5^2)\cdots(t^2 - n^2) & \text{if } n \text{ is odd.}\end{cases}$$

If $\mathfrak{b}$ is an n-dimensional nondegenerate symmetric bilinear form over F, then $p_n(\mathfrak{b}) = 0$ in $W(F)$.

(2) $W(F)$ is integral over $\mathbb{Z}$ if F is formally real and over $\mathbb{Z}/2^r\mathbb{Z}$ for some r if F is not formally real. In particular, the Krull dimension of $W(F)$ is one if F is formally real and zero if not.

(3) Using Exercise 33.14 below show that $W_t(F)$ is 2-primary.

(4) $W(F)$ has no odd-dimensional zero divisors and if $W_t(F) \neq 0$, then we have $\operatorname{zd}(W(F)) = I(F)$. Moreover, $W(F)$ contains no nontrivial idempotents.

(5) If F is formally real, then $W_t(F) = \operatorname{nil}(W(F))$.

31.D. The Local-Global Principle. The main result of this subsection is Theorem 31.22 below due to Pfister (cf. [**110**]).

Let F be a formally real field and $\mathfrak{X}(F)$ the space of orderings. Let $P \in \mathfrak{X}(F)$ and F_P be the real closure of F at P (within a fixed algebraic closure). By Sylvester's Law of Inertia 31.5, the signature map defines an isomorphism $\operatorname{sgn} : W(F_P) \to \mathbb{Z}$. In particular, we have a *signature map* $\operatorname{sgn}_P : W(F) \to \mathbb{Z}$ given by $\operatorname{sgn}_P = \operatorname{sgn} \circ r_{F_P/F}$. This is a ring homomorphism satisfying $W_t(F) \subset \operatorname{Ker}(r_{F_P/F}) = \operatorname{Ker}(\operatorname{sgn}_P)$. We let

$$\mathfrak{P}_P = \operatorname{Ker}(\operatorname{sgn}_P) \text{ in } \operatorname{Spec}(W(F)).$$

Note if $F \subset K \subset F_P$ and $\mathfrak{b}$ is a nondegenerate symmetric bilinear form, then $\operatorname{sgn}_P \mathfrak{b} = \operatorname{sgn}_{F_P^2 \cap K} \mathfrak{b}_K$. In particular, if K is euclidean, then $\operatorname{sgn}_P \mathfrak{b} = \operatorname{sgn} \mathfrak{b}_K$.

Theorem 31.22 (Local-Global Principle). *The sequence*

$$0 \to W_t(F) \to W(F) \xrightarrow{(r_{F_P/F})} \prod_{\mathfrak{X}(F)} W(F_P)$$

is exact.

Proof. We may assume that F is formally real by Proposition 31.4. We saw above that $W_t(F) \subset \operatorname{Ker}(\operatorname{sgn}_P)$ for every ordering $P \in \mathfrak{X}(F)$, so the sequence is a zero sequence. Suppose that $\mathfrak{b} \in W(F)$ is not torsion of 2-power order. By Lemma 31.19, there exists a euclidean field K/F with $\mathfrak{b}_K$ not of 2-power order. As $K^2 \in \mathfrak{X}(K)$, we have $P = K^2 \cap F \in \mathfrak{X}(F)$. Thus $\operatorname{sgn}_P \mathfrak{b} = \operatorname{sgn} \mathfrak{b}_K \neq 0$. The result follows. □

Corollary 31.23. *The map*

$$\mathfrak{X}(F) \to \{\mathfrak{P} \in \operatorname{Spec}(W(F)) \mid W(F)/\mathfrak{P} \simeq \mathbb{Z}\} \quad \text{given by} \quad P \mapsto \mathfrak{P}_P$$

is a bijection.

Proof. Let $\mathfrak{P} \subset W(F)$ be a prime ideal such that $W(F)/\mathfrak{P} \simeq \mathbb{Z}$. As in Proposition 31.20, let $P_{\mathfrak{P}} := \{x \in F^\times \mid \langle\langle x \rangle\rangle \in \mathfrak{P}\} \cup \{0\} \in \mathfrak{X}(F)$.

Claim: $\mathfrak{P} \mapsto P_{\mathfrak{P}}$ is the inverse, i.e., $P = P_{\mathfrak{P}_P}$ and $\mathfrak{P} = \mathfrak{P}_{P_{\mathfrak{P}}}$.

If $P \in \mathfrak{X}(F)$, then certainly $P \subset P_{\mathfrak{P}_P}$, so we must have $P = P_{\mathfrak{P}_P}$ as both are orderings. By definition, we see that the composition $W(F) \to W(F)/\mathfrak{P} \xrightarrow{\sim} \mathbb{Z}$ maps $\langle x \rangle$ to $\operatorname{sgn}_{P_{\mathfrak{P}}} \langle x \rangle$. Hence $\operatorname{Ker}(\operatorname{sgn}_{P_{\mathfrak{P}}}) = \mathfrak{P}$. □

The spectrum of the Witt ring can now be determined. This result is due to Lorenz-Leicht (cf. [**95**]) and Harrison (cf. [**49**]).

Theorem 31.24. $\operatorname{Spec}(W(F))$ *consists of:*

(1) *The fundamental ideal* $I(F)$.
(2) $\mathfrak{P}_P$ *with* $P \in \mathfrak{X}(F)$.
(3) $\mathfrak{P}_{P,p} := \mathfrak{P}_P + pW(F) = \operatorname{sgn}_P^{-1}(p\mathbb{Z})$, *$p$ an odd prime, with* $P \in \mathfrak{X}(F)$.

Moreover, all these ideals are different. The prime ideals in (1) *and* (3) *are the maximal ideals of* $W(F)$. *If* F *is formally real, then the ideals in* (2) *are the minimal primes of* $W(F)$ *and* $\mathfrak{P}_P \subset \mathfrak{P}_{P,p} \cap I(F)$ *for all* $P \in \mathfrak{X}(F)$ *and for all odd primes* p.

PROOF. We may assume that F is formally real by Proposition 31.4. Let $\mathfrak{P}$ be a prime ideal in $W(F)$. Let $a \in F^\times$. As $\langle\langle a, -a\rangle\rangle = 0$ in $W(F)$ either $\langle\langle a\rangle\rangle \in \mathfrak{P}$ or $\langle\langle -a\rangle\rangle \in \mathfrak{P}$. In particular, $\langle a\rangle \equiv \pm\langle 1\rangle \mod \mathfrak{P}$. Hence $W(F)/\mathfrak{P}$ is cyclic generated by $\langle 1\rangle + \mathfrak{P}$, so $W(F)/\mathfrak{P} \simeq \mathbb{Z}$ or $\mathbb{Z}/p\mathbb{Z}$ for p a prime. If $x, y \in F^\times$, then $\langle x\rangle$ and $\langle y\rangle$ are units in $W(F)$, so do not lie in $\mathfrak{P}$. Suppose that $W(F)/\mathfrak{P} \simeq \mathbb{Z}/2\mathbb{Z}$. Then we must have $\langle x, y\rangle \in \mathfrak{P}$ for all $x, y \in F^\times$, hence $\mathfrak{P} = I(F)$. So suppose that $W(F)/\mathfrak{P} \not\simeq \mathbb{Z}/2\mathbb{Z}$. By Proposition 31.20, the set $P = P_{\mathfrak{P}}$ lies in $\mathfrak{X}(F)$. Since $W(F)/\mathfrak{P}_P \simeq \mathbb{Z}$, we have $\mathfrak{P}_P \subset \mathfrak{P}$. Hence $\mathfrak{P} = \mathfrak{P}_P$ or $\mathfrak{P} = \mathfrak{P}_{P,p}$ for a suitable odd prime. As each $P \in \mathfrak{X}(F)$ determines a unique $\mathfrak{P}_P$ and $\mathfrak{P}_{P,p}$ by Corollary 31.20, the result follows. □

Corollary 31.25. *If* F *is formally real, then* $\dim W(F) = 1$ *and the map* $\mathfrak{X}(F) \to \operatorname{Min}\operatorname{Spec}\big(W(F)\big)$ *given by* $P \mapsto \operatorname{Ker}(\operatorname{sgn}_P)$ *is a homeomorphism.*

PROOF. As $\langle\langle 1\rangle\rangle$ does not lie in any minimal prime, for each $a \in F^\times$ either $a \in \mathfrak{P}_P$ or $-a \in \mathfrak{P}_P$ but not both where $P \in \mathfrak{X}(F)$. The sets $H(a) := \{P \mid -a \in P\}$ form a subbase for the topology of $\mathfrak{X}(F)$ (cf. §96). As $a \in P$ for $P \in \mathfrak{X}(F)$ if and only if $\langle\langle a\rangle\rangle \in \mathfrak{P}_P$ if and only if $\mathfrak{P}_P$ lies in the basic open set $\{\mathfrak{P} \mid a \notin \mathfrak{P}$ for $\mathfrak{P} \in \operatorname{Min}\big(\operatorname{Spec} W(F)\big\}$, the result follows. □

The structure of the Witt ring in the formally real case due to Pfister (cf. [**110**]) can now be shown.

Proposition 31.26. *Let* F *be formally real. Then:*

(1) $\operatorname{nil}\big(W(F)\big) = \operatorname{rad}\big(W(F)\big) = W_t(F)$.
(2) $W(F)^\times = \{\mathfrak{b} \mid \operatorname{sgn}_P \mathfrak{b} = \pm 1$ *for all* $P \in \mathfrak{X}(F)\}$
$= \{\langle a\rangle + \mathfrak{c} \mid a \in F^\times$ *and* $\mathfrak{c} \in I^2(F) \cap W_t(F)\}$.
(3) *If* F *is not pythagorean, then* $\operatorname{zd}\big(W(F)\big) = I(F)$.
(4) *If* F *is pythagorean, then* $\operatorname{zd}\big(W(F)\big) = \bigcup_{\mathfrak{X}(F)} \mathfrak{P}_P \subsetneqq I(F)$.
(5) $W(F)$ *is connected, i.e.,* 0 *and* 1 *are the only idempotents in* $W(F)$.
(6) $W(F)$ *is noetherian if and only if* $F^\times/F^{\times 2}$ *is finite.*

PROOF. (1): If $P \in \mathfrak{X}(F)$, then $\mathfrak{P}_P = \bigcap_p \mathfrak{P}_{P,p}$, so $\operatorname{nil}\big(W(F)\big) = \operatorname{rad}\big(W(F)\big)$. By the Local-Global Principle 31.22, we have

$$\begin{aligned} W_t(F) = \operatorname{Ker}\Big(\prod_{P \in \mathfrak{X}(F)} r_{F_P/F}\Big) &= \bigcap_{\mathfrak{X}(F)} \operatorname{Ker}(\operatorname{sgn}_P) \\ &= \bigcap_{\mathfrak{X}(F)} \mathfrak{P}_P = \bigcap_{\mathfrak{X}(F)} \mathfrak{P}_{P,p} = \operatorname{nil}\big(W(F)\big). \end{aligned}$$

(2): We have $\operatorname{sgn}_P\big(W(F)^\times\big) \subset \{\pm 1\}$ for all $P \in \mathfrak{X}(F)$. Let $\mathfrak{b}$ be a nondegenerate symmetric bilinear form satisfying $\operatorname{sgn}_P \mathfrak{b} = \pm 1$ for all $P \in \mathfrak{X}(F)$. Choose $a \in F$ such that $\mathfrak{c} := \mathfrak{b} - \langle a\rangle$ lies in $I^2(F)$ using Proposition 4.13. In particular, $\operatorname{sgn}_P \mathfrak{b} \equiv \operatorname{sgn}_P\langle a\rangle \mod 4$, hence $\operatorname{sgn}_P \mathfrak{b} = \operatorname{sgn}_P\langle a\rangle$ for all $P \in \mathfrak{X}(F)$. Consequently, $\operatorname{sgn}_P \mathfrak{c} = 0$ for all $P \in \mathfrak{X}(F)$ so $\mathfrak{c}$ is torsion by the Local-Global Principle 31.22. By (1), the form $\mathfrak{c}$ is nilpotent hence $\mathfrak{b} \in W(F)^\times$.

(3), (4): As the set of zero divisors is a saturated multiplicative set, it follows by commutative algebra that it is a union of prime ideals.

Suppose that F is not pythagorean. Then $W_t(F) \neq 0$ by Corollary 31.9. In particular, $2^n\mathfrak{b} = 0 \in W(F)$ for some $\mathfrak{b} \neq 0$ in $W(F)$ and $n \geq 1$ by Theorem 31.18. Thus $\langle\langle -1 \rangle\rangle$ is a zero divisor. As $I(F)$ is the only prime ideal containing $\langle\langle -1 \rangle\rangle$, we have $I(F) \subset \mathrm{zd}\big(W(F)\big)$. Since $n\langle 1 \rangle$ is not a zero divisor for any odd integer n by Theorem 31.18, no $\mathfrak{P}_{P,p}$ can lie in $\mathrm{zd}\big(W(F)\big)$. It follows that $\mathrm{zd}\big(W(F)\big) = I(F)$, since $\mathfrak{P}_P \subset I(F)$ for all $P \in \mathfrak{X}(F)$.

Suppose that F is pythagorean. Then $W(F)$ is torsion-free, so $n\langle 1 \rangle$ is not a zero-divisor for any nonzero integer n. In particular, no maximal ideal lies in $\mathrm{zd}\big(W(F)\big)$. Let $P \in \mathfrak{X}(F)$ and $\mathfrak{b} \in \mathfrak{P}_P$. Then $\mathfrak{b}$ is diagonalizable so we have $\mathfrak{b} \simeq \langle a_1, \dots, a_n, b_1, \dots b_n \rangle$ with $a_i, -b_j \in P$ for all i, j. Let $\mathfrak{c} = \langle\langle a_1 b_1, \dots, a_n b_n \rangle\rangle$. Then $\mathfrak{c}$ is nonzero in $W(F)$ as $\mathrm{sgn}_P\, \mathfrak{c} = 2^n$. As $\langle\langle -a_i b_i \rangle\rangle \cdot \mathfrak{c} = 0$ in $W(F)$ for all i, we have $\mathfrak{b} \cdot \mathfrak{c} = 0$, hence $\mathfrak{b} \in \mathrm{zd}\big(W(F)\big)$. Consequently, $\mathfrak{P}_P \subset \mathrm{zd}\big(W(F)\big)$ for all $P \in \mathfrak{X}(F)$, hence $\mathrm{zd}\big(W(F)\big)$ is the union of the minimal primes.

(5): If the result is false, then $1 = e_1 + e_2$ for some nontrivial idempotents e_1, e_2. As $e_1 e_2 = 0$, we have $e_1, e_2 \in \mathrm{zd}\big(W(F)\big) \subset I(F)$ which implies $1 \in I(F)$, a contradiction.

(6): This follows by the same proof for the analogous result in Proposition 31.4. □

Let $I_t(F) := W_t(F) \cap I(F)$.

Proposition 31.27. *If F is formally real, then $W_t(F)$ is generated by $\langle\langle x \rangle\rangle$ with $x \in D\big(\infty\langle 1 \rangle\big)$, i.e., $W_t(F) = I_t(F)$ is generated by torsion 1-fold Pfister forms.*

Proof. Let $\mathfrak{b} \in W_t(F)$. Then $2^n\mathfrak{b} = 0$ for some integer $n > 0$. Thus $\mathfrak{b} \in \mathrm{ann}_{W(F)}\big(2^n\langle 1 \rangle\big)$. By Corollary 6.23, there exist binary forms $\mathfrak{d}_i \in \mathrm{ann}_{W(F)}\big(2^n\langle 1 \rangle\big)$ satisfying $\mathfrak{b} = \mathfrak{d}_1 + \cdots + \mathfrak{d}_m$ in $W(F)$. The result follows. □

31.E. The Witt ring up to isomorphism. Because $I(F)$ is the unique ideal of index two in $W(F)$, we can deduce the following due to Cordes and Harrison (cf. [**26**]):

Theorem 31.28. *Let F and K be two fields. Then $W(F)$ and $W(K)$ are isomorphic as rings if and only if $W(F)/I^3(F)$ and $W(K)/I^3(K)$ are isomorphic as rings.*

Proof. The fundamental ideal is the unique ideal of index two in its Witt ring by Theorem 31.24. Therefore, any ring isomorphism $W(F) \to W(K)$ induces a ring isomorphism $W(F)/I^3(F) \to W(K)/I^3(K)$.

Conversely, let $g : W(F)/I^3(F) \to W(K)/I^3(K)$ be a ring isomorphism. As the fundamental ideal is the unique ideal of index two in its Witt ring by Theorem 31.24, the map g induces an isomorphism $I(F)/I^3(F) \to I(K)/I^3(K)$. From this and Proposition 4.13, it follows easily that g induces an isomorphism $h : F^\times/F^{\times 2} \to K^\times/K^{\times 2}$.

We adopt the following notation. For a coset $\alpha = xK^{\times 2}$, write $\langle \alpha \rangle$ and $\langle\langle \alpha \rangle\rangle$ for the forms $\langle x \rangle$ and $\langle\langle x \rangle\rangle$ in $W(K)$, respectively. We also write $s(a)$ for $h(aF^{\times 2})$. Note that $s(ab) = s(a)s(b)$ for all $a, b \in F^\times$.

By construction,

$$g\big(\langle\langle a \rangle\rangle + I^3(F)\big) \equiv \langle\langle s(a) \rangle\rangle \mod I^2(K)/I^3(K).$$

As $g(1) = 1$, plugging in $a = -1$, we get $\langle s(-1)\rangle = \langle -1\rangle$. In particular,

$$\langle s(1)\rangle + \langle s(-1)\rangle = \langle 1\rangle + \langle -1\rangle = 0 \in W(K). \tag{31.29}$$

Since g is a ring homomorphism, we have

$$\begin{aligned} g\big(\langle\langle a,b\rangle\rangle + I^3(F)\big) &= g\big(\langle\langle a\rangle\rangle + I^3(F)\big)\cdot g\big(\langle\langle b\rangle\rangle + I^3(F)\big) \\ &= \langle\langle s(a)\rangle\rangle \cdot \langle\langle s(b)\rangle\rangle + I^3(K) \\ &= \langle\langle s(a),\ s(b)\rangle\rangle + I^3(K). \end{aligned}$$

for every $a, b \in F^\times$.

If $a+b \neq 0$, we have $\langle\langle a,b\rangle\rangle \simeq \langle\langle a+b, ab(a+b)\rangle\rangle$ by Lemma 4.15(3). Therefore,

$$\langle\langle s(a),\ s(b)\rangle\rangle \equiv \langle\langle s(a+b),\ s\big(ab(a+b)\big)\rangle\rangle \mod I^3(K).$$

By Theorem 6.20, these two 2-fold Pfister forms are equal in $W(K)$. Therefore,

$$\langle s(a)\rangle + \langle s(b)\rangle = \langle s(a+b)\rangle + \langle s(ab(a+b))\rangle \tag{31.30}$$

in $W(K)$.

Let $\mathcal{F}$ be the free abelian group with basis the set of isomorphism classes of 1-dimensional forms $\langle a\rangle$ over F. It follows from Theorem 4.8 and equations (31.29) and (31.30) that the map $\mathcal{F} \to W(K)$ taking $\langle a\rangle$ to $\langle s(a)\rangle$ gives rise to a homomorphism $s : W(F) \to W(K)$. Interchanging the roles of F and K, we have, in a similar fashion, a homomorphism $W(K) \to W(F)$ which is the inverse of s. $\square$

32. Addendum on torsion

We know by Corollary 6.26 that if $\mathfrak{b} \in \operatorname{ann}_{W(F)}\big(2\langle 1\rangle\big)$, i.e., if $2\mathfrak{b} = 0$ in $W(F)$ that $\mathfrak{b} \simeq \mathfrak{d}_1 \perp \cdots \perp \mathfrak{d}_n$ where each $\mathfrak{b}_i$ is a binary form annihilated by 2. In particular, if $\mathfrak{b}$ is an anisotropic bilinear Pfister form such that $2\mathfrak{b} = 0$ in $W(F)$, then the pure subform $\mathfrak{b}'$ of $\mathfrak{b}$ satisfies $D(\mathfrak{b}') \cap D\big(2\langle -1\rangle\big) \neq \emptyset$. In general, if $2^n\mathfrak{b} = 0$ in $W(F)$ with $n > 1$, then $\mathfrak{b}$ is not isometric to binary forms annihilated by 2^n nor does the pure subform of a torsion bilinear Pfister form represent a totally negative element. In this addendum, we construct a counterexample due to Arason and Pfister (cf. [**11**]). We use the following variant of the Cassels-Pfister Theorem 17.3.

Lemma 32.1. *Let* $\operatorname{char} F \neq 2$. *Let* $\varphi = \langle a_1, \dots, a_n\rangle_q$ *be anisotropic over* $F(t)$ *with* $a_1, \dots, a_n \in F[t]$ *all satisfying* $\deg a_i \leq 1$. *Suppose that* $0 \neq q \in D(\varphi_{F(t)}) \cap F[t]$. *Then there exist polynomials* $f_1, \dots, f_n \in F[t]$ *such that* $q = \varphi(f_1, \dots, f_n)$, *i.e.,* $F[t] \otimes_F \varphi$ *represents* q.

Proof. Let $\psi \simeq \langle -q\rangle \perp \varphi$ and let

$$Q := \big\{f = (f_0, \dots, f_n) \in F[t]^{n+1} \mid \mathfrak{b}_\psi(f,f) = 0\big\}.$$

Choose $f \in Q$ such that $\deg f_0$ is minimal. Assume that the result is false. Then $\deg f_0 > 0$. Write $f_i = f_0 g_i + r_i$ with $r_i = 0$ or $\deg r_i < \deg f_0$ for each i using the Euclidean Algorithm. So $\deg r_i^2 \leq 2\deg f_0 - 2$ for all i. Let $g = (1, g_1, \dots, g_n)$ and define $h \in F[t]^{n+1}$ by $h = cf + dg$ with $c = \mathfrak{b}_\psi(g,g)$ and $d = -2\mathfrak{b}_\psi(f,g)$. We have

$$\mathfrak{b}_\psi(cf + dg, cf + dg) = c^2\mathfrak{b}_\psi(f,f) + 2cd\mathfrak{b}_\psi(f,g) + d^2\mathfrak{b}_\psi(g,g) = 0,$$

so $h \in Q$. Therefore,

$$h_0 = \mathfrak{b}_\psi(g,g)f_0 - 2\mathfrak{b}_\psi(f,g)g_0 = \mathfrak{b}_\psi(f_0 g - 2f, g) = -\mathfrak{b}_\psi(f + r, g),$$

so

$$f_0h_0 = -f_0\mathfrak{b}_\psi(f+r,g) = -\mathfrak{b}_\psi(f+r,f-r) = \mathfrak{b}_\psi(r,r) = \sum_{i=1}^{n} a_i r_i,$$

which is not zero as φ is anisotropic. Consequently,

$$\deg h_0 + \deg f_0 \le \max_i\{\deg a_i\} + 2\deg f_0 - 2 \le 2\deg f_0 - 1$$

as $\deg a_i \le 1$ for all i. This is a contradiction. □

Lemma 32.2. *Let F be a formally real field and $x, y \in D\bigl(\infty\langle 1\rangle\bigr)$. Let $\mathfrak{b} = \langle\langle -t, x+ty\rangle\rangle$ be a 2-fold Pfister form over $F(t)$. If $\mathfrak{b} \simeq \mathfrak{b}_1 \perp \mathfrak{b}_2$ over $F(t)$ with $\mathfrak{b}_1$ and $\mathfrak{b}_2$ binary torsion forms over $F(t)$, then there exists a $z \in D\bigl(\infty\langle 1\rangle\bigr)$ such that $x, y \in D\bigl(\langle\langle -z\rangle\rangle\bigr)$.*

PROOF. If x (respectively, y or xy) is a square, let $z = y$ (respectively, $z = x$) to finish. So we may assume they are not squares. As $\mathfrak{b}$ is round, we may also assume that $\mathfrak{d}_1 \simeq \langle\langle w\rangle\rangle$ with $w \in D\bigl(\infty\langle 1\rangle\bigr)$ by Corollary 6.6. In particular, $D(\mathfrak{b}'_F) \cap D\bigl(\infty\langle -1\rangle\bigr) \neq \emptyset$ by Lemma 6.11. Thus, there exists a positive integer n such that $\mathfrak{b}' \perp n\langle 1\rangle$ is isotropic. Let $\mathfrak{c} = \langle t, -(x+yt)\rangle \perp n\langle 1\rangle$. We have $t(x+yt) \in D(\mathfrak{c})$. The form $\langle 1, -y\rangle$ is anisotropic as is $n\langle 1\rangle$, since F is formally real. If $\mathfrak{c}$ is isotropic, then we would have an equation $-tf^2 = \sum g_i^2 - (x+yt)h^2$ in $F[t]$ for some $f, g_i, h \in F[t]$. Comparing leading terms implies that y is a square. So $\mathfrak{c}$ is anisotropic. By Lemma 32.1, there exist $c, d, f_i \in F[t]$ satisfying

$$f_1^2 + \cdots + f_n^2 + tc^2 - (x+yt)d^2 = t(x+yt).$$

Since $\langle 1, -y\rangle$ and $n\langle 1\rangle$ are anisotropic and t^2 occurs on the right hand side, we must have c, d are constants and $\deg f_i \le 1$ for all i. Write $f_i = a_i + b_i t$ with $a_i, b_i \in F$ for $1 \le i \le n$. Then

$$\sum_{i=1}^{n} a_i^2 = xd^2, \quad 2\sum_{i=1}^{n} a_i b_i = -c^2 + x + yd^2, \quad \text{and} \quad \sum_{i=1}^{n} b_i^2 = y.$$

If $d = 0$, then $a_i = 0$ for all i and $x = c^2$ is a square which was excluded. So $d \neq 0$. Let

$$z = 4\sum_{i=1}^{n} a_i^2 \cdot \sum_{i=1}^{n} b_i^2 - 4\Bigl(\sum_{i=1}^{n} a_i b_i\Bigr)^2 = 4xyd^2 - (x - c^2 + yd^2)^2.$$

Applying the Cauchy-Schwarz Inequality in each real closure of F, we see that z is nonnegative in every ordering, so $z \in \widetilde{D}\bigl(\infty\langle 1\rangle\bigr)$. As xy is not a square, $z \neq 0$. As $d \neq 0$, we have $xy \in D\bigl(\langle\langle -z\rangle\rangle\bigr)$. Expanding, one checks

$$z = 4xyd^2 - (x - c^2 + yd^2)^2 = 4xc^2 - (x - yd^2 + c^2)^2.$$

Thus $x \in D\bigl(\langle\langle -z\rangle\rangle\bigr)$. As $\langle\langle -z\rangle\rangle$ is round, $y \in D\bigl(\langle\langle -z\rangle\rangle\bigr)$ also. □

Lemma 32.3. *Let F_0 be a formally real field and $u, y \in D(\infty\langle 1\rangle_{F_0})$. Let $x = u + t^2$ in $F = F_0(t)$. If there exists a $z \in D(\infty\langle 1\rangle_F)$ such that $x, y \in D\bigl(\langle\langle -z\rangle\rangle\bigr)$, then $y \in D\bigl(\langle\langle -u\rangle\rangle\bigr)$.*

PROOF. We may assume that y is not a square. By assumption, we may write

$$z = (u+t^2)f_1^2 - g_1^2 = yf_2^2 - g_2^2 \quad \text{for some} \quad f_1, f_2, g_1, g_2 \in F_0(t).$$

Multiplying this equation by an appropriate square in $F_0(t)$, we may assume that $z \in F[t]$ and that $f_1, g_1, f_2, g_2 \in F_0[t]$ have no common nontrivial factor. As z is totally positive, i.e., lies in $D\big(\infty\langle 1\rangle\big)$, its leading term must be totally positive in F_0. Consequently,

$$\deg g_1 \le 1 + \deg f_1 \quad \text{and} \quad \deg g_2 \le \deg f_2.$$

It follows that $\frac{1}{2}\deg z \le 1 + \deg f_1$. We have $\frac{1}{2}\deg z = \deg f_2$, otherwise $y \in F^2$, a contradiction. Thus, we have

$$\deg f_2 \le 1 + \deg f_1 \quad \text{and} \quad \deg(g_1 \pm g_2) \le 1 + \deg f_1.$$

If $\deg\big((u+t^2)f_1^2 - yf_2^2\big) < 2\deg f_1 + 2$, then y would be a square in F_0, a contradiction. So

$$\deg\big((u+t^2)f_1^2 - yf_2^2\big) = 2 + 2\deg f_1.$$

As $(u+t^2)f_1^2 - yf_2^2 = g_1^2 - g_2^2$, we have $\deg(g_1 \pm g_2) = 1 + \deg f_1$. It follows that either f_1 or $g_1 - g_2$ has a prime factor p of odd degree. Let $\overline{F} = F_0[t]/(p)$ and $^- : F_0[t] \to \overline{F}$ the canonical map. Suppose that $\overline{f}_1 = 0$. Then $\overline{z} = -\overline{g}_1^2$ in $\overline{F}$. As z is a sum of squares in $F_0[t]$ (possibly zero), we must also have $\overline{z}$ is a sum of squares in $\overline{F}$. But $[\overline{F} : F_0]$ is odd, hence $\overline{F_0}$ is still formally real by Theorem 95.3 or Springer's Theorem 18.5. Consequently, we must have $\overline{z} = \overline{g}_1 = 0$. This implies that $y\overline{f}_2^2 = \overline{g}_2^2$. As y cannot be a square in the odd degree extension $\overline{F}$ of F_0 by Springer's Theorem 18.5, we must have $\overline{f}_2 = 0 = \overline{g}_2$. But there exist no prime p dividing f_1, f_2, g_1, and g_2. Thus $p \nmid f_1$ in $F_0[t]$. It follows that $\overline{g}_1 = \overline{g}_2$ which in turn implies that $(u+\overline{t}^2)\overline{f}_1^2 - y\overline{f}_2^2 = 0$. As $\overline{f}_1 \neq 0$, we have $\overline{f}_2 \neq 0$, so we conclude that $\langle u, 1, -y\rangle_{\overline{F}}$ is isotropic. As $[\overline{F} : F_0]$ is odd, $\langle u, 1, -y\rangle$ is isotropic over F_0 by Springer's Theorem 18.5, i.e., $y \in D\big(\langle\langle -u\rangle\rangle\big)$ as needed. □

We now construct the counterexample.

Example 32.4. We apply the above lemmas in the following case. Let $F_0 = \mathbb{Q}(t_1)$ and $u = 1$ and $y = 3$. The element y is a sum of three but not two squares in F_0 by the Substitution Principle 17.7. Let $K = F_0(t_2)$ and $\mathfrak{b} = \langle\langle -t_2, 1 + t_1^2 + 3t_2\rangle\rangle$ over K. Then the Pfister form $4\mathfrak{b}$ is isotropic, hence metabolic so $4\mathfrak{b} = 0$ in $W(K)$. As $1, 3t_2^2 \in D\big(\langle\langle -3t_2^2\rangle\rangle_K\big)$ and $3 \notin D(2\langle 1\rangle_{\mathbb{Q}(t_1)})$, the lemmas imply that $\mathfrak{b}$ is not isometric to an orthogonal sum of binary torsion forms. In particular, it also follows that the form $\mathfrak{b}$ has the property $D(\mathfrak{b}') \cap D(\infty\langle -1\rangle_K) = \emptyset$.

33. The total signature

We saw when F is a formally real field, the torsion in the Witt ring $W(F)$ is determined by the signatures at the orderings on F. In this section, we view the relationship between bilinear forms over a formally real field F and the totality of continuous functions on the topological space $\mathfrak{X}(F)$ of orderings on F with integer values.

We shall use results in Appendices §95 and §96. Let F be a formally real field. The space of orderings $\mathfrak{X}(F)$ is a boolean space, i.e., a totally disconnected compact Hausdorff space with a subbase the collection of sets

$$H(a) = H_F(a) := \big\{P \in \mathfrak{X}(F) \mid -a \in P\big\}. \tag{33.1}$$

Let $\mathfrak{b}$ be a nondegenerate symmetric bilinear form over F. Then we define the *total signature of* $\mathfrak{b}$ to be the map

(33.2) $$\operatorname{sgn}\mathfrak{b} : \mathfrak{X}(F) \to \mathbb{Z} \quad \text{given by} \quad \operatorname{sgn}\mathfrak{b}(P) = \operatorname{sgn}_P \mathfrak{b}.$$

Theorem 33.3. *Let F be formally real. Then*

$$\operatorname{sgn}\mathfrak{b} : \mathfrak{X}(F) \to \mathbb{Z}$$

is continuous with respect to the discrete topology on $\mathbb{Z}$. The topology on $\mathfrak{X}(F)$ is the coarsest topology such that $\operatorname{sgn}\mathfrak{b}$ *is continuous for all* $\mathfrak{b}$.

PROOF. As $\mathbb{Z}$ is a topological group, addition of continuous functions is continuous. As any nondegenerate symmetric bilinear form is diagonalizable over a formally real field, we need only prove the result for $\mathfrak{b} = \langle a \rangle$, $a \in F^\times$. But

$$\big(\operatorname{sgn}\langle a \rangle\big)^{-1}(n) = \begin{cases} \emptyset & \text{if } n \neq \pm 1, \\ H(a) & \text{if } n = -1, \\ H(-a) & \text{if } n = 1. \end{cases}$$

The result follows easily as the $H(a)$ form a subbase. □

Let $C\big(\mathfrak{X}(F), \mathbb{Z}\big)$ be the ring of continuous functions $f : \mathfrak{X}(F) \to \mathbb{Z}$ where $\mathbb{Z}$ has the discrete topology. By the theorem, we have a map

(33.4) $$\operatorname{sgn} : W(F) \to C\big(\mathfrak{X}(F), \mathbb{Z}\big) \quad \text{given by} \quad \mathfrak{b} \mapsto \operatorname{sgn}\mathfrak{b}$$

called the *total signature map*. It is a ring homomorphism. The Local-Global Theorem 31.22 in this terminology states

$$W_t(F) = \operatorname{Ker}(\operatorname{sgn}).$$

We turn to the cokernel of $\operatorname{sgn} : W(F) \to C\big(\mathfrak{X}(F), \mathbb{Z}\big)$. We shall show that it too is a 2-primary torsion group. This generalizes the two observations that $C\big(\mathfrak{X}(F), \mathbb{Z}\big) = 0$ if F is not formally real and $\operatorname{sgn} : W(F) \to C\big(\mathfrak{X}(F), \mathbb{Z}\big)$ is an isomorphism if F is euclidean. We use and generalize the approach and results from [**33**] that we shall need in §41.

If $A \subset \mathfrak{X}(F)$, write χ_A for the characteristic function of A. In particular, $\chi_A \in C\big(\mathfrak{X}(F), \mathbb{Z}\big)$ if A is clopen. Let $f \in C\big(\mathfrak{X}(F), \mathbb{Z}\big)$. Then $A_n = f^{-1}(n)$ is a clopen set. As $\{A_n \mid n \in \mathbb{Z}\}$ partition the compact space $\mathfrak{X}(F)$, only finitely many A_n are nonempty. In particular, $f = \sum n\chi_{A_n}$ is a finite sum. This shows that $C\big(\mathfrak{X}(F), \mathbb{Z}\big)$ is additively generated by χ_A, as A varies over the clopen sets in the boolean space $\mathfrak{X}(F)$.

The finite intersections of the subbase elements (33.1),

(33.5) $$H(a_1, \dots, a_n) := H(a_1) \cap \dots \cap H(a_n) \quad \text{with} \quad a_1, \dots, a_n \in F^\times$$

form a base for the topology of $\mathfrak{X}(F)$. As

$$H(a_1, \dots, a_n) = \operatorname{supp}\big(\langle\langle a_1, \dots, a_n \rangle\rangle\big),$$

where $\operatorname{supp}\mathfrak{b} := \{P \in \mathfrak{X}(F) \mid \operatorname{sgn}_P \mathfrak{b} \neq 0\}$ is the *support* of $\mathfrak{b}$, this base is none other than the collection of clopen sets

(33.6) $$\{\operatorname{supp}(\mathfrak{b}) \mid \mathfrak{b} \text{ is a bilinear Pfister form}\}.$$

We also have

(33.7) $$\operatorname{sgn}\mathfrak{b} = 2^n \chi_{\operatorname{supp}(\mathfrak{b})} \quad \text{if } \mathfrak{b} \text{ is a bilinear } n\text{-fold Pfister form.}$$

Theorem 33.8. *The cokernel of* $\operatorname{sgn} : W(F) \to C\big(\mathfrak{X}(F), \mathbb{Z}\big)$ *is 2-primary torsion.*

PROOF. It suffices to prove for each clopen set $A \subset \mathfrak{X}(F)$ that $2^n\chi_A \in \operatorname{im}(\operatorname{sgn})$ for some $n \geq 0$. As $\mathfrak{X}(F)$ is compact, A is a finite union of clopen sets of the form (33.6) whose characteristic functions lie in im(sgn) by (33.7). By induction, it suffices to show that if A and B are clopen sets in $\mathfrak{X}(F)$ with $2^n\chi_A$ and $2^m\chi_B$ lying in im(sgn) for some integers m and n, then $2^s\chi_{A\cup B}$ lies in im(sgn) for some s. But

$$\chi_{A\cup B} = \chi_A + \chi_B - \chi_A \cdot \chi_B, \tag{33.9}$$

so

$$2^{m+n}\chi_{A\cup B} = 2^m(2^n\chi_A) + 2^n(2^m\chi_B) - (2^n\chi_A)\cdot(2^m\chi_B) \tag{33.10}$$

lies in im(sgn) as needed. □

Define the *reduced stability index* $\operatorname{st}_r(F)$ of F to be n if 2^n is the exponent of the cokernel of the signature map (or infinity if this exponent is not finite).

Refining the argument in the last theorem, we establish:

Lemma 33.11. *Let $C \subset \mathfrak{X}(F)$ be clopen. Then there exists an integer $n > 0$ and a $\mathfrak{b} \in I^n(F)$ satisfying* $\operatorname{sgn}\mathfrak{b} = 2^n\chi_C$. *More precisely, there exists an integer $n > 0$, bilinear n-fold Pfister forms $\mathfrak{b}_i$ satisfying* $\operatorname{supp}(\mathfrak{b}_i) \subset C$, *and integers k_i such that* $\sum k_i \operatorname{sgn}\mathfrak{b}_i = 2^n\chi_C$.

PROOF. As $\mathfrak{X}(F)$ is compact and (33.6) is a base for the topology, there exists an $r \geq 1$ such that $C = A_1 \cup \cdots \cup A_r$ with $A_i = \operatorname{supp}(\mathfrak{b}_i)$ for some m_i-fold Pfister forms $\mathfrak{b}_i$, $i \in [1, r]$. We induct on r. If $r = 1$ the result follows by (33.7), so assume that $r > 1$. Let $A = A_1$, $\mathfrak{b} = \mathfrak{b}_1$, and $B = A_2 \cup \cdots \cup A_r$. By induction, there exists an $m \geq 1$ and a $\mathfrak{c} \in I^m(F)$, a sum (and difference) of Pfister forms with the desired properties with $\operatorname{sgn}\mathfrak{c} = 2^m\chi_B$. Multiplying by a suitable power of 2, we may assume that $m = m_1$. Let $\mathfrak{d} = 2^m(\mathfrak{b} \perp \mathfrak{c}) \perp (-\mathfrak{b}) \otimes \mathfrak{c}$. Then $\mathfrak{d}$ is a sum (and difference) of Pfister forms whose supports all lie in C as $\operatorname{supp}(\mathfrak{a}) = \operatorname{supp}(2\mathfrak{a})$ for any bilinear form $\mathfrak{a}$. By equations (33.9) and (33.10), we have

$$\begin{aligned} 2^{2m}\chi_{A\cup B} &= 2^{2m}\chi_A + 2^{2m}\chi_B - 2^m\chi_A \cdot 2^m\chi_B \\ &= 2^m(\operatorname{sgn}\mathfrak{b} + \operatorname{sgn}\mathfrak{c}) - \operatorname{sgn}\mathfrak{b}\cdot\operatorname{sgn}\mathfrak{c} = \operatorname{sgn}\mathfrak{d}, \end{aligned}$$

so the result follows. □

Using the lemma, we can establish two useful results. The first is:

Theorem 33.12 (Normality Theorem). *Let A and B be disjoint closed subsets of $\mathfrak{X}(F)$. Then there exists an integer $n > 0$ and $\mathfrak{b} \in I^n(F)$ satisfying*

$$\operatorname{sgn}_P \mathfrak{b} = \begin{cases} 2^n & \text{if } P \in A, \\ 0 & \text{if } P \in B. \end{cases}$$

PROOF. The complement $\mathfrak{X}(F) \setminus B$ is a union of clopen sets. As the closed set A is covered by this union of clopen sets and $\mathfrak{X}(F)$ is compact, there exists a finite cover $\{C_1, \ldots, C_r\}$ of A for some clopen sets C_i, $i \in [1, r]$ lying in $\mathfrak{X}(F) \setminus B$. As $C_i \setminus \big(C_i \cap (\bigcup_{j\neq i} C_j)\big)$ is clopen for $i \in [1, r]$, we may assume this is a disjoint union. By Lemma 33.11, there exist $\mathfrak{b}_i \in I^{m_i}(F)$, some m_i, such that $\operatorname{sgn}\mathfrak{b}_i = 2^{m_i}\chi_{C_i}$. Let $n = \max\{m_i \mid 1 \leq i \leq r\}$. Then $\mathfrak{b} = \sum_i 2^{n-m_i}\mathfrak{b}_i$ lies in $I^n(F)$ and satisfies $\mathfrak{b} = 2^n\chi_{\cup_i C_i}$. Since $A \subset \cup_i C_i$, the result follows. □

We now investigate the relationship between elements in $f \in C(\mathfrak{X}(F), 2^m\mathbb{Z})$ and bilinear forms $\mathfrak{b}$ satisfying $2^m \mid \operatorname{sgn}_P \mathfrak{b}$ for all $P \in \mathfrak{X}(F)$. We first need a useful trick.

If $\varepsilon = (\varepsilon_1, \dots, \varepsilon_n) \in \{\pm 1\}^n$ and $\mathfrak{b} = \langle\langle a_1, \dots, a_n \rangle\rangle$ with $a_i \in F^\times$, let

$$\mathfrak{b}_\varepsilon = \langle\langle \varepsilon_1 a_1, \dots, \varepsilon_n a_n \rangle\rangle.$$

Then $\operatorname{supp}(\mathfrak{b}_\varepsilon) \cap \operatorname{supp}(\mathfrak{b}_{\varepsilon'}) = \emptyset$ unless $\varepsilon = \varepsilon'$.

Lemma 33.13. *Let $\mathfrak{b}$ be a bilinear n-fold Pfister form over an arbitrary field F. Then $2^n\langle 1 \rangle = \sum_\varepsilon \mathfrak{b}_\varepsilon$ in $W(F)$, where the sum runs over all $\varepsilon \in \{\pm 1\}^n$.*

PROOF. Let $\mathfrak{b} = \langle\langle a_1, \dots, a_n \rangle\rangle$ and $\mathfrak{c} = \langle\langle a_1, \dots, a_{n-1} \rangle\rangle$ with $a_i \in F^\times$. As $\langle\langle -1 \rangle\rangle = \langle\langle a \rangle\rangle + \langle\langle -a \rangle\rangle$ in $W(F)$ for all $a \in F^\times$, we have

$$\sum_\varepsilon \mathfrak{b}_\varepsilon = \sum_{\varepsilon'} \mathfrak{c}_{\varepsilon'} \langle\langle a_n \rangle\rangle + \sum_{\varepsilon'} \mathfrak{c}_{\varepsilon'} \langle\langle -a_n \rangle\rangle = 2 \sum_{\varepsilon'} \mathfrak{c}_{\varepsilon'}$$

where the ε' run over all $\{\pm 1\}^{n-1}$. The result follows by induction on n. □

The generalization of Lemma 33.13 given in the following exercise is useful. These identities were first observed by Witt, who used them to give a simple proof that $W_t(F)$ is 2-primary (cf. Exercise 31.21(3)).

Exercise 33.14. Let $a_1, \dots, a_n \in F^\times$ and $\varepsilon = (\varepsilon_1, \dots, \varepsilon_n) \in \{\pm 1\}^n$. Suppose that $\mathfrak{b} = \langle\langle a_1, \dots, a_n \rangle\rangle$, $\mathfrak{c} = \langle -a_1, \dots, -a_n \rangle$, and $\mathfrak{e}_\varepsilon = \langle \varepsilon_1, \dots, \varepsilon_n \rangle$. Then the following are true in $W(F)$:

(1) $\mathfrak{c} \cdot \mathfrak{b}_\varepsilon = \mathfrak{e}_\varepsilon \cdot \mathfrak{b}_\varepsilon = (\sum_i \varepsilon_i)\mathfrak{b}_\varepsilon$.

(2) $2^n \mathfrak{c} = \sum_\varepsilon \mathfrak{e}_\varepsilon \cdot \mathfrak{b}_\varepsilon$.

Using Lemma 33.11, we also establish:

Theorem 33.15. *Let $f \in C(\mathfrak{X}(F), 2^m\mathbb{Z})$. Then there is a positive integer n and a $\mathfrak{b} \in I^{m+n}(F)$ such that $2^n f = \operatorname{sgn} \mathfrak{b}$. More precisely, there exists an integer n such that $2^n f$ can be written as a sum $\sum_{i=1}^r k_i \operatorname{sgn} \mathfrak{b}_i$ for some integers k_i and bilinear $(n+m)$-fold Pfister forms $\mathfrak{b}_i$ such that $\operatorname{supp}(\mathfrak{b}_i) \subset \operatorname{supp}(f)$ for every $i = 1, \dots, r$ and whose supports are pairwise disjoint.*

PROOF. We first show:

Claim: Let $g \in C(\mathfrak{X}(F), \mathbb{Z})$. Then there exists a nonnegative integer n and bilinear n-fold Pfister forms $\mathfrak{c}_i$ such that $2^n g = \sum_{i=1}^r s_i \operatorname{sgn} \mathfrak{c}_i$ for some integers s_i with $\operatorname{supp}(\mathfrak{c}_i) \subset \operatorname{supp}(g)$ for every $i = 1, \dots, r$.

The function g is a finite sum of functions $\sum_i i\chi_{g^{-1}(i)}$ with $i \in \mathbb{Z}$ and each $g^{-1}(i)$ a clopen set. For each nonempty $g^{-1}(i)$, there exist a nonnegative integer n_i, bilinear n_i-fold Pfister forms $\mathfrak{b}_{ij}$ with $\operatorname{supp}(\mathfrak{b}_{ij}) \subset g^{-1}(i)$ and integers k_j satisfying $2^{n_i}\chi_{g^{-1}(i)} = \sum_j k_j \operatorname{sgn} \mathfrak{b}_{ij}$ by Lemma 33.11. Let $n = \max_i\{n_i\}$. Then $2^n g = \sum_{i,j} ik_j \operatorname{sgn}(2^{n-n_i}\mathfrak{b}_{ij})$. This proves the Claim.

Let $g = f/2^m$. By the Claim, $2^n g = \sum_{i=1}^r s_i \operatorname{sgn} \mathfrak{c}_i$ for some n-fold Pfister forms $\mathfrak{c}_i$ whose supports lie in $\operatorname{supp}(g) = \operatorname{supp}(f)$. Thus $2^n f = \sum_{i=1}^r s_i \operatorname{sgn} 2^m \mathfrak{c}_i$ with each $2^m\mathfrak{c}_i$ an $(n+m)$-fold Pfister form. Let $\mathfrak{d} = \mathfrak{c}_1 \otimes \cdots \otimes \mathfrak{c}_r$ be an rn-fold Pfister form. By Lemma 33.13, we have $2^{(n+1)r} f = \sum_\varepsilon \operatorname{sgn}(2^m s_i \mathfrak{c}_i \cdot \mathfrak{d}_\varepsilon)$ in $C(\mathfrak{X}(F), \mathbb{Z})$ where ε runs over all $\{\pm 1\}^{rn}$. For each i and ε, the form $\mathfrak{c}_i \cdot \mathfrak{d}_\varepsilon$ is isometric to either $2^{n+m}\mathfrak{d}_\varepsilon$ or is metabolic by Example 4.16(2) and (3). As the $\mathfrak{d}_\varepsilon$ have pairwise disjoint supports, adding the coefficients of the isometric forms $\mathfrak{c}_i \cdot \mathfrak{d}_\varepsilon$ yields the result. □

Corollary 33.16. *Let $\mathfrak{b}$ be a nondegenerate symmetric bilinear form over F and fix $m > 0$. Then $2^n\mathfrak{b} \in I^{n+m}(F)$ for some $n \geq 0$ if and only if*

$$\operatorname{sgn}\mathfrak{b} \in C\big(\mathfrak{X}(F), 2^m\mathbb{Z}\big).$$

PROOF. We may assume that F is formally real as $2s(F)W(F) = 0$.

$\Rightarrow$: If $\mathfrak{d}$ is a bilinear n-fold Pfister form, then $\operatorname{sgn}\mathfrak{d} \in C\big(\mathfrak{X}(F), 2^n\mathbb{Z}\big)$. It follows that $\operatorname{sgn}\big(I^n(F)\big) \subset C\big(\mathfrak{X}(F), 2^n\mathbb{Z}\big)$. Suppose that $2^n\mathfrak{b} \in I^{n+m}(F)$ for some $n \geq 0$. Then $2^n \operatorname{sgn}\mathfrak{b} \in C\big(\mathfrak{X}(F), 2^{n+m}\mathbb{Z}\big)$, hence $\operatorname{sgn}\mathfrak{b} \in C\big(\mathfrak{X}(F), 2^m\mathbb{Z}\big)$.

$\Leftarrow$: By Theorem 33.15, there exists $\mathfrak{c} \in I^{n+m}(F)$ such that $\operatorname{sgn}\mathfrak{c} = 2^n \operatorname{sgn}\mathfrak{b}$. As $W_t(F) = \operatorname{Ker}(\operatorname{sgn})$ is 2-primary torsion by the Local-Global Principle 31.22, there exists a nonnegative integer k such that $2^{n+k}\mathfrak{b} = 2^k\mathfrak{c} \in I^{n+m+k}(F)$. □

This Corollary 33.16 suggests that if $\mathfrak{b}$ is a nondegenerate symmetric bilinear form over F, the following may be true:

(33.17) $\qquad \operatorname{sgn}\mathfrak{b} \in C\big(\mathfrak{X}(F), 2^n\mathbb{Z}\big) \quad$ if and only if $\quad \mathfrak{b} \in I^n(F) + W_t(F)$.

This question was raised by Lam in [**88**].

In particular, in the case that F is a formally real pythagorean field, this would mean

$$\mathfrak{b} \in I^n(F) \quad \text{if and only if} \quad 2^n \mid \operatorname{sgn}_P(\mathfrak{b}) \text{ for all } P \in \mathfrak{X}(F)$$

as $W(F)$ is then torsion-free which would answer a question of Marshall in [**97**].

Of course, if $\mathfrak{b} \in I^n(F) + W_t(F)$, then $\operatorname{sgn}\mathfrak{b} \in C\big(\mathfrak{X}(F), 2^n\mathbb{Z}\big)$. The converse would follow if

$$2^m\mathfrak{b} \in I^{n+m}(F) \quad \text{always implies that} \quad \mathfrak{b} \in I^n(F) + W_t(F).$$

If F were formally real pythagorean, the converse would follow if

$$2^m\mathfrak{b} \in I^{n+m}(F) \quad \text{always implies that} \quad \mathfrak{b} \in I^n(F).$$

Because the nilradical of $W(F)$ is the torsion $W_t(F)$ when F is formally real, the total signature induces an embedding of the *reduced Witt ring*

$$W_{red}(F) := W(F)/\operatorname{nil}\big(W(F)\big) = W(F)/W_t(F)$$

into $C\big(\mathfrak{X}(F), \mathbb{Z}\big)$. Moreover, since $W_t(F)$ is 2-primary, the images of two nondegenerate bilinear forms $\mathfrak{b}$ and $\mathfrak{c}$ are equal in the reduced Witt ring if and only if there exists a nonnegative integer n such that $2^n\mathfrak{b} = 2^n\mathfrak{c}$ in $W(F)$. Let $^- : W(F) \to W_{red}(F)$ be the canonical ring epimorphism. Then the problem above becomes: If $\mathfrak{b}$ is a nondegenerate symmetric bilinear form over F, then

$$\bar{\mathfrak{b}} \in I^n_{red}(F) \quad \text{if and only if} \quad \operatorname{sgn}\mathfrak{b} \in C\big(\mathfrak{X}(F), 2^n\mathbb{Z}\big).$$

where $I^n_{red}(F)$ is the image of $I^n(F)$ in $W_{red}(F)$.

This is all, in fact, true as we shall see in §41 (Cf. Corollaries 41.9 and 41.10).

34. Bilinear and quadratic forms under quadratic extensions

In this section we develop the relationship between bilinear and quadratic forms over a field F and over a quadratic extension K of F. We know that bilinear and quadratic forms can become isotropic over a quadratic extension and we exploit this. We also investigate the transfer map taking forms over K to forms over F induced by a nontrivial F-linear functional. This leads to useful exact sequences of Witt rings and Witt groups done first in the case of characteristic not 2 in [**37**] and by Arason in [**4**] in the case of characteristic 2 by Baeza in [**15**].

34.A. Bilinear forms under a quadratic extension.

We start with bilinear forms and the Witt ring.

Proposition 34.1. *Let K/F be a quadratic field extension and $s : K \to F$ a nontrivial F-linear functional satisfying $s(1) = 0$. Let $\mathfrak{c}$ be an anisotropic bilinear form over K. Then there exist bilinear forms $\mathfrak{b}$ over F and $\mathfrak{a}$ over K such that $\mathfrak{c} \simeq \mathfrak{b}_K \perp \mathfrak{a}$ with $s_*(\mathfrak{a})$ anisotropic.*

PROOF. We induct on $\dim \mathfrak{c}$. Suppose that $s_*(\mathfrak{c})$ is isotropic. It follows that there is a $b \in D(\mathfrak{c}) \cap F$, i.e., $\mathfrak{c} \simeq \langle b \rangle \perp \mathfrak{c}_1$ for some $\mathfrak{c}_1$. Applying the induction hypothesis to $\mathfrak{c}_1$ completes the proof. □

We need the following generalization of Proposition 34.1.

Lemma 34.2. *Let K/F be a quadratic extension of F and $s : K \to F$ a nontrivial F-linear functional satisfying $s(1) = 0$. Let $\mathfrak{f}$ be a bilinear anisotropic n-fold Pfister form over F and $\mathfrak{c}$ a nondegenerate bilinear form over K such that $\mathfrak{f}_K \otimes \mathfrak{c}$ is anisotropic. Then there exists a bilinear form $\mathfrak{b}$ over F and a bilinear form $\mathfrak{a}$ over K such that $\mathfrak{f}_K \otimes \mathfrak{c} \simeq (\mathfrak{f} \otimes \mathfrak{b})_K \perp \mathfrak{f}_K \otimes \mathfrak{a}$ with $\mathfrak{f} \otimes s_*(\mathfrak{a})$ anisotropic.*

PROOF. Let $\mathfrak{d} = \mathfrak{f}_K \otimes \mathfrak{c}$. We may assume that $s_*(\mathfrak{d})$ is isotropic. Then there exists a $b \in D(\mathfrak{d}) \cap F$. If $\mathfrak{c} \simeq \langle a_1, \dots, a_n \rangle$, there exist $x_i \in \widetilde{D}(\mathfrak{f}_K)$, not all zero satisfying $b = x_1 a_1 + \cdots + x_n a_n$. Let $y_i = x_i$ if $x_i \neq 0$ and $y_i = 1$ otherwise. Then

$$\mathfrak{f}_K \otimes \mathfrak{c} \simeq \mathfrak{f}_K \otimes \langle y_1 a_1, \dots, y_n a_n \rangle \simeq \mathfrak{f}_K \otimes \langle b, z_2, \dots, z_n \rangle$$

for some $z_i \in K^\times$ as $G(\mathfrak{f}_K) = D(\mathfrak{f}_K)$. The result follows easily by induction. □

Corollary 34.3. *Let K/F be a quadratic extension of F and $s : K \to F$ a nontrivial F-linear functional satisfying $s(1) = 0$. Let $\mathfrak{f}$ be a bilinear anisotropic n-fold Pfister form and $\mathfrak{c}$ an anisotropic bilinear form over K satisfying $\mathfrak{f} \otimes s_*(\mathfrak{c})$ is hyperbolic. Then there exists a bilinear form $\mathfrak{b}$ over F such that $\dim \mathfrak{b} = \dim \mathfrak{c}$ and $\mathfrak{f}_K \otimes \mathfrak{c} \simeq (\mathfrak{f} \otimes \mathfrak{b})_K$.*

PROOF. If $\mathfrak{f}_K \otimes \mathfrak{c}$ is anisotropic, the result follows by Lemma 34.2, so we may assume that $\mathfrak{f}_K \otimes \mathfrak{c}$ is isotropic. If $\mathfrak{f}_K$ is isotropic, it is hyperbolic and the result follows easily, so we may assume the Pfister form $\mathfrak{f}_K$ is anisotropic. Using Proposition 6.22, we see that there exist a bilinear form $\mathfrak{d}$ with $\mathfrak{f}_K \otimes \mathfrak{d}$ anisotropic and an integer $n \geq 0$ with $\dim \mathfrak{d} + 2n = \dim \mathfrak{c}$ and $\mathfrak{f}_K \otimes \mathfrak{c} \simeq \mathfrak{f}_K \otimes (\mathfrak{d} \perp n\mathbb{H})$. Replacing $\mathfrak{c}$ by $\mathfrak{d}$, we reduce to the anisotropic case. □

Note that if K/F is a quadratic extension and $s, s' : K \to F$ are F-linear functionals satisfying $s(1) = 0 = s'(1)$ with s nontrivial, then $s'_* = as_*$ for some $a \in F$.

Theorem 34.4. *Let K/F be a quadratic field extension and $s : K \to F$ a nonzero F-linear functional such that $s(1) = 0$. Then the sequence*

$$W(F) \xrightarrow{r_{K/F}} W(K) \xrightarrow{s_*} W(F)$$

is exact.

PROOF. Let $b \in F^\times$ then the binary form $s_*\big(\langle b\rangle_K\big)$ is isotropic, hence metabolic. Thus $s_* \circ r_{K/F} = 0$. Let $\mathfrak{c} \in W(K)$. By Proposition 34.1, there exists a decomposition $\mathfrak{c} \simeq \mathfrak{b}_K \perp \mathfrak{c}_1$ with $\mathfrak{b}$ a bilinear form over F and $\mathfrak{c}_1$ a bilinear form over K satisfying $s_*(\mathfrak{c}_1)$ is anisotropic. In particular, if $s_*(\mathfrak{c}) = 0$, we have $\mathfrak{c} = \mathfrak{b}_K$. This proves exactness. □

If K/F is a quadratic extension, denote the quadratic norm form of the quadratic algebra K by $\mathrm{N}_{K/F}$ (cf. §98.B).

Lemma 34.5. *Let K/F be a quadratic extension and $s : K \to F$ a nontrivial F-linear functional. Let $\mathfrak{b}$ be an anisotropic binary bilinear form over F such that the quadratic form $\mathfrak{b} \otimes \mathrm{N}_{K/F}$ is isotropic. Then $\mathfrak{b} \simeq s_*\big(\langle y\rangle\big)$ for some $y \in K^\times$.*

PROOF. Let $\{1, x\}$ be a basis of K over F. Let $\mathfrak{c}$ be the polar form of $\mathrm{N}_{K/F}$. We have

$$\mathfrak{c}(1, x) = \mathrm{N}_{K/F}(1 + x) - \mathrm{N}_{K/F}(x) - \mathrm{N}_{K/F}(1) = \mathrm{Tr}_{K/F}(x)$$

for every $x \in K$. By assumption there are nonzero vectors $v, w \in V_{\mathfrak{b}}$ such that

$$\begin{aligned} 0 &= (\mathfrak{b} \otimes \mathrm{N}_{K/F})(v \otimes 1 + w \otimes x) \\ &= \mathfrak{b}(v, v)\,\mathrm{N}_{K/F}(1) + \mathfrak{b}(v, w)\mathfrak{c}(1, x) + \mathfrak{b}(w, w)\,\mathrm{N}_{K/F}(x) \\ &= \mathfrak{b}(v, v) + \mathfrak{b}(v, w)\,\mathrm{Tr}_{K/F}(x) + \mathfrak{b}(w, w)\,\mathrm{N}_{K/F}(x) \end{aligned}$$

by the definition of tensor product (8.14). Let $f : K \to F$ be an F-linear functional satisfying $f(1) = \mathfrak{b}(w, w)$ and $f(x) = \mathfrak{b}(v, w)$. By (98.1), we have

$$\begin{aligned} f(x^2) &= f\big(-\mathrm{Tr}_{K/F}(x)x - \mathrm{N}_{K/F}(x)\big) \\ &= -\mathrm{Tr}_{K/F}(x)\mathfrak{b}(v, w) - \mathrm{N}_{K/F}(x)\mathfrak{b}(w, w) = \mathfrak{b}(v, v). \end{aligned}$$

Therefore, the F-linear isomorphism $K \to V_{\mathfrak{b}}$ taking 1 to w and x to v is an isometry between $\mathfrak{c} = f_*\big(\langle 1\rangle\big)$ and $\mathfrak{b}$. As f is the composition of s with the endomorphism of K given by multiplication by some element $y \in K^\times$, we have $\mathfrak{b} \simeq f_*\big(\langle 1\rangle\big) \simeq s_*\big(\langle y\rangle\big)$. □

Proposition 34.6. *Let K/F be a quadratic extension and $s : K \to F$ a nontrivial F-linear functional. Let $\mathfrak{b}$ be an anisotropic bilinear form over F. Then there exist bilinear forms $\mathfrak{c}$ over K and $\mathfrak{d}$ over F such that $\mathfrak{b} \simeq s_*(\mathfrak{c}) \perp \mathfrak{d}$ and $\mathfrak{d} \otimes \mathrm{N}_{K/F}$ is anisotropic.*

PROOF. We induct on $\dim \mathfrak{b}$. Suppose that $\mathfrak{b} \otimes \mathrm{N}_{K/F}$ is isotropic. Then there is a 2-dimensional subspace $W \subset V_{\mathfrak{b}}$ with $(\mathfrak{b}|_W) \otimes \mathrm{N}_{K/F}$ isotropic. By Lemma 34.5, we have $\mathfrak{b}|_W \simeq s_*\big(\langle y\rangle\big)$ for some $y \in K^\times$. Applying the induction hypothesis to the orthogonal complement of W in V completes the proof. □

Theorem 34.7. *Let $K = F(\sqrt{a})$ be a quadratic field extension of F with $a \in F^\times$. Let $s : K \to F$ be a nontrivial F-linear functional such that $s(1) = 0$. Then the sequence*

$$W(K) \xrightarrow{s_*} W(F) \xrightarrow{\langle\langle a \rangle\rangle} W(F)$$

is exact where the last homomorphism is multiplication by $\langle\langle a \rangle\rangle$.

PROOF. For every $\mathfrak{c} \in W(F)$ we have $\langle\langle a \rangle\rangle s_*(\mathfrak{c}) = s_*\big(\langle\langle a \rangle\rangle_K \mathfrak{c}\big) = 0$ as $\langle\langle a \rangle\rangle_K = 0$. Therefore, the composition of the two homomorphisms in the sequence is trivial. Since $\mathrm{N}_{K/F} \simeq \langle\langle a \rangle\rangle_q$, the exactness of the sequence now follows from Proposition 34.6. □

34.B. Quadratic forms under a quadratic extension. We now turn to quadratic forms.

Proposition 34.8. *Let K/F be a separable quadratic field extension and φ an anisotropic quadratic form over F. Then*

$$\varphi \simeq (\mathfrak{b} \otimes \mathrm{N}_{K/F}) \perp \psi$$

with $\mathfrak{b}$ a nondegenerate symmetric bilinear form and ψ a quadratic form satisfying ψ_K is anisotropic.

PROOF. Since K/F is separable, the binary form $\sigma := \mathrm{N}_{K/F}$ is nondegenerate. As $F(\sigma) \simeq K$, the statement follows from Corollary 22.12. □

Theorem 34.9. *Let K/F be a separable quadratic field extension and $s : K \to F$ a nonzero functional such that $s(1) = 0$. Then the sequence*

$$W(F) \xrightarrow{r_{K/F}} W(K) \xrightarrow{s_*} W(F) \xrightarrow{\mathrm{N}_{K/F}} I_q(F) \xrightarrow{r_{K/F}} I_q(K) \xrightarrow{s_*} I_q(F)$$

is exact where the middle homomorphism is multiplication by $\mathrm{N}_{K/F}$.

PROOF. In view of Theorem 34.4 and Propositions 34.6 and 34.8, it suffices to prove exactness at $I_q(K)$. Let $\varphi \in I_q(K)$ be an anisotropic form such that $s_*(\varphi)$ is hyperbolic. We show by induction on $n = \dim_K \varphi$ that $\varphi \in \mathrm{im}(r_{K/F})$. We may assume that $n > 0$. Let $W \subset V_\varphi$ be a totally isotropic F-subspace for the form $s_*(\varphi)$ of dimension n. As $\mathrm{Ker}(s) = F$ we have $\varphi(W) \subset F$.

We claim that the K-space KW properly contains W, in particular,

$$\dim_K KW = \frac{1}{2}\dim_F KW > \frac{1}{2}\dim_F W = \frac{n}{2}. \tag{34.10}$$

To prove the claim choose an element $x \in K$ such that $x^2 \notin F$. Then for every nonzero $w \in W$, we have $\varphi(xw) = x^2\varphi(w) \notin F$, hence $xw \in KW$ but $x \notin W$. It follows from the inequality (34.10) that the restriction of $\mathfrak{b}_\varphi$ on KW and therefore on W is nonzero. Consequently, there is a 2-dimensional F-subspace $U \subset W$ such that $\mathfrak{b}_\varphi|_U$ is nondegenerate. Therefore, the K-space KU is also 2-dimensional and the restriction $\psi = \varphi|_U$ is a nondegenerate binary quadratic form over F satisfying $\psi_K \simeq \varphi|_{KU}$. Applying the induction hypothesis to $(\psi_K)^\perp$, we have $(\psi_K)^\perp \in \mathrm{im}(r_{K/F})$. Therefore, $\varphi = \psi_K + (\psi_K)^\perp \in \mathrm{im}(r_{K/F})$. □

Remark 34.11. In Proposition 34.9, we have $\mathrm{Ker}\big(r_{K/F} : I_q(F) \to I_q(K)\big) = W(F)\langle\langle a]]$ when $K = F_a$.

Application of the above in the case of fields of characteristic not 2 provides a proof of the following result shown in [**37**] and by Arason in [**4**]:

Corollary 34.12. *Suppose that* $\operatorname{char} F \neq 2$ *and* $K = F(\sqrt{a})/F$ *is a quadratic field extension with* $a \in F^\times$. *If* $s : K \to F$ *is a nontrivial* F*-linear functional such that* $s(1) = 0$, *then the triangle*

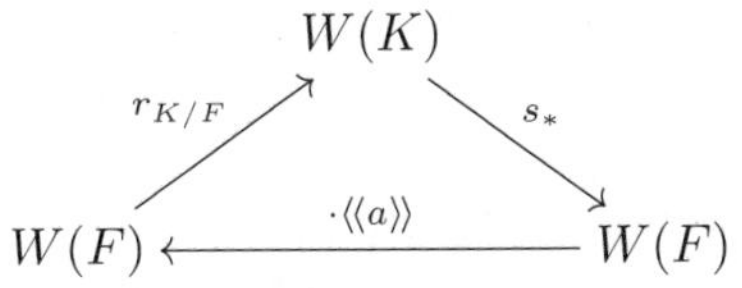

is exact.

PROOF. Since the quadratic norm form $N_{K/F}$ coincides with $\varphi_{\mathfrak{b}}$ where $\mathfrak{b} = \langle\langle a\rangle\rangle$, the map $W(F) \to I_q(F)$ given by multiplication by $N_{K/F}$ is identified with the map $W(F) \to I(F)$ given by multiplication by $\langle\langle a\rangle\rangle$. Note also that $\operatorname{Ker}(r_{K/F}) \subset I(F)$, so the statement follows from Theorem 34.9. □

Remark 34.13. Suppose that $\operatorname{char} F \neq 2$ and $K = F(\sqrt{a})$ is a quadratic extension of F. Let $\mathfrak{b}$ be an anisotropic bilinear form. Then by Proposition 34.8 and Example 9.4, we see that the following are equivalent:

(1) $\mathfrak{b}_K$ is metabolic.

(2) $\mathfrak{b} \in \langle\langle a\rangle\rangle W(F)$.

(3) $\mathfrak{b} \simeq \langle\langle a\rangle\rangle \otimes \mathfrak{c}$ for some symmetric bilinear form $\mathfrak{c}$.

In the case that $\operatorname{char} F = 2$, Theorem 34.9 can be slightly improved.

We need the following computation:

Lemma 34.14. *Let* F *be a field of characteristic* 2 *and* K/F *a quadratic field extension. Let* $s : K \to F$ *be a nonzero* F*-linear functional satisfying* $s(1) = 0$. *Then for every* $x \in K$ *we have*

$$s_*\big(\langle\langle x]]\big) = \begin{cases} 0 & \text{if } x \in F, \\ s(x)\langle\langle \operatorname{Tr}_{K/F}(x)]] & \text{otherwise.} \end{cases}$$

In particular, $s_*\big(\langle\langle x]]\big) \equiv \langle\langle \operatorname{Tr}_{K/F}(x)]]$ *modulo* $I_q^2(F)$.

PROOF. The element x satisfies the quadratic equation $x^2 + ax + b = 0$ for some $a, b \in F$. We have $\operatorname{Tr}_{K/F}(x) = a$ and $s(x^2) = as(x) = s(x)\operatorname{Tr}_{K/F}(x)$. Let $\bar{x} = \operatorname{Tr}_{K/F}(x) - x$. The element $\bar{x}$ satisfies the same quadratic equation and $s(\bar{x}^2) = s(x)\operatorname{Tr}_{K/F}(\bar{x})$.

Let $\{v, w\}$ be the standard basis for the space V of the form $\varphi := \langle\langle x]]$ over K. If $x \in F$, then v and w span the totally isotropic F-subspace of $s_*(\varphi)$, i.e., $s_*(\varphi) = 0$.

Suppose that $x \notin F$. Then $V = W \perp W'$ with $W = Fv \oplus Fxw$ and $W' = F\bar{x}v \oplus Fw$. We have $s_*(\varphi) \simeq s_*(\varphi)|_W \perp s_*(\varphi)|_{W'}$. As $s_*(\varphi)(v) = s(1) = 0$, the form $s_*(\varphi)|_W$ is isotropic and therefore $s_*(\varphi)|_W \simeq \mathbb{H}$. Moreover,

$$s_*(\varphi)(\bar{x}v) = s(\bar{x}^2) = s(x)\operatorname{Tr}_{K/F}(x), \quad s_*(\varphi)(w) = s(x) \quad \text{and}$$

$$s_*\big(\mathfrak{b}_\varphi(\bar{x}v, w)\big) = s(\bar{x}) = s(x),$$

hence $s_*(\varphi)|_{W'} \simeq s(x)\langle\langle \operatorname{Tr}_{K/F}(x)]]$. □

Corollary 34.15. *Suppose that* $\operatorname{char} F = 2$. *Let* K/F *be a separable quadratic field extension and* $s : K \to F$ *a nonzero functional such that* $s(1) = 0$. *Then the sequence*

$$0 \to W(F) \xrightarrow{r_{K/F}} W(K) \xrightarrow{s_*} W(F) \xrightarrow{\cdot N_{K/F}} I_q(F) \xrightarrow{r_{K/F}} I_q(K) \xrightarrow{s_*} I_q(F) \to 0$$

is exact.

PROOF. To prove the injectivity of $r_{K/F}$, it suffices to show that if $\mathfrak{b}$ is an anisotropic bilinear form over F, then $\mathfrak{b}_K$ is also anisotropic. Let $x \in K \setminus F$ be an element satisfying $x^2 + x + a = 0$ for some $a \in F$ and let $\mathfrak{b}_K(v + xw, v + xw) = 0$ for some $v, w \in V_{\mathfrak{b}}$. We have

$$0 = \mathfrak{b}_K(v + xw, v + xw) = \mathfrak{b}(v, v) + a\mathfrak{b}(w, w) + x\mathfrak{b}(w, w),$$

hence $\mathfrak{b}(w, w) = 0 = \mathfrak{b}(v, v)$. Therefore, $v = w = 0$ as $\mathfrak{b}$ is anisotropic.

By Lemma 34.14, we have for every $y \in K$, the form $s_*(\langle\langle y]])$ is similar to $\langle\langle \operatorname{Tr}_{K/F}(y)]]$. As the map s_* is $W(F)$-linear, $I_q(F)$ is generated by the classes of binary forms and the trace map $\operatorname{Tr}_{K/F}$ is surjective, the last homomorphism s_* in the sequence is surjective. □

34.C. The filtrations of the Witt ring and Witt group under quadratic extensions. We turn to the study of relations between the ideals $I^n(F)$, $I^n(K)$ and between the groups $I_q^n(F)$ and $I_q^n(K)$ for a quadratic field extension K/F.

Lemma 34.16. *Let* K/F *be a quadratic extension. Let* $n \geq 1$.

(1) *We have*

$$I^n(K) = I^{n-1}(F)I(K),$$

i.e., $I^n(K)$ *is the* $W(F)$*-module generated by* n*-fold bilinear Pfister forms* $\mathfrak{b}_K \otimes \langle\langle x\rangle\rangle$ *with* $x \in K^\times$ *and* $\mathfrak{b}$ *an* $(n-1)$*-fold bilinear Pfister form over* F.

(2) *If* $\operatorname{char} F = 2$, *then*

$$I_q^n(K) = I^{n-1}(F)I_q(K) + I(K)I_q^{n-1}(F).$$

PROOF. (1): Clearly, to show that $I^n(K) = I^{n-1}(F)I(K)$, it suffices to show this for the case $n = 2$. Let $x, y \in K \setminus F$. As $1, x, y$ are linearly dependent over F, there are $a, b \in F^\times$ such that $ax + by = 1$. Note that the form $\langle\langle ax, by\rangle\rangle$ is isotropic and therefore metabolic. Using the relation

$$\langle\langle uv, w\rangle\rangle = \langle\langle u, w\rangle\rangle + u\langle\langle v, w\rangle\rangle$$

in $W(K)$, we have

$$0 = \langle\langle ax, by\rangle\rangle = \langle\langle x, by\rangle\rangle + a\langle\langle x, by\rangle\rangle = \langle\langle a, b\rangle\rangle + b\langle\langle a, y\rangle\rangle + a\langle\langle x, b\rangle\rangle + ab\langle\langle x, y\rangle\rangle,$$

hence $\langle\langle x, y\rangle\rangle \in I(F)I(K)$.

(2): In view of (1), it is sufficient to consider the case $n = 2$. The group $I_q^2(K)$ is generated by the classes of 2-fold Pfister forms by (9.5). Let $x, y \in K$. If $x \in F$, then $\langle\langle x, y]] \in I(F)I_q(K)$. Otherwise, $y = a + bx$ for some $a, b \in F$. Then, by Lemma 15.1 and Lemma 15.5,

$$\langle\langle x, y]] = \langle\langle x, a]] + \langle\langle x, bx]] = \langle\langle x, a]] + \langle\langle b, bx]] \in I(K)I_q(F) + I(F)I_q(K)$$

since $\langle\langle b, bx]] + \langle\langle x, bx]] = \langle\langle bx, bx]] = 0$. □

Corollary 34.17. *Let K/F be a quadratic extension and $s : K \to F$ a nonzero F-linear functional. Then for every $n \geq 1$:*

(1) $s_*\big(I^n(K)\big) \subset I^n(F)$.
(2) $s_*\big(I_q^n(K)\big) \subset I_q^n(F)$.

PROOF. (1): Clearly, $s_*\big(I(K)\big) \subset I(F)$. It follows from Lemma 34.16 and Frobenius Reciprocity that

$$s_*\big(I^n(K)\big) = s_*\big(I^{n-1}(F)I(K)\big) = I^{n-1}(F)s_*\big(I(K)\big) \subset I^{n-1}(F)I(F) = I^n(F).$$

(2): This follows from (1) if $\operatorname{char} F \neq 2$ and from Lemma 34.16(2) and Frobenius Reciprocity if $\operatorname{char} F = 2$. □

Lemma 34.18. *Let K/F be a quadratic extension and $s, s' : K \to F$ two nonzero F-linear functionals. Let $\mathfrak{b} \in I^n(K)$. Then $s_*(\mathfrak{b}) \equiv s'_*(\mathfrak{b}) \mod I^{n+1}(F)$.*

PROOF. As in the proof of Corollary 20.7, there exists a $c \in K^\times$ such that $s'_*(\mathfrak{c}) = s_*(c\mathfrak{c})$ for all symmetric bilinear forms $\mathfrak{c}$. As $\mathfrak{b} \in I^n(K)$, we have $\langle\langle c\rangle\rangle \cdot \mathfrak{b} \in I^{n+1}(K)$. Consequently, $s_*(\mathfrak{b}) - s'_*(\mathfrak{b}) = s_*\big(\langle\langle c\rangle\rangle \cdot \mathfrak{b}\big)$ lies in $I^{n+1}(F)$. The result follows. □

Corollary 34.19. *Let K/F be a quadratic field extension and $s : K \to F$ a nontrivial F-linear functional. Then $s_*\big(\langle\langle x\rangle\rangle\big) \equiv \langle\langle \mathrm{N}_{K/F}(x)\rangle\rangle$ modulo $I^2(F)$ for every $x \in K^\times$.*

PROOF. By Lemma 34.18, we know that $s_*\big(\langle\langle x\rangle\rangle\big)$ is independent of the nontrivial F-linear functional s modulo $I^2(F)$. Using the functional defined in (20.8), the result follows by Corollary 20.13. □

Let K/F be a separable quadratic field extension and let $s : K \to F$ be a nontrivial F-linear functional such that $s(1) = 0$. It follows from Theorem 34.9 and Corollary 34.17 that we have a well-defined complex
(34.20)

$$I^n(F) \xrightarrow{r_{K/F}} I^n(K) \xrightarrow{s_*} I^n(F) \xrightarrow{\cdot \mathrm{N}_{K/F}} I_q^{n+1}(F) \xrightarrow{r_{K/F}} I_q^{n+1}(K) \xrightarrow{s_*} I_q^{n+1}(F)$$

and this induces (where by, abuse of notation, we label the maps in the same way)
(34.21)

$$\overline{I}^n(F) \xrightarrow{r_{K/F}} \overline{I}^n(K) \xrightarrow{s_*} \overline{I}^n(F) \xrightarrow{\cdot \mathrm{N}_{K/F}} \overline{I}_q^{n+1}(F) \xrightarrow{r_{K/F}} \overline{I}_q^{n+1}(K) \xrightarrow{s_*} \overline{I}_q^{n+1}(F).$$

where $\overline{I}^n(F) := I^n(F)/I^{n+1}(F)$.

By Lemma 34.18, it follows that the homomorphism s_* in (34.21) is independent of the nontrivial F-linear functional $K \to F$ although it is not independent in (34.20).

We show that the complexes (34.20) and (34.21) are exact on bilinear Pfister forms. More precisely, we have

Theorem 34.22. *Let K/F be a separable quadratic field extension and $s : K \to F$ a nontrivial F-linear functional such that $s(1) = 0$.*

(1) *Let $\mathfrak{c}$ be an anisotropic bilinear n-fold Pfister form over K. If $s_*(\mathfrak{c}) \in I^{n+1}(F)$, then there exists a bilinear n-fold Pfister form $\mathfrak{b}$ over F such that $\mathfrak{c} \simeq \mathfrak{b}_K$.*

(2) *Let* $\mathfrak{b}$ *be an anisotropic bilinear* n*-fold Pfister form over* F. *If* $\mathfrak{b}\cdot \mathrm{N}_{K/F} \in I^{n+2}(F)$, *then there exists a bilinear* n*-fold Pfister form* $\mathfrak{c}$ *over* K *such that* $\mathfrak{b} = s_*(\mathfrak{c})$.

(3) *Let* φ *be an anisotropic quadratic* $(n+1)$*-fold Pfister form over* F. *If* $r_{K/F}(\varphi) \in I^{n+2}(K)$, *then there exists a bilinear* n*-fold Pfister form* $\mathfrak{b}$ *over* F *such that* $\varphi \simeq \mathfrak{b}\otimes \mathrm{N}_{K/F}$.

(4) *Let* ψ *be an anisotropic* $(n+1)$*-fold quadratic Pfister form over* K. *If* $s_*(\psi) \in I^{n+2}(F)$, *then there exists a quadratic* $(n+1)$*-fold Pfister form* φ *over* F *such that* $\psi \simeq \varphi_K$.

PROOF. (1): As $\mathfrak{c}$ represents 1, the form $s_*(\mathfrak{c})$ is isotropic and belongs to $I^{n+1}(F)$. It follows from the Hauptsatz 23.7 that $s_*(\mathfrak{c}) = 0$ in $W(F)$. We show by induction on $k \geq 0$ that there is a bilinear k-fold Pfister form $\mathfrak{d}$ over F and a bilinear $(n-k)$-fold Pfister form $\mathfrak{e}$ over K such that $\mathfrak{c}\simeq \mathfrak{d}_K\otimes\mathfrak{e}$. The statement that we need follows when $k=n$.

Suppose we have $\mathfrak{d}$ and $\mathfrak{e}$ for some $k<n$. We have

$$0 = s_*(\mathfrak{c}) = s_*(\mathfrak{d}_K\cdot\mathfrak{e}' \perp \mathfrak{d}_K) = s_*(\mathfrak{d}_K\cdot\mathfrak{e}')$$

in $W(F)$ where, as usual, $\mathfrak{e}'$ denotes the pure subform of $\mathfrak{e}$. In particular, $s_*(\mathfrak{d}_K\otimes\mathfrak{e}')$ is isotropic. Thus there exists $b\in F^\times \cap D(\mathfrak{d}_K\otimes\mathfrak{e}')$. It follows that $\mathfrak{d}_K\otimes\mathfrak{e}\simeq \mathfrak{d}_K\otimes\langle\langle b\rangle\rangle\otimes\mathfrak{f}$ for some Pfister form $\mathfrak{f}$ over K by Theorem 6.15.

(2): By the Hauptsatz 23.7, we have $\mathfrak{b}\otimes\mathrm{N}_{K/F}$ is hyperbolic. We claim that $\mathfrak{b}\simeq\langle\langle a\rangle\rangle\otimes\mathfrak{a}$ for some $a\in \mathrm{N}_{K/F}(K^\times)$ and an $(n-1)$-fold bilinear Pfister form $\mathfrak{a}$ over F. If $\operatorname{char} F\neq 2$, the claim follows from Corollary 6.14. If $\operatorname{char} F = 2$, it follows from Lemma 9.11 that $\mathrm{N}_{K/F}\simeq\langle\langle a]]$ for some $a\in D(\mathfrak{b}')$. Clearly, $a\in\mathrm{N}_{K/F}(K^\times)$ and by Lemma 6.11, the form $\mathfrak{b}$ is divisible by $\langle\langle a\rangle\rangle$. The claim is proven.

As $a\in\mathrm{N}_{K/F}(K^\times)$ there is $y\in K^\times$ such that $s_*\big(\langle\langle y\rangle\rangle\big) = \langle\langle a\rangle\rangle$. It follows that $s_*\big(\langle\langle y\rangle\rangle\cdot\mathfrak{a}\big) = \langle\langle a\rangle\rangle\cdot\mathfrak{a} = \mathfrak{b}$.

(3): By the Hauptsatz 23.7, we have $r_{K/F}(\varphi) = 0$ in $I_q(K)$. The field K is isomorphic to the function field of the 1-fold Pfister form $\mathrm{N}_{K/F}$. The statement now follows from Corollary 23.6.

(4): In the case $\operatorname{char} F\neq 2$, the statement follows from (1). So we may assume that $\operatorname{char} F = 2$. As ψ represents 1, the form $s_*(\psi)$ is isotropic and belongs to $I_q^{n+2}(F)$. It follows from the Hauptsatz 23.7 that $s_*(\psi) = 0\in I_q(F)$. We show by induction for each $k\in[1,\ n]$ that there is a k-fold bilinear Pfister form $\mathfrak{d}$ over F and a quadratic Pfister form ρ over K such that $\psi\simeq\mathfrak{d}_K\otimes\rho$.

Suppose we have $\mathfrak{d}$ and ρ for some $k<n$. As $\dim(\mathfrak{d}_K\otimes\rho') > \frac{1}{2}\dim(\mathfrak{d}_K\otimes\rho)$ with ρ' the pure subform of ρ, the subspace of $s_*(\mathfrak{d}_K\otimes\rho')$ intersects a totally isotropic subspace of $s_*(\mathfrak{d}_K\otimes\rho)$ and therefore is isotropic. Hence there is a $c\in F$ satisfying $c\in D(\mathfrak{d}_K\otimes\rho)\setminus D(\mathfrak{d}_K)$. By Proposition 15.7, $\psi\simeq\mathfrak{d}\otimes\langle\langle c\rangle\rangle_K\otimes\mu$ for some quadratic Pfister form μ.

Applying the statement with $k=n$, we get an n-fold bilinear Pfister form $\mathfrak{b}$ over F such that $\psi\simeq\mathfrak{b}_K\otimes\langle\langle y]]$ for some $y\in K$. As $s_*\big(\langle\langle y]]\big)$ is similar to $\langle\langle\mathrm{Tr}_{K/F}(y)]]$, we have $\mathfrak{b}\otimes\langle\langle\mathrm{Tr}_{K/F}(y)]] = 0\in I_q(F)$. By Corollary 6.14, $\mathrm{Tr}_{K/F}(y) = b+b^2+\mathfrak{b}'(v,v)$ for some $b\in F$ and $v\in V_{\mathfrak{b}'}$. Let $x\in K\setminus F$ be an element such that $x^2+x+a=0$ for some $a\in F$. Set $z = xb+(xb)^2+\mathfrak{b}'_K(xv,xv)\in K$ and $c = y+z$. Since $\mathrm{Tr}_{K/F}(x) = \mathrm{Tr}_{K/F}(x^2) = 1$, we have $\mathrm{Tr}_{K/F}(z) = \mathrm{Tr}_{K/F}(y)$. It follows that $c\in F$.

Again by Corollary 6.14, we see that $\mathfrak{b}_K \otimes \langle\langle z]]$ is hyperbolic and therefore

$$\psi = \mathfrak{b}_K \cdot \langle\langle y]] = \mathfrak{b}_K \cdot \langle\langle y+z]] = \big(\mathfrak{b} \cdot \langle\langle c]]\big)_K. \qquad \square$$

Remark 34.23. Suppose that char $F \neq 2$ and $K = F(\sqrt{a})$ is a quadratic extension of F. Let $\mathfrak{b}$ be an anisotropic bilinear n-fold Pfister form over F. Then $\mathrm{N}_{K/F} = \langle\langle a\rangle\rangle$, so by Theorem 34.22(3), the following are equivalent:

(1) $\mathfrak{b}_K \in I^{n+1}(K)$.

(2) $\mathfrak{b} \in \langle\langle a\rangle\rangle W(F)$.

(3) $\mathfrak{b} \simeq \langle\langle a\rangle\rangle \otimes \mathfrak{c}$ for some $(n-1)$-fold Pfister form $\mathfrak{c}$.

We now consider the case of a purely inseparable quadratic field extension K/F.

Lemma 34.24. *Let K/F be a purely inseparable quadratic field extension and $s : K \to F$ a nonzero F-linear functional satisfying $s(1) = 0$. Let $b \in F^\times$. Then the following conditions are equivalent:*

(1) $b \in \mathrm{N}_{K/F}(K^\times)$.

(2) $\langle\langle b\rangle\rangle_K = 0 \in W(K)$.

(3) $\langle\langle b\rangle\rangle = s_*\big(\langle y\rangle\big)$ *for some* $y \in K^\times$.

PROOF. The equality $\mathrm{N}_{K/F}(K^\times) = K^2 \cap F^\times$ proves (1) $\Leftrightarrow$ (2). For any $y \in F^\times$, it follows by Corollary 34.19 that $s_*\big(\langle y\rangle\big)$ is similar to $\langle\langle \mathrm{N}_{K/F}(y)\rangle\rangle$. This proves that (1) $\Leftrightarrow$ (3). $\square$

Proposition 34.25. *Let K/F be a purely inseparable quadratic field extension and $s : K \to F$ a nontrivial F-linear functional such that $s(1) = 0$. Let $\mathfrak{b}$ be an anisotropic bilinear form over F. Then there exist bilinear forms $\mathfrak{c}$ over K and $\mathfrak{d}$ over F satisfying $\mathfrak{b} \simeq s_*(\mathfrak{c}) \perp \mathfrak{d}$ and $\mathfrak{d}_K$ is anisotropic.*

PROOF. We induct on $\dim \mathfrak{b}$. Suppose that $\mathfrak{b}_K$ is isotropic. Then there is a 2-dimensional subspace $W \subset V_{\mathfrak{b}}$ such that $(\mathfrak{b}|_W)_K$ is isotropic. By Lemma 34.24, we have $\mathfrak{b}|_W \simeq s_*\big(\langle y\rangle\big)$ for some $y \in K^\times$. Applying the induction hypothesis to the orthogonal complement of W in V completes the proof. $\square$

Theorem 34.4 and Proposition 34.25 yield

Corollary 34.26. *Let K/F be a purely inseparable quadratic field extension and $s : K \to F$ a nonzero F-linear functional such that $s(1) = 0$. Then the sequence*

$$W(F) \xrightarrow{r_{K/F}} W(K) \xrightarrow{s_*} W(F) \xrightarrow{r_{K/F}} W(K)$$

is exact.

Let K/F be a purely inseparable quadratic field extension and $s : K \to F$ a nonzero linear functional such that $s(1) = 0$. It follows from Corollaries 34.17 and 34.26 that we have well-defined complexes

$$I^n(F) \xrightarrow{r_{K/F}} I^n(K) \xrightarrow{s_*} I^n(F) \xrightarrow{r_{K/F}} I^n(K) \tag{34.27}$$

and

$$\bar{I}^n(F) \xrightarrow{r_{K/F}} \bar{I}^n(K) \xrightarrow{s_*} \bar{I}^n(F) \xrightarrow{r_{K/F}} \bar{I}^n(K). \tag{34.28}$$

As in the separable case, the homomorphism s_* in (34.28) is independent of the nontrivial F-linear functional $K \to F$ by Lemma 34.18 although it is not independent in (34.27).

We show that the complexes (34.27) and (34.28) are exact on quadratic Pfister forms.

Theorem 34.29. *Let K/F be a purely inseparable quadratic field extension and $s : K \to F$ a nontrivial F-linear functional such that $s(1) = 0$.*

(1) *Let $\mathfrak{c}$ be anisotropic n-fold bilinear Pfister form over K. If $s_*(\mathfrak{c}) \in I^{n+1}(F)$, then there exists an $\mathfrak{b}$ over K such that $\mathfrak{c} \simeq \mathfrak{b}_K$.*

(2) *Let $\mathfrak{b}$ be anisotropic n-fold bilinear Pfister form over F. If $\mathfrak{b}_K \in I^{n+1}(K)$, then there exists an n-fold bilinear Pfister form $\mathfrak{c}$ such that $\mathfrak{b} = s_*(\mathfrak{c})$.*

PROOF. (1): The proof is the same as in Theorem 34.22(1).

(2): By the Hauptsatz 23.7, we have $\mathfrak{b}_K = 0 \in W(K)$. In particular, $\mathfrak{b}_K$ is isotropic and hence there is a 2-dimensional subspace $W \subset V_{\mathfrak{b}}$ with $\mathfrak{b}|_W$ nondegenerate and isotropic over K. Hence $\mathfrak{b}|_W$ is similar to $\langle\langle b\rangle\rangle$ for some $b \in F^\times$. As $\langle\langle b\rangle\rangle_K = 0$, by Lemma 34.24 $\langle\langle b\rangle\rangle = s_*(\langle\langle y\rangle\rangle)$ for some $y \in K^\times$. By Corollary 6.17, we have $\mathfrak{b} \simeq \langle\langle b\rangle\rangle \otimes \mathfrak{d}$ for some bilinear Pfister form $\mathfrak{d}$. Finally,

$$\mathfrak{b} = \langle\langle b\rangle\rangle \cdot \mathfrak{d} = s_*(\langle\langle y\rangle\rangle) \cdot \mathfrak{d} = s_*(\langle\langle y\rangle\rangle \cdot \mathfrak{d}) \in W(F). \qquad \square$$

We shall show in Theorems 40.3, 40.5, and 40.6 that the complexes (34.20), (34.21), (34.27) and (34.28) are exact for any n. Note that the exactness for small n (up to 2) can be shown by elementary means.

34.D. Torsion in the Witt ring under a quadratic extension. We turn to the transfer of the torsion ideal in the Witt ring of a quadratic extension to obtain results found in [**31**]. We need the following lemma.

Lemma 34.30. *Let K/F be a quadratic field extension of F and $\mathfrak{b}$ a bilinear Pfister form over F.*

(1) *If $\mathfrak{c}$ is an anisotropic bilinear form over K such that $\mathfrak{b}_K \otimes \mathfrak{c}$ is defined over F, then there exists a form $\mathfrak{d}$ over F such that $\mathfrak{b}_K \otimes \mathfrak{c} \simeq (\mathfrak{b} \otimes \mathfrak{d})_K$.*

(2) $r_{K/F}(W(F)) \cap \mathfrak{b}_K W(K) = r_{K/F}(\mathfrak{b}W(F))$.

PROOF. (1): Let $\mathfrak{c} = \langle a_1, \dots, a_n\rangle$. We induct on $\dim \mathfrak{c} = n$. By hypothesis, there is a $c \in F^\times \cap D(\mathfrak{b}_K \otimes \mathfrak{c})$. Write $c = a_1b_1 + \cdots + a_nb_n$ with $b_i \in \widetilde{D}(\mathfrak{b}_K)$. Let $c_i = b_i$ if $b_i \neq 0$ and 1 if not. Then $\mathfrak{e} := \langle a_1c_1, \dots, a_nc_n\rangle$ represents c so $\mathfrak{e} \simeq \langle c\rangle \perp \mathfrak{f}$. Since $b_i \in G_K(\mathfrak{b})$, we have

$$\mathfrak{b}_K \otimes \mathfrak{c} \simeq \mathfrak{b}_K \otimes \mathfrak{e} \simeq \mathfrak{b}_K \otimes \langle c\rangle \perp \mathfrak{b}_K \otimes \mathfrak{f}.$$

As $\mathfrak{b}_K \otimes \mathfrak{f} \in \operatorname{im}(r_{K/F})$, its anisotropic part is defined over F by Proposition 34.1 and Theorem 34.4. By induction, there exists a form $\mathfrak{g}$ such that $\mathfrak{b}_K \otimes \mathfrak{f} \simeq \mathfrak{b}_K \otimes \mathfrak{g}_K$. Then $\langle c\rangle \perp \mathfrak{g}$ works.

(2) follows easily from (1). $\square$

Proposition 34.31. *Let $K = F(\sqrt{a})/F$ be a quadratic extension with $a \in F^\times$ and $s : K \to F$ a nontrivial F-linear functional such that $s(1) = 0$. Let $\mathfrak{b}$ be an n-fold bilinear Pfister form. Then*

$$s_*(W(K)) \cap \operatorname{ann}_{W(F)}(\mathfrak{b}) = s_*(\operatorname{ann}_{W(K)}(\mathfrak{b}_K)).$$

PROOF. By Frobenius Reciprocity, we have

$$s_*(\operatorname{ann}_{W(K)}(\mathfrak{b}_K)) \subset s_*(W(K)) \cap \operatorname{ann}_{W(F)}(\mathfrak{b}).$$

Conversely, if $\mathfrak{c} \in s_*\big(W(K)\big) \cap \operatorname{ann}_{W(F)}(\mathfrak{b})$, we can write $\mathfrak{c} = s_*(\mathfrak{d})$ for some form $\mathfrak{d}$ over K. By Theorem 34.4 and Lemma 34.30,

$$\mathfrak{b}_K \otimes \mathfrak{d} \in r_{K/F}\big(W(F)\big) \cap \mathfrak{b}_K W(K) = r_{K/F}\big(\mathfrak{b}W(F)\big).$$

Hence there exists a form $\mathfrak{e}$ over F such that $\mathfrak{b}_K \otimes \mathfrak{d} = (\mathfrak{b} \otimes \mathfrak{e})_K$. Let $\mathfrak{f} = \mathfrak{d} \perp -\mathfrak{e}_K$. Then $\mathfrak{c} = s_*(\mathfrak{d}) = s_*(\mathfrak{f}) \in s_*\big(\operatorname{ann}_{W(K)}(\mathfrak{b}_K)\big)$ as needed. □

The torsion $W_t(F)$ of $W(F)$ is 2-primary. Thus applying the proposition to $\rho = 2^n\langle 1\rangle$ for all n yields

Corollary 34.32. *Let $K = F(\sqrt{a})$ be a quadratic extension of F with $a \in F^\times$ and $s : K \to F$ a nontrivial F-linear functional such that $s(1) = 0$. Then $W_t(F) \cap s_*\big(W(K)\big) = s_*\big(W_t(K)\big)$.*

We also have the following:

Corollary 34.33. *Suppose that F is a field of characteristic different from 2 and $K = F(\sqrt{a})$ a quadratic extension of F. Let $s : K \to F$ be a nontrivial F-linear functional such that $s(1) = 0$. Then*

$$\begin{aligned}\langle\langle a\rangle\rangle W(F) \cap \operatorname{ann}_{W(F)}\big(2\langle 1\rangle\big) &= \operatorname{Ker}(r_{K/F}) \cap s_*\big(W(K)\big)\\ &\subset \operatorname{ann}_{W(F)}\big(2\langle 1\rangle\big) \cap \operatorname{ann}_{W(F)}\big(\langle\langle a\rangle\rangle\big) = s_*\big(\operatorname{ann}_{W(K)}(2\langle 1\rangle)\big).\end{aligned}$$

PROOF. As $\langle\langle a, a\rangle\rangle \simeq \langle\langle -1, a\rangle\rangle$, we have

$$\langle\langle a\rangle\rangle W(F) \cap \operatorname{ann}_{W(F)}\big(2\langle 1\rangle\big) = \langle\langle a\rangle\rangle W(F) \cap \operatorname{ann}_{W(F)}\big(\langle\langle a\rangle\rangle\big),$$

which yields the first equality by Corollary 34.12. As $\langle\langle a\rangle\rangle W(F) \subset \operatorname{ann}_{W(F)}\big(\langle\langle a\rangle\rangle\big)$, we have the inclusion. Finally, $s_*\big(W(K)\big) \cap \operatorname{ann}_{W(F)}\big(2\langle 1\rangle\big) = s_*\big(\operatorname{ann}_{W(K)}(2\langle 1\rangle_K)\big)$ by Proposition 34.31, so Corollary 34.12 yields the second equality. □

Remark 34.34. Suppose that F is a formally real field and K a quadratic extension. Let $s_* : W(K) \to W(F)$ be the transfer induced by a nontrivial F-linear functional such that $s(1) = 0$. Then it follows by Corollaries 34.12 and 34.32 that the maps induced by $r_{K/F}$ and s_* yield an exact sequence

$$0 \to W_{red}(K/F) \to W_{red}(F) \xrightarrow{r_{K/F}} W_{red}(K) \xrightarrow{s_*} W_{red}(F)$$

(again abusing notation for the maps) where

$$W_{red}(K/F) := \operatorname{Ker}\big(W_{red}(F) \to W_{red}(K)\big).$$

By Corollary 33.15, we have a zero sequence

$$0 \to I^n_{red}(K/F) \to I^n_{red}(F) \xrightarrow{r_{K/F}} I^n_{red}(K) \xrightarrow{s_*} I^n_{red}(F)$$

where $I^n_{red}(K/F) := \operatorname{Ker}\big(I^n_{red}(F) \to I^n_{red}(K)\big)$.

In fact, we shall see in §41 that this sequence is also exact.

35. Torsion in $I^n(F)$ and torsion Pfister forms

In this section we study the property that $I(F)$ is nilpotent, i.e., that there exists an n such that $I^n(F) = 0$. For such an n to exist, the field must be nonformally real. In order to study all fields we broaden this investigation to the study of the existence of an n such that $I^n(F)$ is torsion-free. We wish to establish the relationship between this occurring over F and over a quadratic field extension K. This more general case is more difficult, so in this section we look at the simpler property that there are no torsion bilinear n-fold Pfister forms over the field F. This was the approach introduced and used in [**37**] and would be equivalent to $I^n(F)$ being torsion-free if we knew that torsion bilinear n-fold Pfister forms generate the torsion in $I^n(F)$. This is in fact true as shown in [**8**] and which we shall show in §41, but cannot be proven by these elementary methods.

35.A. The property A_n. In this section we begin to study torsion in $I^n(F)$ for a field F. We set

$$I_t^n(F) := W_t(F) \cap I^n(F).$$

Note that the group $I_t^1(F) = I_t(F)$ is generated by torsion binary forms by Proposition 31.27.

It is obvious that

$$I_t^n(F) \supset I^{n-1}(F)I_t(F).$$

Proposition 35.1. $I_t^2(F) = I(F)I_t(F)$.

PROOF. Note that for all $a, a' \in F^\times$ and $w, w' \in D(\infty\langle 1\rangle)$, we have

$$a\langle\langle w\rangle\rangle + a'\langle\langle w'\rangle\rangle = a\langle\langle -aa', w\rangle\rangle + a'w\langle\langle ww'\rangle\rangle,$$

hence

$$a\langle\langle w\rangle\rangle + a'\langle\langle w'\rangle\rangle \equiv a'w\langle\langle ww'\rangle\rangle \mod I(F)I_t(F).$$

Let $\mathfrak{b} \in I_t^2(F)$. By Proposition 31.27, we have $\mathfrak{b}$ is a sum of binary forms $a\langle\langle w\rangle\rangle$ with $a \in F^\times$ and $w \in D(\infty\langle 1\rangle)$. Repeated application of the congruence above shows that $\mathfrak{b}$ is congruent to a binary form $a\langle\langle w\rangle\rangle$ modulo $I(F)I_t(F)$. As $a\langle\langle w\rangle\rangle \in I^2(F)$ we have $a\langle\langle w\rangle\rangle = 0$ and therefore $\mathfrak{b} \in I(F)I_t(F)$. □

We shall prove in §41 that the equality $I_t^n(F) = I^{n-1}(F)I_t(F)$ holds for every n.

It is easy to determine Pfister forms of order 2 (cf. Corollary 6.14).

Lemma 35.2. *Let $\mathfrak{b}$ be a bilinear n-fold Pfister form. Then $2\mathfrak{b} = 0$ in $W(F)$ if and only if either* char $F = 2$ *or $\mathfrak{b} = \langle\langle w\rangle\rangle \otimes \mathfrak{c}$ for some $w \in D(2\langle 1\rangle)$ and $\mathfrak{c}$ is an $(n-1)$-fold Pfister form.*

Proposition 35.3. *Let F be a field and $n \geq 1$ an integer. The following conditions are equivalent:*

(1) *There are no n-fold Pfister forms of order 2 in $W(F)$.*
(2) *There are no anisotropic n-fold Pfister forms of finite order in $W(F)$.*
(3) *For every $m \geq n$, there are no anisotropic m-fold Pfister forms of finite order in $W(F)$.*

PROOF. The implications (3) ⇒ (2) ⇒ (1) are trivial.

(1) ⇒ (3): If char $F = 2$ the statement is clear as $W(F)$ is torsion. Assume that char $F \neq 2$. Let $2^k\mathfrak{b} = 0$ in $W(F)$ for some $k \geq 1$ and $\mathfrak{b}$ an m-fold Pfister form with $m \geq n$. We show by induction on k that $\mathfrak{b} = 0$ in $W(F)$. It follows from Lemma 35.2 that $2^{k-1}\mathfrak{b} \simeq \langle\langle w\rangle\rangle \otimes \mathfrak{c}$ for some $w \in D(2\langle 1\rangle)$ and a $(k+m-2)$-fold Pfister form $\mathfrak{c}$. Let $\mathfrak{d}$ be an $(n-1)$-fold Pfister form dividing $\mathfrak{c}$. Again by Lemma 35.2, the form $2\langle\langle w\rangle\rangle \cdot \mathfrak{d}$ is 0 in $W(F)$, hence by assumption, $\langle\langle w\rangle\rangle \cdot \mathfrak{d} = 0$ in $W(F)$. It follows that $2^{k-1}\mathfrak{b} = \langle\langle w\rangle\rangle \cdot \mathfrak{c} = 0$ in $W(F)$. By the induction hypothesis, $\mathfrak{b} = 0$ in $W(F)$. □

We say that a field F *satisfies* A_n if the equivalent conditions of Proposition 35.3 hold. It follows from the definition that the condition A_n implies A_m for every $m \geq n$. It follows from Proposition 31.9 that F satisfies A_1 if and only if F is pythagorean.

If F is not formally real, the condition A_n is equivalent to $I^n(F) = 0$ as the group $W(F)$ is torsion.

As the group $I_t(F)$ is generated by torsion binary forms, the property A_n implies that $I^{n-1}(F)I_t(F) = 0$.

Exercise 35.4. Suppose that F is a field of characteristic not 2. If K is a quadratic extension of F, let $s^K : K \to F$ be a nontrivial F-linear functional such that $s^K(1) = 0$. Show the following are equivalent:

(1) F satisfies A_{n+1}.

(2) $s_*^{F(\sqrt{w})}\big(P_n(F(\sqrt{w}))\big) = P_n(F)$ for every $w \in D\big(\infty\langle 1\rangle\big)$.

(3) $s_*^{F(\sqrt{w})}\big(I^n(F(\sqrt{w}))\big) = I^n(F)$ for every $w \in D\big(\infty\langle 1\rangle\big)$.

We now study the property A_n under field extensions. The case of fields of characteristic 2 is easy.

Lemma 35.5. *Let K/F be a finite extension of fields of characteristic 2. Then $I^n(F) = 0$ if and only if $I^n(K) = 0$.*

PROOF. The property $I^n(E) = 0$ for a field E is equivalent to $[E : E^2] < 2^n$ by Example 6.5. We have $[K : F] = [K^2 : F^2]$, as the Frobenius map $K \to K^2$ given by $x \to x^2$ is an isomorphism. Hence

$$[K : K^2] = [K : F^2]/[K^2 : F^2] = [K : F^2]/[K : F] = [F : F^2]. \tag{35.6}$$

Thus we have $I^n(K) = 0$ if and only if $I^n(F) = 0$. □

Let F_0 be a formally real field satisfying A_1, i.e., a pythagorean field. Let $F_n = F_0((t_1))\cdots((t_n))$ be the iterated Laurent series field over F_0. Then F_n is also formally real pythagorean (cf. Example 31.6), hence F_n satisfies A_n for all $n \geq 1$. However, $K_n = F_n\big(\sqrt{-1}\big)$ does not satisfy A_n as $\langle\langle t_1, \dots, t_n\rangle\rangle$ is an anisotropic form over the nonformally real field K_n. Thus the property A_n is not preserved under quadratic extensions. Nevertheless, we have

Proposition 35.7. *Suppose that F satisfies A_n. Let $K = F(\sqrt{a})$ be a quadratic extension of F with $a \in F^\times$. Then K satisfies A_n if either of the following two conditions hold:*

(1) *$a \in D\big(\infty\langle 1\rangle\big)$.*

(2) *Every bilinear n-fold Pfister form over F becomes metabolic over K.*

PROOF. If $\operatorname{char} F = 2$, then $I^n(F) = 0$, hence $I^n(K) = 0$ by Lemma 35.5. So we may assume that $\operatorname{char} F \neq 2$. Let $y \in K^\times$ satisfy $y \in D\big(2\langle 1\rangle_K\big)$ and let $\mathfrak{e}$ be an $(n-1)$-fold Pfister form over K. By Lemma 35.2, it suffices to show that $\mathfrak{b} := \langle\langle y\rangle\rangle \otimes \mathfrak{e}$ is trivial in $W(K)$. Let $s_* : W(K) \to W(F)$ be the transfer induced by a nontrivial F-linear functional $s(1) = 0$.

We claim that $s_*(\mathfrak{b}) = 0$. Suppose that $n = 1$. Then $s_*(\mathfrak{b}) \in I_t(F) = 0$. So we may assume that $n \geq 2$. As $I^{n-1}(K)$ is generated by Pfister forms of the form $\langle\langle z\rangle\rangle \otimes \mathfrak{d}_K$ with $z \in K^\times$ and $\mathfrak{d}$ an $(n-2)$-fold Pfister form over F by Lemma 34.16, we may assume that $\mathfrak{b} = \langle\langle y, z\rangle\rangle \otimes \mathfrak{d}_K$.

We have $s_*(\langle\langle y, z\rangle\rangle) \in I_t^2(F) = I(F)I_t(F)$ by Proposition 35.1. So

$$s_*(\langle\langle y, z\rangle\rangle \cdot \mathfrak{d}_K) = s_*(\langle\langle y, z\rangle\rangle) \cdot \mathfrak{d}$$

lies in $I^{n-1}(F)I_t(F)$ which is trivial by A_n. The claim is proven.

It follows that $\mathfrak{b} = \mathfrak{c}_K$ for some n-fold Pfister form $\mathfrak{c}$ over F by Theorem 34.22. Thus we are done if every n-fold Pfister form over F becomes hyperbolic over K. So assume that $a \in D(\infty\langle 1\rangle)$. As $\mathfrak{b}$ is torsion in $W(K)$, there exists an m such that $2^m\mathfrak{b} = 0$ in $W(F)$. Thus $2^m\mathfrak{c}_K$ is hyperbolic so $2^m\mathfrak{c}$ is a sum of binary forms $x\langle\langle ay^2 + x^2\rangle\rangle$ in $W(F)$ for some x, y, z in F by Corollary 34.12. In particular, $2^m\mathfrak{c}$ is torsion and so trivial by A_n for F. The result follows. □

Corollary 35.8. *Suppose that $I^n(F) = 0$ (in particular, F is not formally real). Let K/F be a quadratic extension. Then $I^n(K) = 0$.*

In general, the above corollary does not hold if K/F is not quadratic. For example, let F be the quadratic closure of the rationals, so $I(F) = 0$. There exist algebraic extensions K of F such that $I(K) \neq 0$, e.g., $K = F(\sqrt[3]{2})$. It is true, however, that in this case $I^2(K) = 0$. It is still an unanswered question whether $I^2(K) = 0$ when K/F is finite and F is an arbitrary quadratically closed field, equivalently whether the cohomological 2-dimension of a quadratically closed field is at most one.

If $I^n(F)$ is torsion-free, then F satisfies A_n. Conversely, if F satisfies A_1, then $I(F)$ is torsion-free by Proposition 31.9. If F satisfies A_2, then it follows from Proposition 35.1 that $I^2(F)$ is torsion-free as $I_t(F)$ is generated by torsion binary forms.

Proposition 35.9. *A field F satisfies A_3 if and only if $I^3(F)$ is torsion-free.*

PROOF. The statement is obvious if F is not formally real, so we may assume that $\operatorname{char} F \neq 2$. Let $\mathfrak{b} \in I^3(F)$ be a torsion element. By Proposition 35.1,

$$\mathfrak{b} = \sum_{i=1}^{r} x_i\langle\langle y_i, w_i\rangle\rangle$$

for some $x_i, y_i \in F^\times$ and $w_i \in D(\infty\langle 1\rangle)$. We show by induction on r that $\mathfrak{b} = 0$.

It follows from Proposition 35.7 that $K = F(\sqrt{w})$ with $w = w_r$ satisfies A_3. By the induction hypothesis, we have $\mathfrak{b}_K = 0$. Thus $\mathfrak{b} = \langle\langle w\rangle\rangle \cdot \mathfrak{c}$ for some $\mathfrak{c} \in W(F)$ by Corollary 34.12. Then $\mathfrak{c}$ must be even-dimensional as the determinant of $\mathfrak{c}$ is trivial. Choose $d \in F^\times$ such that $\mathfrak{d} := \mathfrak{c} + \langle\langle d\rangle\rangle \in I^2(F)$.

Thus in $W(F)$,

$$\mathfrak{b} = \langle\langle w\rangle\rangle \cdot \mathfrak{d} - \langle\langle w, d\rangle\rangle.$$

Note that $\langle\langle w\rangle\rangle \cdot \mathfrak{d} = 0$ in $W(F)$ by A_3. Consequently, $\langle\langle w, d\rangle\rangle \in I^3(F)$, so it is zero in $W(F)$ by the Hauptsatz 23.7. This shows $\mathfrak{b} = 0$. □

We shall show in Corollary 41.5 below that $I^n(F)$ is torsion-free if and only if F satisfies A_n for every $n \geq 1$.

We have an application for quadratic forms first shown in [**36**].

Theorem 35.10 (Classification Theorem). *Let F be a field.*

(1) *Dimension and total signature classify the isometry classes of nondegenerate quadratic forms over F if and only if $I_q(F)$ is torsion-free, i.e., F is*

pythagorean. In particular, if F is not formally real, then dimension classifies the isometry classes of forms over F if and only if F is quadratically closed.

(2) *Dimension, discriminant and total signature classify the isometry classes of nondegenerate even-dimensional quadratic forms over F if and only if $I_q^2(F)$ is torsion-free. In particular, if F is not formally real, then dimension and discriminant classify the isometry classes of forms over F if and only if $I_q^2(F) = 0$.*

(3) *Dimension, discriminant, Clifford invariant, and total signature classify the isometry classes of nondegenerate even-dimensional quadratic forms over F if and only if $I_q^3(F)$ is torsion-free. In particular, if F is not formally real, then dimension, discriminant, and Clifford invariant classify the isometry classes of forms over F if and only if $I_q^3(F) = 0$.*

PROOF. We prove (3) as the others are similar (and easier). If $I_q^3(F)$ is not torsion-free, then there exists an anisotropic torsion form $\varphi \in P_3(F)$ by Proposition 35.9 if F is formally real and trivially if F is not formally real as then $I_q(F)$ is torsion. As φ and $4\mathbb{H}$ have the same dimension, discriminant, Clifford invariant, and total signature but are not isometric, these invariants do not classify.

Conversely, assume that $I_q^3(F)$ is torsion-free. Let φ and ψ be nondegenerate even-dimensional quadratic forms having the same dimension, discriminant, Clifford invariant, and total signature. Then by Theorem 13.7, the form $\theta := \varphi \perp -\psi$ lies in $I_q^2(F)$ and is torsion. As φ and ψ have the same dimension, it suffices to show that θ is hyperbolic. Thus the result is equivalent to showing that if a torsion form $\theta \in I_q^2(F)$ has trivial Clifford invariant and $I_q^3(F)$ is torsion-free, then θ is hyperbolic.

The case $\operatorname{char} F = 2$ follows from Theorem 16.3. So we may assume that $\operatorname{char} F \neq 2$. By Proposition 35.1, we can write $\theta = \sum_{i=1}^r a_i\langle\langle b_i, c_i\rangle\rangle$ in $I_q(F)$ with $\langle\langle c_i\rangle\rangle$ torsion forms. We prove that θ is hyperbolic by induction on r.

Let $K = F(\sqrt{c})$ with $c = c_r$. Clearly, $\theta_K \in I_q^2(K)$ is torsion and has trivial Clifford invariant. By Proposition 35.7 and Corollary 35.9, we have $I_q^3(K)$ is torsion-free. By the induction hypothesis, θ_K is hyperbolic. By Corollary 23.6, we conclude that $\theta = \psi \cdot \langle\langle c\rangle\rangle$ in $I_q(F)$ for some quadratic form ψ. As $\operatorname{disc}(\theta)$ is trivial, $\dim\psi$ is even. Choose $d \in F^\times$ such that $\tau := \psi + \langle\langle d\rangle\rangle \in I^2(F)$. Then

$$\theta = \tau \cdot \langle\langle c\rangle\rangle - \langle\langle d, c\rangle\rangle$$

in $W(F)$.

As the torsion form $\tau \otimes \langle\langle c\rangle\rangle$ belongs to $I_q^3(F)$, it is hyperbolic. As the Clifford invariant of θ is trivial, it follows that the Clifford invariant of $\langle\langle d, c\rangle\rangle$ must also be trivial. By Corollary 12.5, $\langle\langle d, c\rangle\rangle$ is hyperbolic and hence θ is hyperbolic. □

Remark 35.11. The Stiefel-Whitney classes introduced in (5.4) are defined on nondegenerate bilinear forms. If $\mathfrak{b}$ is such a form then the $w_i(\mathfrak{b})$ determine $\operatorname{sgn}\mathfrak{b}$ for every $P \in \mathfrak{X}(F)$ by Remark 5.8 and Example 5.13. We also have $w_i = e_i$ for $i = 1, 2$ by Corollary 5.9.

Let $\mathfrak{b}$ and $\mathfrak{b}'$ be two nondegenerate symmetric bilinear forms of the same dimension. Suppose that $w(\mathfrak{b}) = w(\mathfrak{b}')$, then $w\big([\mathfrak{b}] - [\mathfrak{b}']\big) = 1$, where $[\ ]$ is the class of a form in the Witt-Grothendieck ring $\widehat{W}(F)$. It follows that $[\mathfrak{b}] - [\mathfrak{b}']$ lies in $\widehat{I}^3(F)$ by (5.11), hence $\mathfrak{b} - \mathfrak{b}'$ lies in $I^3(F)$. As the w_i determine the total signature of a form, we have $\mathfrak{b} - \mathfrak{b}'$ is torsion by the Local-Global Principle 31.22. It follows that the dimension and total Stiefel-Whitney class determine the isometry class of anisotropic bilinear forms if and only if $I^3(F)$ is torsion-free.

Suppose that $\operatorname{char} F \neq 2$. Then all metabolic forms are hyperbolic, so in this case the dimension and total Stiefel-Whitney class determine the isometry class of nondegenerate symmetric bilinear forms if and only if $I^3(F)$ is torsion-free. In addition, we can define another Stiefel-Whitney map

$$\hat{w} : \widehat{W}(F) \to H^*(F)[[t]]^\times$$

to be the composition of w and the map $k_*(F)[[t]] \to H^*(F)[[t]]$ induced by the norm residue homomorphism $h_F^* : k_*(F) \to H^*(F)$ in §101.5. Then dimension and $\hat{w}$ classify the isometry classes of nondegenerate bilinear forms if and only if $I^3(F)$ is torsion-free by Theorem 35.10 as h_* is an isomorphism if F is a real closed field and $\tilde{w}_2$ is the classical Hasse invariant so determines the Clifford invariant.

We turn to the question on whether the property A_n goes down. This was first done in [**37**]. We use the proof given in [**31**].

Theorem 35.12. *Let K/F be a finite normal extension. If K satisfies A_n so does F.*

PROOF. Let $G = \operatorname{Gal}(K/F)$ and let H be a Sylow 2-subgroup of G. Set $E = K^H$, $L = K^G$. The field extension L/F is purely inseparable, so $[L : F]$ is either odd or L/F is a tower of successive quadratic extensions. The extension K/E is a tower of successive quadratic extensions and $[E : L]$ is odd. Thus we may assume that $[K : F]$ is either 2 or odd. Springer's Theorem 18.5 solves the case of odd degree. Hence we may assume that K/F is a quadratic extension.

The case $\operatorname{char} F = 2$ follows from Lemma 35.5. Thus we may assume that the characteristic of F is different from 2 and therefore $K = F(\sqrt{a})$ with $a \in F^\times$. Let $s : K \to F$ be a nontrivial F-linear functional with $s(1) = 0$.

Let $\mathfrak{b}$ be a 2-torsion bilinear n-fold Pfister form. We must show that $\mathfrak{b} = 0$ in $W(F)$. As $\mathfrak{b}_K = 0$ we have

$$\mathfrak{b} \in \langle\langle a\rangle\rangle W(F) \cap \operatorname{ann}_{W(F)}\big(2\langle 1\rangle\big)$$

by Corollary 34.12. As $\langle\langle a, a\rangle\rangle = \langle\langle a, -1\rangle\rangle$, it follows that $\langle\langle a\rangle\rangle \cdot \mathfrak{b} = 0$ in $W(F)$, hence by Corollary 6.14, we can write $\mathfrak{b} \simeq \langle\langle b\rangle\rangle \otimes \mathfrak{c}$ for some $(n-1)$-fold Pfister form $\mathfrak{c}$ and $b \in D\big(\langle\langle a\rangle\rangle\big)$. Choose $x \in K^\times$ such that $s_*\big(\langle x\rangle\big) = \langle\langle b\rangle\rangle$ and let $\mathfrak{d} = x\mathfrak{c}_K$. Then

$$s_*(\mathfrak{d}) = s_*\big(\langle x\rangle\big)\mathfrak{c} = \langle\langle b\rangle\rangle\mathfrak{c} = \mathfrak{b}.$$

If $\mathfrak{d} = 0$, then $\mathfrak{b} = 0$ and we are done. So we may assume that $\mathfrak{d}$ and therefore $\mathfrak{c}_K$ is anisotropic.

We have $s_*(2\mathfrak{d}) = 2\mathfrak{b} = 0$ in $W(F)$, hence the form $s_*(2\mathfrak{d})$ is isotropic. Therefore, $2\mathfrak{d}$ represents an element $c \in F^\times$, hence there exist $u, v \in \widetilde{D}(\mathfrak{c}_K)$ satisfying $x(u+v) = c$. But the form $\langle\langle u+v\rangle\rangle \otimes \mathfrak{c}$ is 2-torsion and K satisfies A_n. Consequently, $u + v \in D(\mathfrak{c}_K) = G(\mathfrak{c}_K)$ as $\mathfrak{c}_K$ is anisotropic. We have

$$\mathfrak{d} = x\mathfrak{c}_K \simeq x(u+v)\mathfrak{c}_K = c\mathfrak{c}_K.$$

Therefore, $0 = s_*(\mathfrak{d}) = \mathfrak{b}$ in $W(F)$ as needed. □

Corollary 35.13. *Let K/F be a finite normal extension with F not formally real. If $I^n(K) = 0$ for some n, then $I^n(F) = 0$.*

The above corollary is, in fact, true without the condition that K/F be normal. We have already shown this to be the case for fields of characteristic 2 in Lemma 35.5 and will show it to be true for characteristic not 2 in Corollary 43.9 below. In fact, it follows from §43 below that if K/F a finitely generated field extension of transcendence degree m, then $I^n(K) = 0$ implies $n \geq m$ and $I^{n-m}(F) = 0$. But we cannot do this by the elementary means employed here.

Corollary 35.14. *Let K/F be a quadratic extension.*

(1) *Suppose that $I^n(K) = 0$. Then L satisfies A_n for every extension L/F such that $[L : F] \leq 2$.*
(2) *Suppose that $I^n(K) = 0$. Then $I^n(F) = \langle\langle -w\rangle\rangle I^{n-1}(F)$ for every $w \in D(\infty\langle 1\rangle)$.*
(3) *Suppose that $I^n(F) = \langle\langle -w\rangle\rangle I^{n-1}(F)$ for some $w \in F^\times$. Then both F and K satisfy A_{n+1} and if* $\operatorname{char} F \neq 2$, *then $w \in D(\infty\langle 1\rangle)$.*

PROOF. (1), (2): By Corollaries 35.8 and 35.13 if F is not formally real, then $I^n(F) = 0$ if and only if $I^n(L) = 0$ for any quadratic extension L/F. In particular, (1) and (2) follow if F is not formally real. So suppose that F is formally real. We may assume that $K = F(\sqrt{a})$ with $a \in F^\times$. Then $I^n\big(L(\sqrt{a})\big) = 0$ by Proposition 35.7, hence $I^n(L)$ satisfies A_n by Theorem 35.12. This establishes (1).

Let $w \in D(\infty\langle 1\rangle)$. Then $F(\sqrt{-w})$ is not formally real. By (1), the field $F(\sqrt{-w})$ satisfies A_n, hence $I^n\big(F(\sqrt{-w})\big) = 0$. In particular, if $\mathfrak{b}$ is a bilinear n-fold Pfister form, then $\mathfrak{b}_{F(\sqrt{-w})}$ is metabolic. Thus $\mathfrak{b} \simeq \langle\langle -w\rangle\rangle \otimes \mathfrak{c}$ for some $(n-1)$-fold Pfister form $\mathfrak{c}$ over F by Remark 34.23 and (2) follows.

(3): If $\operatorname{char} F = 2$, then $I^n(F) = 0$, hence $I^n(K) = 0$ by Corollary 35.8. So we may assume that $\operatorname{char} F \neq 2$. By Remark 34.23, we have $2^n\langle 1\rangle \simeq \langle\langle -w\rangle\rangle \otimes \mathfrak{b}$ for some bilinear $(n-1)$-fold Pfister form $\mathfrak{b}$. As $2^n\langle 1\rangle$ only represents elements in $\widetilde{D}(\infty\langle 1\rangle)$, we have $w \in D(\infty\langle 1\rangle)$.

To show the first statement, it suffices to show that $L = F(\sqrt{-w})$ satisfies A_{n+1} by (1) and (2). Since $I^{n+1}(L)$ is generated by Pfister forms of the type $\langle\langle x\rangle\rangle \otimes \mathfrak{c}_L$ where $x \in L^\times$ and $\mathfrak{c}$ is an n-fold Pfister form over F by Lemma 34.16, we have $I^{n+1}(L) \subset \langle\langle -w\rangle\rangle I^n(L) = \{0\}$. □

If F is the field of 2-adic numbers, then $I^2(F) = 2I(F)$ and K satisfies $I^3(K) = 0$ for all finite extensions K/F but no such K satisfies $I^2(K) = 0$. In particular, statement (3) of Corollary 35.14 is the best possible.

Corollary 35.15. *Let F be a field extension of transcendence degree n over a real closed field. Then $D\big(2^n\langle 1\rangle\big) = D\big(\infty\langle 1\rangle\big)$.*

PROOF. As $F\big(\sqrt{-1}\big)$ is a C_n-field by Theorem 97.7 below, we have

$$I^n\big(F(\sqrt{-1})\big) = 0.$$

Therefore, F satisfies A_n by Corollary 35.14. □

35.B. Torsion-freeness and $I^n\big(F(\sqrt{-1})\big) = 0$. We intend to prove a result of Krüskemper 35.26 (cf. [**87**]) showing F satisfying A_n and $I^n(F)$ torsion-free are equivalent when $I^n\big(F(\sqrt{-1})\big) = 0$. We follow Arason's notes (cf. [**3**]) generalizing our congruence relations of Pfister forms developed in §24.

Let $\mathfrak{b}$ be a bilinear Pfister form. For simplicity, we set

$$I_{\mathfrak{b}}(F) = \big\{\mathfrak{c} \in I(F) \mid \mathfrak{b}\cdot\mathfrak{c} = 0 \in W(F)\big\} = I(F) \cap \operatorname{ann}_{W(F)}(\mathfrak{b}) \subset I(F).$$

We note if $\mathfrak{b}$ is metabolic, then $I_{\mathfrak{b}}(F) = I(F)$. We tacitly assume that $\mathfrak{b}$ is anisotropic below.

Lemma 35.16. *Let $\mathfrak{c}$ be a bilinear $(n-1)$-fold Pfister form and $d \in D_F(\mathfrak{b} \otimes \mathfrak{c})$. Then $\langle\langle d\rangle\rangle \cdot \mathfrak{c} \in I^{n-1}(F)I_{\mathfrak{b}}(F)$.*

PROOF. We induct on n. The hypothesis implies that $\mathfrak{b} \cdot \langle\langle d\rangle\rangle \cdot \mathfrak{c} = 0$ in $W(F)$, hence $\langle 1, -d\rangle \cdot \mathfrak{c} \in I_{\mathfrak{b}}(F)$. In particular, the case $n = 1$ is trivial. So assume that $n > 1$ and that the lemma holds for $(n-2)$-fold Pfister forms. Write $\mathfrak{c} = \langle\langle a\rangle\rangle \otimes \mathfrak{d}$ where $\mathfrak{d}$ is an $(n-2)$-fold Pfister form. Then $d = e_1 - ae_2$, where $e_1, e_2 \in \widetilde{D}(\mathfrak{b} \otimes \mathfrak{d})$. If $e_2 = 0$, then we are done by the induction hypothesis. So assume that $e_2 \neq 0$. Then $d = e_2(e - a)$, where $e = e_1/e_2 \in \widetilde{D}(\mathfrak{b} \otimes \mathfrak{d})$. By the induction hypothesis, we have

$$\begin{aligned}\langle\langle d\rangle\rangle \cdot \mathfrak{c} &= \langle\langle e_2(e-a)\rangle\rangle \cdot \mathfrak{c} = \langle\langle e-a\rangle\rangle \cdot \mathfrak{c} + \langle\langle e-a, e_2\rangle\rangle \cdot \mathfrak{c} \\ &\equiv \langle\langle e-a\rangle\rangle \cdot \mathfrak{c} \mod I^{n-1}(F)I_{\mathfrak{b}}(F).\end{aligned}$$

It follows that we may assume that $e_2 = 1$, hence that $d = e - a$. But then

$$\langle\langle d, a\rangle\rangle = \langle\langle e-a, a\rangle\rangle = \langle\langle e, a'\rangle\rangle$$

for some $a' \neq 0$ by Lemma 4.15 , hence

$$\langle\langle d\rangle\rangle \cdot \mathfrak{c} = \langle\langle d, a\rangle\rangle \cdot \mathfrak{d} = \langle\langle e, a'\rangle\rangle \cdot \mathfrak{d}.$$

By the induction hypothesis, it follows that $\langle\langle d\rangle\rangle \cdot \mathfrak{c} \in I^{n-1}(F)I_{\mathfrak{b}}(F)$. □

Lemma 35.17. *Let $\mathfrak{e}$ be a bilinear n-fold Pfister form and $b \in D(\mathfrak{b} \otimes \mathfrak{e}')$. Then there is a bilinear $(n-1)$-fold Pfister form $\mathfrak{f}$ such that $\mathfrak{e} \equiv \langle\langle b\rangle\rangle \cdot \mathfrak{f} \mod I^{n-1}(F)I_{\mathfrak{b}}(F)$.*

PROOF. We induct on n. If $n = 1$, then $\mathfrak{e}' = \langle\langle a\rangle\rangle$ and $b = ax$ for some $x \in D(-\mathfrak{b})$. It follows that

$$\langle\langle b\rangle\rangle = \langle\langle ax\rangle\rangle = \langle\langle a\rangle\rangle + a\langle\langle x\rangle\rangle \equiv \langle\langle a\rangle\rangle \mod I_{\mathfrak{b}}(F).$$

Now assume that $n > 1$ and that the lemma holds for $(n-1)$-fold Pfister forms. Write $\mathfrak{e} = \langle\langle a\rangle\rangle \otimes \mathfrak{d}$ with $\mathfrak{d}$ an $(n-1)$-fold Pfister form. Then $b = c + ad$, where $c \in \widetilde{D}(\mathfrak{b} \otimes \mathfrak{d}')$ and $d \in \widetilde{D}(\mathfrak{b} \otimes \mathfrak{d})$. If $d = 0$, then we are done by the induction hypothesis. So assume that $d \neq 0$. Then

$$\begin{aligned}\langle\langle ad\rangle\rangle \cdot \mathfrak{d} &= \langle\langle a\rangle\rangle \cdot \mathfrak{d} + a\langle\langle d\rangle\rangle \cdot \mathfrak{d} \\ &\equiv \langle\langle a\rangle\rangle \cdot \mathfrak{d} \mod I^{n-1}(F)I_{\mathfrak{b}}(F)\end{aligned}$$

by Lemma 35.16. It follows that we may assume that $d = 1$, hence $b = c + a$. If $c = 0$, then $b = a$ and there is nothing to prove. So assume that $c \neq 0$. By the induction hypothesis, we can write

$$\mathfrak{d} \equiv \langle\langle c\rangle\rangle \cdot \mathfrak{g} \mod I^{n-2}(F)I_{\mathfrak{b}}(F)$$

with $\mathfrak{g}$ an $(n-2)$-fold Pfister form. As

$$\langle\langle a, c\rangle\rangle = \langle\langle b-c, c\rangle\rangle \simeq \langle\langle b, c'\rangle\rangle$$

for some $c' \neq 0$ by Lemma 4.15, it follows that

$$\begin{aligned}\mathfrak{e} = \langle\langle a\rangle\rangle \cdot \mathfrak{d} &\equiv \langle\langle a, c\rangle\rangle \otimes \mathfrak{g} \\ &= \langle\langle b, c'\rangle\rangle \cdot \mathfrak{g} \mod I^{n-1}(F)I_{\mathfrak{b}}(F)\end{aligned}$$

as needed. □

Lemma 35.18. *Let $\mathfrak{e}$ be a bilinear n-fold Pfister form and $\mathfrak{h}$ a bilinear form over F.*

(1) *If $\mathfrak{e} \in I_{\mathfrak{b}}(F)$, then $\mathfrak{e} \in I^{n-1}(F)I_{\mathfrak{b}}(F)$.*
(2) *If $\mathfrak{h} \cdot \mathfrak{e} \in I_{\mathfrak{b}}(F)$, then $\mathfrak{h} \cdot \mathfrak{e} \in I^{n-1}(F)I_{\mathfrak{b}}(F)$.*

PROOF. (1): The hypothesis implies that $\mathfrak{b} \cdot \mathfrak{e} = 0$ in $W(F)$. In particular, $\mathfrak{b} \otimes \mathfrak{e} = \mathfrak{b} \perp \mathfrak{b} \otimes \mathfrak{e}'$ is isotropic. It follows that there exists an element $b \in D_F(\mathfrak{b}) \cap D_F(\mathfrak{b} \otimes \mathfrak{e}')$. By Lemma 35.17,

$$\mathfrak{e} \equiv \langle\langle b \rangle\rangle \cdot \mathfrak{f} \equiv 0 \mod I^{n-1}(F)I_{\mathfrak{b}}(F).$$

(2): The hypothesis implies that $\mathfrak{h} \cdot \mathfrak{b} \cdot \mathfrak{e} = 0$ in $W(F)$. If $\mathfrak{b} \cdot \mathfrak{e} = 0$ in $W(F)$ then, by (1), we have $\mathfrak{e} \in I^{n-1}(F)I_{\mathfrak{b}}(F)$ and we are done. Otherwise, we have $\mathfrak{h} \in I_{\mathfrak{b}\otimes\mathfrak{e}}(F)$, which is generated by the forms $\langle\langle x \rangle\rangle$, with $x \in D(\mathfrak{b} \otimes \mathfrak{e})$. It therefore suffices to prove the claim in the case $\mathfrak{h} = \langle\langle x \rangle\rangle$. But then, by (1), we even have $\mathfrak{h} \cdot \mathfrak{e} \in I^n(F)I_{\mathfrak{b}}(F)$. □

Lemma 35.19. *Suppose the bilinear n-fold Pfister forms $\mathfrak{e}, \mathfrak{f}$ satisfy*

$$a\mathfrak{e} \equiv b\mathfrak{f} \mod I_{\mathfrak{b}}(F)$$

with $a, b \in F^\times$. Then

$$a\mathfrak{e} \equiv b\mathfrak{f} \mod I^{n-1}(F)I_{\mathfrak{b}}(F).$$

PROOF. We induct on n. As the case $n = 1$ is trivial, we may assume that $n > 1$ and that the claim holds for $(n-1)$-fold Pfister forms. The hypothesis implies that $a\mathfrak{b} \otimes \mathfrak{e} \simeq b\mathfrak{b} \otimes \mathfrak{f}$, in particular, $b/a \in D_F(\mathfrak{b} \otimes \mathfrak{e})$. By Lemma 35.16, we therefore have

$$a\mathfrak{e} \equiv b\mathfrak{f} \mod I^{n-1}(F)I_{\mathfrak{b}}(F)$$

(actually, $\mod I^n(F)I_{\mathfrak{b}}(F)$). Hence we may assume that $a = b$. Dividing by a, we may even assume that $a = b = 1$. Write

$$\mathfrak{e} = \langle\langle c \rangle\rangle \otimes \mathfrak{g} \quad \text{and} \quad \mathfrak{f} = \langle\langle d \rangle\rangle \otimes \mathfrak{h}$$

with $\mathfrak{g}, \mathfrak{h}$ being $(n-1)$-fold Pfister forms. The hypothesis now implies that $\mathfrak{b} \otimes \mathfrak{e}' \simeq \mathfrak{b} \otimes \mathfrak{f}'$. In particular, $d \in D(\mathfrak{b} \otimes \mathfrak{e}')$. By Lemma 35.17, we can write $\mathfrak{e} \equiv \langle\langle d \rangle\rangle \cdot \mathfrak{g}_1 \mod I^{n-1}(F)I_{\mathfrak{b}}(F)$ with $\mathfrak{g}_1$ an $(n-1)$-fold Pfister form. It follows that we may assume that $c = d$. By the induction hypothesis, we then have $\mathfrak{g} \equiv \mathfrak{h} \mod I^{n-2}(F)I_{\mathfrak{b}\otimes\langle\langle d \rangle\rangle}(F)$, hence

$$\langle\langle d \rangle\rangle \cdot \mathfrak{g} \equiv \langle\langle d \rangle\rangle \cdot \mathfrak{h} \mod \langle\langle d \rangle\rangle I^{n-2}(F)I_{\mathfrak{b}\otimes\langle\langle d \rangle\rangle}(F).$$

We are therefore finished if we can show that $\langle\langle d \rangle\rangle I_{\mathfrak{b}\otimes\langle\langle d \rangle\rangle}(F) \subseteq I(F)I_{\mathfrak{b}}(F)$. Now, $I_{\mathfrak{b}\otimes\langle\langle d \rangle\rangle}(F)$ is generated by the $\langle\langle x \rangle\rangle$, with $x \in D\big(\mathfrak{b} \otimes \langle\langle d \rangle\rangle\big)$. For such a generator $\langle\langle x \rangle\rangle$, we have $\mathfrak{b} \cdot \langle\langle d, x \rangle\rangle = 0$ in $W(F)$, hence, by Lemma 35.18, the form $\langle\langle d, x \rangle\rangle$ lies in $I(F)I_{\mathfrak{b}}(F)$. □

Proposition 35.20. *Let $\mathfrak{e}, \mathfrak{f}, \mathfrak{g}$ be bilinear n-fold Pfister forms. Assume that*

$$a\mathfrak{e} \equiv b\mathfrak{f} + c\mathfrak{g} \mod I_{\mathfrak{b}}(F).$$

Then

$$a\mathfrak{e} \equiv b\mathfrak{f} + c\mathfrak{g} \mod I^{n-1}(F)I_{\mathfrak{b}}(F).$$

PROOF. The hypothesis implies that $a\mathfrak{b}\cdot\mathfrak{e} = b\mathfrak{b}\cdot\mathfrak{f} + c\mathfrak{b}\cdot\mathfrak{g}$ in $W(F)$. In particular, the form $b\mathfrak{b}\otimes\mathfrak{f} \perp c\mathfrak{b}\otimes\mathfrak{g}$ is isotropic. It follows that there exists $d \in D(b\mathfrak{b}\otimes\mathfrak{f}) \cap D(-c\mathfrak{b}\otimes\mathfrak{g})$. By Lemma 35.16, we then have

$$b\mathfrak{f} \equiv d\mathfrak{f} \mod I^{n-1}(F)I_{\mathfrak{b}}(F) \quad \text{and} \quad c\mathfrak{g} \equiv -d\mathfrak{g} \mod I^{n-1}(F)I_{\mathfrak{b}}(F)$$

(actually, $\mod I^n(F)I_{\mathfrak{b}}(F)$). Hence we may assume that $c = -b$. Dividing by b, we may even assume that $b = 1$ and $c = -1$. Then the hypothesis implies that $a\mathfrak{b}\cdot\mathfrak{e} = \mathfrak{b}\cdot\mathfrak{f} - \mathfrak{b}\cdot\mathfrak{g}$ in $W(F)$ and we have to prove that $a\mathfrak{e} \equiv \mathfrak{f} - \mathfrak{g} \mod I^{n-1}(F)I_{\mathfrak{b}}(F)$.

As $a\mathfrak{b}\cdot\mathfrak{e} = \mathfrak{b}\cdot\mathfrak{f} - \mathfrak{b}\cdot\mathfrak{g}$ in $W(F)$, it follows that $\mathfrak{b}\otimes\mathfrak{f}$ and $\mathfrak{b}\otimes\mathfrak{g}$ are linked using Proposition 6.21 and with $\mathfrak{b}$ dividing the linkage. Hence there exists an $(n-1)$-fold Pfister form $\mathfrak{d}$ and elements $b', c' \neq 0$ such that $\mathfrak{b}\otimes\mathfrak{f} \simeq \mathfrak{b}\otimes\mathfrak{d}\otimes\langle\langle b'\rangle\rangle$ and $\mathfrak{b}\otimes\mathfrak{g} \simeq \mathfrak{b}\otimes\mathfrak{d}\otimes\langle\langle c'\rangle\rangle$ (and hence $\mathfrak{b}\otimes\mathfrak{e} \simeq \mathfrak{b}\otimes\mathfrak{d}\otimes\langle\langle b'c'\rangle\rangle$). By Lemma 35.19, we then have

$$\mathfrak{f} \equiv \mathfrak{d}\cdot\langle\langle b'\rangle\rangle \quad \text{and also} \quad \mathfrak{g} \equiv \mathfrak{d}\cdot\langle\langle c'\rangle\rangle \mod I^{n-1}(F)I_{\mathfrak{b}}(F).$$

We may therefore assume that $\mathfrak{f} = \mathfrak{d}\otimes\langle\langle b'\rangle\rangle$ and $\mathfrak{g} = \mathfrak{d}\otimes\langle\langle c'\rangle\rangle$. Then $\mathfrak{f} - \mathfrak{g} = \mathfrak{d}\cdot\langle -b', c'\rangle = -b'\mathfrak{d}\cdot\langle\langle b'c'\rangle\rangle$ in $W(F)$. The result now follows from Lemma 35.19. □

Remark 35.21. From Lemmas 35.16-35.19 and Proposition 35.20, we easily see that the corresponding results hold for the torsion part $I_t(F)$ of $I(F)$ instead of $I_{\mathfrak{b}}(F)$. Indeed, in each case, we only have to use our result for $\mathfrak{b} = 2^k\langle 1\rangle$ for some $k \geq 0$.

We always have $2I^n(F) \subset I^{n+1}(F)$ for a field F. For some interesting fields, we have equality, i.e., $2I^n(F) = I^{n+1}(F)$ for some positive integer n. In particular, we shall see in Lemma 41.1 below that this is true for some n for any field of finite transcendence degree over its prime field. (This is easy if the field has positive characteristic but depends on Fact 16.2 when the characteristic of F is 0.) We shall now investigate when this phenomenon holds for a field.

Proposition 35.22. *Let F be a field. Then $2I^n(F) = I^{n+1}(F)$ if and only if every anisotropic bilinear $(n+1)$-fold Pfister form $\mathfrak{b}$ is divisible by $2\langle 1\rangle$, i.e., $\mathfrak{b} \simeq 2\mathfrak{c}$ for some n-fold Pfister form $\mathfrak{c}$.*

PROOF. If $2\langle 1\rangle$ is metabolic, the result is trivial so assume not. In particular, we may assume that char $F \neq 2$. Suppose $2I^n(F) = I^{n+1}(F)$ and $\mathfrak{b}$ is an anisotropic bilinear $(n+1)$-fold Pfister form. By assumption, there exist $\mathfrak{d} \in I^n(F)$ such that $\mathfrak{b} = 2\mathfrak{d}$ in $W(F)$. By Remark 34.23, we have $\mathfrak{b} \simeq 2\mathfrak{c}$ for some n-fold Pfister form $\mathfrak{c}$. □

It is also useful to study a variant of the property that $2I^n(F) = I^{n+1}(F)$ introduced by Bröcker in [**19**]. Recall that $I^n_{red}(F)$ is the image of $I^n(F)$ under the canonical homomorphism $W(F) \to W_{red}(F) = W(F)/W_t(F)$. We investigate the case that $2I^n_{red}(F) = I^{n+1}_{red}(F)$ for some positive integer n. Of course, if $2I^n(F) = I^{n+1}(F)$, then $2I^n_{red}(F) = I^{n+1}_{red}(F)$. We shall show that the above proposition generalizes. Further, we shall show this property is characterized by the cokernel of the signature map

$$\operatorname{sgn} : W(F) \to C(\mathfrak{X}(F), \mathbb{Z}).$$

Recall that the cokernel of this map is a 2-primary group by Theorem 33.8.

Proposition 35.23. *Suppose the reduced stability $\operatorname{st}_r(F)$ is finite and equals n. Then n is the least nonnegative integer such that $2I^n_{red}(F) = I^{n+1}_{red}(F)$. Moreover,* *for any bilinear $(n+1)$-fold Pfister form $\mathfrak{b}$, there exists an n-fold Pfister form $\mathfrak{c}$ such that* $\mathfrak{b} \equiv 2\mathfrak{c} \mod W_t(F)$.

PROOF. Let $\mathfrak{b}$ be an anisotropic bilinear $(n+1)$-fold Pfister form. In particular, $\operatorname{sgn} \mathfrak{b} \in C(\mathfrak{X}(F), 2^{n+1}\mathbb{Z})$. By assumption, there exists a bilinear form $\mathfrak{d}$ satisfying $\operatorname{sgn} \mathfrak{d} = \frac{1}{2} \operatorname{sgn} \mathfrak{b}$. Thus $\mathfrak{b} - 2\mathfrak{d} \in W_t(F)$, hence there exists an integer m such that $2^m\mathfrak{b} = 2^{m+1}\mathfrak{d}$ in $W(F)$ by Theorem 31.18. If $2^m\mathfrak{b}$ is metabolic the result is trivial, so we may assume it is anisotropic. By Proposition 6.22, there exists $\mathfrak{f}$ such that $2^m\mathfrak{b} \simeq 2^{m+1}\mathfrak{f}$. Therefore, $2^m\mathfrak{b} \simeq 2^{m+1}\mathfrak{c}$ for some bilinear n-fold Pfister form $\mathfrak{c}$ by Corollary 6.17. Hence $2I^n_{red}(F) = I^{n+1}_{red}(F)$.

Conversely, suppose that $2I^n_{red}(F) = I^{n+1}_{red}(F)$. Let $f \in C(\mathfrak{X}(F), \mathbb{Z})$. It suffices to show that there exists a bilinear form $\mathfrak{b}$ satisfying $\operatorname{sgn} \mathfrak{b} = 2^n f$. By Theorem 33.15, there exists an integer m and a bilinear form $\mathfrak{b} \in I^m(F)$ satisfying $\operatorname{sgn} \mathfrak{b} = 2^m f$. So we are done if $m \le n$. If $m > n$, then there exists $\mathfrak{c} \in I^n(F)$ such that $\operatorname{sgn} \mathfrak{b} = \operatorname{sgn} 2^{m-n}\mathfrak{c}$ and $2^n f = \operatorname{sgn} \mathfrak{c}$. □

Remark 35.24. If $2I^n_{red}(F) = I^{n+1}_{red}(F)$, then for any bilinear $(n+m)$-fold Pfister form $\mathfrak{b}$ there exists an n-fold Pfister form $\mathfrak{c}$ such that $\mathfrak{b} \equiv 2^m\mathfrak{c} \mod I_t(F)$ and $I^{n+m}_{red}(F) = 2^m I^n_{red}(F)$. Similarly, if $2I^n(F) = I^{n+1}(F)$, then for any bilinear $(n+m)$-fold Pfister form $\mathfrak{b}$ there exists an n-fold Pfister form $\mathfrak{c}$ such that $\mathfrak{b} \simeq 2^m\mathfrak{c}$ and $I^{n+m}(F) = 2^m I^n(F)$.

Suppose that $2I^n_{red}(F) = I^{n+1}_{red}(F)$. Let $\mathfrak{b}$ be an n-fold Pfister form over F and let $d \in F^\times$. Write

$$\langle\langle d \rangle\rangle \cdot \mathfrak{b} \equiv 2\mathfrak{e} \mod I_t(F) \quad \text{and} \quad \langle\langle -d \rangle\rangle \cdot \mathfrak{b} \equiv 2\mathfrak{f} \mod I_t(F)$$

for some n-fold Pfister forms $\mathfrak{e}$ and $\mathfrak{f}$ over F. By adding, we then get $2\mathfrak{b} \equiv 2\mathfrak{e} + 2\mathfrak{f} \mod I_t(F)$, hence also $\mathfrak{b} \equiv \mathfrak{e} + \mathfrak{f} \mod I_t(F)$. By Proposition 35.20, it follows that we even have $\mathfrak{b} \equiv \mathfrak{e} + \mathfrak{f} \mod I^{n-1}(F)I_t(F)$.

We generalize this as follows:

Lemma 35.25. *Suppose that $2I^n_{red}(F) = I^{n+1}_{red}(F)$. Let $\mathfrak{b}$ be a bilinear n-fold Pfister form and let $d_1, \dots, d_m \in F^\times$. Write*

$$\langle\langle \varepsilon_1 d_1, \dots, \varepsilon_m d_m \rangle\rangle \cdot \mathfrak{b} \equiv 2^m \mathfrak{c}_\varepsilon \mod I_t(F)$$

with $\mathfrak{c}_\varepsilon$ a bilinear n-fold Pfister form for every $\varepsilon = (\varepsilon_1, \dots, \varepsilon_m) \in \{\pm 1\}^m$. Then

$$\mathfrak{b} \equiv \sum_\varepsilon \mathfrak{c}_\varepsilon \mod I^{n-1}(F)I_t(F).$$

PROOF. We induct on m. The case $m = 1$ is done above. So assume that $m > 1$. Write $\langle\langle \varepsilon_2 d_2, \dots, \varepsilon_m d_m \rangle\rangle \cdot \mathfrak{b} \equiv 2^{m-1}\mathfrak{d}_{\varepsilon'} \mod I_t(F)$ with $\mathfrak{d}_{\varepsilon'}$ a bilinear n-fold Pfister form for every $\varepsilon' = (\varepsilon_2, \dots, \varepsilon_m) \in \{\pm 1\}^{m-1}$. By the induction hypothesis, we then have $\mathfrak{b} \equiv \sum_{\varepsilon'} \mathfrak{d}_{\varepsilon'} \mod I^{n-1}(F)I_t(F)$. It therefore suffices to show that

$$\mathfrak{d}_{\varepsilon'} \equiv \mathfrak{c}_{(+1,\varepsilon')} + \mathfrak{c}_{(-1,\varepsilon')} \mod I^{n-1}(F)I_t(F)$$

for every ε'. Since

$$\begin{aligned} 2^m \mathfrak{d}_{\varepsilon'} &\equiv 2\langle\langle \varepsilon_2 d_2, \dots, \varepsilon_m d_m \rangle\rangle \cdot \mathfrak{c} = \big(\langle\langle d \rangle\rangle + \langle\langle -d \rangle\rangle\big) \cdot \langle\langle \varepsilon_2 d_2, \dots, \varepsilon_m d_m \rangle\rangle \cdot \mathfrak{e} \\ &\equiv 2^m \mathfrak{c}_{(+1,\varepsilon')} + 2^m \mathfrak{c}_{(-1,\varepsilon')} \mod I_t(F) \end{aligned}$$

in $W(F)$, hence also $\mathfrak{d}_{\varepsilon'} \equiv \mathfrak{c}_{(+1,\varepsilon')} + \mathfrak{c}_{(-1,\varepsilon')} \mod I_t(F)$. By Proposition 35.20, it follows that $\mathfrak{d}_{\varepsilon'} \equiv \mathfrak{c}_{(+1,\varepsilon')} + \mathfrak{c}_{(-1,\varepsilon')} \mod I^{n-1}(F)I_t(F)$. □

Krüskemper's main result in [**87**] is:

Theorem 35.26. *Let* $2I^n_{red}(F) = I^{n+1}_{red}(F)$. *Then*

$$I^n_t(F) = I^{n-1}(F)I_t(F).$$

PROOF. Suppose that $\sum_{i=1}^r a_i\mathfrak{b}_i \in I_t(F)$, where $\mathfrak{b}_1, \dots, \mathfrak{b}_r$ are bilinear n-fold Pfister forms and $a_i \in F^\times$. We prove by induction on r that this implies that $\sum_{i=1}^r a_i\mathfrak{b}_i \in I^{n-1}(F)I_t(F)$. The case $r = 1$ is simply Lemma 35.18, so assume that $r > 1$.

Write $\mathfrak{b}_i = \langle\langle a_{i1}, \dots, a_{in}\rangle\rangle$ for $i = 1, \dots, r$ and let $m = rn$ and

$$(d_1, \dots, d_m) = (a_{11}, \dots, a_{1n}, a_{21}, \dots, a_{2n}, \dots, a_{r1}, \dots, a_{rn}).$$

Write

$$\langle\langle \varepsilon_1 d_1, \dots, \varepsilon_m d_m\rangle\rangle \cdot \mathfrak{b}_i \equiv 2^m \mathfrak{c}_{i\varepsilon} \mod I_t(F)$$

with $\mathfrak{c}_{i\varepsilon}$ bilinear n-fold Pfister forms for every $i = 1, \dots, r$ and every

$$\varepsilon = (\varepsilon_1, \dots, \varepsilon_m) \in \{\pm 1\}^m.$$

By Lemma 35.25,

$$\sum_{i=1}^r a_i\mathfrak{b}_i \equiv \sum_{\varepsilon}\sum_{i=1}^r a_i\mathfrak{c}_{i\varepsilon} \mod I^{n-1}(F)I_t(F).$$

If $\varepsilon^{(1)} \neq \varepsilon^{(2)}$ in $\{\pm 1\}^m$, then $\operatorname{sgn}\langle\langle \varepsilon_1^{(1)} d_1, \dots, \varepsilon_m^{(1)} d_m\rangle\rangle$ and $\operatorname{sgn}\langle\langle \varepsilon_1^{(2)} d_1, \dots, \varepsilon_m^{(2)} d_m\rangle\rangle$ have disjoint supports on $\mathfrak{X}(F)$, hence the same holds for $\operatorname{sgn}\mathfrak{c}_{i\varepsilon^{(1)}}$ and $\operatorname{sgn}\mathfrak{c}_{j\varepsilon^{(2)}}$. It therefore follows from the hypothesis that

$$\sum_{i=1}^r a_i\mathfrak{c}_{i\varepsilon} \equiv 0 \mod I_t(F) \quad \text{for each } \varepsilon.$$

Clearly, it suffices to show that $\sum_{i=1}^r a_i\mathfrak{c}_{i\varepsilon} \equiv 0 \mod I^{n-1}(F)I_t(F)$ for each ε.

Fix ε in $\{\pm 1\}^m$. Suppose that $\varepsilon \neq (1, \dots, 1)$. If -1 occurs as the coordinate of ε corresponding to a_{ji} for some $j \in [1,\ r]$ and $i \in [1,\ n]$, then $\langle\langle \varepsilon_1 d_1, \dots, \varepsilon_m d_m\rangle\rangle \cdot \mathfrak{b}_j = 0$ in $W(F)$ and we may assume for all such j that $\mathfrak{c}_{j\varepsilon} = 0$ in $W(F)$. In particular, if $\varepsilon \neq (1, \dots, 1)$, then

$$\sum_{i=1}^r a_i\mathfrak{c}_{i\varepsilon} = \sum_{\substack{i=1\\ i\neq j}}^r a_i\mathfrak{c}_{i\varepsilon} \equiv 0 \mod I^{n-1}(F)I_t(F)$$

by the induction hypothesis. So we may assume that $\varepsilon = (1, \dots, 1)$. Then

$$\langle\langle \varepsilon_1 d_1, \dots, \varepsilon_m d_m\rangle\rangle \otimes \mathfrak{b}_i \simeq \langle\langle d_1, \dots, d_m\rangle\rangle \otimes \mathfrak{b}_i \simeq 2^n\langle\langle d_1, \dots, d_m\rangle\rangle$$

is independent of i. We therefore may assume that $\mathfrak{c}_{i\varepsilon}$, for $i \in [1,\ r]$, are all equal to a single $\mathfrak{c}$. Let $\mathfrak{d} = \langle a_1, \dots, a_r\rangle$, then

$$\mathfrak{d} \cdot \mathfrak{c} = \sum_{i=1}^r a_i\mathfrak{c}_{i\varepsilon} \equiv 0 \mod I_t(F).$$

By Lemma 35.18, we conclude that $\mathfrak{d}\cdot\mathfrak{c} \in I^{n-1}(F)I_t(F)$ and the theorem follows. □

Corollary 35.27. *The following are equivalent for a field* F *of characteristic different from 2:*

(1) $I^{n+1}\big(F(\sqrt{-1})\big) = 0$.
(2) F *satisfies* A_{n+1} *and* $2I^n(F) = I^{n+1}(F)$.

(3) *F satisfies A_{n+1} and $2I^n_{red}(F) = I^{n+1}_{red}(F)$.*
(4) *$I^{n+1}(F)$ is torsion-free and $2I^n(F) = I^{n+1}(F)$.*

PROOF. (1) $\Rightarrow$ (2): By Theorem 35.12, F satisfies A_{n+1}. Theorem 34.22 applied to the quadratic extension $F(\sqrt{-1})/F$ gives $2I^n(F) = I^{n+1}(F)$.

(2) $\Rightarrow$ (3) is trivial as $2I^n_{red}(F) = I^{n+1}_{red}(F)$ if $2I^n(F) = I^{n+1}(F)$.

(3) $\Rightarrow$ (4): As the torsion $(n+1)$-fold Pfister forms generate the torsion in $I^{n+1}(F)$ by Theorem 35.26, we have $I^{n+1}(F)$ is torsion-free. Suppose that $\mathfrak{b}$ is an $(n+1)$-fold Pfister form. Then there exist $\mathfrak{c} \in I^n(F)$ and $\mathfrak{d} \in W_t(F)$ such that $\mathfrak{b} = 2\mathfrak{c} + \mathfrak{d}$ in $W(F)$. Hence for some N, we have $2^N\mathfrak{b} = 2^{N+1}\mathfrak{c}$. As $I^{n+1}(F)$ is torsion-free, we have $\mathfrak{b} = 2\mathfrak{c}$ in $W(F)$, hence $\mathfrak{b}_{F(\sqrt{-1})}$ is hyperbolic. By Theorem 34.22, there exists an n-fold Pfister form $\mathfrak{f}$ such that $\mathfrak{b} \simeq 2\mathfrak{f}$. It follows that $2I^n(F) = I^{n+1}(F)$.

(4) $\Rightarrow$ (1) follows from Theorem 34.22 for the quadratic extension $F(\sqrt{-1})/F$ as forms in $W(K)$ transfer to torsion forms in $W(F)$. □

Remark 35.28. By Corollary 35.14, the condition that $I^{n+1}(F(\sqrt{-1})) = 0$ is equivalent to $I^{n+1}(K) = 0$ for some quadratic extension K/F. In particular, if $I^{n+1}(K) = 0$ for some quadratic extension K/F, then $I^{n+1}(F)$ is torsion-free. Much more is true. If K/F is a field extension of transcendence degree m, then $I^n(K)$ torsion-free implies $n \geq m$ and $I^{n-m}(F)$ is torsion-free. We shall prove this in Corollary 43.9 below.

Corollary 35.29. *Let F be a real closed field and K/F a finitely generated extension of transcendence degree n. Then $I^{n+1}(K)$ is torsion-free and $2I^n(K) = I^{n+1}(K)$.*

PROOF. As $K(\sqrt{-1})$ is a C_n-field by Theorem 97.7, we have $I^{n+1}(K(\sqrt{-1})) = 0$ and hence $2I^n(K) = I^{n+1}(K)$ by Corollary 35.27 applied to the field K. □

Corollary 35.30. *Let F be a field satisfying $I^{n+1}(F) = 2I^n(F)$. Then $I^{n+2}(F)$ is torsion-free.*

PROOF. If $-1 \in F^2$, then $I^{n+1}(F) = 0$ and the result follows. In particular, we may assume that $\operatorname{char} F \neq 2$. By Theorem 35.26, it suffices to show that F satisfies A_{n+2}. Let $\mathfrak{b}$ be an $(n+2)$-fold Pfister form such that $2\mathfrak{b} = 0$ in $W(F)$. By Lemma 35.2, we can write $\mathfrak{b} = \langle\langle w\rangle\rangle \cdot \mathfrak{c}$ in $W(F)$ with $\mathfrak{c}$ an $(n+1)$-fold Pfister form and $w \in D(2\langle 1\rangle)$. By assumption, $\mathfrak{c} = 2\mathfrak{d}$ in $W(F)$ for some n-fold Pfister form $\mathfrak{d}$. Hence $\mathfrak{b} = 2\langle\langle w\rangle\rangle \cdot \mathfrak{d} = 0$ in $W(F)$. □

Remark 35.31. Any local field F satisfies $I^3(F) = 0$ (cf. [**89**, Cor. VI.2.15]). Let $\mathbb{Q}_2$ be the field of 2-adic numbers. Then, up to isomorphism, $\left(\frac{-1,-1}{\mathbb{Q}_2}\right)$ is the unique quaternion algebra (cf. [**89**, Cor. VI.2.24]), hence $I^2(\mathbb{Q}_2) = 2I(\mathbb{Q}_2) = \{0, 4\langle 1\rangle\} \neq 0$. Thus, in general, $I^{n+2}(F)$ cannot be replaced by $I^{n+1}(F)$ in the corollary above.

We shall return to these matters in §41.

CHAPTER VI

u-invariants

36. The $\bar{u}$-invariant

Given a field F, it is interesting to see if there exists a uniform bound on the dimension of anisotropic forms over F, i.e., if there exists an integer n such that every quadratic form over F of dimension greater than n is isotropic and if such an n exists, what is the minimum. For example, a consequence of the Chevalley-Warning Theorem (cf. [**123**, I.2, Th. 3]), is that over a finite field every 3-dimensional quadratic form is isotropic and a consequence of the Lang-Nagata Theorem (cf. Theorem 97.7 below) is that every (2^n+1)-dimensional form over a field of transcendence degree n over an algebraically closed field is isotropic. When a field is of characteristic different from 2, anisotropic forms are nondegenerate. This is no longer the case for fields of characteristic 2. This leads to the consideration of two invariants. If F is a formally real field, then $n\langle 1\rangle$ can never be isotropic. To obtain meaningful arithmetic data about formally real fields, we shall strengthen the condition on our forms. Although this makes computation more delicate, it is a useful generalization. In this section, we shall, for the most part, look at the simpler case of fields that are not formally real.

Let F be a field. We call a quadratic form φ over F *locally hyperbolic* if φ_{F_P} is hyperbolic at each real closure F_P of F (if any). If F is formally real, then the dimension of every locally hyperbolic form is even. If F is not formally real, every form is locally hyperbolic. For any field F, a locally hyperbolic form is one that is torsion in the Witt ring $W(F)$.

Define the *u-invariant* of F to be the smallest integer $u(F) \geq 0$ such that every nondegenerate locally hyperbolic quadratic form over F of dimension $> u(F)$ is isotropic (or infinity if no such integer exists) and the *$\bar{u}$-invariant* of F to be the smallest integer $\bar{u}(F) \geq 0$ such that every locally hyperbolic quadratic form over F of dimension $> \bar{u}(F)$ is isotropic (or infinity if no such integer exists). The u-invariant above was first defined in [**34**] and the $\bar{u}$-invariant by Baeza in [**16**].

Remark 36.1. (1) We have $\bar{u}(F) \geq u(F)$.

(2) If $\operatorname{char} F \neq 2$, every anisotropic form is nondegenerate, hence $\bar{u}(F) = u(F)$.

(3) If F is formally real, the integer $\bar{u}(F) = u(F)$ is even.

(4) As any (nondegenerate) quadratic form contains (nondegenerate) subforms of all smaller dimensions, if F is not formally real, we have $u(F) \leq n$ if and only if every nondegenerate quadratic form of dimension $n+1$ is isotropic and $\bar{u}(F) \leq n$ if and only if every quadratic form of dimension $n+1$ is isotropic.

Example 36.2. (1) If F is a formally real field, then $\bar{u}(F) = 0$ if and only if F is pythagorean.

(2) Suppose that F is a quadratically closed field. If $\operatorname{char} F \neq 2$, then $\bar{u}(F) = 1$ as every form is diagonalizable. If $\operatorname{char} F = 2$, then $\bar{u}(F) \leq 2$ with equality if F is not separably closed by Example 7.33.

(3) If F is a finite field, then $\bar{u}(F) = 2$.

(4) Let F be a field of characteristic not 2 and ρ a quadratic form over the Laurent series field $F((t))$. As ρ is diagonalizable, $\rho \cong \varphi \perp t\psi$ for some quadratic forms φ and ψ over F. Using Lemma 19.5, we deduce that ρ is anisotropic if and only if both φ and ψ are. Moreover, one can check that ρ is torsion if and only if both φ and ψ are torsion. If $\bar{u}(F)$ is finite, it follows that $\bar{u}\big(F((t))\big) = 2\bar{u}(F)$. This equality is also true if $\operatorname{char} F = 2$ (cf. [**16**]).

(5) If F is a C_n-field, then $\bar{u}(F) \leq 2^n$.

(6) If F is a local field, then $\bar{u}(F) = 4$. If $\operatorname{char} F > 0$, then $u(F) = 4$ by Example (4). If $\operatorname{char} F = 0$ this follows as in the argument for Example (4) using also Lemma 19.4 (cf. also [**111**]).

(7) If F is a global field, then $\bar{u}(F) = 4$. If $\operatorname{char} F = 0$, this follows from the Hasse-Minkowski Theorem [**89**, VI.3.1]. If $\operatorname{char} F > 0$, then F is a C_2-field by Theorem 97.7.

Proposition 36.3. *Let F be a field with $I_q^3(F) = 0$. If $1 < u(F) < \infty$, then $u(F)$ is even.*

PROOF. The hypothesis implies that F is not formally real. Suppose that $u(F) > 1$ is odd and let φ be a nondegenerate anisotropic quadratic form with $\dim \varphi = u(F)$. We claim that $\varphi \simeq \psi \perp \langle -a \rangle$ for some $\psi \in I_q^2(F)$ and $a \in F^\times$. If $\operatorname{char} F \neq 2$, then $\varphi \perp \langle a \rangle \in I_q^2(F)$ for some $a \in F^\times$ by Proposition 4.13. This form is isotropic, hence $\varphi \perp \langle a \rangle \simeq \psi \perp \mathbb{H}$ for some $\psi \in I_q^2(F)$ and therefore $\varphi \simeq \psi \perp \langle -a \rangle$. If $\operatorname{char} F = 2$, write $\varphi \simeq \mu \perp \langle a \rangle$ for some form μ and $a \in F^\times$. Choose $b \in F$ such that the discriminant of the form $\mu \perp [a, b]$ is trivial, i.e., $\mu \perp [a, b] \in I_q^2(F)$. By assumption the form $\mu \perp [a, b]$ is isotropic, i.e., $\mu \perp [a, b] \simeq \psi \perp \mathbb{H}$ for a form $\psi \in I_q^2(F)$. It follows from (8.7) that

$$\varphi \simeq \mu \perp \langle a \rangle \sim \mu \perp [a, b] \perp \langle a \rangle \sim \psi \perp \langle a \rangle,$$

hence $\varphi \simeq \psi \perp \langle a \rangle$ as these forms have the same dimension. This proves the claim.

Let $b \in D(\psi)$. As $\langle\langle ab \rangle\rangle \otimes \psi \in I_q^3(F) = 0$, we have $ab \in G(\psi)$. Therefore, $a = ab/b \in D(\psi)$ and hence the form φ is isotropic, a contradiction. □

Corollary 36.4. *The u-invariant of a field is not equal to $3, 5$ or 7.*

Let $r > 0$ be an integer. Define the $\bar{u}_r$-invariant of F to be the smallest integer $\bar{u}_r(F) \geq 0$ such that every set of r quadratic forms on a vector space over F of dimension $> \bar{u}_r(F)$ has common nontrivial zero.

In particular, if $\bar{u}_r(F)$ is finite, then F is not a formally real field. We also have $\bar{u}_1(F) = \bar{u}(F)$ when F is not formally real. The following is due to Leep (cf. [**93**]):

Theorem 36.5. *Let F be a field. Then for every $r > 1$, we have*

$$\bar{u}_r(F) \leq r\bar{u}_1(F) + \bar{u}_{r-1}(F).$$

PROOF. We may assume that $\bar{u}_{r-1}(F)$ is finite. Let $\varphi_1, \dots, \varphi_r$ be quadratic forms on a vector space V over F of dimension $n > r\bar{u}_1(F) + \bar{u}_{r-1}(F)$. We shall show that the forms have an isotropic vector in V. Let W be a totally isotropic subspace of V of the forms $\varphi_1, \dots, \varphi_{r-1}$ of the largest dimension d. Let V_i be the orthogonal complement of W in V relative to φ_i for each $i \in [1,\ r-1]$. We have $\dim V_i \geq n - d$.

Let $U = V_1 \cap \cdots \cap V_{r-1}$. Then $W \subset U$ and $\dim U \geq n - (r-1)d$. Choose a subspace $U' \subset U$ such that $U = W \oplus U'$. We have

$$\dim U' \geq n - rd > r\big(\bar{u}_1(F) - d\big) + \bar{u}_{r-1}(F).$$

If $d \leq \bar{u}_1(F)$, then $\dim U' > \bar{u}_{r-1}(F)$, hence the forms $\varphi_1, \dots, \varphi_{r-1}$ have an isotropic vector $u \in U'$. Then the subspace $W \oplus Fu$ is totally isotropic for these forms, contradicting the maximality of W.

It follows that $d > \bar{u}_1(F)$. The form φ_r therefore has an isotropic vector in U' which is isotropic for all the φ_i's. □

Corollary 36.6. *If F is not formally real, then $\bar{u}_r(F) \leq \frac{1}{2}r(r+1)\bar{u}(F)$.*

This improves the linear bound in [**31**] to:

Corollary 36.7. *Let K/F be a finite field extension of degree r. If F is not formally real, then $\bar{u}(K) \leq \frac{1}{2}(r+1)\bar{u}(F)$.*

Proof. Let $s_1, s_2, \dots, s_r$ be a basis for the space of F-linear functionals on K. Let φ be a quadratic form over K of dimension $n > \frac{1}{2}(r+1)\bar{u}(F)$. As $\dim(s_i)_*(\varphi) = rn > \frac{1}{2}r(r+1)\bar{u}(F)$ for each $i \in [1,\ r]$, by Corollary 36.6, the forms $(s_i)_*(\varphi)$ have a common isotropic vector which is, then an isotropic vector for φ. □

Let K/F be a finite extension with F not formally real. We shall show that if $\bar{u}(K)$ is finite, then so is $\bar{u}(F)$. We begin with the case that F is a field of characteristic 2 with an observation by Mammone, Moresi, and Wadsworth (cf. [**96**]).

Lemma 36.8. *Let F be a field of characteristic 2. Let φ be an even-dimensional nondegenerate quadratic form over F and ψ a totally singular quadratic form over F. If $\varphi \perp \psi$ is anisotropic, then*

$$\frac{1}{2}\dim\varphi + \dim\psi \leq [F : F^2].$$

Proof. Let $\varphi \simeq [a_1, b_1] \perp \cdots \perp [a_m, b_m]$ with $a_i, b_i \in F$ and $\psi \simeq \langle c_1, \dots, c_n\rangle$ with $c_i \in F^\times$. For each $i \in [1,\ m]$ let $d_i \in D\big([a_i, b_i]\big)$. Then $\{c_1, \dots, c_n, d_1, \dots, d_m\}$ is F^2-linearly independent. The result follows. □

Proposition 36.9. *Let F be a field of characteristic 2 and K/F a finite extension. Then*

$$\bar{u}(F) \leq 2\bar{u}(K) \leq 4\bar{u}(F).$$

Proof. If $c_1, \dots, c_n$ are F^2-linearly independent, then the form $\langle c_1, \dots, c_n\rangle$ is anisotropic. By the lemma, it follows that we have

$$[F : F^2] \leq \bar{u}(F) \leq 2[F : F^2].$$

As $[F : F^2] = [K : K^2]$ (cf. (35.6)), we have

$$\bar{u}(F) \leq 2[F : F^2] = 2[K : K^2] \leq 2\bar{u}(K) \leq 4[K : K^2] = 4[F : F^2] \leq 4\bar{u}(F). \quad \square$$

Remark 36.10. Let F be a field of characteristic 2. The proof above shows that every anisotropic totally singular quadratic form has dimension at most $[F : F^2]$ and if $[F : F^2]$ is finite, then there exists an anisotropic totally singular quadratic form of dimension $[F : F^2]$.

Remark 36.11. Let F be a field of characteristic 2 such that $[F : F^2]$ is infinite but F separably closed. Then $\bar{u}(F)$ is infinite but $u(F) = 1$ by Exercise 7.34.

We now look at finiteness of $\bar{u}$ coming down from a quadratic extension following the ideas in [**34**].

Proposition 36.12. *Let K/F be a quadratic extension with F not formally real. If $\bar{u}(K)$ is finite, then $\bar{u}(F) < 4\bar{u}(K)$.*

PROOF. If $\operatorname{char} F = 2$, then $\bar{u}(F) \leq 2\bar{u}(K)$ by Proposition 36.9, so we may assume that $\operatorname{char} F \neq 2$. We first show that $\bar{u}(F)$ is finite. Let φ be an anisotropic quadratic form over F. By Proposition 34.8, there exist quadratic forms φ_1 and μ_0 over F with $(\mu_0)_K$ anisotropic satisfying

$$\varphi \simeq \langle\langle a\rangle\rangle \otimes \varphi_1 \perp \mu_0$$

where $a \in F^\times$ satisfies $K = F(\sqrt{a})$. In particular, $\dim(\mu_0) \leq \bar{u}(K)$. Analogously, there exist quadratic forms φ_2 and μ_1 over F with $(\mu_1)_K$ anisotropic satisfying

$$\varphi_1 \simeq \langle\langle a\rangle\rangle \otimes \varphi_2 \perp \mu_1.$$

Hence

$$\varphi \simeq \langle\langle a\rangle\rangle \otimes \big(\langle\langle a\rangle\rangle \otimes \varphi_2 \perp \mu_1\big) \perp \mu_0 \simeq 2\langle\langle a\rangle\rangle \otimes \varphi_2 \perp \langle\langle a\rangle\rangle \otimes \mu_1 \perp \mu_0$$

as $\langle\langle a, a\rangle\rangle = 2\langle\langle a\rangle\rangle$. Continuing in this way, we see that

$$\varphi \simeq 2^{k-1}\langle\langle a\rangle\rangle \otimes \varphi_k \perp 2^{k-2}\langle\langle a\rangle\rangle \otimes \mu_{k-1} \perp \cdots \perp \langle\langle a\rangle\rangle \otimes \mu_1 \perp \mu_0$$

for some forms φ_k and μ_i over F satisfying $\dim \mu_i \leq \bar{u}(K)$ for all i. By Proposition 31.4, there exists an integer n such that $2^n\langle\langle a\rangle\rangle = 0$ in $W(F)$. It follows that

$$\dim \varphi \leq (2^n + \cdots + 2 + 1)\bar{u}(K) \leq 2^{n+1}\bar{u}(K),$$

hence $\dim \varphi$ is finite.

We now show that $\bar{u}(F) < 4\bar{u}(K)$. As $\bar{u}(F)$ is finite, there exists an anisotropic form φ over F of dimension $\bar{u}(F)$. Let $s : K \to F$ be a nontrivial F-linear functional satisfying $s(1) = 0$. We can write

$$\varphi \simeq \mu \perp s_*(\psi)$$

with quadratic forms ψ over K and μ over F satisfying $\mu \otimes N_{K/F}$ is anisotropic by Proposition 34.6. Then

$$\dim s_*(\psi) \leq 2\bar{u}(K) \quad \text{and} \quad \dim \mu \leq \bar{u}(F)/2.$$

If $\dim s_*(\psi) = 2\bar{u}(K)$, then ψ is a $\bar{u}(K)$-dimensional form over K, hence universal as every $\big(\bar{u}(K)+1\big)$-dimensional form is isotropic over the nonformally real field K. In particular, $\psi \simeq \langle x\rangle_K \perp \psi_1$ for some $x \in F^\times$. Thus $s_*(\psi) = s_*(\psi_1)$ in $W(F)$, so $s_*(\psi)$ is isotropic, a contradiction. Therefore, we have $\dim s_*(\psi) < 2\bar{u}(K)$, hence

$$2\bar{u}(K) > \dim s_*(\psi) = \dim \varphi - \dim \mu \geq \bar{u}(F) - \bar{u}(F)/2 \geq \bar{u}(F)/2.$$

The result follows. $\square$

Proposition 36.13. *Let K/F be a finite extension with F not formally real. Then $\bar{u}(F)$ is finite if and only if $\bar{u}(K)$ is finite.*

PROOF. If $\operatorname{char} F = 2$, the result follows by Proposition 36.9, so we may assume that $\operatorname{char} F \neq 2$. By Corollary 36.7, we need only show if $\bar{u}(K)$ is finite, then $\bar{u}(F)$ is also finite. Let L be the normal closure of K/F and E_0 the fixed field of the Galois group of L/F. Then E_0/F is of odd degree as $\operatorname{char} F \neq 2$. Let E be the fixed field of a Sylow 2-subgroup of the Galois group of L/F. Then E/F is also of odd degree. Therefore, if $\bar{u}(E)$ is finite so is $\bar{u}(F)$ by Springer's Theorem 18.5. Hence we may assume that $E = F$. By Corollary 36.7, we have $\bar{u}(L)$ is finite, so we may assume that $L = K$ and K/F is a Galois 2-extension. By induction on $[K : F]$, we may assume that K/F is a quadratic extension, the case established in Proposition 36.12. □

Let K/F be a normal extension of degree $2^m r$ with r odd and F not formally real. If $\bar{u}(K)$ is finite, the argument in Proposition 36.13 and the bound in Proposition 36.12 shows that $\bar{u}(F) \leq 4^r \bar{u}(K)$. We shall improve this bound in Remark 37.6 below.

37. The u-invariant for formally real fields

If F is formally real and K/F finite, then $\bar{u}(K)$ can be infinite and $\bar{u}(F)$ finite. Indeed, let F_0 be the euclidean field of real constructible numbers. Then there exist extensions E_r/F_0 of degree r none of which are both pythagorean and formally real. In particular, $\bar{u}(E_r) > 0$. It is easy to see that $\bar{u}(E_r) \leq 4$. (In fact, it can be shown that $\bar{u}(E_r) \leq 2$.) For example, E_2 is the quadratic closure of the rational numbers. Let $F = F_0((t_1))\cdots((t_n))\cdots$ denote the field of iterated Laurent series in infinitely many variables over F_0. Then F is pythagorean by Example 36.2(1) so $\bar{u}(F) = 0$. However, the field $E_r((t_1))\cdots((t_n))\cdots$ has infinite u-invariant by Example 36.2(4). In fact, in [**42**] for each positive integer n, formally real fields F_n are constructed with $\bar{u}(F_n) = 2^n$ and having a formally real quadratic extension K/F_n with $\bar{u}(K) = \infty$ and formally fields F'_n are constructed with $\bar{u}(F'_n) = 2^n$ and such that every finite nonformally real extension L of F has infinite $\bar{u}$-invariant.

However, we can determine when finiteness of the $\bar{u}$-invariant persists when going up a quadratic extension and when coming down one. Since we already know this when the base field is not formally real, we shall mostly be interested in the formally real case. In particular, we shall assume that the fields in this section are of characteristic different from 2 and hence the $\bar{u}$-invariant and u-invariant are identical.

We need some preliminaries.

Lemma 37.1. *Let F be a field of characteristic different from 2 and $K = F(\sqrt{a})$ a quadratic extension of F. Let $b \in F^\times \setminus F^{\times 2}$ and $\varphi \in \operatorname{ann}_{W(F)}(\langle\langle b\rangle\rangle)$ be anisotropic. Then $\varphi = \varphi_1 + \varphi_2$ in $W(F)$ for some forms φ_1 and φ_2 over F satisfying:*

(1) *$\varphi_1 \in \langle\langle a\rangle\rangle W(F) \cap \operatorname{ann}_{W(F)}(\langle\langle b\rangle\rangle)$ is anisotropic.*
(2) *$\varphi_2 \in \operatorname{ann}_{W(F)}(\langle\langle b\rangle\rangle)$.*
(3) *$(\varphi_2)_K$ is anisotropic.*

PROOF. By Corollary 6.23 the dimension of φ is even. We induct on $\dim \varphi$. If φ_K is hyperbolic, then $\varphi = \varphi_1$ works by Corollary 34.12 and if φ_K is anisotropic, then $\varphi = \varphi_2$ works. So we may assume that φ_K is isotropic but not hyperbolic. In particular, $\dim \varphi \geq 4$. By Proposition 34.8, we can write

$$\varphi \simeq x\langle\langle a\rangle\rangle \perp \mu$$

for some $x \in F^\times$ and even-dimensional form μ over F. As $\varphi \in \operatorname{ann}_{W(F)}\big(\langle\langle b\rangle\rangle\big)$, we have $\langle\langle b\rangle\rangle \cdot \mu = -x\langle\langle b, a\rangle\rangle$ in $W(F)$, so $\dim\big(\langle\langle b\rangle\rangle \otimes \mu\big)_{an} = 0$ or 4. Therefore, by Proposition 6.25, we can write

$$\mu \simeq \mu_1 \perp y\langle\langle c\rangle\rangle$$

for some $y, c \in F^\times$ and even-dimensional form $\mu_1 \in \operatorname{ann}_{W(F)}\big(\langle\langle b\rangle\rangle\big)$. Substituting in the previous isometry and taking determinants, we see that $ac \in D\big(\langle\langle b\rangle\rangle\big)$ by Proposition 6.25. Thus $c = az$ for some $z \in D\big(\langle\langle b\rangle\rangle\big)$. Consequently,

$$\varphi \simeq x\langle\langle a\rangle\rangle \perp y\langle\langle az\rangle\rangle \perp \mu_1 = x\langle\langle a, -xyz\rangle\rangle + y\langle\langle z\rangle\rangle + \mu_1$$

in $W(F)$. Let $\mu_2 \simeq \big(y\langle\langle z\rangle\rangle \perp \mu_1\big)_{an}$. As $y\langle\langle z\rangle\rangle$ lies in $\operatorname{ann}_{W(F)}\big(\langle\langle b\rangle\rangle\big)$, so does μ_2 and hence also $x\langle\langle a, -xyz\rangle\rangle$. By induction on $\dim \varphi$, we can write $\mu_2 = \widetilde{\varphi}_1 + \widetilde{\varphi}_2$ in $W(F)$ where $\widetilde{\varphi}_1$ satisfies condition (1) and $\widetilde{\varphi}_2$ satisfies conditions (2) and (3). It follows that

$$\varphi_1 \simeq \big(\langle\langle a, -xyz\rangle\rangle \perp \widetilde{\varphi}_1\big)_{an} \quad \text{and} \quad \varphi_2 \simeq \widetilde{\varphi}_2$$

work. □

Exercise 37.2. Let φ and ψ be 2-fold Pfister forms over a field of characteristic not 2. Prove that the group $\varphi W(F) \cap \operatorname{ann}_{W(F)}(\psi) \cap I^2(F)$ is generated by 2-fold Pfister forms ρ in $\operatorname{ann}_{W(F)}(\psi)$ that are divisible by φ. This exercise generalizes. (Cf. Exercise 41.8 below.)

As we are interested in the case of fields of characteristic not 2, we shall show that to test finiteness of the u-invariant, it suffices to look at $\operatorname{ann}_{W(F)}\big(2\langle 1\rangle\big)$. For a field of characteristic not 2, let

$$u'(F) := \max\big\{\dim \varphi \mid \varphi \text{ is an anisotropic form over } F \text{ and } 2\varphi = 0 \text{ in } W(F)\big\}$$

or ∞ if no such maximum exists.

Lemma 37.3. *Let F be a field of characteristic not 2. Then $u'(F)$ is finite if and only if $u(F)$ is finite. Moreover, if $u(F)$ is finite, then $u(F) = u'(F) = 0$ or $u'(F) \leq u(F) < 2u'(F)$.*

PROOF. We may assume that $u'(F) > 0$, i.e., that F is not a formally real pythagorean field. Let φ be an n-dimensional anisotropic form over F. Suppose that $n \geq 2u'(F)$. By Proposition 6.25, we can write $\varphi \simeq \mu_1 \perp \varphi_1$ with $\mu_1 \in \operatorname{ann}_{W(F)}\big(2\langle 1\rangle\big)$ and $2\varphi_1$ anisotropic. By assumption, $\dim \mu_1 \leq u'(F)$. Thus

$$2u'(F) \leq \dim \varphi = \dim \mu_1 + \dim \varphi_1 \leq u'(F) + \dim \varphi_1,$$

hence $2u'(F) \leq \dim 2\varphi_1$. As $(2\varphi)_{an} \simeq 2\varphi_1$, we have $\dim(2\varphi)_{an} \geq 2u'(F)$. Repeating the argument, we see inductively that $\dim(2^m\varphi)_{an} \geq 2u'(F)$ for all m. In particular, φ is not torsion. The result follows. □

Hoffmann has shown that there exist fields F satisfying $u'(F) < u(F)$. (Cf. [**54**].)

Let K/F be a quadratic extension. As it is not true that $u(F)$ is finite if and only if $u(K)$ is when F is formally real, we need a further condition for this to be true. This condition is given by a relative u-invariant.

Let F be a field of characteristic not 2 and L/F a field extension. The *relative u-invariant* of L/F is defined as

$$u(L/F) := \max\{\dim(\varphi_L)_{an} \mid \varphi \text{ a quadratic form over } F \text{ with } \varphi_L \text{ torsion in } W(L)\}$$

or ∞ if no such integer exists.

The introduction of this invariant and its study come from Lee in [**91**] and [**40**]. We shall prove

Theorem 37.4. *Let F be a field of characteristic different than 2 and K a quadratic extension of F. Then $u(F)$ and $u(K/F)$ are both finite if and only if $u(K)$ is finite. Moreover, we have:*

1. *If $u(F)$ and $u(K/F)$ are both finite, then $u(K) \le u(F) + u(K/F)$. If, in addition, K is not formally real, then $u(K) \le \frac{1}{2}u(F) + u(K/F)$.*
2. *If $u(K)$ is finite, then $u(K/F) \le u(K)$ and $u(F) < 6u(K)$ or $u(F) = u(K) = 0$. If, in addition, K is not formally real, then $u(F) < 4u(K)$.*

PROOF. Let $K = F(\sqrt{a})$ and $s_* : W(K) \to W(F)$ be the transfer induced by the F-linear functional defined by $s(1) = 0$ and $s(\sqrt{a}) = 1$.

Claim: Let φ be an anisotropic quadratic form over K such that $s_*(\varphi)$ is torsion in $W(F)$. Then there exist a form σ over F and a form ψ over K satisfying:

(a) $\dim \sigma = \dim \varphi$.
(b) ψ is a torsion form in $W(K)$.
(c) $\dim \psi \le 2 \dim \varphi$ and $\varphi \simeq (\sigma_K \perp \psi)_{an}$.
(d) If $s_*(\varphi)$ is anisotropic over F, then $\dim \varphi \le \dim \psi$.

In particular, if $u(F)$ is finite and $s_*(\varphi)$ anisotropic and torsion, then $\dim \varphi \le \frac{1}{2}u(F)$ and $\dim \psi \le u(F)$.

Let $2^n s_*(\varphi) = 0$ in $W(F)$ for some integer n. By Corollary 34.3 with $\rho = 2^n\langle 1\rangle$, there exists a form σ over F such that $\dim \sigma = \dim \varphi$ and $2^n\varphi \simeq 2^n\sigma_K$. Let $\psi \simeq (\varphi \perp (-\sigma))_{an}$. Then ψ is a torsion form in $W(K)$ as it has trivial total signature. Condition (c) holds by construction and (d) holds as $s_*(\psi) = s_*(\varphi)$ in $W(F)$.

We now prove (1). Suppose that both $u(F)$ and $u(K/F)$ are finite. Let τ be an anisotropic torsion form over K. By Proposition 34.1, there exists an isometry $\tau \simeq \varphi \perp \mu_K$ for some form τ over K satisfying $s_*(\varphi)$ is anisotropic and some form μ over F. As $s_*(\varphi) = s_*(\tau)$ is torsion, we can apply the claim to φ. Let σ over F and ψ over K be forms as in the claim. By the last statement of the claim, we have $\dim \varphi \le \frac{1}{2}u(F)$. In particular, we have $\dim \psi \le 2\dim \varphi \le u(F)$ and $\varphi = \psi + \sigma_K$ in $W(K)$. Since τ and ψ are torsion, so is $(\sigma + \mu)_K$. As $\tau = \psi + \big((\sigma \perp \mu)_K\big)_{an}$ in $W(K)$, it follows that $\dim \tau \le u(F) + u(K/F)$ as needed.

Finally, if K is not formally real, then as above, we have $\tau \simeq \varphi \perp \mu_K$ with $\dim \varphi \le \frac{1}{2}u(F)$. As every F-form is torsion in $W(K)$, we have $\dim \mu_K \le u(K/F)$ and the proof of (1) is complete.

We now prove (2). Suppose that $u(K)$ is finite. Certainly $u(K/F) \le u(K)$. We show the rest of the first statement. By Lemma 37.3, it suffices to show that $u'(F) \le 3u'(K)$. Let $\varphi \in \operatorname{ann}_{W(F)}\big(2\langle 1\rangle\big)$ be anisotropic. By Lemma 37.1 and

Corollary 34.33, we can decompose $\varphi = \varphi_1 + \varphi_2$ in $W(F)$ with $\varphi_2 \in \operatorname{ann}_{W(F)}(2\langle 1\rangle)$ satisfying $(\varphi_2)_K$ is anisotropic and φ_1 is anisotropic over F and lies in

$$\langle\langle a\rangle\rangle W(F) \cap \operatorname{ann}_{W(F)}(2\langle 1\rangle) \subset \operatorname{ann}_{W(F)}(\langle\langle a\rangle\rangle) \cap \operatorname{ann}_{W(F)}(2\langle 1\rangle)$$

using Lemma 34.33. In particular, $(\varphi_2)_K \in \operatorname{ann}_{W(K)}(2\langle 1\rangle)$ so $\dim \varphi_2 \leq u'(K)$. Consequently, to show that $u'(F) \leq 3u'(K)$, it suffices to show $\dim \varphi_1 \leq 2u'(K)$. This follows from (i) of the following (with $\sigma = \varphi_1$):

Claim: Let σ be a nondegenerate quadratic form over F.

(i) If $\sigma \in \operatorname{ann}_{W(F)}(\langle\langle a\rangle\rangle) \cap \operatorname{ann}_{W(F)}(2\langle 1\rangle)$, then $\dim \sigma_{an} \leq 2u'(K)$.

(ii) If $\sigma \in \operatorname{ann}_{W(F)}(\langle\langle a\rangle\rangle) \cap W_t(F)$, then $\dim \sigma_{an} \leq 2u(K)$ with inequality if K is not formally real.

By Corollary 34.33, in the situation of (i), there exists $\tau \in \operatorname{ann}_{W(K)}(2\langle 1\rangle)$ such that $\sigma = s_*(\tau)$. Then $\dim \sigma_{an} \leq \dim s_*(\tau_{an}) \leq 2 \dim \tau_{an} \leq 2u'(K)$ as needed.

We turn to the proof of (ii) which implies the the bound on $u(F)$ in statement (2) for arbitrary K (with $\sigma = \varphi_1$). In the situation of (ii), we have $\dim \sigma_{an} \leq u(K)$ by Corollaries 34.12 and 34.32. If K is not formally real, then any $u(K)$-dimensional form τ over K is universal. In particular, $D(\tau) \cap F^\times \neq \emptyset$ and (ii) follows.

Now assume that K is not formally real. Let φ be an anisotropic torsion form over F of dimension $u(F)$. As $\operatorname{im}(s_*) = \operatorname{ann}_{W(F)}(\langle\langle a\rangle\rangle)$ by Corollary 34.12, using Proposition 6.25, we have a decomposition $\varphi \simeq \varphi_3 \perp \varphi_4$ with φ_4 a form over F satisfying $\langle\langle a\rangle\rangle \otimes \varphi_4$ is anisotropic and $\varphi_3 \simeq s_*(\tau)$ for some form τ over K. Since φ_3 lies in

$$s_*\big(W(K)\big) = s_*\big(W_t(K)\big) = \operatorname{ann}_{W(F)}(\langle\langle a\rangle\rangle) \cap W_t(F)$$

by Corollaries 34.12 and 34.32, we have $\dim \varphi_3 < 2u(K)$ by the claim above. As $\langle\langle a\rangle\rangle \cdot \varphi_4 = \langle\langle a\rangle\rangle \cdot \varphi$ in $W(F)$, hence is torsion, we have $\dim \varphi_4 \leq u(F)/2$. Therefore, $2u(K) > \dim \varphi_3 = \dim \varphi - \dim \varphi_4 \geq u(F) - u(F)/2$ and $u(F) < 4u(K)$. □

Of course, by Theorem 36.6 if F is not formally real and $K = F(\sqrt{a})$ is a quadratic extension, then $u(K) \leq \frac{3}{2}u(F)$.

Corollary 37.5. *Let F be a field of transcendence degree n over a real closed field. Then $u(F) < 2^{n+2}$.*

PROOF. $F(\sqrt{-1})$ is a C_n-field by Corollary 97.7. □

Remark 37.6. Let F be a field of characteristic not 2 and K/F a finite normal extension. Suppose that $u(K)$ is finite. If K/F is quadratic, then the proof of Theorem 37.4 shows that $u'(F) \leq 3u'(K)$. If K/F is of degree $2^r m$ with m odd, arguing as in Proposition 36.13, shows that $u'(F) \leq 3^r u'(K)$, hence $u(F) \leq 2 \cdot 3^r u(K)$.

One case where the bound in the remark can be sharpened is the following which generalizes the case of a pythagorean field of characteristic different from 2.

Proposition 37.7. *Let F be a field of characteristic different from 2 and K/F a finite normal extension. If $u(K) \leq 2$, then $u(F) \leq 2$.*

PROOF. By Proposition 35.1, we know for a field E that $I^2(E)$ is torsion-free if and only if E satisfies A_2, i.e., there are no anisotropic 2-fold torsion Pfister forms. In particular, as $u(K) \leq 2$, we have $I^2(K)$ is torsion-free. Arguing as in Proposition 36.13, we reduce to the case that $K = F(\sqrt{a})$ is a quadratic extension of F, hence

$I^2(F)$ is also torsion-free by Theorem 35.12. It follows that every torsion element ρ in $I(F)$ lies in $\mathrm{ann}_{W(F)}(2\langle 1\rangle)$. In particular, by Proposition 6.25, we can write $\rho \simeq \langle\langle w\rangle\rangle \mod I^2(F)$ for some $w \in D(2\langle 1\rangle)$, hence $\rho \simeq \langle\langle w\rangle\rangle$ some $w \in D(2\langle 1\rangle)$ and is universal. Consequently, every even-dimensional anisotropic torsion form over F is of dimension at most two. Suppose that there exists an odd-dimensional anisotropic torsion form φ over F. Then F is not formally real, hence all forms are torsion. As every 2-dimensional form over F is universal by the above, we must have $\dim \varphi = 1$. The result follows. □

Corollary 37.8. *Let F be a field of transcendence degree one over a real closed field. Then $u(F) \leq 2$.*

Exercise 37.9. Let F be a field of arbitrary characteristic and $a \in F^\times$ totally positive. If $K = F(\sqrt{a})$, then $u(K) \leq 2u(F)$.

We next show if K/F is a quadratic extension with K not formally real, then the relative u-invariant already determines finiteness. We note

Remark 37.10. Suppose that $\operatorname{char} F \neq 2$ and $K = F(\sqrt{a})$ is a quadratic extension of F that is not formally real. If φ is a nondegenerate quadratic form over F, then, by Proposition 34.8, there exist forms φ_1 and ψ such that $\varphi \simeq \langle\langle a\rangle\rangle \otimes \psi \perp \varphi_1$ with $\dim \varphi_1 \leq u(K/F)$.

We need the following simple lemma.

Lemma 37.11. *Let F be a field of characteristic different from 2 and $K = F(\sqrt{a})$ a quadratic extension of F that is not formally real. Suppose that $u(K/F) < 2^m$. Then $I^{m+1}(F)$ is torsion-free, $I^{m+1}(K) = 0$, and the exponent of $W_t(F)$ is at most 2^{m+1}.*

Proof. If $\rho \in P_m(F)$, then $r^*_{K/F}(\rho) = 0$ as K is not formally real. So $I^m(F) = \langle\langle a\rangle\rangle I^{m-1}(F)$ by Theorem 34.22. It follows that $I^{m+1}(K) = 0$ by Lemma 34.16. Hence $I^{m+1}(F(\sqrt{-1})) = 0$ by Corollary 35.14. The result follows by Corollary 35.27. □

If F is a local field in the above, then one can show that $u(K/F) = 2$ for any quadratic extension K of F, but neither $I^2(F)$ nor $I^2(K)$ is torsion-free.

Theorem 37.12. *Let F be a field of characteristic different from 2. Suppose that K is a quadratic extension of F with K not formally real. Then $u(K/F)$ is finite if and only if $u(K)$ is finite.*

Proof. By Theorem 37.4, we may assume that $u(K/F)$ is finite and must show that $u(F)$ is also finite. Let φ be an anisotropic form over F satisfying $2\varphi = 0$ in $W(F)$. By the lemma, $I^{n-1}(F)$ is torsion-free for some $n \geq 1$. We apply the Remark 37.10 iteratively. In particular, if $\dim \varphi$ is large, then $\varphi \simeq x\rho \perp \psi$ for some $\rho \in P_n(F)$ (cf. the proof of Proposition 36.12). Indeed, computation shows that if $u(K/F) < 2^m$ and $\dim \varphi > 2^m(2^{m+2} - 1)$, then $n = m + 2$ works. As ρ is an anisotropic Pfister form and $I^n(F)$ is torsion-free, 2ρ is also anisotropic. Scaling φ, we may assume that $x = 1$. Write $\psi \simeq \varphi_1 \perp \varphi_2$ with $2\varphi_1 = 0$ in $W(F)$ and $2\varphi_2$ anisotropic. Then we have $2\rho \simeq 2(-\varphi_2)$. If $b \in D(-\varphi_2)$, then $2\langle\langle b\rangle\rangle \cdot \rho$ is isotropic, hence is zero in $W(F)$. As $I^n(F)$ is torsion-free, $\langle\langle b\rangle\rangle \cdot \rho = 0$ in $W(F)$ and $b \in D(\rho)$. It follows that φ cannot be anisotropic if $\dim \varphi > 2^m(2^{m+2} - 1)$. By Lemma 37.3, it follows that $u(F) \leq 2^{m+1}(2^{m+2} - 1)$ and the result follows by Theorem 37.4. □

The bounds in the proof can be improved but are still very weak. The theorem does not generalize to the case when K is formally real. Indeed, let F_0 be a formally real subfield of the algebraic closure of the rationals having square classes represented by $\pm 1, \pm w$ where w is a sum of (two) squares. Let $F = F_0((t_1))((t_2))\cdots$ and $K = F(\sqrt{w})$. Then, using Corollary 34.12, we see that $u(K/F) = 0$ but both $u(F)$ and $u(K)$ are infinite.

Corollary 37.13. *Let F be a field of characteristic different from 2. Then $u(K)$ is finite for all finite extensions of F if and only if $u\big(F(\sqrt{-1})\big)$ is finite if and only if $u(F(\sqrt{-1})/F)$ is finite.*

Remark 37.14. Using the theory of abstract Witt rings, Schubert proved that if F is a formally real field, then $u(K)$ is finite for all finite extensions of F if and only if $u(F)$ is finite and $I_{red}^{n+1}(F) = 2I_{red}^{n}(F)$ for some n if and only if $u\big(F(\sqrt{-1})\big)$ is finite (cf. [**122**]). Recall that the condition on the reduced Witt ring is equivalent to the cokernel of the total signature map having finite exponent by Proposition 35.23.

38. Construction of fields with even u-invariant

In 1953 Kaplansky conjectured in [**68**] that $u(F)$ if finite was always a power of two. This was shown to be false in [**101**] where it was shown there exist fields having u-invariant six and afterwards in [**102**] having u-invariant any even integer. Subsequently, Izhboldin constructed fields having u-invariant 9 in [**64**] and Vishik has constructed fields having u-invariant $2^r + 1$ for every $r \geq 3$ in [**135**].

By taking iterated Laurent series fields over the complex numbers, we can construct fields whose u-invariant is 2^n for any $n \geq 0$. (We also know that formally real pythagorean fields have u-invariant zero.) In this section, given any even integer $m > 0$, we construct fields whose u-invariant is m. This construction was first done in [**102**].

Lemma 38.1. *Let $\varphi \in I_q^2(F)$ be a form of dimension $2n \geq 2$. Then φ is a sum of $n-1$ general quadratic 2-fold Pfister forms in $I_q^2(F)$ and* $\operatorname{ind}\operatorname{clif}(\varphi) \leq 2^{n-1}$.

Proof. We induct on n. If $n = 1$, we have $\varphi = 0$ and the statement is clear. If $n = 2$, φ is a general 2-fold Pfister form and by Proposition 12.4, we have $\operatorname{clif}(\varphi) = [Q]$, where Q is a quaternion algebra such that Nrd_Q is similar to φ. Hence $\operatorname{ind}\operatorname{clif}(\varphi) \leq 2$.

In the case $n \geq 3$ write $\varphi = \sigma \perp \psi$ where σ is a binary form. Choose $a \in F^\times$ such that the form $a\sigma \perp \psi$ is isotropic, i.e., $a\sigma \perp \psi \simeq \mathbb{H} \perp \mu$ for some form μ of dimension $2n-2$. We have in $I_q(F)$:

$$\varphi = \sigma + \psi = \langle\langle a\rangle\rangle\sigma + \mu$$

and therefore $\operatorname{clif}(\varphi) = \operatorname{clif}\big(\langle\langle a\rangle\rangle\sigma\big)\cdot\operatorname{clif}(\mu)$ by Lemma 14.2. Applying the induction hypothesis to μ, we have φ is a sum of $n-1$ general quadratic 2-fold Pfister forms and

$$\operatorname{ind}\operatorname{clif}(\varphi) \leq \operatorname{ind}\operatorname{clif}\big(\langle\langle a\rangle\rangle\sigma\big)\cdot\operatorname{ind}\operatorname{clif}(\mu) \leq 2\cdot 2^{n-2} = 2^{n-1}. \qquad \square$$

Corollary 38.2. *In the condition of the lemma assume that* $\operatorname{ind}\operatorname{clif}(\varphi) = 2^{n-1}$. *Then φ is anisotropic.*

PROOF. Suppose φ is isotropic, i.e., $\varphi \simeq \mathbb{H} \perp \psi$ for some ψ of dimension $2n-2$. Applying Lemma 38.1 to ψ, we have $\operatorname{ind}\operatorname{clif}(\varphi) = \operatorname{ind}\operatorname{clif}(\psi) \leq 2^{n-2}$, a contradiction. □

Lemma 38.3. *Let D be a tensor product of $n-1$ quaternion algebras ($n \geq 1$). Then there is a $\varphi \in I_q^2(F)$ of dimension $2n$ such that $\operatorname{clif}(\varphi) = [D]$ in $\operatorname{Br}(F)$.*

PROOF. We induct on n. The case $n = 1$ follows from Proposition 12.4. If $n \geq 2$ write $D = Q \otimes B$, where Q is a quaternion algebra and B is a tensor product of $n-2$ quaternion algebras. By the induction hypothesis, there is $\psi \in I_q^2(F)$ of dimension $2n-2$ such that $\operatorname{clif}(\psi) = [B]$. Choose a quadratic 2-fold Pfister form σ with $\operatorname{clif}(\sigma) = [Q]$ and an element $a \in F^\times$ such that $a\sigma \perp \psi$ is isotropic, i.e., $a\sigma \perp \psi \simeq \mathbb{H} \perp \varphi$ for some φ of dimension $2n$. Then φ works as $\operatorname{clif}(\varphi) = \operatorname{clif}(\sigma) \cdot \operatorname{clif}(\psi) = [Q] \cdot [B] = [D]$. □

Let $\mathfrak{A}$ be a set (of isometry classes) of irreducible quadratic forms. For any finite subset $S \subset \mathfrak{A}$ let X_S be the product of all the quadrics X_φ with $\varphi \in S$. If $S \subset T$ are two subsets of $\mathfrak{A}$ we have the dominant projection $X_T \to X_S$ and therefore the inclusion of function fields $F(X_S) \to F(X_T)$. Set $F_{\mathfrak{A}} = \operatorname{colim} F_S$ over all finite subsets $S \subset \mathfrak{A}$. By construction, all quadratic forms $\varphi \in \mathfrak{A}$ are isotropic over the field extension $F_{\mathfrak{A}}/F$. A field is called 2-*special* if every finite extension of it has degree a power of 2 (cf. §101.B).

Theorem 38.4. *Let F be a field and $n \geq 1$ an integer. Then there is a field extension E of F satisfying:*

(1) $u(E) = 2n$.
(2) $I_q^3(E) = 0$.
(3) *E is 2-special.*

PROOF. To every field L, we associate three fields $L^{(1)}$, $L^{(2)}$, and $L^{(3)}$ as follows:

Let $\mathfrak{A}$ be the set of isometry classes of all nondegenerate quadratic forms over L of dimension $2n+1$. We set $L^{(1)} = L_{\mathfrak{A}}$. Every nondegenerate quadratic form over L of dimension $2n+1$ is isotropic over $L^{(1)}$.

Let $\mathfrak{B}$ be the set of isometry classes of all quadratic 3-fold Pfister forms over L. We set $L^{(2)} = L_{\mathfrak{B}}$. By construction, every quadratic 3-fold Pfister form over L is isotropic over $L^{(2)}$.

Finally, let $L^{(3)}$ be a 2-special closure of L (cf. §101.B).

Let D be a central division L-algebra of degree 2^{n-1}. By Corollaries 30.9, 30.11, and 98.5, the algebra D remains a division algebra when extended to $L^{(1)}$, $L^{(2)}$, or $L^{(3)}$.

Let L be a field extension of F such that there is a central division algebra D over L that is a tensor product of $n-1$ quaternion algebras (cf. Proposition 98.18). By Lemma 38.3, there is a $\varphi \in I_q^2(L)$ of dimension $2n$ such that $\operatorname{clif}(\varphi) = [D]$ in $\operatorname{Br}(L)$.

We construct a tower of field extensions $E_0 \subset E_1 \subset E_2 \subset \dots$ by induction. We set $E_0 = L$. If E_i is defined we set $E_{i+1} = (((E_i)^{(1)})^{(2)})^{(3)}$. Note that the field E_{i+1} is 2-special and all nondegenerate quadratic forms of dimension $2n+1$ and all 3-fold Pfister forms over E_i are isotropic over E_{i+1}. Moreover, the algebra D remains a division algebra over E_{i+1}.

Now set $E = \bigcup E_i$. Clearly, E has the following properties:

(i) All $(2n+1)$-dimensional Pfister forms over E are isotropic. In particular, $u(E) \leq 2n$.
(ii) The field E is 2-special.
(iii) All quadratic 3-fold Pfister forms over E are isotropic. In particular, $I_q^3(E) = 0$.
(iv) The algebra D_E is a division algebra.

As $\operatorname{clif}(\varphi_E) = [D_E]$, it follows from Corollary 38.2 that φ_E is anisotropic. In particular, $u(E) = 2n$. □

39. Addendum: Linked fields and the Hasse number

Theorem 39.1. *Let F be a field. Then the following conditions are equivalent:*

(1) *Every pair of quadratic 2-fold Pfister forms over F are linked.*
(2) *Every 6-dimensional form in $I_q^2(F)$ is isotropic.*
(3) *The tensor product of two quaternion algebras over F is not a division algebra.*
(4) *Every two division quaternion algebras over F have isomorphic separable quadratic subfields.*
(5) *Every two division quaternion algebras over F have isomorphic quadratic subfields.*
(6) *The classes of quaternion algebras in $\operatorname{Br}(F)$ form a subgroup.*

Proof. (1) $\Rightarrow$ (2): Let ψ be a 6-dimensional form in $I_q^2(F)$. By Lemma 38.1, we have $\psi = \varphi_1 + \varphi_2$, where φ_1 and φ_2 are general quadratic 2-fold Pfister forms. By assumption, φ_1 and φ_2 are linked. Therefore, the class of ψ in $I_q^2(F)$ is represented by a form of dimension 4, hence ψ is isotropic.

(2) $\Rightarrow$ (1): Let φ_1 and φ_2 be two quadratic 2-fold Pfister forms over F. Then $\varphi_1 - \varphi_2 = \psi$ for some 6-dimensional form $\psi \in I_q^2(F)$. As ψ is isotropic, we have $\mathfrak{i}_0(\varphi_1 \perp -\varphi_2) \geq 2$, i.e., φ_1 and φ_2 are linked.

(1) $\Rightarrow$ (4): Let Q_1 and Q_2 be division quaternion algebras over F. Let φ_1 and φ_2 be the reduced norm quadratic forms of Q_1 and Q_2, respectively. By assumption, φ_1 and φ_2 are linked. In particular, φ_1 and φ_2 are split by a separable quadratic field extension L/F. Hence L splits Q_1 and Q_2 and therefore L is isomorphic to subfields of Q_1 and Q_2.

(3) $\Leftrightarrow$ (4) $\Leftrightarrow$ (5) is proven in Theorem 98.19.

(3) $\Leftrightarrow$ (6) is obvious.

(4) $\Rightarrow$ (1): Let φ_1 and φ_2 be two anisotropic 2-fold Pfister forms over F. Let Q_1 and Q_2 be two division quaternion algebras with the reduced norm forms φ_1 and φ_2, respectively. By assumption, Q_1 and Q_2 have quadratic subfields isomorphic to a separable quadratic extension L/F. By Example 9.7, the forms φ_1 and φ_2 are divisible by the norm form of L/F and hence are linked. □

A field F is called *linked* if F satisfies the conditions of Theorem 39.1.

For a formally real field F, the u-invariant can be thought of as a weak Hasse Principle, i.e., every locally hyperbolic form of dimension $> u(F)$ is isotropic. A variant of the u-invariant introduced in [**31**] naturally arises. We call a quadratic form φ over F *locally isotropic* or *totally indefinite* if φ_{F_P} is isotropic at each real closure F_P of F (if any), i.e., φ is indefinite at each real closure of F (if any). The *Hasse number* of a field F is defined to be

$$\widetilde{u}(F) := \max\{\dim\varphi \mid \varphi \text{ is a locally isotropic anisotropic form over } F\}$$

or ∞ if no such maximum exists. For fields that are not formally real this coincides with the $\bar{u}$-invariant. If a field is formally real, finiteness of its $\widetilde{u}$-invariant is a very strong condition and is a form of a strong Hasse Principle. For example, if F is a global field, then $\widetilde{u}(F) = 4$ by Meyer's Theorem [**104**] (a forerunner of the Hasse-Minkowski Principle [**121**]) and if F is the function field of a real curve, then $\widetilde{u}(F) = 2$ (cf. Example 39.11 below), but if $F/\mathbb{R}$ is formally real and finitely generated of transcendence degree > 1, then, although $u(F)$ is finite, its Hasse number $\widetilde{u}(F)$ is infinite.

Exercise 39.2. Show, if the Hasse number is finite, then it cannot be $3, 5$, or 7.

We establish another characterization of $\widetilde{u}(F)$ using a concept introduced in [**38**]. We say F satisfies Property H_n with $n > 1$ if there exist no anisotropic, locally isotropic forms of dimension n. Thus if $\widetilde{u}(F)$ is finite, then

$$\widetilde{u}(F) + 1 = \min\{n \mid F \text{ satisfies } H_m \text{ for all } m \geq n\}.$$

Remark 39.3. Every 6-dimensional form in $I_q^2(F)$ is locally isotropic, since every element in $I^2(F)$ has signature divisible by 4 at every ordering. Hence if $\widetilde{u}(F) \leq 4$, then F is linked by Theorem 39.1.

Lemma 39.4. *Let F be a linked field of characteristic not 2. Then*

(1) *Any pair of n-fold Pfister forms are linked for $n \geq 2$.*
(2) *If $\varphi \in P_n(F)$, then $\varphi \simeq \langle\langle -w_1, x\rangle\rangle$ if $n = 2$ and $\varphi = 2^{n-3}\langle\langle -w_1, -w_2, x\rangle\rangle$ if $n \geq 3$ for some $w_1, w_2 \in D(3\langle 1\rangle)$ and $x \in F^\times$.*
(3) *For every $n \geq 0$ and $\varphi \in I^n(F)$, there exists an integer m and $\rho_i \in GP_i(F)$ with $n \leq i \leq m$ satisfying $\varphi = \sum_{i=n}^m \rho_i$ in $W(F)$. Moreover, if φ is a torsion element, then each ρ_i is torsion.*
(4) *$I^4(F)$ is torsion-free.*

PROOF. (1), (2): Any pair of n-fold Pfister forms is easily seen to be linked by induction, so (1) is true. As any 2-fold Pfister form is linked to $4\langle 1\rangle$, statement (2) holds for $n = 2$. Let $\rho = \langle\langle a, b, c\rangle\rangle$ be a 3-fold Pfister form, then applying the $n = 2$ case gives $\rho = \langle\langle w_1, x, y\rangle\rangle = \langle\langle w_1, w_2, z\rangle\rangle$ for some $x, y, z \in F^\times$ and $w_1, w_2 \in D(3\langle 1\rangle)$. This establishes the $n = 3$ case. Let $\rho = \langle\langle a, b, c, d\rangle\rangle$ be a 4-fold Pfister form. By assumption, there exist $x, y, z \in F^\times$ such that $\langle\langle a, b\rangle\rangle \simeq \langle\langle x, y\rangle\rangle$ and $\langle\langle c, d\rangle\rangle \simeq \langle\langle x, z\rangle\rangle$. Thus

$$\rho = \langle\langle a, b, c, d\rangle\rangle \simeq \langle\langle x, y, x, z\rangle\rangle \simeq \langle\langle -1, y, x, z\rangle\rangle = 2\langle\langle y, x, z\rangle\rangle. \tag{39.5}$$

Statement (2) follows.

(3): Let ψ and τ be n-fold Pfister forms. As they are linked, $\psi - \tau = a\langle\langle b\rangle\rangle \cdot \mu$ in $W(F)$ for some $(n-1)$-fold Pfister form μ and $a, b \in F^\times$. Then

$$x\psi + y\tau = x\psi - x\tau + x\tau + y\tau = ax\langle\langle b\rangle\rangle \cdot \mu + x\langle\langle -xy\rangle\rangle \cdot \tau.$$

The first part now follows by repeating this argument. If φ is torsion, then inductively, each ρ_i is torsion by the Hauptsatz 23.7, so the second statement follows.

(4): By (3), it suffices to show there are no anisotropic torsion n-Pfister forms with $n > 3$. By Proposition 35.3, it suffices to show if $\rho \in P_4(F)$ satisfies $2\rho = 0$

in $W(F)$, then $\rho = 0$ in $W(F)$. By Lemma 35.2, we can write $\rho \simeq \langle\langle a,b,c,w\rangle\rangle$ with $w \in D(2\langle 1\rangle)$ and $a,b,c \in F^\times$. Applying equation (39.5) with $d = w$, we have $\rho \simeq 2\langle\langle y,x,z\rangle\rangle \simeq 2\langle\langle y,c,w\rangle\rangle$ which is hyperbolic. The result follows. □

Lemma 39.6. *Let* $\operatorname{char} F \neq 2$ *and* $n \geq 2$. *If* F *is linked and* F *satisfies* H_n, *then it satisfies* H_{n+1}.

PROOF. Let φ be an $(n+1)$-dimensional anisotropic quadratic form with $n \geq 2$. Replacing φ by $x\varphi$ for an appropriate $x \in F^\times$, we may assume that $\varphi = \langle w, b, wb\rangle \perp \varphi_1$ for some $w, b \in F^\times$ and a form φ_1 over F and by Lemma 39.4 that $w \in D(3\langle 1\rangle)$. Let $\varphi_2 = \langle w, b\rangle \perp \varphi_1$. As $\operatorname{sgn}_P\langle b\rangle = \operatorname{sgn}_P\langle wb\rangle$ for all $P \in \mathfrak{X}(F)$, the form φ is locally isotropic if and only if φ_2 is. The result follows by induction. □

Remark 39.7. If $\operatorname{char} F \neq 2$ and $n \geq 4$, then F satisfies Property H_{n+1} if it satisfies Property H_n. However, in general, H_3 does not imply H_4 (cf. [**38**]).

Exercise 39.8. Let F be a formally real pythagorean field. Then $\widetilde{u}(F)$ is finite if and only if $I^2(F) = 2I(F)$. Moreover, if this is the case, then $\widetilde{u}(F) = 0$.

The following theorem from [**31**] strengthens the result in [**35**].

Theorem 39.9. *Let* F *be a linked field of characteristic not* 2. *Then* $u(F) = \widetilde{u}(F)$ *and* $\widetilde{u}(F) = 0, 1, 2, 4,$ *or* 8.

PROOF. We first show that $\widetilde{u}(F) = 0, 1, 2, 4,$ or 8. We know that $I^4(F)$ is torsion-free by Lemma 39.4. We first show that F satisfies H_9, hence $\widetilde{u}(F) \leq 8$ by Lemma 39.6. Let φ be a 9-dimensional locally isotropic form over F. Replacing φ by $x\varphi$ for an appropriate $x \in F^\times$, we can assume that $\varphi = \langle 1\rangle + \varphi_1$ in $W(F)$ with $\varphi_1 \in I^2(F)$ using Proposition 4.13. By Lemma 39.4, we have a congruence

$$\varphi \equiv \langle 1\rangle + \rho_2 - \rho_3 \mod I^4(F) \tag{39.10}$$

for some $\rho_i \in P_i(F)$ with $i = 2, 3$. Write $\rho_2 \simeq \langle\langle a,b\rangle\rangle$ and $\rho_3 \simeq \langle\langle c,d,e\rangle\rangle$. As F is linked, we may assume that $e = b$ and $-d \in D_F(\rho_2')$. Thus we have

$$\begin{aligned}\varphi &\equiv \langle 1\rangle + \langle\langle a,b\rangle\rangle - \langle\langle c,d,b\rangle\rangle\\ &\equiv \langle 1\rangle - d\big(\langle\langle a,b\rangle\rangle - \langle\langle c,b\rangle\rangle\big) - \langle\langle c,b\rangle\rangle\\ &\equiv -cd\langle\langle ac,b\rangle\rangle - \langle\langle c,b\rangle\rangle' \mod I^4(F).\end{aligned}$$

Let $\mu = \varphi \perp cd\langle\langle ac,b\rangle\rangle \perp \langle\langle c,b\rangle\rangle'$ be a locally isotropic form over F lying in $I^4(F)$. In particular, for all $P \in \mathfrak{X}(F)$, we have $16 \mid \operatorname{sgn}_P \mu$. As the locally isotropic form μ is 16-dimensional, $|\operatorname{sgn}_P \mu| < 16$ for all $P \in \mathfrak{X}(F)$, so $\operatorname{sgn}_P \mu = 0$ for all $P \in \mathfrak{X}(F)$ and $\mu \in I_t^4(F) = 0$. Consequently, $\varphi = -cd\langle\langle ac,b\rangle\rangle \perp \big(-\langle\langle c,b\rangle\rangle'\big)$ in $W(F)$, so φ is isotropic and $\widetilde{u}(F) \leq 8$.

Suppose that $\widetilde{u}(F) < 8$. Then there are no anisotropic torsion 3-fold Pfister forms over F. It follows that $I^3(F)$ is torsion-free by Lemma 39.4. We show $\widetilde{u}(F) \leq 4$. To do this it suffices to show that F satisfies H_5 by Lemma 39.6. Let φ be a 5-dimensional, locally isotropic space over F. Arguing as above but going mod $I^3(F)$, we may assume that

$$\varphi \equiv \langle 1\rangle - \langle\langle a,b\rangle\rangle = -\langle\langle a,b\rangle\rangle' \mod I^3(F).$$

Let $\mu = \varphi \perp \langle\langle a,b\rangle\rangle'$ be an 8-dimensional, locally isotropic form over F lying in $I^3(F)$. As above, it follows that μ is locally hyperbolic, hence $\mu \in I_t^3(F) = 0$. Thus $\varphi = -\langle\langle a,b\rangle\rangle'$ in $W(F)$ so it is isotropic and $\widetilde{u}(F) < 4$. In a similar way, we see that

$\widetilde{u}(F) = 0, 1, 2$ are the only other possibilities. This shows that $\widetilde{u}(F) = 0, 1, 2, 4, 8$. The argument above and Lemma 39.4 show that $u(F) = \widetilde{u}(F)$. □

Note the proof shows that if F is linked and $I^n(F)$ is torsion-free, then $\widetilde{u}(F) \leq 2^{n-1}$.

It is still unknown whether Theorem 39.9 is true if $\operatorname{char} F = 2$.

Example 39.11. (1) If $F(\sqrt{-1})$ is a C_1 field, then $I^2(F(\sqrt{-1})) = 0$. It follows that $I^2(F) = 2I(F)$ and is torsion-free by Corollary 35.14 and Proposition 35.1 (or Corollary 35.27). In particular, F is linked and $\widetilde{u}(F) \leq 2$.

(2) If F is a local or global field, then $\widetilde{u}(F) = 4$.

(3) Let F_0 be a local field and $F = F_0((t))$ be a Laurent series field. As $u(F_0) = 4$, and F is not formally real, we have $\widetilde{u}(F) = u(F) = 8$. If $\operatorname{char} F \neq 2$, this field F is linked by the following exercise:

Exercise 39.12. Let F be a field of characteristic not 2. Show that the field $F((t))$ is linked if and only if there exist no 4-dimensional anisotropic quadratic forms over F_0 of nontrivial discriminant.

There exist linked formally real fields with Hasse number 8, but the construction of such fields is more delicate (cf. [**42**]).

Remark 39.13. Let F be a formally real field. Then it can be shown that $\widetilde{u}(F)$ is finite if and only if $u(F)$ is finite and $I^2(F_{py}) = 2I^n(F_{py})$ (cf. [**42**]). If both of these invariants are finite, they may be different (cf. [**112**].)

Remark 39.14. We say that a formally real field F satisfies condition S_n if the map $s_* : I^n_{red}(F(\sqrt{w})) \to I^n_{red}(F)$ induced by $s : F(\sqrt{w}) \to F$ with $s(1) = 0$ for all $w \in D(\infty\langle 1\rangle)$ is surjective (cf. Exercise 35.4). In [**42**], it is shown that $\widetilde{u}(F)$ is finite if and only if $u(F)$ is finite, F satisfies S_1, and $\operatorname{st}_r(F) \leq 1$. Moreover, it is also shown in [**42**] that $\operatorname{st}_r(F_{py}) \leq n$ if and only if $\operatorname{st}_r(F) \leq n$ and F satisfies S_n. This last statement is the real analog of the first three conditions of Corollary 35.27.

CHAPTER VII

Applications of the Milnor Conjecture

40. Exact sequences for quadratic extensions

In this section, we derive the first consequences of the validity of the Milnor Conjecture for fields of characteristic different from 2. In particular, we show that the infinite complexes (34.20) and (34.21) of the powers of I and $\bar{I}$ arising from a quadratic extension of a field of characteristic different from 2 are in fact exact as shown in the paper [**107**] of Orlov, Vishik, and Voevodsky. For fields of characteristic 2, we also show this to be true for separable quadratic extensions as well as proving the exactness of the corresponding complexes (34.27) and (34.28) for purely inseparable quadratic extensions. In addition, we show that for all fields, the ideals $I_q^n(F)$ coincide with the ideals $J_n(F)$ defined in terms of the splitting patterns of quadratic forms.

We need the following lemmas.

Lemma 40.1. *Let K/F be a quadratic field extension and $s : K \to F$ a nonzero F-linear functional. Then for every $n \geq 0$, the diagram*

$$\begin{array}{ccc} k_n(K) & \xrightarrow{c_{K/F}} & k_n(F) \\ {\scriptstyle f_n}\downarrow & & \downarrow{\scriptstyle f_n} \\ \bar{I}^n(K) & \xrightarrow{s_*} & \bar{I}^n(F) \end{array}$$

commutes where the vertical homomorphisms are defined in (5.1).

PROOF. As all the maps in the diagram are $K_*(F)$-linear, in view of Lemma 34.16, it is sufficient to check commutativity only when $n = 1$. The statement follows now from Corollary 34.19 and Fact 100.8. □

Lemma 40.2. *Suppose that F is a field of characteristic 2 and K/F a quadratic field extension. Let $s : K \to F$ be a nonzero F-linear functional. Then the diagram*

$$\begin{array}{ccc} \bar{I}_q^n(K) & \xrightarrow{s_*} & \bar{I}_q^n(F) \\ {\scriptstyle \bar{e}_n}\downarrow & & \downarrow{\scriptstyle \bar{e}_n} \\ H^n(K) & \xrightarrow{c_{K/F}} & H^n(F) \end{array}$$

is commutative.

PROOF. It follows from Lemma 34.16 that it is sufficient to prove the statement in the case $n = 1$. This follows from Lemmas 34.14 and 34.18, since the corestriction map $c_{K/F} : H^1(K) \to H^1(F)$ is induced by the trace map $\mathrm{Tr}_{K/F} : K \to F$ (cf. Example 101.2). □

We set $I^n(F) = W(F)$ if $n \leq 0$.

We first consider the case $\operatorname{char} F \neq 2$.

Theorem 40.3. *Suppose that F is a field of characteristic different from 2 and $K = F(\sqrt{a})/F$ a quadratic extension with $a \in F^\times$. Let $s : K \to F$ be an F-linear functional such that $s(1) = 0$. Then the following infinite sequences are exact:*

$$\cdots \xrightarrow{s_*} I^{n-1}(F) \xrightarrow{\cdot\langle\langle a\rangle\rangle} I^n(F) \xrightarrow{r_{K/F}} I^n(K) \xrightarrow{s_*} I^n(F) \xrightarrow{\cdot\langle\langle a\rangle\rangle} I^{n+1}(F) \to \cdots,$$

$$\cdots \xrightarrow{s_*} \bar{I}^{n-1}(F) \xrightarrow{\cdot\langle\langle a\rangle\rangle} \bar{I}^n(F) \xrightarrow{r_{K/F}} \bar{I}^n(K) \xrightarrow{s_*} \bar{I}^n(F) \xrightarrow{\cdot\langle\langle a\rangle\rangle} \bar{I}^{n+1}(F) \to \cdots.$$

PROOF. Consider the diagram

$$\begin{array}{ccccccccc}
k_{n-1}(F) & \xrightarrow{\cdot\{a\}} & k_n(F) & \xrightarrow{r_{K/F}} & k_n(K) & \xrightarrow{c_{K/F}} & k_n(F) & \xrightarrow{\cdot\{a\}} & k_{n+1}(F) \\
\downarrow & & \downarrow & & \downarrow & & \downarrow & & \downarrow \\
\bar{I}^{n-1}(F) & \xrightarrow{\cdot\langle\langle a\rangle\rangle} & \bar{I}^n(F) & \xrightarrow{r_{K/F}} & \bar{I}^n(K) & \xrightarrow{s_*} & \bar{I}^n(F) & \xrightarrow{\cdot\langle\langle a\rangle\rangle} & \bar{I}^{n+1}(F)
\end{array}$$

where the vertical homomorphisms are defined in (5.1). It follows from Lemma 40.1 that the diagram is commutative. By Fact 5.15, the vertical maps in the diagram are isomorphisms. The top sequence in the diagram is exact by Proposition 101.10. Therefore, the bottom sequence is also exact.

To prove exactness of the first sequence in the statement consider the commutative diagram

$$\begin{array}{cccccc}
 & I^{n+1}(F) \to & I^{n+1}(K) \to & I^{n+1}(F) \to & I^{n+2}(F) \to & I^{n+2}(K) \\
 & \downarrow & \downarrow & \downarrow & \downarrow & \downarrow \\
I^{n-1}(F) \to & I^n(F) \to & I^n(K) \to & I^n(F) \to & I^{n+1}(F) \to & I^{n+1}(K) \\
\downarrow & \downarrow & \downarrow & \downarrow & \downarrow & \\
\bar{I}^{n-1}(F) \to & \bar{I}^n(F) \to & \bar{I}^n(K) \to & \bar{I}^n(F) \to & \bar{I}^{n+1}(F) &
\end{array}$$

with the horizontal sequences considered above and natural vertical maps. By the first part of the proof the bottom sequence is exact. Therefore, exactness of the middle sequence implies exactness of the top one. Thus the statement follows by induction on n (with the base of the induction given by Corollary 34.12). □

Remark 40.4. Let $\operatorname{char} F \neq 2$. Then the second exact sequence in Theorem 40.3 can be rewritten as

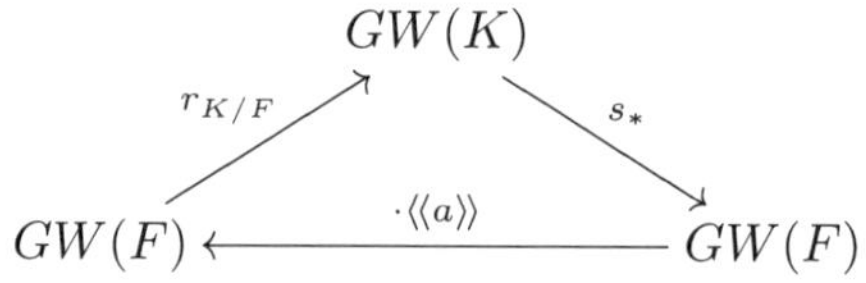

is exact (cf. Corollary 34.12).

Now consider the case of fields of characteristic 2. We consider separately the cases of separable and purely inseparable quadratic field extensions.

Theorem 40.5. *Suppose that F is a field of characteristic 2 and K/F a separable quadratic field extension. Let $s : K \to F$ be a nonzero F-linear functional such that $s(1) = 0$. Then the following sequences are exact:*

$$0 \to I^n(F) \xrightarrow{r_{K/F}} I^n(K) \xrightarrow{s_*} I^n(F) \xrightarrow{\cdot \mathrm{N}_{K/F}} I_q^{n+1}(F) \xrightarrow{r_{K/F}} I_q^{n+1}(K) \xrightarrow{s_*} I_q^{n+1}(F) \to 0,$$

$$0 \to \bar{I}^n(F) \xrightarrow{r_{K/F}} \bar{I}^n(K) \xrightarrow{s_*} \bar{I}^n(F) \xrightarrow{\cdot \mathrm{N}_{K/F}} \bar{I}_q^{n+1}(F) \xrightarrow{r_{K/F}} \bar{I}_q^{n+1}(K) \xrightarrow{s_*} \bar{I}_q^{n+1}(F) \to 0.$$

PROOF. Consider the diagram

$$\begin{array}{ccccccc}
0 \to k_n(F) & \xrightarrow{r_{K/F}} & k_n(K) & \xrightarrow{c_{K/F}} & k_n(F) \\
\downarrow & & \downarrow & & \downarrow \\
0 \to \bar{I}^n(F) & \xrightarrow{r_{K/F}} & \bar{I}^n(K) & \xrightarrow{s_*} & \bar{I}^n(F)
\end{array}$$

$$\begin{array}{ccccccc}
k_n(F) & \xrightarrow{\cdot[K]} & H^{n+1}(F) & \xrightarrow{r_{K/F}} & H^{n+1}(K) & \xrightarrow{c_{K/F}} & H^{n+1}(F) \to 0 \\
\downarrow & & \uparrow & & \uparrow & & \uparrow \\
\bar{I}^n(F) & \xrightarrow{\cdot \mathrm{N}_{K/F}} & \bar{I}_q^{n+1}(F) & \xrightarrow{r_{K/F}} & \bar{I}_q^{n+1}(K) & \xrightarrow{s_*} & \bar{I}_q^{n+1}(F) \to 0
\end{array}$$

where the vertical homomorphisms are defined in (5.1) and Fact 16.2. The middle map in the top row is the multiplication by the class $[K] \in H^1(F)$. By Proposition 101.12, the top sequence is exact. By Facts 5.15 and 16.2, the vertical maps are isomorphisms. Therefore, the bottom sequence is exact.

Exactness of the second sequence in the statement follows by induction on n from the first part of the proof and commutativity of the diagram

$$\begin{array}{ccccccc}
0 \to I^{n+1}(F) \to & I^{n+1}(K) \to & I^{n+1}(F) \to & I_q^{n+2}(F) \to & I_q^{n+2}(K) \to & I_q^{n+2}(F) \to 0 \\
\downarrow & \downarrow & \downarrow & \downarrow & \downarrow & \downarrow \\
0 \to I^n(F) \to & I^n(K) \to & I^n(F) \to & I_q^{n+1}(F) \to & I_q^{n+1}(K) \to & I_q^{n+1}(F) \to 0 \\
\downarrow & \downarrow & \downarrow & \downarrow & \downarrow & \downarrow \\
0 \to \bar{I}^n(F) \to & \bar{I}^n(K) \to & \bar{I}^n(F) \to & \bar{I}_q^{n+1}(F) \to & \bar{I}_q^{n+1}(K) \to & \bar{I}_q^{n+1}(F) \to 0.
\end{array}$$

The base of the induction follows from Corollary 34.15. □

Theorem 40.6. *Suppose that F is a field of characteristic 2 and K/F a purely inseparable quadratic field extension. Let $s : K \to F$ be an F-linear functional such that $s(1) = 0$. Then the following sequences are exact:*

$$\cdots \xrightarrow{s_*} I^n(F) \xrightarrow{r_{K/F}} I^n(K) \xrightarrow{s_*} I^n(F) \xrightarrow{r_{K/F}} I^n(K) \xrightarrow{s_*} \cdots,$$

$$\cdots \xrightarrow{s_*} \bar{I}^n(F) \xrightarrow{r_{K/F}} \bar{I}^n(K) \xrightarrow{s_*} \bar{I}^n(F) \xrightarrow{r_{K/F}} \bar{I}^n(K) \xrightarrow{s_*} \cdots.$$

PROOF. Consider the diagram

$$\begin{array}{ccccccc} k_n(F) & \xrightarrow{r_{K/F}} & k_n(K) & \xrightarrow{c_{K/F}} & k_n(F) & \xrightarrow{r_{K/F}} & k_n(K) \\ \downarrow & & \downarrow & & \downarrow & & \downarrow \\ \overline{I}^n(F) & \xrightarrow{r_{K/F}} & \overline{I}^n(K) & \xrightarrow{s_*} & \overline{I}^n(F) & \xrightarrow{r_{K/F}} & \overline{I}^n(K) \end{array}$$

where the vertical homomorphisms are defined in (5.1). The diagram is commutative by Lemma 40.1. By Fact 5.15, the vertical maps in the diagram are isomorphisms. The top sequence in the diagram is exact by Proposition 100.12. Therefore, the bottom sequence is also exact. The proof of exactness of the second sequence in the statement of the theorem is similar to the one in Theorems 40.3 and 40.5. □

For further applications, we shall also need the following result which follows immediately from the validity of the Milnor Conjecture proven by Voevodsky in [**136**] and Orlov, Vishik, and Voevodsky in [**107**, Th. 2.1]:

Fact 40.7. *Let* $\operatorname{char} F \neq 2$ *and let* ρ *be a quadratic* n*-fold Pfister form over* F*. Then the sequence*

$$\coprod H^*(L) \xrightarrow{\sum c_{L/F}} H^*(F) \xrightarrow{\cup e_n(\rho)} H^{*+n}(F) \xrightarrow{r_{F(\rho)/F}} H^{*+n}\big(F(\rho)\big),$$

is exact, where the direct sum is taken over all quadratic field extensions L/F *such that* ρ_L *is isotropic.*

Fact 40.8 (cf. [**13**, Th. 5.4]). *Let* $\operatorname{char} F = 2$ *and let* ρ *be a quadratic* n*-fold Pfister form over* F*. Then the kernel of* $r_{F(\rho)/F} : H^n(F) \to H^n\big(F(\rho)\big)$ *coincides with* $\{0, e_n(\rho)\}$.

Corollary 40.9. *Let* ρ *be a quadratic* n*-fold Pfister form over an arbitrary field* F*. Then the kernel of the natural homomorphism* $\overline{I}_q^n(F) \to \overline{I}_q^n\big(F(\rho)\big)$ *coincides with* $\{0, \bar{\rho}\}$.

PROOF. Under the isomorphism $\overline{I}_q^n(F) \xrightarrow{\sim} H^n(F)$ (cf. Fact 16.2), the homomorphism in the statement is identified with $H^n(F) \to H^n\big(F(\rho)\big)$. The statement now follows from Fact 40.7 if $\operatorname{char} F \neq 2$ and Fact 40.8 if $\operatorname{char} F = 2$. □

The following statement generalizes Proposition 25.13 and was first proven by Orlov, Vishik, and Voevodsky in [**107**].

Theorem 40.10. *If* F *is a field, then* $J_n(F) = I_q^n(F)$ *for every* $n \geq 1$.

PROOF. By Corollary 25.12, we have the inclusion $I_q^n(F) \subset J_n(F)$. Let $\varphi \in J_n(F)$. We show by induction on n that $\varphi \in I_q^n(F)$. As $\varphi \in J_{n-1}(F)$, by the induction hypothesis, we have $\varphi \in I_q^{n-1}(F)$. The form φ is a sum of m general $(n-1)$-fold Pfister forms in $I_q^{n-1}(F)$ for some m. Let ρ be one of them. Let $K = F(\rho)$. Since φ_K is a sum of $m-1$ general $(n-1)$-Pfister forms in $I_q^{n-1}(K)$, by induction on m we have $\varphi_K \in I_q^n(K)$. By Corollary 40.9, we have either $\varphi \in I_q^n(F)$ or $\varphi \equiv \rho$ modulo $I_q^n(F)$. But the latter case does not occur as $\varphi \in J_n(F)$ and $\rho \notin J_n(F)$. □

41. Annihilators of Pfister forms

The main purpose of this section is to establish the generalization of Corollary 6.23 and Theorem 9.12 on the annihilators of bilinear and quadratic Pfister forms and show these annihilators respect the grading induced by the fundamental ideal. We even show that if α is a bilinear or quadratic Pfister form, then the annihilator $\operatorname{ann}_{W(F)}(\alpha) \cap I^n(F)$ is not only generated by bilinear Pfister forms annihilated by α but is, in fact, generated by bilinear n-fold Pfister forms of the type $\mathfrak{b} \otimes \mathfrak{c}$ with $\mathfrak{b} \in \operatorname{ann}_{W(F)}(\alpha)$ a 1-fold bilinear Pfister form and $\mathfrak{c}$ a bilinear $(n-1)$-fold Pfister form. In particular, Pfister forms of the type $\langle\langle w, a_2, \dots, a_n \rangle\rangle$ with $w \in D\big(\infty\langle 1 \rangle\big)$ and $a_i \in F^\times$ generate $I_t^n(F)$, thus solving the problems raised at the end of §33. These results first appeared in [**8**].

Let F be a field. The smallest integer n such that $I^{n+1}(F) = 2I^n(F)$ and $I^{n+1}(F)$ is torsion-free is called the *stable range of* F and is denoted by $\operatorname{st}(F)$. We say that F has *finite stable range* if such an n exists and write $\operatorname{st}(F) = \infty$ if such an n does not exist. By Corollary 35.30, a field F has finite stable range if and only if $I^{n+1}(F) = 2I^n(F)$ for some integer n. If F is not formally real, then $\operatorname{st}(F)$ is the smallest integer n such that $I^{n+1}(F) = 0$. If F is formally real, then it follows from Corollary 35.27 that $\operatorname{st}(F) = \operatorname{st}\big(F(\sqrt{-1})\big)$, i.e., $\operatorname{st}(F)$ is the smallest integer n such that $I^{n+1}\big(F(\sqrt{-1})\big) = 0$.

Lemma 41.1. *Suppose that F has finite transcendence degree n over its prime subfield. Then* $\operatorname{st}(F) \le n+2$ *if* $\operatorname{char} F = 0$ *and* $\operatorname{st}(F) \le n+1$ *if* $\operatorname{char} F > 0$.

Proof. If the characteristic of F is positive, then F is a C_{n+1}-field (cf. 97.7) as finite fields are C_1 fields by the Chevalley-Warning Theorem (cf. [**123**, I.2, Th. 3]) and, therefore, every $(n+2)$-fold Pfister form is isotropic, so $I^{n+2}(F) = 0$, i.e., $\operatorname{st}(F) \le n+1$. If the characteristic of F is 0, then the cohomological 2-dimension of $F(\sqrt{-1})$ is at most $n+2$ by [**124**, II.4.1, Prop. 10 and II.4.2, Prop. 11]. By Fact 16.2 and the Hauptsatz 23.7, we have $I^{n+3}\big(F(\sqrt{-1})\big) = 0$. Thus $\operatorname{st}(F) \le n+2$. □

As many problems in a field F reduce to finitely many elements over its prime field, we can often reduce to a problem over a given field to another over a field having finite stable range. We then can try to solve the problem when the stable range is finite. We shall use this idea repeatedly below.

Exercise 41.2. Let K/F be a finite simple extension of degree r. If $I^n(F) = 0$, then $I^{n+r}(K) = 0$. In particular, if a field has finite stable range, then any finite extension also has finite stable range.

Next we study graded annihilators.

Let $\mathfrak{b}$ be a bilinear n-fold Pfister form. For any $m \ge 0$ set

$$\operatorname{ann}_m(\mathfrak{b}) = \{\mathfrak{a} \in I^m(F) \mid \mathfrak{a} \cdot \mathfrak{b} = 0 \in W(F)\},$$
$$\overline{\operatorname{ann}}_m(\mathfrak{b}) = \{\bar{\mathfrak{a}} \in \bar{I}^m(F) \mid \bar{\mathfrak{a}} \cdot \bar{\mathfrak{b}} = 0 \in GW(F)\}.$$

Similarly, for a quadratic n-fold Pfister form ρ and any $m \ge 0$ set

$$\operatorname{ann}_m(\rho) = \{\mathfrak{a} \in I^m(F) \mid \mathfrak{a} \cdot \rho = 0 \in I_q(F)\},$$
$$\overline{\operatorname{ann}}_m(\rho) = \{\bar{\mathfrak{a}} \in \bar{I}^m(F) \mid \bar{\mathfrak{a}} \cdot \bar{\rho} = 0 \in \bar{I}_q(F)\}.$$

It follows from Corollary 6.23 and Theorem 9.12 that $\text{ann}_1(\mathfrak{b})$ and $\text{ann}_1(\rho)$ are generated by the binary forms in them. Thus the following theorem determines completely the graded annihilators.

Theorem 41.3. *Let $\mathfrak{b}$ and ρ be bilinear and quadratic n-fold Pfister forms respectively. Then for any $m \geq 1$, we have*

$$\text{ann}_m(\mathfrak{b}) = I^{m-1}(F) \cdot \text{ann}_1(\mathfrak{b}), \qquad \overline{\text{ann}}_m(\mathfrak{b}) = \overline{I}^{m-1}(F) \cdot \overline{\text{ann}}_1(\mathfrak{b}),$$

$$\text{ann}_m(\rho) = I^{m-1}(F) \cdot \text{ann}_1(\rho), \qquad \overline{\text{ann}}_m(\rho) = \overline{I}^{m-1}(F) \cdot \overline{\text{ann}}_1(\rho).$$

PROOF. The case $\text{char}\, F = 2$ is proven in [**14**, Th.1.1 and 1.2]. We assume that $\text{char}\, F \neq 2$. It is sufficient to consider the case of the bilinear form $\mathfrak{b}$.

It follows from Fact 40.7 and Fact 16.2 that the sequence

$$\coprod \overline{I}^m(L) \xrightarrow{\sum s_*} \overline{I}^m(F) \xrightarrow{\cdot \mathfrak{b}} \overline{I}^{n+m}(F),$$

is exact where the direct sum is taken over all quadratic field extensions L/F with $\mathfrak{b}_L$ isotropic. By Lemma 34.16, we have $I^m(L) = I^{m-1}(F)I(L)$, hence the image of $s_* : I^m(L) \to I^m(F)$ is contained in $I^{m-1}(F) \cdot \text{ann}_1(\mathfrak{b})$. Therefore, the kernel of the second map in the sequence coincides with the image of $I^{m-1}(F) \cdot \text{ann}_1(\mathfrak{b})$ in $\overline{I}^m(F)$. This proves $\overline{\text{ann}}_m(\mathfrak{b}) = \overline{I}^{m-1}(F) \cdot \overline{\text{ann}}_1(\mathfrak{b})$.

Let $\mathfrak{c} \in \text{ann}_m(\mathfrak{b})$. We need to show that $\mathfrak{c} \in I^{m-1}(F) \cdot \text{ann}_1(\mathfrak{b})$. We may assume that F is finitely generated over its prime field and hence F has finite stable range by Lemma 41.1. Let k be an integer such that $k + m > \text{st}(F)$. Repeatedly applying exactness of the sequence above, we see that $\mathfrak{c}$ is congruent to an element $\mathfrak{a} \in I^{k+m}(F)$ modulo $I^{m-1}(F) \cdot \text{ann}_1(\mathfrak{b})$. Replacing $\mathfrak{c}$ by $\mathfrak{a}$ we may assume that $m > \text{st}(F)$.

We claim that it suffices to prove the result for $\mathfrak{c}$ an m-fold Pfister form. By Theorem 33.15, for any $\mathfrak{c} \in I^m(F)$, there is an integer n such that

$$2^n \,\text{sgn}\, \mathfrak{c} = \sum_{i=1}^{r} k_i \cdot \text{sgn}\, \mathfrak{c}_i,$$

with $k_i \in \mathbb{Z}$ and $(n+m)$-fold Pfister forms $\mathfrak{c}_i$ with pairwise disjoint supports. As $m > \text{st}(F)$, it follows from Proposition 35.22 that $\mathfrak{c}_i \simeq 2^n \mathfrak{d}_i$ for some m-fold Pfister forms $\mathfrak{d}_i$. Since $I^m(F)$ is torsion-free, we have

$$\mathfrak{c} = \sum_{i=1}^{r} k_i \cdot \mathfrak{d}_i$$

in $I^m(F)$ and the supports of the $\mathfrak{d}_i$'s are pairwise disjoint. In particular, if $\mathfrak{b} \otimes \mathfrak{c}$ is hyperbolic, then $\text{supp}(\mathfrak{b}) \cap \text{supp}(\mathfrak{c}) = \emptyset$, so $\text{supp}(\mathfrak{b}) \cap \text{supp}(\mathfrak{d}_i) = \emptyset$ for every i. As $I^m(F)$ is torsion-free, this would mean that $\mathfrak{b} \otimes \mathfrak{c}_i$ is hyperbolic for every i and establishes the claim. Therefore, we may assume that $\mathfrak{c}$ is a Pfister form.

The result now follows from Lemma 35.18(1). □

We turn to the generators for $I^n_t(F)$, the torsion in $I^n(F)$.

Theorem 41.4. *For any field F we have $I^n_t(F) = I^{n-1}(F)I_t(F)$.*

PROOF. Let $\mathfrak{c} \in I^n_t(F)$. Then $2^m \mathfrak{c} = 0$ for some m. Applying Theorem 41.3 to the Pfister form $\mathfrak{b} = 2^m\langle 1\rangle$, we have

$$\mathfrak{c} \in \text{ann}_n(\mathfrak{b}) = I^{n-1}(F) \cdot \text{ann}_1(\mathfrak{b}) \subset I^{n-1}(F)I_t(F). \qquad \square$$

Recall that by Proposition 31.27, the group $I_t(F)$ is generated by binary torsion forms. Hence Theorem 41.4 yields

Corollary 41.5. *A field F satisfies property A_n if and only if $I^n(F)$ is torsion-free.*

Remark 41.6. By Theorem 41.4, every torsion bilinear n-fold Pfister form $\mathfrak{b}$ can be written as a $\mathbb{Z}$-linear combination of the (torsion) forms $\langle\langle a_1, a_2, \dots, a_n\rangle\rangle$ with $a_1 \in D\big(\infty\langle 1\rangle\big)$. Note that $\mathfrak{b}$ itself may not be isometric to a form like this (cf. Example 32.4).

Theorem 41.7. *Let $\mathfrak{b}$ and ρ be bilinear and quadratic n-fold Pfister forms respectively. Then for any $m \geq 0$, we have*

$$W(F)\mathfrak{b} \cap I^{n+m}(F) = I^m(F)\mathfrak{b},$$

$$W(F)\rho \cap I_q^{n+m}(F) = I^m(F)\rho.$$

Proof. We prove the first equality (the second being similar). Let $\mathfrak{c} \in W(F)\mathfrak{b} \cap I^{n+m}(F)$. We show by induction on m that $\mathfrak{c} \in I^m(F)\mathfrak{b}$. Suppose that $\mathfrak{c} = \mathfrak{a} \cdot \mathfrak{b}$ in $W(F)$ for some $\mathfrak{a} \in I^{m-1}(F)$, i.e., $\bar{\mathfrak{a}} \in \overline{\operatorname{ann}}_{m-1}(\mathfrak{b})$. By Theorem 41.3, we have $\bar{\mathfrak{a}} = \bar{\mathfrak{d}}\bar{\mathfrak{e}}$ for some $\mathfrak{d} \in I^{m-2}(F)$ and $\mathfrak{e} \in W(F)$ satisfying $\bar{\mathfrak{e}} \in \overline{\operatorname{ann}}_1(\mathfrak{b})$. Let $\mathfrak{f}$ be a binary bilinear form congruent to $\mathfrak{e}$ modulo $I^2(F)$. As $\bar{\mathfrak{f}}\bar{\mathfrak{b}} = \bar{\mathfrak{e}}\bar{\mathfrak{b}} = 0 \in \bar{I}^{n+1}(F)$, the general $(n+1)$-fold Pfister form $\mathfrak{f} \otimes \mathfrak{b}$ belongs to $I^{n+2}(F)$. By the Hauptsatz 23.7, we have $\mathfrak{f} \cdot \mathfrak{b} = 0$ in $W(F)$. Since $\mathfrak{a} \equiv \mathfrak{d}\mathfrak{f}$ modulo $I^m(F)$ it follows that $\mathfrak{c} = \mathfrak{a}\mathfrak{b} \in I^m(F)\mathfrak{b}$. □

Exercise 41.8. Let $\mathfrak{b}$ and $\mathfrak{c}$ be bilinear k-fold and n-fold Pfister forms, respectively, over a field F of characteristic not 2. Prove that for any $m \geq 1$ the group

$$W(F)\mathfrak{c} \cap \operatorname{ann}_{W(F)}(\mathfrak{b}) \cap I^{m+n}(F)$$

is generated by $(m+n)$-fold Pfister forms $\mathfrak{d}$ in $\operatorname{ann}_{W(F)}(\mathfrak{b})$ that are divisible by $\mathfrak{c}$.

The theorem allows us to answer the problems at the end of §33. The solution to Lam's problem in [**88**] was proven in [**8**] and independently by Dickmann and Miraglia (cf. [**29**]).

Corollary 41.9. *Let $\mathfrak{b}$ be a form over F. If $2^n\mathfrak{b} \in I^{n+m}(F)$, then $\mathfrak{b} \in I^m(F) + W_t(F)$. In particular,*

$$\operatorname{sgn}(\mathfrak{b}) \in C\big(\mathfrak{X}(F), 2^m\mathbb{Z}\big) \quad \textit{if and only if} \quad \mathfrak{b} \in I^m(F) + W_t(F).$$

Proof. Suppose that $\operatorname{sgn}\mathfrak{b} \in C\big(\mathfrak{X}(F), 2^m\mathbb{Z}\big)$. By Theorem 33.15, there exists a form $\mathfrak{a} \in I^{n+m}(F)$ such that $2^n \operatorname{sgn}\mathfrak{b} = \operatorname{sgn}\mathfrak{a}$ for some n. In particular, $2^n\mathfrak{b} - \mathfrak{a} \in W_t(F)$. Therefore, $2^{k+n}\mathfrak{b} = 2^k\mathfrak{a}$ for some k. By Theorem 41.7, applied to the form $2^{k+n}\langle 1\rangle$, we may write $2^k\mathfrak{a} = 2^{k+n}\mathfrak{c}$ for some $\mathfrak{c} \in I^n(F)$. Then $\mathfrak{b} - \mathfrak{c}$ lies in $W_t(F)$ as needed. □

As a consequence, we obtain a solution to Marshall's question in [**97**]. It was first proved by Dickmann and Miraglia in [**28**].

Corollary 41.10. *Let F be a formally real pythagorean field. Let $\mathfrak{b}$ be a form over F. If $2^n\mathfrak{b} \in I^{n+m}(F)$, then $\mathfrak{b} \in I^m(F)$. In particular, $\operatorname{sgn}\big(I^m(F)\big) = C\big(\mathfrak{X}(F), 2^m\mathbb{Z}\big)$.*

If F is a formally real field, let $GC(\mathfrak{X}(F),\mathbb{Z})$ be the graded ring

$$GC(\mathfrak{X}(F),\mathbb{Z}) := \coprod 2^nC(\mathfrak{X}(F),\mathbb{Z})/2^{n+1}C(\mathfrak{X}(F),\mathbb{Z}) = \coprod C(\mathfrak{X}(F), 2^n\mathbb{Z}/2^{n+1}\mathbb{Z})$$

and $GW_t(F)$ the graded ideal in $GW(F)$ induced by $I_t(F)$. Then Corollary 41.9 implies that the signature induces an exact sequence

$$0 \to GW_t(F) \to GW(F) \to GC(\mathfrak{X}(F),\mathbb{Z})$$

and Corollary 41.10 says if F is a formally real pythagorean field, then the signature induces an isomorphism $GW(F) \to GC(\mathfrak{X}(F),\mathbb{Z})$.

We interpret this result in terms of the reduced Witt ring and prove the result mentioned at the end of §34.

Theorem 41.11. *Let K be a quadratic extension and $s : K \to F$ a nonzero F-linear functional such that $s(1) = 0$. Then the sequence*

$$0 \to I^n_{red}(K/F) \to I^n_{red}(F) \xrightarrow{r_{K/F}} I^n_{red}(K) \xrightarrow{s_*} I^n_{red}(F)$$

is exact.

PROOF. We need only to show exactness at $I^n_{red}(K)$. Let $\mathfrak{c} \in I^n_{red}(K)$ satisfy $s_*(\mathfrak{c})$ is trivial in $I^n_{red}(F)$, i.e., the form $s_*(\mathfrak{c})$ is torsion. By Theorem 41.4, we have $s_*(\mathfrak{c}) = \sum \mathfrak{a}_i\mathfrak{b}_i$ with $\mathfrak{a}_i \in I^{n-1}(F)$ and $\mathfrak{b}_i \in I_t(F)$. It follows by Corollary 34.32 that $\mathfrak{b}_i = s_*(\mathfrak{d}_i)$ for some torsion forms $\mathfrak{d}_i \in I(K)$. Therefore, the form $\mathfrak{e} := \mathfrak{c} - \sum(\mathfrak{a}_i)_K\mathfrak{d}_i$ belongs to the kernel of $s_* : I^n(K) \to I^n(F)$. It follows from Theorems 40.3 and 40.5 that $\mathfrak{e} = r_{K/F}(\mathfrak{f})$ for some $\mathfrak{f} \in I^n(F)$. Therefore, $\mathfrak{c} \equiv r_{K/F}(\mathfrak{f})$ modulo torsion. □

42. Presentation of $I^n(F)$

In this section, using the validity of the Milnor Conjecture, we show that the presentation established for $I^2(F)$ in Theorem 4.22 generalizes to a presentation for $I^n(F)$. This was first proven in [**8**] in the case of characteristic not 2. The characteristic 2 case was also proven independently by Arason and Baeza from that given below in [**6**].

Let $n \geq 2$ and let $\underline{I}_n(F)$ be the abelian group generated by all the isometry classes $[\mathfrak{b}]$ of bilinear n-fold Pfister forms $\mathfrak{b}$ subject to the generating relations:

(1) $[\langle\langle 1, 1, \dots, 1\rangle\rangle] = 0$.
(2) $[\langle\langle ab, c\rangle\rangle \otimes \mathfrak{d}] + [\langle\langle a, b\rangle\rangle \otimes \mathfrak{d}] = [\langle\langle a, bc\rangle\rangle \otimes \mathfrak{d}] + [\langle\langle b, c\rangle\rangle \otimes \mathfrak{d}]$ for all $a, b, c \in F^\times$ and bilinear $(n-2)$-fold Pfister forms $\mathfrak{d}$.

Note that the group $\underline{I}_2(F)$ was defined earlier in Section §4.C.

There is a natural surjective group homomorphism $g_n : \underline{I}_n(F) \to I^n(F)$ taking the class $[\mathfrak{b}]$ of a bilinear n-fold Pfister form $\mathfrak{b}$ to $\mathfrak{b}$ in $I^n(F)$. The map g_2 is an isomorphism by Theorem 4.22.

As in the proof of Lemma 4.18, applying both relations repeatedly, we find that $[\langle\langle a_1, a_2, \dots, a_n\rangle\rangle] = 0$ if $a_1 = 1$. It follows that for any bilinear m-fold Pfister form $\mathfrak{b}$, the assignment $\mathfrak{a} \mapsto \mathfrak{a} \otimes \mathfrak{b}$ gives rise to a well-defined homomorphism

$$\underline{I}_n(F) \to \underline{I}_{n+m}(F)$$

taking $[\mathfrak{a}]$ to $[\mathfrak{a} \otimes \mathfrak{b}]$.

Lemma 42.1. *Let $\mathfrak{b}$ be a metabolic bilinear n-fold Pfister form. Then $[\mathfrak{b}] = 0$ in $\underline{I}_n(F)$.*

PROOF. We prove the statement by induction on n. Since g_2 is an isomorphism, the statement is true if $n = 2$. In the general case, we write $\mathfrak{b} = \langle\langle a \rangle\rangle \otimes \mathfrak{c}$ for some $a \in F^\times$ and bilinear $(n-1)$-fold Pfister form $\mathfrak{c}$. We may assume by induction that $\mathfrak{c}$ is anisotropic. It follows from Corollary 6.14 that $\mathfrak{c} \simeq \langle\langle b \rangle\rangle \otimes \mathfrak{d}$ for some $b \in F^\times$ and bilinear $(n-2)$-fold Pfister form $\mathfrak{d}$ such that $\langle\langle a, b \rangle\rangle$ is metabolic. By the case $n = 2$, we have $\big[\langle\langle a, b \rangle\rangle\big] = 0$ in $\underline{I}_2(F)$, hence $[\mathfrak{b}] = [\langle\langle a, b \rangle\rangle \otimes \mathfrak{d}] = 0$ in $\underline{I}_n(F)$. $\square$

For each n, let $\alpha_n : \underline{I}_{n+1}(F) \to \underline{I}_n(F)$ be the map given by

$$[\langle\langle a, b \rangle\rangle \otimes \mathfrak{c}] \mapsto [\langle\langle a \rangle\rangle \otimes \mathfrak{c}] + [\langle\langle b \rangle\rangle \otimes \mathfrak{c}] - [\langle\langle ab \rangle\rangle \otimes \mathfrak{c}].$$

We show that this map is well-defined. Let $\langle\langle a, b \rangle\rangle \otimes \mathfrak{c}$ and $\langle\langle a', b' \rangle\rangle \otimes \mathfrak{c}'$ be isometric bilinear n-fold Pfister forms. We need to show that

$$(42.2)\quad [\langle\langle a \rangle\rangle \otimes \mathfrak{c}] + [\langle\langle b \rangle\rangle \otimes \mathfrak{c}] - [\langle\langle ab \rangle\rangle \otimes \mathfrak{c}] = [\langle\langle a' \rangle\rangle \otimes \mathfrak{c}'] + [\langle\langle b' \rangle\rangle \otimes \mathfrak{c}'] - [\langle\langle a'b' \rangle\rangle \otimes \mathfrak{c}']$$

in $\underline{I}_n(F)$. By Theorem 6.10, the forms $\langle\langle a, b \rangle\rangle \otimes \mathfrak{c}$ and $\langle\langle a', b' \rangle\rangle \otimes \mathfrak{c}'$ are chain p-equivalent. Thus we may assume that one of the following cases hold:

(1) $a = a'$, $b = b'$ and $\mathfrak{c} \simeq \mathfrak{c}'$.
(2) $\langle\langle a, b \rangle\rangle \simeq \langle\langle a', b' \rangle\rangle$ and $\mathfrak{c} = \mathfrak{c}'$.
(3) $a = a'$, $\mathfrak{c} = \langle\langle c \rangle\rangle \otimes \mathfrak{d}$, and $\mathfrak{c}' = \langle\langle c' \rangle\rangle \otimes \mathfrak{d}$ for some $c \in F^\times$ and bilinear $(n-2)$-fold Pfister form $\mathfrak{d}$ and $\langle\langle b, c \rangle\rangle \simeq \langle\langle b', c' \rangle\rangle$.

It follows that it is sufficient to prove the statement in the case $n = 2$. The equality (42.2) holds if we compose the morphism α_2 with the homomorphism $g_2 : \underline{I}_2(F) \to I^2(F)$. But g_2 is an isomorphism, hence α_n is well-defined. It is easy to check that α_n is a homomorphism.

The homomorphism α_n fits in the commutative diagram

$$\begin{array}{ccc} \underline{I}_{n+1}(F) & \xrightarrow{\alpha_n} & \underline{I}_n(F) \\ \downarrow{\scriptstyle g_{n+1}} & & \downarrow{\scriptstyle g_n} \\ I^{n+1}(F) & \longrightarrow & I^n(F) \end{array}$$

with the bottom map the inclusion.

Lemma 42.3. *The natural homomorphism*

$$\gamma : \operatorname{Coker}(\alpha_n) \to \bar{I}^n(F)$$

is an isomorphism.

PROOF. Consider the map

$$\tau : (F^\times)^n \to \operatorname{Coker}(\alpha_n)$$

given by $(a_1, a_2, \dots, a_n) \mapsto \big[\langle\langle a_1, a_2, \dots, a_n \rangle\rangle\big] + \operatorname{im}(\alpha_n)$. Clearly τ is symmetric with respect to permutations of the a_i's.

By definition of α_n we have

$$\big[\langle\langle a \rangle\rangle \otimes \mathfrak{c}\big] + \big[\langle\langle b \rangle\rangle \otimes \mathfrak{c}\big] \equiv \big[\langle\langle ab \rangle\rangle \otimes \mathfrak{c}\big] \mod \operatorname{im}(\alpha_n)$$

for any bilinear $(n-1)$-fold Pfister form $\mathfrak{c}$. It follows that τ is multilinear.

The map τ also satisfies the Steinberg relation. Indeed, if $a_1 + a_2 = 1$, then $\big[\langle\langle a_1, a_2 \rangle\rangle\big] = 0$ in $\underline{I}_2(F)$ as g_2 is an isomorphism and therefore $\big[\langle\langle a_1, a_2, \dots, a_n \rangle\rangle\big] = 0$ in $\underline{I}_n(F)$.

As the group $\operatorname{Coker}(\alpha_n)$ has exponent 2, the map τ induces a group homomorphism

$$k_n(F) = K_n(F)/2K_n(F) \to \operatorname{Coker}(\alpha_n)$$

which we also denote by τ. The composition $\gamma \circ \tau$ takes a symbol $\{a_1, a_2, \dots, a_n\}$ to $\langle\langle a_1, a_2, \dots, a_n \rangle\rangle + I^{n+1}(F)$. By Fact 5.15, the map $\gamma \circ \tau$ is an isomorphism. As τ is surjective, we have γ is an isomorphism. $\square$

It follows from Lemma 42.3 that we have a commutative diagram

$$\begin{array}{ccccccccc}
 & & \underline{I}_{n+1}(F) & \xrightarrow{\alpha_n} & \underline{I}_n(F) & \longrightarrow & \overline{I}^n(F) & \longrightarrow & 0 \\
 & & {\scriptstyle g_{n+1}}\downarrow & & {\scriptstyle g_n}\downarrow & & \| & & \\
0 & \longrightarrow & I^{n+1}(F) & \longrightarrow & I^n(F) & \longrightarrow & \overline{I}^n(F) & \longrightarrow & 0
\end{array}$$

with exact rows. It follows that if g_{n+1} is an isomorphism, then g_n is also an isomorphism.

Theorem 42.4. *If $n \geq 2$, the abelian group $I^n(F)$ is generated by the isometry classes of bilinear n-fold Pfister forms subject to the generating relations*

(1) $\langle\langle 1, 1, \dots, 1 \rangle\rangle = 0$.

(2) $\langle\langle ab, c \rangle\rangle \cdot \mathfrak{d} + \langle\langle a, b \rangle\rangle \cdot \mathfrak{d} = \langle\langle a, bc \rangle\rangle \cdot \mathfrak{d} + \langle\langle b, c \rangle\rangle \cdot \mathfrak{d}$ *for all $a, b, c \in F^\times$ and bilinear $(n-2)$-fold Pfister forms $\mathfrak{d}$.*

PROOF. We shall show that the surjective map $g_n : \underline{I}_n(F) \to I^n(F)$ is an isomorphism. Any element in the kernel of $g_n = g_{n,F}$ belongs to the image of the natural map $g_{n,F'} \to g_{n,F}$ where F' is a subfield of F, finitely generated over the prime subfield of F. Thus we may assume that F is finitely generated. It follows from Lemma 41.1 that F has finite stable range. The discussion preceding the theorem shows that we may also assume that $n > \operatorname{st}(F)$.

If F is not formally real, then $I^n(F) = 0$, i.e., every bilinear n-fold Pfister form is metabolic. By Lemma 42.1, the group $\underline{I}_n(F)$ is trivial and we are done.

In what follows we may assume that F is formally real, in particular, $\operatorname{char} F \neq 2$.

We let M be the abelian group given by generators $\{\mathfrak{b}\}$, the isometry classes of bilinear n-fold Pfister forms $\mathfrak{b}$ over F, and relations $\{\mathfrak{b}\} = \{\mathfrak{c}\} + \{\mathfrak{d}\}$ where the bilinear n-fold Pfister forms $\mathfrak{b}, \mathfrak{c}$ and $\mathfrak{d}$ satisfy $\mathfrak{b} = \mathfrak{c} + \mathfrak{d}$ in $W(F)$. In particular, $\{\mathfrak{b}\} = 0$ in M if $\mathfrak{b} = 0$ in $W(F)$.

We claim that the homomorphism

$$\delta : M \to \underline{I}_n(F) \quad \text{given by} \quad \{\mathfrak{b}\} \mapsto [\mathfrak{b}]$$

is well-defined. To see this, it suffices to check that if $\mathfrak{b}, \mathfrak{c}$ and $\mathfrak{d}$ satisfy $\mathfrak{b} = \mathfrak{c} + \mathfrak{d}$ in $W(F)$, then $[\mathfrak{b}] = [\mathfrak{c}] + [\mathfrak{d}]$ in $\underline{I}_n(F)$. As $\operatorname{char} F \neq 2$, it follows from Proposition 24.5 that there are $c, d \in F^\times$ and a bilinear $(n-1)$-fold Pfister form $\mathfrak{a}$ such that

$$\mathfrak{c} \simeq \langle\langle c \rangle\rangle \otimes \mathfrak{a}, \quad \mathfrak{d} \simeq \langle\langle d \rangle\rangle \otimes \mathfrak{a}, \quad \mathfrak{b} \simeq \langle\langle cd \rangle\rangle \otimes \mathfrak{a}.$$

The equality $\mathfrak{b} = \mathfrak{c} + \mathfrak{d}$ implies that $\langle\langle c, d \rangle\rangle \cdot \mathfrak{a} = 0$ in $W(F)$. Therefore,

$$0 = \alpha_n\big([\langle\langle c, d \rangle\rangle \otimes \mathfrak{a}]\big) = [\mathfrak{c}] + [\mathfrak{d}] - [\mathfrak{b}]$$

in $\underline{I}_n(F)$, hence the claim.

Let $\mathfrak{b}$ be a bilinear n-fold Pfister form and $d \in F^\times$. As $I^{n+1}(F) = 2I^n(F)$, we can write $\langle\langle d \rangle\rangle \cdot \mathfrak{b} = 2\mathfrak{c}$ and $\langle\langle -d \rangle\rangle \cdot \mathfrak{b} = 2\mathfrak{d}$ in $W(F)$ with $\mathfrak{c}$, $\mathfrak{d}$ bilinear n-fold Pfister forms. Adding, we then get $2\mathfrak{b} = 2\mathfrak{c} + 2\mathfrak{d}$ in $W(F)$, hence $\mathfrak{b} = \mathfrak{c} + \mathfrak{d}$ since $I^n(F)$ is torsion-free. It follows that $[\mathfrak{b}] = [\mathfrak{c}] + [\mathfrak{d}]$ in M. We generalize this as follows:

Lemma 42.5. *Let F be a formally real field having finite stable range. Suppose that n is a positive integer $\geq \operatorname{st}(F)$. Let $\mathfrak{b} \in P_n(F)$ and $d_1, \dots, d_m \in F^\times$. For every $\varepsilon = (\varepsilon_1, \dots, \varepsilon_m) \in \{\pm 1\}^m$ write $\langle\langle \varepsilon_1 d_1, \dots, \varepsilon_m d_m \rangle\rangle \otimes \mathfrak{b} \simeq 2^m \mathfrak{c}_\varepsilon$ with $\mathfrak{c}_\varepsilon \in P_n(F)$. Then $[\mathfrak{b}] = \sum_\varepsilon [\mathfrak{c}_\varepsilon]$ in M.*

PROOF. We induct on m: The case $m = 1$ was done above. So we assume that $m > 1$. For every $\varepsilon' = (\varepsilon_2, \dots, \varepsilon_m) \in \{\pm 1\}^{m-1}$ write

$$\langle\langle \varepsilon_2 d_2, \dots, \varepsilon_m d_m \rangle\rangle \otimes \mathfrak{b} \simeq 2^{m-1} \mathfrak{d}_{\varepsilon'}$$

with $\mathfrak{d}_{\varepsilon'} \in P_n(F)$. By the induction hypothesis, we then have $[\mathfrak{b}] = \sum_{\varepsilon'} [\mathfrak{d}_{\varepsilon'}]$ in M. It therefore suffices to show that $[\mathfrak{d}_{\varepsilon'}] = [\mathfrak{c}_{(1,\varepsilon')}] + [\mathfrak{c}_{(-1,\varepsilon')}]$ for every ε'. But

$$\begin{aligned} 2^m \mathfrak{d}_{\varepsilon'} &= 2\langle\langle \varepsilon_2 d_2, \dots, \varepsilon_m d_m \rangle\rangle \cdot \mathfrak{b} \\ &= \big(\langle\langle d_1 \rangle\rangle + \langle\langle -d_1 \rangle\rangle\big) \cdot \langle\langle \varepsilon_2 d_2, \dots, \varepsilon_m d_m \rangle\rangle \cdot \mathfrak{b} \\ &= 2^m \mathfrak{c}_{(1,\varepsilon')} + 2^m \mathfrak{c}_{(-1,\varepsilon')} \end{aligned}$$

in $W(F)$, hence $\mathfrak{d}_{\varepsilon'} = \mathfrak{c}_{(1,\varepsilon')} + \mathfrak{c}_{(-1,\varepsilon')}$ in $W(F)$. Consequently, $[\mathfrak{d}_{\varepsilon'}] = [\mathfrak{c}_{(1,\varepsilon')}] + [\mathfrak{c}_{(-1,\varepsilon')}]$ in M. □

Proposition 42.6. *Let F be a formally real field having finite stable range. Suppose that n is a positive integer $\geq \operatorname{st}(F)$. Then every element in M can be written as a $\mathbb{Z}$-linear combination $\sum_{j=1}^s l_j \cdot [\mathfrak{c}_j]$ with forms $\mathfrak{c}_1, \dots, \mathfrak{c}_s \in P_n(F)$ having pairwise disjoint supports in $\mathfrak{X}(F)$.*

PROOF. Let $\mathfrak{a} = \sum_{i=1}^r k_i \cdot [\mathfrak{b}_i] \in M$. Write $\mathfrak{b}_i \simeq \langle\langle a_{i1}, \dots, a_{in} \rangle\rangle$ for $i = 1, \dots, r$. For every matrix $\varepsilon = (\varepsilon_{ik})_{i=1,k=1}^{r,n}$ in $\{\pm 1\}^{r \times n}$ let $\mathfrak{f}_\varepsilon \simeq \bigotimes_{j=1}^r \bigotimes_{l=1}^n \langle\langle \varepsilon_{jl} a_{jl} \rangle\rangle$ and write $\mathfrak{f}_\varepsilon \otimes \mathfrak{b}_i \simeq 2^{rn} \mathfrak{c}_{i,\varepsilon}$ with $\mathfrak{c}_{i,\varepsilon}$ bilinear n-fold Pfister forms for $i \in [1,\ r]$. By Lemma 42.5, we have $[\mathfrak{b}_i] = \sum_\varepsilon [\mathfrak{c}_{i,\varepsilon}]$ in M for $i \in [1,\ r]$, hence

$$\mathfrak{a} = \sum_{i=1}^r k_i \cdot [\mathfrak{b}_i] = \sum_{i=1}^r k_i \cdot \sum_\varepsilon [\mathfrak{c}_{i,\varepsilon}] = \sum_\varepsilon \sum_{i=1}^r k_i \cdot [\mathfrak{c}_{i,\varepsilon}]$$

in M.

For each ε, write $\mathfrak{f}_\varepsilon \simeq 2^{nr-n} \mathfrak{d}_\varepsilon$ with $\mathfrak{d}_\varepsilon$ a bilinear n-fold Pfister form. Clearly, the $\mathfrak{f}_\varepsilon$ have pairwise disjoint supports, hence also the $\mathfrak{d}_\varepsilon$. Now look at a pair (i, ε). If all the ε_{ik}, $k \in [1,\ r]$, are 1, then $\mathfrak{f}_\varepsilon \otimes \mathfrak{b}_i = 2^n \mathfrak{f}_\varepsilon = 2^{nr} \mathfrak{d}_\varepsilon$, hence $\mathfrak{c}_{i,\varepsilon} = \mathfrak{d}_\varepsilon$. If, however, some ε_{ik}, $k \in [1,\ r]$, is -1, then $\mathfrak{f}_\varepsilon \otimes \mathfrak{b}_i = 0$, hence $\mathfrak{c}_{i,\varepsilon} = 0$. It follows that for each ε we have $\sum_{i=1}^r k_i \cdot [\mathfrak{c}_{i,\varepsilon}] = l_\varepsilon \cdot \mathfrak{d}_\varepsilon$ for some integer l_ε. Consequently,

$$\mathfrak{a} = \sum_\varepsilon \sum_{i=1}^r k_i \cdot [\mathfrak{c}_{i,\varepsilon}] = \sum_\varepsilon l_\varepsilon \cdot \mathfrak{d}_\varepsilon. \qquad \square$$

Applying Proposition 42.6 to an element in the kernel of the composition

$$M \xrightarrow{\delta} \underline{I}_n(F) \xrightarrow{g_n} I^n(F) \xrightarrow{\text{sgn}} C\big(\mathfrak{X}(F), \mathbb{Z}\big),$$

we see that all the coefficients l_j are 0. Hence the composition is injective. Since δ is surjective, it follows that g_n is injective and therefore an isomorphism. The proof of Theorem 42.4 is complete. □

Remark 42.7. In [**6**], Arason and Baeza also establish a presentation of $I_q^n(F)$. They show that as an additive group $I_q^n(F)$ for $n \geq 2$ is isomorphic to the equivalence classes $\{\varphi\}$ of n-fold quadratic Pfister forms φ over F subject to the relations:

(1_q) $\{\alpha\otimes[1, d_1+d_2]\} = \{\alpha\otimes[1, d_1]\}+\{\alpha\otimes[1, d_2]\}$ for any $(n-1)$-fold bilinear Pfister form α and $d_1, d_2 \in F$.

(2_q) $\{\langle\langle a\rangle\rangle \otimes \rho\} + \{\langle\langle b\rangle\rangle \otimes \rho\} = \{\langle\langle a+b\rangle\rangle \otimes \rho\} + \{\langle\langle ab(a+b)\rangle\rangle \otimes \rho\}$ for any $(n-1)$-fold quadratic Pfister form ρ and $a, b \in F^\times$ satisfying $a+b \neq 0$.

43. Going down and torsion-freeness

We show in this section that if K/F is a finite extension with $I^n(K)$ torsion-free, then $I^n(F)$ is torsion-free. Since we already know this to be true if $\operatorname{char} F = 2$ by Lemma 35.5, we need only show this when $\operatorname{char} F \neq 2$. In this case, we show that the solution of the Milnor Conjecture that the norm residue map is an isomorphism implies the result as shown in [**7**] and, in fact, leads to the more general Corollary 43.10 below.

Let F be a field of characteristic not 2. For any integer $k, n \geq 0$ consider Galois cohomology groups (cf. §101)

$$H^n(F, k) := H^{n,n-1}(F, \mathbb{Z}/2^k\mathbb{Z}).$$

In particular, $H^n(F, 1) = H^n(F)$.

By Corollary 101.7, there is an exact sequence

$$0 \to H^n(F, r) \to H^n(F, r+s) \to H^n(F, s).$$

For a field extension L/F set

$$H^n(L/F, k) := \operatorname{Ker}\big(H^n(F, k) \xrightarrow{r_{L/F}} H^n(L, k)\big).$$

For all $r, s \geq 0$, we have an exact sequence

$$0 \to H^n(L/F, r) \to H^n(L/F, r+s) \to H^n(L/F, s). \tag{43.1}$$

Proposition 43.2. *Let* $\operatorname{char} F \neq 2$. *Suppose* $I_t^n(F) = 0$. *Then* $H^n(F_{py}/F, k) = 0$ *for all* k.

Proof. Let $\alpha \in H^n(F_{py}/F)$. As F_{py} is the union of admissible extensions over F (cf. Definition 31.13), there is an admissible sub-extension L/F of F_{py}/F such that $\alpha_L = 0$. We induct on the degree $[L : F]$ to show $\alpha_L = 0$. Let E be a subfield of L such that E/F is admissible and $L = E(\sqrt{d})$ with $d \in D(2\langle 1\rangle_E)$. It follows from the exactness of the cohomology sequence (cf. Theorem 99.13) for the quadratic extension L/E that $\alpha_E \in H^{n-1}(E) \cup (d)$. By Proposition 35.7, the field E satisfies A_n. Hence all the torsion Pfister forms $\langle\langle a_1, \dots, a_{n-1}, d\rangle\rangle$ over E are trivial, consequently, $H^{n-1}(E) \cup (d) = 0$ by Fact 16.2 and therefore $\alpha_E = 0$. By the induction hypothesis, $\alpha = 0$.

We have shown that $H^n(F_{py}/F) = 0$. Triviality of the group $H^n(F_{py}/F, k)$ follows then by induction on k from exactness of the sequence (43.1). □

Exercise 43.3. Let $\operatorname{char} F \neq 2$. Show that if $H^n(F_{py}/F) = 0$, then $I^n(F)$ is torsion-free.

Lemma 43.4. *A field* F *of characteristic different from* 2 *is pythagorean if and only if* F *has no cyclic extensions of degree* 4.

PROOF. Consider the exact sequence

$$H^1(F,2) \xrightarrow{g} H^1(F) \xrightarrow{b} H^2(F),$$

where b is the Bockstein homomorphism, $b\big((a)\big) = (a) \cup (-1)$ (cf. Proposition 101.14). The field F is not pythagorean if and only if there is nonsquare $a \in F^\times$ such that $a \in D\big(2\langle 1\rangle\big)$. The latter is equivalent to $(a)\cup(-1) = 0$ in $H^2(F) = \operatorname{Br}_2(F)$ which in its turn is equivalent to $(a) \in \operatorname{im}(g)$, i.e., the quadratic extension $F(\sqrt{a})/F$ can be embedded into a cyclic extension of degree 4. □

Let F be a field of characteristic different from 2 such that $\mu_{2^n} \subset F$ with $n > 1$ and $m \leq n$. Then Kummer theory implies that the natural map

$$F^\times/F^{\times 2^n} = H^1(F,n) \to H^1(F,m) = F^\times/F^{\times 2^m} \tag{43.5}$$

is surjective.

Lemma 43.6. *Let F be a pythagorean field of characteristic different from 2. Then*

$$c_{F(\sqrt{-1})/F} : H^1\big(F(\sqrt{-1}), s\big) \to H^1(F,s)$$

is trivial for every s.

PROOF. If F is nonreal, then it is quadratically closed, so $H^1(F,s) = 0$. Therefore, we may assume that F is formally real. In particular, $F\big(\sqrt{-1}\big) \neq F$.

Let $\beta \in H^1(F, s+1) = \operatorname{Hom}_{\text{cont}}(\Gamma_F, \mathbb{Z}/2^{s+1}\mathbb{Z})$. Then the kernel of β is an open subgroup U of Γ_F with Γ_F/U cyclic of 2-power order. As F is pythagorean, F has no cyclic extensions of a 2-power order greater than 2 by Lemma 43.4. It follows that $[\Gamma_F : U] \leq 2$, hence β lies in the image of $H^1(F) \to H^1(F, s+1)$. Consequently, β lies in the kernel of $H^1(F, s+1) \to H^1(F, s)$. This shows that the natural map $H^1(F, s+1) \to H^1(F, s)$ is trivial. The statement now follows from the commutativity of the diagram

$$\begin{array}{ccc}
H^1\big(F(\sqrt{-1}), s+1\big) & \xrightarrow{c_{F(\sqrt{-1})/F)}} & H^1(F, s+1) \\
\downarrow & & \downarrow{\scriptstyle 0} \\
H^1\big(F(\sqrt{-1}), s\big) & \xrightarrow{c_{F(\sqrt{-1})/F)}} & H^1(F, s)
\end{array}$$

together with the surjectivity of $H^1\big(F(\sqrt{-1}), s+1\big) \to H^1\big(F(\sqrt{-1}), s\big)$ which holds by (43.5) as $\mu_{2^\infty} \subset \mathbb{Q}_{py}\big(\sqrt{-1}\big) \subset F\big(\sqrt{-1}\big)$. □

Lemma 43.7. *Let F be a field of characteristic different from 2 satisfying $\mu_{2^s} \subset F(\sqrt{-1})$. Then for every $d \in D\big(2\langle 1\rangle\big)$ the class (d) belongs to the image of the natural map $H^1(F_{py}/F, s) \to H^1(F_{py}/F)$.*

PROOF. By (43.5), the natural map $g : H^1\big(F(\sqrt{-1}), s\big) \to H^1\big(F(\sqrt{-1})\big)$ is surjective. As $d \in N_{F(\sqrt{-1})/F}\big(F(\sqrt{-1})\big)$, there exists a $\gamma \in H^1\big(F(\sqrt{-1}), s\big)$ satisfying $(d) = g\big(c_{F(\sqrt{-1})/F}(\gamma)\big)$. By Lemma 43.6, we have $c_{F(\sqrt{-1})/F}(\gamma) \in H^1(F_{py}/F, s)$ and the image of $c_{F(\sqrt{-1})/F}(\gamma)$ in $H^1(F_{py}/F)$ coincides with (d). □

Theorem 43.8. *Let $\operatorname{char} F \neq 2$ and K/F a finite field extension. If $I^n(K)$ is torsion-free for some n, then $I^n(F)$ is also torsion-free.*

PROOF. Let 2^r be the largest power of 2 dividing $[K : F]$. Suppose first that the field $F(\sqrt{-1})$ contains $\mu_{2^{r+1}}$.

By Corollary 41.5, we know that $I^n(F)$ is torsion-free if and only if F satisfies A_n. By Lemma 35.2, it suffices to show any bilinear n-fold Pfister form $\langle\langle a_1, \dots, a_{n-1}, d\rangle\rangle$ with $a_1, \dots, a_{n-1} \in F^\times$ and $d \in D(2\langle 1\rangle)$ is hyperbolic. By Lemma 43.7, there is an $\alpha \in H^1(F_{py}/F, r+1)$ such that the natural map

$$H^1(F_{py}/F, r+1) \to H^1(F_{py}/F)$$

takes α to (d).

Recall that the graded group $H^*(F_{py}/F, r+1)$ has the natural structure of a module over the Milnor ring $K_*(F)$ (cf. 101.5). Consider the element

$$\beta = \{a_1, \dots, a_{n-1}\} \cdot \alpha \in H^n(F_{py}/F, r+1).$$

As $I_t^n(K) = 0$, we have $H^n(K_{py}/K, r+1) = 0$ by Proposition 43.2. Therefore,

$$[K : F] \cdot \beta = c_{K/F} \circ r_{K/F}(\beta) = 0,$$

hence $2^r\beta = 0$. The composition

$$H^n(F, r+1) \to H^n(F) \to H^n(F, r+1)$$

coincides with the multiplication by 2^r. Since the second homomorphism is injective by (43.1), the image $\{a_1, \dots, a_{n-1}\} \cdot (d) = (a_1, \dots, a_{n-1}, d)$ of β in $H^n(F)$ is trivial. Therefore, $\langle\langle a_1, \dots, a_{n-1}, d\rangle\rangle$ is hyperbolic by Fact 16.2.

Consider the general case. As $\mu_{2^\infty} \subset F_{py}(\sqrt{-1})$ there is a subfield $E \subset F_{py}$ such that $\mu_{2^{r+1}} \subset E(\sqrt{-1})$ and E/F is an admissible extension. Then $L := KE$ is an admissible extension of K. In particular, $I^n(L)$ is torsion-free by Proposition 35.7 and Corollary 41.5. Note also that the degree $[L : E]$ divides $[K : F]$. By the first part of the proof applied to the extension L/E we have $I_t^n(E) = 0$. It follows from Theorem 35.12 and Corollary 41.5 that $I_t^n(F) = 0$. □

Corollary 43.9. *Let K be a finite extension of a nonformally real field F. If $I^n(K) = 0$, then $I^n(F) = 0$.*

PROOF. If $\operatorname{char} F = 2$, this was shown in Lemma 35.5. If $\operatorname{char} F \neq 2$, this follows from Theorem 43.8. □

Define the *torsion-free index* of a field F to be

$$\nu(F) := \min\{n \mid I^n(F) \text{ is torsion-free}\}$$

(or infinity if no such integer exists). Then we have

Corollary 43.10. *Let K/F a finitely generated field extension. Then*

$$\nu(F) + \operatorname{tr.deg}(K/F) \le \nu(K).$$

PROOF. If K/F is finite, then $\nu(F) \le \nu(K)$ by Theorem 43.8. It follows from this and induction that we may assume that $K = F(t)$. If $\mathfrak{b} \in I_t^n(F)$ where $\nu(K) \le n+1$, then $\langle\langle t\rangle\rangle \cdot \mathfrak{b} = 0$ in $W(K)$ as it is torsion in $I^{n+1}(K)$. It follows by Example 19.13 that $\mathfrak{b} = 0$ in $W(F)$. □

Remark 43.11. Let F be a field of characteristic not 2.

(1) Define the absolute stability index $\operatorname{st}_a(F)$ to be the minimum n (if it exists) such that $I^{n+1}(F) = 2I^n(F)$. Corollary 35.27 implies that

$$\nu(F(\sqrt{-1})) \le n+1 \quad \text{if and only if} \quad \nu(F) \le n+1 \text{ and } \operatorname{st}_a(F) \le n.$$

(which, by Remark 35.28, is equivalent to $I^{n+1}(L) = 0$ for some quadratic extension L/F.) By Fact 16.2 this is equivalent to $H^{n+1}\big(F(\sqrt{-1})\big) = 0$.

(2) Let K/F be a finitely generated field extension of transcendental degree m. In [**41**] it is shown that if $m > 0$, then $H^{n+1}\big(F(\sqrt{-1})\big) = 0$ is equivalent to $\mathrm{cd}_2\big(K(\sqrt{-1})\big) \leq n + 1$. This is not true for K/F algebraic, e.g., let F be the quadratic closure of $\mathbb{Q}$ and $K = F\big(\sqrt[3]{2}\big)$.

(3) The "going up" problem of the relationship of $\nu(F)$ and $\nu(K)$ even when K/F is finite is unclear. In [**37**] it was shown that $\nu(K) \leq \nu(F) + [K : F] - 1$ when F is not formally real and that $\nu(K) \geq 1 + \nu(F)$ is possible, but no uniform bound is known even when F is quadratically closed. Leep (unpublished) has improved this bound to $\nu(K) \leq \nu(F) + [\log_2([K : F]/3)] + 1$ (for $F \neq K$). Similarly, it can be shown that $\mathrm{st}_a(K) \leq \mathrm{st}_a(F) + [K : F] - 1$, but again no uniform bound is known. This compares to the relative case where Bröcker showed in [**19**] using valuation theory that the reduced stability satisfied $st_r(F) + m \leq st_r K \leq st_r(F) + \mathrm{tr.deg}(K/F) + 1$.

CHAPTER VIII

On the Norm Residue Homomorphism of Degree Two

44. The main theorem

In this chapter we prove the degree two case of the Milnor Conjecture (cf. Fact 101.6).

Theorem 44.1. *For every field F of characteristic not 2, the norm residue homomorphism*

$$h_F := h_F^2 : K_2(F)/2K_2(F) \to \operatorname{Br}_2(F),$$

taking $\{a,b\} + 2K_2(F)$ to the class of the quaternion algebra $\left(\frac{a,b}{F}\right)$ is an isomorphism.

Corollary 44.2. *Let F be a field of characteristic not 2. Then*

(1) *The group $\operatorname{Br}_2(F)$ is generated by the classes of quaternion algebras.*
(2) *The following is the list of the defining relations between classes of quaternion algebras:*
 (i) $\left(\frac{aa',b}{F}\right) = \left(\frac{a,b}{F}\right)\cdot\left(\frac{a',b}{F}\right)$; $\left(\frac{a,bb'}{F}\right) = \left(\frac{a,b}{F}\right)\cdot\left(\frac{a,b'}{F}\right)$,
 (ii) $\left(\frac{a,b}{F}\right)\cdot\left(\frac{a,b}{F}\right) = 1$,
 (iii) $\left(\frac{a,b}{F}\right) = 1$ *if* $a + b = 1$.

The main idea of the proof is to compare the norm residue homomorphisms h_F and $h_{F(C)}$, where C is a smooth conic curve over F. The function field $F(C)$ is a generic splitting field for a symbol in $k_2(F) := K_2(F)/2K_2(F)$, so passing from F to $F(C)$ allows us to carry out inductive arguments.

Theorem 44.1 was originally proven in [**100**]. The proof used a specialization argument reducing the problem to the study of the function field of a conic curve and a comparison theorem of Suslin [**128**] on the behavior of the norm residue homomorphism over the function field of a conic curve.

The "elementary" proof presented in this chapter relies neither on a specialization argument nor on higher K-theory. The key point of the proof is Theorem 46.1. It is also a consequence of Quillen's computation of higher K-theory of a conic curve [**113**, §8, Th. 4.1] and a theorem of Rehmann and Stuhler on the group K_2 of a quaternion algebra given in [**114**].

Other "elementary" proofs of the bijectivity of h_F, avoiding higher K-theory, but still using a specialization argument, were given by Arason in [**5**] and Wadsworth in [**138**].

45. Geometry of conic curves

In this section we establish interrelations between projective conic curves and their corresponding quaternion algebras over a field F of arbitrary characteristic.

45.A. Quaternion algebras and conic curves. Let Q be a quaternion algebra over a field F. Recall (cf. Appendix 98.E) that Q carries a *canonical involution* $a \mapsto \bar{a}$, the *reduced trace* linear map

$$\operatorname{Trd} : Q \to F, \quad a \mapsto a + \bar{a}$$

and the *reduced norm* quadratic map

$$\operatorname{Nrd} : Q \to F, \quad a \mapsto a\bar{a}.$$

Every element $a \in Q$ satisfies the quadratic equation

$$a^2 - \operatorname{Trd}(a)a + \operatorname{Nrd}(a) = 0.$$

Set

$$V_Q := \operatorname{Ker}(\operatorname{Trd}) = \{a \in Q \mid \bar{a} = -a\},$$

so V_Q is a 3-dimensional subspace of Q. Note that $x^2 = -\operatorname{Nrd}(x) \in F$ for any $x \in V_Q$, and the map $\varphi_Q : V_Q \to F$ given by $\varphi_Q(x) = x^2$ is a quadratic form on V_Q. The space V_Q is the orthogonal complement to 1 in Q with respect to the nondegenerate bilinear form on Q:

$$(a, b) \mapsto \operatorname{Trd}(ab).$$

The quadric C_Q of the form $\varphi_Q(x)$ in the projective plane $\mathbb{P}(V_Q)$ is a smooth projective *conic curve.* In fact, every smooth projective conic curve (1-dimensional quadric) is of the form C_Q for some quaternion algebra Q (cf. Exercise 12.6).

Proposition 45.1. *The following conditions are equivalent:*

(1) *Q is split.*
(2) *C_Q is isomorphic to the projective line $\mathbb{P}^1$.*
(3) *C_Q has a rational point.*

PROOF. (1) $\Rightarrow$ (2): The algebra Q is isomorphic to the matrix algebra $\mathbb{M}_2(F)$. Hence V_Q is the space of trace 0 matrices and C_Q is given by the equation $t_0^2 + t_1 t_2 = 0$. The morphism $C_Q \to \mathbb{P}^1$, given by $[t_0 : t_1 : t_2] \mapsto [t_0 : t_1] = [-t_2 : t_0]$ is an isomorphism.

(2) $\Rightarrow$ (3) is obvious.

(3) $\Rightarrow$ (1): There is a nonzero element $x \in Q$ such that $x^2 = 0$. In particular, Q is not a division algebra and therefore Q is split. □

Corollary 45.2. *Let Q be a quaternion algebra. Then*

(1) *Every divisor on C_Q of degree zero is principal.*
(2) *If Q is a division algebra, the degree of every closed point of C_Q is even.*

PROOF. (1): Let w be a divisor on C_Q of degree zero. Let K/F be a separable quadratic splitting field of Q and σ the nontrivial automorphism of K over F. By Proposition 45.1, the conic C_Q is isomorphic to the projective line over K. As $\deg \sigma(w_K) = \deg w_K$, there is a function $f \in K(C_Q)^\times$ with $\operatorname{div}(f) = w_K$. Let $s = f \cdot \sigma(f^{-1})$. Since $\operatorname{div}(s) = w_K - \sigma(w_K) = 0$ we have $s \in K^\times$ and $\operatorname{N}_{K/F}(s) = s \cdot \sigma(s) = 1$. By the Hilbert Theorem 90, there is $t \in K^\times$ with $s = \sigma(t) \cdot t^{-1}$. Setting $g = tf$ we have $\sigma(g) = g$, i.e., $g \in F(C_Q)^\times$ and $\operatorname{div}(g) = w$.

(2): Let $x \in C_Q$ be a closed point. As C_Q has a rational point over $F(x)$, by Proposition 45.1, the algebra Q is split over $F(x)$. It follows from Corollary 98.5 that the degree $[F(x) : F]$ is even. □

Example 45.3. If $\operatorname{char} F \neq 2$, there is a basis $1, i, j, k$ of Q such that $a := i^2 \in F^\times$, $b := j^2 \in F^\times$, $k = ij = -ji$ (cf. Example 98.10). Then $V_Q = Fi \oplus Fj \oplus Fk$ and C_Q is given by the equation $at_0^2 + bt_1^2 - abt_2^2 = 0$.

Example 45.4. If $\operatorname{char} F = 2$, there is a basis $1, i, j, k$ of Q such that $a := i^2 \in F$, $b := j^2 \in F$, $k = ij = ji + 1$ (cf. Example 98.11). Then $V_Q = F1 \oplus Fi \oplus Fj$ and C_Q is given by the equation $t_0^2 + at_1^2 + bt_2^2 + t_1 t_2 = 0$.

For every $a \in Q$ define the F-linear function l_a on V_Q by the formula

$$l_a(x) = \operatorname{Trd}(ax).$$

The reduced trace form Trd is a nondegenerate bilinear form on Q. Indeed, to see this it is sufficient to check over a splitting field of Q where Q is isomorphic to a matrix algebra, and it is clear in this case. Hence every F-linear function on V_Q is equal to l_a for some $a \in Q$.

Lemma 45.5. *Let $a, b \in Q$ and $\alpha, \beta \in F$. Then:*

(1) $l_a = l_b$ *if and only if* $a - b \in F$.
(2) $l_{\alpha a + \beta b} = \alpha l_a + \beta l_b$.
(3) $l_{\bar{a}} = -l_a$.
(4) $l_{a^{-1}} = -(\operatorname{Nrd} a)^{-1} \cdot l_a$ *if a is invertible.*

PROOF. (1): This follows from the fact that V_Q is orthogonal to F with respect to the bilinear form Trd.

(2) is obvious.

(3): For any $x \in V_Q$, we have

$$l_{\bar{a}}(x) = \operatorname{Trd}(\bar{a}x) = \operatorname{Trd}(\overline{\bar{a}x}) = \operatorname{Trd}(\bar{x}a) = -\operatorname{Trd}(xa) = -\operatorname{Trd}(ax) = -l_a(x).$$

(4): It follows from (2) and (3) that $(\operatorname{Nrd} a) l_{a^{-1}} = l_{\bar{a}} = -l_a$. □

Every element $a \in Q \setminus F$ generates a quadratic subalgebra $F[a] = F \oplus Fa$ of Q. Conversely, every quadratic subalgebra K of Q is of the form $F[a]$ for any $a \in K \setminus F$. By Lemma 45.5, the linear form l_a on V_Q is independent, up to a multiple in $F^\times$, of the choice of $a \in K \setminus F$. Hence the line in $\mathbb{P}(V_Q)$ given by the equation $l_a(x) = 0$ is determined by K. The intersection of this line with the conic C_Q is an effective divisor on C_Q of degree two. Thus, we have the following maps:

Quadratic subalgebras of Q	$\to$	Rational points in $\mathbb{P}(V_Q^*)$	$=$	Lines in $\mathbb{P}(V_Q)$	$\to$	Degree 2 effective divisors on C_Q

Proposition 45.6. *The two maps above are bijections.*

PROOF. The first map is a bijection since every line in $\mathbb{P}(V_Q)$ is given by an equation $l_a = 0$ for some $a \in Q \setminus F$ and a generates a quadratic subalgebra of Q. The last map is a bijection since the embedding of C_Q as a closed subscheme of $\mathbb{P}(V_Q)$ is given by a complete linear system. □

Remark 45.7. Effective divisors on C_Q of degree 2 are rational points of the symmetric square $S^2(C_Q)$. Proposition 45.6 essentially asserts that $S^2(C_Q)$ is isomorphic to the projective plane $\mathbb{P}(V_Q^*)$.

Suppose Q is a division algebra. Then the conic curve C_Q has no rational points and quadratic subalgebras of Q are quadratic (maximal) subfields of Q. An effective cycle on C_Q of degree 2 is a closed point of degree 2. Thus, by Proposition 45.6, we have bijections

$$\boxed{\begin{array}{c}\text{Quadratic}\\ \text{subfields of } Q\end{array}} \xrightarrow{\sim} \boxed{\begin{array}{c}\text{Rational points}\\ \text{in } \mathbb{P}(V_Q^*)\end{array}} = \boxed{\begin{array}{c}\text{Lines}\\ \text{in } \mathbb{P}(V_Q)\end{array}} \xrightarrow{\sim} \boxed{\begin{array}{c}\text{Points of}\\ \text{degree 2 in } C_Q\end{array}}$$

In what follows, we shall frequently use the bijection constructed above between the set of quadratic subfields of Q and the set of degree 2 closed points of C_Q.

45.B. Key identity. In the following proposition, we write a multiple of the quadratic form φ_Q on V_Q as a degree two polynomial of linear forms.

Proposition 45.8. *Let Q be a quaternion algebra over F. For any $a, b, c \in Q$,*

$$l_{a\bar{b}} \cdot l_c + l_{b\bar{c}} \cdot l_a + l_{c\bar{a}} \cdot l_b = \big(\operatorname{Trd}(cba) - \operatorname{Trd}(abc)\big) \cdot \varphi_Q.$$

PROOF. We write T for Trd in this proof. For every $x \in V_Q$ we have:

$$\begin{aligned}
l_{a\bar{b}}(x) \cdot l_c(x) &= T(a\bar{b}x)T(cx)\\
&= T\big(a(T(b) - b)x\big)T(cx)\\
&= T(ax)T(b)T(cx) - T(abx)T(cx)\\
&= T(ax)T(b)T(cx) - T\big(abT(cx)x\big)\\
&= T(ax)T(b)T(cx) - T(abc)x^2 + T(abx\bar{c}x),
\end{aligned}$$

$$\begin{aligned}
l_{b\bar{c}}(x) \cdot l_a(x) &= T(b\bar{c}x)T(ax)\\
&= T\big((T(b) - \bar{b})\bar{c}x\big)T(ax)\\
&= T(\bar{c}x)T(b)T(ax) - T(\bar{b}\bar{c}x)T(ax)\\
&= -T(ax)T(b)T(cx) - T\big(\bar{b}\bar{c}x(ax + \bar{x}\bar{a})\big)\\
&= -T(ax)T(b)T(cx) - T(\bar{b}\bar{c}xax) + T(\bar{b}\bar{c}\bar{a})x^2\\
&= -T(ax)T(b)T(cx) - T(ax\bar{b}\bar{c}x) + T(cba)x^2
\end{aligned}$$

$$\begin{aligned}
l_{c\bar{a}}(x) \cdot l_b(x) &= T(c\bar{a}x)T(bx)\\
&= -T(a\bar{c}x)T(bx)\\
&= -T\big(aT(bx)\bar{c}x\big)\\
&= -T(abx\bar{c}x) + T(ax\bar{b}\bar{c}x).
\end{aligned}$$

Adding the equalities yields the result. $\square$

45.C. Residue fields of points of C_Q and quadratic subfields of Q. Suppose that the quaternion algebra Q is a division algebra. Recall that quadratic subfields of Q correspond bijectively to degree 2 points of C_Q. We shall show how to identify a quadratic subfield of Q with the residue field of the corresponding point in C_Q of degree 2.

Choose a quadratic subfield $K \subset Q$. For every $a \in Q \setminus K$, one has $Q = K \oplus aK$. We define a map

$$\mu_a : V_Q^* \to K$$

by the rule: if $c = u + av$ for $u, v \in K$, then $\mu_a(l_c) = v$. Clearly,

$$\mu_a(l_c) = 0 \qquad \text{if and only if} \qquad c \in K.$$

By Lemma 45.5, the map μ_a is well-defined and F-linear. If $b \in Q \setminus K$ is another element, we have

$$\mu_b(l_c) = \mu_b(l_a)\mu_a(l_c), \tag{45.9}$$

hence the maps μ_a and μ_b differ by the multiple $\mu_b(l_a) \in K^\times$. The map μ_a extends to an F-algebra homomorphism

$$\mu_a : S^\bullet(V_Q^*) \to K$$

in the usual way (where $S^\bullet$ denotes the symmetric algebra).

Let $x \in C_Q \subset \mathbb{P}(V_Q)$ be the point of degree 2 corresponding to the quadratic subfield K. The local ring $\mathcal{O}_{\mathbb{P}(V_Q),x}$ is the subring of the quotient field of the symmetric algebra $S^\bullet(V_Q^*)$ generated by the fractions l_c/l_d for all $c \in Q$ and $d \in Q \setminus K$.

Fix an element $a \in Q \setminus F$. We define an F-algebra homomorphism

$$\mu : \mathcal{O}_{\mathbb{P}(V_Q),x} \to K$$

by the formula

$$\mu(l_c/l_d) = \mu_a(l_c)/\mu_a(l_d).$$

Note that $\mu_a(l_d) \neq 0$ since $d \notin K$ and the map μ is independent of the choice of $a \in Q \setminus K$ by (45.9).

We claim that the map μ vanishes on the quadratic form φ_Q defining C_Q in $\mathbb{P}(V_Q)$. Proposition 45.8 gives a formula for a multiple of the quadratic form φ_Q with the coefficient $\alpha := \operatorname{Trd}(cba) - \operatorname{Trd}(abc)$.

Lemma 45.10. *There exist $a \in Q \setminus K$, $b \in K$, and $c \in Q$ such that $\alpha \neq 0$.*

PROOF. Pick any $b \in K \setminus F$ and any $a \in Q$ such that $ab \neq ba$. Clearly, $a \in Q \setminus K$. Then $\alpha = \operatorname{Trd}\big((ba - ab)c\big)$ is nonzero for some $c \in Q$ since the bilinear form Trd is nondegenerate on Q. □

Choose a, b and c as in Lemma 45.10. We have $\mu_a(l_b) = 0$ since $b \in K$. Also, $\mu_a(l_a) = 1$ and $\mu_a(l_{a\bar{b}}) = \bar{b}$. Write $c = u + av$ for $u, v \in K$, then $\mu_a(l_c) = v$. As

$$b\bar{c} = b\bar{u} + b\bar{v}\bar{a} = b\bar{u} + \operatorname{Trd}(b\bar{v}\bar{a}) - av\bar{b},$$

we have $\mu_a(l_{b\bar{c}}) = -v\bar{b}$ and by Proposition 45.8,

$$\alpha\mu(\varphi_Q) = \mu_a(l_{a\bar{b}})\mu_a(l_c) + \mu_a(l_{b\bar{c}})\mu_a(l_a) + \mu_a(l_{c\bar{a}})\mu_a(l_b) = \bar{b}v - v\bar{b} = 0.$$

Since Q is a division algebra and $\alpha \neq 0$, we have $\mu(\varphi_Q) = 0$ as claimed.

The local ring $\mathcal{O}_{C_Q,x}$ coincides with the factor ring

$$\mathcal{O}_{\mathbb{P}(V_Q),x}/\varphi_Q\mathcal{O}_{\mathbb{P}(V_Q),x}.$$

Therefore, μ factors through an F-algebra homomorphism

$$\mu : \mathcal{O}_{C_Q,x} \to K.$$

Let $e \in K \setminus F$. The function l_e/l_a is a local parameter of the local ring $\mathcal{O}_{C_Q,x}$, i.e., it generates the maximal ideal of $\mathcal{O}_{C_Q,x}$. Since $\mu(l_e/l_a) = 0$, the map μ induces a field isomorphism

$$F(x) \xrightarrow{\sim} K \tag{45.11}$$

of degree 2 field extensions of F. We have proved

Proposition 45.12. *Let Q be a division quaternion algebra. Let $K \subset Q$ be a quadratic subfield and $x \in C_Q$ the corresponding point of degree 2. Then the residue field $F(x)$ is canonically isomorphic to K over F. Let $a \in Q$ and $b \in Q \setminus K$. Write $a = u + bv$ for unique $u, v \in K$. Then the value $(l_a/l_b)(x)$ in $F(x)$ of the function l_a/l_b at the point x corresponds to the element $v \in K$ under the isomorphism* (45.11).

46. Key exact sequence

In this section we prove exactness of a sequence comparing the groups $K_2(F)$ and $K_2\big(F(C)\big)$ over a field F of arbitrary characteristic.

Let C be a smooth curve over a field F. For every (closed) point $x \in C$, there is a residue homomorphism

$$\partial_x : K_2\big(F(C)\big) \to K_1\big(F(x)\big) = F(x)^\times$$

induced by the discrete valuation of the local ring $\mathcal{O}_{C,x}$ (cf. §49.A).

In this section we prove the following:

Theorem 46.1. *Let C be a conic curve over a field F. Then the sequence*

$$K_2(F) \to K_2\big(F(C)\big) \xrightarrow{\partial} \coprod_{x \in C} F(x)^\times \xrightarrow{c} F^\times,$$

with $\partial = (\partial_x)$ and $c = (\mathrm{N}_{F(x)/F})$, is exact.

46.A. Filtration on $K_2\big(F(C)\big)$. Let C be a conic over F. If C has a rational point, i.e., $C \simeq \mathbb{P}^1_F$, the statement of Theorem 46.1 is Milnor's computation of $K_2\big(F(t)\big)$ given in Theorem 100.7. So we may (and will) assume that C has no rational point. By Corollary 45.2, the degree of every closed point of C is even.

Fix a closed point $x_0 \in C$ of degree 2. As in §29, for any $n \in \mathbb{Z}$ let L_n be the F-subspace

$$\{f \in F(C)^\times \mid \operatorname{div}(f) + nx_0 \geq 0\} \cup \{0\}$$

of $F(C)$. Clearly, $L_n = 0$ if $n < 0$. Recall that $L_0 = F$ and $L_n \cdot L_m \subset L_{n+m}$. It follows from Lemma 29.7 that $\dim L_n = 2n + 1$ if $n \geq 0$.

We write $L_n^\times$ for $L_n \setminus \{0\}$. Note that the value $g(x)$ in $F(x)$ is defined for every $g \in L_n^\times$ and point $x \neq x_0$.

Since any divisor on C of degree zero is principal by Corollary 45.2, for every point $x \in C$ of degree $2n$ we can choose a function $p_x \in L_n^\times$ satisfying $\operatorname{div}(p_x) = x - nx_0$. In particular, $p_{x_0} \in F^\times$. Note that p_x is uniquely determined up to a scalar multiple. Clearly, $p_x(x) = 0$ if $x \neq x_0$. Every function in $L_n^\times$ can be written as the product of a nonzero constant and finitely many p_x for some points x of degree at most $2n$.

Lemma 46.2. *Let $x \in C$ be a point of degree $2n$ different from x_0. If $g \in L_m$ satisfies $g(x) = 0$, then $g = p_x q$ for some $q \in L_{m-n}$. In particular, $g = 0$ if $m < n$.*

PROOF. Consider the F-linear map

$$e_x : L_m \to F(x), \quad e_x(g) = g(x).$$

If $m < n$, the map e_x is injective since x does not belong to the support of the divisor of a function in $L_m^\times$. Suppose that $m = n$ and $g \in \operatorname{Ker} e_x$. Then $\operatorname{div}(g) = x - nx_0$ and hence g is a multiple of p_x. Thus, the kernel of e_x is 1-dimensional. By dimension count, e_x is surjective.

Therefore, for arbitrary $m \geq n$, the map e_x is surjective and

$$\dim \operatorname{Ker}(e_x) = \dim L_m - \deg(x) = 2m + 1 - 2n.$$

The image of the injective linear map $L_{m-n} \to L_m$ given by multiplication by p_x is contained in $\operatorname{Ker}(e_x)$ and of dimension $\dim L_{m-n} = 2m + 1 - 2n$. Therefore, $\operatorname{Ker}(e_x) = p_x L_{m-n}$. □

For every $n \in \mathbb{Z}$, let M_n be the subgroup of $K_2\big(F(C)\big)$ generated by the symbols $\{f, g\}$ with $f, g \in L_n^\times$, i.e., $M_n = \{L_n^\times, L_n^\times\}$. We have the following filtration:

$$0 = M_{-1} \subset M_0 \subset M_1 \subset \cdots \subset K_2\big(F(C)\big). \tag{46.3}$$

Note that M_0 coincides with the image of the homomorphism $K_2(F) \to K_2\big(F(C)\big)$ and $K_2\big(F(C)\big)$ is the union of all the M_n as the group $F(C)^\times$ is the union of the subsets $L_n^\times$.

If $f \in L_n^\times$, the degree of every point in the support of $\operatorname{div}(f)$ is at most $2n$. In particular, $\partial_x(M_{n-1}) = 0$ for every point x of degree $2n$. Therefore, for every $n \geq 0$, we have a well-defined homomorphism

$$\partial_n : M_n/M_{n-1} \to \coprod_{\deg x = 2n} F(x)^\times$$

induced by ∂_x over all points $x \in C$ of degree $2n$.

We refine the filtration (46.3) by adding an extra term M' between M_0 and M_1. Set $M' := \{L_1^\times, L_0^\times\} = \{L_1^\times, F^\times\}$, so the group M' is generated by M_0 and symbols of the form $\{p_x, \alpha\}$ for all points $x \in C$ of degree 2 and all $\alpha \in F^\times$.

Denote by A' the subgroup of $\coprod_{\deg x = 2} F(x)^\times$ consisting of all families (α_x) such that $\alpha_x \in F^\times$ for all x and $\prod_x \alpha_x = 1$. Clearly, $\partial_1(M'/M_0) \subset A'$.

Theorem 46.1 is a consequence of the following three propositions.

Proposition 46.4. *If $n \geq 2$, the map*

$$\partial_n : M_n/M_{n-1} \to \coprod_{\deg x = 2n} F(x)^\times$$

is an isomorphism.

Proposition 46.5. *The restriction $\partial' : M'/M_0 \to A'$ of ∂_1 is an isomorphism.*

Proposition 46.6. *The sequence*

$$0 \to M_1/M' \xrightarrow{\partial_1} \left(\coprod_{\deg x = 2} F(x)^\times \right) / A' \xrightarrow{c} F^\times$$

is exact.

PROOF OF THEOREM 46.1. Since $K_2(F(C))$ is the union of M_n, it is sufficient to prove that the sequence

$$0 \to M_n/M_0 \xrightarrow{\partial} \coprod_{\deg x \le 2n} F(x)^\times \xrightarrow{c} F^\times$$

is exact for every $n \ge 1$. We induct on n. The case $n = 1$ follows from Propositions 46.5 and 46.6. The induction step is guaranteed by Proposition 46.4. □

46.B. Proof of Proposition 46.4. To effect the proof, we shall construct the inverse map of ∂_n. We need the following two lemmas.

Lemma 46.7. *Let $x \in C$ be a point of degree $2n > 2$. Then for every $u \in F(x)^\times$, there exist $f \in L_{n-1}^\times$ and $h \in L_1^\times$ such that $(f/h)(x) = u$.*

PROOF. The F-linear map $e_x : L_{n-1} \to F(x)$, $e_x(f) = f(x)$, is injective by Lemma 46.2. Hence

$$\dim \operatorname{Coker}(e_x) = \deg(x) - \dim L_{n-1} = 2n - (2n-1) = 1.$$

Consider the F-linear map

$$g : L_1 \to \operatorname{Coker}(e_x), \quad g(h) = u \cdot h(x) + \operatorname{Im}(e_x).$$

Since $\dim L_1 = 3$, the kernel of g contains a nonzero function $h \in L_1^\times$. We have $u \cdot h(x) = f(x)$ for some $f \in L_{n-1}^\times$. Since $\deg x > 2$ the value $h(x)$ is nonzero. Hence $u = (f/h)(x)$. □

Let $x \in C$ be a point of degree $2n > 2$. Define a map

$$\psi_x : F(x)^\times \to M_n/M_{n-1}$$

as follows: By Lemma 46.7, for each element $u \in F(x)^\times$, we can choose $f \in L_{n-1}^\times$ and $h \in L_1^\times$ such that $(f/h)(x) = u$. We set

$$\psi_x(u) = \left\{p_x, \frac{f}{h}\right\} + M_{n-1}.$$

Lemma 46.8. *The map ψ_x is a well-defined homomorphism.*

PROOF. Let $f' \in L_{n-1}^\times$ and $h' \in L_1^\times$ be two functions with $(f'/h')(x) = u$. Then $f'h - fh' \in L_n$ and $(f'h - fh')(x) = 0$. By Lemma 46.2, we have $f'h - fh' = \lambda p_x$ for some $\lambda \in F$. If $\lambda = 0$, then $f/h = f'/h'$.

Suppose $\lambda \ne 0$. Since $(\lambda p_x)/(f'h) + (fh')/(f'h) = 1$, we have

$$0 = \left\{\frac{\lambda p_x}{f'h}, \frac{fh'}{f'h}\right\} \equiv \left\{p_x, \frac{f}{h}\right\} - \left\{p_x, \frac{f'}{h'}\right\} \quad \bmod M_{n-1}.$$

Hence

$$\left\{p_x, \frac{f}{h}\right\} + M_{n-1} = \left\{p_x, \frac{f'}{h'}\right\} + M_{n-1},$$

so that the map ψ is well-defined.

Let $u_3 = u_1 u_2 \in F(x)^\times$. Choose $f_i \in L_{n-1}^\times$ and $h_i \in L_1^\times$ satisfying $(f_i/h_i)(x) = u_i$ for $i = 1, 2, 3$. The function $f_1 f_2 h_3 - f_3 h_1 h_2$ belongs to L_{2n-1} and has zero value at x. We have $f_1 f_2 h_3 - f_3 h_1 h_2 = p_x q$ for some $q \in L_{n-1}$ by Lemma 46.2. Since $(p_x q)/(f_1 f_2 h_3) + (f_3 h_1 h_2)/(f_1 f_2 h_3) = 1$,

$$0 = \left\{\frac{p_x q}{f_1 f_2 h_3}, \frac{f_3 h_1 h_2}{f_1 f_2 h_3}\right\} \equiv \left\{p_x, \frac{f_3}{h_3}\right\} - \left\{p_x, \frac{f_1}{h_1}\right\} - \left\{p_x, \frac{f_2}{h_2}\right\} \quad \bmod M_{n-1}.$$

Thus, $\psi_x(u_3) = \psi_x(u_1) + \psi_x(u_2)$. □

By Lemma 46.8, we have a homomorphism

$$\psi_n = \sum \psi_x : \coprod_{\deg x = 2n} F(x)^\times \to M_n/M_{n-1}.$$

We claim that ∂_n and ψ_n are isomorphisms inverse to each other. If x is a point of degree $2n > 2$ and $u \in F(x)^\times$, choose $f \in L_{n-1}^\times$ and $h \in L_1^\times$ such that $(f/h)(x) = u$. We have

$$\partial_x\Big(\Big\{p_x, \frac{f}{h}\Big\}\Big) = \Big(\frac{f}{h}\Big)(x) = u$$

and the symbol $\{p_x, f/h\}$ has no nontrivial residues at other points of degree $2n$. Therefore, $\partial_n \circ \psi_n$ is the identity.

To finish the proof of Proposition 46.4, it suffices to show that ψ_n is surjective. The group M_n/M_{n-1} is generated by classes of the form $\{p_x, g\} + M_{n-1}$ and $\{p_x, p_y\} + M_{n-1}$, where $g \in L_{n-1}^\times$ and x, y are distinct points of degree $2n$. Clearly,

$$\{p_x, g\} + M_{n-1} = \psi_x\big(g(x)\big),$$

hence $\{p_x, g\} + M_{n-1} \in \operatorname{Im} \psi_n$.

By Lemma 46.7, there are elements $f \in L_{n-1}^\times$ and $h \in L_1^\times$ such that $p_x(y) = (f/h)(y)$. The function $p_x h - f$ belongs to $L_{n+1}^\times$ and has zero value at y. Therefore, $p_x h - f = p_y q$ for some $q \in L_1^\times$ by Lemma 46.2. Since $(p_y q)/(p_x h) + f/(p_x h) = 1$ we have

$$0 = \Big\{\frac{p_y q}{p_x h}, \frac{f}{p_x h}\Big\} \equiv \{p_x, p_y\} \mod \operatorname{Im}(\psi_n). \quad \square$$

46.C. Proof of Proposition 46.5. Define a homomorphism

$$\rho : A' \to M'/M_0$$

by the rule

$$\rho\Big(\coprod \alpha_x\Big) = \sum_{\deg x = 2} \{p_x, \alpha_x\} + M_0.$$

Since $\partial_x\{p_x, \alpha\} = \alpha$ and $\partial_{x_0}\{p_x, \alpha\} = \alpha^{-1}$ for every $x \neq x_0$ and the product of all α_x is equal to 1, the composition $\partial' \circ \rho$ is the identity. Clearly, ρ is surjective. $\square$

46.D. Generators and relations of $A(Q)/A'$. It remains to prove Proposition 46.6. Let Q be a quaternion division algebra satisfying $C \xrightarrow{\sim} C_Q$. By Proposition 45.12, the norm homomorphism

$$\coprod_{\deg x = 2} F(x)^\times \to F^\times$$

is canonically isomorphic to the norm homomorphism

$$\coprod K^\times \to F^\times, \tag{46.9}$$

where the coproduct is taken over all quadratic subfields $K \subset Q$. Note that the norm map $N_{K/F} : K^\times \to F^\times$ is the restriction of the reduced norm Nrd on K. Let $A(Q)$ be the kernel of the norm homomorphism (46.9). Under the above canonical isomorphism the subgroup A' of $\coprod F(x)^\times$ corresponds to the subgroup of $A(Q)$ (that we still denote by A') consisting of all families (a_K) satisfying $a_K \in F^\times$ and $\prod a_K = 1$, i.e., A' is the intersection of $A(Q)$ and $\coprod F^\times$. Therefore, Proposition 46.6 asserts that the canonical homomorphism

$$\partial_1 : M_1/M' \to A(Q)/A' \tag{46.10}$$

is an isomorphism. In the proof of Proposition 46.6, we shall construct the inverse isomorphism. In order to do so, it is convenient to have a presentation of the group $A(Q)/A'$ by generators and relations.

We define a map (not a homomorphism!)

$$Q^\times \to \Big(\coprod K^\times\Big)/A', \quad a \mapsto \widetilde{a}$$

as follows: If $a \in Q^\times$ is not a scalar, it is contained in a unique quadratic subfield K of Q. Therefore, a defines an element of the coproduct $\coprod K^\times$. We denote by $\widetilde{a}$ the corresponding class in $\big(\coprod K^\times\big)/A'$. If $a \in F^\times$, of course, a belongs to all quadratic subfields. Nevertheless, a defines a unique element $\widetilde{a}$ of the factor group $\big(\coprod K^\times\big)/A'$ (we place a in any quadratic subfield). Clearly,

$$\widetilde{(ab)} = \widetilde{a} \cdot \widetilde{b} \quad \text{if } a \text{ and } b \text{ commute.} \tag{46.11}$$

(Note that we are using multiplicative notation for the operation in the factor group.) Obviously, the group $\big(\coprod K^\times\big)/A'$ is an abelian group generated by the $\widetilde{a}$ for all $a \in Q^\times$ with the set of defining relations given by (46.11).

The group $A(Q)/A'$ is generated (as an abelian group) by the products

$$\widetilde{a}_1\widetilde{a}_2\cdots\widetilde{a}_n \quad \text{with } a_i \in Q^\times \text{ and } \operatorname{Nrd}(a_1a_2\cdots a_n) = 1,$$

with the following set of defining relations:

(1) $(\widetilde{a}_1\widetilde{a}_2\cdots\widetilde{a}_n)\cdot(\widetilde{a}_{n+1}\widetilde{a}_{n+2}\cdots\widetilde{a}_{n+m}) = (\widetilde{a}_1\widetilde{a}_2\cdots\widetilde{a}_{n+m})$.

(2) $\widetilde{a}\widetilde{b}\big(\widetilde{a^{-1}}\big)\big(\widetilde{b^{-1}}\big) = 1$.

(3) If a_{i-1} and a_i commute, then $\widetilde{a}_1\cdots\widetilde{a}_{i-1}\widetilde{a}_i\cdots\widetilde{a}_n = \widetilde{a}_1\cdots\widetilde{a_{i-1}a_i}\cdots\widetilde{a}_n$.

The set of generators is too large for our purposes. In the next subsection, we shall find another presentation of $A(Q)/A'$ (cf. Corollary 46.27). More precisely, we shall define an abstract group G by generators and relations (with a "better" set of generators) and prove that G is isomorphic to $A(Q)/A'$.

46.E. The group G. Let Q be a division quaternion algebra over a field F. Consider the abelian group G defined by generators and relations as follows: The sign $*$ will denote the operation in G with 1 its identity element.

Generators: Symbols (a, b, c) for all ordered triples with a, b, c elements of $Q^\times$ satisfying $abc = 1$. Note that if (a, b, c) is a generator of G, then so are the cyclic permutations (b, c, a) and (c, a, b).

Relations:

$(R1)$: $(a, b, cd) * (ab, c, d) = (b, c, da) * \big(bc, d, a\big)$ for all $a, b, c, d \in Q^\times$ satisfying $abcd = 1$;

$(R2)$: $(a, b, c) = 1$ if a and b commute.

For an (ordered) sequence $a_1, a_2, \ldots, a_n$ $(n \geq 1)$ of elements in $Q^\times$ satisfying $a_1a_2\ldots a_n = 1$, we define a symbol

$$(a_1, a_2, \ldots, a_n) \in G$$

by induction on n as follows. The symbol is trivial if $n = 1$ or 2. If $n \geq 3$, we set

$$(a_1, a_2, \ldots, a_n) := (a_1, a_2, \ldots, a_{n-2}, a_{n-1}a_n) * (a_1a_2\cdots a_{n-2}, a_{n-1}, a_n).$$

Note that if $a_1a_2\ldots a_n = 1$, then $a_2\ldots a_na_1 = 1$.

Lemma 46.12. *The symbols do not change under cyclic permutations, i.e.,* $(a_1, a_2, \ldots, a_n) = (a_2, \ldots, a_n, a_1)$ *if* $a_1a_2\ldots a_n = 1$.

PROOF. We induct on n. The statement is clear if $n = 1$ or 2. If $n = 3$,

$$\begin{aligned}(a_1, a_2, a_3) &= (a_1, a_2, a_3) * (a_1 a_2, a_3, 1) \text{ (relation } R2)\\ &= (a_2, a_3, a_1) * (a_2 a_3, 1, a_1) \text{ (relation } R1)\\ &= (a_2, a_3, a_1) \text{ (relation } R2).\end{aligned}$$

Suppose that $n \geq 4$. We have

$$\begin{aligned}(a_1, a_2, \dots, a_n) &= (a_1, \dots, a_{n-2}, a_{n-1}a_n) * (a_1 a_2 \cdots a_{n-2}, a_{n-1}, a_n) \text{ (definition)}\\ &= (a_2, \dots, a_{n-2}, a_{n-1}a_n, a_1) * (a_1 a_2 \cdots a_{n-2}, a_{n-1}, a_n) \text{ (induction)}\\ &= (a_2, \dots, a_{n-2}, a_{n-1}a_n a_1) * (a_2 a_3 \cdots a_{n-2}, a_{n-1}a_n, a_1)\\ &\quad * (a_1 a_2 \cdots a_{n-2}, a_{n-1}, a_n) \text{ (definition)}\\ &= (a_2, \dots, a_{n-2}, a_{n-1}a_n a_1) * (a_1, a_2 a_3 \cdots a_{n-2}, a_{n-1}a_n)\\ &\quad * (a_1 a_2 \cdots a_{n-2}, a_{n-1}, a_n) \text{ (case } n = 3)\\ &= (a_2, \dots, a_{n-2}, a_{n-1}a_n a_1) * (a_2 a_3 \cdots a_{n-2}, a_{n-1}, a_n a_1)\\ &\quad * (a_2 a_3 \cdots a_{n-1}, a_n, a_1) \text{ (relation } R1)\\ &= (a_2, \dots, a_{n-2}, a_{n-1}, a_n a_1) * (a_2 a_3 \cdots a_{n-1}, a_n, a_1) \text{ (definition)}\\ &= (a_2, \dots, a_n, a_1) \text{ (definition).}\end{aligned}$$

□

Lemma 46.13. *If $a_1 a_2 \dots a_n = 1$ and a_{i-1} commutes with a_i for some i, then*

$$(a_1, \dots, a_{i-1}, a_i, \dots, a_n) = (a_1, \dots, a_{i-1}a_i, \dots, a_n).$$

PROOF. We may assume that $n \geq 3$ and $i = n$ by Lemma 46.12. We have

$$\begin{aligned}&(a_1, \dots, a_{n-2}, a_{n-1}, a_n)\\ &\quad = (a_1, \dots, a_{n-2}, a_{n-1}a_n) * (a_1 a_2 \cdots a_{n-2}, a_{n-1}, a_n) \text{ (definition)}\\ &\quad = (a_1, \dots, a_{n-2}, a_{n-1}a_n) \text{ (relation } R2).\end{aligned}$$

□

Lemma 46.14. $(a_1, \dots, a_n) * (b_1, \dots, b_m) = (a_1, \dots, a_n, b_1, \dots, b_m)$.

PROOF. We induct on m. By Lemma 46.13, we may assume that $m \geq 3$. We have

$$\begin{aligned}L.H.S. &= (a_1, \dots, a_n) * (b_1, \dots, b_{m-1}b_m) * (b_1 b_2 \cdots b_{m-2}, b_{m-1}, b_m) \text{ (definition)}\\ &= (a_1, \dots, a_n, b_1, \dots, b_{m-1}b_m) * (b_1 b_2 \cdots b_{m-2}, b_{m-1}, b_m) \text{ (induction)}\\ &= (a_1, \dots, a_n, b_1, \dots, b_m) \text{ (definition).}\end{aligned}$$

□

As usual, we write $[a, b]$ for the commutator $aba^{-1}b^{-1}$.

Lemma 46.15. *Let $a, b \in Q^\times$.*

(1) *For every nonzero $b' \in Fb + Fba$, one has $[a, b] = [a, b']$. Similarly, $[a, b] = [a', b]$ for every nonzero $a' \in Fa + Fab$.*

(2) *For every nonzero $b' \in Fb + Fba + Fbab$ there exists $a' \in Q^\times$ such that $[a, b] = [a', b] = [a', b']$.*

PROOF. (1): We have $b' = bx$, where $x \in F + Fa$. Hence x commutes with a, so $[a, b] = [a, b']$. The proof of the second statement is similar.

(2): There is nonzero $a' \in Fa + Fab$ such that $b' \in Fb + Fba'$. By the first part, $[a, b] = [a', b] = [a', b']$. □

Corollary 46.16. (1) *Let* $[a, b] = [c, d]$. *Then there are* $a', b' \in Q^\times$ *such that* $[a, b] = [a', b] = [a', b'] = [c, b'] = [c, d]$.

(2) *Every pair of commutators in* $Q^\times$ *can be written in the form* $[a, b]$ *and* $[c, d]$ *with* $b = c$.

PROOF. (1): If $[a, b] = 1 = [c, d]$, we can take $a' = b' = 1$. Otherwise, both sets $\{b, ba, bab\}$ and $\{d, dc\}$ are linearly independent. Let b' be a nonzero element in the intersection of the subspaces $Fb + Fba + Fbab$ and $Fd + Fdc$. The statement follows from Lemma 46.15.

(2): Let $[a, b]$ and $[c, d]$ be two commutators. We may clearly assume that $[a, b] \neq 1 \neq [c, d]$, so that both sets $\{b, ba, bab\}$ and $\{c, cd\}$ are linearly independent. Choose a nonzero element b' in the intersection of $Fb + Fba + Fbab$ and $Fc + Fcd$. By Lemma 46.15, we have $[a, b] = [a', b']$ for some $a' \in Q^\times$ and $[c, d] = [b', d]$. □

Lemma 46.17. *Let* $h \in Q^\times$. *The following conditions are equivalent:*

(1) $h = [a, b]$ *for some* $a, b \in Q^\times$.
(2) $h \in [Q^\times, Q^\times]$.
(3) $\operatorname{Nrd}(h) = 1$.

PROOF. The implications (1) ⇒ (2) ⇒ (3) are obvious.

(3) ⇒ (1): Let K be a separable quadratic subfield containing h. (If h is purely inseparable, then $h^2 \in F$ and therefore $h = 1$.) Since $N_{K/F}(h) = \operatorname{Nrd}(h) = 1$, by the Hilbert Theorem 90, we have $h = \bar{b}b^{-1}$ for some $b \in K^\times$. By the Noether-Skolem Theorem, $\bar{b} = aba^{-1}$ for some $a \in Q^\times$, hence $h = [a, b]$. □

Let $h \in Q^\times$ satisfy $\operatorname{Nrd}(h) = 1$. Then $h = [a, b] = aba^{-1}b^{-1}$ for some $a, b \in Q^\times$ by Lemma 46.17. Consider the following element

$$\widehat{h} = (b, a, b^{-1}, a^{-1}, h) \in G.$$

Lemma 46.18. *The element* $\widehat{h}$ *does not depend on the choice of* a *and* b.

PROOF. Let $h = [a, b] = [c, d]$. By Corollary 46.16(1), we may assume that either $a = c$ or $b = d$. Consider the first case (the latter case is similar). We can write $d = bx$, where x commutes with a. We have

$$\begin{aligned}
(d, a, d^{-1}, a^{-1}, h) &= (bx, a, x^{-1}b^{-1}, a^{-1}, h) \\
&= (bx, x^{-1}, b^{-1}) * (b, x, a, x^{-1}b^{-1}, a^{-1}, h) \text{ (Lemmas 46.13, 46.14)} \\
&= (bx, x^{-1}, b^{-1}) * (a^{-1}, h, b, x, a, x^{-1}b^{-1}) \text{ (Lemma 46.12)} \\
&= (a^{-1}, h, b, x, a, x^{-1}b^{-1}, bx, x^{-1}, b^{-1}) \text{ (Lemma 46.14)} \\
&= (a^{-1}, h, b, a, b^{-1}) \text{ (Lemma 46.13)} \\
&= (b, a, b^{-1}, a^{-1}, h) \text{ (Lemma 46.12).}
\end{aligned}$$

□

Lemma 46.19. *For every* $h_1, h_2 \in [Q^\times, Q^\times]$ *we have*

$$\widehat{h_1 h_2} = \widehat{h_1} * \widehat{h_2} * (h_1 h_2, h_2^{-1}, h_1^{-1}).$$

PROOF. By Corollary 46.16(2), we have $h_1 = [a_1, c]$ and $h_2 = [c, b_2]$ for some $a_1, b_2, c \in Q^\times$. Then $h_1h_2 = [a_1b_2^{-1}, b_2cb_2^{-1}]$ and

$$\begin{aligned}
&\widehat{h_1} * \widehat{h_2} * (h_1h_2, h_2^{-1}, h_1^{-1})\\
&\qquad = (c, a_1, c^{-1}, a_1^{-1}, h_1, h_2, b_2, c, b_2^{-1}, c^{-1}) * (h_1h_2, h_2^{-1}, h_1^{-1})\\
&\qquad = (b_2, c, b_2^{-1}, c^{-1}, c, a_1, c^{-1}, a_1^{-1}, h_1, h_2) * (h_2^{-1}, h_1^{-1}h_1h_2)\\
&\qquad = (b_2, c, b_2^{-1}, a_1, c^{-1}, a_1^{-1}, h_1h_2)\\
&\qquad = (b_2, c, b_2^{-1}, b_2c^{-1}b_2^{-1}) * (b_2cb_2^{-1}, a_1, c^{-1}, a_1^{-1}, h_1h_2)\\
&\qquad = (b_2^{-1}, b_2c^{-1}b_2^{-1}, b_2, c) * (c^{-1}, a_1^{-1}, h_1h_2, b_2cb_2^{-1}, a_1)\\
&\qquad = (b_2^{-1}, b_2c^{-1}b_2^{-1}, b_2, c, c^{-1}, a_1^{-1}, h_1h_2, b_2cb_2^{-1}, a_1)\\
&\qquad = (b_2^{-1}, b_2c^{-1}b_2^{-1}, b_2, a_1^{-1}, h_1h_2, b_2cb_2^{-1}, a_1b_2^{-1}, b_2a_1^{-1}, a_1)\\
&\qquad = (b_2, a_1^{-1}, h_1h_2, b_2cb_2^{-1}, a_1b_2^{-1}, b_2c^{-1}b_2^{-1}) * (b_2a_1^{-1}, a_1, b_2^{-1})\\
&\qquad = (b_2a_1^{-1}, a_1, b_2^{-1}, b_2, a_1^{-1}, h_1h_2, b_2cb_2^{-1}, a_1b_2^{-1}, b_2c^{-1}b_2^{-1})\\
&\qquad = (b_2a_1^{-1}, h_1h_2, b_2cb_2^{-1}, a_1b_2^{-1}, b_2c^{-1}b_2^{-1})\\
&\qquad = (b_2cb_2^{-1}, a_1b_2^{-1}, b_2c^{-1}b_2^{-1}, b_2a_1^{-1}, h_1h_2)\\
&\qquad = \widehat{h_1h_2}. \qquad \square
\end{aligned}$$

Let $a_1, a_2, \dots, a_n \in Q^\times$ satisfying $\operatorname{Nrd}(h) = 1$ where $h = a_1a_2 \dots a_n$. We set

$$((a_1, a_2, \dots, a_n)) := (a_1, a_2, \dots, a_n, h^{-1}) * \widehat{h} \in G. \tag{46.20}$$

Lemma 46.21. $((a_1, a_2, \dots, a_n)) * ((b_1, b_2, \dots, b_m)) = ((a_1, \dots, a_n, b_1, \dots, b_m))$.

PROOF. Set $h := a_1 \cdots a_n$ and $h' := b_1 \cdots b_m$. We have

$$\begin{aligned}
L.H.S. &= (a_1, a_2, \dots, a_n, h^{-1}) * (b_1, b_2, \dots, b_m, (h')^{-1}) * \widehat{h} * \widehat{h'}\\
&= (a_1, a_2, \dots, a_n, b_1, b_2, \dots, b_m, (h')^{-1}, h^{-1}) * \widehat{h} * \widehat{h'}\\
&= (a_1, a_2, \dots, a_n, b_1, b_2, \dots, b_m, (hh')^{-1}) * (hh', (h')^{-1}, h) * \widehat{h} * \widehat{h'}\\
&= (a_1, a_2, \dots, a_n, b_1, b_2, \dots, b_m, (hh')^{-1}) * \widehat{hh'} \text{ (Lemma 46.19)}\\
&= R.H.S. \qquad \square
\end{aligned}$$

The following lemma is a consequence of the definition (46.20) and Lemma 46.13.

Lemma 46.22. *If a_{i-1} commutes with a_i for some i, then*

$$((a_1, \dots, a_{i-1}, a_i, \dots, a_n)) = ((a_1, \dots, a_{i-1}a_i, \dots, a_n)).$$

Lemma 46.23. $((a, b, a^{-1}, b^{-1})) = 1$.

PROOF. Set $h = [a, b]$. We have

$$L.H.S. = (a, b, a^{-1}, b^{-1}, h^{-1}) * \widehat{h} = (a, b, a^{-1}, b^{-1}, h^{-1}) * (b, a, b^{-1}, a^{-1}, h) = 1. \qquad \square$$

We want to establish an isomorphism between G and $A(Q)/A'$. To do so we define a map $\pi : G \to A(Q)/A'$ by the formula

$$\pi\big((a, b, c)\big) = \tilde{a}\tilde{b}\tilde{c} \in A(Q)/A',$$

where $a, b, c \in Q^\times$ satisfy $abc = 1$. Clearly, π is well-defined.

Let $a_1, a_2, \dots, a_n \in Q^\times$ with $a_1 a_2 \cdots a_n = 1$. By induction on n, we have

$$\pi\big((a_1, a_2, \dots, a_n)\big) = \widetilde{a}_1 \widetilde{a}_2 \cdots \widetilde{a}_n \in A(Q)/A'.$$

Hence π is a homomorphism by Lemma 46.14.

Let $h \in [Q^\times, Q^\times]$. Write $h = [a, b]$ for $a, b \in Q^\times$. We have

$$\pi(\widehat{h}) = \pi\big((b, a, b^{-1}, a^{-1}, h)\big) = \widetilde{h}.$$

If $a_1, a_2, \dots, a_n \in Q^\times$ satisfies $\mathrm{Nrd}(h) = 1$ with $h = a_1 a_2 \cdots a_n$, then

$$\pi\big(((a_1, a_2, \dots, a_n))\big) = \pi\big((a_1, a_2, \dots, a_n, h^{-1})\big) * \pi(\widehat{h}) = \widetilde{a}_1 \widetilde{a}_2 \cdots \widetilde{a}_n. \tag{46.24}$$

Define a homomorphism $\theta : A(Q)/A' \to G$ as follows: Let $a_1, a_2, \dots, a_n \in Q^\times$ satisfy $\mathrm{Nrd}(a_1 a_2 \cdots a_n) = 1$. Define θ by

$$\theta(\widetilde{a}_1 \widetilde{a}_2 \cdots \widetilde{a}_n) = ((a_1, a_2, \dots, a_n)). \tag{46.25}$$

The relation at the end of subsection 46.D and Lemmas 46.21, 46.22, and 46.23 show that θ is a well-defined homomorphism. Formulas (46.24) and (46.25) yield

Proposition 46.26. *The maps π and θ are isomorphisms inverse to each other.*

Corollary 46.27. *The group $A(Q)/A'$ is generated by products $\tilde{a}\tilde{b}\tilde{c}$ for all ordered triples a, b, c of elements in $Q^\times$ satisfying $abc = 1$ with the following set of defining relations:*

($R1'$): $\big(\tilde{a}\tilde{b}(\widetilde{cd})\big) \cdot \big((\widetilde{ab})\tilde{c}\tilde{d}\big) = \big(\tilde{b}\tilde{c}(\widetilde{da})\big) \cdot \big((\widetilde{da})\tilde{b}\tilde{c}\big)$ *for all $a, b, c, d \in Q^\times$ such that $abcd = 1$;*

($R2'$): $\tilde{a}\tilde{b}\tilde{c} = 1$ *if a and b commute.*

46.F. Proof of Proposition 46.6. To prove Proposition 46.6, we need to prove that the homomorphism ∂_1 in (46.10) is an isomorphism.

We shall view the fraction l_a/l_b for $a, b \in Q \setminus F$ as a nonzero rational function on C, i.e., $l_a/l_b \in F(C)^\times$.

Lemma 46.28. *Let K_0 be the quadratic subfield of Q corresponding to the fixed closed point x_0 of degree 2 on C and let $b \in K_0 \setminus F$. Then the space L_1 consists of all the fractions l_a/l_b with $a \in Q$.*

Proof. Obviously, $l_a/l_b \in L_1$. It follows from Lemma 45.5 that the space of all fractions l_a/l_b is 3-dimensional. On the other hand, $\dim L_1 = 3$. □

By Lemma 46.28, the group M' is generated by symbols of the form $\{l_a/l_b, \alpha\}$ for all $a, b \in Q \setminus F$ and $\alpha \in F^\times$ and the group M_1 is generated by symbols $\{l_a/l_b, l_c/l_d\}$ for all $a, b, c, d \in Q \setminus F$.

Let $a, b, c \in Q$ satisfy $abc = 1$. We define an element

$$[a, b, c] \in M_1/M'$$

as follows: If at least one of a, b and c belongs to $F^\times$, we set $[a, b, c] = 0$. Otherwise, the linear forms l_a, l_b and l_c are nonzero and we set

$$[a, b, c] := \left\{\frac{l_a}{l_c}, \frac{l_b}{l_c}\right\} + M'.$$

Lemma 45.5 and the equality $\{u, -u\} = 0$ in $K_2\big(F(C)\big)$ yield:

Lemma 46.29. *Let $a, b, c \in Q^\times$ be such that $abc = 1$ and let $\alpha \in F^\times$. Then*

(1) $[a, b, c] = [b, c, a]$*;*

(2) $[\alpha a, \alpha^{-1}b, c] = [a, b, c]$;
(3) $[a, b, c] + [c^{-1}, b^{-1}, a^{-1}] = 0$;
(4) *If* a *and* b *commute, then* $[a, b, c] = 0$.

Lemma 46.30. $\partial_1([a, b, c]) = \widetilde{\tilde{a}\tilde{b}\tilde{c}}$.

PROOF. We may assume that none of a, b, or c is a constant. Let x, y, and z be the points of C of degree 2 corresponding to quadratic subfields $F[a]$, $F[b]$, and $F[c]$ that we identify with $F(x)$, $F(y)$ and $F(z)$, respectively.

Consider the following element in the class $[a, b, c]$:

$$w = \Big\{\frac{l_a}{l_c}, \frac{l_b}{l_c}\Big\} + \Big\{\frac{l_b}{l_c}, \operatorname{Nrd}(a)\Big\} + \Big\{\frac{l_b}{l_a}, -\operatorname{Nrd}(b)\Big\}.$$

By Proposition 45.12 (identifying residue fields with the corresponding quadratic extensions) and Lemma 45.5,

$$\begin{aligned}
\partial_x(w) &= \frac{l_b}{l_c}(x)\big(-\operatorname{Nrd}(b)\big)^{-1} = -\operatorname{Nrd}(b)\frac{l_{b^{-1}}}{l_{b^{-1}a^{-1}}}(x)\big(-\operatorname{Nrd}(b)\big)^{-1} = a,\\
\partial_y(w) &= \frac{l_c}{l_a}(y)\big(-\operatorname{Nrd}(ab)\big) = -\operatorname{Nrd}(a)^{-1}\frac{l_{b^{-1}a^{-1}}}{l_{a^{-1}}}(x)\big(-\operatorname{Nrd}(ab)\big)\\
&= -\operatorname{Nrd}(a)^{-1}\bar{b}^{-1}\big(-\operatorname{Nrd}(ab)\big) = b,\\
\partial_z(w) &= -\frac{l_a}{l_b}(z)\operatorname{Nrd}(a)^{-1} = \operatorname{Nrd}(a)\frac{l_{bc}}{l_b}(x)\operatorname{Nrd}(a)^{-1} = c.
\end{aligned}$$

□

Lemma 46.31. *Let* $a, b, c, d \in Q \setminus F$ *be such that* $cd, da \notin F$ *and* $abcd = 1$. *Then*

$$\Big\{\frac{l_a l_c}{l_{cd}l_{da}}, \frac{l_b l_d}{l_{cd}l_{da}}\Big\} \in M'.$$

PROOF. In Proposition 45.8 plugging in the elements c^{-1}, ab and b for a, b and c, respectively, and using Lemma 45.5, we get elements $\alpha, \beta, \gamma \in F^\times$ such that on the conic C,

$$\alpha l_a l_c + \beta l_b l_d + \gamma l_{cd} l_{da} = 0.$$

Then

$$-\frac{\alpha l_a l_c}{\gamma l_{cd} l_{da}} - \frac{\beta l_b l_d}{\gamma l_{cd} l_{da}} = 1$$

and

$$0 = \Big\{-\frac{\alpha l_a l_c}{\gamma l_{cd} l_{da}}, -\frac{\beta l_b l_d}{\gamma l_{cd} l_{da}}\Big\} \equiv \Big\{\frac{l_a l_c}{l_{cd} l_{da}}, \frac{l_b l_d}{l_{cd} l_{da}}\Big\} \mod M'.$$

□

Proposition 46.32. *Let* $a, b, c, d \in Q^\times$ *be such that* $abcd = 1$. *Then*

$$[a, b, cd] + [ab, c, d] = [b, c, da] + [bc, d, a].$$

PROOF. We first note that if one of the elements a, b, ab, c, d, cd belongs to $F^\times$, the equality holds. For example, if $a \in F^\times$, then the equality reads $[ab, c, d] = [b, c, da]$ and follows from Lemma 46.29 and if $\alpha = ab \in F^\times$, then again by Lemma 46.29,

$$L.H.S. = 0 = [b, c, da] + [(da)^{-1}, \alpha^{-1}c^{-1}, \alpha b^{-1}] = R.H.S.$$

So we may assume that none of the elements belong to $F^\times$. It follows from Lemma 45.5(4) that l_{cd}/l_{ab} and l_{da}/l_{bc} belong to $F^\times$. By Lemmas 46.29 and 46.31, we have in M_1/M':

$$\begin{aligned}
0 &= \left\{\frac{l_a l_c}{l_{cd} l_{da}}, \frac{l_b l_d}{l_{cd} l_{da}}\right\} + M' \\
&= \left\{\frac{l_a}{l_{cd}}, \frac{l_b}{l_{cd}}\right\} + \left\{\frac{l_c}{l_{da}}, \frac{l_b}{l_{da}}\right\} + \left\{\frac{l_a}{l_{cd}}, \frac{l_d}{l_{da}}\right\} + \left\{\frac{l_c}{l_{da}}, \frac{l_d}{l_{cd}}\right\} + M' \\
&= [a,b,cd] - [b,c,da] + \left(\left\{\frac{l_a}{l_{da}}, \frac{l_d}{l_{da}}\right\} + \left\{\frac{l_{da}}{l_{cd}}, \frac{l_d}{l_{da}}\right\}\right) + \left\{\frac{l_c}{l_{da}}, \frac{l_d}{l_{cd}}\right\} + M' \\
&= [a,b,cd] - [b,c,da] + \left\{\frac{l_a}{l_{bc}}, \frac{l_d}{l_{bc}}\right\} + \left(\left\{\frac{l_{da}}{l_{cd}}, \frac{l_d}{l_{cd}}\right\} + \left\{\frac{l_c}{l_{da}}, \frac{l_d}{l_{cd}}\right\}\right) + M' \\
&= [a,b,cd] - [b,c,da] - [bc,d,a] + \left\{\frac{l_c}{l_{cd}}, \frac{l_d}{l_{cd}}\right\} + M' \\
&= [a,b,cd] - [b,c,da] - [bc,d,a] + \left\{\frac{l_c}{l_{ab}}, \frac{l_d}{l_{ab}}\right\} + M' \\
&= [a,b,cd] - [b,c,da] - [bc,d,a] + [ab,c,d]. \qquad \square
\end{aligned}$$

We shall use the presentation of the group $A(Q)/A'$ by generators and relations given in Corollary 46.27. We define a homomorphism

$$\mu : A(Q)/A' \to M_1/M'$$

by the formula

$$\mu(\widetilde{abc}) := [a,b,c]$$

for all $a, b, c \in Q$ such that $abc = 1$. It follows from Lemma 46.29(4) and Proposition 46.32 that μ is well-defined. Lemma 46.30 implies that $\partial_1 \circ \mu$ is the identity.

To show that μ is the inverse of ∂_1 it suffices to prove that μ is surjective.

The group M_1/M' is generated by elements of the form $w = \{l_{a'}/l_{c'}, l_{b'}/l_{c'}\} + M'$ for $a', b', c' \in Q \backslash F$. We may assume that $1, a', b'$ and c' are linearly independent (otherwise, $w = 0$). In particular, $1, a', b'$ and $a'b'$ form a basis of Q, hence

$$c' = \alpha + \beta a' + \gamma b' + \delta a'b'$$

for some $\alpha, \beta, \gamma, \delta \in F$ with $\delta \neq 0$. We have

$$(\gamma\delta^{-1} + a')(\beta + \delta b') = \varepsilon + c'$$

for $\varepsilon = \beta\gamma\delta^{-1} - \alpha$. Set

$$a := \gamma\delta^{-1} + a', \quad b := \beta + \delta b', \quad c := (\varepsilon + c')^{-1}.$$

We have $abc = 1$. It follows from Lemma 45.5 that

$$w = \left\{\frac{l_{a'}}{l_{c'}}, \frac{l_{b'}}{l_{c'}}\right\} + M' = \left\{\frac{l_a}{l_c}, \frac{l_b}{l_c}\right\} + M' = [a,b,c].$$

By definition of μ, we have $\mu(\widetilde{abc}) = [a,b,c] = w$, hence μ is surjective. The proof of Proposition 46.6 is complete. $\square$

47. Hilbert Theorem 90 for K_2

In this section we prove the K_2-analog of the classical Hilbert Theorem 90.

Throughout this section let L/F be a Galois quadratic field extension with the Galois group $G = \{1, \sigma\}$. For every field extension E/F linearly disjoint from L/F, the field $LE = L \otimes_F E$ is a quadratic Galois extension of E with Galois group isomorphic to G. The group G acts naturally on $K_2(LE)$. We write $(1-\sigma)u$ for $u-\sigma(u)$ with $u \in K_2(LE)$ and set

$$V(E) := K_2(LE)/(1-\sigma)K_2(LE).$$

If $E \to E'$ is a homomorphism of field extensions of F linearly disjoint from L/F, there is a natural homomorphism

$$V(E) \to V(E').$$

Proposition 47.1. *Let C be a conic curve over F and L/F a Galois quadratic field extension such that C has a rational point over L. Then the natural homomorphism $V(F) \to V\big(F(C)\big)$ is injective.*

PROOF. Let $u \in K_2(L)$ satisfy $u_{L(C)} = (1-\sigma)v$ for some $v \in K_2\big(L(C)\big)$. For a closed point $x \in C$ the L-algebra $L(x) := L \otimes_F F(x)$ is isomorphic to the direct product of residue fields $L(y)$ for all closed points $y \in C_L$ over $x \in C$. We denote the product of all the $\partial_y(v) \in L(y)^\times$ with y over x by $\partial_x(v) \in L(x)^\times$.

Set $a_x := \partial_x(v) \in L(x)^\times$. Then

$$a_x/\sigma(a_x) = \partial_x(v)/\sigma\big(\partial_x(v)\big) = \partial_x\big((1-\sigma)v\big) = \partial_x\big(u_{L(C)}\big) = 1,$$

i.e., $a_x \in F(x)^\times$. Applying Theorem 46.1 to C_L, we have

$$\prod_{x \in C} \mathrm{N}_{F(x)/F}(a_x) = \mathrm{N}_{L/F}\Big(\prod_{y \in C_L} \mathrm{N}_{L(y)/L}(a_y)\Big) = \mathrm{N}_{L/F}\Big(\prod_{y \in C_L} \mathrm{N}_{L(y)/L}(\partial_y(v))\Big) = 1.$$

Hence applying Theorem 46.1 to C, there is a $w \in K_2\big(F(C)\big)$ satisfying $\partial_x(w) = a_x$ for all $x \in C$. Set $v' := v - w_{L(X)} \in K_2\big(L(C)\big)$. As

$$\partial_x(v') = \partial_x(v)\partial_x(w)^{-1} = a_x a_x^{-1} = 1,$$

applying Theorem 46.1 to C_L, there exists an $s \in K_2(L)$ with $s_{L(C)} = v'$. We have

$$(1-\sigma)s_{L(C)} = (1-\sigma)v' = (1-\sigma)v = u_{L(C)},$$

i.e., $(1-\sigma)s-u$ splits over $L(C)$. Since $L(C)/L$ is a purely transcendental extension, we must have $(1-\sigma)s - u = 0$ (cf. Example 100.6), hence

$$u = (1-\sigma)s \in \mathrm{Im}(1-\sigma). \qquad \square$$

Corollary 47.2. *For any finitely generated subgroup $H \subset F^\times$, there is a field extension F'/F linearly disjoint from L/F such that the natural homomorphism $V(F) \to V(F')$ is injective and $H \subset \mathrm{N}_{L'/F'}(L'^\times)$ where $L' = LF'$.*

PROOF. By induction it suffices to assume that H is generated by one element b. Set $F' = F(C)$, where $C = C_Q$ is the conic curve associated with the quaternion algebra $Q = (L/F, b)$ (cf. §98.E). Since Q splits over F', we have $b \in \mathrm{N}_{L'/F'}(L'^\times)$ by Facts 98.13(4) and 98.14(5). The conic C has a rational point over L, therefore, the homomorphism $V(F) \to V(F')$ is injective by Proposition 47.1. $\square$

For any two elements $x, y \in L^\times$, we write $\langle x, y\rangle$ for the class of the symbol $\{x, y\}$ in $V(F)$. Let f be the group homomorphism

$$f = f_F : \mathrm{N}_{L/F}(L^\times) \otimes F^\times \to V(F), \quad f\big(\mathrm{N}_{L/F}(x) \otimes a\big) = \langle x, a\rangle.$$

The map f is well-defined. Indeed, if $\mathrm{N}_{L/F}(x) = \mathrm{N}_{L/F}(y)$ for $x, y \in L^\times$, then $y = xz\sigma(z)^{-1}$ for some $z \in L^\times$ by the classical Hilbert Theorem 90. Hence $\{y, a\} = \{x, a\} + (1-\sigma)\{z, a\}$ and consequently $\langle y, a\rangle = \langle x, a\rangle$.

Lemma 47.3. *Let* $b \in \mathrm{N}_{L/F}(L^\times)$. *Then* $f\big(b \otimes (1-b)\big) = 0$.

PROOF. If $b = d^2$ for some $d \in F^\times$, then

$$f\big(b \otimes (1-b)\big) = \langle d, 1-d^2 \rangle = \langle d, 1-d \rangle + \langle d, 1+d \rangle = \langle -1, 1+d \rangle = 0$$

since $-1 = z\sigma(z)^{-1}$ for some $z \in L^\times$.

So we may assume that b is not a square in F. Set

$$F' = F[t]/(t^2 - b), \quad L' = L[t]/(t^2 - b).$$

Note that L' is either a field or product of two copies of the field F'. Let $u \in F'$ be the class of t, so that $u^2 = b$. Choose $x \in L^\times$ with $\mathrm{N}_{L/F}(x) = b$. Note that $\mathrm{N}_{L'/F'}(x/u) = b/u^2 = 1$ and $\mathrm{N}_{L'/L}(1-u) = 1-b$.

The automorphism σ extends to an automorphism of L' over F'. Applying the classical Hilbert Theorem 90 to the extension L'/F', there is a $v \in L'^\times$ such that $v\sigma(v)^{-1} = x/u$. We have

$$\begin{aligned} f\big(b\otimes(1-b)\big) = \langle x, 1-b \rangle = \langle x, \mathrm{N}_{L'/L}(1-u) \rangle = c_{F'/F}\langle x, 1-u \rangle = c_{F'/F}(\langle x/u, 1-u \rangle) \\ = c_{F'/F}(\langle v\sigma(v)^{-1}, 1-u \rangle) = (1-\sigma)c_{F'/F}(\langle v, 1-u \rangle) = 0, \end{aligned}$$

where $c_{F'/F} : V(F') \to V(F)$ is induced by the norm map $c_{L'/L} : K_2(L') \to K_2(L)$. □

Theorem 47.4 (Hilbert Theorem 90 for K_2). *Let L/F be a Galois quadratic extension and σ the generator of* $\mathrm{Gal}(L/F)$. *Then the sequence*

$$K_2(L) \xrightarrow{1-\sigma} K_2(L) \xrightarrow{c_{L/F}} K_2(F)$$

is exact.

PROOF. Let $u \in K_2(L)$ satisfy $c_{L/F}(u) = 0$. By Proposition 100.2, the group $K_2(L)$ is generated by symbols of the form $\{x, a\}$ with $x \in L^\times$ and $a \in F^\times$. Therefore, we can write

$$u = \sum_{j=1}^{m} \{x_j, a_j\}$$

for some $x_j \in L^\times$ and $a_j \in F^\times$, and

$$c_{L/F}(u) = \sum_{j=1}^{m} \{\mathrm{N}_{L/F}(x_j), a_j\} = 0.$$

Hence by definition of $K_2(F)$, we have in $F^\times \otimes F^\times$:

$$\sum_{j=1}^{m} \mathrm{N}_{L/F}(x_j) \otimes a_j = \sum_{i=1}^{n} \pm \big(b_i \otimes (1-b_i)\big) \tag{47.5}$$

for some $b_i \in F^\times$. Clearly, the equality (47.5) holds in $H \otimes F^\times$ for some finitely generated subgroup $H \subset F^\times$ containing all the $\mathrm{N}_{L/F}(x_j)$ and b_i.

By Corollary 47.2, there is a field extension F'/F linearly disjoint from L/F such that the natural homomorphism $V(F) \to V(F')$ is injective and

$$H \subset \mathrm{N}_{L'/F'}(L'^\times)$$

with $L' = LF'$. The equality (47.5) then holds in $\mathrm{N}_{L'/F'}(L'^{\times}) \otimes F'^{\times}$. Now apply the map $f_{F'}$ to both sides of (47.5). By Lemma 47.3, the class of $u_{L'}$ in $V(F')$ is equal to

$$\sum_{j=1}^{m} \langle x_j, a_j \rangle = f_{F'}\Big(\sum_{j=1}^{m} \mathrm{N}_{L/F}(x_j) \otimes a_j \Big) = \sum_{i=1}^{n} \pm f_{F'}\big(b_i \otimes (1 - b_i)\big) = 0,$$

i.e., $u_{L'} \in (1-\sigma)K_2(L')$. Since the map $V(F) \to V(F')$ is injective, we conclude that $u \in (1-\sigma)K_2(L)$. □

Theorem 47.6. *Let $u \in K_2(F)$ satisfy $2u = 0$. Then $u = \{-1, a\}$ for some $a \in F^{\times}$. In particular, $u = 0$ if $\mathrm{char}(F) = 2$.*

PROOF. Let $G = \{1, \sigma\}$ be a cyclic group of order two. Consider a G-action on the field $L = F((t))$ of a Laurent power series defined by

$$\sigma(t) = \begin{cases} -t & \text{if } \mathrm{char}\, F \neq 2, \\ t/(1+t) & \text{if } \mathrm{char}\, F = 2. \end{cases}$$

We have a Galois quadratic extension L/E with $E = L^G$.

Consider the diagram

$$\begin{array}{ccc} K_2(L) & \xrightarrow{1-\sigma} & K_2(L) \\ \partial \big\downarrow & & \big\downarrow s \\ F^{\times} & \xrightarrow{\{-1\}} & K_2(F), \end{array}$$

where ∂ is the residue homomorphism of the canonical discrete valuation of L, the map $s = s_t$ is the specialization homomorphism of the parameter t (cf. §100.D), and the bottom homomorphism is multiplication by $\{-1\}$. We claim that the diagram is commutative. The group $K_2(L)$ is generated by elements of the form $\{f, g\}$ and $\{t, g\}$ with f and g in $F[[t]]$ having nonzero constant term. If $\mathrm{char}\, F \neq 2$, we have

$$\begin{aligned} s \circ (1-\sigma)(\{f,g\}) &= s(\{f,g\} - \{\sigma(f), \sigma(g)\}) \\ &= \{f(0), g(0)\} - \{\sigma(f)(0), \sigma(g)(0)\} \\ &= 0 = \{-1\} \cdot \partial(\{f,g\}) \end{aligned}$$

and

$$\begin{aligned} s \circ (1-\sigma)(\{t,g\}) &= s(\{t,g\} - \{-t, \sigma(g)\}) \\ &= \{-1, g(0)\} \\ &= \{-1\} \cdot \partial(\{t,g\}). \end{aligned}$$

If $\mathrm{char}\, F = 2$, we obviously have $s(u) = s\big(\sigma(u)\big)$ for every $u \in K_2(L)$, hence $s \circ (1-\sigma) = 0$.

Since $c_{L/E}(u_L) = 2u_E = 0$, by Theorem 47.4, we have $u = (1-\sigma)v$ for some $v \in K_2(L)$. The commutativity of the diagram yields

$$u = s(u_L) = s\big((1-\sigma)v\big) = \{-1, \partial(v)\}.$$ □

48. Proof of the main theorem

In this section we prove Theorem 44.1. Let F be a field of characteristic different from 2.

48.A. Injectivity of h_F. Suppose that $h_F\big(u + 2K_2(F)\big) = 1$ for an element $u \in K_2(F)$. Let u be a sum of n symbols. We prove by induction on n that $u \in 2K_2(F)$.

First consider the case $n = 1$, i.e., $u = \{a, b\}$ with $a, b \in F^\times$. Since $\left(\frac{a,b}{F}\right)$ is a split quaternion algebra, there are $x, y \in F$ such that $b = x^2 - ay^2$. It follows from Lemma 100.3 that $\{a, b\} \in 2K_2(F)$.

Next consider the case $n = 2$, i.e. $u = \{a, b\} + \{c, d\}$. By assumption, the algebra $\left(\frac{a,b}{F}\right) \otimes \left(\frac{c,d}{F}\right)$ is split, or equivalently, $\left(\frac{a,b}{F}\right)$ and $\left(\frac{c,d}{F}\right)$ are isomorphic. By the Chain Lemma 98.15, we may assume that $a = c$ and hence $u = \{a, bd\}$, so the statement follows from the case $n = 1$.

Now consider the general case. Write u in the form $u = \{a, b\} + v$ for $a, b \in F^\times$ and an element $v \in K_2F$ that is a sum of $n - 1$ symbols. We may assume that $\{a, b\} \notin 2K_2(F)$, i.e., the quaternion algebra $Q := \left(\frac{a,b}{F}\right)$ does not split. Let $C = C_Q$ be the conic curve over F corresponding to Q and set $L = F(C)$. The conic C is given by the equation

$$at_0^2 + bt_1^2 = abt_2^2$$

in projective coordinates (cf. Example 45.3). Set $x = t_0/t_2$ and $y = t_1/t_2$. As $b^{-1}x^2 + a^{-1}y^2 = 1$, we have

$$0 = \{b^{-1}x^2, a^{-1}y^2\} = 2\{x, a^{-1}y^2\} - 2\{b, y\} - \{a, b\}$$

and therefore $\{a, b\} = 2r$ in $K_2(L)$ with $r = \{x, a^{-1}y^2\} - \{b, y\}$. Let $p \in C$ be the point of degree 2 given by $Z = 0$. The element r has only one nontrivial residue at the point p and $\partial_p(r) = -1$.

Since the quaternion algebra $\left(\frac{a,b}{F}\right)$ is split over L, we have $h_L\big(v_L + 2K_2L\big) = 1$. By induction, $v_L = 2w$ for some element $w \in K_2(L)$.

Set $c_x := \partial_x(w)$ for every point $x \in C$. Since

$$c_x^2 = \partial_x(2w) = \partial_x(v_L) = 1,$$

we have $c_x = (-1)^{n_x}$ for $n_x = 0$ or 1. By Corollary 45.2(2), the degree of every point of C is even, hence

$$\sum_{x \in C} n_x \deg(x) = 2m$$

for some $m \in \mathbb{Z}$. As every degree zero divisor on C is principal by Corollary 45.2(1), there is a function $f \in L^\times$ with $\operatorname{div}(f) = \sum n_x x - mp$. Set

$$w' := w + \{-1, f\} + kr \in K_2(L)$$

where $k = m + n_p$. If $x \in C$ is a point different from p, then

$$\partial_x(w') = \partial_x(w) \cdot (-1)^{n_x} = 1.$$

Since, also,

$$\partial_p(w') = \partial_p(w) \cdot (-1)^m \cdot (-1)^k = (-1)^{n_p + m + k} = 1,$$

we have $\partial_x(w') = 1$ for all $x \in C$. By Theorem 46.1, it follows that $w' = s_L$ for some $s \in K_2(F)$. Thus

$$v_L = 2w = 2w' - 2kr = 2s_L - \{a^k, b\}_L.$$

Set $v' := v - 2s + \{a^k, b\} \in K_2(F)$; we have $v'_L = 0$. The conic C has a rational point over the quadratic extension $E = F(\sqrt{a})$. Since the field extension $E(C)/E$

is purely transcendental and $v'_{E(C)} = 0$, we have $v'_E = 0$ by Example 100.6 and therefore $2v' = N_{E/F}(v'_E) = 0$. It follows from Theorem 47.6 that $v' = \{-1, d\}$ for some $d \in F^\times$. Hence modulo $2K_2(F)$, the element v is the sum of two symbols $\{a^k, b\}$ and $\{-1, d\}$. Consequently, we are reduced to the case $n = 2$ that has already been considered.

48.B. Surjectivity of h_F. Recall that we are writing $k_2(F)$ for $K_2F/2K_2(F)$.

Proposition 48.1. *Let L/F be a quadratic extension. Then the sequence*

$$k_2(F) \xrightarrow{r_{L/F}} k_2(L) \xrightarrow{c_{L/F}} k_2(F)$$

is exact.

PROOF. Let $u \in K_2(L)$ satisfy $c_{L/F}(u) = 2v$ for some $v \in K_2(F)$. Then

$$c_{L/F}(u - v_L) = 2v - 2v = 0$$

and, by Theorem 47.4, we have $u - v_L = (1 - \sigma)w$ for some $w \in K_2(L)$. Hence

$$u = v_L + (1 - \sigma)w = \big(v + c_{L/F}(w)\big)_L - 2\sigma w. \qquad \square$$

We now finish the proof of Theorem 44.1. Let $s \in \mathrm{Br}_2(F)$. First suppose that the field F is 2-special (cf. §101.B). By induction on the index of s, we prove that $s \in \mathrm{Im}(h_F)$. By Proposition 101.15, there exists a quadratic extension L/F with $\mathrm{ind}(s_L) < \mathrm{ind}(s)$. By induction, $s_L = h_L(u)$ for some $u \in k_2(L)$. It follows from Proposition 101.9 that

$$h_F\big(c_{L/F}(u)\big) = c_{L/F}\big(h_L(u)\big) = c_{L/F}(s_L) = 1.$$

The injectivity of h_F implies that $c_{L/F}(u) = 0$ so by Proposition 48.1, we have $u = v_L$ for some $v \in k_2(F)$. Therefore,

$$h_F(v)_L = h_L(v_L) = h_L(u) = s_L,$$

hence $s - h_F(v)$ splits over L and must be the class of a quaternion algebra. Consequently, $s - h_F(v) = h_F(w)$ for some symbol w in $k_2(F)$, so $s = h_F(v+w) \in \mathrm{Im}(h_F)$.

In the general case, applying the first part of the proof to a maximal odd degree extension of F (cf. §101.B and Proposition 101.16), we see that there exists an odd degree extension E/F such that $s_E = h_E(v)$ for some $v \in k_2(E)$. Then again by Proposition 101.9,

$$s = c_{E/F}(s_E) = c_{E/F}\big(h_E(v)\big) = h_F\big(c_{E/F}(v)\big).$$

Part 2

Algebraic cycles

CHAPTER IX

Homology and Cohomology

The word "*scheme*" in this book (with the exception of §49) will mean a separated scheme of finite type over a field and a "*variety*" will mean an integral scheme.

In this chapter we develop the K-homology and K-cohomology theories of schemes over a field that generalizes the theory of Chow groups. We follow the approach of [**117**] given by Rost. There are two advantages of having such general theories rather than just that of Chow groups. First we have a long (infinite) localization exact sequence. This tool together with the 5-lemma allows us to give simple proofs of some basic results in the theory such as the Homotopy Invariance and Projective Bundle Theorems. Secondly, the construction of the deformation map (called the specialization homomorphism in [**45**]), used in the definition of the pull-back homomorphisms, is much easier; it does not require intersections with Cartier divisors.

We view the K-homology as a covariant functor from the category of schemes of finite type over a field to the category of abelian groups and the K-cohomology as a contravariant functor from the category of smooth schemes of finite type over a field. The fact that K-homology groups for smooth schemes coincide with K-cohomology groups should be viewed as Poincaré duality.

49. The complex $C_*(X)$

The purpose of this section is to construct complexes $C_*(X)$ giving the homology and cohomology theories that we need.

Throughout this section (and only in this section), we need to extend the class of separated schemes of finite type over a field and consider the class of *excellent schemes* (cf. [**48**, §7.8]). The class of excellent schemes contains:

1. Schemes of finite type over a field.
2. Closed and open subschemes of excellent schemes.
3. $\operatorname{Spec}(\mathcal{O}_{X,x})$ where x is a point of a scheme X of finite type over a field.
4. $\operatorname{Spec}(R)$ where R is a complete noetherian local ring.

We shall use the following properties of excellent schemes:

A. If X is excellent integral, then the normalization morphism $\widetilde{X} \to X$ is finite and $\widetilde{X}$ is excellent.

B. An excellent scheme X is catenary, i.e., given irreducible closed subschemes $Z \subset Y \subset X$, all maximal chains of closed irreducible subsets between Z and Y have the same length.

If x is a point of a scheme X, we write $\kappa(x)$ for the residue field of x (and we shall use the standard notation $F(x)$ when X is a scheme over a field F). We write

$\dim x$ for the dimension of the closure $\overline{\{x\}}$ and $X_{(p)}$ for the set of point of X of dimension p.

The word "*scheme*" in this section will mean an excellent scheme of finite dimension.

49.A. Residue homomorphism for local rings. Let R be a 1-dimensional local excellent domain with quotient field L and residue field E. Let $\widetilde{R}$ denote the integral closure of R in L. The ring $\widetilde{R}$ is semilocal, 1-dimensional, and finite as R-algebra. Let $M_1, M_2, \dots, M_n$ be all of the maximal ideals of $\widetilde{R}$. Each localization $\widetilde{R}_{M_i}$ is integrally closed, noetherian and 1-dimensional, hence a DVR. Let v_i denote the discrete valuation of $\widetilde{R}_{M_i}$ and E_i its residue field. The field extension E_i/E is finite. Let K_* denote the Milnor K-groups (cf. §100) and define the *residue homomorphism*

$$\partial_R : K_*(L) \to K_{*-1}(E),$$

by the formula

$$\partial_R = \sum_{i=1}^{n} c_{E_i/E} \circ \partial_{v_i},$$

where

$$\partial_{v_i} : K_*(L) \to K_{*-1}(E_i)$$

is the residue homomorphism associated with the discrete valuation v_i on L (cf. §100.B) and

$$c_{E_i/E} : K_{*-1}(E_i) \to K_{*-1}(E)$$

is the norm homomorphism (cf. §100.E).

Let X be a scheme. For every pair of points $x, x' \in X$, we define a homomorphism

$$\partial^x_{x'} : K_*\big(\kappa(x)\big) \to K_{*-1}\big(\kappa(x')\big)$$

as follows: Let Z be the closure of $\{x\}$ in X considered as an integral closed subscheme (subvariety) of X. If $x' \in Z$ (in this case we say that x' is a *specialization of* x) and $\dim x = \dim x' + 1$, then the local ring $R = \mathcal{O}_{Z,x'}$ is a 1-dimensional excellent local domain with quotient field $\kappa(x)$ and residue field $\kappa(x')$. We set $\partial^x_{x'} = \partial_R$. Otherwise, set $\partial^x_{x'} = 0$.

Lemma 49.1. *Let X be a scheme. For each $x \in X$ and every $\alpha \in K_*\big(\kappa(x)\big)$ the residue $\partial^x_{x'}(\alpha)$ is nontrivial for only finitely many points $x' \in X$.*

PROOF. We may assume that $X = \operatorname{Spec}(A)$ where A is an integrally closed noetherian domain, x the generic point of X and α the symbol $\{a_1, a_2, \dots, a_n\}$ with nonzero $a_i \in A$. For every point $x' \in X$ of codimension 1, let $v_{x'}$ be the corresponding discrete valuation of the quotient field of A. For each i, there is a bijection between the set of all x' satisfying $v_{x'}(a_i) \neq 0$ and the set of minimal prime ideals of the (noetherian) ring A/a_iA and hence this set is finite. Thus, for all but finitely many x' we have $v_{x'}(a_i) = 0$ for all i and therefore $\partial^x_{x'}(\alpha) = 0$. $\square$

It follows from Lemma 49.1 that for a scheme X there is a well-defined endomorphism $d = d_X$ of the direct sum

$$C(X) := \coprod_{x \in X} K_*\big(\kappa(x)\big)$$

such that the (x, x')-component of d is equal to $\partial^x_{x'}$.

Example 49.2. Let $Z \subset X$ be a closed subscheme and let $z_1, z_2, \ldots$ be all of the generic points of Z. We set

$$[Z] := \sum m_i z_i \in \coprod_{x \in X} K_0\big(\kappa(x)\big) \subset C(X),$$

where $m_i = l(\mathcal{O}_{Z,z_i})$ is the length of the local ring $\mathcal{O}_{Z,z_i}$. The element $[Z]$ is called the *cycle of* Z *on* X.

Example 49.3. Let X be a scheme, $x \in X$, and $f \in \kappa(x)^\times$. We view f as an element of $K_1\big(\kappa(x)\big) \subset C(X)$. Then the element

$$d_X(f) \in \coprod_{x \in X} K_0\big(\kappa(x)\big) \subset C(X)$$

is called the *divisor of* f and is denoted by $\operatorname{div}(f)$.

The group $C(X)$ is graded: we write for any $p \geq 0$,

$$C_p(X) := \coprod_{x \in X_{(p)}} K_*\big(\kappa(x)\big).$$

The endomorphism d_X of $C_*(X)$ has degree -1 with respect to this grading. We also set

$$C_{p,n}(X) := \coprod_{x \in X_{(p)}} K_{p+n}\big(\kappa(x)\big),$$

hence $C_p(X)$ is the coproduct of $C_{p,n}(X)$ over all n. Note that the graded group $C_{*,n}(X)$ is invariant under d_X for every n.

Let X be a scheme over a field F. Then the group $C_p(X)$ has a natural structure of a left and right $K_*(F)$-module for all p and d_X is a homomorphism of right $K_*(F)$-modules.

If X is the disjoint union of two schemes X_1 and X_2, we have

(49.4) $$C_*(X) = C_*(X_1) \oplus C_*(X_2)$$

and $d_X = d_{X_1} \oplus d_{X_2}$.

49.B. Multiplication with an invertible function. Let a be an invertible regular function on a scheme X. For every $\alpha \in C_*(X)$, we write $\{a\} \cdot \alpha$ for the element of $C_*(X)$ satisfying

$$(\{a\} \cdot \alpha)_x = \{a(x)\} \cdot \alpha_x$$

for every $x \in X$. We denote by $\{a\}$ the endomorphism of $C_*(X)$ given by $\alpha \mapsto \{a\} \cdot \alpha$.

The product $\alpha \cdot \{a\}$ is defined similarly.

Let $a_1, a_2, \ldots, a_n$ be invertible regular functions on a scheme X. We write $\{a_1, a_2, \ldots, a_n\} \cdot \alpha$ for the product $\{a_1\} \cdot \{a_2\} \cdot \ldots \cdot \{a_n\} \cdot \alpha$ and $\{a_1, a_2, \ldots, a_n\}$ for the endomorphism of $C_*(X)$ given by $\alpha \mapsto \{a_1, a_2, \ldots, a_n\} \cdot \alpha$.

Proposition 49.5. *Let a be an invertible function on a scheme X and $\alpha \in C_*(X)$. Then*

$$d_X\big(\alpha \cdot \{a\}\big) = d_X(\alpha) \cdot \{a\} \quad \textit{and} \quad d_X\big(\{a\} \cdot \alpha\big) = -\{a\} \cdot d_X(\alpha).$$

PROOF. The statement follows from Fact 100.4(1) and the projection formula for the norm map in Proposition 100.8(3). □

By Fact 100.1, it follows that

$$\{a_1, a_2\} = -\{a_2, a_1\} \quad \text{and} \quad \{a_1, a_2\} = 0 \quad \text{if} \quad a_1 + a_2 = 1.$$

49.C. Push-forward homomorphisms. Let $f : X \to Y$ be a morphism of schemes. We define the *push-forward homomorphism*

$$f_* : C_*(X) \to C_*(Y)$$

as follows: Let $x \in X$ and $y \in Y$. If $y = f(x) \in Y$ and the field extension $\kappa(x)/\kappa(y)$ is finite, we set

$$(f_*)^x_y := c_{\kappa(x)/\kappa(y)} : K_*\big(\kappa(x)\big) \to K_*\big(\kappa(y)\big)$$

and set $(f_*)^x_y = 0$ otherwise. It follows from transitivity of the norm map that if $g : Y \to Z$ is another morphism, then $(g \circ f)_* = g_* \circ f_*$.

If either

(1) f is a morphism of schemes of finite type over a field or
(2) f is a finite morphism,

the push-forward f_* is a graded homomorphism of degree 0. Indeed, if $y = f(x)$, then $\dim y = \dim x$ if and only if $\kappa(x)/\kappa(y)$ is a finite extension for all $x \in X$.

If f is a morphism of schemes over a field F, then f_* is a homomorphism of left and right $K_*(F)$-modules.

Example 49.6. If $f : X \to Y$ is a closed embedding, then f_* is a monomorphism satisfying $f_* \circ d_X = d_Y \circ f_*$. Moreover, if in addition f is a bijection on points (e.g., if f is the canonical morphism $Y_{red} \to Y$), then f_* is an isomorphism.

Remark 49.7. Let X be a localization of a scheme Y (e.g., X is an open subscheme of Y) and $f : X \to Y$ the natural morphism. For every point $x \in X$, the natural ring homomorphism $\mathcal{O}_{Y,f(x)} \to \mathcal{O}_{X,x}$ is an isomorphism. It follows from the definitions that for any $x, x' \in X$, we have

$$(f_* \circ d_X)^x_{y'} = (f_*)^{x'}_{y'} \circ (d_X)^x_{x'} = (d_Y)^y_{y'} \circ (f_*)^x_y = (d_Y \circ f_*)^x_{y'}$$

where $y = f(x)$ and $y' = f(x')$. Note that if $y'' \in Y$ does not belong to the image of f, then $(f_* \circ d_X)^x_{y''} = 0$, but in general $(d_Y \circ f_*)^x_{y''}$ may be nonzero.

The following rule is a consequence of the projection formula for Milnor's K-groups (cf. Fact 100.8(3)).

Proposition 49.8. *Let $f : X \to Y$ be a morphism of schemes and a an invertible regular function on Y. Then*

$$f_* \circ \{a'\} = \{a\} \circ f_*$$

where $a' = f^*(a) = a \circ f$.

Proposition 49.9. *Let $f : X \to Y$ be either*

(1) *a proper morphism of schemes of finite type over a field or*
(2) *a finite morphism.*

Then the diagram

$$\begin{array}{ccc} C_p(X) & \xrightarrow{d_X} & C_{p-1}(X) \\ {\scriptstyle f_*}\downarrow & & \downarrow{\scriptstyle f_*} \\ C_p(Y) & \xrightarrow{d_Y} & C_{p-1}(Y) \end{array}$$

is commutative.

PROOF. Let $x \in X_{(p)}$ and $y' \in Y_{(p-1)}$. The (x, y')-component of both compositions in the diagram can be nontrivial only if y' belongs to the closure of the point $y = f(x)$, i.e., if y' is a specialization of y. We have

$$p = \dim x \geq \dim y \geq \dim y' = p - 1,$$

therefore, $\dim y$ can be either equal to p or $p-1$. Note that if f is finite, then $\dim y = p$.

Case 1: $\dim(y) = p$.

In this case, the field extension $\kappa(x)/\kappa(y)$ is finite. Replacing X by the closure of $\{x\}$ and Y by the closure of $\{y\}$ we may assume that x and y are the generic points of X and Y respectively.

First suppose that X and Y are normal. Since the morphism f is proper, the points $x' \in X_{p-1}$ satisfying $f(x') = y'$ are in a bijective correspondence with the extensions of the valuation $v_{y'}$ of the field $\kappa(y)$ to the field $\kappa(x)$. Hence by Fact 100.8(4),

$$\begin{aligned}
(d_Y \circ f_*)^x_{y'} &= \partial^y_{y'} \circ c_{\kappa(x)/\kappa(y)} \\
&= \sum_{f(x')=y'} c_{\kappa(x')/\kappa(y')} \circ \partial^x_{x'} \\
&= \sum_{f(x')=y'} (f_*)^{x'}_{y'} \circ (d_X)^x_{x'} \\
&= (f_* \circ d_X)^x_{y'}.
\end{aligned}$$

In the general case let $g : \widetilde{X} \to X$ and $h : \widetilde{Y} \to Y$ be the normalizations and let $\tilde{x}$ and $\tilde{y}$ be the generic points of $\widetilde{X}$ and $\widetilde{Y}$ respectively. Note that $\kappa(\tilde{x}) \simeq \kappa(x)$ and $\kappa(\tilde{y}) \simeq \kappa(y)$. There is a natural morphism $\tilde{f} : \widetilde{X} \to \widetilde{Y}$ over f.

Consider the following diagram:

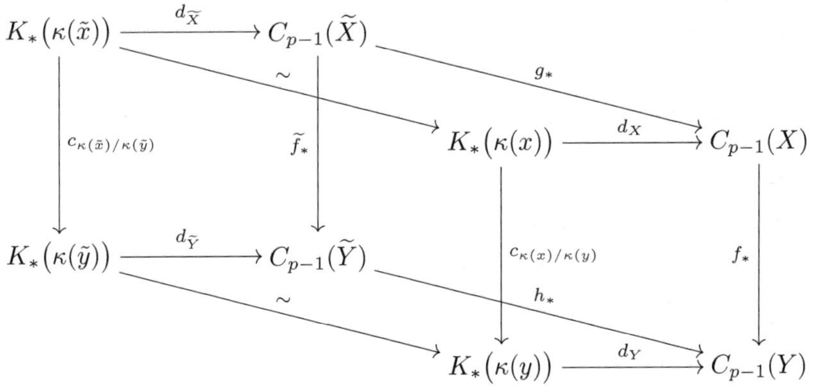

By the first part of the proof, the back face of the diagram is commutative. The left face is obviously commutative. The right face is commutative by functoriality of the push-forward. The upper and the bottom faces are commutative by definition of the maps d_X and d_Y. Hence the front face is also commutative, i.e., the (x, y')-components of the compositions $f_* \circ d_X$ and $d_Y \circ f_*$ coincide.

Note that we have proved the proposition in the case when f is finite. Before proceeding to Case 2, as a corollary we also deduce

Theorem 49.10 (Weil's Reciprocity Law). *Let X be a complete integral curve over a field F. Then the composition*

$$K_{*+1}\big(F(X)\big) \xrightarrow{d_X} \coprod_{x \in X_{(0)}} K_*\big(F(x)\big) \xrightarrow{\sum c_{\kappa(x)/F}} K_*(F)$$

is trivial.

PROOF. The case $X = \mathbb{P}^1_F$ follows from Theorem 100.7. The general case can be reduced to the case of the projective line as follows. Let f be a nonconstant rational function on X. We view f as a finite morphism $f : X \to \mathbb{P}^1_F$ over F. By the first case of the proof of Proposition 49.9, the left square of the diagram

$$\begin{array}{ccccc} K_{*+1}\big(F(X)\big) & \xrightarrow{d_X} & \coprod_{x \in X_{(0)}} K_*\big(F(x)\big) & \xrightarrow{\sum c_{\kappa(x)/F}} & K_*(F) \\ \downarrow{\scriptstyle f_*} & & \downarrow{\scriptstyle f_*} & & \| \\ K_{*+1}\big(F(\mathbb{P}^1)\big) & \xrightarrow{d_{\mathbb{P}^1}} & \coprod_{y \in \mathbb{P}^1_{(0)}} K_*\big(F(y)\big) & \xrightarrow{\sum c_{\kappa(y)/F}} & K_*(F) \end{array}$$

is commutative. The right square is commutative by the transitivity property of the norm map (cf. Fact 100.8(1)). Finally, the statement of the theorem follows from the commutativity of the diagram. □

Weil's Reciprocity Law can be reformulated as follows:

Corollary 49.11. *Proposition* 49.9 *holds for the structure morphism* $X \to \operatorname{Spec}(F)$.

We return to the proof of Proposition 49.9.

Case 2: $\dim(y) = p - 1$.

In this case $y' = y$. We replace Y by $\operatorname{Spec}\big(\kappa(y)\big)$ and X by the fiber $X \times_Y \operatorname{Spec}\big(\kappa(y)\big)$ of f over y. We can further replace X by the closure of x in X. Thus, X is a proper integral curve over the field $\kappa(y)$ and the result follows from Corollary 49.11. □

49.D. Pull-back homomorphisms. Let $g : Y \to X$ be a flat morphism of schemes. We say that g is of *relative dimension* d if for every $x \in X$ in the image of g and for every generic point y of $g^{-1}(\overline{\{x\}})$ we have $\dim y = \dim x + d$.

In this book *all flat morphisms will be assumed of constant relative dimension.*

Let $g : Y \to X$ be a flat morphism of relative dimension d. For every point $x \in X$, let Y_x denote the fiber scheme

$$Y \times_X \operatorname{Spec}\big(\kappa(x)\big)$$

over $\kappa(x)$. We identify the underlying topological space of Y_x with a subspace of Y.

The following statement is a direct consequence of the going-down theorem [**99**, Ch. 1, Th. 4].

Lemma 49.12. *For every $x \in X$ we have:*

(1) $\dim y \leq \dim x + d$ *for every* $y \in Y_x$.
(2) *A point* $y \in Y_x$ *is generic in* Y_x *if and only if* $\dim y = \dim x + d$.

If y is a generic point of Y_x, the local ring $\mathcal{O}_{Y_x,y}$ is noetherian 0-dimensional, hence is artinian. We define the *ramification index of* y by

$$e_y(g) := l(\mathcal{O}_{Y_x,y}),$$

where l denotes the length (cf. §102).

The *pull-back homomorphism*

$$g^* : C_*(X) \to C_{*+d}(Y)$$

is defined as follows: Let $x \in X$ and $y \in Y$. If $g(y) = x$ and y is a generic point of Y_x, we set

$$(g^*)^x_y := e_y(g) \cdot r_{\kappa(y)/\kappa(x)} : K_*\big(\kappa(x)\big) \to K_*\big(\kappa(y)\big)$$

where $r_{\kappa(y)/\kappa(x)}$ is the restriction homomorphism (cf. §100.A) and $(g_*)^x_y = 0$ otherwise.

Suppose that Z is of pure dimension d over a field F. The structure morphism $p : Z \to \operatorname{Spec}(F)$ is flat of relative dimension d. The image of the identity under the composition

$$p^* : \mathbb{Z} = K_0(F) = C_{0,0}(\operatorname{Spec} F) \xrightarrow{p^*} C_{d,-d}(Z) \xrightarrow{i_*} C_{d,-d}(X),$$

where $i : Z \to X$ is the closed embedding, is equal to the cycle $[Z]$ of Z.

Example 49.13. Let $p : E \to X$ be a vector bundle of rank r. Then p is a flat morphism of relative dimension r and $p^*\big([X]\big) = [E]$.

Example 49.14. Let X be a scheme of finite type over F and let L/F be an arbitrary field extension. The natural morphism $g : X_L \to X$ is flat of relative dimension 0. The pull-back homomorphism

$$g^* : C_p(X) \to C_p(X_L)$$

is called the *change of field homomorphism*.

Example 49.15. An open embedding $j : U \to X$ is a flat morphism of relative dimension 0. The pull-back homomorphism

$$j^* : C_p(X) \to C_p(U)$$

is called the *restriction homomorphism*.

The following proposition is an immediate consequence of the definitions.

Proposition 49.16. *Let $g : Y \to X$ be a flat morphism and a an invertible function on X. Then*

$$g^* \circ \{a\} = \{a'\} \circ g^*,$$

where $a' = g^*(a) = a \circ g$.

Let g be a morphism of schemes over a field F. It follows from Proposition 49.16 that g^* is a homomorphism of left and right $K_*(F)$-modules.

Let $g : Y \to X$ and $h : Z \to Y$ be flat morphisms. Let $z \in Z$ and $y = h(z)$, $x = g(y)$. It follows from Lemma 49.12 that z is a generic point of Z_x if and only if z is a generic point of Z_y and y is a generic point of Y_x.

Lemma 49.17. *Let z be a generic point of Z_x. Then $e_z(g \circ h) = e_z(h) \cdot e_y(g)$.*

PROOF. The statement follows from Corollary 102.2 with $B = \mathcal{O}_{Y_x,y}$ and $C = \mathcal{O}_{Z_x,z}$. Note that $C/QC = \mathcal{O}_{Z_y,z}$ where Q is the maximal ideal of B. □

Proposition 49.18. *Let $g : Y \to X$ and $h : Z \to Y$ be flat morphisms of constant relative dimension. Then $(g \circ h)^* = h^* \circ g^*$.*

PROOF. Let $x \in X$ and $z \in Z$. We compute the (z, x)-components of both sides of the equality. We may assume that $x = (g \circ h)(z)$. Let $y = h(z)$. By Lemma 49.17, we have

$$\begin{aligned}\big((g \circ h)^*\big)_z^x &= e_z(g \circ h) \cdot r_{\kappa(z)/\kappa(x)} \\ &= e_z(h) \cdot e_y(g) \cdot r_{\kappa(z)/\kappa(y)} \circ r_{\kappa(y)/\kappa(x)} \\ &= (h^*)_z^y \circ (g^*)_y^x \\ &= (h^* \circ g^*)_z^x.\end{aligned}$$

□

Next consider the fiber product diagram

(49.19)
$$\begin{array}{ccc} X' & \xrightarrow{g'} & X \\ {\scriptstyle f'}\downarrow & & \downarrow{\scriptstyle f} \\ Y' & \xrightarrow{g} & Y. \end{array}$$

Proposition 49.20. *Let g and g' in (49.19) be flat morphisms of relative dimension d. Suppose that either*

(1) *f is a morphism of schemes of finite type over a field or*
(2) *f is a finite morphism.*

Then the diagram

$$\begin{array}{ccc} C_p(X) & \xrightarrow{g'^*} & C_{p+d}(X') \\ {\scriptstyle f_*}\downarrow & & \downarrow{\scriptstyle f'_*} \\ C_p(Y) & \xrightarrow{g^*} & C_{p+d}(Y') \end{array}$$

is commutative.

PROOF. Let $x \in X_{(p)}$ and $y' \in Y'_{(p+d)}$. We shall compare the (x, y')-components of both compositions in the diagram. These components are trivial unless $g(y') = f(x)$. Denote this point by y. By Lemma 49.12,

$$p + d = \dim y' \le \dim y + d \le \dim x + d = p + d,$$

hence $\dim y = \dim x = p$ and y' is a generic point of Y'_y. In particular, the field extension $\kappa(x)/\kappa(y)$ is finite.

Let S be the set of all $x' \in X'$ satisfying $f'(x') = y'$ and $g'(x') = x$. Again by Lemma 49.12,

$$p + d = \dim y' \le \dim x' \le \dim x + d = p + d,$$

hence $\dim x' = \dim y' = p+d$ and x' is a generic point of X'_x. In particular, the field extension $\kappa(x')/\kappa(y')$ is finite. The set S is in a natural bijective correspondence with the finite set $\operatorname{Spec}\big(\kappa(y') \otimes_{\kappa(y)} \kappa(x)\big)$.

The local ring $C = \mathcal{O}_{X'_x, x'}$ is a localization of the ring $\mathcal{O}_{Y'_y, y'} \otimes_{\kappa(y)} \kappa(x)$ and hence is flat over $B = \mathcal{O}_{Y'_y, y'}$. Let Q be the maximal ideal of B. The factor ring C/QC is the localization of the tensor product $\kappa(y') \otimes_{\kappa(y)} \kappa(x)$ at the prime ideal corresponding to x'. Denote by $l_{x'}$ the length of C/QC.

By Corollary 102.2,

$$e_{x'}(g') = l_{x'} \cdot e_{y'}(g) \tag{49.21}$$

for every $x' \in S$. It follows from (49.21) and Fact 100.8(5) that

$$\begin{aligned}
(f'_* \circ g'^*)^x_{y'} &= \sum_{x' \in S} (f'_*)^{x'}_{y'} \circ (g'^*)^x_{x'} \\
&= \sum_{x' \in S} e_{x'}(g') \cdot c_{\kappa(x')/\kappa(y')} \circ r_{\kappa(x')/\kappa(x)} \\
&= e_{y'}(g) \cdot \sum_{x' \in S} l_{x'} \cdot c_{\kappa(x')/\kappa(y')} \circ r_{\kappa(x')/\kappa(x)} \\
&= e_{y'}(g) \cdot r_{\kappa(y')/\kappa(y)} \circ c_{\kappa(x)/\kappa(y)} \\
&= (g^*)^y_{y'} \circ (f_*)^x_y \\
&= (g^* \circ f_*)^x_{y'}.
\end{aligned}$$

□

Remark 49.22. It follows from the definitions that Proposition 49.20 holds also for arbitrary f if Y' is a localization of Y (cf. Remark 49.7).

Proposition 49.23. *Let $g : Y \to X$ be a flat morphism of relative dimension d. Then the diagram*

$$\begin{array}{ccc}
C_p(X) & \xrightarrow{d_X} & C_{p-1}(X) \\
{\scriptstyle g^*}\downarrow & & \downarrow{\scriptstyle g^*} \\
C_{p+d}(Y) & \xrightarrow{d_Y} & C_{p+d-1}(Y)
\end{array}$$

is commutative.

PROOF. Let $x \in X_{(p)}$ and $y' \in Y_{(p+d-1)}$. We compare the (x, y')-components of both compositions in the diagram. Let $y_1, \dots, y_k$ be all generic points of $Y_x \subset Y$ satisfying $y' \in \overline{\{y_i\}}$. We have

$$(d_Y \circ g^*)^x_{y'} = \sum_{i=1}^k (d_Y)^{y_i}_{y'} \circ (g^*)^x_{y_i} = \sum_{i=1}^k e_{y_i}(g) \cdot (d_Y)^{y_i}_{y'} \circ r_{\kappa(y_i)/\kappa(x)}. \tag{49.24}$$

Set $x' = g(y')$. If $x' \notin \overline{\{x\}}$, then both components $(g^* \circ d_X)^x_{y'}$ and $(d_Y \circ g^*)^x_{y'}$ are trivial.

Suppose $x' \in \overline{\{x\}}$. We have

$$p = \dim x \geq \dim x' \geq \dim y' - d = p - 1.$$

Therefore, $\dim x'$ is either p or $p-1$.

Case 1: $\dim(x') = p$, i.e., $x' = x$.

The component $(g^* \circ d_X)^x_{y'}$ is trivial since $(g^*)^{\tilde{x}}_{y'} = 0$ for every $\tilde{x} \neq x'$. By assumption, every discrete valuation of $\kappa(y_i)$ with center y' is trivial on $\kappa(x)$. Therefore, the map $(d_Y)^{y_i}_{y'}$ is trivial on the image of $r_{\kappa(y_i)/\kappa(x)}$. It follows from formula (49.24) that $(d_Y \circ g^*)^x_{y'} = 0$.

Case 2: $\dim(x') = p - 1$.

We have y' is a generic point of $Y_{x'}$ and

$$(g^* \circ d_X)^x_{y'} = (g^*)^{x'}_{y'} \circ (d_X)^x_{x'} = e_{y'}(g) \cdot r_{\kappa(y')/\kappa(x')} \circ \partial^x_{x'}. \tag{49.25}$$

Replacing X by $\overline{\{x\}}$ and Y by $g^{-1}(\overline{\{x\}})$, we may assume that $X = \overline{\{x\}}$. By Propositions 49.9 and 49.20, we can replace X by its normalization $\widetilde{X}$ and Y by the fiber product $Y \times_X \widetilde{X}$, so we may assume that X is normal.

Let $Y_1, \dots, Y_k$ be all irreducible components of Y containing y', so that y_i is the generic point of Y_i for all i. Let $\widetilde{Y}_i$ be the normalization of Y_i and let $\tilde{y}_i$ be the generic points of $\widetilde{Y}_i$. We have $\kappa(\tilde{y}_i) = \kappa(y_i)$. Let t be a prime element of the discrete valuation ring $R = \mathcal{O}_{X,x'}$.

The local ring $A = \mathcal{O}_{Y,y'}$ is 1-dimensional; its minimal prime ideals are in bijective correspondence with the set of points $y_1, \dots, y_k$.

Fix $i \in [1,\ k]$. We write A_i for the factor ring of A by the corresponding minimal prime ideal. Since A is flat over R, the prime element t is not a zero divisor in A, hence the image of t in A_i is not zero for every i. Let $\widetilde{A}_i$ be the normalization of the ring A_i.

Let S_i be the set of all points $w \in \widetilde{Y}_i$ such that $g(w) = x'$. There is a natural bijection between S_i and the set of all maximal ideals of $\widetilde{A}_i$. Moreover, if Q is a maximal ideal of $\widetilde{A}_i$ corresponding to a point $w \in S_i$, then the local ring $\mathcal{O}_{\widetilde{Y}_i,w}$ coincides with the localization of $\widetilde{A}_i$ with respect to Q.

Denote by $l_{i,w}$ the length of the ring $\mathcal{O}_{\widetilde{Y}_i,w}/t\mathcal{O}_{\widetilde{Y}_i,w}$. Applying Lemma 102.3 to the A-algebra $\widetilde{A}_i$ and $M = \widetilde{A}_i/t\widetilde{A}_i$, we have

$$l_A\big(\widetilde{A}_i/t\widetilde{A}_i\big) = \sum_{w \in S_i} l_{i,w} \cdot \big[\kappa(w) : \kappa(y')\big]. \tag{49.26}$$

On the other hand, $l_{i,w}$ is the ramification index of the discrete valuation ring $\mathcal{O}_{\widetilde{Y}_i,w}$ over R. It follows from Fact 100.4(2) that

$$\partial_w^{\tilde{y}_i} \circ r_{\kappa(y_i)/\kappa(x)} = l_{i,w} \cdot r_{\kappa(w)/\kappa(x')} \circ \partial_{x'}^{x} \tag{49.27}$$

for every $w \in S_i$.

By (49.26), (49.27), and Fact 100.8(3), for every i, we have

$$\begin{aligned}
(d_Y)_{y'}^{y_i} \circ r_{\kappa(y_i)/\kappa(x)} &= \sum c_{\kappa(w)/\kappa(y')} \circ \partial_w^{\tilde{y}_i} \circ r_{\kappa(y_i)/\kappa(x)} \\
&= \sum c_{\kappa(w)/\kappa(y')} \cdot l_{i,w} \cdot r_{\kappa(w)/\kappa(x')} \circ \partial_{x'}^{x} \\
&= \sum l_{i,w} \cdot c_{\kappa(w)/\kappa(y')} \circ r_{\kappa(w)/\kappa(y')} \circ r_{\kappa(y')/\kappa(x')} \circ \partial_{x'}^{x} \\
&= \sum l_{i,w} \cdot \big[\kappa(w) : \kappa(y')\big] \cdot r_{\kappa(y')/\kappa(x')} \circ \partial_{x'}^{x} \\
&= l_A\big(\widetilde{A}_i/t\widetilde{A}_i\big) \cdot r_{\kappa(y')/\kappa(x')} \circ \partial_{x'}^{x}
\end{aligned}$$

(where all summations are taken over all $w \in S_i$).

The factor A-module $\widetilde{A}_i/A_i$ is of finite length, hence by Lemma 102.4, we have $h(t, A_i) = h(t, \widetilde{A}_i)$ where h is the Herbrand index. Since t is not a zero divisor in either A_i or in $\widetilde{A}_i$, we have $l_A(\widetilde{A}_i/t\widetilde{A}_i) = l_A(A_i/tA_i) = l(A_i/tA_i)$. Therefore,

$$(d_Y)_{y'}^{y_i} \circ r_{\kappa(y_i)/\kappa(x)} = l(A_i/tA_i) \cdot r_{\kappa(y')/\kappa(x')} \circ \partial_{x'}^{x}. \tag{49.28}$$

The local ring $\mathcal{O}_{Y_x,y_i} = \mathcal{O}_{Y,y_i}$ is the localization of A with respect to the minimal prime ideal corresponding to y_i. The ring $\mathcal{O}_{Y_{x'},y'}$ is canonically isomorphic to A/tA.

Applying Lemma 102.5 to the ring A and the module $M = A$ we get the equality

$$e_{y'}(g) = h(t, A) = \sum_{i=1}^{k} l(\mathcal{O}_{Y_x, y_i}) \cdot l(A_i/tA_i) = \sum_{i=1}^{k} e_{y_i}(g) \cdot l(A_i/tA_i). \tag{49.29}$$

It follows from (49.25), (49.28), and (49.29) that

$$\begin{aligned}(d_Y \circ g^*)^x_{y'} &= \sum_{i=1}^{k} e_{y_i}(g) \cdot (d_Y)^{y_i}_{y'} \circ r_{\kappa(y_i)/\kappa(x)} \\ &= \sum_{i=1}^{k} e_{y_i}(g) \cdot l(A_i/tA_i) \cdot r_{\kappa(y')/\kappa(x')} \circ \partial^x_{x'} \\ &= e_{y'}(g) \cdot r_{\kappa(y')/\kappa(x')} \circ \partial^x_{x'} \\ &= (g^* \circ d_X)^x_{y'}.\end{aligned}$$

□

The following proposition was proven by Kato in [**79**].

Proposition 49.30. *For every scheme X over a field, the map d_X is a differential of $C_*(X)$, i.e., $(d_X)^2 = 0$.*

PROOF. We will prove the statement in several steps.

Step 1: $X = \operatorname{Spec}(R)$, where $R = F[[s,t]]$ and F is a field.

A polynomial $t^n + a_1 t^{n-1} + a_2 t^{n-2} + \cdots + a_n$ over the ring $F[[s]]$ is called *marked* if $a_i \in sF[[s]]$ for all i. We shall use the following properties of marked polynomials derived from the Weierstrass Preparation Theorem [**18**, Ch.VII, §3, no. 8]:

A. Every height 1 ideal of the ring R is either equal to sR or is generated by a unique marked polynomial.

B. A marked polynomial f is irreducible in R if and only if f is irreducible in $F((s))[t]$.

It follows that the multiplicative group $F((s,t))^\times$ is generated by $R^\times, s, t$ and the set H of all power series of the form $t^{-n} \cdot f$ where f is a marked polynomial of degree n.

If $r \in R^\times$ and $\alpha \in K_*\big(F((s,t))\big)$, then by Proposition 49.5,

$$(d_X)^2(\{r\} \cdot \alpha) = -d_X\big(\{\bar{r}\} \cdot d_X(\alpha)\big) = \{\bar{r}\} \cdot (d_X)^2(\alpha),$$

where $\bar{r} \in F$ is the residue of r. Thus it suffices to prove the following:

(i) $(d_X)^2\big(\{s,t\}\big) = 0$,

(ii) $(d_X)^2\big(\{f, g_1, \dots, g_n\}\big) = 0$ where $f \in H$ and all g_i belong to the subgroup generated by s, t and H.

For every point $x \in X_{(1)}$ set $\partial_x = \partial^y_x$, where y is the generic point of X and $\partial^x = \partial^x_z$, where z is the closed point of X. Thus,

$$\big((d_X)^2\big)^y_z = \sum_{x \in X_{(1)}} \partial^x \circ \partial_x : K_*\big(F((s,t))\big) \to K_{*-2}(F).$$

To prove (i) let x_s and x_t be the points of $X_{(1)}$ given by the ideals sR and tR, respectively. We have

$$\sum_{x \in X_{(1)}} \partial^x \circ \partial_x\big(\{s,t\}\big) = \partial^{x_s}\big(\{t\}\big) - \partial^{x_t}\big(\{s\}\big) = 1 - 1 = 0.$$

To prove (ii) consider the field $L = F((s))$ and the natural morphism

$$h : X' = \operatorname{Spec}\big(R[s^{-1}]\big) \to \operatorname{Spec}\big(L[t]\big) = \mathbb{A}^1_L.$$

By the properties of marked polynomials, the map h identifies the set $X'_{(0)} = X_{(1)} - \{x_s\}$ with the subset of the closed points of $\mathbb{A}^1_L$ given by irreducible marked polynomials. For every $x \in X'$ we write $\bar{x}$ for the point $h(x) \in \mathbb{A}^1_L$. Note that for $x \in X'_{(0)} = X_{(1)} - \{x_s\}$, the residue fields $\kappa(x)$ and $L(\bar{x})$ are canonically isomorphic. In particular, the field $\kappa(x)$ can be viewed as a finite extension of L. By Fact 100.8(4), we have $\partial^x = \partial \circ c_{\kappa(x)/L}$, where $\partial : K_*(L) \to K_{*-1}(F)$ is given by the canonical discrete valuation of L.

Let $x \in X'_{(0)} = X_{(1)} - \{x_s\}$. We write $\partial_{\bar{x}}$ for $\partial^{\bar{y}}_{\bar{x}}$. Under the identification of $\kappa(x)$ with $L(\bar{x})$, we have $\partial_{\bar{x}} = \partial_x \circ i$ where $i : K_*\big(L(t)\big) \to K_*\big(F((s,t))\big)$ is the canonical homomorphism. Therefore,

$$\begin{aligned}\sum_{x \in X_{(1)}} \partial^x \circ \partial_x \circ i &= \partial^{x_s} \circ \partial_{x_s} \circ i + \partial \circ \sum_{x \in X'_{(0)}} c_{\kappa(x)/L} \circ \partial_x \circ i \\ &= \partial^{x_s} \circ \partial_{x_s} \circ i + \partial \circ \sum_{x \in X'_{(0)}} c_{L(\bar{x})/L} \circ \partial_{\bar{x}}.\end{aligned}$$

Let $\alpha = \{f, g_1, \dots, g_n\} \in K_{n+1}\big(L(t)\big)$ with f and g_i as in (ii). Note that the divisors in $\mathbb{A}^1_L$ of the functions f and g_i are supported in the image of h. Hence $\partial_p(\alpha) = 0$ for every closed point $\mathbb{A}^1_L$ not in the image of h. Moreover, for the point q of $\mathbb{P}^1_L$ at infinity, $f(q) = 1$ and therefore, $\partial_q(\alpha) = 0$. Hence, by Weil's Reciprocity Law 49.10, applied to $\mathbb{P}^1_L$,

$$\sum_{x \in X'_{(0)}} c_{L(\bar{x})/L} \circ \partial_{\bar{x}}(\alpha) = \sum_{p \in \mathbb{P}^1_L} c_{L(p)/L} \circ \partial_p(\alpha) = 0.$$

Notice also that $f(x_s) = 1$, hence $\partial_{x_s} \circ i(\alpha) = 0$ and therefore,

$$(d_X)^2\big(\{f, g_1, \dots, g_n\}\big) = \sum_{x \in X_{(1)}} \partial^x \circ \partial_x \circ i(\alpha) = 0.$$

Step 2: $X = \operatorname{Spec}(S)$, where S is a (noetherian) local complete 2-dimensional ring containing a field.

Let $M \subset S$ be the maximal ideal. By Cohen's theorem [**140**, Ch. VIII, Th.27], there is a subfield $F \subset S$ such that the natural ring homomorphism $F \to S/M$ is an isomorphism.

Choose local parameters $s, t \in M$ and consider the subring $R = F[[s,t]] \subset S$. Denote by P the maximal ideal of R. There is an integer r such that $M^r \subset PS$. We claim that the R-algebra S is finite. Indeed, note that

$$\bigcap_{n>0} P^n S \subset \bigcap_{n>0} M^n = 0.$$

Since S/M^r is of finite length and there is a natural surjection $S/M^r \to S/PS$, the ring S/PS is a finitely generated R/P-module. As the ring R is complete, S is a finitely generated R-module as claimed.

It follows from the claim that the natural morphism $f : X \to Y = \operatorname{Spec}(R)$ is finite. By Proposition 49.9 and Step 1,

$$f_* \circ (d_X)^2 = (d_Y)^2 \circ f_* = 0.$$

The rings R and S have isomorphic residue fields, hence $(d_X)^2 = 0$.

Step 3: $X = \operatorname{Spec}(S)$ where S is a 2-dimensional (noetherian) local ring containing a field.

Let $\widehat{S}$ be the completion of S. The natural morphism $f : Y = \operatorname{Spec}(\widehat{S}) \to X$ is flat of relative dimension 0. By Proposition 49.23 and Step 2,

$$g^* \circ (d_X)^2 = (d_Y)^2 \circ g^* = 0.$$

The rings $\widehat{S}$ and S have isomorphic residue fields, hence $(d_X)^2 = 0$.

Step 4: X is a scheme over a field.

Let x and x' be two points of X such x' is of codimension 2 in $\overline{\{x\}}$. We need to show that the (x, x')-component of $(d_X)^2$ is trivial. We may assume that $X = \overline{\{x\}}$. The ring $S = \mathcal{O}_{X,x'}$ is local 2-dimensional. The natural morphism $f : Y = \operatorname{Spec}(S) \to X$ is flat of constant relative dimension. By Proposition 49.23 and Step 3,

$$f^* \circ (d_X)^2 = (d_Y)^2 \circ f^* = 0.$$

The field $\kappa(x')$ and the residue field of S are isomorphic, therefore, the (x, x')-component of $(d_X)^2$ is trivial. □

Definition 49.31. Let X be a scheme. The complex $\big(C_*(X), d_X\big)$ is called the *Rost complex of* X.

49.E. Boundary map. Let X be a scheme of finite type over a field and $Z \subset X$ a closed subscheme. Set $U = X \setminus Z$. For every $p \geq 0$, the set $X_{(p)}$ is the disjoint union of $Z_{(p)}$ and $U_{(p)}$, hence

$$C_p(X) = C_p(Z) \oplus C_p(U).$$

Consider the closed embedding $i : Z \to X$ and the open immersion $j : U \to X$. The sequence of complexes

$$0 \to C_*(Z) \xrightarrow{i_*} C_*(X) \xrightarrow{j^*} C_*(U) \to 0$$

is exact. This sequence is not split in general as a sequence of complexes, but it splits canonically termwise. Let $v : C_*(U) \to C_*(X)$ and $w : C_*(X) \to C_*(Z)$ be the canonical inclusion and projection. Note that v and w do not commute with the differentials in general. We have $j^* \circ v = \mathrm{id}$ and $w \circ i_* = \mathrm{id}$.

We define the *boundary map*

$$\partial_Z^U : C_p(U) \to C_{p-1}(Z)$$

by $\partial_Z^U := w \circ d_X \circ v$.

Example 49.32. Let $X = \mathbb{A}_F^1$, $Z = \{0\}$, and $U = \mathbb{G}_m := \mathbb{A}_F^1 \setminus \{0\}$. Then

$$\partial_Z^U\big(\{t\} \cdot [U]\big) = [Z],$$

where t is the coordinate function on $\mathbb{A}_F^1$.

Proposition 49.33. *Let X be a scheme and $Z \subset X$ a closed subscheme. Set $U = X \setminus Z$. Then $d_Z \circ \partial_Z^U = -\partial_Z^U \circ d_U$.*

PROOF. By the definition of $\partial = \partial_Z^U$, we have $i_* \circ \partial = d_X \circ v - v \circ d_U$. Hence by Propositions 49.9 and 49.30,

$$\begin{aligned} i_* \circ d_Z \circ \partial &= d_X \circ i_* \circ \partial \\ &= d_X \circ (d_X \circ v - v \circ d_U) \\ &= -\, d_X \circ v \circ d_U \\ &= (v \circ d_U - d_X \circ v) \circ d_U \\ &= -\, i_* \circ \partial \circ d_U . \end{aligned}$$

Since i_* is injective, we have $d_Z \circ \partial = -\partial \circ d_U$. □

Proposition 49.34. *Let a be an invertible function on X and let a', a'' be the restrictions of a on U and Z, respectively. Then*

$$\partial_Z^U\big(\alpha \cdot \{a'\}\big) = \partial_Z^U(\alpha) \cdot \{a''\} \qquad \textit{and} \qquad \partial_Z^U\big(\{a'\} \cdot \alpha\big) = -\{a''\} \cdot \partial_Z^U(\alpha)$$

for every $\alpha \in C_*(U)$.

PROOF. The homomorphisms v and w commute with the products. The statement follows from Proposition 49.5. □

Let

$$\begin{array}{ccccc} Z' & \xrightarrow{\ i'\ } & X' & \xleftarrow{\ j'\ } & U' \\ {\scriptstyle g}\downarrow & & {\scriptstyle f}\downarrow & & {\scriptstyle h}\downarrow \\ Z & \xrightarrow{\ i\ } & X & \xleftarrow{\ j\ } & U \end{array} \tag{49.35}$$

be a commutative diagram with i and i' closed embeddings, j and j' open embeddings and $U = X \setminus Z,\ U' = X' \setminus Z'$.

Proposition 49.36. *Suppose that we have the diagram* (49.35).

(1) *If f, g and h are proper morphisms of schemes of finite type over a field, then the diagram*

$$\begin{array}{ccc} C_p(U') & \xrightarrow{\ \partial_{Z'}^{U'}\ } & C_{p-1}(Z') \\ {\scriptstyle h_*}\downarrow & & \downarrow{\scriptstyle g_*} \\ C_p(U) & \xrightarrow{\ \partial_Z^U\ } & C_{p-1}(Z) \end{array}$$

is commutative.

(2) *Suppose that both squares in the diagram* (49.35) *are fiber squares. If f is flat of constant relative dimension d, then so are g and h and the diagram*

$$\begin{array}{ccc} C_p(U) & \xrightarrow{\ \partial_Z^U\ } & C_{p-1}(Z) \\ {\scriptstyle h^*}\downarrow & & \downarrow{\scriptstyle g^*} \\ C_{p+d}(U') & \xrightarrow{\ \partial_{Z'}^{U'}\ } & C_{p+d-1}(Z') \end{array}$$

is commutative.

PROOF. (1) Consider the diagram

$$\begin{array}{ccccccc}
C_p(U') & \xrightarrow{v'} & C_p(X') & \xrightarrow{d_{X'}} & C_{p-1}(X') & \xrightarrow{w'} & C_{p-1}(Z') \\
{\scriptstyle h_*}\downarrow & & {\scriptstyle f_*}\downarrow & & \downarrow{\scriptstyle f_*} & & \downarrow{\scriptstyle g_*} \\
C_p(U) & \xrightarrow{v} & C_p(X) & \xrightarrow{d_X} & C_{p-1}(X) & \xrightarrow{w} & C_{p-1}(Z).
\end{array}$$

The left and the right squares are commutative by the local nature of the definition of the push-forward homomorphisms. The middle square is commutative by Proposition 49.9.

The proof of (2) is similar — one uses Proposition 49.23. As both squares of the diagram are fiber squares, for any point $z \in Z$ (respectively, $u \in U$), the fibers Z'_z and $X'_{i(z)}$ (respectively, U'_u and $X'_{j(u)}$) are naturally isomorphic. □

Let Z_1 and Z_2 be closed subschemes of a scheme X. Set

$$T_1 = Z_1 \setminus Z_2, \quad T_2 = Z_2 \setminus Z_1, \quad U_i = X \setminus Z_i, \quad U = U_1 \cap U_2, \quad Z = Z_1 \cap Z_2.$$

We have the following fiber product diagram of open and closed embeddings:

$$\begin{array}{ccccc}
Z & \longrightarrow & Z_2 & \longleftarrow & T_2 \\
\downarrow & & \downarrow & & \downarrow \\
Z_1 & \longrightarrow & X & \longleftarrow & U_1 \\
\uparrow & & \uparrow & & \uparrow \\
T_1 & \longrightarrow & U_2 & \longleftarrow & U.
\end{array}$$

Denote by $\partial_t, \partial_b, \partial_l, \partial_r$ the boundary homomorphisms for the top, bottom, left and right triples of the diagram respectively.

Proposition 49.37. *The morphism*

$$\partial_l \circ \partial_b + \partial_t \circ \partial_r : C_*(U) \to C_{*-2}(Z)$$

is homotopic to zero.

PROOF. The differential of $C_*(X)$ relative to the decomposition

$$C_*(X) = C_*(U) \oplus C_*(T_1) \oplus C_*(T_2) \oplus C_*(Z)$$

is given by the matrix

$$d_X = \begin{pmatrix} d_U & * & * & * \\ \partial_b & * & * & * \\ \partial_r & * & * & * \\ h & \partial_l & \partial_t & d_Z \end{pmatrix}$$

where $h : C_*(U) \to C_{*-1}(Z)$ is some morphism. The equality $(d_X)^2 = 0$ gives

$$h \circ d_U + d_Z \circ h + \partial_l \circ \partial_b + \partial_t \circ \partial_r = 0.$$

In other words, $-h$ is a contracting homotopy for $\partial_l \circ \partial_b + \partial_t \circ \partial_r$. □

50. External products

From now on the word "scheme" means a *separated scheme of finite type over a field.*

Let X and Y be two schemes over F. We define the *external product*

$$C_p(X) \times C_q(Y) \to C_{p+q}(X \times Y), \quad (\alpha, \beta) \mapsto \alpha \times \beta$$

as follows: For a point $v \in (X \times Y)_{(p+q)}$, we set $(\alpha \times \beta)_v = 0$ unless the point v projects to a point x in $X_{(p)}$ and y in $Y_{(q)}$. In the latter case

$$(\alpha \times \beta)_v = l_v \cdot r_{F(v)/F(x)}(\alpha_x) \cdot r_{F(v)/F(y)}(\beta_y),$$

where l_v is the length of the artinian local ring of v on $\operatorname{Spec}(F(x)) \times \operatorname{Spec}(F(y))$.

The external product is graded symmetric with respect to X and Y. More precisely, if $\alpha \in C_{p,n}(X)$ and $\beta \in C_{q,m}(Y)$, then

$$\beta \times \alpha = (-1)^{(p+n)(q+m)}(\alpha \times \beta). \tag{50.1}$$

For every point $x \in X$, we write Y_x for $Y \times \operatorname{Spec}(F(x))$ and h_x for the canonical flat morphism $Y_x \to Y$ of relative dimension 0. Note that Y_x is a scheme over $F(x)$, in particular, $C_*(Y_x)$ is a module over $K_*(F(x))$. Denote by $i_x : Y_x \to X \times Y$ the canonical morphism. Let $\alpha \in C_p(X)$ and $\beta \in C_q(Y)$. Unfolding the definitions, we see that

$$\alpha \times \beta = \sum_{x \in X_{(p)}} (i_x)_*\big(\alpha_x \cdot (h_x)^*(\beta)\big).$$

Symmetrically, for every point $y \in Y$, we write X_y for $X \times \operatorname{Spec}(F(y))$ and k_y for the canonical flat morphism $X_y \to X$ of relative dimension 0. Note that X_y is a scheme over $F(y)$, in particular, $C_*(X_y)$ is a module over $K_*(F(y))$. Denote by $j_y : X_y \to X \times Y$ the canonical morphism. Let $\alpha \in C_p(X)$ and $\beta \in C_q(Y)$. Then

$$\alpha \times \beta = \sum_{y \in Y_{(q)}} (j_y)_*\big((k_y)^*(\alpha) \cdot \beta_y\big).$$

Proposition 50.2. *For every $\alpha \in C_*(X)$, $\beta \in C_*(Y)$, and $\gamma \in C_*(Z)$, we have*

$$(\alpha \times \beta) \times \gamma = \alpha \times (\beta \times \gamma).$$

PROOF. It is sufficient to show that for every point $w \in (X \times Y \times Z)_{(p+q+r)}$ projecting to $x \in X_{(p)}$, $y \in Y_{(q)}$ and $z \in Z_{(r)}$ respectively, the w-components of both sides of the equality are equal to

$$r_{F(w)/F(x)}(\alpha_x) \cdot r_{F(w)/F(y)}(\beta_y) \cdot r_{F(w)/F(z)}(\gamma_z)$$

times the multiplicity that is the length of the local ring C of the point w on $\operatorname{Spec}(F(x)) \times \operatorname{Spec}(F(y)) \times \operatorname{Spec}(F(z))$. Let $v \in (X \times Y)_{(p+q)}$ be the projection of w. The multiplicity of the v-component of $\alpha \times \beta$ is equal to the length of the local ring B of the point v on $\operatorname{Spec}(F(x)) \times \operatorname{Spec}(F(y))$. Clearly, C is flat over B. Let Q be the maximal ideal of B. The factor ring C/QC is the local ring of w on $\operatorname{Spec}(F(v)) \times \operatorname{Spec}(F(z))$. Then the multiplicity of the w-component of the left hand side of the equality is equal to $l(B) \cdot l(C/QC)$. By Corollary 102.2, the latter number is equal to $l(C)$. The multiplicity of the right hand side of the equality can be computed similarly. □

Proposition 50.3. *For every $\alpha \in C_{p,n}(X)$ and $\beta \in C_{q,m}(Y)$ we have*

$$d_{X \times Y}(\alpha \times \beta) = d_X(\alpha) \times \beta + (-1)^{p+n} \alpha \times d_Y(\beta).$$

PROOF. We may assume that $\alpha \in K_{p+n}\big(F(x)\big)$ and $\beta \in K_{q+m}\big(F(y)\big)$ for some points $x \in X_{(p)}$ and $y \in Y_{(q)}$. For a point $z \in (X \times Y)_{(p+q-1)}$ the z-components of all three terms in the formula are trivial unless the projections of z to X and Y are specializations of x and y respectively. By dimension count, z projects either to x or to y.

Consider the first case. We have $\big(d_X(\alpha) \times \beta\big)_z = 0$. The point z belongs to the image of i_x and the morphism i_x factors as $Y_x \to \overline{\{x\}} \times Y \hookrightarrow X \times Y$. The scheme Y_x is a localization of $\overline{\{x\}} \times Y$. By Remark 49.7 and Proposition 49.9, the z-components of $d_{X\times Y} \circ (i_x)_*$ and $(i_x)_* \circ d_{Y_x}$ are equal.

By Propositions 49.5 and 49.23, we have

$$\begin{aligned}
\big[d_{X\times Y}(\alpha \times \beta)\big]_z &= \big[d_{X\times Y} \circ (i_x)_*\big(\alpha \cdot (h_x)^*(\beta)\big)\big]_z \\
&= \big[(i_x)_* \circ d_{Y_x}\big(\alpha \cdot (h_x)^*(\beta)\big)\big]_z \\
&= (-1)^{p+n}\big[(i_x)_*\big(\alpha \cdot d_{Y_x} \circ (h_x)^*(\beta)\big)\big]_z \\
&= (-1)^{p+n}\big[(i_x)_*\big(\alpha \cdot (h_x)^*(d_Y\beta)\big)\big]_z \\
&= (-1)^{p+n}\big[\alpha \times d_Y(\beta)\big]_z.
\end{aligned}$$

In the second case, symmetrically, we have $\big(\alpha \times d_Y(\beta)\big)_z = 0$ and

$$d_{X\times Y}(\alpha \times \beta)_z = (d_X(\alpha) \times \beta)_z. \qquad \square$$

Proposition 50.4. *Let $f : X \to X'$ and $g : Y \to Y'$ be morphisms. Then for every $\alpha \in C_p(X)$ and $\beta \in C_q(Y)$ we have*

$$(f \times g)_*(\alpha \times \beta) = f_*(\alpha) \times g_*(\beta).$$

PROOF. Clearly, it suffices to consider the case that f is the identity of X. Let $x \in X_{(p)}$ and let $i'_x : Y'_x \to X \times Y'$, $h'_x : Y'_x \to Y'$, and $g_x : Y_x \to Y'_x$ be canonical morphisms. We have

$$(1_X \times g) \circ i_x = i'_x \circ g_x \quad \text{and} \quad g \circ h_x = h'_x \circ g_x.$$

By Propositions 49.8 and 49.20, we have

$$\begin{aligned}
(1_X \times g)_*(\alpha \times \beta) &= (1_X \times g)_* \circ \sum (i_x)_*\big(\alpha_x \cdot (h_x)^*(\beta)\big) \\
&= \sum (i'_x)_* \circ (g_x)_*\big(\alpha_x \cdot (h_x)^*(\beta)\big) \\
&= \sum (i'_x)_*\big(\alpha_x \cdot (g_x)_*(h_x)^*(\beta)\big) \\
&= \sum (i'_x)_*\big(\alpha_x \cdot (h'_x)^* g_*(\beta)\big) \\
&= \alpha \times g_*(\beta).
\end{aligned} \qquad \square$$

Proposition 50.5. *Let $f : X' \to X$ and $g : Y' \to Y$ be flat morphisms. Then for every $\alpha \in C_p(X)$ and $\beta \in C_q(Y)$ we have*

$$(f \times g)^*(\alpha \times \beta) = f^*(\alpha) \times g^*(\beta).$$

PROOF. Clearly, it suffices to consider the case that f is the identity of X. Let $x \in X_{(p)}$ and let $i'_x : Y'_x \to X \times Y'$, $h'_x : Y'_x \to Y'$ and $g_x : Y'_x \to Y_x$ be canonical morphisms. We have

$$(1_X \times g) \circ i'_x = i_x \circ g_x \quad \text{and} \quad g \circ h'_x = h_x \circ g_x.$$

Note that the scheme Y_x is a localization of $\overline{\{x\}} \times Y$. By Proposition 49.20 and Remark 49.22,

$$(1_X \times g)^* \circ (i_x)_* = (i'_x)_* \circ (g_x)^*.$$

By Propositions 49.16 and 49.18, we have

$$\begin{aligned}(1_X \times g)^*(\alpha \times \beta) &= (1_X \times g)^* \circ \sum (i_x)_*\big(\alpha_x \cdot (h_x)^*(\beta)\big) \\ &= \sum (i'_x)_* \circ (g_x)^*\big(\alpha_x \cdot (h_x)^*(\beta)\big) \\ &= \sum (i'_x)_*\big(\alpha_x \cdot (g_x)^*(h_x)^*(\beta)\big) \\ &= \sum (i'_x)_*\big(\alpha_x \cdot (h'_x)^* g^*(\beta)\big) \\ &= \alpha \times g^*(\beta).\end{aligned}$$

□

Corollary 50.6. *Let $f : X \times Y \to X$ be the projection. Then for every $\alpha \in C_*(X)$, we have $f^*(\alpha) = \alpha \times [Y]$.*

PROOF. We apply Proposition 50.5 and Example 49.2 to $f = 1_X \times g$, where $g : Y \to \operatorname{Spec}(F)$ is the structure morphism. □

Proposition 50.7. *Let X and Y be schemes over F. Let $Z \subset X$ be a closed subscheme and $U = X \setminus Z$. Then for every $\alpha \in C_p(U)$ and $\beta \in C_q(Y)$, we have*

$$\partial_Z^U(\alpha) \times \beta = \partial_{Z\times Y}^{U\times Y}(\alpha \times \beta).$$

PROOF. We may assume that $\beta \in K_*\big(F(y)\big)$ for some $y \in Y$. By Propositions 49.36(1) and 50.4 we may also assume that $Y = \overline{\{y\}}$. For any scheme V denote by $k^V : V_y \to V$ and $j^V : V_y \to V \times Y$ the canonical morphisms. Let $v \in (Z \times Y)_{(p+q-1)}$. The v-component of both sides of the equality are trivial unless v belongs to the image of j^Z. By Remark 49.7, the v-component of $j_*^Z \circ \partial_{Z_y}^{U_y}$ and $\partial_{Z\times Y}^{U\times Y} \circ j_*^U$ are equal. It follows from Propositions 49.34 and 49.36(2) that

$$\begin{aligned}\big[\partial_Z^U(\alpha) \times \beta\big]_v &= \big[j_*^Z\big((k^Z)^*(\partial_Z^U \alpha) \cdot \beta\big)\big]_v \\ &= \big[j_*^Z\big(\partial_{Z_y}^{U_y}(k^U)^*(\alpha) \cdot \beta\big)\big]_v \\ &= \big[j_*^Z \circ \partial_{Z_y}^{U_y}\big((k^U)^*(\alpha) \cdot \beta\big)\big]_v \\ &= \big[\partial_{Z\times Y}^{U\times Y} \circ j_*^U\big((k^U)^*(\alpha) \cdot \beta\big)\big]_v \\ &= \big[\partial_{Z\times Y}^{U\times Y}(\alpha \times \beta)\big]_v.\end{aligned}$$

□

Proposition 50.8. *Let X and Y be two schemes and let a be an invertible regular function on X. Then for every $\alpha \in C_p(X)$ and $\beta \in C_q(Y)$ we have*

$$\big(\{a\} \cdot \alpha\big) \times \beta = \{a'\} \cdot (\alpha \times \beta),$$

where a' is the pull-back of a on $X \times Y$.

PROOF. Let $\bar{a}$ be the pull-back of a on X_y. It follows from Propositions 49.8 and 49.16 that

$$\begin{aligned}(\{a\}\cdot\alpha)\times\beta &= \sum(j_y)_*\big((k_y)^*(\{a\}\alpha)\cdot\beta_y\big)\\ &= \sum(j_y)_*\big(\{\bar{a}\}(k_y)^*(\alpha)\cdot\beta_y\big)\\ &= \sum\{a'\}(j_y)_*\big((k_y)^*(\alpha)\cdot\beta_y\big)\\ &= \{a'\}\cdot(\alpha\times\beta).\end{aligned}$$

□

51. Deformation homomorphisms

In this section we construct deformation homomorphisms . We shall use them later to define Gysin and pull-back homomorphisms. We follow Rost's approach in [**117**] for the definition of deformation homomorphisms. (Deformation homomorphisms are called *specialization homomorphism* in [**45**].) Recall that we only consider separated schemes of finite type over a field.

Let $f : Y \to X$ be a closed embedding, D_f the deformation scheme and C_f the normal cone of f. Recall that $D_f \setminus C_f$ is canonically isomorphic to $\mathbb{G}_m \times X$ (cf. §104.E). We define the *deformation homomorphism* as the composition

$$\sigma_f : C_*(X) \xrightarrow{p^*} C_{*+1}(\mathbb{G}_m \times X) \xrightarrow{\{t\}} C_{*+1}(\mathbb{G}_m \times X) \xrightarrow{\partial} C_*(C_f)$$

where $p : \mathbb{G}_m \times X \to X$ is the projection, the coordinate t of $\mathbb{G}_m$ is considered as an invertible function on $\mathbb{G}_m \times X$ and $\partial := \partial_{C_f}^{\mathbb{G}_m\times X}$ is taken with respect to the open and closed subsets of the deformation scheme D_f.

Example 51.1. Let $f = 1_X$ for a scheme X. Then $D_f = \mathbb{A}^1 \times X$ and $C_f = X$. We claim that σ_f is the identity. Indeed, it suffices to prove that the composition

$$C_*(X) \xrightarrow{p^*} C_{*+1}(\mathbb{G}_m \times X) \xrightarrow{\{t\}} C_{*+1}(\mathbb{G}_m \times X) \xrightarrow{\partial} C_*(X)$$

is the identity. By Propositions 50.5, 50.7, 50.8, and Example 49.32, for every $\alpha \in C_*(X)$, we have

$$\begin{aligned}\partial\big(\{t\}\cdot p^*(\alpha)\big) &= \partial\big(\{t\}\cdot([\mathbb{G}_m]\times\alpha)\big)\\ &= \partial\big((\{t\}\cdot[\mathbb{G}_m])\times\alpha\big)\\ &= \partial\big(\{t\}\cdot[\mathbb{G}_m]\big)\times\alpha\\ &= \big[\{0\}\big]\times\alpha\\ &= \alpha.\end{aligned}$$

The following statement is a consequence of Propositions 49.5, 49.23 and 49.33.

Proposition 51.2. *Let $f : Y \to X$ be a closed embedding. Then $\sigma_f \circ d_X = d_{C_f} \circ \sigma_f$.*

Consider the fiber product diagram

$$\begin{array}{ccc} Y' & \xrightarrow{f'} & X' \\ {\scriptstyle g}\downarrow & & \downarrow{\scriptstyle h} \\ Y & \xrightarrow{f} & X \end{array} \qquad (51.3)$$

with f and f' closed embeddings. Then we have the fiber product diagram (cf. §104.E)

$$\begin{array}{ccccc} C_{f'} & \longrightarrow & D_{f'} & \longleftarrow & \mathbb{G}_m \times X' \\ {\scriptstyle k}\downarrow & & {\scriptstyle l}\downarrow & & \downarrow{\scriptstyle 1\times h} \\ C_f & \longrightarrow & D_f & \longleftarrow & \mathbb{G}_m \times X. \end{array} \tag{51.4}$$

Proposition 51.5. *If the morphism h in diagram (51.3) is flat of relative dimension d then the morphism k in (51.4) is flat of relative dimension d and the diagram*

$$\begin{array}{ccc} C_p(X) & \xrightarrow{\sigma_f} & C_p(C_f) \\ {\scriptstyle h^*}\downarrow & & \downarrow{\scriptstyle k^*} \\ C_{p+d}(X') & \xrightarrow{\sigma_{f'}} & C_{p+d}(C_{f'}) \end{array}$$

is commutative.

PROOF. By Proposition 104.23, we have $D_{f'} = D_f \times_X X'$, hence the morphisms l and k in the diagram (51.4) are flat of relative dimension d. It follows from Propositions 49.16, 49.18, and 49.36(2) that the diagram

$$\begin{array}{ccccccc} C_*(X) & \xrightarrow{p^*} & C_{*+1}(\mathbb{G}_m \times X) & \xrightarrow{\{t\}} & C_{*+1}(\mathbb{G}_m \times X) & \xrightarrow{\partial} & C_*(C_f) \\ {\scriptstyle h^*}\downarrow & & {\scriptstyle (1\times h)^*}\downarrow & & {\scriptstyle (1\times h)^*}\downarrow & & \downarrow{\scriptstyle k^*} \\ C_{*+d}(X') & \xrightarrow{p^*} & C_{*+d+1}(\mathbb{G}_m \times X') & \xrightarrow{\{t\}} & C_{*+d+1}(\mathbb{G}_m \times X') & \xrightarrow{\partial} & C_{*+d}(C_{f'}) \end{array}$$

is commutative. □

Proposition 51.6. *If the morphism h in (51.3) is a proper morphism, then the diagram*

$$\begin{array}{ccc} C_*(X') & \xrightarrow{\sigma_{f'}} & C_*(C_{f'}) \\ {\scriptstyle h_*}\downarrow & & \downarrow{\scriptstyle k_*} \\ C_*(X) & \xrightarrow{\sigma_f} & C_*(C_f) \end{array}$$

is commutative.

PROOF. The natural morphism $D_{f'} \to D_f \times_X X'$ is a closed embedding by Proposition 104.23, hence the morphism l in the diagram (51.4) is proper. It follows from Propositions 49.8, 49.20, and 49.36(1) that the diagram

$$\begin{array}{ccccccc} C_*(X') & \xrightarrow{p^*} & C_{*+1}(\mathbb{G}_m \times X') & \xrightarrow{\{t\}} & C_{*+1}(\mathbb{G}_m \times X') & \xrightarrow{\partial} & C_*(C_{f'}) \\ {\scriptstyle h_*}\downarrow & & {\scriptstyle (1\times h)_*}\downarrow & & {\scriptstyle (1\times h)_*}\downarrow & & \downarrow{\scriptstyle k_*} \\ C_*(X) & \xrightarrow{p^*} & C_{*+1}(\mathbb{G}_m \times X) & \xrightarrow{\{t\}} & C_{*+1}(\mathbb{G}_m \times X) & \xrightarrow{\partial} & C_*(C_f) \end{array}$$

is commutative. □

Corollary 51.7. *Let $f : Y \to X$ be a closed embedding. Then the composition $\sigma_f \circ f_*$ coincides with the push-forward map $C_*(Y) \to C_*(C_f)$ for the zero section $Y \to C_f$.*

PROOF. The statement follows from Proposition 51.6, applied to the fiber product square

$$\begin{array}{ccc} Y & = & Y \\ \| & & \downarrow f \\ Y & \xrightarrow{f} & X \end{array}$$

and Example 51.1. □

Lemma 51.8. *Let $f : X \to \mathbb{A}^1 \times W$ be a morphism. Suppose that the composition $X \to \mathbb{A}^1 \times W \to \mathbb{A}^1$ and the restriction of f on both $f^{-1}(\mathbb{G}_m \times W)$ and $f^{-1}(\{0\} \times W)$ is flat. Then f is flat.*

PROOF. Let $x \in X$, $y = f(x)$, and $z \in \mathbb{A}^1$ the projection of y. Set $A = \mathcal{O}_{\mathbb{A}^1,z}$, $B = \mathcal{O}_{\mathbb{A}^1 \times W, y}$, and $C = \mathcal{O}_{X,x}$. We need to show that C is flat over B. If $z \neq 0$, this follows from the flatness of the restriction of f on $f^{-1}(\mathbb{G}_m \times W)$.

Suppose that $z = 0$. Let M be the maximal ideal of A. The rings B/MB and C/MC are the local rings of y on $\{0\} \times W$ and of x on $f^{-1}(\{0\} \times W)$ respectively. By assumption, C/MC is flat over B/MB and C is flat over A. It follows from [**99**, 20G] that C is flat over B. □

Lemma 51.9. *Let $f : U \to V$ be a closed embedding and $g : V \to W$ a flat morphism. Suppose that the composition*

$$q : C_f \to U \xrightarrow{f} V \xrightarrow{g} W$$

is flat. Then $\sigma_f \circ g^* = q^*$.

PROOF. Consider the composition $u : D_f \to \mathbb{A}^1 \times V \xrightarrow{1 \times g} \mathbb{A}^1 \times W$. The restriction of u on $u^{-1}(\mathbb{G}_m \times W)$ is isomorphic to $1 \times g : \mathbb{G}_m \times V \to \mathbb{G}_m \times W$ and is therefore flat. The restriction of u on $u^{-1}(W \times \{0\})$ coincides with q and is flat by assumption. The projection $D_f \to \mathbb{A}^1$ is also flat. It follows from Lemma 51.8 that the morphism u is flat.

Consider the fiber product diagram

$$\begin{array}{ccccc} C_f & \longrightarrow & D_f & \longleftarrow & \mathbb{G}_m \times V \\ {\scriptstyle q}\downarrow & & {\scriptstyle u}\downarrow & & \downarrow{\scriptstyle 1\times g} \\ W & \longrightarrow & \mathbb{A}^1 \times W & \longleftarrow & \mathbb{G}_m \times W. \end{array}$$

By Propositions 49.16, 49.18, and 49.36(2), the following diagram is commutative:

$$\begin{array}{ccccccc} C_*(W) & \xrightarrow{p^*} & C_{*+1}(\mathbb{G}_m \times W) & \xrightarrow{\{t\}} & C_{*+1}(\mathbb{G}_m \times W) & \xrightarrow{\partial} & C_*(W) \\ {\scriptstyle g^*}\downarrow & & \downarrow{\scriptstyle 1\times g^*} & & \downarrow{\scriptstyle 1\times g^*} & & \downarrow{\scriptstyle q^*} \\ C_{*+d}(V) & \longrightarrow & C_{*+d+1}(\mathbb{G}_m \times V) & \xrightarrow{\{t\}} & C_{*+d+1}(\mathbb{G}_m \times V) & \xrightarrow{\partial} & C_{*+d}(C_f), \end{array}$$

where d is the relative dimension of g. As the composition in the top row of the diagram is the identity by Example 51.1, the result follows. □

If $f : Y \to X$ is a regular closed embedding, we shall write N_f for the normal bundle C_f.

Let $g : Z \to Y$ and $f : Y \to X$ be regular closed embeddings. Then $f \circ g : Z \to X$ is also a regular closed embedding by Proposition 104.15. The normal bundles of the regular closed embeddings $i : N_f|_Z \to N_f$ and $j : N_g \to N_{f \circ g}$ are canonically isomorphic; we denote them by N (cf. §104.E).

Lemma 51.10. *In the setup above, the morphisms of complexes $\sigma_i \circ \sigma_f$ and $\sigma_j \circ \sigma_{f\circ g} : C_*(X) \to C_*(N)$ are homotopic.*

PROOF. Let $D = D_{f,g}$ be the double deformation scheme (cf. §104.F). We have the following fiber product diagram of open and closed embeddings:

$$\begin{array}{ccccc} N & \longrightarrow & D_i & \longleftarrow & N_f \times \mathbb{G}_m \\ \downarrow & & \downarrow & & \downarrow \\ D_j & \longrightarrow & D & \longleftarrow & D_f \times \mathbb{G}_m \\ \uparrow & & \uparrow & & \uparrow \\ \mathbb{G}_m \times N_{f\circ g} & \longrightarrow & \mathbb{G}_m \times D_{fg} & \longleftarrow & \mathbb{G}_m \times X \times \mathbb{G}_m. \end{array}$$

We shall use the notation $\partial_t, \partial_b, \partial_l, \partial_r$ for the boundary morphisms as in §49.E. For every scheme V, denote by p_V either of the projections $V \times \mathbb{G}_m \to V$ or $\mathbb{G}_m \times V \to V$. Write p for the projection $\mathbb{G}_m \times X \times \mathbb{G}_m \to X$.

By Proposition 49.34 and 51.5, we have

$$\begin{aligned} \sigma_i \circ \sigma_f &= \partial_t \circ \{s\} \circ p_{N_f}^* \circ \sigma_f \\ &= \partial_t \circ \{s\} \circ \sigma_{f \times \mathbb{G}_m} \circ p_X^* \\ &= \partial_t \circ \{s\} \circ \partial_r \circ \{t\} \circ p^* \\ &= -\partial_t \circ \partial_r \circ \{s, t\} \circ p^*, \end{aligned}$$

and similarly,

$$\begin{aligned} \sigma_j \circ \sigma_{fg} &= \partial_l \circ \{t\} \circ p_{N_{f\circ g}}^* \circ \sigma_{fg} \\ &= \partial_l \circ \{t\} \circ \sigma_{\mathbb{G}_m \times f\circ g} \circ p_X^* \\ &= \partial_l \circ \{t\} \circ \partial_b \circ \{s\} \circ p^* \\ &= -\partial_l \circ \partial_b \circ \{t, s\} \circ p^*. \end{aligned}$$

As $\{s,t\} = -\{t,s\}$ (cf. §49.B) and the compositions $\partial_t \circ \partial_r$ and $-\partial_l \circ \partial_b$ are homotopic by Proposition 49.37, the result follows. □

52. K-homology groups

Let X be a separated scheme of finite type over a field F. The complex $C_*(X)$ is the coproduct of complexes $C_{*q}(X)$ over all $q \in \mathbb{Z}$. Denote the pth homology group of the complex $C_{*q}(X)$ by $A_p(X, K_q)$ and call it the *K-homology group of X*. In other words, $A_p(X, K_q)$ is the homology group of the complex

$$\coprod_{\dim x = p+1} K_{p+q+1}\big(F(x)\big) \xrightarrow{d_X} \coprod_{\dim x = p} K_{p+q}\big(F(x)\big) \xrightarrow{d_X} \coprod_{\dim x = p-1} K_{p+q-1}\big(F(x)\big).$$

It follows from the definition that $A_p(X, K_q) = 0$ if $p+q < 0$, $p < 0$, or $p > \dim X$.

The group $A_p(X, K_{-p})$ is a factor group of $\coprod_{\dim x = p} K_0\big(F(x)\big)$. If $Z \subset X$ is a closed subscheme, the coset of the cycle $[Z]$ of Z in $A_p(X, K_{-p})$ (cf. Example 49.2) will be also denoted by $[Z]$.

If X is the disjoint union of two schemes X_1 and X_2, then by (49.4), we have

$$A_p(X, K_q) = A_p(X_1, K_q) \oplus A_p(X_2, K_q).$$

Example 52.1. We have

$$A_p\big(\mathrm{Spec}(F), K_q\big) = \begin{cases} K_q(F) & \text{if } p = 0, \\ 0 & \text{otherwise.} \end{cases}$$

It follows from Fact 100.5 that

$$A_p(\mathbb{A}^1_F, K_q) = \begin{cases} K_{q+1}(F) & \text{if } p = 1, \\ 0 & \text{otherwise.} \end{cases}$$

52.A. Push-forward homomorphisms. If $f : X \to Y$ is a proper morphism of schemes, the push-forward homomorphism $f_* : C_{*q}(X) \to C_{*q}(Y)$ is a morphism of complexes by Proposition 49.9. This induces the *push-forward homomorphism* of K-homology groups

(52.2) $$f_* : A_p(X, K_q) \to A_p(Y, K_q).$$

Thus, the assignment $X \mapsto A_*(X, K_*)$ gives rise to a functor from the category of schemes and proper morphisms to the category of bigraded abelian groups and bigraded homomorphisms.

Example 52.3. Let $f : X \to Y$ be a closed embedding such that f is a bijection on points. It follows from Example 49.6 that the push-forward homomorphism f_* in (52.2) is an isomorphism.

52.B. Pull-back homomorphism. If $g : Y \to X$ is a flat morphism of relative dimension d, the pull-back homomorphism $g^* : C_{*q}(X) \to C_{*+d,q-d}(Y)$ is a morphism of complexes by Proposition 49.23. This induces the *pull-back homomorphism* of the K-homology groups

$$g^* : A_p(X, K_q) \to A_{p+d}(Y, K_{q-d}).$$

The assignment $X \mapsto A_*(X, K_*)$ gives rise to a contravariant functor from the category of schemes and flat morphisms to the category of abelian groups.

Example 52.4. If X is a variety of dimension d over F, then the flat structure morphism $p : X \to \mathrm{Spec}(F)$ of relative dimension d induces a natural pull-back homomorphism

$$p^* : K_q F = A_0\big(\mathrm{Spec}(F), K_q\big) \to A_d(X, K_{q-d})$$

giving $A_*(X, K_*)$ a structure of a $K_*(F)$-module.

Example 52.5. It follows from Example 52.1 that the pull-back homomorphism

$$f^* : A_p\big(\mathrm{Spec}(F), K_q\big) \to A_{p+1}(\mathbb{A}^1_F, K_{q-1})$$

given by the flat structure morphism $f : \mathbb{A}^1_F \to \mathrm{Spec}(F)$ is an isomorphism.

52.C. Product. Let X and Y be two schemes over F. It follows from Proposition 50.3 that there is a well-defined pairing

$$A_p(X, K_n) \otimes A_q(Y, K_m) \to A_{p+q}(X \times Y, K_{n+m})$$

taking the classes of cycles α and β to the class of the external product $\alpha \times \beta$ (cf. §50).

52.D. Localization. Let X be a scheme and $Z \subset X$ a closed subscheme. Set $U := X \setminus Z$ and consider the closed embedding $i : Z \to X$ and the open immersion $j : U \to X$. The exact sequence of complexes

$$0 \to C_*(Z) \xrightarrow{i_*} C_*(X) \xrightarrow{j^*} C_*(U) \to 0$$

induces the long *localization exact sequence* of K-homology groups

$$(52.6) \qquad \dots \xrightarrow{\delta} A_p(Z, K_q) \xrightarrow{i_*} A_p(X, K_q) \xrightarrow{j^*} A_p(U, K_q) \xrightarrow{\delta} A_{p-1}(Z, K_q) \xrightarrow{i_*} \dots$$

The map δ is called the *connecting homomorphism*. It is induced by the boundary map of complexes $\partial_Z^U : C_*(U) \to C_{*-1}(Z)$ (cf. Proposition 49.33).

52.E. Deformation. Let $f : Y \to X$ be a closed embedding. It follows from Proposition 51.2 that the deformation homomorphism σ_f of complexes induces the *deformation homomorphism* of homology groups

$$\sigma_f : A_p(X, K_q) \to A_p(C_f, K_q),$$

where C_f is the normal cone of f.

Proposition 52.7. *Let Z be a closed equidimensional subscheme of a scheme X and $g : f^{-1}(Z) \to Z$ the restriction of f. Then $\sigma_f([Z]) = h_*([C_g])$, where $h : C_g \to C_f$ is the closed embedding of cones.*

PROOF. Let $i : Z \to X$ be the closed embedding and $q : Z \to \operatorname{Spec}(F)$, $r : C_f \to \operatorname{Spec}(F)$ the structure morphisms. Consider the diagram

$$\begin{array}{ccccc} A_0(\operatorname{Spec}(F), K_0) & \xrightarrow{q^*} & A_d(Z, K_{-d}) & \xrightarrow{i_*} & A_d(X, K_{-d}) \\ \| & & \downarrow{\scriptstyle \sigma_g} & & \downarrow{\scriptstyle \sigma_f} \\ A_0(\operatorname{Spec}(F), K_0) & \xrightarrow{r^*} & A_d(C_g, K_{-d}) & \xrightarrow{h_*} & A_d(C_f, K_{-d}), \end{array}$$

where $d = \dim Z$. The left square is commutative by Lemma 51.9 and the right one by Proposition 51.6. Consequently, $\sigma_f([Z]) = \sigma_f \circ i_* \circ q^*(1) = h_* \circ r^*(1) = h_*([C_g])$. □

52.F. Continuity. Let X be a variety of dimension n and $f : Y \to X$ a dominant morphism. Let x denote the generic point of X and Y_x the generic fiber of f. For every nonempty open subscheme $U \subset X$, the natural flat morphism $g_U : Y_x \to f^{-1}(U)$ is of relative dimension $-n$. Hence we have the pull-back homomorphism

$$(52.8) \qquad g_U^* : C_*\big(f^{-1}(U)\big) \to C_{*-n}(Y_x).$$

The following proposition is a straightforward consequence of the definition of the complexes C_*.

Proposition 52.9. *The pull-back homomorphism g_U^* of (52.8) induces isomorphisms*

$$\operatorname{colim} C_p\big(f^{-1}(U)\big) \xrightarrow{\sim} C_{p-n}(Y_x) \quad \text{and} \quad \operatorname{colim} A_p\big(f^{-1}(U), K_q\big) \xrightarrow{\sim} A_{p-n}(Y_x, K_{q+n})$$

for all p and q, where the colimits are taken over all nonempty open subschemes U of X.

52.G. Homotopy invariance. Let $g: Y \to X$ be a morphism of schemes over F. Recall that for every $x \in X$, we denote by Y_x the fiber scheme $g^{-1}(x) = Y \times_X \operatorname{Spec}\big(F(x)\big)$ over the field $F(x)$.

Proposition 52.10. *Let $g: Y \to X$ be a flat morphism of relative dimension d. Suppose that for every $x \in X$, the pull-back homomorphism*

$$A_p\big(\operatorname{Spec}\big(F(x)\big), K_q\big) \to A_{p+d}(Y_x, K_{q-d})$$

is an isomorphism for every p and q. Then the pull-back homomorphism

$$g^*: A_p(X, K_q) \to A_{p+d}(Y, K_{q-d})$$

is an isomorphism for every p and q.

PROOF. *Step 1: X is a variety.*

We induct on $n = \dim X$. The case $n = 0$ is obvious. In general, let $U \subset X$ be a nonempty open subset and $Z = X \setminus U$ with the structure of a reduced scheme. Set $V = g^{-1}(U)$ and $T = g^{-1}(Z)$. We have closed embeddings $i: Z \to X$, $k: T \to Y$ and open immersions $j: U \to X$, $l: V \to Y$. By induction, the pull-back homomorphism $(g|_T)^*$ in the diagram

$$\begin{array}{ccccc}
A_{p+1}(U, K_q) & \xrightarrow{\delta} & A_p(Z, K_q) & \xrightarrow{i_*} & A_p(X, K_q) \\
\downarrow{\scriptstyle (g|_V)^*} & & \downarrow{\scriptstyle (g|_T)^*} & & \downarrow{\scriptstyle g^*} \\
A_{p+d+1}(V, K_{q-d}) & \xrightarrow{\delta} & A_{p+d}(T, K_{q-d}) & \xrightarrow{k_*} & A_{p+d}(Y, K_{q-d})
\end{array}$$

$$\begin{array}{ccccc}
A_p(X, K_q) & \xrightarrow{j^*} & A_p(U, K_q) & \xrightarrow{\delta} & A_{p-1}(Z, K_q) \\
\downarrow{\scriptstyle g^*} & & \downarrow{\scriptstyle (g|_V)^*} & & \downarrow{\scriptstyle (g|_T)^*} \\
A_{p+d}(Y, K_{q-d}) & \xrightarrow{l^*} & A_{p+d}(V, K_{q-d}) & \xrightarrow{\delta} & A_{p+d-1}(T, K_{q-d})
\end{array}$$

is an isomorphism. The diagram is commutative by Propositions 49.18, 49.20, and 49.36(2).

Let $x \in X$ be the generic point. By Proposition 52.9, the colimit of the homomorphisms

$$(g|_V)^*: A_p(U, K_q) \to A_{p+d}(V, K_{q-d})$$

over all nonempty open subschemes U of X is isomorphic to the pull-back homomorphism

$$A_{p-n}\big(\operatorname{Spec}\big(F(x)\big), K_{q+n}\big) \to A_{p-n+d}(Y_x, K_{q+n-d}).$$

By assumption, it is an isomorphism. Taking the colimits of all terms of the diagram, we conclude by the 5-lemma that g^* is an isomorphism.

Step 2: X is reduced.

We induct on the number m of irreducible components of X. The case $m = 1$ is Step 1. Let Z be a (reduced) irreducible component of X and let $U = X \setminus Z$. Consider the commutative diagram as in Step 1. By Step 1, we have $(g|_T)^*$ is an isomorphism. The pull-back $(g|_V)^*$ is also an isomorphism by the induction hypothesis. By the 5-lemma, g^* is an isomorphism.

Step 3: X is an arbitrary scheme.

Let X' be the reduced scheme X_{red}. Consider the fiber product diagram

$$\begin{array}{ccc} Y' & \xrightarrow{g'} & X' \\ {\scriptstyle f}\downarrow & & \downarrow{\scriptstyle h} \\ Y & \xrightarrow{g} & X, \end{array}$$

with f and h closed embeddings. By Proposition 49.20, we have $g^* \circ h_* = f_* \circ g'^*$. It follows from Example 52.3 that the maps f_* and h_* are isomorphisms. Finally, g'^* is an isomorphism by Step 2, hence we conclude that g^* is also an isomorphism. $\square$

Corollary 52.11. *The pull-back homomorphism*

$$g^* : A_p(X, K_q) \to A_{p+d}(X \times \mathbb{A}^d_F, K_{q-d})$$

induced by the projection $g : X \times \mathbb{A}^d_F \to X$ is an isomorphism. In particular,

$$A_p(\mathbb{A}^d, K_q) = \begin{cases} K_{q+d}(F) & \text{if } p = d, \\ 0 & \text{otherwise.} \end{cases}$$

PROOF. Example 52.5 and Proposition 52.10 yield the statement in the case $d = 1$. The general case follows by induction. $\square$

A morphism $g : Y \to X$ is called an *affine bundle of rank d* if g is flat and the fiber of g over any point $x \in X$ is isomorphic to the affine space $\mathbb{A}^d_{F(x)}$. For example, a vector bundle of rank d is an affine bundle of rank d.

The following statement is a useful criterion establish a morphism is an affine bundle. We shall use it repeatedly in the sequel.

Lemma 52.12. *A morphism $f : Y \to X$ over F is an affine bundle of rank d if for any local commutative F-algebra R and any morphism $\operatorname{Spec}(R) \to X$ over F, the fiber product $Y \times_X \operatorname{Spec}(R)$ is isomorphic to $\mathbb{A}^d_R$ over R.*

PROOF. Applying the condition to the local ring $R = \mathcal{O}_{X,x}$ for each $x \in X$, we see that f is flat and the fiber of f over x is the affine space $\mathbb{A}^d_{F(x)}$. $\square$

The following theorem implies that affine spaces are essentially negligible for K-homology computation.

Theorem 52.13 (Homotopy Invariance). *Let $g : Y \to X$ be an affine bundle of rank d. Then the pull-back homomorphism*

$$g^* : A_p(X, K_q) \to A_{p+d}(Y, K_{q-d})$$

is an isomorphism for every p and q.

PROOF. For every $x \in X$, we have $Y_x \simeq \mathbb{A}^d_{F(x)}$. Applying Corollary 52.11 to $X = \operatorname{Spec}\big(F(x)\big)$, we see that the pull-back homomorphism

$$A_p\big(\operatorname{Spec}\big(F(x)\big), K_q\big) \to A_{p+d}(Y_x, K_{q-d})$$

is an isomorphism for every p and q. By Proposition 52.10, the map g^* is an isomorphism. $\square$

Corollary 52.14. *Let $f : E \to X$ be a vector bundle of rank d. Then the pull-back homomorphism*

$$f^* : A_p(X, K_*) \to A_{p+d}(E, K_{*-d})$$

is an isomorphism for every p.

53. Euler classes and projective bundle theorem

In this section, we compute the K-homology for projective spaces and more generally for projective bundles.

53.A. Euler class. Let $p : E \to X$ be a vector bundle of rank r. Let $s : X \to E$ denote the zero section. Note that p is a flat morphism of relative dimension r and s is a closed embedding. By Corollary 52.14, the pull-back homomorphism p^* is an isomorphism. We call the composition

$$e(E) = (p^*)^{-1} \circ s_* : A_*(X, K_*) \to A_{*-r}(X, K_{*+r})$$

the *Euler class of* E. Note that isomorphic vector bundles over X have equal Euler classes.

Proposition 53.1. *Let* $0 \to E' \xrightarrow{f} E \xrightarrow{g} E'' \to 0$ *be an exact sequence of vector bundles over* X. *Then* $e(E) = e(E'') \circ e(E')$.

PROOF. Consider the fiber product diagram

$$\begin{array}{ccc} E' & \xrightarrow{f} & E \\ {\scriptstyle p'}\downarrow & & \downarrow{\scriptstyle g} \\ X & \xrightarrow{s''} & E''. \end{array}$$

By Proposition 49.20, we have $g^* \circ s''_* = f_* \circ p'^*$, hence

$$\begin{aligned} e(E'') \circ e(E') &= (p''^*)^{-1} \circ s''_* \circ (p'^*)^{-1} \circ s'_* \\ &= (p''^*)^{-1} \circ g^{*-1} \circ f_* \circ s'_* \\ &= (p'' \circ g)^{*-1} \circ (f \circ s')_* \\ &= p^{*-1} \circ s_* \\ &= e(E). \end{aligned}$$ □

Corollary 53.2. *The Euler classes of any two vector bundles* E *and* E' *over* X *commute:* $e(E') \circ e(E) = e(E) \circ e(E')$.

PROOF. By Proposition 53.1, we have

$$e(E') \circ e(E) = e(E' \oplus E) = e(E \oplus E') = e(E) \circ e(E').$$ □

Proposition 53.3. *Let* $f : Y \to X$ *be a morphism and* E *a vector bundle over* X. *Then the pull-back* $E' = f^*(E)$ *is a vector bundle over* Y *and*

(1) *If* f *is proper, then* $e(E) \circ f_* = f_* \circ e(E')$.
(2) *If* f *is flat, then* $f^* \circ e(E) = e(E') \circ f^*$.

PROOF. We have two fiber product diagrams

$$\begin{array}{ccc} E' & \xrightarrow{g} & E \\ {\scriptstyle q}\downarrow & & \downarrow{\scriptstyle p} \\ Y & \xrightarrow{f} & X \end{array} \quad \text{and} \quad \begin{array}{ccc} Y & \xrightarrow{f} & X \\ {\scriptstyle j}\downarrow & & \downarrow{\scriptstyle i} \\ E' & \xrightarrow{g} & E \end{array}$$

where p and q are the natural morphisms and i and j are the zero sections.

(1): By Proposition 49.20, we have $p^* \circ f_* = g_* \circ q^*$. Hence

$$\begin{aligned} e(E) \circ f_* &= (p^*)^{-1} \circ i_* \circ f_* \\ &= (p^*)^{-1} \circ g_* \circ j_* \\ &= f_* \circ (q^*)^{-1} \circ j_* \\ &= f_* \circ e(E'). \end{aligned}$$

(2): Again by Proposition 49.20, we have $g^* \circ i_* = j_* \circ f^*$. Hence

$$\begin{aligned} f^* \circ e(E) &= f^* \circ (p^*)^{-1} \circ i_* \\ &= (q^*)^{-1} \circ g^* \circ i_* \\ &= (q^*)^{-1} \circ j_* \circ f^* \\ &= e(E') \circ f^*. \end{aligned}$$

□

Proposition 53.4. *Let $p : E \to X$ and $p' : E' \to X'$ be vector bundles. Then*

$$e(E \times E')(\alpha \times \alpha') = e(E)(\alpha) \times e(E')(\alpha')$$

for every $\alpha \in A_*(X, K_*)$ and $\alpha' \in A_*(X', K_*)$.

Proof. Let $s : X \to E$ and $s' : X' \to E'$ be zero sections. It follows from Propositions 50.4 and 50.5 that

$$\begin{aligned} e(E \times E')(\alpha \times \alpha') &= (p \times p')^{*^{-1}} \circ (s \times s')_*(\alpha \times \alpha') \\ &= (p^{*^{-1}} \times p'^{*^{-1}}) \circ (s_* \times s'_*)(\alpha \times \alpha') \\ &= \big(p^{*^{-1}} \circ s_*(\alpha)\big) \times \big(p'^{*^{-1}} \circ s'_*(\alpha')\big) \\ &= e(E)(\alpha) \times e(E')(\alpha'). \end{aligned}$$

□

Proposition 53.5. *The Euler class $e(\mathbb{1})$ is trivial.*

Proof. It suffices to proof that the push-forward homomorphism s_* for the zero section $s : X \to \mathbb{A}^1 \times X$ is trivial. Let t be the coordinate function on $\mathbb{A}^1$. We view $\{t\}$ as an element of $C_1(\mathbb{A}^1) = K_1\big(F(\mathbb{A}^1)\big)$. Clearly, $d_{\mathbb{A}^1}\big(\{t\}\big) = \operatorname{div}(t) = [0]$. It follows from Proposition 50.3 that for every $\alpha \in A^*(X, K_*)$, one has

$$s_*(\alpha) = [0] \times \alpha = d_{\mathbb{A}^1}\big(\{t\}\big) \times \alpha = d_{\mathbb{A}^1 \times X}\big(\{t\} \times \alpha\big) = 0$$

in $A^*(\mathbb{A}^1 \times X, K_*)$ □

53.B. K-homology of projective spaces. Let V be a vector space of dimension $d+1$ over F, X the projective space $\mathbb{P}_F(V)$ and V_p a subspace of V of dimension $p+1$ for $p = 0, \dots, d$. We view $\mathbb{P}(V_p)$ as a subvariety of X. Let $x_p \in X$ be the generic point of $\mathbb{P}(V_p)$. Consider the generator 1_p of $K_0\big(F(x_p)\big) = \mathbb{Z}$ viewed as a subgroup of $C_{p,-p}(X)$. We claim that the class l_p of the generator 1_p in $A_p(X, K_{-p})$ does not depend on the choice of V_p.

The statement is trivial if $p = d$. Suppose $p < d$ and let V'_p be another subspace of dimension $p+1$. We may assume that V_p and V'_p are subspaces of a subspace $W \subset V$ of dimension $p+2$. Let h and h' be linear forms on W satisfying $\operatorname{Ker}(h) = V_p$ and $\operatorname{Ker}(h') = V'_p$. View the ratio $f = h/h'$ as a rational function on $\mathbb{P}_F(W)$, so $\operatorname{div}(f) = 1_p - 1_{p'}$. By definition of the K-homology group $A_p(X, K_{-p})$, the classes l_p and $l_{p'}$ of 1_p and $1_{p'}$ respectively in $A_p(X, K_{-p})$ coincide.

Let X be a scheme over F and set $\mathbb{P}^d_X = \mathbb{P}^d_F \times X$. For every $i = 0, \dots, d$ consider the external product homomorphism

$$A_{*-i}(X, K_{*+i}) \to A_*\big(\mathbb{P}^d_X, K_*\big), \quad \alpha \mapsto l_i \times \alpha.$$

The following proposition computes the K-homology of the projective space $\mathbb{P}^d_X$.

Proposition 53.6. *For any scheme X, the homomorphism*

$$\coprod_{i=0}^{d} A_{*-i}(X, K_{*+i}) \to A_*\big(\mathbb{P}^d_X, K_*\big)$$

taking $\sum \alpha_i$ to $\sum l_i \times \alpha_i$ is an isomorphism.

PROOF. We induct on d. The case $d = 0$ is obvious since $\mathbb{P}^d_X = X$. If $d > 0$ we view $\mathbb{P}^{d-1}_X$ as a closed subscheme of $\mathbb{P}^d_X$ with open complement $\mathbb{A}^d_X$. Consider the closed and open embeddings $f : \mathbb{P}^{d-1}_X \to \mathbb{P}^d_X$ and $g : \mathbb{A}^d_X \to \mathbb{P}^d_X$. In the diagram

$$\begin{array}{ccccc}
0 & \longrightarrow & \coprod_{i=0}^{d-1} A_{*-i}(X, K_{*+i}) & \longrightarrow & \coprod_{i=0}^{d} A_{*-i}(X, K_{*+i}) \\
 & & \big\downarrow & & \big\downarrow \\
\dots & \xrightarrow{\delta} & A_*\big(\mathbb{P}^{d-1}_X, K_*\big) & \xrightarrow{f_*} & A_*\big(\mathbb{P}^d_X, K_*\big)
\end{array}$$

$$\begin{array}{ccccc}
\coprod_{i=0}^{d} A_{*-i}(X, K_{*+i}) & \longrightarrow & A_{*-d}(X, K_{*+d}) & \longrightarrow & 0 \\
\big\downarrow & & \big\downarrow{\scriptstyle h^*} & & \\
A_*\big(\mathbb{P}^d_X, K_*\big) & \xrightarrow{g^*} & A_*\big(\mathbb{A}^d_X, K_*\big) & \xrightarrow{\delta} & \dots
\end{array}$$

the bottom row is the localization exact sequence and $h : \mathbb{A}^d_X \to X$ is the canonical morphism. The left square is commutative by Proposition 50.4 and the right square by Proposition 50.5.

Let $q : \mathbb{P}^d_X \to X$ be the projection. Since $h = q \circ g$, we have $h^* = g^* \circ q^*$. By Corollary 52.11, we have h^* is an isomorphism, hence g^* is surjective. Therefore, all connecting homomorphisms δ in the bottom localization exact sequence are trivial. It follows that the map f_* is injective, i.e., the bottom sequence of the two maps f_* and g^* is short exact. By the induction hypothesis, the left vertical homomorphism is an isomorphism. By the 5-lemma so is the middle one. □

Corollary 53.7. $A_p(\mathbb{P}^d_F, K_q) = \begin{cases} K_{p+q}(F) \cdot l_p & \textit{if } 0 \le p \le d, \\ 0 & \textit{otherwise.} \end{cases}$

Example 53.8. Let L be the canonical line bundle over $X = \mathbb{P}^d_F$. We claim that $e(L)(l_p) = l_{p-1}$ for every $p = 1, \dots, d$. First consider the case $p = d$. By §104.C, we have $L = \mathbb{P}^{d+1} \setminus \{0\}$, where $0 = [0 : \dots 0 : 1]$ and the morphism $f : L \to X$ takes $[S_0 : \dots : S_n : S_{n+1}]$ to $[S_0 : \dots : S_n]$. The image Z of the zero section $s : X \to L$ is given by $S_{n+1} = 0$. Let $H \subset X$ be the hyperplane given by $S_0 = 0$. We have $\operatorname{div}(S_{n+1}/S_0) = [Z] - [f^{-1}(H)]$ and therefore, in $A_{d-1}(X, K_{1-d})$:

$$e(L)(l_d) = (f^*)^{-1}\big(s_*([X])\big) = (f^*)^{-1}\big([Z]\big) = (f^*)^{-1}\big([f^{-1}(H)]\big) = [H] = l_{d-1}.$$

In the general case consider a linear closed embedding $g : \mathbb{P}^p_F \to \mathbb{P}^d_F$. The pull-back $L' := g^*(L)$ is the canonical bundle over $\mathbb{P}^p_F$. By the first part of the proof and Proposition 53.3(1),

$$e(L)(l_p) = e(L)\big(g_*(l_p)\big) = g_*\big(e(L')(l_p)\big) = g_*(l_{p-1}) = l_{p-1}.$$

Example 53.9. Let L' be the tautological line bundle over $X = \mathbb{P}^d_F$. Similar to Example 53.8, we get $e(L')(l_p) = -l_{p-1}$ for every $p \in [1,\ d]$.

53.C. Projective bundle theorem. Let $E \to X$ be a vector bundle of rank $r > 0$. Consider the associated projective bundle morphism $q : \mathbb{P}(E) \to X$. Note that q is a flat morphism of relative dimension $r - 1$. Let $L \to \mathbb{P}(E)$ be either the canonical or the tautological line bundle and e the Euler class of L.

Theorem 53.10 (Projective Bundle Theorem). *Let $E \to X$ be a vector bundle of rank $r > 0$. Then the homomorphism*

$$\varphi(E) := \coprod_{i=1}^{r} e^{r-i} \circ q^* : \coprod_{i=1}^{r} A_{*-i+1}(X, K_{*+i-1}) \to A_*\big(\mathbb{P}(E), K_*\big)$$

is an isomorphism. In other words, every $\alpha \in A_*\big(\mathbb{P}(E), K_*\big)$ can be written in the form

$$\alpha = \sum_{i=1}^{r} e^{r-i}\big(q^*(\alpha_i)\big)$$

for uniquely determined elements $\alpha_i \in A_{*-i+1}(X, K_{*+i-1})$.

PROOF. Suppose that L is the canonical line bundle; the case of the tautological bundle is treated similarly. If E is a trivial vector bundle, we have $\mathbb{P}(E) = X \times \mathbb{P}^{r-1}_F$. Let L' be the canonical line bundle over $\mathbb{P}^{r-1}_F$. It follows from Example 53.8 that

$$\begin{aligned} e(L)^{r-i}\big(q^*(\alpha)\big) &= e(L)^{r-i}(l_{r-1} \times \alpha) \\ &= e(L')^{r-i}(l_{r-1}) \times \alpha \\ &= l_{i-1} \times \alpha. \end{aligned}$$

Hence the map $\varphi(E)$ coincides with the one in Proposition 53.6, consequently is an isomorphism.

In general, we induct on $d = \dim X$. If $d = 0$, the vector bundle is trivial. If $d > 0$ choose an open subscheme $U \subset X$ such that dimension of $Z = X \setminus U$ is less than d and the vector bundle $E|_U$ is trivial. In the diagram

$$\begin{array}{ccccc} \dots \longrightarrow & \coprod_{i=1}^{r} A_{*-i+1}(Z, K_{*+i-1}) & \longrightarrow & \coprod_{i=1}^{r} A_{*-i+1}(X, K_{*+i-1}) \\ & \Big\downarrow{\scriptstyle \varphi(E|_Z)} & & \Big\downarrow{\scriptstyle \varphi(E)} \\ \dots \longrightarrow & A_*\big(\mathbb{P}(E|_Z), K_*\big) & \longrightarrow & A_*\big(\mathbb{P}(E), K_*\big) \end{array}$$

$$\begin{array}{ccccc} \coprod_{i=1}^{r} A_{*-i+1}(X, K_{*+i-1}) & \longrightarrow & \coprod_{i=1}^{r} A_{*-i+1}(U, K_{*+i-1}) & \longrightarrow \dots \\ \Big\downarrow{\scriptstyle \varphi(E)} & & \Big\downarrow{\scriptstyle \varphi(E|_U)} \\ A_*\big(\mathbb{P}(E), K_*\big) & \longrightarrow & A_*\big(\mathbb{P}(E|_U), K_*\big) & \longrightarrow \dots \end{array}$$

with rows the localization long exact sequences, the homomorphisms $\varphi(E|_Z)$ are isomorphisms by the induction hypothesis and $\varphi(E|_U)$ are isomorphisms as $E|_U$ is trivial. The diagram is commutative by Proposition 53.3. The statement follows by the 5-lemma. □

Remark 53.11. It follows from Propositions 49.18, 49.20, and 53.3 that the isomorphisms $\varphi(E)$ are natural with respect to push-forward homomorphisms for proper morphisms of the base schemes and with respect to pull-back homomorphisms for flat morphisms.

Corollary 53.12. *The pull-back homomorphism*

$$q^* : A_{*-r+1}(X, K_{*+r-1}) \to A_*\big(\mathbb{P}(E), K_*\big)$$

is a split injection.

Proposition 53.13 (Splitting Principle). *Let $E \to X$ be a vector bundle. Then there is a flat morphism $f : Y \to X$ of constant relative dimension, say d, such that:*

(1) *The pull-back homomorphism $f^* : A_*(X, K_*) \to A_{*+d}(Y, K_{*-d})$ is injective.*
(2) *The vector bundle $f^*(E)$ has a filtration by subbundles with quotients line bundles.*

PROOF. We induct on the rank r of E. Consider the projective bundle $q : \mathbb{P}(E) \to X$. The pull-back homomorphism q^* is injective by Corollary 53.12. The tautological line bundle L over $\mathbb{P}(E)$ is a sub-bundle of the vector bundle $q^*(E)$. Applying the induction hypothesis to the factor bundle $q^*(E)/L$ over $\mathbb{P}(E)$, we find a flat morphism $g : Y \to \mathbb{P}(E)$ of constant relative dimension satisfying the conditions (1) and (2). The composition $f = q \circ g$ works. □

To prove various relations between K-homology classes, the splitting principle allows us to assume that all the vector bundles involved have filtration by sub-bundles with line factors.

54. Chern classes

In this section we construct Chern classes of vector bundles as operations on K-homology.

Let $E \to X$ be a vector bundle of rank $r > 0$ and $q : \mathbb{P}(E) \to X$ the associated projective bundle. By Theorem 53.10, for every $\alpha \in A_*(X, K_*)$ there exist unique $\alpha_i \in A_{*-i}(X, K_{*+i})$, $i = 0, \dots, r$ such that

$$-e^r\big(q^*(\alpha)\big) = \sum_{i=1}^{r} (-1)^i e^{r-i}\big(q^*(\alpha_i)\big),$$

where e is the Euler class of the tautological line bundle L over $\mathbb{P}(E)$, i.e.,

$$\sum_{i=0}^{r} (-1)^i e^{r-i}\big(q^*(\alpha_i)\big) = 0, \tag{54.1}$$

where $\alpha_0 = \alpha$. Thus we have obtained group homomorphisms

$$c_i(E) : A_*(X, K_*) \to A_{*-i}(X, K_{*+i}), \quad \alpha \mapsto \alpha_i = c_i(E)(\alpha) \tag{54.2}$$

for every $i = 0, \dots, r$. These are called the *Chern classes of* E. By definition, $c_0(E)$ is the identity. We also set $c_i = 0$ if $i > r$ or $i < 0$ and define the *total Chern class of* E by

$$c(E) := c_0(E) + c_1(E) + \cdots + c_r(E)$$

viewed as an endomorphism of $A_*(X, K_*)$. If E is the zero bundle (of rank 0), we set $c_0(E) = 1$ and $c_i(E) = 0$ if $i \neq 0$.

Proposition 54.3. *If E is a line bundle then $c_1(E) = e(E)$.*

PROOF. We have $\mathbb{P}(E) = X$ and $L = E$ by Example 104.19. Therefore the equality (54.1) reads $e(E)(\alpha) - \alpha_1 = 0$, hence $c_1(E)(\alpha) = \alpha_1 = e(E)(\alpha)$. □

Example 54.4. If L is a line bundle, then $c(L) = 1 + e(L)$. In particular, $c(\mathbb{1}) = 1$ by Proposition 53.5.

Proposition 54.5. *Let $f : Y \to X$ be a morphism and E a vector bundle over X. Set $E' = f^*(E)$. Then*

(1) *If f is proper then $c(E) \circ f_* = f_* \circ c(E')$.*
(2) *If f is flat then $f^* \circ c(E) = c(E') \circ f^*$.*

PROOF. Let $\operatorname{rank} E = r$. Consider the fiber product diagram

$$\begin{array}{ccc} \mathbb{P}(E') & \xrightarrow{\ h\ } & \mathbb{P}(E) \\ {\scriptstyle q'}\downarrow & & \downarrow{\scriptstyle q} \\ Y & \xrightarrow{\ f\ } & X \end{array}$$

with flat morphisms q and q' of constant relative dimension $r - 1$. Denote by e and e' the Euler classes of the tautological line bundle L over $\mathbb{P}(E)$ and L' over $\mathbb{P}(E')$, respectively. Note that $L' = h^*(L)$.

(1): By Proposition 49.20, we have $h_* \circ (q')^* = q^* \circ f_*$. By the definition of Chern classes, for every $\alpha' \in A_*(Y, K_*)$ and $\alpha'_i = c_i(E')(\alpha')$, we have:

$$\sum_{i=0}^{r} (-1)^i (e')^{r-i} \big(q'^*(\alpha'_i)\big) = 0.$$

Applying h_*, by Propositions 49.20 and 53.3(1), we have

$$\begin{aligned} 0 &= h_*\left(\sum_{i=0}^{r} (-1)^i (e')^{r-i} \big(q'^*(\alpha'_i)\big)\right) \\ &= \sum_{i=0}^{r} (-1)^i e^{r-i} \big(h_* q'^*(\alpha'_i)\big) \\ &= \sum_{i=0}^{r} (-1)^i e^{r-i} \big(q^* f_*(\alpha'_i)\big). \end{aligned}$$

Hence $c_i(E)\big(f_*(\alpha')\big) = f_*(\alpha'_i) = f_* c_i(E')(\alpha')$.

(2): By the definition of the Chern classes, for every $\alpha \in A_*(X, K_*)$ and $\alpha_i = c_i(E)(\alpha)$ we have

$$\sum_{i=0}^{r} (-1)^i e^{r-i} (q^* \alpha_i) = 0.$$

Applying h^*, by Proposition 53.3(2), we have

$$\begin{aligned} 0 &= h^*\Big(\sum_{i=0}^{r}(-1)^i e^{r-i}\big(q^*(\alpha_i)\big)\Big) \\ &= \sum_{i=0}^{r}(-1)^i (e')^{r-i}\big(h^* q^*(\alpha_i)\big) \\ &= \sum_{i=0}^{r}(-1)^i (e')^{r-i}\big(q'^* f^*(\alpha_i)\big). \end{aligned}$$

Hence $c_i(E')\big(f^*(\alpha)\big) = f^*(\alpha_i) = f^* c_i(E)(\alpha)$. □

Proposition 54.6. *Let E be a vector bundle over X possessing a filtration by subbundles with factors line bundles $L_1, L_2, \dots, L_r$. Then for every $i = 1, \dots, r$, we have*

$$c_i(E) = \sigma_i\big(e(L_1), \dots, e(L_r)\big)$$

where σ_i is the ith elementary symmetric function, i.e.,

$$c(E) = \prod_{i=1}^{r}\big(1 + e(L_i)\big) = \prod_{i=1}^{r} c(L_i).$$

PROOF. Let $q : \mathbb{P}(E) \to X$ be the canonical morphism and let e be the Euler class of the tautological line bundle L over $\mathbb{P}(E)$. It follows from formula (54.1) and Proposition 54.5 that it suffices to prove that

$$\prod_{i=1}^{r}\big(e - e(q^* L_i)\big) = 0$$

as an operation on $A_*(\mathbb{P}(E), K_*)$. We induct on r. The case $r = 1$ follows from the fact that the tautological bundle L coincides with E over $\mathbb{P}(E) = X$ (cf. Example 104.19). In the general case, let E' be a subbundle of E having a filtration by subbundles with factors line bundles $L_1, L_2, \dots, L_{r-1}$ and with $E/E' \simeq L_r$. Consider the natural morphism $f : U = \mathbb{P}(E) \setminus \mathbb{P}(E') \to \mathbb{P}(L_r)$. Under the identification of $\mathbb{P}(L_r)$ with X, the bundle L_r is the tautological line bundle over $\mathbb{P}(L_r)$. Hence $f^*(L_r)$ is isomorphic to the restriction of L to U. In other words, $L|_U \simeq q^*(L_r)|_U$ and therefore $e(L|_U) = e(q^*(L_r)|_U)$. It follows from Proposition 53.3 that for every $\alpha \in A_*(\mathbb{P}(E), K_*)$, we have

$$\big(e - e(q^* L_r)\big)(\alpha)|_U = \big(e(L|_U) - e(q^*(L_r)|_U)\big)(\alpha|_U) = 0.$$

By the exactness of the localization sequence (52.6), there is an element $\beta \in A_*(\mathbb{P}(E'), K_*)$ satisfying

$$i_*(\beta) = \big(e - e(q^* L_r)\big)(\alpha),$$

where $i : \mathbb{P}(E') \to \mathbb{P}(E)$ is the closed embedding. Let L' be the the tautological line bundle over $\mathbb{P}(E')$ and $q' : \mathbb{P}(E') \to X$ the canonical morphism. We have $q' = q \circ i$.

By induction and Proposition 53.3,

$$\begin{aligned}\prod_{i=1}^{r}\big(e - e(q^*L_i)\big)(\alpha) &= \prod_{i=1}^{r-1}\big(e - e(q^*L_i)\big)(i_*\beta)\\ &= i_*\left(\prod_{i=1}^{r-1}\big(e(L') - e(i^*q^*L_i)\big)(\beta)\right)\\ &= i_*\left(\prod_{i=1}^{r-1}\big(e(L') - e(q'^*L_i)\big)(\beta)\right)\\ &= 0.\end{aligned}$$

□

Proposition 54.7 (Whitney Sum Formula). *Let* $0 \to E' \xrightarrow{f} E \xrightarrow{g} E'' \to 0$ *be an exact sequence of vector bundles over* X. *Then* $c(E) = c(E') \circ c(E'')$, *i.e.,*

$$c_n(E) = \sum_{i+j=n} c_i(E') \circ c_j(E'')$$

for every n.

PROOF. By the Splitting Principle 53.13 and Proposition 54.5(2), we may assume that E' and E'' have filtrations by subbundles with quotients line bundles $L'_1, \dots, L'_r$ and $L''_1, \dots, L''_s$, respectively. Consequently, E has a filtration with factors $L'_1, \dots, L'_r, L''_1, \dots, L''_s$. It follows from Proposition 54.6 that

$$c(E') \circ c(E'') = \prod_{i=1}^{r} c(L'_i) \circ \prod_{j=1}^{s} c(L''_i) = c(E).$$

□

The same proof as for Corollary 53.2 yields:

Corollary 54.8. *The Chern classes of any two vector bundles* E *and* E' *over* X *commute:* $c(E') \circ c(E) = c(E) \circ c(E')$.

By Example 54.4, we have

Corollary 54.9. *If* E *is a vector bundle over* X, *then* $c(E \oplus \mathbb{1}) = c(E)$. *In particular, if* E *is a trivial vector bundle, then* $c(E) = 1$.

The Whitney Sum Formula 54.7 allows us to define Chern classes not only for vector bundles over a scheme X but also for elements of the Grothendieck group $K_0(X)$ of the category of vector bundles over X. Note that for a vector bundle E over X the endomorphisms $c_i(E)$ are nilpotent for $i > 0$; therefore, the total Chern class $c(E)$ is an invertible endomorphism. By the Whitney Sum Formula, the assignment $E \mapsto c(E) \in \operatorname{Aut}\big(A_*(X, K_*)\big)$ gives rise to the total Chern class homomorphism

$$c : K_0(X) \to \operatorname{Aut}\big(A_*(X, K_*)\big).$$

55. Gysin and pull-back homomorphisms

In this section we consider contravariant properties of K-homology.

55.A. Gysin homomorphisms. Let $f : Y \to X$ be a regular closed embedding of codimension r and let $p_f : N_f \to Y$ be the canonical morphism (cf. §104.B). We define the *Gysin homomorphism* as the composition

$$f^\star : A_*(X, K_*) \xrightarrow{\sigma_f} A_*(N_f, K_*) \xrightarrow{(p_f^*)^{-1}} A_{*-r}(Y, K_{*+r}).$$

Proposition 55.1. *Let $Z \xrightarrow{g} Y \xrightarrow{f} X$ be regular closed embeddings. Then $(f \circ g)^\star = g^\star \circ f^\star$.*

PROOF. The normal bundles of the regular closed embeddings $i : N_f|_Z \to N_f$ and $j : N_g \to N_{f\circ g}$ are canonically isomorphic, denote them by N. Consider the diagram

$$\begin{array}{ccccc}
C_*(X) & \xrightarrow{\sigma_{f\circ g}} & C_*(N_{f\circ g}) & \xleftarrow{p^*_{fg}} & C_*(Z) \\
{\scriptstyle\sigma_f}\downarrow & & {\scriptstyle\sigma_j}\downarrow & & \Vert \\
C_*(N_f) & \xrightarrow{\sigma_i} & C_*(N) & \xleftarrow{(p_g p_j)^*} & C_*(Z) \\
{\scriptstyle p_f^*}\uparrow & & {\scriptstyle p_j^*}\uparrow & & \Vert \\
C_*(Y) & \xrightarrow{\sigma_g} & C_*(N_g) & \xleftarrow{p_g^*} & C_*(Z).
\end{array}$$

The bottom right square is commutative by Proposition 49.18. The bottom left and upper right squares are commutative by Proposition 51.5 and Lemma 51.9, respectively. The upper left square is commutative up to homotopy by Lemma 51.10. The statement follows from commutativity of the diagram up to homotopy. □

Let

$$(55.2)\qquad \begin{array}{ccc}
Y' & \xrightarrow{f'} & X' \\
{\scriptstyle g}\downarrow & & \downarrow{\scriptstyle h} \\
Y & \xrightarrow{f} & X
\end{array}$$

be a fiber product diagram with f and f' regular closed embeddings. The natural morphisms $i : N_{f'} \to g^*(N_f)$ of normal bundles over Y' are closed embeddings. The factor bundle $E = g^*(N_f)/N_{f'}$ over Y' is called the *excess vector bundle*.

Proposition 55.3 (Excess Formula). *Let h be a proper morphism. Then in the notation of diagram* (55.2),

$$f^\star \circ h_* = g_* \circ e(E) \circ f'^\star.$$

PROOF. Let

$$p : N_f \to Y, \quad p' : N_{f'} \to Y', \quad i : N_{f'} \to g^*(N_f),$$

$$r : g^*(N_f) \to N_f \quad \text{and} \quad t : g^*(N_f) \to Y'$$

be the canonical morphisms. It suffices to prove that the diagram

$$\begin{array}{ccccccc}
C_*(X') & \xrightarrow{\sigma_{f'}} & C_*(N_{f'}) & \xleftarrow{p'^*} & C_*(Y') & & \\
 & & & \searrow^{i_*} & & \searrow^{e(E)} & \\
\Big\downarrow h_* & & \Big\downarrow (ri)_* & & C_*\big(g^*(N_f)\big) & \xleftarrow{t^*} & C_*(Y') \\
 & & & \swarrow_{r_*} & & \swarrow_{g_*} & \\
C_*(X) & \xrightarrow{\sigma_f} & C_*(N_f) & \xleftarrow{p^*} & C_*(Y) & &
\end{array}$$

commutes.

The commutativity everywhere but at the top parallelogram follows by Propositions 49.20 and 51.6. Hence it suffices to show that $t^* \circ e(E) = i_* \circ p'^*$. Consider the fiber product diagram

$$\begin{array}{ccc}
N_{f'} & \xrightarrow{i} & g^*(N_f) \\
{\scriptstyle p'}\Big\downarrow & & \Big\downarrow{\scriptstyle j} \\
X & \xrightarrow{s} & E,
\end{array}$$

where j is the natural morphism of vector bundles and s is the zero section. Let $q : E \to Y'$ be the natural morphism. It follows from the equality $q \circ j = t$ and Proposition 49.20 that

$$t^* \circ e(E) = t^* \circ q^{*-1} \circ s_* = j^* \circ s_* = i_* \circ p'^*. \qquad \square$$

Corollary 55.4. *Suppose under the conditions of Proposition 55.3 that f and f' are regular closed embeddings of the same codimension. Then $f^\bigstar \circ h_* = g_* \circ f'^\bigstar$.*

PROOF. In this case, $E = 0$ so $e(E)$ is the identity. $\square$

A consequence of Propositions 49.18 and 51.5 is the following:

Proposition 55.5. *Suppose in the diagram (55.2) that h is a flat morphism. Then the diagram*

$$\begin{array}{ccc}
A_*(X, K_*) & \xrightarrow{f^\bigstar} & A_*(Y, K_*) \\
{\scriptstyle h^*}\Big\downarrow & & \Big\downarrow{\scriptstyle g^*} \\
A_*(X', K_*) & \xrightarrow{f'^\bigstar} & A_*(Y', K_*)
\end{array}$$

is commutative.

Proposition 55.6. *Let $f : Y \to X$ be a regular closed embedding of equidimensional schemes. Then $f^\bigstar([X]) = [Y]$.*

PROOF. By Example 49.13 and Proposition 52.7,

$$f^\bigstar([X]) = (p_f^*)^{-1} \circ \sigma_f([X]) = (p_f^*)^{-1}([N_f]) = [Y]. \qquad \square$$

Lemma 55.7. *Let $i : U \to V$ and $g : V \to W$ be a regular closed embedding and a flat morphism, respectively, and let $h = g \circ i$. If h is flat, then $h^* = i^\bigstar \circ g^*$.*

PROOF. Let $p : N_i \to U$ be the canonical morphism. By Lemma 51.9, we have $\sigma_i \circ g^* = (h \circ p)^* = p^* \circ h^*$, hence $i^\bigstar \circ g^* = (p^*)^{-1} \circ \sigma_i \circ g^* = h^*$. $\square$

We turn to the study of the functorial behavior of Euler and Chern classes under Gysin homomorphisms. The next proposition is a consequence of Corollary 55.4 and Proposition 55.5 (cf. the proof of Proposition 53.3).

Proposition 55.8. *Let $f : Y \to X$ be a regular closed embedding and L a line bundle over X. Set $L' = f^*(L)$. Then $f^\star \circ e(L) = e(L') \circ f^\star$.*

As is in the proof of Proposition 54.5, we get

Proposition 55.9. *Let $f : Y \to X$ be a regular closed embedding and E a vector bundle over X. Set $E' = f^*(E)$. Then $f^\star \circ c(E) = c(E') \circ f^\star$.*

Proposition 55.10. *Let $f : Y \to X$ be a regular closed embedding. Then $f^\star \circ f_* = e(N_f)$.*

PROOF. Let $p : N_f \to Y$ and $s : Y \to N_f$ be the canonical morphism and the zero section of the normal bundle respectively. By Corollary 51.7,

$$f^\star \circ f_* = (p^*)^{-1} \circ \sigma_f \circ f_* = (p^*)^{-1} \circ s_* = e(N_f). \qquad \square$$

Proposition 55.11. *Let $f : Y \to X$ be a closed embedding given by a sheaf of locally principal ideals $I \subset \mathcal{O}_X$. Let $f' : Y' \to X$ be the closed embedding given by the sheaf of ideals I^n for some $n > 0$ and $g : Y \to Y'$ the canonical morphism. Then*

$$f'^\star = n(g_* \circ f^\star).$$

PROOF. We define a natural finite morphism $h : D_f \to D_{f'}$ of deformation schemes as follows: We may assume that X is affine, $X = \operatorname{Spec}(A)$, and $Y = \operatorname{Spec}(A/I)$. Then $D_f = \operatorname{Spec}(\widetilde{A})$ and $D_{f'} = \operatorname{Spec}(\widetilde{A}')$ where

$$\widetilde{A} = \coprod_{k \in \mathbb{Z}} I^{-k} t^k, \qquad \widetilde{A}' = \coprod_{k \in \mathbb{Z}} I^{-kn} (t')^k$$

(cf. §104.E).

The morphism h is induced by the ring homomorphism $\widetilde{A}' \to \widetilde{A}$ taking a component $I^{-kn} t'^k$ identically to $I^{-kn} t^{kn}$. In particular, the image of t' is equal to t^n.

The morphism h yields a commutative diagram (cf. §104.E)

$$\begin{array}{ccccc} N_f & \longrightarrow & D_f & \longleftarrow & \mathbb{G}_m \times X \\ \downarrow{\scriptstyle r} & & \downarrow{\scriptstyle h} & & \downarrow{\scriptstyle q} \\ N_{f'} & \longrightarrow & D_{f'} & \longleftarrow & \mathbb{G}_m \times X, \end{array}$$

where q is the identity on X and the nth power morphism on $\mathbb{G}_m$. Let ∂ (respectively, ∂') be the boundary map with respect to the top row (respectively, the bottom row) of the diagram. It follows from Proposition 49.36(1) that

$$r_* \circ \partial = \partial' \circ q_*. \tag{55.12}$$

For any $\alpha \in C_*(X)$ we have

$$q_*\big(\{t\} \cdot ([\mathbb{G}_m] \times \alpha)\big) = \{\pm t'\} \cdot ([\mathbb{G}_m] \times \alpha) \tag{55.13}$$

since the norm of t in the field extension $F(t)/F(t')$ is equal to $\pm t'$. By (55.12) and (55.13), we have

$$\begin{aligned}(r_* \circ \sigma_f)(\alpha) &= (r_* \circ \partial)\big(\{t\} \cdot ([\mathbb{G}_m] \times \alpha)\big) \\ &= (\partial' \circ q_*)\big(\{t\} \cdot ([\mathbb{G}_m] \times \alpha)\big) \\ &= \partial'\big(\{\pm t'\} \cdot ([\mathbb{G}_m] \times \alpha)\big) \\ &= \sigma_{f'}(\alpha),\end{aligned}$$

hence

$$r_* \circ \sigma_f = \sigma_{f'}. \tag{55.14}$$

The morphism p factors into the composition of morphisms in the first row of the commutative diagram

$$\begin{array}{ccccc} N_f & \xrightarrow{i} & (N_f)^{\otimes n} & \xrightarrow{j} & N_{f'} \\ {\scriptstyle p}\downarrow & & {\scriptstyle s}\downarrow & & {\scriptstyle p'}\downarrow \\ Y & = & Y & \xrightarrow{g} & Y' \end{array}$$

of vector bundles. As the morphism i is finite flat of degree n, the composition $i_* \circ i^*$ is multiplication by n. The right square of the diagram is a fiber square. Hence by Proposition 49.20, we have

$$r_* \circ p^* = j_* \circ i_* \circ i^* \circ s^* = n(j_* \circ s^*) = n(p'^* \circ g_*).$$

It follows from (55.14) that

$$f'^{\bigstar} = (p'^*)^{-1} \circ \sigma_{f'} = (p'^*)^{-1} \circ r_* \circ \sigma_f = n\big(g_* \circ (p^*)^{-1} \circ \sigma_f\big) = n(g_* \circ f^{\bigstar})$$

as needed. □

55.B. The pull-back homomorphisms. Let $f : Y \to X$ be a morphism of equidimensional schemes with X smooth. By Corollary 104.14, the morphism

$$i = (1_Y, f) : Y \to Y \times X$$

is a regular closed embedding of codimension $d_X = \dim X$ with the normal bundle $N_i = f^*(T_X)$, where T_X is the tangent bundle of X (cf. Corollary 104.14). The projection $p : Y \times X \to X$ is a flat morphism of relative dimension d_Y. Set $d = d_X - d_Y$. We define the *pull-back homomorphism*

$$f^* : A_*(X, K_*) \to A_{*-d}(Y, K_{*+d}) \tag{55.15}$$

as the composition $i^{\bigstar} \circ p^*$.

We use the same notation for the pull-back homomorphism just defined and the flat pull-back. The following proposition justifies this notation.

Proposition 55.16. *Let $f : Y \to X$ be a flat morphism of equidimensional schemes and let X be smooth. Then the pull-back f^* in* (55.15) *coincides with the flat pull-back homomorphism.*

PROOF. This follows by applying Lemma 55.7 to the closed embedding $i = (1_Y, f) : Y \to Y \times X$ and to the projection $g : Y \times X \to X$. □

We have the following two propositions about the compositions of the pull-back maps.

Proposition 55.17. *Let $Z \xrightarrow{g} Y \xrightarrow{f} X$ be morphisms of equidimensional schemes with X smooth and g flat. Then $(f \circ g)^* = g^* \circ f^*$.*

PROOF. Consider the fiber product diagram

$$\begin{array}{ccc} Z & \xrightarrow{g} & Y \\ {\scriptstyle i_Z}\downarrow & & \downarrow{\scriptstyle i_Y} \\ Z \times X & \xrightarrow{h} & Y \times X, \end{array}$$

where $i_Y = (1_Y, f)$, $i_Z = (1_Z, fg)$, $h = (g, 1_X)$ and two projections $p_Y : Y \times X \to X$ and $p_Z : Z \times X \to X$. We have $p_Z = p_Y \circ h$. By Propositions 49.18 and 55.5,

$$\begin{aligned} (f \circ g)^* &= i_Z^{\bigstar} \circ p_Z^* \\ &= i_Z^{\bigstar} \circ h^* \circ p_Y^* \\ &= g^* \circ i_Y^{\bigstar} \circ p_Y^* \\ &= g^* \circ f^*. \end{aligned}$$

□

Proposition 55.18. *Let $Z \xrightarrow{g} Y \xrightarrow{f} X$ be morphisms of equidimensional schemes with both Y and X smooth. Then $(f \circ g)^* = g^* \circ f^*$.*

PROOF. Consider the commutative diagram

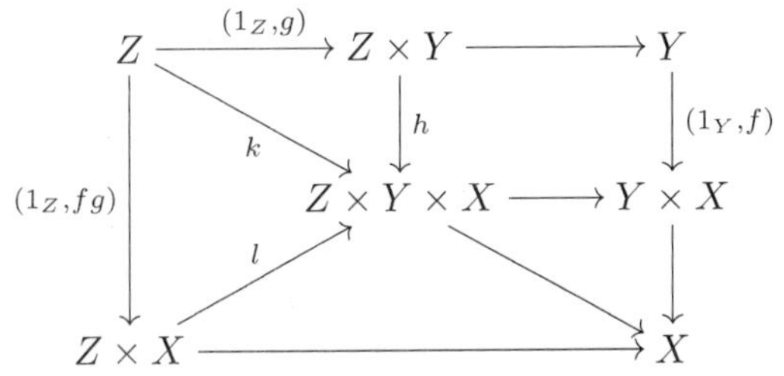

where

$$k = (1_Z, g, fg), \quad h(z, y) = \big(z, y, f(y)\big), \quad l(z, x) = \big(z, g(z), x\big),$$

and all unmarked arrows are the projections. Applying the Gysin homomorphisms or the flat pull-backs for all arrows in the diagram, we get a diagram of homomorphisms of the K-homology groups that is commutative by Propositions 49.18, 55.1, 55.5, and Lemma 55.7. □

We next show that the pull-back homomorphism for a regular closed embedding coincides with the Gysin homomorphism:

Proposition 55.19. *Let $f : Y \to X$ be a regular closed embedding of equidimensional schemes with X smooth. Then $f^* = f^{\bigstar}$.*

PROOF. The commutative diagram

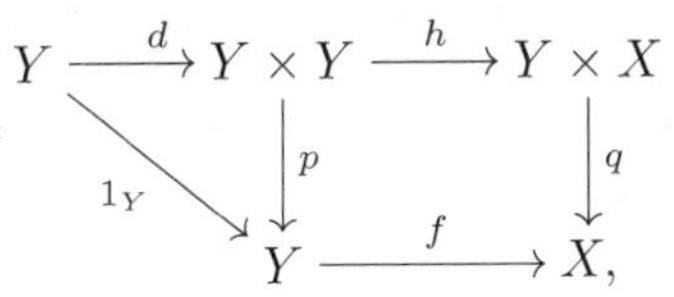

with d the diagonal embedding and $h = 1_Y \times f$ gives rise to the diagram

$$\begin{array}{ccccc}
A_*(X, K_*) & \xrightarrow{f^\bigstar} & A_*(Y, K_*) & & \\
\downarrow{\scriptstyle q^*} & & \downarrow{\scriptstyle p^*} & \searrow^{1} & \\
A_*(Y \times X, K_*) & \xrightarrow{h^\bigstar} & A_*(Y \times Y, K_*) & \xrightarrow{d^\bigstar} & A_*(Y, K_*).
\end{array}$$

The square is commutative by Proposition 55.5 and the triangle by Lemma 55.7. Let $g = h \circ d$. Then

$$f^* = g^\bigstar \circ q^* = d^\bigstar \circ h^\bigstar \circ q^* = f^\bigstar. \qquad \square$$

The pull-back homomorphisms commute with external products:

Proposition 55.20. *Let $f : X' \to X$ and $g : Y' \to Y$ be morphisms of equidimensional schemes with X and Y smooth. Then for every $\alpha \in C_*(X)$ and $\beta \in C_*(Y)$, we have*

$$(f \times g)^*(\alpha \times \beta) = f^*(\alpha) \times g^*(\beta).$$

PROOF. It suffices to do the case with $g = 1_Y$ and by Proposition 50.5 with f a regular closed embedding. Denote by $q_X : \mathbb{G}_m \times X \to X$ and $p_f : N_f \to X'$ the canonical morphisms. Note that $N_{f \times 1_Y} = N_f \times Y$. Consider the diagram

$$\begin{array}{ccccc}
C_*(X) & \xrightarrow{q_X^*} & C_*(\mathbb{G}_m \times X) & \xrightarrow{\{t\}} & C_*(\mathbb{G}_m \times X) \\
\downarrow & & \downarrow & & \downarrow \\
C_*(X \times Y) & \xrightarrow{q_{X \times Y}^*} & C_*(\mathbb{G}_m \times X \times Y) & \xrightarrow{\{t\}} & C_*(\mathbb{G}_m \times X \times Y)
\end{array}$$

$$\begin{array}{ccccc}
C_*(\mathbb{G}_m \times X) & \xrightarrow{\partial} & C_*(N_f) & \xleftarrow{p_f^*} & C_*(X') \\
\downarrow & & \downarrow & & \downarrow \\
C_*(\mathbb{G}_m \times X \times Y) & \xrightarrow{\partial} & C_*(N_{f \times 1_Y}) & \xleftarrow{(p_f \times 1_Y)^*} & C_*(X' \times Y)
\end{array}$$

where all vertical homomorphisms are given by the external product with β. The commutativity of all squares follow from Propositions 50.5, 50.7, and 50.8. $\square$

Proposition 55.21. *Let $f : Y \to X$ be a morphism of equidimensional schemes with X smooth. Then $f^*([X]) = [Y]$.*

PROOF. Let $i = (1_Y, f) : Y \to Y \times X$ be the graph of f and let $p : Y \times X \to X$ be the projection. It follows from Corollary 50.6 and Proposition 55.6 that

$$f^*([X]) = i^\bigstar \circ p^*([X]) = i^\bigstar([Y \times X]) = [Y]. \qquad \square$$

The following statement on the commutativity of the pull-back maps and the Chern classes is a consequence of Propositions 54.5(2) and 55.9.

Proposition 55.22. *Let $f : Y \to X$ be a morphism of equidimensional schemes with X smooth and E a vector bundle over X. Set $E' = f^*(E)$. Then $f^* \circ c(E) = c(E') \circ f^*$.*

56. K-cohomology ring of smooth schemes

We now consider the case that our scheme X is smooth. This allows us to introduce the *K-cohomology groups* $A^*(X, K_*)$ which we do as follows: If X is irreducible of dimension d, we set

$$A^p(X, K_q) := A_{d-p}(X, K_{q-d}).$$

In the general case, let $X_1, X_2, \dots, X_s$ be (disjoint) irreducible components of X. We set

$$A^p(X, K_q) := \coprod_{i=1}^{s} A^p(X_i, K_q).$$

In particular, if X is an equidimensional smooth scheme of dimension d, then $A^p(X, K_q) = A_{d-p}(X, K_{q-d})$.

Let $f : Y \to X$ be a morphism of smooth schemes. We define the pull-back homomorphism

$$f^* : A^p(X, K_q) \to A^p(Y, K_q)$$

as follows: If X and Y are both irreducible and of dimension d_X and d_Y, respectively, we define f^* as in §55.B:

$$f^* : A^p(X, K_q) = A_{d_X-p}(X, K_{q-d_X}) \xrightarrow{f^*} A_{d_Y-p}(X, K_{q-d_Y}) = A^p(Y, K_q).$$

If just Y is irreducible, we have $f(Y) \subset X_i$ for an irreducible component X_i of X. We define the pull-back as the composition

$$A^p(X, K_q) \to A^p(X_i, K_q) \xrightarrow{f^*} A^p(Y, K_q),$$

where the first map is the canonical projection. Finally, in the general case, we define f^* as the direct sum of the homomorphisms $A^p(X, K_q) \to A^p(Y_j, K_q)$ over all irreducible components Y_j of Y.

It follows from Proposition 55.18 that if $Z \xrightarrow{g} Y \xrightarrow{f} X$ are morphisms of smooth schemes, then $(f \circ g)^* = g^* \circ f^*$.

Let X be a smooth scheme. Denote by

$$d = d_X : X \to X \times X$$

the diagonal closed embedding. The composition

$$A^p(X, K_q) \otimes A^{p'}(X, K_{q'}) \xrightarrow{\times} A^{p+p'}(X \times X, K_{q+q'}) \xrightarrow{d^*} A^{p+p'}(X, K_{q+q'}) \tag{56.1}$$

defines a product on $A^*(X, K_*)$.

Remark 56.2. If $X = X_1 \coprod X_2$, then $A^*(X, K_*) = A^*(X_1, K_*) \oplus A^*(X_2, K_*)$. Since the image of the diagonal morphism d_X does not intersect $X_1 \times X_2$, the product of two classes from $A^*(X_1, K_*)$ and $A^*(X_2, K_*)$ is zero.

Proposition 56.3. *The product in* (56.1) *is associative.*

Proof. Let $\alpha, \beta, \gamma \in A^*(X, K_*)$. By Proposition 55.20, we have

$$\begin{aligned}(\alpha \times \beta) \times \gamma &= d^*\big(d^*(\alpha \times \beta) \times \gamma\big)\\ &= d^* \circ (d \times 1_X)^*(\alpha \times \beta \times \gamma)\\ &= \big((d \times 1_X) \circ d\big)^*(\alpha \times \beta \times \gamma)\\ &= c^*(\alpha \times \beta \times \gamma),\end{aligned}$$

where $c : X \to X \times X \times X$ is the diagonal embedding. Similarly, $\alpha \times (\beta \times \gamma) = c^*(\alpha \times \beta \times \gamma)$. $\square$

Proposition 56.4. *For every smooth scheme X, the product in $A^*(X, K_*)$ is bigraded commutative, i.e., if $\alpha \in A^p(X, K_q)$ and $\alpha' \in A^{p'}(X, K_{q'})$, then*

$$\alpha \cdot \alpha' = (-1)^{(p+q)(p'+q')} \alpha' \cdot \alpha.$$

PROOF. It follows from (50.1) that

$$\alpha \cdot \alpha' = d^*(\alpha \times \alpha') = (-1)^{(p+q)(p'+q')} d^*(\alpha' \times \alpha) = (-1)^{(p+q)(p'+q')} \alpha' \cdot \alpha. \quad \square$$

Let X be a smooth scheme and let $X_1, X_2, \dots$ be the irreducible components of X. Then $[X] = \sum [X_i]$ in $A^0(X, K_0)$.

Proposition 56.5. *The class $[X]$ is the identity in $A^*(X, K_*)$ under the product.*

PROOF. We may assume that X is irreducible. Let $f : X \times X \to X$ be the first projection. Since $f \circ d = 1_X$, it follows from Corollary 50.6 and Proposition 55.16 that

$$\alpha \cdot [X] = d^*\big(\alpha \times [X]\big) = d^* f^*(\alpha) = \alpha. \quad \square$$

We have proven:

Theorem 56.6. *Let X be a smooth scheme. Then $A^*(X, K_*)$ is a bigraded commutative associative ring with the identity $[X]$.*

Remark 56.7. If $X_1, \dots, X_n$ are the irreducible components of a smooth scheme X, the ring $A^*(X, K_*)$ is the product of the rings $A^*(X_1, K_*), \dots, A^*(X_n, K_*)$.

Proposition 56.8. *Let $f : Y \to X$ be a morphism of smooth schemes. Then*

$$f^*(\alpha \cdot \beta) = f^*(\alpha) \cdot f^*(\beta)$$

for all $\alpha, \beta \in A^*(X, K_*)$ and $f^*\big([X]\big) = [Y]$.

PROOF. Since $(f \times f) \circ d_Y = d_X \circ f$, it follows from Propositions 55.18 and 55.20 that

$$\begin{aligned} f^*(\alpha \cdot \beta) &= f^* \circ d_X^*(\alpha \times \beta) \\ &= d_Y^* \circ (f \times f)^*(\alpha \times \beta) \\ &= d_Y^*\big(f(\alpha) \times f(\beta)\big) \\ &= f^*(\alpha) \cdot f^*(\beta). \end{aligned}$$

The second equality follows from Proposition 55.21. $\square$

It follows from Proposition 56.8 that the correspondence $X \mapsto A^*(X, K_*)$ gives rise to a cofunctor from the category of smooth schemes and arbitrary morphisms to the category of bigraded rings and homomorphisms of bigraded rings.

Proposition 56.9 (Projection Formula). *Let $f : Y \to X$ be a proper morphism of smooth schemes. Then*

$$f_*\big(\alpha \cdot f^*(\beta)\big) = f_*(\alpha) \cdot \beta$$

for every $\alpha \in A^*(Y, K_*)$ and $\beta \in A^*(X, K_*)$.

Proof. Let $g = (1_Y \times f) \circ d_Y$. Then we have the fiber product diagram

$$\begin{array}{ccc} Y & \xrightarrow{g} & Y \times X \\ {\scriptstyle f}\downarrow & & \downarrow{\scriptstyle f\times 1_X} \\ X & \xrightarrow{d_X} & X \times X. \end{array} \tag{56.10}$$

It follows from Propositions 49.20 and 55.20 that

$$\begin{aligned} f_*\big(\alpha \cdot f^*(\beta)\big) &= f_* \circ d_Y^*\big(\alpha \times f^*(\beta)\big) \\ &= f_* \circ d_Y^* \circ (1_Y \times f)^*(\alpha \times \beta) \\ &= f_* \circ g^*(\alpha \times \beta) \\ &= d_X^* \circ (f \times 1_Y)_*(\alpha \times \beta) \\ &= d_X^*(f_*(\alpha) \times \beta) \\ &= f_*(\alpha) \cdot \beta. \end{aligned}$$

□

The projection formula asserts that the push-forward homomorphism f_* is $A^*(X, K_*)$-linear if we view $A^*(Y, K_*)$ as a $A^*(X, K_*)$-module via f^*.

The following statement is an analog of the projection formula.

Proposition 56.11. *Let $f : Y \to X$ be a morphism of equidimensional schemes with X smooth. Then*

$$f_*\big(f^*(\beta)\big) = f_*\big([Y]\big) \cdot \beta$$

for every $\beta \in A^*(X, K_*)$.

Proof. The closed embeddings g and d_X in the diagram (56.10) are regular of the same codimension (cf. Corollary 104.14). Let $p : X \times X \to X$ be the second projection. Then the composition $q = p \circ (f \times 1_X) : Y \times X \to X$ is also the projection. By Propositions 50.4, 50.5, 55.17, 55.21, and Corollaries 50.6, 55.4, we have

$$\begin{aligned} f_*\big(f^*(\beta)\big) &= f_* \circ g^\star \circ q^*(\beta) \\ &= f_* \circ g^\star \circ (f \times 1_X)^* \circ p^*(\beta) \\ &= d_X^\star \circ (f \times 1_X)_* \circ (f \times 1_X)^* \circ p^*(\beta) \\ &= d_X^\star \circ (f_* \times \mathrm{id}) \circ (f^* \times \mathrm{id})\big([X] \times \beta\big) \\ &= d_X^\star\big(f_* \circ f^*\big([X]\big) \times \beta\big) \\ &= d_X^\star\big(f_*([Y]) \times \beta\big) \\ &= f_*\big([Y]\big) \cdot \beta. \end{aligned}$$

□

CHAPTER X

Chow Groups

In this chapter we study Chow groups as special cases of K-homology and K-cohomology theories, so we can apply results from the previous chapter. Chow groups will remain the main tool in the rest of the book. We also develop the theory of Segre classes that will be used in the chapter on the Steenrod operations that follows.

57. Definition of Chow groups

Recall that a *scheme* is a separated scheme of finite type over a field and a *variety* is an integral scheme.

57.A. Two equivalent definitions of Chow groups.

Definition 57.1. Let X be a scheme over F and $p \in \mathbb{Z}$. We call the group

$$\mathrm{CH}_p(X) := A_p(X, K_{-p})$$

the *Chow group of dimension p classes of cycles on* X.

By definition,

$$\mathrm{CH}_p(X) = \mathrm{Coker}\left(\coprod_{x \in X_{(p+1)}} K_1\big(F(x)\big) \xrightarrow{d_X} \coprod_{x \in X_{(p)}} K_0\big(F(x)\big) \right).$$

Note that $K_1\big(F(x)\big) = F(x)^\times$ and $K_0\big(F(x)\big) = \mathbb{Z}$. Thus the Chow group $\mathrm{CH}_p(X)$ is the factor group of the free abelian group

$$\mathrm{Z}_p(X) = \coprod_{x \in X_{(p)}} \mathbb{Z},$$

called the *group of p-dimensional cycles on* X, by the subgroup generated by the divisors $d_X(f) = \mathrm{div}(f)$ for all $f \in F(x)^\times$ and $x \in X_{(p+1)}$.

A point $x \in X$ of dimension p gives rise to a *prime cycle* in $\mathrm{Z}_p(X)$, denoted by $[x]$. Thus, an element of $\mathrm{Z}_p(X)$ is a finite formal linear combination $\sum n_x[x]$ with $n_x \in \mathbb{Z}$ and $\dim x = p$. We will often write $\overline{\{x\}}$ instead of x, so that an element of $\mathrm{Z}_p(X)$ is a finite formal linear combination $\sum n_Z[Z]$ where the sum is taken over closed subvarieties $Z \subset X$ of dimension p. We will use the same notation for the classes of cycles in $\mathrm{CH}_p(X)$. Note that a closed subscheme $W \subset X$ (not necessarily integral) defines the cycle $[W] \in \mathrm{Z}(X)$ (cf. Example 49.2).

Example 57.2. Let X be a scheme of dimension d. The group $\mathrm{CH}_d(X) = \mathrm{Z}_d(X)$ is free with basis the classes of irreducible components (generic points) of X of dimension d.

The divisor of a function can be computed in a simpler way. Let R be a 1-dimensional Noetherian local domain with quotient field L. We define the *order homomorphism*

$$\mathrm{ord}_R : L^\times \to \mathbb{Z}$$

by the formula $\mathrm{ord}_R(r) = l(R/rR)$ for every nonzero $r \in R$.

Let Z be a variety over F of dimension d. For any point $x \in Z$ of dimension $d-1$, the local ring $\mathcal{O}_{Z,x}$ is 1-dimensional. Hence the order homomorphism

$$\mathrm{ord}_x = \mathrm{ord}_{\mathcal{O}_{Z,x}} : F(Z)^\times \to \mathbb{Z}$$

is well-defined.

Proposition 57.3. *Let Z be a variety over F of dimension d and $f \in F(Z)^\times$. Then $\mathrm{div}(f) = \sum \mathrm{ord}_x(f) \cdot x$, where the sum is taken over all points $x \in Z$ of dimension $d-1$.*

PROOF. Let R be the local ring $\mathcal{O}_{Z,x}$, where x is a point of dimension $d-1$. Let $\widetilde{R}$ denote the integral closure of R in $F(Z)$. For every nonzero $f \in R$, the x-component of $\mathrm{div}(f)$ is equal to

$$\sum l(\widetilde{R}_Q/f\widetilde{R}_Q) \cdot [\widetilde{R}/Q : F(x)],$$

where the sum is taken over all maximal ideals Q of $\widetilde{R}$. Applying Lemma 102.3 to the $\widetilde{R}$-module $M = \widetilde{R}/f\widetilde{R}$, we have the x-component equals $l_R(\widetilde{R}/f\widetilde{R})$. Since $\widetilde{R}/R$ is an R-module of finite length, $l_R(\widetilde{R}/f\widetilde{R}) = l_R(R/fR) = \mathrm{ord}_x(f)$. □

We wish to give an equivalent definition of Chow groups. To do so we need some preliminaries.

Let Z be a variety over F of dimension d and $f : Z \to \mathbb{P}^1$ a dominant morphism. Thus f is a flat morphism of relative dimension $d-1$. For any rational point $a \in \mathbb{P}^1$, the pull-back scheme $f^{-1}(a)$ is an equidimensional subscheme of Z of dimension $d-1$. Note that we can view f as a rational function on Z.

Lemma 57.4. *Let f be as above. Then $\mathrm{div}(f) = [f^{-1}(0)] - [f^{-1}(\infty)]$ on Z.*

PROOF. Let $x \in Z$ be a point of dimension $d-1$ with the x-component of $\mathrm{div}(f)$ nontrivial. Then $f(x) = 0$ or $f(x) = \infty$.

Consider the first case, so $f \in \mathcal{O}_{Z,x}$. By Proposition 57.3, the x-component of $\mathrm{div}(f)$ is equal to $\mathrm{ord}_x(f)$. The local ring $\mathcal{O}_{f^{-1}(0),x}$ coincides with $\mathcal{O}_{Z,x}/f\mathcal{O}_{Z,x}$, therefore, the x-component of $[f^{-1}(0)]$ is equal to

$$l(\mathcal{O}_{f^{-1}(0),x}) = l(\mathcal{O}_{Z,x}/f\mathcal{O}_{Z,x}) = \mathrm{ord}_x(f).$$

Similarly (applying the above argument to the function f^{-1}), we see that in the second case the x-component of $[f^{-1}(\infty)]$ is equal to $\mathrm{ord}_x(f^{-1}) = -\mathrm{ord}_x(f)$. □

Let X be a scheme and $Z \subset X \times \mathbb{P}^1$ a closed subvariety of dimension d with Z dominant over $\mathbb{P}^1$. Hence the projection $f : Z \to \mathbb{P}^1$ is flat of relative dimension $d-1$. For every rational point $a \in \mathbb{P}^1$, the projection $p : X \times \mathbb{P}^1 \to X$ maps the subscheme $f^{-1}(a)$ isomorphically onto a closed subscheme of X that we denote by $Z(a)$. It follows from Lemma 57.4 that

$$p_*\big(\mathrm{div}(f)\big) = [Z(0)] - [Z(\infty)]. \tag{57.5}$$

In particular, the classes of $[Z(0)]$ and $[Z(\infty)]$ coincide in $\mathrm{CH}(X)$.

Let $\mathrm{Z}(X;\mathbb{P}^1)$ denote the subgroup of $\mathrm{Z}(X\times\mathbb{P}^1)$ generated by the classes of closed subvarieties of $X \times \mathbb{P}^1$ that are dominant over $\mathbb{P}^1$. For any cycle $\beta \in \mathrm{Z}(X;\mathbb{P}^1)$ and any rational point $a \in \mathbb{P}^1$, the cycle $\beta(a) \in \mathrm{Z}(X)$ is well-defined.

If $\alpha = \sum n_Z[Z] \in \mathrm{Z}(X)$, we write $\alpha\times[\mathbb{P}^1]$ for the cycle $\sum n_Z[Z\times\mathbb{P}^1] \in \mathrm{Z}(X;\mathbb{P}^1)$. Clearly, $\big(\alpha \times [\mathbb{P}^1]\big)(a) = \alpha$ for any rational point $a \in \mathbb{P}^1$.

Let α and α' be two cycles on a scheme X. We say that α and α' are *rationally equivalent* if the classes of α and α' are equal in $\mathrm{CH}(X)$.

Proposition 57.6. *Two cycles α and α' on a scheme X are rationally equivalent if and only if there is a cycle $\beta \in \mathrm{Z}(X;\mathbb{P}^1)$ satisfying $\alpha = \beta(0)$ and $\alpha' = \beta(\infty)$.*

PROOF. It was shown in (57.5) that the classes of the cycles $\beta(0)$ and $\beta(\infty)$ are equal in $\mathrm{CH}(X)$. Conversely, suppose that the classes of α and α' are equal in $\mathrm{CH}(X)$. By the definition of the Chow group, there are closed subvarieties $Z_i \subset X$ and nonconstant rational functions g_i on Z_i such that

$$\alpha - \alpha' = \sum \operatorname{div}(g_i).$$

Let V_i be closure of the graph of g_i in $Z_i \times \mathbb{P}^1 \subset X \times \mathbb{P}^1$ and let $f_i : V_i \to \mathbb{P}^1$ be the induced morphism. Since g_i is nonconstant, the morphism f_i is dominant and $[V_i] \in \mathrm{Z}(X;\mathbb{P}^1)$.

The projection $p : X \times \mathbb{P}^1 \to X$ maps V_i birationally onto Z_i, hence by Proposition 49.9,

$$\operatorname{div}(g_i) = \operatorname{div}\big(p_*(f_i)\big) = p_* \operatorname{div}(f_i) = [V_i(0)] - [V_i(\infty)].$$

Let $\beta' = \sum[V_i] \in \mathrm{Z}(X;\mathbb{P}^1)$. We have

$$\alpha - \alpha' = \beta'(0) - \beta'(\infty).$$

Consider the cycle

$$\gamma = \alpha - \beta'(0) = \alpha' - \beta'(\infty)$$

and set $\beta'' = \gamma \times [\mathbb{P}^1]$ and $\beta = \beta' + \beta''$. Then $\beta(0) = \beta'(0) + \beta''(0) = \beta'(0) + \gamma = \alpha$ and similarly $\beta(\infty) = \alpha'$. □

It follows from the above that an equivalent definition of the Chow group $\mathrm{CH}(X)$ is given as the factor group of the the group of cycles $\mathrm{Z}(X)$ modulo the subgroup of cycles of the form $\beta(0) - \beta(\infty)$ for all $\beta \in \mathrm{Z}(X;\mathbb{P}^1)$.

57.B. Functorial properties of the Chow groups. We now specialize the functorial properties developed in the previous chapter to Chow groups.

A proper morphism $f : X \to Y$ gives rise to the *push-forward homomorphism*

$$f_* : \mathrm{CH}_p(X) \to \mathrm{CH}_p(Y).$$

Example 57.7. Let X be a complete scheme over F. The push-forward homomorphism $\deg : \mathrm{CH}(X) \to \mathrm{CH}(\operatorname{Spec} F) = \mathbb{Z}$ induced by the structure morphism $X \to \operatorname{Spec}(F)$ is called the *degree homomorphism*. For any $x \in X$, we have

$$\deg([x]) = \begin{cases} \deg(x) = [F(x) : F] & \text{if } x \text{ is a closed point,} \\ 0 & \text{otherwise.} \end{cases}$$

A flat morphism $g : Y \to X$ of relative dimension d defines the *pull-back homomorphism*

$$g^* : \mathrm{CH}_p(X) \to \mathrm{CH}_{p+d}(Y).$$

Proposition 57.8. *Let $g : Y \to X$ be a flat morphism of schemes over F of relative dimension d and $W \subset X$ a closed subscheme of pure dimension k. Then $g^*([W]) = [g^{-1}(W)]$ in $Z_{d+k}(Y)$.*

PROOF. Consider the fiber product diagram of natural morphisms

$$\begin{array}{ccccc} g^{-1}(W) & \xrightarrow{f} & W & \xrightarrow{p} & \mathrm{Spec}(F) \\ {\scriptstyle j}\downarrow & & \downarrow{\scriptstyle i} & & \\ Y & \xrightarrow{g} & X. & & \end{array}$$

By Propositions 49.18 and 49.20,

$$g^*([W]) = g^* \circ i_* \circ p^*(1) = j_* \circ f^* \circ p^*(1) = j_* \circ (p \circ f)^*(1) = [g^{-1}(W)]. \quad \square$$

The localization property (cf. §52.D) yields:

Proposition 57.9 (Localization Sequence). *Let X be a scheme, $Z \subset X$ a closed subscheme and $U = X \setminus Z$. Let $i : Z \to X$ and $j : U \to X$ be the closed embedding and the open immersion respectively. Then the sequence*

$$\mathrm{CH}_p(Z) \xrightarrow{i_*} \mathrm{CH}_p(X) \xrightarrow{j^*} \mathrm{CH}_p(U) \to 0$$

is exact.

The following proposition shows that the restriction on the generic fiber of a morphism is surjective on Chow groups.

Proposition 57.10. *Let X be a variety of dimension n and $f : Y \to X$ a dominant morphism. Let x denote the generic point of X and Y_x the generic fiber of f. Then the pull-back homomorphism $\mathrm{CH}_p(Y) \to \mathrm{CH}_{p-n}(Y_x)$ is surjective.*

PROOF. By the continuity property (cf. Proposition 52.9), the pull-back homomorphism $\mathrm{CH}_p(Y) \to \mathrm{CH}_{p-n}(Y_x)$ is the colimit of surjective restriction homomorphisms $\mathrm{CH}_p(Y) \to \mathrm{CH}_p(f^{-1}(U))$ over all nonempty open subschemes U of X and therefore is surjective. $\square$

Corollary 57.11. *For every variety X of dimension n and scheme Y over F, the pull-back homomorphism $\mathrm{CH}_p(X \times Y) \to \mathrm{CH}_{p-n}(Y_{F(X)})$ is surjective.*

Let X and Y be two schemes. It follows from §52.C that there is a product map of Chow groups

$$\mathrm{CH}_p(X) \otimes \mathrm{CH}_q(Y) \to \mathrm{CH}_{p+q}(X \times Y).$$

Proposition 57.12. *Let $Z \subset X$ and $W \subset Y$ be two closed equidimensional subschemes of dimensions d and e, respectively. Then*

$$[Z \times W] = [Z] \times [W] \quad \text{in} \quad Z_{d+e}(X \times Y).$$

PROOF. Let $p : Z \to \mathrm{Spec}(F)$ and $q : W \to \mathrm{Spec}(F)$ be the structure morphisms and $i : Z \to X$ and $j : W \to Y$ the closed embeddings. By Example 49.2 and Propositions 50.4, 50.5, we have

$$[Z \times W] = (i \times j)_* \circ (p \times q)^*(1) = (i_* \circ p^*(1)) \times (j_* \circ q^*(1)) = [Z] \times [W]. \quad \square$$

Theorem 57.13 (Homotopy Invariance; cf. Theorem 52.13). *Let $g : Y \to X$ be a flat morphism of schemes over F of relative dimension d. Suppose that for every $x \in X$, the fiber Y_x is isomorphic to the affine space $\mathbb{A}^d_{F(x)}$. Then the pull-back homomorphism*

$$g^* : \mathrm{CH}_p(X) \to \mathrm{CH}_{p+d}(Y)$$

is an isomorphism for every p.

Theorem 57.14 (Projective Bundle Theorem; cf. Theorem 53.10). *Let $E \to X$ be a vector bundle of rank r, $q : \mathbb{P}(E) \to X$ the associated projective bundle morphism, and e the Euler class of the canonical or tautological line bundle over $\mathbb{P}(E)$. Then the homomorphism*

$$\coprod_{i=1}^{r} e^{r-i} \circ q^* : \coprod_{i=1}^{r} \mathrm{CH}_{*-i+1}(X) \to \mathrm{CH}_*\big(\mathbb{P}(E)\big)$$

is an isomorphism, i.e., every $\alpha \in \mathrm{CH}_*\big(\mathbb{P}(E)\big)$ can be written in the form

$$\alpha = \sum_{i=1}^{r} e^{r-i}\big(q^*(\alpha_i)\big)$$

for uniquely determined elements $\alpha_i \in \mathrm{CH}_{*-i+1}(X)$.

Example 57.15. Let $X = \mathbb{P}(V)$, where V is a vector space of dimension $d+1$ over F. For every $p \in [0,\ d]$, let $l_p \in \mathrm{CH}_p\big(\mathbb{P}(V)\big)$ be the class of the subscheme $\mathbb{P}(V_p)$ of X, where V_p is a subspace of V of dimension $p+1$. By Corollary 53.7,

$$\mathrm{CH}_p\big(\mathbb{P}(V)\big) = \begin{cases} \mathbb{Z} \cdot l_p & \text{if} \quad 0 \le p \le d, \\ 0 & \text{otherwise.} \end{cases}$$

Let $f : Y \to X$ be a regular closed embedding of codimension r. As usual we write N_f for the normal bundle of f. The *Gysin homomorphism*

$$f^\star : \mathrm{CH}_*(X) \to \mathrm{CH}_{*-r}(Y)$$

is defined by the formula $f^\star = (p^*)^{-1} \circ \sigma_f$, where $p : N_f \to Y$ is the canonical morphism and σ_f is the deformation homomorphism.

Let $Z \subset X$ be a closed subscheme of pure dimension k and set $W = f^{-1}(Z)$. The cone C_g of the restriction $g : W \to Z$ of f is of pure dimension k. Proposition 52.7 yields:

Corollary 57.16. *Under the conditions of Proposition 52.7, we have $f^\star\big([Z]\big) = (p^*)^{-1} h_*\big([C_g]\big)$.*

Lemma 57.17. *Let C' be an irreducible component of C_g. Then C' is an integral cone over a closed subvariety $W' \subset W$ with $\dim W' \ge k - r$.*

Proof. Let N' be the restriction of the normal bundle N_f on W'. Since C' is a closed subvariety of N' of dimension k (cf. Example 104.3), we have

$$k = \dim C' \le \dim N' = \dim W' + r. \qquad \square$$

Corollary 57.18. *Let $V \subset W$ be an irreducible component. Then there is an irreducible component of C_g that is a cone over V. In particular, $\dim V \ge k - r$.*

PROOF. Let $v \in V$ be the generic point. Since the canonical morphism $q : C_g \to W$ is surjective (that is, split by the zero section), there is an irreducible component $C' \subset C_g$ such that $v \in \operatorname{Im} q$. Clearly, $\operatorname{Im} q = V$, i.e., C' is a cone over V. □

We say that the scheme Z has a *proper inverse image with respect to* f if every irreducible component of $W = f^{-1}(Z)$ has dimension $k - r$.

Proposition 57.19. *Let $f : Y \to X$ be a regular closed embedding of schemes over F of codimension r and $Z \subset X$ a closed equidimensional subscheme having proper inverse image with respect to f. Let $V_1, V_2, \dots, V_s$ be all the irreducible components of $W = f^{-1}(Z)$, so $[W] = \sum n_i[V_i]$ for some $n_i > 0$. Then*

$$f^\star([Z]) = \sum_{i=1}^{s} m_i[V_i],$$

for some integers m_i with $1 \le m_i \le n_i$.

PROOF. Let $g : W \to Z$ be the restriction of f and C_i the restriction of the cone C_g on V_i. Let N_i be the restriction to V_i of the normal cone N_f. As N_i is a vector bundle of rank r over the variety V_i of dimension $k - r$, the variety N_i is of dimension k. Moreover, the N_i are all of the irreducible components of the restriction N of N_f to W and $[N] = \sum n_i[N_i]$.

The cone C_g is a closed subscheme of N of pure dimension k. Hence C_i is a closed subscheme of N_i of pure dimension k for each i. Since N_i is a variety of dimension k, the closed embedding of C_i into N_i is an isomorphism. In particular, the C_i are all of the irreducible components of C_g, so $[C_g] = \sum m_i[C_i]$ with $m_i = l(\mathcal{O}_{C_g, x_i})$ and where $x_i \in C_g$ is the generic point of C_i. In view of Example 49.13, we have

$$h_*\big([C_i]\big) = [N_i] = p^*\big([V_i]\big)$$

and by Corollary 57.16,

$$f^\star\big([Z]\big) = (p^*)^{-1} h_*\big([C_g]\big) = (p^*)^{-1} \sum m_i h_*\big([C_i]\big) = \sum m_i[V_i].$$

Finally, the closed embedding $h : C_g \to N$ induces a surjective ring homomorphism $\mathcal{O}_{N, y_i} \to \mathcal{O}_{C_g, x_i}$, where $y_i \in N$ is the generic point of N_i. Therefore,

$$1 \le m_i = l(\mathcal{O}_{C_g, x_i}) \le l(\mathcal{O}_{N, y_i}) = n_i. \qquad \square$$

Corollary 57.20. *Suppose the conditions of Proposition 57.19 hold and, in addition, the scheme W is reduced. Then $f^\star\big([Z]\big) = \sum [V_i]$, i.e., all the $m_i = 1$.*

PROOF. Indeed, all $n_i = 1$, hence all $m_i = 1$. □

Let Y and Z be closed subvarieties of a smooth scheme X of codimensions p and q, respectively. We say that Y *and* Z *intersect properly* if every component of $Y \cap Z$ has codimension $p + q$.

Applying Proposition 57.19 to the regular diagonal embedding $X \to X \times X$ and the subscheme $Y \times Z$ yields the following:

Proposition 57.21. *Let Y and Z be two closed subvarieties of a smooth scheme X that intersect properly. Let $V_1, V_2, \dots, V_s$ be all irreducible components of $W = Y \cap Z$*

and $[W] = \sum n_i[V_i]$ *for some* $n_i > 0$. *Then*

$$[Y] \cdot [Z] = \sum_{i=1}^{s} m_i[V_i],$$

for some integers m_i *with* $1 \le m_i \le n_i$.

Corollary 57.22. *Suppose the conditions of Proposition* 57.21 *hold and in addition the scheme* W *is reduced. Then* $[Y] \cdot [Z] = \sum [V_i]$, *i.e., all the* $m_i = 1$.

If X is smooth, we write $\mathrm{CH}^p(X)$ for the group $A^p(X, K_p)$ and call it the *Chow group of codimension* p *classes of cycles on* X. We now apply the results from §56 to this group. The graded group $\mathrm{CH}^*(X)$ has the structure of a commutative associative ring with the identity 1_X. A morphism $f : Y \to X$ of smooth schemes induces a pull-back graded ring homomorphism $f^* : \mathrm{CH}^*(X) \to \mathrm{CH}^*(Y)$.

Example 57.23. Let $h \in \mathrm{CH}^1\big(\mathbb{P}(V)\big)$ be the class of a hyperplane of the projective space $\mathbb{P}(V)$, where V is a vector space of dimension $d+1$, and $l_p \in \mathrm{CH}_p\big(\mathbb{P}(V)\big)$ as in Example 57.15. Then

$$h \cdot l_p = l_{p-1}$$

for all $p \in [1, d]$. Indeed, $h = [\mathbb{P}(U)]$ and $l_p = [\mathbb{P}(V_p)]$ where U and V_p are subspaces of V of dimensions n and $p+1$ respectively. We can choose these subspaces so that the subspace $V_{p-1} = U \cap V_p$ has dimension p. Then $\mathbb{P}(U) \cap \mathbb{P}(V_p) = \mathbb{P}(V_{p-1})$ and we have equality by Corollary 57.22. It follows that $\mathrm{CH}^p\big(\mathbb{P}(V)\big) = \mathbb{Z} \cdot h^p$ for $p \in [0, d]$. In particular, the ring $\mathrm{CH}^*\big(\mathbb{P}(V)\big)$ is generated by h with the one relation $h^{d+1} = 0$.

57.C. Cartier divisors and the Euler class. Let D be a Cartier divisor on a variety X of dimension d and $L(D)$ its associated line bundle over X. Let $[D]$ denote the associated divisor in $Z_{d-1}(X)$. Recall that if D is a principal Cartier divisor given by a nonzero rational function f on X, then $[D] = \mathrm{div}(f)$.

Lemma 57.24. *In the notation above,* $e\big(L(D)\big)\big([X]\big) = [D]$ *in* $\mathrm{CH}_{d-1}(X)$.

Proof. Let $p : L(D) \to X$ and $s : X \to L(D)$ be the canonical morphism and the zero section respectively. Let $X = \bigcup U_i$ be an open covering and f_i rational functions on U_i giving the Cartier divisor D. Let $\mathcal{L}(D)$ be the locally free sheaf of sections of $L(D)$. The group of sections $\mathcal{L}(D)(U_i)$ consists of all rational functions f on X such that $f \cdot f_i$ is regular on U_i. Thus we can view f_i as a section of the dual bundle $L(D)^\vee$ over U_i. The line bundle $L(D)$ is the spectrum of the symmetric algebra

$$\mathcal{O}_X \oplus \mathcal{L}(D)^\vee \cdot t \oplus \big(\mathcal{L}(D)^\vee\big)^{\otimes 2} \cdot t^2 \oplus \dots$$

of the sheaf $\mathcal{L}(D)^\vee$. The rational functions $(f_i \cdot t)/f_i$ on $p^{-1}(U_i)$ agree on intersections, so they give a well-defined rational function on $L(D)$. We denote this function by t.

We claim that $\mathrm{div}(t) = s_*\big([X]\big) - p^*\big([D]\big)$. The statement is of a local nature, so we may assume that X is affine, say $X = \mathrm{Spec}(A)$ and D is a principal Cartier divisor given by a rational function f on X. We have $L(D) = \mathrm{Spec}\big(A[ft]\big)$ and by Proposition 49.23,

$$\mathrm{div}(t) = \mathrm{div}(ft) - \mathrm{div}(p^*f) = s_*\big([X]\big) - p^*\big(\mathrm{div}(f)\big) = s_*\big([X]\big) - p^*\big([D]\big)$$

proving the claim. By the claim, the classes $s_*\big([X]\big)$ and $p^*\big([D]\big)$ are equal in $\mathrm{CH}_d\big(L(D)\big)$. Hence, $e\big(L(D)\big)\big([X]\big) = (p^*)^{-1} \circ s_*\big([X]\big) = [D]$. □

Example 57.25. Let $C = \operatorname{Spec}(S^\bullet)$ be an integral cone with $S^\bullet$ a sheaf of graded $\mathcal{O}_X$-algebras as in §104.A. Consider the cone $C \oplus \mathbb{1} = \operatorname{Spec}\big(S^\bullet[t]\big)$. The family of functions t/s with $s \in S^1$ on the principal open subscheme $D(s)$ of the projective bundle $\mathbb{P}(C \oplus \mathbb{1})$ gives rise to a Cartier divisor D on $\mathbb{P}(C \oplus \mathbb{1})$ with $L(D)$ the canonical line bundle. The associated divisor $\widetilde{D}$ coincides with $\mathbb{P}(C)$. It follows from Lemma 57.24 that $e\big(L(D)\big)\big([\mathbb{P}(C \oplus \mathbb{1})]\big) = [\mathbb{P}(C)]$.

Proposition 57.26. *Let L and L' be line bundles over a scheme X. Then $e(L \otimes L') = e(L) + e(L')$ on $\operatorname{CH}(X)$.*

PROOF. It suffices to prove that both sides of the equality coincide on the class $[Z]$ of a closed subvariety Z in X. Denote by $i : Z \to X$ the closed embedding. Choose Cartier divisors D and D' on Z so that $L|_Z \simeq L(D)$ and $L'|_Z \simeq L(D')$. Then $L|_Z \otimes L'|_Z \simeq L(D + D')$. By Proposition 53.3(1),

$$\begin{aligned} e(L \otimes L')\big([Z]\big) &= i_* \circ e(L|_Z \otimes L'|_Z)\big([Z]\big) \\ &= i_* \circ e\big(L(D + D')\big)\big([Z]\big) \\ &= i_*[\widetilde{D + D'}] \\ &= i_*[\widetilde{D}] + i_*[\widetilde{D}'] \\ &= i_* \circ e\big(L(D)\big) + i_* \circ e\big(L(D')\big) \\ &= e(L)\big([Z]\big) + e(L')\big([Z]\big). \end{aligned}$$

□

Corollary 57.27. *For any line bundle L over X, we have $e(L^\vee) = -e(L)$.*

58. Segre and Chern classes

In this section, we define Segre classes and consider their relations with Chern classes. The Segre class for a vector bundle is the inverse of the Chern class. The advantage of Segre classes is that they can be defined for arbitrary cones (not just for vector bundles like Chern classes). We follow the book [**45**] for the definition of Segre classes.

58.A. Segre classes. Let $C = \operatorname{Spec}(S^\bullet)$ be a cone over X (cf. §104.A). Let $q : \mathbb{P}(C \oplus \mathbb{1}) \to X$ be the natural morphism and L the canonical line bundle over $\mathbb{P}(C \oplus \mathbb{1})$ (cf. §104). Let $e(L)^\bullet$ denote the *total Euler class* $\sum_{k \geq 0} e(L)^k$ viewed as an operation on $\operatorname{CH}\big(\mathbb{P}(C \oplus \mathbb{1})\big)$.

We define the *Segre homomorphism*

$$\operatorname{sg}^C : \operatorname{CH}\big(\mathbb{P}(C \oplus \mathbb{1})\big) \to \operatorname{CH}(X) \quad \text{by}$$

$$\operatorname{sg}^C = q_* \circ e(L)^\bullet.$$

The class $\operatorname{Sg}(C) := \operatorname{sg}^C\big([\mathbb{P}(C \oplus \mathbb{1})]\big)$ in $\operatorname{CH}(X)$ is known as the *total Segre class of C*.

Proposition 58.1. *If C is a cone over X, then $\operatorname{Sg}(C \oplus \mathbb{1}) = \operatorname{Sg}(C)$.*

PROOF. If $[C] = \sum m_i[C_i]$, where the C_i are the irreducible components of C, then

$$[\mathbb{P}(C \oplus \mathbb{1}^k)] = \sum m_i[\mathbb{P}(C_i \oplus \mathbb{1}^k)]$$

for $k \geq 1$. Therefore, we may assume that C is a variety. Let L and L' be canonical line bundles over $\mathbb{P}(C \oplus \mathbb{1}^2)$ and $\mathbb{P}(C \oplus \mathbb{1})$, respectively. We have $L' = i^*(L)$, where $i : \mathbb{P}(C \oplus \mathbb{1}) \to \mathbb{P}(C \oplus \mathbb{1}^2)$ is the closed embedding. By Example 57.25, we have

$$e(L)\big([\mathbb{P}(C \oplus \mathbb{1}^2)]\big) = [\mathbb{P}(C \oplus \mathbb{1})],$$

Let $q : \mathbb{P}(C \oplus \mathbb{1}^2) \to X$ be the canonical morphism. It follows from Proposition 53.3(1) that

$$\begin{aligned}\mathrm{Sg}(C \oplus \mathbb{1}) &= q_* \circ e(L)^\bullet([\mathbb{P}(C \oplus \mathbb{1}^2)]) \\ &= q_* \circ e(L)^\bullet\big(i_*([\mathbb{P}(C \oplus \mathbb{1})])\big) \\ &= q_* i_* \circ e\big(i^*(L)\big)^\bullet([\mathbb{P}(C \oplus \mathbb{1})]) \\ &= (q \circ i)_* \circ e(L')^\bullet([\mathbb{P}(C \oplus \mathbb{1})]) \\ &= \mathrm{Sg}(C). \end{aligned}$$

□

Proposition 58.2. *Let C be a cone over a scheme X over F and $i : Z \to X$ a closed embedding. Let D be a closed subcone of the restriction of C on Z. Then the diagram*

$$\begin{array}{ccc} \mathrm{CH}\big(\mathbb{P}(D \oplus \mathbb{1})\big) & \xrightarrow{\mathrm{sg}^D} & \mathrm{CH}(Z) \\ {\scriptstyle j_*}\downarrow & & \downarrow{\scriptstyle i_*} \\ \mathrm{CH}\big(\mathbb{P}(C \oplus \mathbb{1})\big) & \xrightarrow{\mathrm{sg}^C} & \mathrm{CH}(X) \end{array}$$

is commutative, where $j : \mathbb{P}(D \oplus \mathbb{1}) \to \mathbb{P}(C \oplus \mathbb{1})$ is the closed embedding. In particular, $i_*\big(\mathrm{Sg}(D)\big) = \mathrm{sg}^C\big(\mathbb{P}(D \oplus \mathbb{1})\big)$.

PROOF. The canonical line bundle L_D over $\mathbb{P}(D \oplus \mathbb{1})$ is the pull-back $j^*(L_C)$. It follows from the projection formula (cf. Proposition 53.3(1)) that

$$\begin{aligned}\mathrm{sg}^C \circ j_* &= (q_C)_* \circ e(L_C)^\bullet \circ j_* \\ &= (q_C)_* \circ j_* \circ e\big(j^*(L_C)\big)^\bullet \\ &= i_* \circ (q_D)_* \circ e(L_D)^\bullet \\ &= i_* \circ \mathrm{sg}^D. \end{aligned}$$

□

If $C = E$ is a vector bundle over X, the projection q is a flat morphism of relative dimension $r = \operatorname{rank} E$, and we define the *total Segre operation* $s(E)$ on $\mathrm{CH}(X)$ by

$$s(E) : \mathrm{CH}(X) \to \mathrm{CH}(X), \quad s(E) = \mathrm{sg}^E \circ q^* = q_* \circ e(L)^\bullet \circ q^*.$$

In particular, $\mathrm{Sg}(E) = s(E)\big([X]\big)$.

For every $k \in \mathbb{Z}$ denote the degree k component of the operation $s(E)$ by $s_k(E)$, so it is the operation

$$s_k(E) : \mathrm{CH}_n(X) \to \mathrm{CH}_{n-k}(X) \quad \text{given by}$$

$$s_k(E) = q_* \circ e(L)^{k+r} \circ q^*. \tag{58.3}$$

Proposition 58.4. *Let $f : Y \to X$ be a morphism of schemes over F and E a vector bundle over X. Set $E' = f^*(E)$. Then*

(1) *If f is proper, then $s(E) \circ f_* = f_* \circ s(E')$.*
(2) *If f is flat, then $f^* \circ s(E) = s(E') \circ f^*$.*

PROOF. Consider the fiber product diagram

$$\begin{array}{ccc} \mathbb{P}(E') & \xrightarrow{h} & \mathbb{P}(E) \\ {\scriptstyle q'}\downarrow & & \downarrow{\scriptstyle q} \\ Y & \xrightarrow{f} & X \end{array}$$

with flat morphisms q and q' of constant relative dimension $r-1$ where $r = \operatorname{rank} E$. Denote by e and e' the Euler classes of the canonical line bundle L over $\mathbb{P}(E)$ and L' over $\mathbb{P}(E')$, respectively. Note that $L' = h^*(L)$.

By Propositions 49.20 and 53.3, we have

$$\begin{aligned} s(E) \circ f_* &= q_* \circ e(L)^\bullet \circ q^* \circ f_* \\ &= q_* \circ e(L)^\bullet \circ h_* \circ q'^* \\ &= q_* \circ h_* \circ e(L')^\bullet \circ q'^* \\ &= f_* \circ q'_* \circ e(L')^\bullet \circ q'^* \\ &= f_* \circ s(E'), \end{aligned}$$

and

$$\begin{aligned} f^* \circ s(E) &= f^* \circ q_* \circ e(L)^\bullet \circ q^* \\ &= q'_* \circ h^* \circ e(L)^\bullet \circ q^* \\ &= q'_* \circ e(L')^\bullet \circ h^* \circ q^* \\ &= q'_* \circ e(L')^\bullet \circ q'^* \circ f^* \\ &= s(E') \circ f^*. \end{aligned}$$

□

Proposition 58.5. *Let E be a vector bundle over a scheme X over F. Then*

$$s_i(E) = \begin{cases} 0 & \text{if } i < 0, \\ \mathrm{id} & \text{if } i = 0. \end{cases}$$

PROOF. Let $\alpha \in \mathrm{CH}(X)$. We need to prove that $s_i(E)(\alpha) = 0$ if $i < 0$ and $s_0(E)(\alpha) = \alpha$. We may assume that $\alpha = [Z]$, where $Z \subset X$ is a closed subvariety. Let $i : Z \to X$ be the closed embedding. By Proposition 58.4(1), we have

$$s(E)(\alpha) = s(E) \circ i_*\big([Z]\big) = i_* \circ s(E')\big([Z]\big),$$

where $E' = i^*(E)$. Hence it is sufficient to prove the statement for the vector bundle E' over Z and the cycle $[Z]$. Therefore, we may assume that X is a variety of dimension d and $\alpha = [X]$ in $\mathrm{CH}_d(X)$. Since $s_i(E)(\alpha) \in \mathrm{CH}_{d-i}(X)$, by dimension count, $s_i(E)(\alpha) = 0$ if $i < 0$.

To prove the second identity, by Proposition 58.4(2), we may replace X by an open subscheme. Therefore, we can assume that E is a trivial vector bundle, i.e., $\mathbb{P}(E) = X \times \mathbb{P}^{r-1}$. Applying Example 53.8 and Proposition 53.3(2) to the projection $X \times \mathbb{P}^{r-1} \to \mathbb{P}^{r-1}$, we have

$$\begin{aligned} s_0(E)\big([X]\big) &= q_* \circ e(L)^{r-1} \circ \; q^*\big([X]\big) \\ &= q_* \circ e(L)^{r-1}\big([X] \times \mathbb{P}^{r-1}\big) = q_*\big([X] \times \mathbb{P}^0\big) = [X]. \end{aligned}$$

□

Let $E \to X$ be a vector bundle of rank r. The restriction of Chern classes defined in §54 on Chow groups provides operations

$$c_i(E) : \mathrm{CH}_*(X) \to \mathrm{CH}_{*-i}(X), \quad \alpha \mapsto \alpha_i = c_i(E)(\alpha)$$

Example 58.6. In view of Examples 53.8 and 57.23, the class $e(L)$ of the canonical line bundle L over $\mathbb{P}^d$ acts on $\mathrm{CH}(\mathbb{P}^d) = \mathbb{Z}[h]/(h^{d+1})$ by multiplication by the class h of a hyperplane in $\mathbb{P}^d$.

By Example 104.20, the class of the tangent bundle of the projective space $\mathbb{P}^d$ in $K_0(\mathbb{P}^d)$ is equal to $(d+1)[L]-1$, hence $c(T_{\mathbb{P}^d})$ is multiplication by $(1+h)^{d+1}$.

Example 58.7. For a vector bundle E, we have $c_i(E^\vee) = (-1)^i c_i(E)$. Indeed, by the Splitting Principle 53.13, we may assume that E has a filtration by subbundles with factors line bundles $L_1, L_2, \dots, L_r$. The dual bundle $E^\vee$ then has a filtration by subbundles with factors line bundles $L_1^\vee, L_2^\vee, \dots, L_r^\vee$. As $e(L_k^\vee) = -e(L_k)$ by Corollary 57.27, it follows from Proposition 54.6 that

$$c_i(E^\vee) = \sigma_i\big(e(L_1^\vee), \dots, e(L_r^\vee)\big) = (-1)^i \sigma_i\big(e(L_1), \dots, e(L_r)\big) = (-1)^i c_i(E)$$

where σ_i is the ith elementary symmetric function.

Let e and $\tilde{e}$ be the Euler classes of the tautological and the canonical line bundles over $\mathbb{P}(E)$, respectively. By Corollary 57.27, we have $\tilde{e} = -e$. Therefore, the formula (54.1) can be rewritten as

$$\sum_{i=0}^{r} \tilde{e}^{\,r-i} \circ q^* \circ c_i(E) = 0, \tag{58.8}$$

where $q : \mathbb{P}(E) \to X$ is the canonical morphism.

Proposition 58.9. *Let E be a vector bundle over X. Then $s(E) = c(E)^{-1}$.*

PROOF. Applying $q_* \circ \tilde{e}^{\,k-1}$ to the equality (58.8) for the vector bundle $E \oplus \mathbb{1}$ of rank $r+1$, we get for every $k \geq 1$:

$$0 = \sum_{i \geq 0} q_* \circ \tilde{e}^{\,r+k-i} \circ q^* \circ c_i(E \oplus \mathbb{1}) = \sum_{i \geq 0} s_{k-i}(E) \circ c_i(E \oplus \mathbb{1})$$

in view of (58.3). By Corollary 54.9, we have $c_i(E \oplus \mathbb{1}) = c_i(E)$. As $s_0(E) = 1$ and $s_i(E) = 0$ if $i < 0$ by Proposition 58.5, we have $s(E) \circ c(E) = 1$. □

Proposition 58.10. *Let $E \to X$ be a vector bundle and $E' \subset E$ a subbundle of corank r. Then*

$$[\mathbb{P}(E')] = \sum_{i=0}^{r} \tilde{e}^{r-i} \circ q^* \circ c_i(E/E')\big([X]\big) \tag{58.11}$$

in $\mathrm{CH}\big(\mathbb{P}(E)\big)$.

PROOF. Applying (58.8) to the factor bundle E/E', we have

$$\sum_{i=0}^{r} e'^{\,r-i} \circ q'^* \circ c_i(E/E') = 0,$$

where $q' : \mathbb{P}(E/E') \to X$ is the canonical morphism and e' is the Euler class of the canonical line bundle over $\mathbb{P}(E/E')$. Applying the pull-back homomorphism with respect to the canonical morphism $\mathbb{P}(E) \setminus \mathbb{P}(E') \to \mathbb{P}(E/E')$, we see that the restriction of the right hand side of the formula in (58.11) to $\mathbb{P}(E) \setminus \mathbb{P}(E')$ is trivial. By the localization property (cf. §52.D), the right hand side in (58.11) is equal to $k[\mathbb{P}(E')]$ for some $k \in \mathbb{Z}$.

To determine k, we can replace X by an open subscheme of X and assume that E and E' are trivial vector bundles of rank n and $n-r$, respectively. The right hand side in (58.11) is then equal to

$$\tilde{e}^r \circ q^*\big([X]\big) = \tilde{e}^r\big([\mathbb{P}^{n-1} \times X]\big) = [\mathbb{P}^{n-r-1} \times X] = [\mathbb{P}(E')],$$

therefore $k = 1$. □

Proposition 58.12. *Let E and E' be vector bundles over schemes X and X', respectively. Then*

$$c(E \times E')(\alpha \times \alpha') = c(E)(\alpha) \times c(E')(\alpha')$$

for any $\alpha \in \mathrm{CH}(X)$ and $\alpha' \in \mathrm{CH}(X')$.

PROOF. Let p and p' be the projections of $X \times X'$ to X and X', respectively. We claim that for any $\beta \in \mathrm{CH}(X)$ and $\beta' \in \mathrm{CH}(X')$, we have

$$c\big(p^*(E)\big)(\beta \times \beta') = c(E)(\beta) \times \beta', \tag{58.13}$$

$$c\big(p'^*(E')\big)(\beta \times \alpha') = \beta \times c(E')(\beta'). \tag{58.14}$$

To prove the claim, by Proposition 54.5, we may assume that $\beta = [X]$ and $\beta' = [X']$. Then (58.13) and (58.14) follow from Proposition 54.5(2).

Since $E \times E' = p^*(E) \oplus p'^*(E')$, by the Whitney Sum Formula 54.7 and by (58.13), (58.14), we have

$$\begin{aligned} c(E \times E')(\alpha \times \alpha') &= c\big(p^*(E) \oplus p'^*(E')\big)(\alpha \times \alpha') \\ &= c\big(p^*(E)\big) \circ c\big(p'^*(E')\big)(\alpha \times \alpha') \\ &= c\big(p^*(E)\big)\big(\alpha \times c(E')(\alpha')\big) \\ &= c(E)(\alpha) \times c(E')(\alpha'). \end{aligned}$$

□

Proposition 58.15. *Let E be a vector bundle over a smooth scheme X. Then $c(E)(\alpha) = c(E)\big([X]\big) \cdot \alpha$ for every $\alpha \in \mathrm{CH}(X)$.*

PROOF. Consider the vector bundle $E' = E \times X$ over $X \times X$. Let $d : X \to X \times X$ be the diagonal embedding. We have $E = d^*(E')$. By Propositions 58.12 and 55.9,

$$\begin{aligned} c(E)(\alpha) &= c\big(d^*(E')\big)\big(d^\star([X] \times \alpha)\big) \\ &= d^\star c(E \times X)\big([X] \times \alpha\big) \\ &= d^\star\big(c(E)([X]) \times \alpha\big) \\ &= c(E)\big([X]\big) \cdot \alpha. \end{aligned}$$

□

Proposition 58.15 shows that for a vector bundle E over a smooth scheme X, the Chern class operation $c(E)$ is multiplication by the class $\beta = c(E)\big([X]\big)$. We shall sometimes write $c(E) = \beta$ to mean that $c(E)$ is multiplication by β.

Let $f : Y \to X$ be a morphism of schemes, i.e., Y is a scheme over X. Assume that X is a smooth variety. We shall see that $\mathrm{CH}(Y)$ has a natural structure of a module over the ring $\mathrm{CH}(X)$. Indeed, as we saw in §55.B, the morphism

$$i = (1_Y, f) : Y \to Y \times X$$

is a regular closed embedding of codimension $\dim X$. For every $\alpha \in \mathrm{CH}(Y)$ and $\beta \in \mathrm{CH}(X)$, we set

$$\alpha \cdot \beta = i^\star(\alpha \times \beta). \tag{58.16}$$

Proposition 58.17. *Let X be a smooth variety and Y a scheme over X. Then* $\mathrm{CH}(Y)$ *is a module over* $\mathrm{CH}(X)$ *under the product defined in* (58.16). *Let $g : Y' \to Y$ be a proper (respectively flat) morphism of schemes over X. Then the homomorphism g_* (respectively g^*) is* $\mathrm{CH}(X)$*-linear.*

PROOF. The composition of i and the projection $p : Y \times X \to Y$ is the identity on Y. It follows from Lemma 55.7 that $\alpha \cdot [X] = i^\star\big(\alpha \times [X]\big) = i^\star \circ p^*(\alpha) = 1_Y^*(\alpha) = \alpha$, i.e., the identity $[X]$ of $\mathrm{CH}(X)$ acts on $\mathrm{CH}(Y)$ trivially.

Consider the fiber product diagram

$$\begin{array}{ccc} Y & \xrightarrow{i} & Y \times X \\ {\scriptstyle i}\downarrow & & \downarrow{\scriptstyle h} \\ Y \times X & \xrightarrow{k} & Y \times X \times X, \end{array}$$

where $k = 1_Y \times d_X$ and $h = i \times 1_X$. It follows from Corollary 55.4 that for any $\alpha \in \mathrm{CH}(Y)$ and $\beta, \gamma \in \mathrm{CH}(X)$, we have

$$\begin{aligned} \alpha \cdot (\beta \cdot \gamma) &= i^\star\big(\alpha \times (\beta \cdot \gamma)\big) = i^\star k^\star(\alpha \times \beta \times \gamma) \\ &= i^\star h^\star(\alpha \times \beta \times \gamma) = i^\star\big((\alpha \cdot \beta) \times \gamma\big) = (\alpha \cdot \beta) \cdot \gamma. \end{aligned}$$

Consider the fiber product diagram

$$\begin{array}{ccc} Y' & \xrightarrow{i'} & Y' \times X \\ {\scriptstyle g}\downarrow & & \downarrow{\scriptstyle g \times 1_X} \\ Y & \xrightarrow{i} & Y \times X. \end{array}$$

Suppose first that the morphism g is proper. By Corollary 55.4,

$$g_*(\alpha' \cdot \beta) = g_* \circ i'^\star(\alpha' \times \beta) = i^\star(g \times 1_X)_*(\alpha' \times \beta) = i'^\star\big(g_*(\alpha') \times \beta\big) = g_*(\alpha') \cdot \beta$$

for all $\alpha' \in \mathrm{CH}(Y')$ and $\beta \in \mathrm{CH}(X)$.

If g is flat, it follows from Proposition 55.5 that

$$g^*(\alpha \cdot \beta) = g^* \circ i^\star(\alpha \times \beta) = i'^\star \circ (g \times 1_X)^*(\alpha \times \beta) = i'^\star\big(g^*(\alpha) \times \beta\big) = g^*(\alpha) \cdot \beta$$

for all $\alpha \in \mathrm{CH}(Y)$ and $\beta \in \mathrm{CH}(X)$. □

Proposition 58.18. *Let $f : Y \to X$ be a morphism of schemes with X smooth and $g : Y \to Y'$ a flat morphism. Suppose that for every point $y' \in Y'$, the pull-back homomorphism* $\mathrm{CH}(X) \to \mathrm{CH}(Y_{y'})$ *induced by the natural morphism of the fiber $Y_{y'}$ to X is surjective. Then the homomorphism*

$$h : \mathrm{CH}(Y') \otimes \mathrm{CH}(X) \to \mathrm{CH}(Y), \quad \alpha \otimes \beta \mapsto g^*(\alpha) \cdot \beta$$

is surjective.

PROOF. The proof is similar to the one given for Proposition 52.10. Obviously, we may assume that Y' is reduced.

Step 1: Y' is a variety.

We induct on $n = \dim Y'$. The case $n = 0$ is obvious. In general, let $U' \subset Y'$ be a nonempty open subset and $Z' = Y' \setminus U'$ have the structure of a reduced scheme. Set $U = g^{-1}(U')$ and $Z = g^{-1}(Z')$. We have closed embeddings $i : Z \to Y$,

$i' : Z' \to Y'$ and open immersions $j : U \to Y$, $j' : U' \to Y'$. By induction, the homomorphism h_Z is surjective in the diagram

$$\begin{array}{ccccccc} \mathrm{CH}(Z') \otimes \mathrm{CH}(X) & \xrightarrow{i'_* \otimes 1} & \mathrm{CH}(Y') \otimes \mathrm{CH}(X) & \xrightarrow{j^* \otimes 1} & \mathrm{CH}(U') \otimes \mathrm{CH}(X) & \longrightarrow & 0 \\ \downarrow{\scriptstyle h_Z} & & \downarrow{\scriptstyle h_Y} & & \downarrow{\scriptstyle h_U} & & \\ \mathrm{CH}(Z) & \xrightarrow{i_*} & \mathrm{CH}(Y) & \xrightarrow{j'^*} & \mathrm{CH}(U) & \longrightarrow & 0 \end{array}$$

The diagram is commutative by Proposition 58.17.

Let $y' \in Y'$ be the generic point. By Proposition 52.9, the colimit of the homomorphisms

$$(h_U)^* : \mathrm{CH}(U') \otimes \mathrm{CH}(X) \to \mathrm{CH}(U)$$

over all nonempty open subschemes U' of Y' is isomorphic to the pull-back homomorphism $\mathrm{CH}(X) \to \mathrm{CH}(Y_{y'})$ which is surjective by assumption. Taking the colimits of all terms of the diagram, we conclude by the 5-lemma that h_Y is also surjective.

Step 2: Y' *is an arbitrary scheme.*

We induct on the number m of irreducible components of Y'. The case $m = 1$ is Step 1. Let Z' be a (reduced) irreducible component of Y' and let $U' = Y' \setminus Z'$. Consider the commutative diagram as in Step 1. By Step 1, the map h_Z is surjective. The map h_U is also surjective by the induction hypothesis. By the 5-lemma, h_Y is surjective. □

Proposition 58.19. *Let* C *and* C' *be cones over schemes* X *and* X'*, respectively. Then*

$$\mathrm{Sg}(C \times C') = \mathrm{Sg}(C) \times \mathrm{Sg}(C') \in \mathrm{CH}(X \times X').$$

PROOF. Set $\widetilde{C} = C \oplus \mathbb{1}$ and $\widetilde{C}' = C' \oplus \mathbb{1}$. Let L and L' be the tautological line bundles over $\mathbb{P}(\widetilde{C})$ and $\mathbb{P}(\widetilde{C}')$, respectively (cf. §104.D). We view $L \times L'$ as a vector bundle over $\mathbb{P}(\widetilde{C}) \times \mathbb{P}(\widetilde{C}')$. The canonical morphism $L \times L' \to \widetilde{C} \times \widetilde{C}'$ induces a morphism

$$f : \mathbb{P}(L \times L') \to \mathbb{P}(\widetilde{C} \times \widetilde{C}').$$

If D is a cone, we write D° for the complement of the zero section in D. By §104.C, we have $L^\circ = \widetilde{C}^\circ$ and $L'^\circ = \widetilde{C}'^\circ$. The open subsets $\widetilde{C}^\circ \times \widetilde{C}'^\circ$ in $\widetilde{C} \times \widetilde{C}'$ and $L^\circ \times L'^\circ$ in $L \times L'$ are dense. Hence f maps any irreducible component of $\mathbb{P}(L \times L')$ birationally onto an irreducible component of $\mathbb{P}(\widetilde{C} \times \widetilde{C}')$. In particular,

$$f_*\big([\mathbb{P}(L \times L')]\big) = [\mathbb{P}(\widetilde{C} \times \widetilde{C}')].$$

Let $\widetilde{L}$ be the canonical line bundle over $\mathbb{P}(\widetilde{C} \times \widetilde{C}')$. Then $f^*(\widetilde{L})$ is the canonical line bundle over $\mathbb{P}(L \times L')$. Let $q : \mathbb{P}(\widetilde{C} \times \widetilde{C}') \to X \times X'$ be the natural morphism. By Proposition 58.1 and the Projection Formula 56.9, we have

$$\begin{aligned} \mathrm{Sg}(C \times C') &= \mathrm{Sg}\big((C \times C') \oplus \mathbb{1}\big) \\ &= q_* \circ e(\widetilde{L})^\bullet\big([\mathbb{P}(\widetilde{C} \times \widetilde{C}')]\big) \\ &= q_* \circ e(\widetilde{L})^\bullet f_*\big([\mathbb{P}(L \times L')]\big) \\ &= q_* \circ f_* \circ e(f^*\widetilde{L})^\bullet\big([\mathbb{P}(L \times L')]\big). \end{aligned}$$

The normal bundle N of the closed embedding $\mathbb{P}(\widetilde{C}) \times \mathbb{P}(\widetilde{C}') \to L \times L'$, given by the zero section, coincides with $L \times L'$. By definition of the Segre class and the Segre operation, we have

$$p_* \circ e(f^*\widetilde{L})^\bullet\big([\mathbb{P}(L \times L')]\big) = \mathrm{Sg}(N) = s(N)\big([\mathbb{P}(\widetilde{C}) \times \mathbb{P}(\widetilde{C}')]\big),$$

where $p : \mathbb{P}(L \times L') \to \mathbb{P}(\widetilde{C}) \times \mathbb{P}(\widetilde{C}')$ is the natural morphism. By Propositions 58.12 and 58.9,

$$s(N)\big([\mathbb{P}(\widetilde{C}) \times \mathbb{P}(\widetilde{C}')]\big) = s(L)\big([\mathbb{P}(\widetilde{C})]\big) \times s(L')\big([\mathbb{P}(\widetilde{C}')]\big).$$

Let $g : \mathbb{P}(\widetilde{C}) \to X$ and $g' : \mathbb{P}(\widetilde{C}') \to X'$ be the natural morphisms and set $h = g \times g'$. By Proposition 50.4,

$$\begin{aligned} h_* \circ s(L)\big([\mathbb{P}(\widetilde{C})] \times s(L')([\mathbb{P}(\widetilde{C}')])\big) \\ = \big(g_* \circ s(L)([\mathbb{P}(\widetilde{C})])\big) \times \big(g'_* \circ s(L')([\mathbb{P}(\widetilde{C}')])\big) = \mathrm{Sg}(C) \times \mathrm{Sg}(C'). \end{aligned}$$

To finish the proof it is sufficient to notice that $q \circ f = h \circ p$ and therefore $q_* \circ f_* = h_* \circ p_*$. □

Exercise 58.20. (Strong Splitting Principle) Let E be a vector bundle over X. Prove that there is a flat morphism $f : Y \to X$ such that the pull-back homomorphism $f^* : \mathrm{CH}_*(X) \to \mathrm{CH}_*(Y)$ is injective and $f^*(E)$ is a direct sum of line bundles.

Exercise 58.21. Let E be a vector bundle of rank r. Prove that $e(E) = c_r(E)$.

CHAPTER XI

Steenrod Operations

In this chapter we develop Steenrod operations on Chow groups modulo 2. There are two reasons why we do not consider the operations modulo an arbitrary prime integer. First, this case is sufficient for our applications as the number 2 is the only "critical" prime for projective quadrics. Second, our approach does not immediately generalize to the case of an arbitrary prime integer.

Unfortunately, we need to assume that the characteristic of the base field is different from 2 in this chapter as we do not know how to define Steenrod operations in characteristic 2.

In this chapter, the word *scheme* means a quasi-projective scheme over a field F of characteristic not 2. We write $\mathrm{Ch}(X)$ for $\mathrm{CH}(X)/2\,\mathrm{CH}(X)$.

Let X be a scheme. Consider the homomorphism $\mathrm{Z}(X) \to \mathrm{Ch}(X)$ taking the class $[Z]$ of a closed subvariety $Z \subset X$ to $j_*\big(\mathrm{Sg}(T_Z)\big)$ modulo 2, where Sg is the total Segre class (cf. §58.A), T_Z is the tangent cone over Z (cf. Example 104.5) and $j : Z \to X$ is the closed embedding. We shall prove that this map factors through rational equivalence yielding the *Steenrod operation modulo* 2 *of* X *(of homological type)*

$$\mathrm{Sq}^X : \mathrm{Ch}(X) \to \mathrm{Ch}(X).$$

Thus we shall have

$$\mathrm{Sq}^X\big([Z]\big) = j_*\big(\mathrm{Sg}(T_Z)\big)$$

modulo 2. We shall see that the operation Sq^X commutes with the push-forward homomorphisms, so it can be viewed as a functor from the category of schemes to the category of abelian groups.

For a smooth scheme X, we can then define the *Steenrod operation modulo* 2 *of* X *(of cohomological type)* by the formula

$$\mathrm{Sq}_X := c(T_X) \circ \mathrm{Sq}^X .$$

This formula can be viewed as a Riemann-Roch type relation between the operations Sq^X and Sq_X. We shall show that the operation Sq_X commutes with pull-back homomorphisms, so it can be viewed as a contravariant functor from the category of smooth schemes to the category of abelian groups.

In this chapter, we shall also prove the standard properties of the Steenrod operations.

Steenrod operations for motivic cohomology modulo a prime integer p of a smooth scheme X were originally constructed by Voevodsky in [**137**]. The reduced power operations (but not the Bockstein operation) restrict to the Chow groups of X. An "elementary" construction of the reduced power operations modulo p on Chow groups (requiring equivariant Chow groups) was given by Brosnan in [**20**].

59. Definition of the Steenrod operations

Consider a cyclic group $G = \{1, \sigma\}$ of order 2. Let Y be a G-scheme such that $p : Y \to Y/G$ is a G-torsor over a field F (cf. §105.A). In particular, p is flat of relative dimension 0 by Proposition 105.5(1). For a point $z \in Y/G$, let Y_z be the fiber of p over z. By Example 105.6, we have $Y_z = \operatorname{Spec}(K)$, where K is an étale quadratic algebra over $F(z)$. Then either K is a field (and the fiber Y_z has only one point y) or $K = F(z) \times F(z)$ (and the fiber has two points y_1 and y_2). In any case, every point in Y_z is unramified (cf. Proposition 105.5(2)). It follows that for the pull-back homomorphism $p^* : \mathrm{Z}(Y/G) \to \mathrm{Z}(Y)$, we have

$$p^*([z]) = \begin{cases} [y] & \text{if } K \text{ is a field,} \\ [y_1]+[y_2] & \text{otherwise.} \end{cases}$$

Similarly, for a point y in the fiber Y_z, we have:

$$p_*([y]) = \begin{cases} 2[z] & \text{if } K \text{ is a field,} \\ [z] & \text{otherwise.} \end{cases}$$

In particular, $p_* \circ p^*$ is multiplication by 2.

We have $\sigma(y) = y$ if K is a field and $\sigma(y_1) = y_2$ otherwise. In particular,

$$p^* \circ p_* = 1 + \sigma^*, \tag{59.1}$$

where $\sigma^* : \mathrm{Z}(Y) \to \mathrm{Z}(Y)$ is the (pull-back) isomorphism induces by σ.

The cycles $[y]$ and $[y_1] + [y_2]$ generate the group $\mathrm{Z}(Y)^G$ of G-invariant cycles. We have proved

Proposition 59.2. *Let $p : Y \to Y/G$ be a G-torsor. Then the pull-back homomorphism*

$$p^* : \mathrm{Z}(Y/G) \to \mathrm{Z}(Y)^G$$

is an isomorphism.

Suppose $\operatorname{char} F \neq 2$. Let X be a scheme over F. The group G acts on $X^2 \times \mathbb{A}^1 = X \times X \times \mathbb{A}^1$ by $\sigma(x, x', t) = (x', x, -t)$. We have $(X^2 \times \mathbb{A}^1)^G = \Delta(X) \times \{0\}$, where $\Delta : X \to X^2$ is the diagonal morphism (cf. §105.B). Set

$$U_X := (X^2 \times \mathbb{A}^1) \setminus \big(\Delta(X) \times \{0\}\big). \tag{59.3}$$

The group G acts naturally on U_X. The morphism $p : U_X \to U_X/G$ is a G-torsor (cf. Example 105.4).

Let $\alpha \in \mathrm{Z}(X)$ be a cycle. The cycle $\alpha^2 \times \mathbb{A}^1 := \alpha \times \alpha \times \mathbb{A}^1$ in $\mathrm{Z}(X^2 \times \mathbb{A}^1)$ is invariant under the above G-action as is the restriction of the cycle $\alpha^2 \times \mathbb{A}^1$ on U_X. It follows from Proposition 59.2 that the pull-back homomorphism p^* identifies $\mathrm{Z}(U_X/G)$ with $\mathrm{Z}(U_X)^G$. Let $\alpha_G^2 \in \mathrm{Z}(U_X/G)$ denote the cycle satisfying $p^*(\alpha_G^2) = (\alpha^2 \times \mathbb{A}^1)|_{U_X}$.

We then have a map

$$\mathrm{Z}(X) \to \mathrm{Z}(U_X/G), \quad \alpha \mapsto \alpha_G^2. \tag{59.4}$$

Lemma 59.5. *If α and α' are rationally equivalent cycles in $\mathrm{Z}(X)$, then α_G^2 and ${\alpha'}_G^2$ are rationally equivalent cycles in $\mathrm{Z}(U_X/G)$.*

PROOF. As in §57.A, let $\mathrm{Z}(X; \mathbb{P}^1)$ denote the subgroup of $\mathrm{Z}(X \times \mathbb{P}^1)$ generated by the classes of closed subvarieties in $X \times \mathbb{P}^1$ dominant over $\mathbb{P}^1$. Let $W \subset X \times \mathbb{P}^1$ and $W' \subset X' \times \mathbb{P}^1$ be two closed subvarieties dominant over $\mathbb{P}^1$. The projections $W \to \mathbb{P}^1$ and $W' \to \mathbb{P}^1$ are flat and hence so is the fiber product $W \times_{\mathbb{P}^1} W' \to \mathbb{P}^1$.

Therefore, every irreducible component of $W \times_{\mathbb{P}^1} W'$ is dominant over $\mathbb{P}^1$, i.e., the cycle $[W \times_{\mathbb{P}^1} W']$ belongs to $\mathrm{Z}(X \times X'; \mathbb{P}^1)$. By linearity, this construction extends to an external product over $\mathbb{P}^1$ given by

$$\mathrm{Z}(X;\mathbb{P}^1) \times \mathrm{Z}(X';\mathbb{P}^1) \to \mathrm{Z}(X \times X';\mathbb{P}^1), \quad (\beta, \beta') \mapsto \beta \times_{\mathbb{P}^1} \beta'.$$

By Proposition 57.12,

$$[(W \times_{\mathbb{P}^1} W')(a)] = [W(a) \times W'(a)] = [W(a)] \times [W'(a)] \tag{59.6}$$

for any rational point a of $\mathbb{P}^1$. If $X' = X$ and $\beta' = \beta$, write $\tilde{\beta}^2$ for $\beta \times_{\mathbb{P}^1} \beta$.

By Proposition 57.6, there is a cycle $\beta \in \mathrm{Z}(X;\mathbb{P}^1)$ such that $\alpha = \beta(0)$ and $\alpha' = \beta(\infty)$. Consider the cycle $\tilde{\beta}^2 \times [\mathbb{A}^1] \in \mathrm{Z}(X^2 \times \mathbb{A}^1; \mathbb{P}^1)$.

Let G act on $X^2 \times \mathbb{A}^1 \times \mathbb{P}^1$ by $\sigma(x, x', t, s) = (x', x, -t, s)$. The cycle $\tilde{\beta}^2 \times [\mathbb{A}^1]$ is G-invariant. Since $U_X \times \mathbb{P}^1$ is a G-torsor over $(U_X/G) \times \mathbb{P}^1$, the restriction of the cycle $\tilde{\beta}^2 \times [\mathbb{A}^1]$ on $U_X \times \mathbb{P}^1$ gives rise to a well-defined cycle

$$\tilde{\beta}^2_G \in \mathrm{Z}(U_X/G; \mathbb{P}^1)$$

satisfying

$$q^*(\tilde{\beta}^2_G) = (\tilde{\beta}^2 \times [\mathbb{A}^1])|_{U_X \times \mathbb{P}^1}, \tag{59.7}$$

where $q : U_X \times \mathbb{P}^1 \to (U_X/G) \times \mathbb{P}^1$ is the canonical morphism.

Let $Z \subset (U_X/G) \times \mathbb{P}^1$ be a closed subvariety dominant over $\mathbb{P}^1$. We have $p^{-1}(Z(a)) = q^{-1}(Z)(a)$ for any rational point a of $\mathbb{P}^1$, where $p : U_X \to U_X/G$ is the canonical morphism. It follows from Proposition 57.8 that

$$p^*(\gamma(a)) = (q^*(\gamma))(a) \tag{59.8}$$

for every cycle $\gamma \in \mathrm{Z}(U_X/G; \mathbb{P}^1)$.

Let $\beta = \sum n_i [W_i]$. Then applying (59.8) to $\gamma = \tilde{\beta}^2_G$, we see by (59.6) and (59.7) that

$$\begin{aligned} p^*(\tilde{\beta}^2_G(a)) &= (q^*(\tilde{\beta}^2_G))(a) = (\tilde{\beta}^2 \times [\mathbb{A}^1])|_{U_X \times \mathbb{P}^1}(a) \\ &= \sum n_i n_j [W_i \times_{\mathbb{P}^1} W_j \times \mathbb{A}^1]|_{U_X \times \mathbb{P}^1}(a) \\ &= \sum n_i n_j [W_i(a) \times W_j(a) \times \mathbb{A}^1]|_{U_X} \\ &= p^*(\beta(a)^2_G) \end{aligned}$$

in $\mathrm{Z}(U_X)$. It follows that $\tilde{\beta}^2_G(a) = \beta(a)^2_G$ in $\mathrm{Z}(U_X/G)$ since p^* is injective on cycles. In particular, $\tilde{\beta}^2_G(0) = \beta(0)^2_G = \alpha^2_G$ and $\tilde{\beta}^2_G(\infty) = \beta(\infty)^2_G = {\alpha'}^2_G$, i.e., the cycles α^2_G and ${\alpha'}^2_G$ are rationally equivalent by Proposition 57.6. □

By Lemma 59.5, we have a well-defined map (but not a homomorphism!)

$$v_X : \mathrm{CH}(X) \to \mathrm{CH}(U_X/G), \quad [\alpha] \mapsto [\alpha^2_G]. \tag{59.9}$$

Consider the blowup B_X of $X^2 \times \mathbb{A}^1$ along $\Delta(X) \times \{0\}$. The exceptional divisor is the projective cone $\mathbb{P}(T_X \oplus \mathbb{1})$, where T_X is the tangent cone of X. The open complement $B_X \setminus \mathbb{P}(T_X \oplus \mathbb{1})$ is naturally isomorphic to U_X and the group $G = \{1, \sigma\}$ acts naturally on B_X (cf. §105.B).

By Proposition 105.7, the composition

$$i : \mathbb{P}(T_X \oplus \mathbb{1}) \hookrightarrow B_X \to B_X/G$$

is a locally principal divisor with normal line bundle $L^{\otimes 2}$ where L is the canonical line bundle over $\mathbb{P}(T_X \oplus \mathbb{1})$.

We define a map

$$u_X : \mathrm{Ch}(U_X/G) \to \mathrm{Ch}\big(\mathbb{P}(T_X \oplus \mathbb{1})\big)$$

as follows: Let $\delta \in \mathrm{Ch}(U_X/G)$. By the localization property (Proposition 57.9), there is $\beta \in \mathrm{Ch}(B_X/G)$ such that $\beta|_{(U_X/G)} = \delta$. We set

$$u_X(\delta) = i^\star(\beta).$$

We claim that the result is independent of the choice of β. Indeed, if $\beta' \in \mathrm{Ch}(B_X/G)$ is another element with $\beta'|_{(U_X/G)} = \delta$, then by the localization, $\beta' = \beta + i_*(\gamma)$ for some $\gamma \in \mathrm{Ch}(T_X \oplus \mathbb{1})$. By Proposition 55.10, we have $(i^\star \circ i_*)(\gamma) = e(L^{\otimes 2})(\gamma) = 2e(L)(\gamma) = 0$ modulo 2, hence

$$i^\star(\beta') = i^\star(\beta) + (i^\star \circ i_*)(\gamma) = i^\star(\beta)$$

as needed.

Let $q : B_X \to B_X/G$ be the projection.

Lemma 59.10. *The composition $i^\star \circ q_* : \mathrm{Ch}(B_X) \to \mathrm{Ch}\big(\mathbb{P}(T_X \oplus \mathbb{1})\big)$ is zero.*

PROOF. The scheme $Y := q^{-1}\big(\mathbb{P}(T_X \oplus \mathbb{1})\big)$ is a locally principal closed subscheme of B_X. The sheaf of ideals in $\mathcal{O}_{B_X}$ defining Y is the square of the sheaf of ideals of $\mathbb{P}(T_X \oplus \mathbb{1})$ as a subscheme of B_X. Let $j : Y \to B_X$ be the closed embedding and $p : Y \to \mathbb{P}(T_X \oplus \mathbb{1})$ the natural morphism. By Corollary 55.4, we have $i^\star \circ q_* = p_* \circ j^\star$. It follows from Proposition 55.11 that $j^\star$ is trivial modulo 2. □

Proposition 59.11. *For every scheme X, the map u_X is a homomorphism.*

PROOF. Let $p : U_X \to U_X/G$ be the projection. For any two cycles $\alpha = \sum n_i[Z_i]$ and $\alpha' = \sum n'_i[Z_i]$ on X, we have

$$p^*(\alpha + \alpha')^2_G - p^*(\alpha^2_G) - p^*({\alpha'_G}^2) = (1 + \sigma^*)(\gamma),$$

where

$$\gamma = \sum_{i<j} n_i n'_j [Z_i \times Z_j \times \mathbb{A}^1]|_{U_X} \in \mathrm{Z}(U_X).$$

Since $p^* \circ p_* = 1 + \sigma^*$ by (59.1) and p^* is injective on cycles by Proposition 59.2, we have

$$(\alpha + \alpha')^2_G - \alpha^2_G - {\alpha'_G}^2 = p_*(\gamma).$$

Let β, β', $\beta'' \in \mathrm{Ch}(B_X/G)$ and $\delta \in \mathrm{Ch}(U_X)$ be cycles restricting to α, α', $\alpha + \alpha'$ and γ, respectively, satisfying

$$\beta'' - \beta - \beta' = q_*(\delta).$$

By Lemma 59.10,

$$u_X(\alpha + \alpha') - u_X(\alpha) - u_X(\alpha') = i^\star(\beta'') - i^\star(\beta) - i^\star(\beta') = (i^\star \circ q_*)(\delta) = 0. \quad \square$$

Definition 59.12. Let X be a scheme. We define the *Steenrod operation of homological type* as the composition

$$\mathrm{Sq}^X : \mathrm{Ch}(X) \xrightarrow{v_X} \mathrm{Ch}(U_X/G) \xrightarrow{u_X} \mathrm{Ch}\big(\mathbb{P}(T_X \oplus \mathbb{1})\big) \xrightarrow{\mathrm{sg}^{T_X}} \mathrm{Ch}(X),$$

where sg^{T_X} is the Segre homomorphism defined in §58.A.

For every integer k we write

$$\mathrm{Sq}_k^X : \mathrm{Ch}_*(X) \to \mathrm{Ch}_{*-k}(X),$$

for the component of Sq^X decreasing dimension by k.

Proposition 59.13. *Let Z be a closed subvariety of a scheme X. Then $\mathrm{Sq}^X([Z]) = j_*\big(\mathrm{Sg}(T_Z)\big)$, where $j : Z \to X$ is the closed embedding and Sg is the total Segre class.*

PROOF. Let $\alpha = [Z] \in \mathrm{CH}(X)$. We have $v_X(\alpha) = \alpha_G^2 = \big[U_Z/G\big]$. Set $\beta = \big[B_Z/G\big] \in \mathrm{CH}(B_Z/G)$. By Proposition 55.6, we have $i_Z^\star(\beta) = \big[\mathbb{P}(T_Z \oplus \mathbb{1})\big]$, where $i_Z : \mathbb{P}(T_Z \oplus \mathbb{1}) \to B_Z/G$ is the closed embedding.

Consider the diagram

$$\begin{array}{ccccc} \mathrm{Ch}(B_Z/G) & \xrightarrow{i_Z^\star} & \mathrm{Ch}\big(\mathbb{P}(T_Z \oplus \mathbb{1})\big) & \xrightarrow{\mathrm{sg}^{T_Z}} & \mathrm{Ch}(Z) \\ {\scriptstyle k_*}\downarrow & & \downarrow & & \downarrow{\scriptstyle j_*} \\ \mathrm{Ch}(B_X/G) & \xrightarrow{i_X^\star} & \mathrm{Ch}\big(\mathbb{P}(T_X \oplus \mathbb{1})\big) & \xrightarrow{\mathrm{sg}^{T_X}} & \mathrm{Ch}(X) \end{array}$$

with vertical maps the push-forward homomorphisms. The diagram is commutative by Corollary 55.4 and Proposition 58.2. The commutativity yields

$$\begin{aligned} \mathrm{Sq}^X([Z]) &= (\mathrm{sg}^{T_X} \circ i_X^\star)\big(k_*(\beta)\big) \\ &= (j_* \circ \mathrm{sg}^{T_Z} \circ i_Z^\star)(\beta) \\ &= (j_* \circ \mathrm{sg}^{T_Z})\big([\mathbb{P}(T_Z \oplus \mathbb{1})]\big) \\ &= j_*\big(\mathrm{Sg}(T_Z)\big). \end{aligned}$$

□

Remark 59.14. The maps v_X, u_X and sg^{T_X} commute with arbitrary field extensions, hence so do Steenrod operations. More precisely, if L/F is a field extension, then the diagram

$$\begin{array}{ccc} \mathrm{Ch}(X) & \xrightarrow{\mathrm{Sq}^X} & \mathrm{Ch}(X) \\ \downarrow & & \downarrow \\ \mathrm{Ch}(X_L) & \xrightarrow{\mathrm{Sq}^{X_L}} & \mathrm{Ch}(X_L) \end{array}$$

commutes.

60. Properties of the Steenrod operations

In this section, we establish the standard properties of Steenrod operations of homological type.

60.A. Formula for a smooth cycle. Let Z be a smooth closed subvariety of a scheme X. By Proposition 58.9, the total Segre class $\mathrm{Sg}(T_Z)$ coincides with $s(T_Z)\big([Z]\big) = c(T_Z)^{-1}\big([Z]\big) = c(-T_Z)\big([Z]\big)$, where c is the total Chern class. Hence by Proposition 59.13,

(60.1) $$\mathrm{Sq}^X\big([Z]\big) = j_* \circ c(-T_Z)\big([Z]\big),$$

where $j : Z \to X$ is the closed embedding.

60.B. External products.

Theorem 60.2. *Let X and Y be two schemes over a field F of characteristic not 2. Then* $\mathrm{Sq}^{X\times Y}(\alpha\times\beta)=\mathrm{Sq}^X(\alpha)\times\mathrm{Sq}^Y(\beta)$ *for any* $\alpha\in\mathrm{Ch}(X)$ *and* $\beta\in\mathrm{Ch}(Y)$. *Equivalently,*

$$\mathrm{Sq}_n^{X\times Y}(\alpha\times\beta)=\sum_{k+m=n}\mathrm{Sq}_k^X(\alpha)\times\mathrm{Sq}_m^Y(\beta)$$

for all n.

PROOF. We may assume that $\alpha=[V]$ and $\beta=[W]$ where V and W are closed subvarieties of X and Y, respectively. Let $i:V\to X$ and $j:W\to Y$ be the closed embeddings. By Propositions 50.4, 58.19, and Corollary 104.8,

$$\begin{aligned}\mathrm{Sq}^{X\times Y}(\alpha\times\beta)&=(i\times j)_*\circ\mathrm{Sg}(T_{V\times W})\\&=(i_*\times j_*)\circ\mathrm{Sg}(T_V\times T_W)\\&=(i_*\times j_*)\circ\big(\mathrm{Sg}(T_V)\times\mathrm{Sg}(T_W)\big)\\&=i_*\circ\mathrm{Sg}(T_V)\times j_*\circ\mathrm{Sg}(T_W)\\&=\mathrm{Sq}^X(\alpha)\times\mathrm{Sq}^Y(\beta).\end{aligned}$$ □

60.C. Functoriality of Sq^X.

Lemma 60.3. *Let $i:Y\to X$ be a closed embedding. Then* $i_*\circ\mathrm{Sq}^Y=\mathrm{Sq}^X\circ i_*$.

PROOF. Let $Z\subset Y$ be a closed subscheme and let $j:Z\to Y$ be the closed embedding. By Proposition 59.13, we have

$$i_*\circ\mathrm{Sq}^Y\big([Z]\big)=i_*\circ j_*\circ\mathrm{Sg}(T_Z)=(ij)_*\circ\mathrm{Sg}(T_Z)=\mathrm{Sq}^X\big(i_*[Z]\big).$$ □

Lemma 60.4. *Let $p:\mathbb{P}^r\times X\to X$ be the projection. Then* $p_*\circ\mathrm{Sq}^{\mathbb{P}^r\times X}=\mathrm{Sq}^X\circ p_*$.

PROOF. The group $\mathrm{CH}(\mathbb{P}^r\times X)$ is generated by cycles $\alpha=[\mathbb{P}^k\times Z]$ for all closed subvarieties $Z\subset X$ and $k\le r$ by Theorem 57.14. It follows from Lemma 60.3 that we may assume $Z=X$ and $k=r$. The statement is obvious if $r=0$, so we may assume that $r>0$. Since $p_*(\alpha)=0$, we need to prove that $p_*\,\mathrm{Sq}^{\mathbb{P}^r\times X}(\alpha)=0$.

By Theorem 60.2, we have

$$\mathrm{Sq}^{\mathbb{P}^r\times X}(\alpha)=\mathrm{Sq}^{\mathbb{P}^r}\big([\mathbb{P}^r]\big)\times\mathrm{Sq}^X\big([X]\big).$$

It follows from Example 58.6 and (60.1) that

$$\mathrm{Sq}^{\mathbb{P}^r}\big([\mathbb{P}^r]\big)=c(T_{\mathbb{P}^r})^{-1}\big([\mathbb{P}^r]\big)=(1+h)^{-r-1},$$

where $h=c_1(L)$ is the class of a hyperplane in $\mathbb{P}^r$. By Proposition 50.4,

$$p_*\big(\mathrm{Sq}^{\mathbb{P}^r\times X}(\alpha)\big)=\deg(1+h)^{-r-1}\cdot\mathrm{Sq}^X\big([X]\big).$$

We have

$$\deg(1+h)^{-r-1}=\binom{-r-1}{r}=(-1)^r\binom{2r}{r}$$

and the latter binomial coefficient is even if $r>0$. □

Theorem 60.5. *Let $f : Y \to X$ be a projective morphism. Then the diagram*

$$\begin{array}{ccc} \mathrm{Ch}(Y) & \xrightarrow{\mathrm{Sq}^Y} & \mathrm{Ch}(Y) \\ {\scriptstyle f_*}\downarrow & & {\scriptstyle f_*}\downarrow \\ \mathrm{Ch}(X) & \xrightarrow{\mathrm{Sq}^X} & \mathrm{Ch}(X) \end{array}$$

is commutative.

PROOF. The projective morphism f factors as the composition of a closed embedding $Y \to \mathbb{P}^r \times X$ and the projection $\mathbb{P}^r \times X \to X$, so the statement follows from Lemmas 60.3 and 60.4. □

Theorem 60.6. $\mathrm{Sq}_k^X = 0$ *if* $k < 0$ *and* Sq_0^X *is the identity.*

PROOF. First suppose that X is a variety of dimension d. By dimension count, the class $\mathrm{Sq}_k^X([X]) = \mathrm{Sg}_{d-k}(T_X)$ is trivial if $k < 0$. To compute $\mathrm{Sq}_0^X([X])$, we can extend the base field to a perfect one and replace X by a smooth open subscheme. Then by (60.1),

$$\mathrm{Sq}_0^X([X]) = c_0(-T_X)([X]) = [X],$$

i.e., Sq_0^X is the identity on $\mathrm{Ch}_d(X)$.

In general, let $Z \subset X$ be a closed subvariety and let $j : Z \to X$ be the closed embedding. Then by Lemma 60.3 and the first part of the proof, the class $\mathrm{Sq}_k^X([Z]) = j_*\big(\mathrm{Sq}_k^Z([Z])\big)$ is trivial for $k < 0$ and is equal to $[Z] \in \mathrm{Ch}(X)$ if $k = 0$. □

61. Steenrod operations for smooth schemes

In this section, we define Steenrod operations of cohomological type and prove their basic properties.

Lemma 61.1. *Let $f : Y \to X$ be a regular closed embedding of schemes of codimension r and $g : U_Y/G \to U_X/G$ the closed embedding induced by f. Then g is a regular closed embedding of codimension $2r$ and the following diagram is commutative:*

$$\begin{array}{ccc} \mathrm{CH}(X) & \xrightarrow{\upsilon_X} & \mathrm{CH}(U_X/G) \\ {\scriptstyle f^\star}\downarrow & & \downarrow{\scriptstyle g^\star} \\ \mathrm{CH}(Y) & \xrightarrow{\upsilon_Y} & \mathrm{CH}(U_Y/G) \end{array}$$

PROOF. The closed embedding $U_Y \to U_X$ is regular of codimension $2r$ and the morphism $U_X \to U_X/G$ is faithfully flat. Hence g is also a regular closed embedding by Proposition 104.11. Let $p : N \to Y$ be the normal bundle of f. The Gysin homomorphism $f^\star$ is the composition of the deformation homomorphism $\sigma_f : \mathrm{CH}(X) \to \mathrm{CH}(N)$ and the inverse to the pull-back isomorphism $p_f^* : \mathrm{CH}(Y) \to \mathrm{CH}(N)$ (cf. §55.A).

The normal bundle N_h of the closed embedding $h : U_Y \to U_X$ is the restriction of the vector bundle $N^2 \times \mathbb{A}^1$ on U_Y.

Consider the diagram

$$
\begin{array}{ccccccc}
\mathrm{Z}(X) & \longrightarrow & \mathrm{Z}(X^2 \times \mathbb{A}^1)^G & \longrightarrow & \mathrm{Z}(U_X)^G & = & \mathrm{Z}(U_X/G) \\
\downarrow{\scriptstyle \sigma_f} & & \downarrow{\scriptstyle \sigma_{f^2 \times 1}} & & \downarrow{\scriptstyle \sigma_h} & & \downarrow{\scriptstyle \sigma_g} \\
\mathrm{Z}(N) & \longrightarrow & \mathrm{Z}(N^2 \times \mathbb{A}^1)^G & \longrightarrow & \mathrm{Z}(N_h)^G & = & \mathrm{Z}(N_h/G) \\
\uparrow & & \uparrow & & \uparrow & & \uparrow \\
\mathrm{Z}(Y) & \longrightarrow & \mathrm{Z}(Y^2 \times \mathbb{A}^1)^G & \longrightarrow & \mathrm{Z}(U_Y)^G & = & \mathrm{Z}(U_Y/G)
\end{array}
\tag{61.2}
$$

where the first homomorphism in every row takes a cycle α to $\alpha^2 \times [\mathbb{A}^1]$, the other unmarked maps are pull-back homomorphisms with respect to flat morphisms and the equalities follow from Proposition 59.2.

The deformation homomorphism is defined by $\sigma_f\big(\sum n_i[Z_i]\big) = [C_{k_i}]$, where $k_i : Y \cap Z_i \to Z_i$ is the restriction of f by Proposition 52.7, so the commutativity of the upper left square follows from the equality of cycles $[C_{k_i} \times C_{k_j}] = [C_{k_i \times k_j}]$ (cf. Proposition 104.7). The two other top squares are commutative by Proposition 51.5. The commutativity of the left bottom square follows from Propositions 57.8 and 57.12. The two other squares are commutative by Proposition 49.18.

The normal bundle N_h is an open subscheme of U_N and of $N^2 \times \mathbb{A}^1$. Let $j : N_h \to U_N$ and $l : N_h/G \to U_N/G$ be the open embeddings. The following diagram of the pull-back homomorphisms

$$
\begin{array}{ccccccc}
\mathrm{Z}(N) & \longrightarrow & \mathrm{Z}(N^2 \times \mathbb{A}^1)^G & \longrightarrow & \mathrm{Z}(U_N)^G & = & \mathrm{Z}(U_N/G) \\
\Vert & & \Vert & & {\scriptstyle j^*}\downarrow & & \downarrow{\scriptstyle l^*} \\
\mathrm{Z}(N) & \longrightarrow & \mathrm{Z}(N^2 \times \mathbb{A}^1)^G & \longrightarrow & \mathrm{Z}(N_h)^G & = & \mathrm{Z}(N_h/G)
\end{array}
$$

is commutative by Proposition 49.18. It follows from Lemma 59.5 that the composition in the top row factors through the rational equivalence, hence so does the composition in the bottom row and then in the middle row of the diagram (61.2). Therefore the diagram (61.2) yields a commutative diagram:

$$
\begin{array}{ccccc}
\mathrm{CH}(X) & \xrightarrow{v_X} & \mathrm{CH}(U_X/G) & & \\
{\scriptstyle \sigma_f}\downarrow & & & \searrow{\scriptstyle \sigma_g} & \\
\mathrm{CH}(N) & \longrightarrow & \mathrm{CH}(U_N/G) & \xrightarrow{l_*} & \mathrm{CH}(N_h/G) \\
{\scriptstyle p_f^*}\uparrow\wr & & & \nearrow{\scriptstyle p_g^*}\,\sim & \\
\mathrm{CH}(Y) & \xrightarrow{v_Y} & \mathrm{CH}(U_Y/G) & &
\end{array}
$$

The lemma follows from the commutativity of this diagram. $\square$

Let $f : Y \to X$ be a closed embedding of smooth schemes with the normal bundle $N \to Y$. Consider the diagram

$$
\begin{array}{ccc}
\mathbb{P}(T_Y \oplus \mathbb{1}) & \xrightarrow{j} & \mathbb{P}(T_X \oplus \mathbb{1}) \\
{\scriptstyle p}\downarrow & & \downarrow{\scriptstyle q} \\
Y & \xrightarrow{f} & X.
\end{array}
$$

Lemma 61.3. *We have* $c(N) \circ f^* \circ \operatorname{sg}^{T_X} = \operatorname{sg}^{T_Y} \circ j^*$.

PROOF. By the Projective Bundle Theorem 57.14, the group $\operatorname{CH}(\mathbb{P}(T_X \oplus \mathbb{1}))$ is generated by the elements $\beta = e(L_X)^k(q^*(\alpha))$ with $k \geq 0$ and $\alpha \in \operatorname{CH}(X)$. We have

$$e(L_X)^\bullet(\beta) = e(L_X)^\bullet(q^*(\alpha)). \tag{61.4}$$

Since $j^*(L_X) = L_Y$ and $j^* \circ q^* = p^* \circ f^*$, we have $j^*\beta = e(L_{T_Y})^k(p^*(f^*(\alpha)))$ by Proposition 53.3(2), and therefore,

$$e(L_Y)^\bullet(j^*(\beta)) = e(L_Y)^\bullet \circ p^*(f^*(\alpha)). \tag{61.5}$$

By Proposition 104.16, $c(N) \circ s(f^*(T_X)) = c(N) \circ c(f^*(T_X))^{-1} = c(T_Y)^{-1} = s(T_Y)$. It follows from (61.4), (61.5), Propositions 54.7 and 58.4(2) that

$$\begin{aligned}
c(N) \circ f^* \circ \operatorname{sg}^{T_X}(\beta) &= c(N) \circ f^* \circ q^* \circ e(L_X)^\bullet(\beta) \\
&= c(N) \circ f^* \circ q^* \circ e(L_X)^\bullet(q^*(\alpha)) \\
&= c(N) \circ f^* \circ s(T_X)(\alpha) \\
&= c(N) \circ s(f^*T_X)(f^*(\alpha)) \\
&= s(T_Y)(f^*(\alpha)) \\
&= p^* \circ e(L_Y)^\bullet \circ p^*(f^*(\alpha)) \\
&= p^* \circ e(L_Y)^\bullet(j^*(\beta)) \\
&= \operatorname{sg}^{T_Y} \circ j^*(\beta).
\end{aligned}$$

□

Proposition 61.6. *Let* $f : Y \to X$ *be a closed embedding of smooth schemes with the normal bundle* N. *Then* $c(N) \circ f^* \circ \operatorname{Sq}^X = \operatorname{Sq}^Y \circ f^*$.

PROOF. By Proposition 105.8, the schemes B_Y/G and B_X/G are smooth. Let

$$j : \mathbb{P}(T_Y \oplus \mathbb{1}) \to \mathbb{P}(T_X \oplus \mathbb{1}) \quad \text{and} \quad h : B_Y/G \to B_X/G$$

be the closed embeddings induced by f. Let $\alpha \in \operatorname{Ch}(X)$. Choose $\beta \in \operatorname{Ch}(B_X/G)$ satisfying $\beta|_{(U_X/G)} = \alpha_G^2$ (cf. (59.4)). It follows from Proposition 55.19 and Lemma 61.1 that

$$(h^*(\beta))|_{U_Y/G} = (f^*(\alpha))_G^2.$$

By Proposition 55.19 and Lemma 61.3,

$$\begin{aligned}
c(N) \circ f^* \circ \operatorname{Sq}^X(\alpha) &= c(N) \circ f^* \circ \operatorname{sg}^{T_X} \circ i_X^\star(\beta) \\
&= \operatorname{sg}^{T_Y} \circ j^* \circ i_X^\star(\beta) \\
&= \operatorname{sg}^{T_Y} \circ i_Y^\star \circ h^*(\beta) \\
&= \operatorname{Sq}^Y \circ f^*(\alpha).
\end{aligned}$$

□

Definition 61.7. Let X be a smooth scheme. We define the *Steenrod operations of cohomological type* by the formula

$$\operatorname{Sq}_X = c(T_X) \circ \operatorname{Sq}^X.$$

We write Sq_X^k for kth homogeneous part of Sq_X. Thus Sq_X^k is an operation

$$\operatorname{Sq}_X^k : \operatorname{Ch}^*(X) \to \operatorname{Ch}^{*+k}(X).$$

Proposition 61.8 (Wu Formula). *Let Z be a smooth closed subscheme of a smooth scheme X. Then $\mathrm{Sq}_X([Z]) = j_* \circ c(N)([Z])$, where N is the normal bundle of the closed embedding $j : Z \to X$.*

PROOF. By Proposition 54.5 and (60.1),

$$\begin{aligned}\mathrm{Sq}_X([Z]) &= c(T_X) \circ \mathrm{Sq}^X([Z]) \\ &= c(T_X) \circ j_* \circ c(-T_Z)([Z]) \\ &= j_* \circ c(i^*(T_X)) \circ c(-T_Z)([Z]) \\ &= j_* \circ c(N)([Z])\end{aligned}$$

since $c(T_Z) \circ c(N) = c(j^*(T_X))$. □

Theorem 61.9. *Let $f : Y \to X$ be a morphism of smooth schemes. Then the diagram*

$$\begin{array}{ccc} \mathrm{Ch}(X) & \xrightarrow{\mathrm{Sq}_X} & \mathrm{Ch}(X) \\ {\scriptstyle f^*}\downarrow & & \downarrow{\scriptstyle f^*} \\ \mathrm{Ch}(Y) & \xrightarrow{\mathrm{Sq}_Y} & \mathrm{Ch}(Y) \end{array}$$

is commutative.

PROOF. Suppose first that f is a closed embedding with normal bundle N. It follows from Propositions 54.5(2) and 61.6 that

$$\begin{aligned}f^* \circ \mathrm{Sq}_X &= f^* \circ c(T_X) \circ \mathrm{Sq}^X \\ &= c(f^*(T_X)) \circ f^* \circ \mathrm{Sq}^X \\ &= c(T_Y) \circ c(N) \circ f^* \circ \mathrm{Sq}^X \\ &= c(T_Y) \circ \mathrm{Sq}^Y \circ f^* \\ &= \mathrm{Sq}_Y \circ f^*.\end{aligned}$$

Second, consider the case of the projection $g : Y \times X \to X$. Let $Z \subset X$ be a closed subvariety. By (60.1), Propositions 58.12, 57.12, Corollary 104.8, and Theorem 60.2, we have $g^*([Z]) = [Y \times Z] = [Y] \times [Z]$ and

$$\begin{aligned}\mathrm{Sq}_{Y\times X}(g^*([Z])) &= c(T_{Y\times X}) \circ \mathrm{Sq}^{Y\times X}([Y \times Z]) \\ &= \big(c(T_Y) \times c(T_X)\big)\big(\mathrm{Sq}^Y([Y]) \times \mathrm{Sq}^X([Z])\big) \\ &= \big(c(T_Y) \circ \mathrm{Sq}^Y([Y])\big) \times \big(c(T_X) \circ \mathrm{Sq}^X([Z])\big) \\ &= [Y] \times \mathrm{Sq}_X([Z]) \\ &= g^*\big(\mathrm{Sq}_X([Z])\big).\end{aligned}$$

In the general case, write $f = g \circ h$ where $h = (\mathrm{id}_X, f) : Y \to Y \times X$ is the closed embedding and $g : Y \times X \to X$ is the projection. Then by the above,

$$f^* \circ \mathrm{Sq}_X = h^* \circ g^* \circ \mathrm{Sq}_X = h^* \circ \mathrm{Sq}_{Y\times X} \circ g^* = \mathrm{Sq}_Y \circ h^* \circ g^* = \mathrm{Sq}_Y \circ f^*. \quad □$$

Proposition 61.10. *Let $f : Y \to X$ be a smooth projective morphism of smooth schemes. Then*

$$\mathrm{Sq}_X \circ f_* = f_* \circ c(-T_f) \circ \mathrm{Sq}_Y,$$

where T_f is the relative tangent bundle of f.

PROOF. It follows from the exactness of the sequence

$$0 \to T_f \to T_Y \to f^*(T_X) \to 0$$

that $c(T_Y) = c(T_f) \circ c(f^*T_X)$. By Proposition 54.5(1) and Theorem 60.5,

$$\begin{aligned} \mathrm{Sq}_X \circ f_* &= c(T_X) \circ \mathrm{Sq}^X \circ f_* \\ &= c(T_X) \circ f_* \circ \mathrm{Sq}^Y \\ &= c(T_X) \circ f_* \circ c(-T_Y) \circ \mathrm{Sq}_Y \\ &= f_* \circ c(f^*T_X) \circ c(-T_Y) \circ \mathrm{Sq}_Y \\ &= f_* \circ c(-T_f) \circ \mathrm{Sq}_Y . \end{aligned}$$

□

Let X be a smooth variety of dimension d and let $Z \subset X$ be a closed subvariety. Consider the closed embedding $j : \mathbb{P}(T_Z \oplus \mathbb{1}) \to \mathbb{P}(T_X \oplus \mathbb{1})$. By the Projective Bundle Theorem 57.14 applied to the vector bundle $T_X \oplus \mathbb{1}$ over X of rank $d+1$, there are unique elements $\alpha_0, \alpha_1, \dots, \alpha_d \in \mathrm{Ch}(X)$ such that

$$j_*\big([\mathbb{P}(T_Z \oplus \mathbb{1})]\big) = \sum_{k=0}^{d} e(L)^k\big(q^*(\alpha_k)\big)$$

in $\mathrm{Ch}\big(\mathbb{P}(T_X \oplus \mathbb{1})\big)$, where L is the canonical line bundle over $\mathbb{P}(T_X \oplus \mathbb{1})$ and $q : \mathbb{P}(T_X \oplus \mathbb{1}) \to X$ is the natural morphism. We set $\alpha := \alpha_0 + \alpha_1 + \cdots + \alpha_d \in \mathrm{Ch}(X)$.

Lemma 61.11. $\mathrm{Sq}^X\big([Z]\big) = s(T_X)(\alpha)$.

PROOF. Let $p : \mathbb{P}(T_Z \oplus \mathbb{1}) \to Z$ be the projection and $i : Z \to X$ the closed embedding, so that $i \circ p = q \circ j$. The canonical line bundle L' over $\mathbb{P}(T_Z \oplus \mathbb{1})$ coincides with $j^*(L)$ and by Proposition 53.3,

$$\begin{aligned} \mathrm{Sq}^X\big([Z]\big) &= i_* \,\mathrm{Sg}(T_Z) \\ &= i_* \circ p_* \circ e(L')^\bullet\big([\mathbb{P}(T_Z \oplus \mathbb{1})]\big) \\ &= q_* \circ j_* \circ e\big(j^*(L)\big)^\bullet\big([\mathbb{P}(T_Z \oplus \mathbb{1})]\big) \\ &= q_* \circ e(L)^\bullet \circ j_*\big([\mathbb{P}(T_Z \oplus \mathbb{1})]\big) \\ &= q_* \circ e(L)^\bullet \circ \sum_{k=0}^{d} e(L)^k\big(q^*(\alpha_k)\big) \\ &= q_* \circ e(L)^\bullet \circ q^*(\alpha) \\ &= s(T_X)(\alpha). \end{aligned}$$

□

Corollary 61.12. $\mathrm{Sq}_X\big([Z]\big) = \alpha$ *in* $\mathrm{Ch}(X)$.

PROOF. By Lemma 61.11 and Proposition 58.9,

$$\mathrm{Sq}_X\big([Z]\big) = c(T_X)\big(\mathrm{Sq}^X([Z])\big) = c(T_X)s(T_X)(\alpha) = \alpha.$$

□

Theorem 61.13. *Let X be a smooth scheme. Then for any $\beta \in \mathrm{Ch}^k(X)$,*

$$\mathrm{Sq}_X^r(\beta) = \begin{cases} \beta & \text{if } r = 0, \\ \beta^2 & \text{if } r = k, \\ 0 & \text{if } r < 0 \text{ or } r > k. \end{cases}$$

PROOF. By definition and Theorem 60.6, we have $\mathrm{Sq}_X^k = 0$ if $k < 0$ and Sq_X^0 is the identity operation.

We may assume that X is a variety and $\beta = [Z]$ where $Z \subset X$ is a closed subvariety of codimension k. Since $\alpha_i \in \mathrm{Ch}^{2k-i}(X)$, we have $\mathrm{Sq}_X^r(\beta) = \alpha_{k-r}$ by Corollary 61.12. Therefore, $\mathrm{Sq}_X^r(\beta) = 0$ if $r > k$.

As $\mathrm{Sq}_X^k(\beta) = \alpha_0$, it remains to prove that $\beta^2 = \alpha_0$. Consider the diagonal embedding $d : X \to X^2$ and the closed embedding $h : T_Z \to T_X$. By the definition of the product in $\mathrm{Ch}(X)$ and Proposition 52.7,

$$p^*(\beta^2) = p^* \circ d_X^*\big([Z^2]\big) = \sigma_d\big([Z^2]\big) = h_*\big([T_Z]\big) \in \mathrm{Ch}(T_X),$$

where $p : T_X \to X$ is the canonical morphism. Let $j : T_X \to \mathbb{P}(T_X \oplus \mathbb{1})$ be the open embedding. Since the pull-back $j^*(L)$ of the canonical line bundle L over $\mathbb{P}(T_X \oplus \mathbb{1})$ is a trivial line bundle over T_X, we have

$$j^* \circ e(L)^s\big(q^*(\alpha)\big) = e\big(j^*(L)\big)^s\big(j^* \circ q^*(\alpha)\big) = \begin{cases} p^*(\alpha) & \text{if } s = 0, \\ 0 & \text{if } s > 0 \end{cases}$$

for every $\alpha \in \mathrm{Ch}(X)$. Hence

$$p^*(\beta^2) = \big[T_Z\big] = j^*\big(\big[\mathbb{P}(T_Z \oplus \mathbb{1})\big]\big) = p^*(\alpha_0),$$

therefore, $\beta^2 = \alpha_0$ since p^* is an isomorphism. □

Theorem 61.14. *Let X and Y be two smooth schemes. Then*

$$\mathrm{Sq}_{X\times Y} = \mathrm{Sq}_X \times \mathrm{Sq}_Y .$$

PROOF. By Corollary 104.8, we have $T_{X\times Y} = T_X \times T_Y$. It follows from Theorem 60.2 and Proposition 58.12 that

$$\begin{aligned} \mathrm{Sq}_{X\times Y} &= c(T_{X\times Y}) \circ \mathrm{Sq}^{X\times Y} \\ &= \big(c(T_X) \circ \mathrm{Sq}^X\big) \times \big(c(T_Y) \circ \mathrm{Sq}^Y\big) \\ &= \mathrm{Sq}_X \times \mathrm{Sq}_Y . \end{aligned}$$

□

Corollary 61.15 (Cartan Formula). *Let X be a smooth scheme. Then $\mathrm{Sq}_X(\alpha \cdot \beta) = \mathrm{Sq}_X(\alpha) \cdot \mathrm{Sq}_X(\beta)$ for all $\alpha, \beta \in \mathrm{Ch}(X)$. Equivalently,*

$$\mathrm{Sq}_X^n(\alpha \cdot \beta) = \sum_{k+m=n} \mathrm{Sq}_X^k(\alpha) \cdot \mathrm{Sq}_X^m(\beta)$$

for all n.

PROOF. Let $i : X \to X \times X$ be the diagonal embedding. Then by Theorems 61.9 and 61.14,

$$\begin{aligned} \mathrm{Sq}_X(\alpha \cdot \beta) &= \mathrm{Sq}_X\big(i^*(\alpha \times \beta)\big) \\ &= i^* \,\mathrm{Sq}_{X\times Y}(\alpha \times \beta) \\ &= i^*\big(\mathrm{Sq}_X(\alpha) \times \mathrm{Sq}_X(\beta)\big) \\ &= \mathrm{Sq}_X(\alpha) \cdot \mathrm{Sq}_X(\beta). \end{aligned}$$

□

Example 61.16. Let $X = \mathbb{P}^d$ be projective space and $h \in \mathrm{Ch}^1(X)$ the class of a hyperplane. By Theorem 61.13, we have $\mathrm{Sq}_X(h) = h + h^2 = h(1+h)$. It follows from Corollary 61.15 that

$$\mathrm{Sq}_X(h^i) = h^i(1+h)^i, \quad \mathrm{Sq}_X^r(h^i) = \binom{i}{r} h^{i+r}.$$

By Example 104.20, the class of the tangent bundle T_X is equal to $(d+1)[L] - 1$, where L is the canonical line bundle over X. Hence

$$c(T_X) = c(L)^{d+1} = (1+h)^{d+1}$$

and

$$\mathrm{Sq}^X(h^i) = c(T_X)^{-1} \circ \mathrm{Sq}_X(h^i) = h^i(1+h)^{i-d-1}.$$

CHAPTER XII

Category of Chow Motives

In this chapter we study Chow motives. The notion of a Chow motive is due to Grothendieck. Many (co)homology theories defined on the category $\mathrm{Sm}(F)$ of smooth complete schemes, such as Chow groups and more generally K-(co)homology groups take values in the category of abelian groups. But the category $\mathrm{Sm}(F)$ itself does not have the structure of an additive category as we cannot add morphisms of schemes.

In this chapter, for an arbitrary commutative ring Λ, we shall construct the additive categories of correspondences $\mathrm{CR}(F,\Lambda)$, $\mathrm{CR}_*(F,\Lambda)$ and motives $\mathrm{CM}(F,\Lambda)$, $\mathrm{CM}_*(F,\Lambda)$ together with functors

$$\begin{array}{ccccc} \mathrm{Sm}(F) & \longrightarrow & \mathrm{CR}(F,\Lambda) & \longrightarrow & \mathrm{CM}(F,\Lambda) \\ & & \downarrow & & \downarrow \\ & & \mathrm{CR}_*(F,\Lambda) & \longrightarrow & \mathrm{CM}_*(F,\Lambda) \end{array}$$

so that the theories with values in the category of abelian groups mentioned above factor through them. All of these the new categories do have the additional structure of additive category. This makes them easier to work with than the category $\mathrm{Sm}(F)$. Applications of these categories can be found in Chapter XVII later in this book.

Some classical theorems also have motivic analogs. For example, the Projective Bundle Theorem 53.10 has such an analog (cf. Theorem 63.10) below. We shall see that the motive of a projective bundle splits into a direct sum of certain motives already in the category of correspondences $\mathrm{CR}(F,\Lambda)$. From this the classical Projective Bundle Theorem 53.10 can be obtained by applying an appropriate functor to the decomposition in $\mathrm{CR}(F,\Lambda)$.

In this chapter, scheme means a separated scheme of finite type over a field.

62. Correspondences

Definition 62.1. Let X and Y be two schemes over F. A *correspondence* between X and Y is an element of $\mathrm{CH}(X \times Y)$.

For example, the graph of a morphism between X and Y is a correspondence. In this section we study functorial properties of correspondences.

For a scheme Y over F, we have two canonical morphisms: the structure morphism $p_Y : Y \to \mathrm{Spec}(F)$ and the diagonal closed embedding $d_Y : Y \to Y \times Y$. If Y is complete, the map p_Y is proper, and if Y is smooth, the closed embedding d_Y is regular.

Let X, Y and Z be schemes over F with Y complete and smooth. We consider morphisms

$$ {}^{X}p_Y^{Z} := 1_X \times p_Y \times 1_Z : X \times Y \times Z \to X \times Z$$

and

$$ {}^{X}\!d^{Z}_{Y} := 1_X \times d_Y \times 1_Z : X \times Y \times Z \to X \times Y \times Y \times Z. $$

If $X = \operatorname{Spec}(F)$, we shall simply write p^{Z}_{Y} and d^{Z}_{Y}.

Define a bilinear pairing of K-homology groups (cf. §52)

$$ A_*(Y \times Z, K_*) \times A_*(X \times Y, K_*) \to A_*(X \times Z, K_*) $$

by

$$ (\beta, \alpha) \mapsto \beta \circ \alpha := \big({}^{X}\!p^{Z}_{Y}\big)_* \circ \big({}^{X}\!d^{Z}_{Y}\big)^{\star}(\alpha \times \beta). \tag{62.2} $$

For an element $\alpha \in A_*(X \times Y, K_*)$ we write α^t for its image in $A_*(Y \times X, K_*)$ under the *exchange isomorphism* $X \times Y \simeq Y \times X$. The element α^t is called the *transpose of* α. By the definition of the pairing,

$$ (\beta \circ \alpha)^t = \alpha^t \circ \beta^t. $$

Proposition 62.3. *The pairing* (62.2) *is associative. More precisely, for any four schemes X, Y, Z, T over F with Y and Z smooth and complete and any $\alpha \in A_*(X \times Y, K_*)$, $\beta \in A_*(Y \times Z, K_*)$, and $\gamma \in A_*(Z \times T, K_*)$, we have*

$$ (\gamma \circ \beta) \circ \alpha = \big({}^{X}\!p^{T}_{Y\times Z}\big)_* \circ \big({}^{X}\!d^{T}_{Y\times Z}\big)^{\star}(\alpha \times \beta \times \gamma) = \gamma \circ (\beta \circ \alpha). $$

PROOF. We prove the first equality. It follows from Corollary 55.4 that

$$ \big({}^{X}\!d^{T}_{Y}\big)^{\star} \circ \big({}^{X\times Y\times Y}\!p^{T}_{Z}\big)_* = \big({}^{X\times Y}\!p^{T}_{Z}\big)_* \circ \big({}^{X}\!d^{Z\times T}_{Y}\big)^{\star}. $$

By Propositions 50.4, 50.5, and 55.1, we have

$$ \begin{aligned} (\gamma \circ \beta) \circ \alpha &= \big({}^{X}\!p^{T}_{Y}\big)_* \circ \big({}^{X}\!d^{T}_{Y}\big)^{\star}\Big(\alpha \times \big({}^{Y}\!p^{T}_{Z}\big)_*\big({}^{Y}\!d^{T}_{Z}\big)^{\star}(\beta \times \gamma)\Big) \\ &= \big({}^{X}\!p^{T}_{Y}\big)_* \circ \big({}^{X}\!d^{T}_{Y}\big)^{\star} \circ \big({}^{X\times Y\times Y}\!p^{T}_{Z}\big)_* \circ \big({}^{X\times Y\times Y}\!d^{T}_{Z}\big)^{\star}(\alpha \times \beta \times \gamma) \\ &= \big({}^{X}\!p^{T}_{Y}\big)_* \circ \big({}^{X\times Y}\!p^{T}_{Z}\big)_* \circ \big({}^{X}\!d^{Z\times T}_{Y}\big)^{\star} \circ \big({}^{X\times Y\times Y}\!d^{T}_{Z}\big)^{\star}(\alpha \times \beta \times \gamma) \\ &= \big({}^{X}\!p^{T}_{Y\times Z}\big)_* \circ \big({}^{X}\!d^{T}_{Y\times Z}\big)^{\star}(\alpha \times \beta \times \gamma). \end{aligned} $$

□

Let $f : X \to Y$ be a morphism of schemes. The isomorphic image of X under the closed embedding $(1_X, f) : X \to X \times Y$ is called the *graph of* f and is denoted by Γ_f. Thus, Γ_f is a closed subscheme of $X \times Y$ isomorphic to X under the projection $X \times Y \to X$. The class $[\Gamma_f]$ belongs to $\operatorname{CH}(X \times Y)$.

Proposition 62.4. *Let X, Y, Z be schemes over F with Y smooth and complete.*

(1) *For every morphism $g : Y \to Z$ and $\alpha \in A_*(X \times Y, K_*)$,*

$$ [\Gamma_g] \circ \alpha = (1_X \times g)_*(\alpha). $$

(2) *For every morphism $f : X \to Y$ and $\beta \in A_*(Y \times Z, K_*)$,*

$$ \beta \circ [\Gamma_f] = (f \times 1_Z)^*(\beta). $$

PROOF. (1): Consider the commutative diagram

$$ \begin{array}{ccccc} X \times Y & \xleftarrow{\;{}^{X\times Y}\!p_Y\;} & X \times Y \times Y & \xleftarrow{\;{}^{X}\!d_Y\;} & X \times Y \\ & & \Big\downarrow{\scriptstyle r} & & \Big\downarrow{\scriptstyle t} \\ & & X \times Y \times Y \times Z & \xleftarrow{\;{}^{X}\!d^{Z}_{Y}\;} & X \times Y \times Z & \xrightarrow{\;{}^{X}\!p^{Z}_{Y}\;} & X \times Z \end{array} $$

where $r = 1_{X\times Y} \times (1_Y, g)$ and $t = 1_X \times (1_Y, g)$.

The composition ${}^{X\times Y}p_Y \circ {}^X d_Y$ is the identity on $X \times Y$ and ${}^X p_Y^Z \circ t = 1_X \times g$. It follows from Corollary 55.4 that $\big({}^X d_Y^Z\big)^\bigstar \circ r_* = t_* \circ \big({}^X d_Y\big)^\bigstar$. We have

$$\begin{aligned}
[\Gamma_g] \circ \alpha &= \big({}^X p_Y^Z\big)_* \circ \big({}^X d_Y^Z\big)^\bigstar (\alpha \times [\Gamma_g]) \\
&= \big({}^X p_Y^Z\big)_* \circ \big({}^X d_Y^Z\big)^\bigstar \circ r_*(\alpha \times [Y]) \\
&= \big({}^X p_Y^Z\big)_* \circ \big({}^X d_Y^Z\big)^\bigstar \circ r_* \circ \big({}^{X\times Y}p_Y\big)^*(\alpha) \\
&= \big({}^X p_Y^Z\big)_* \circ t_* \circ \big({}^X d_Y\big)^\bigstar \circ \big({}^{X\times Y}p_Y\big)^*(\alpha) \\
&= (1_X \times g)_*(\alpha).
\end{aligned}$$

(2): Consider the commutative diagram

$$\begin{array}{ccccc}
Y \times Z & \xleftarrow{p_X^{Y\times Z}} & X \times Y \times Z & \xleftarrow{\ v\ } & X \times Z \\
& & u\big\downarrow & & v\big\downarrow \\
& & X \times Y \times Y \times Z & \xleftarrow{{}^X d_Y^Z} & X \times Y \times Z & \xrightarrow{{}^X p_Y^Z} & X \times Z
\end{array}$$

where $u = (1_X, f) \times 1_{Y\times Z}$ and $v = (1_X, f) \times 1_Z$.

The composition ${}^X p_Y^Z \circ v$ is the identity on $X \times Z$ and $p_X^{Y\times Z} \circ v = f \times 1_Z$. It follows from Corollary 55.4 that $\big({}^X d_Y^Z\big)^\bigstar \circ u_* = v_* \circ v^*$. We have

$$\begin{aligned}
\beta \circ [\Gamma_f] &= \big({}^X p_Y^Z\big)_* \circ \big({}^X d_Y^Z\big)^\bigstar \big([\Gamma_f] \times \beta\big) \\
&= \big({}^X p_Y^Z\big)_* \circ \big({}^X d_Y^Z\big)^\bigstar \circ u_*\big([X] \times \beta\big) \\
&= \big({}^X p_Y^Z\big)_* \circ \big({}^X d_Y^Z\big)^\bigstar \circ u_* \circ \big(p_X^{Y\times Z}\big)^*(\beta) \\
&= \big({}^X p_Y^Z\big)_* \circ v_* \circ v^* \circ \big(p_X^{Y\times Z}\big)^*(\beta) \\
&= (f \times 1_Z)^*(\beta).
\end{aligned}$$

□

Corollary 62.5. *Let X and Y be schemes over F and $\alpha \in A_*(X \times Y, K_*)$. Then*

(1) *If Y is smooth and complete, then $[\Gamma_{1_Y}] \circ \alpha = \alpha$.*
(2) *If X is smooth and complete, then $\alpha \circ [\Gamma_{1_X}] = \alpha$.*

Corollary 62.6. *Let $f : X \to Y$ and $g : Y \to Z$ be two morphisms. If Y is smooth and complete, then $[\Gamma_g] \circ [\Gamma_f] = [\Gamma_{gf}]$.*

PROOF. By Proposition 62.4(1),

$$\begin{aligned}
[\Gamma_g] \circ [\Gamma_f] &= (1_X \times g)_*\big([\Gamma_f]\big) \\
&= (1_X \times g)_*(1_X, f)_*\big([X]\big) \\
&= (1_X, gf)_*\big([X]\big) \\
&= [\Gamma_{gf}].
\end{aligned}$$

□

Let X, Y and Z be arbitrary schemes and $\alpha \in A_*(X \times Y, K_*)$. If X is smooth and complete, we have a well-defined homomorphism

$$\alpha_* : A_*(Z \times X, K_*) \to A_*(Z \times Y, K_*), \quad \beta \mapsto \alpha \circ \beta.$$

If $\alpha = [\Gamma_f]$ with $f : X \to Y$ a morphism, it follows from Proposition 62.4(1) that $\alpha_* = (1_Z \times f)_*$.

If X is smooth and $Z = \operatorname{Spec}(F)$, we get a homomorphism $\alpha_* : A_*(X, K_*) \to A_*(Y, K_*)$. We have simpler formula for α_* in the following case:

Proposition 62.7. *Let $\alpha = [T]$ with $T \subset X \times Y$ a closed subscheme with X smooth and complete. Then $\alpha_* = q_* \circ p^*$, where $p : T \to X$ and $q : T \to Y$ are the projections.*

PROOF. Let $r : X \times Y \to Y$ be the projection, $i : T \to X \times Y$ the closed embedding, and $f : T \to X \times T$ the graph of the projection p. Consider the commutative diagram

$$\begin{array}{ccccc} X \times T & \xleftarrow{f} & T & \xrightarrow{q} & Y \\ {\scriptstyle 1_X \times i}\downarrow & & {\scriptstyle i}\downarrow & & \| \\ X \times X \times Y & \xleftarrow{d_X^Y} & X \times Y & \xrightarrow{r} & Y. \end{array}$$

It follows from Corollary 55.4 that $i_* \circ f^* = (d_X^Y)^\bigstar \circ (1_X \times i)_*$. Therefore, for every $\beta \in A_*(X, K_*)$, we have

$$\begin{aligned} \alpha_*(\beta) &= r_* \circ (d_X^Y)^\bigstar (\beta \times \alpha) \\ &= r_* \circ (d_X^Y)^\bigstar \circ (1_X \times i)_* \big(\beta \times [T]\big) \\ &= r_* \circ i_* \circ f^* \big(\beta \times [T]\big) \\ &= q_* \circ f^* \big(\beta \times [T]\big) \\ &= q_* \circ p^*(\beta). \end{aligned}$$ □

If Y is smooth and complete, we have a well-defined homomorphism

$$\alpha^* : A_*(Y \times Z, K_*) \to A_*(X \times Z, K_*), \quad \beta \mapsto \beta \circ \alpha.$$

If $\alpha = [\Gamma_f]$ for a flat morphism $f : X \to Y$, it follows from Proposition 62.4(2) that $\alpha^* = (f \times 1_Z)^*$.

Let X, Y and Z be arbitrary schemes, $\alpha \in A_*(X \times Y, K_*)$ and $g : Y \to Z$ a proper morphism. We define the composition of g and α by

$$g \circ \alpha := (1_X \times g)_*(\alpha) \in A_*(X \times Z, K_*).$$

If $g \circ \alpha = [\Gamma_h]$ for some morphism $h : X \to Z$, we abuse notation and just write $g \circ \alpha = h$. If Y is smooth and complete, we have $g \circ \alpha = [\Gamma_g] \circ \alpha$ by Proposition 62.4(1).

Similarly, if $\beta \in A_*(Y \times Z, K_*)$ and $f : X \to Y$ is a flat morphism, we define the composition of β and f by

$$\beta \circ f := (f \times 1_Z)^*(\beta) \in A_*(X \times Z, K_*).$$

If Y is smooth and complete, we have $\beta \circ f = \beta \circ [\Gamma_f]$ by Proposition 62.4(2).

The following statement is an analogue of Proposition 62.3 with fewer assumptions on the schemes.

Proposition 62.8. *Let X, Y, Z and T be arbitrary schemes.*

(1) *Let $\alpha \in A_*(X \times Y, K_*)$, $\gamma \in A_*(T \times X, K_*)$, and $g : Y \to Z$ a proper morphism. If X is smooth and complete, then*

$$(g \circ \alpha) \circ \gamma = g \circ (\alpha \circ \gamma), \quad \text{i.e.,} \quad (g \circ \alpha)_* = g_* \circ \alpha_*.$$

(2) *Let* $\beta \in A_*(Y \times Z, K_*)$, $\delta \in A_*(Z \times T, K_*)$, *and* $f : X \to Y$ *a flat morphism. If* Z *is smooth and complete, then*

$$\delta \circ (\beta \circ f) = (\delta \circ \beta) \circ f, \quad i.e., \quad (\beta \circ f)^* = f^* \circ \beta^*.$$

PROOF. (1): Consider the commutative diagram with fiber product squares

$$\begin{array}{ccccc}
T \times X \times X \times Y & \xleftarrow{{}^{T}d^{Y}_{X}} & T \times X \times Y & \xrightarrow{{}^{T}p^{Y}_{X}} & T \times Y \\
\Big\downarrow {\scriptstyle 1_{T\times X\times X}\times g} & & \Big\downarrow {\scriptstyle 1_{T\times X}\times g} & & \Big\downarrow {\scriptstyle 1_T \times g} \\
T \times X \times X \times Z & \xleftarrow{{}^{T}d^{Z}_{X}} & T \times X \times Z & \xrightarrow{{}^{T}p^{Z}_{X}} & T \times Z.
\end{array}$$

It follows from Proposition 50.4 and Corollary 55.4 that

$$\begin{aligned}
g \circ (\alpha \circ \gamma) &= (1_T \times g)_*(\alpha \circ \gamma) \\
&= (1_T \times g)_* \circ \big({}^{T}p^{Y}_{X}\big)_* \circ \big({}^{T}d^{Y}_{X}\big)^{\bigstar}(\gamma \times \alpha) \\
&= \big({}^{T}p^{Z}_{X}\big)_* \circ \big({}^{T}d^{Z}_{X}\big)^{\bigstar} \circ (1_{T\times X\times X} \times g)_*(\gamma \times \alpha) \\
&= (1_X \times g)_*(\alpha) \circ \gamma \\
&= (g \circ \alpha) \circ \gamma.
\end{aligned}$$

(2): The proof is similar using Propositions 49.20, 50.5 and 55.5 instead. □

If $\gamma \in A_*(Y \times X, K_*)$ and $g : Y \to Z$ is a proper morphism, we write $\gamma \circ g^t$ for $(g \circ \gamma^t)^t \in A_*(Z \times X, K_*)$. Similarly, if $\delta \in A_*(Z \times Y, K_*)$ and $f : X \to Y$ is a flat morphism, we define the composition $f^t \circ \delta \in A_*(Z \times X, K_*)$ to be $(\delta^t \circ f)^t$.

63. Categories of correspondences

Let Λ be a commutative ring. For a scheme Z, we write $\mathrm{CH}(Z;\Lambda)$ for the Λ-module $\mathrm{CH}(Z) \otimes \Lambda$.

Let X and Y be smooth complete schemes over F. Let $X_1, X_2, \dots, X_n$ be the irreducible components of X of dimension $d_1, d_2, \dots, d_n$, respectively. For every $i \in \mathbb{Z}$, we set

$$\mathrm{Corr}_i(X, Y; \Lambda) := \coprod_{k=1}^{n} \mathrm{CH}_{i+d_k}(X_k \times Y; \Lambda).$$

An element $\alpha \in \mathrm{Corr}_i(X, Y; \Lambda)$ is called a *correspondence between* X *and* Y *of degree* i *with coefficients in* Λ. We write $\alpha : X \rightsquigarrow Y$.

Let Z be another smooth complete scheme. By Proposition 62.3, the bilinear pairing $(\beta, \alpha) \mapsto \beta \circ \alpha$ on Chow groups yields an associative pairing (composition)

(63.1) $$\mathrm{Corr}_i(Y, Z; \Lambda) \times \mathrm{Corr}_j(X, Y; \Lambda) \to \mathrm{Corr}_{i+j}(X, Z; \Lambda).$$

The following proposition gives an alternative formula for this composition involving only projection morphisms.

Proposition 63.2. $\beta \circ \alpha = \big({}^{X}p^{Z}_{Y}\big)_*\Big(\big({}^{X\times Y}p_Z\big)^*(\alpha) \cdot \big(p^{Y\times Z}_{X}\big)^*(\beta)\Big).$

PROOF. Let $f : X \times Y \times Y \times Z \to X \times Y \times Z \times X \times Y \times Z$ be defined by $f(x,y,y',z) = (x,y,z,x,y',z)$. We have $f \circ {}^X d_Y^Z = d_{X\times Y\times Z}$, therefore,

$$\begin{aligned}\beta \circ \alpha &= \left({}^X p_Y^Z\right)_* \circ \left({}^X d_Y^Z\right)^\bigstar (\alpha \times \beta)\\ &= \left({}^X p_Y^Z\right)_* \circ \left({}^X d_Y^Z\right)^\bigstar \circ f^*\big(\alpha \times [Z] \times [X] \times \beta\big)\\ &= \left({}^X p_Y^Z\right)_* \circ (d_{X\times Y\times Z})^\bigstar \Big(\left({}^{X\times Y}p_Z\right)^*(\alpha) \times \left(p_X^{Y\times Z}\right)^*(\beta)\Big)\\ &= \left({}^X p_Y^Z\right)_* \Big(\left({}^{X\times Y}p_Z\right)^*(\alpha) \cdot \left(p_X^{Y\times Z}\right)^*(\beta)\Big). \qquad \square\end{aligned}$$

Define the category $\mathrm{CR}_*(F,\Lambda)$ *of correspondences with coefficients in* Λ over F as follows: Objects of $\mathrm{CR}_*(F,\Lambda)$ are smooth complete schemes over F. A morphism between X and Y is an element of the graded group

$$\coprod_{k\in\mathbb{Z}} \mathrm{Corr}_k(X,Y;\Lambda).$$

Composition of morphisms is given by (63.1). The identity morphism of X in $\mathrm{CR}_*(F,\Lambda)$ is $\Gamma_{1_X} \otimes 1$, where Γ_{1_X} is the class of the graph of the identity morphism 1_X (cf. Corollary 62.5). The direct sum in $\mathrm{CR}_*(F,\Lambda)$ is given by the disjoint union of schemes. As the composition law in $\mathrm{CR}_*(F,\Lambda)$ is bilinear and associative by Proposition 62.3, the category $\mathrm{CR}_*(F,\Lambda)$ is additive. Abusing notation, we write Λ for the object $\mathrm{Spec}(F)$ in this category.

An object of $\mathrm{CR}_*(F,\Lambda)$ is called a *Chow motive* or simply a *motive*. If X is a smooth complete scheme, we write $M(X)$ for it as an object in $\mathrm{CR}_*(F,\Lambda)$.

We define another category $\mathrm{C}(F,\Lambda)$ as follows. Objects of $\mathrm{C}(F,\Lambda)$ are pairs (X,i), where X is a smooth complete scheme over F and $i \in \mathbb{Z}$. A morphism between (X,i) and (Y,j) is an element of $\mathrm{Corr}_{i-j}(X,Y;\Lambda)$. The composition of morphisms is given by (63.1). The morphisms between two objects form an abelian group and the composition is bilinear and associative by Proposition 62.3, therefore, $\mathrm{C}(F,\Lambda)$ is a preadditive category.

There is an additive functor $\mathrm{C}(F,\Lambda) \to \mathrm{CR}_*(F,\Lambda)$ taking an object (X,i) to X, and that is the natural inclusion on morphisms.

Let $\mathcal{A}$ be a preadditive category. Define the *additive completion* of $\mathcal{A}$ to be the following category $\widetilde{\mathcal{A}}$: its objects are finite sequences of objects $A_1,\dots,A_n$ of $\mathcal{A}$ written in the form $\coprod_{i=1}^n A_i$. A morphism between $\coprod_{i=1}^n A_i$ and $\coprod_{j=1}^m B_j$ is given by an $n\times m$-matrix of morphisms $A_i \to B_j$. The composition of morphisms is the matrix multiplication. The category $\widetilde{\mathcal{A}}$ has finite products and coproducts and is therefore an additive category. The category $\mathcal{A}$ is a full subcategory of $\widetilde{\mathcal{A}}$.

Let $\mathrm{CR}(F,\Lambda)$ denote the additive completion of $\mathrm{C}(F,\Lambda)$. We call it the *category of graded correspondences with coefficients in* Λ over F. An object of $\mathrm{CR}(F,\Lambda)$ is also called a *Chow motive* or simply a *motive*. We will write $M(X)(i)$ for (X,i) and simply $M(X)$ for $(X,0)$. The functor $\mathrm{C}(F,\Lambda) \to \mathrm{CR}_*(F,\Lambda)$ extends naturally to an additive functor

$$\mathrm{CR}(F,\Lambda) \to \mathrm{CR}_*(F,\Lambda). \tag{63.3}$$

taking $M(X)(i)$ to $M(X)$. Write $\Lambda(i)$ for the motive $(\mathrm{Spec}\,F, i)$ in $\mathrm{CR}(F,\Lambda)$. The motives $\Lambda(i)$ in $\mathrm{CR}(F,\Lambda)$ and Λ in $\mathrm{CR}_*(F,\Lambda)$ are called the *Tate motives*.

Functor (63.3) is faithful but not full. Nevertheless, it has the following nice property.

Proposition 63.4. *Let f be a morphism in $\mathrm{CR}(F,\Lambda)$. If the image of f in the category $\mathrm{CR}_*(F,\Lambda)$ is an isomorphism, then f itself is an isomorphism.*

PROOF. Let f be a morphism between the objects

$$\coprod_{i=1}^{n} X_i(a_i) \quad \text{and} \quad \coprod_{j=1}^{m} Y_j(b_j).$$

Thus f is given by an $n \times m$ matrix $A = (f_{ij})$ with $f_{ij} \in \mathrm{Corr}_{a_j - b_i}(X_j, Y_i; \Lambda)$. Let $B = (g_{kl})$ be the matrix of the inverse of f in $\mathrm{CR}_*(F,\Lambda)$, so that $g_{kl} \in \mathrm{Corr}_*(Y_l, X_k; \Lambda)$. Let $\bar{g}_{kl}$ be the homogeneous component of g_{kl} of degree $b_l - a_k$ and $\overline{B} = (\bar{g}_{kl})$. As $AB = A\overline{B}$ and $\overline{B}A = BA$ are the identity matrices, we have $\overline{B} = B = A^{-1}$. Therefore, B is the matrix of the inverse of f in $\mathrm{CR}(F,\Lambda)$. □

A ring homomorphism $\Lambda \to \Lambda'$ gives rise to natural functors $\mathrm{CR}_*(F,\Lambda) \to \mathrm{CR}_*(F,\Lambda')$ and $\mathrm{CR}(F,\Lambda) \to \mathrm{CR}(F,\Lambda')$ that are identical on objects. We simply write $\mathrm{CR}_*(F)$ for $\mathrm{CR}_*(F,\mathbb{Z})$ and $\mathrm{CR}(F)$ for $\mathrm{CR}(F,\mathbb{Z})$.

It follows from Corollary 62.6 that there is a functor

$$\mathrm{Sm}(F) \to \mathrm{CR}(F,\Lambda)$$

taking a smooth complete scheme X to $M(X)$ and a morphism $f : X \to Y$ to $[\Gamma_f] \otimes 1$ in $\mathrm{Corr}_0(X,Y;\Lambda) = \mathrm{Mor}_{\mathrm{CR}(F,\Lambda)}\big(M(X), M(Y)\big)$, where Γ_f is the graph of f.

Let X and Y be smooth complete schemes and $i, j \in \mathbb{Z}$. We have

$$\mathrm{Hom}_{\mathrm{CR}(F)}\big(M(X)(i), M(Y)(j)\big) = \mathrm{Corr}_{i-j}(X,Y;\Lambda).$$

In particular,

$$\mathrm{Hom}_{\mathrm{CR}(F,\Lambda)}\big(\Lambda(i), M(X)\big) = \mathrm{CH}_i(X;\Lambda), \tag{63.5}$$

$$\mathrm{Hom}_{\mathrm{CR}(F,\Lambda)}\big(M(X), \Lambda(i)\big) = \mathrm{CH}^i(X;\Lambda). \tag{63.6}$$

We define Chow groups with coefficients in Λ for an arbitrary motive M as follows:

$$\mathrm{CH}_i(M;\Lambda) := \mathrm{Hom}_{\mathrm{CR}(F,\Lambda)}\big(\Lambda(i), M\big), \quad \mathrm{CH}^i(M;\Lambda) := \mathrm{Hom}_{\mathrm{CR}(F,\Lambda)}\big(M, \Lambda(i)\big).$$

The category $\mathrm{CR}(F,\Lambda)$ has the structure of a *tensor category* given by

$$M(X)(i) \otimes M(Y)(j) := M(X \times Y)(i+j).$$

In particular,

$$M(X)(i) \otimes \Lambda(j) = M(X)(i+j).$$

Let Y be a smooth variety of dimension d. By the definition of a morphism in $\mathrm{CR}(F)$, the equality

$$\mathrm{Hom}_{\mathrm{CR}(F,\Lambda)}\big(M(Y)(i), N\big) = \mathrm{CH}_{d+i}\big(M(Y) \otimes N;\Lambda\big) \tag{63.7}$$

holds for every N of the form $M(X)(j)$, where X is a smooth complete scheme; and, therefore, by additivity it holds for all motives N. Similarly,

$$\mathrm{Hom}_{\mathrm{CR}(F,\Lambda)}\big(N, M(Y)(i)\big) = \mathrm{CH}^{d+i}\big(N \otimes M(Y);\Lambda\big). \tag{63.8}$$

The following statement is a variant of the Yoneda Lemma.

Lemma 63.9. *Let $\alpha : N \to P$ be a morphism in $\mathrm{CR}(F,\Lambda)$. Then the following conditions are equivalent:*

(1) *α is an isomorphism.*
(2) *For every smooth complete scheme Y, the homomorphism*

$$(1_Y \otimes \alpha)_* : \mathrm{CH}_*\big(M(Y) \otimes N; \Lambda\big) \to \mathrm{CH}_*\big(M(Y) \otimes P; \Lambda\big)$$

is an isomorphism.
(3) *For every smooth complete scheme X, the homomorphism*

$$(1_Y \otimes \alpha)^* : \mathrm{CH}^*\big(M(Y) \otimes P; \Lambda\big) \to \mathrm{CH}^*\big(M(Y) \otimes N; \Lambda\big)$$

is an isomorphism.

PROOF. Clearly (1) $\Rightarrow$ (2) and (1) $\Rightarrow$ (3). We prove that (2) implies (1) (the proof of the implication (3) $\Rightarrow$ (1) is similar and left to the reader). It follows from (63.7) that the natural homomorphism

$$\mathrm{Hom}_{\mathrm{CR}(F,\Lambda)}(M, N) \to \mathrm{Hom}_{\mathrm{CR}(F,\Lambda)}(M, P)$$

is an isomorphism if $M = M(Y)(i)$ for any smooth complete variety Y. By additivity, it is isomorphism for all motives M. The statement now follows from the (usual) Yoneda Lemma. □

The following statement is the motivic version of the Projective Bundle Theorem 53.10.

Theorem 63.10. *Let $E \to X$ be a vector bundle of rank r over a smooth complete scheme X. Then the motives $M\big(\mathbb{P}(E)\big)$ and $\coprod_{i=0}^{r-1} M(X)(i)$ are naturally isomorphic in $\mathrm{CR}(F, \Lambda)$.*

PROOF. Let Y be a smooth complete scheme over F. Applying the Projective Bundle Theorem 53.10 to the vector bundle $E \times Y \to X \times Y$, we see that the Chow groups of $\coprod_{i=0}^{r-1} M(X \times Y)(i)$ and $M\big(\mathbb{P}(E) \times Y\big)$ are isomorphic. Moreover, in view of Remark 53.11, this isomorphism is natural in Y with respect to morphisms in the category $\mathrm{CR}(F, \Lambda)$. In other words, the functors on $\mathrm{CR}(F, \Lambda)$ represented by the objects $\coprod_{i=0}^{r-1} M(X)(i)$ and $M\big(\mathbb{P}(E)\big)$ are isomorphic. By the Yoneda Lemma, the objects are isomorphic in $\mathrm{CR}(F, \Lambda)$. □

Corollary 63.11. *The motive $M\big(\mathbb{P}(E)\big)$ is isomorphic to the direct sum $M(X)^r$ of r copies of $M(X)$ in the category $\mathrm{CR}_*(F, \Lambda)$.*

64. Category of Chow motives

Let $\mathcal{A}$ be an additive category. An idempotent $e : A \to A$ in $\mathcal{A}$ is called *split* if there is an isomorphism $f : A \xrightarrow{\sim} B \oplus C$ such that e coincides with the composition $A \xrightarrow{f} B \oplus C \xrightarrow{p} B \xrightarrow{i} B \oplus C \xrightarrow{f^{-1}} A$, where p and i are the canonical morphisms.

The *idempotent completion of an additive category* $\mathcal{A}$ is the category $\overline{\mathcal{A}}$ defined as follows. Objects of $\overline{\mathcal{A}}$ are the pairs (A, e), where A is an object of $\mathcal{A}$ and $e : A \to A$ is an idempotent. The group of morphisms between (A, e) and (B, f) is $f \circ \mathrm{Hom}_{\mathcal{A}}(A, B) \circ e$. Every idempotent in $\overline{\mathcal{A}}$ is split.

The assignment $A \mapsto (A, 1_A)$ defines a full and faithful functor from $\mathcal{A}$ to $\overline{\mathcal{A}}$. We identify $\mathcal{A}$ with a full subcategory of $\overline{\mathcal{A}}$.

Let Λ be a commutative ring. The idempotent completion of the category $\mathrm{CR}(F, \Lambda)$ is called the *category of graded Chow motives with coefficients in* Λ and is denoted by $\mathrm{CM}(F, \Lambda)$. By definition, every object of $\mathrm{CM}(F, \Lambda)$ is a direct summand of a finite direct sum of motives of the form $M(X)(i)$, where X is a smooth complete scheme over F. We write $\mathrm{CM}(F)$ for $\mathrm{CM}(F,\mathbb{Z})$.

Similarly, the idempotent completion $\mathrm{CM}_*(F,\Lambda)$ of $\mathrm{CR}_*(F,\Lambda)$ is called the *category of Chow motives with coefficients in* Λ. Note that Proposition 63.4 holds for the natural functor $\mathrm{CM}(F,\Lambda) \to \mathrm{CM}_*(F,\Lambda)$.

We have functors

$$\mathrm{Sm}(F) \to \mathrm{CR}(F,\Lambda) \to \mathrm{CM}(F,\Lambda).$$

The second functor is full and faithful, so we can view $\mathrm{CR}(F,\Lambda)$ as a full subcategory of $\mathrm{CM}(F,\Lambda)$ which we do. Note that $\mathrm{CM}(F,\Lambda)$ inherits the structure of a tensor category.

An object of $\mathrm{CM}(F,\Lambda)$ is also called a *motive*. We will keep the same notation $M(X)(i)$, $\Lambda(i)$, etc. for the corresponding motives in $\mathrm{CM}(F,\Lambda)$. The motives $\Lambda(i)$ and Λ are called the *Tate motives*.

We use formulas (63.5) and (63.6) in order to define Chow groups with coefficients in Λ for an arbitrary motive M:

$$\mathrm{CH}_i(M;\Lambda) := \mathrm{Hom}_{\mathrm{CM}(F,\Lambda)}\big(\Lambda(i), M\big), \quad \mathrm{CH}^i(M;\Lambda) := \mathrm{Hom}_{\mathrm{CM}(F,\Lambda)}\big(M, \Lambda(i)\big).$$

The functor from $\mathrm{CM}(F,\Lambda)$ to the category of Λ-modules, taking a motive M to $\mathrm{CH}_i(M;\Lambda)$ (respectively the cofunctor $M \mapsto \mathrm{CH}^i(M;\Lambda)$) is then represented (respectively corepresented) by $\Lambda(i)$.

Let Y be a smooth variety of dimension d. It follows from (63.7) and (63.8) that

$$\mathrm{Hom}_{\mathrm{CM}(F,\Lambda)}\big(M(Y)(i), N\big) = \mathrm{CH}_{d+i}\big(M(Y)\otimes N;\Lambda\big), \tag{64.1}$$

$$\mathrm{Hom}_{\mathrm{CM}(F,\Lambda)}\big(N, M(Y)(i)\big) = \mathrm{CH}^{d+i}\big(N\otimes M(Y);\Lambda\big).$$

for all motives N in $\mathrm{CM}(F,\Lambda)$.

Let M and N be objects in $\mathrm{CM}(F)$. The tensor product of two morphisms $M \to \Lambda(i)$ and $N \to \Lambda(j)$ defines a pairing

$$\mathrm{CH}^*(M;\Lambda)\otimes \mathrm{CH}^*(N;\Lambda) \to \mathrm{CH}^*(M\otimes N;\Lambda). \tag{64.2}$$

Note that this is an isomorphism whenever M (or N) is a Tate motive.

We say that an object M of $\mathrm{CR}(F,\Lambda)$ is *split* if M is isomorphic to a (finite) coproduct of Tate motives. The additivity property of the pairing yields

Proposition 64.3. *Let M (or N) be a split motive. Then the homomorphism* (64.2) *is an isomorphism.*

65. Duality

There is an additive *duality functor* $* : \mathrm{CM}(F,\Lambda)^{op} \to \mathrm{CM}(F,\Lambda)$ uniquely determined by the rule $M(X)(i)^* = M(X)(-d-i)$ where X is a smooth complete variety of dimension d and $\alpha^* = \alpha^t$ for any correspondence α. In particular, $\Lambda(i)^* = \Lambda(-i)$. The composition $*\circ *$ is the identity functor.

It follows from the definition of the duality functor that

$$\mathrm{Hom}_{\mathrm{CM}(F,\Lambda)}(M^*, N^*) = \mathrm{Hom}_{\mathrm{CM}(F,\Lambda)}(N, M)$$

for every two motives M and N. In particular, setting $N = \Lambda(i)$, we get

$$\mathrm{CH}^i(M^*;\Lambda) = \mathrm{CH}_{-i}(M;\Lambda).$$

The equality (64.1) reads as follows:

$$\operatorname{Hom}_{\mathrm{CM}(F,\Lambda)}\big(M(Y)(i), N\big) = \mathrm{CH}_0\big(M(Y)(i)^* \otimes N; \Lambda\big) \tag{65.1}$$

for every smooth complete scheme Y. Set

$$\underline{\operatorname{Hom}}(M, N) := M^* \otimes N$$

for every two motives M and N. By additivity, the equality (65.1) yields

$$\operatorname{Hom}_{\mathrm{CM}(F,\Lambda)}(M, N) = \mathrm{CH}_0\big(\underline{\operatorname{Hom}}(M, N); \Lambda\big).$$

Since the duality functor commutes with the tensor product, the definition of $\underline{\operatorname{Hom}}$ satisfies the *associativity law*

$$\underline{\operatorname{Hom}}(M \otimes N, P) = \underline{\operatorname{Hom}}\big(M, \underline{\operatorname{Hom}}(N, P)\big)$$

for all motives M, N and P. Applying CH_0 we get

$$\operatorname{Hom}_{\mathrm{CM}(F,\Lambda)}(M \otimes N, P) = \operatorname{Hom}_{\mathrm{CM}(F,\Lambda)}\big(M, \underline{\operatorname{Hom}}(N, P)\big).$$

66. Motives of cellular schemes

Motives of cellular schemes were considered in [**70**]. Recall that a morphism $p : U \to Y$ over F is an affine bundle of rank d if f is flat and the fiber of p over any point $y \in Y$ is isomorphic to the affine space $\mathbb{A}^d_{F(y)}$.

A scheme X over F is called *(relatively) cellular* if there is given a filtration by closed subschemes

$$\emptyset = X_{-1} \subset X_0 \subset X_1 \subset \cdots \subset X_n = X \tag{66.1}$$

together with affine bundles $p_i : U_i := X_i \setminus X_{i-1} \to Y_i$ of rank d_i for all $i \in [0, n]$, where Y_i is a smooth complete scheme for all $i \in [1,\ n]$ (we don't assume that Y_0 is either smooth or complete). The scheme U_i is called a *cell* of X and Y_i is the *base* of the cell U_i.

The graph Γ_{p_i} of the morphism p_i is a subscheme of $U_i \times Y_i$. Let α_i in $\mathrm{CH}(X_i \times Y_i)$ be the class of the closure of Γ_{p_i} in $X_i \times Y_i$. We have $\alpha_0 = \Gamma_{p_0}$.

We view α_i as a correspondence $X_i \rightsquigarrow Y_i$ of degree 0. Let $f_i : X_i \to X$ be the closed embedding. The correspondence $f_i \circ \alpha_i^t \in \mathrm{CH}(Y_i \times X)$ between Y_i and X is well-defined and of degree d_i for all $i \geq 1$.

Let Z be a scheme over F. If h is a morphism of schemes, we still write h for the morphism $1_Z \times h$. We define homomorphisms

$$a_i : \mathrm{CH}_{*-d_i}(Z \times Y_i) \to \mathrm{CH}_*(Z \times X)$$

for all $i \in [0, n]$ as follows: We set $a_0 = (f_0)_* \circ (p_0^*)^{-1}$ and $a_i = (f_i \circ \alpha_i^t)_*$ if $i \geq 1$. If Y_0 is smooth and complete, then $a_0 = (f_0 \circ \alpha_0^t)_*$.

Theorem 66.2. *Let X be a cellular scheme with filtration* (66.1). *Then for every scheme Z over F, the homomorphism*

$$\sum_{i=0}^{n} a_i : \coprod_{i=0}^{n} \mathrm{CH}_{*-d_i}(Z \times Y_i) \to \mathrm{CH}_*(Z \times X) \tag{66.3}$$

is an isomorphism.

PROOF. Let $g_i : U_i \to X_i$ denote the open embedding. By the definition of α_i, we have $\alpha_i \circ g_i = p_i$. It follows from Proposition 62.8(2) that for every scheme Z, the composition

$$A(Y_i \times Z, K_*) \xrightarrow{\alpha_i^*} A(X_i \times Z, K_*) \xrightarrow{g_i^*} A(U_i \times Z, K_*)$$

coincides with the pull-back homomorphism p_i^* for $i \geq 1$. By Theorem 52.13, the map p_i^* is an isomorphism. Hence g_i^* is a split surjection. Therefore, in the localization exact sequence (cf. §52.D),

$$A_{k+1}(X_i \times Z, K_{-k}) \xrightarrow{g_i^*} A_{k+1}(U_i \times Z, K_{-k}) \xrightarrow{\delta}$$
$$\mathrm{CH}_k(X_{i-1} \times Z) \to \mathrm{CH}_k(X_i \times Z) \xrightarrow{g_i^*} \mathrm{CH}_k(U_i \times Z) \to 0,$$

the connecting homomorphism δ is trivial. Consequently, we have a short exact sequence

$$0 \to \mathrm{CH}(X_{i-1} \times Z) \to \mathrm{CH}(X_i \times Z) \xrightarrow{s_i} \mathrm{CH}(Y_i \times Z) \to 0$$

where $s_i = p_i^{*-1} \circ g_i^*$ and s_i is split by $\alpha_i^* : \mathrm{CH}(Y_i \times Z) \to \mathrm{CH}(X_i \times Z)$. In particular, $\mathrm{CH}(X_i \times Z)$ is isomorphic to $\mathrm{CH}(X_{i-1} \times Z) \oplus \mathrm{CH}(Y_i \times Z)$. Iterating, we see that $\mathrm{CH}(X \times Z)$ is isomorphic to the coproduct of the $\mathrm{CH}(Y_i \times Z)$ over all $i \in [0,\ n]$. The inclusion of $\mathrm{CH}(Y_i \times Z)$ into $\mathrm{CH}(X \times Z)$ coincides with the composition

$$\mathrm{CH}(Y_i \times Z) \xrightarrow{\alpha_i^*} \mathrm{CH}(X_i \times Z) \xrightarrow{(f_i)_*} \mathrm{CH}(X \times Z),$$

where α_0^* is understood to be p_0^*. If $i \geq 1$, we have $a_i = (f_i)_* \circ (\alpha_i^t)_*$ by Proposition 62.8(1). Under the identification of $\mathrm{CH}(Y_i \times Z)$ with $\mathrm{CH}(Z \times Y_i)$, we have $(\alpha_i^t)_* = \alpha_i^*$, hence $a_i = (f_i)_* \circ \alpha_i^*$. It follows that the homomorphism (66.3) is an isomorphism. □

Lemma 63.9 yields

Corollary 66.4. *Let X be a smooth complete cellular scheme with filtration* (66.1) *and with all Y_i smooth complete. Then the morphism*

$$\coprod_{i=0}^{n} M(Y_i)(d_i) \to M(X),$$

defined by the sequence of correspondences $f_i \circ \alpha_i^t$ is an isomorphism in the category of correspondences $\mathrm{CR}(F)$.

Example 66.5. Let $X = \mathbb{P}^n$. Consider the filtration given by $X_i = \mathbb{P}^i$, $i \in [0,\ n]$. We have $U_i = \mathbb{A}^i$ and $Y_i = \mathrm{Spec}(F)$. By Corollary 66.4,

$$M(\mathbb{P}^n) = \mathbb{Z} \oplus \mathbb{Z}(1) \oplus \cdots \oplus \mathbb{Z}(n).$$

Example 66.6. Let φ be a nondegenerate quadratic form on V and X the associated quadric of say dimension d. Consider the following filtration on $X \times X$: X_0 is the image of the diagonal embedding of X into $X \times X$, X_1 consists of all pairs of orthogonal isotropic lines (L_1, L_2), and $X_2 = X \times X$. We also set $Y_0 = X$ (with the identity projection of X_0 on Y_0), $Y_2 = X$, and Y_1 is the flag variety Fl of pairs (L, P), where L and P are a totally isotropic line and plane, respectively, satisfying $L \subset P$.

We claim that the morphism $p_1 : U_1 \to Y_1$ taking a pair (L_1, L_2) to $(L_1, L_1 + L_2)$ is an affine bundle. To do this we use the criterion of Lemma 52.12. Let R be a local commutative F-algebra. An R-point of Y_1 is a pair (L, P), where L and P are totally isotropic direct summands of the R-module $V_R := V \otimes_F R$ of rank 1 and 2, respectively, with $L \subset P$. Let $\{e, f\}$ be an R-basis of P such that $L = Re$. Then the morphism $\mathbb{A}^1_R \to \operatorname{Spec}(R) \times_{Y_1} U_1$ taking an a to the point $\big(L, R(ae + f)\big)$ of the fiber product is an isomorphism. It follows from Lemma 52.12 that p_1 is an affine bundle.

We claim that the first projection $p_2 : U_2 \to Y_2$ is an affine bundle of rank d. We again apply the criterion of Lemma 52.12. Let R be a local commutative F-algebra. An R-point of Y_2 is a totally isotropic direct summand $L \subset V_R$ of rank 1. Choose a basis of V_R so that φ_R is given by a polynomial $t_0t_1 + \psi(T')$, where ψ is a quadratic form in the variables $T' = (t_2, \dots, t_{d+1})$ over R, and the orthogonal complement $L^\perp$ is given by $t_0 = 0$. Then the fiber product $\operatorname{Spec}(R) \times_{Y_2} U_2$ is given by the equation $\frac{t_1}{t_0} + \psi(\frac{T'}{t_0}) = 0$ and therefore is isomorphic to $\mathbb{A}^d_R$. It follows by Lemma 52.12 that p_2 is an affine bundle.

By Corollary 66.4, we conclude that

$$M(X \times X) \simeq M(X) \oplus M(\mathrm{Fl})(1) \oplus M(X)(d).$$

Example 66.7. Assume that the quadric X in Example 66.6 is isotropic. The cellular structure on $X \times X$ is a structure "over X" in the sense that $X \times X$ itself as well as the bases Y_i of the cells have morphisms to X with affine bundles of the cellular structure morphisms over X. Making the base change of the cellular structure with respect to an F-point $\operatorname{Spec}(F) \to X$ of the isotropic quadric X corresponding to an isotropic line L, we get a cellular structure on X given by the filtration $X'_0 \subset X'_1 \subset X'_2 = X$, where $X'_0 = \{L\}$ and X'_1 consists of all isotropic lines orthogonal to L. We have $Y'_0 = \operatorname{Spec}(F)$, Y'_1 is the quadric given by the quadratic form on $L^\perp/L$ induced by φ, and $Y'_2 = \operatorname{Spec}(F)$. The quadric Y'_1 is isomorphic to a projective quadric Y of dimension $d-2$, given by a quadratic form Witt equivalent to φ. By Corollary 66.4,

$$M(X) \simeq \mathbb{Z} \oplus M(Y)(1) \oplus \mathbb{Z}(d).$$

67. Nilpotence Theorem

Let Λ be a commutative ring and Y a smooth complete scheme over F. For every scheme X and elements $\alpha \in \operatorname{CH}(Y \times Y; \Lambda)$ and $\beta \in \operatorname{CH}(X \times Y; \Lambda)$, the compositions $\alpha^k = \alpha \circ \cdots \circ \alpha$ in $\operatorname{CH}(Y \times Y; \Lambda)$ and $\alpha^k \circ \beta$ in $\operatorname{CH}(X \times Y; \Lambda)$ are defined.

Theorem 67.1 (Nilpotence Theorem). *Let Y be a smooth complete scheme and X a scheme of dimension d over F. Let $\alpha \in \operatorname{CH}(Y \times Y; \Lambda)$ be an element satisfying $\alpha \circ \operatorname{CH}\big(Y_{F(x)}; \Lambda\big) = 0$ for every $x \in X$. Then*

$$\alpha^{d+1} \circ \operatorname{CH}(X \times Y; \Lambda) = 0.$$

PROOF. Consider the filtration

$$0 = C_{-1} \subset C_0 \subset \cdots \subset C_d = \operatorname{CH}(X \times Y; \Lambda),$$

where C_i is the Λ-submodule of $\operatorname{CH}(X \times Y; \Lambda)$ generated by the images of the push-forward homomorphisms

$$\operatorname{CH}(W \times Y; \Lambda) \to \operatorname{CH}(X \times Y; \Lambda),$$

for all closed subvarieties $W \subset X$ of dimension at most k. It suffices to prove that $\alpha \circ C_k \subset C_{k-1}$ for all $k \in [0,\ d]$.

Let W be a closed subvariety of X of dimension k. Denote by $i : W \to X$ the closed embedding and by w the generic point of W. Pick any element $\beta \in \mathrm{CH}(W \times Y; \Lambda)$. We shall prove that $\alpha \circ \big(i_*(\beta)\big) \in C_{k-1}$. Let β_w be the pull-back of β under the canonical morphism $Y_{F(w)} \to W \times Y$. By assumption, $\alpha \circ \beta_w = 0$. By the Continuity Property 52.9, there is a nonempty open subscheme U (a neighborhood of w) in W satisfying $\alpha \circ (\beta|_{U \times Y}) = 0$. It follows by Proposition 62.8(2) that

$$(\alpha \circ \beta)|_{U \times Y} = \alpha \circ (\beta|_{U \times Y}) = 0.$$

The complement V of U in W is a closed subscheme of W of dimension less than k. It follows from the exactness of the localization sequence (cf. Proposition 57.9)

$$\mathrm{CH}(V \times Y; \Lambda) \to \mathrm{CH}(W \times Y; \Lambda) \to \mathrm{CH}(U \times Y; \Lambda) \to 0$$

that $\alpha \circ \beta$ belongs to the image of the first map in this sequence. Therefore, the push-forward of the element $\alpha \circ \beta$ in $\mathrm{CH}(X \times Y; \Lambda)$ lies in the image of the push-forward homomorphism

$$\mathrm{CH}(V \times Y; \Lambda) \to \mathrm{CH}(X \times Y; \Lambda).$$

Consequently, $\alpha \circ (i_* \beta) = \alpha \circ (\beta \circ i^t) = (\alpha \circ \beta) \circ i^t = (i \times 1_Y)_*(\alpha \circ \beta) \in C_{k-1}$. □

The Nilpotence Theorem 67.1 was originally proven by Rost using the cycle modules technique.

Part 3

Quadratic forms and algebraic cycles

CHAPTER XIII

Cycles on Powers of Quadrics

Throughout this chapter, F is a field (of an arbitrary characteristic) and, in all sections but §70 and §71, X is a smooth projective quadric over F of even dimension $D = 2d \geq 0$ or of odd dimension $D = 2d+1 \geq 1$ given by a nondegenerate quadratic form φ on a vector space V over F (of dimension $D + 2$). For any integer $r \geq 1$, we write X^r for the direct product $X \times \cdots \times X$ (over F) of r copies of X.

In this chapter, we study algebraic cycles on X^r for all r. We also obtain many results for the special case $r = 2$. In order to make their statements and proofs more accessible, we introduce certain diagrams of cycles on X^2 (cf. §73).

68. Split quadrics

In this section the quadric X will be *split*, i.e., the Witt index $\mathrm{i}_0(X)$ has the maximal value $d + 1$.

Let V be the underlying vector space of φ. Fix a maximal totally isotropic subspace $W \subset V$. We write $\mathbb{P}(V)$ for the projective space of V; this is the projective space in which the quadric X lies as a hypersurface. Note that the subspace $\mathbb{P}(W)$ of $\mathbb{P}(V)$ is contained in X.

Proposition 68.1. *Let* $h \in \mathrm{CH}^1(X)$ *be the pull-back of the hyperplane class in* $\mathrm{CH}^1\big(\mathbb{P}(V)\big)$. *For any integer* $i \in [0,\ d]$, *let* $l_i \in \mathrm{CH}_i(X)$ *be the class of an* i*-dimensional subspace of* $\mathbb{P}(W)$. *Then the total Chow group* $\mathrm{CH}(X)$ *is free with basis* $\{h^i, l_i |\ i \in [0,\ d]\}$. *Moreover, the following multiplication rule holds in the ring* $\mathrm{CH}(X)$:

$$h \cdot l_i = l_{i-1} \quad \textit{for any} \quad i \in [1, d].$$

PROOF. Let $W^\perp$ be the orthogonal complement of W in V (clearly, $W^\perp = W$ if D is even; otherwise, $W^\perp$ contains W as a hyperplane). The quotient map $V \to V/W^\perp$ induces a morphism $X \setminus \mathbb{P}(W) \to \mathbb{P}(V/W^\perp)$, which is an affine bundle of rank $D - d$. Therefore, by Theorem 66.2,

$$\mathrm{CH}_i(X) \simeq \mathrm{CH}_i\big(\mathbb{P}(W)\big) \oplus \mathrm{CH}_{i-D+d}\big(\mathbb{P}(V/W^\perp)\big)$$

for any i, where the injection $\mathrm{CH}_*\big(\mathbb{P}(W)\big) \hookrightarrow \mathrm{CH}_*(X)$ is the push-forward with respect to the embedding $\mathbb{P}(W) \hookrightarrow X$.

To better understand the second summand in the decomposition of $\mathrm{CH}(X)$, we note that the reduced intersection of $\mathbb{P}(W^\perp)$ with X in $\mathbb{P}(V)$ is $\mathbb{P}(W)$, and that the affine bundle $X \setminus \mathbb{P}(W) \to \mathbb{P}(V/W^\perp)$ above is the composite of the closed embedding $X \setminus \mathbb{P}(W) \hookrightarrow \mathbb{P}(V) \setminus \mathbb{P}(W^\perp)$ with the evident vector bundle $\mathbb{P}(V) \setminus \mathbb{P}(W^\perp) \to \mathbb{P}(V/W^\perp)$. It follows that for any $i \leq d$, the image of $\mathrm{CH}^i\big(\mathbb{P}(V/W^\perp)\big)$ in $\mathrm{CH}^i(X)$ coincides with the image of the pull-back $\mathrm{CH}^i\big(\mathbb{P}(V)\big) \to \mathrm{CH}^i(X)$ (which is generated by h^i).

To check the multiplication formula, we consider the closed embeddings $f : \mathbb{P}(W) \hookrightarrow X$ and $g : X \hookrightarrow \mathbb{P}(V)$. Write L_i for the class in $\mathrm{CH}(\mathbb{P}(W))$ of an i-dimensional linear subspace of $\mathbb{P}(W)$, and H for the hyperplane class in $\mathrm{CH}(\mathbb{P}(V))$. Since $h = g^*(H)$ and $l_i = f_*(L_i)$, we have by the projection formula (Proposition 56.9) and functoriality of the pull-back (Proposition 55.18),

$$h \cdot l_i = g^*(H) \cdot f_*(L_i) = f_*\big((g \circ f)^*(H) \cdot L_i\big).$$

By Corollary 57.20 (together with Propositions 104.16 and 55.19), we see that $(g \circ f)^*(H)$ is the hyperplane class in $\mathrm{CH}(\mathbb{P}(W))$; hence $(g \circ f)^*(H) \cdot L_i = L_{i-1}$ by Example 57.23. □

Proposition 68.2. *For each $i \in [0,\ D/2)$, the i-dimensional subspaces of $\mathbb{P}(V)$ lying inside of X have the same class in $\mathrm{CH}_i(X)$. If D is even, there are precisely two different classes of d-dimensional subspaces and the sum of these two classes is equal to h^d.*

PROOF. By Proposition 68.1, the push-forward homomorphism $\mathrm{CH}_i(X) \to \mathrm{CH}_i(\mathbb{P}(V))$ is injective (even bijective) if $i \in [0,\ D/2)$. Since the i-dimensional linear subspaces of $\mathbb{P}(V)$ have the same class in $\mathrm{CH}(\mathbb{P}(V))$, the first statement of Proposition 68.2 follows.

Assume that D is even. Then $\{h^d, l_d\}$ is a basis for the group $\mathrm{CH}_d(X)$, where l_d is the class of the special linear subspace $\mathbb{P}(W) \subset X$. Let $l'_d \in \mathrm{CH}_d(X)$ be the class of an arbitrary d-dimensional linear subspace of X. Since l_d and l'_d have the same image under the push-forward homomorphism $\mathrm{CH}_d(X) \to \mathrm{CH}_d(\mathbb{P}(V))$ whose kernel is generated by $h^d - 2l_d$, one has $l'_d = l_d + n(h^d - 2l_d)$ for some $n \in \mathbb{Z}$. Since there exists a linear automorphism of X moving l_d to l'_d, and h^d is, of course, invariant with respect to any linear automorphism, h^d and l'_d also form a basis for $\mathrm{CH}_d(X)$; consequently, the determinant of the matrix

$$\begin{pmatrix} 1 & n \\ 0 & 1-2n \end{pmatrix}$$

is ± 1, i.e., n is 0 or 1 and l'_d is l_d or $h^d - l_d$. So there are at most two different rational equivalence classes of d-dimensional linear subspaces of X and the sum of two different classes (if they exist) is equal to h^d.

Now let U be a d-codimensional subspace of V containing W (as a hyperplane). The orthogonal complement $U^\perp$ has codimension 1 in $W^\perp = W$; therefore $\operatorname{codim}_U U^\perp = 2$. The induced 2-dimensional quadratic form on $U/U^\perp$ is a hyperbolic plane. The corresponding quadric consists of two points $W/U^\perp$ and $W'/U^\perp$ for a uniquely determined maximal totally isotropic subspace $W' \subset V$. Moreover, the intersection $X \cap \mathbb{P}(U)$ is reduced and its irreducible components are $\mathbb{P}(W)$ and $\mathbb{P}(W')$. Therefore, $h^d = [X \cap \mathbb{P}(U)] = [\mathbb{P}(W)] + [\mathbb{P}(W')]$ and it follows that $[\mathbb{P}(W)] \neq [\mathbb{P}(W')]$. □

Exercise 68.3. Determine a complete multiplication table for $\mathrm{CH}(X)$ by showing that

(1) $h^{d+1} = 2l_{D-d-1}$;

(2) if D is not divisible by 4, then $l_d^2 = 0$;

(3) if D is divisible by 4, then $l_d^2 = l_0$.

Exercise 68.4. Assume that D is even and let $l_d, l'_d \in \mathrm{CH}(X)$ be two different classes of d-dimensional subspaces. Let f be the automorphism of $\mathrm{CH}(X)$ induced by a reflection. Show that $f(l_d) = l'_d$.

If D is even, an *orientation* of the quadric is the choice of one of two classes of d-dimensional linear subspaces in $\mathrm{CH}(X)$. We denote this class by l_d. An even-dimensional quadric with an orientation is called *oriented.*

Proposition 68.5. *For any $r \geq 1$, the Chow group $\mathrm{CH}(X^r)$ is free with basis given by the external products of the basis elements $\{h^i, l_i\}$, $i \in [0,\ d]$, of $\mathrm{CH}(X)$.*

PROOF. The cellular structure on X, constructed in the proof of Proposition 68.1, together with the calculation of the Chow motive of a projective space (cf. Example 66.5) show by Corollary 66.4 that the motive of X is split. Therefore, the homomorphism $\mathrm{CH}(X)^{\otimes r} \to \mathrm{CH}(X^r)$, given by the external product of cycles is an isomorphism by Proposition 64.3. $\square$

69. Isomorphisms of quadrics

Let φ and ψ be two quadratic forms. A *similitude between φ and ψ (with multiplier $a \in F^\times$)* is an isomorphism $f : V_\varphi \to V_\psi$ such that $\varphi(v) = a\psi\big(f(v)\big)$ for all $v \in V_\varphi$. A similitude between φ and ψ induces an isomorphism of projective spaces $\mathbb{P}(V_\varphi) \xrightarrow{\sim} \mathbb{P}(V_\psi)$ and projective quadrics $X_\varphi \xrightarrow{\sim} X_\psi$.

Let $i : X_\varphi \to \mathbb{P}(V_\varphi)$ be the embedding. We consider the locally free sheaves

$$\mathcal{O}_{X_\varphi}(s) := i^*\big(\mathcal{O}_{\mathbb{P}(V_\varphi)}(s)\big)$$

over X_φ for every $s \in \mathbb{Z}$.

Lemma 69.1. *Let φ be a nonzero quadratic form of dimension at least 2. Then $H^0\big(X_\varphi, \mathcal{O}_{X_\varphi}(-1)\big) = 0$ and $H^0\big(X_\varphi, \mathcal{O}_{X_\varphi}(1)\big)$ is canonically isomorphic to V_φ^*.*

PROOF. We have $H^0\big(\mathbb{P}(V_\varphi), \mathcal{O}_{\mathbb{P}(V_\varphi)}(-1)\big) = 0$, $H^0\big(\mathbb{P}(V_\varphi), \mathcal{O}_{\mathbb{P}(V_\varphi)}(1)\big) \simeq V_\varphi^*$ and $H^1\big(\mathbb{P}(V_\varphi), \mathcal{O}_{\mathbb{P}(V_\varphi)}(s)\big) = 0$ for any s (see [**50**, Ch. III, Th. 5.1]). The statements follow from exactness of the cohomology sequence for the short exact sequence

$$0 \to \mathcal{O}_{\mathbb{P}(V_\varphi)}(s-2) \xrightarrow{\varphi} \mathcal{O}_{\mathbb{P}(V_\varphi)}(s) \to i_*\mathcal{O}_{X_\varphi}(s) \to 0. \qquad \square$$

Lemma 69.2. *Let $\alpha : X_\varphi \xrightarrow{\sim} X_\psi$ be an isomorphism of smooth projective quadrics. Then $\alpha^*\big(\mathcal{O}_{X_\psi}(1)\big) \simeq \mathcal{O}_{X_\varphi}(1)$.*

PROOF. In the case $\dim\varphi = 2$ the sheaves $\mathcal{O}_{X_\varphi}(1)$ and $\mathcal{O}_{X_\psi}(1)$ are free and the statement is obvious.

We may assume that $\dim\varphi > 2$. As the Picard group of smooth projective varieties injects under field extensions, we also may assume that both forms are split. We identify the groups $\mathrm{Pic}(X_\varphi)$ and $\mathrm{CH}^1(X_\varphi)$. The class of the sheaf $\mathcal{O}_{X_\varphi}(1)$ corresponds to the class $h \in \mathrm{CH}^1(X_\varphi)$ of a hyperplane section. It is sufficient to show that $\alpha^*(h) = \pm h$, since the class $-h$ cannot occur as the sheaf $\mathcal{O}_{X_\varphi}(-1)$ has no nontrivial global sections by Lemma 69.1.

If $\dim\varphi > 4$ then, by Proposition 68.1, the group $\mathrm{CH}^1(X_\varphi)$ is infinite cyclic and generated by h. Thus $\alpha^*(h) = \pm h$.

If $\dim\varphi = 3$, then h is twice the generator l_0 of the infinite cyclic group $\mathrm{CH}^1(X_\varphi)$ and the result follows in a similar fashion.

Finally, if $\dim \varphi = 4$, then the group $\mathrm{CH}^1(X_\varphi)$ is a free abelian group with two generators l_1 and l'_1 satisfying $l_1 + l'_1 = h$ (cf. the proof of Proposition 68.2). Using the fact that the pull-back map $\alpha^* : \mathrm{CH}(X_\varphi) \to \mathrm{CH}(X_\varphi)$ is a ring homomorphism, one concludes that $\alpha^*(l_1 + l'_1) = \pm(l_1 + l'_1)$. □

Theorem 69.3. *Every isomorphism between smooth projective quadrics X_φ and X_ψ is induced by a similitude between φ and ψ.*

PROOF. Let $\alpha : X_\varphi \xrightarrow{\sim} X_\psi$ be an isomorphism. By Lemma 69.2, $\alpha^*\big(\mathcal{O}_{X_\psi}(1)\big) \simeq \mathcal{O}_{X_\varphi}(1)$. Lemma 69.1 therefore gives an isomorphism of vector spaces

$$V_\psi^* = H^0\big(X_\psi, \mathcal{O}_{X_\psi}(1)\big) \xrightarrow{\sim} H^0\big(X_\varphi, \mathcal{O}_{X_\varphi}(1)\big) = V_\varphi^*. \tag{69.4}$$

Thus α is given by the induced graded ring isomorphism $S^\bullet(V_\psi^*) \to S^\bullet(V_\varphi^*)$ which must take the ideal (ψ) to (φ); i.e., it takes ψ to a multiple of φ. In other words, the linear isomorphism $f : V_\varphi \to V_\psi$ dual to (69.4) is a similitude between φ and ψ inducing α. □

Corollary 69.5. *Let φ and ψ be nondegenerate quadratic forms. The quadrics X_φ and X_ψ are isomorphic if and only if φ and ψ are similar.*

For a quadratic form φ, all similitudes $V_\varphi \to V_\varphi$ form the *group of similitudes* $\mathrm{GO}(\varphi)$. For every $a \in F^\times$, the endomorphism of V_φ given by the product with a is a similitude. Therefore $F^\times$ can be identified with a subgroup of $\mathrm{GO}(\varphi)$. The factor group $\mathrm{PGO}(\varphi) := \mathrm{GO}(\varphi)/F^\times$ is called the *group of projective similitudes*. Every projective similitude induces an automorphism of the quadric X_φ, so we have a group homomorphism $\mathrm{PGO}(\varphi) \to \mathrm{Aut}(X_\varphi)$.

Corollary 69.6. *Let φ be a nondegenerate quadratic form. Then the map*

$$\mathrm{PGO}(\varphi) \to \mathrm{Aut}(X_\varphi)$$

is an isomorphism.

70. Isotropic quadrics

The motive of a smooth isotropic quadric is computed in terms of a smooth quadric of smaller dimension in Example 66.7 as follows:

Proposition 70.1. *Let X be a smooth isotropic projective quadric given by a quadratic form φ with $\dim \varphi \geq 2$ and let Y be a (smooth) projective quadric given by a quadratic form of dimension $\dim \varphi - 2$ that is Witt equivalent to φ (we assume that $Y = \emptyset$ if $\dim X \leq 1$). Then*

$$M(X) \simeq \mathbb{Z} \oplus M(Y)(1) \oplus \mathbb{Z}(D)$$

with $D = \dim X$. In particular,

$$\mathrm{CH}_*(X) \simeq \mathrm{CH}_*(\mathbb{Z}) \oplus \mathrm{CH}_{*-1}(Y) \oplus \mathrm{CH}_{*-D}(\mathbb{Z}).$$

The motivic decomposition of Proposition 70.1 was originally observed by Rost.

In most of this book we are interested only in *smooth* quadrics. Nevertheless, we make some observations concerning the general (not necessarily smooth) case. Let φ be a quadratic form on a vector space V over F and let X be the associated projective quadric. Suppose a rational point x on X is singular. Then φ is nonzero and x is determined by a nonzero vector $v \in V$ in the radical of φ (cf. proof of Proposition 22.1). Let Y be the quadric given by the quadratic form on V/Fv

induced by φ, so $\dim Y = \dim X - 1$ and $X \setminus \{x\}$ is an affine bundle over Y of rank 1. Since the first homomorphism in the exact localization sequence

$$\mathrm{CH}(\{x\}) \to \mathrm{CH}(X) \to \mathrm{CH}(Y) \to 0$$

is a split injection (with the splitting given by the degree homomorphism $\mathrm{CH}(X) \to \mathrm{CH}(\operatorname{Spec} F)$), we have

Lemma 70.2. *Let x be a singular rational point of a quadric X and let Y be as above. Then $\mathrm{CH}_0(X)$ is (the infinite cyclic group) generated by $[x]$ and $\mathrm{CH}_i(X)$ is isomorphic to $\mathrm{CH}_{i-1}(Y)$ for all $i > 0$.*

Now assume that φ is nonzero and that $X = X_\varphi$ has a nonsingular rational point $x \in X$ (although X itself may not be smooth). The point x is given by a nonzero isotropic vector $v \in V$. The projective quadric Y, given by the restriction of φ on the orthogonal complement $v^\perp$ of v in V, is a closed subscheme of X containing x. Since $\varphi \neq 0$, the vector v lies outside of the radical of φ, so that $v^\perp$ is a hyperplane of V. If φ restricted to $v^\perp$ is zero, then Y is a projective space; otherwise x is a *singular* point of Y (and $\operatorname{codim}_X Y = 1$). The difference $X \setminus Y$ is isomorphic to an affine space (cf. proof of Proposition 22.9). Thus we can apply Theorem 66.2 to X with the one term filtration $Y \subset X$. Theorem 66.2 and Lemma 70.2 yield (we also include the trivial case of $\varphi = 0$):

Lemma 70.3. *Let x be a nonsingular rational point of a quadric X and Y be as above. Then the push-forward homomorphism $\mathrm{CH}_i(Y) \to \mathrm{CH}_i(X)$ is an isomorphism for any $i < \dim X$. In particular, if $\dim X > 0$, the group $\mathrm{CH}_0(X)$ is generated by $[x]$.*

Combining Lemma 70.2 with Lemma 70.3 and adding the trivial case of $\varphi = 0$ yields

Corollary 70.4. *Let X be an isotropic (not necessarily smooth) projective quadric X. Then the degree homomorphism $\deg : \mathrm{CH}_0(X) \to \mathbb{Z}$ is an isomorphism unless φ is a hyperbolic plane, i.e., X is a disjoint union of two copies of $\operatorname{Spec} F$.*

71. The Chow group of dimension 0 cycles on quadrics

Recall that for every $p \in [0, n]$, the group $\mathrm{CH}^p(\mathbb{P}^n_F)$ is infinite cyclic generated by the class h^p where $h \in \mathrm{CH}^1(\mathbb{P}^n_F)$ is the class of a hyperplane in $\mathbb{P}^n_F$ (cf. Example 57.23). Thus for every $p \in [0, n]$ and $\alpha \in \mathrm{CH}^p(\mathbb{P}^n_F)$, we have $\alpha = mh^p$ for a uniquely determined integer m. We call m the *degree of* α and write $m = \deg(\alpha)$. We have $\deg(\alpha\beta) = \deg(\alpha)\deg(\beta)$ for all homogeneous cycles $\alpha \in \mathrm{CH}^p(\mathbb{P}^n_F)$ and $\beta \in \mathrm{CH}^q(\mathbb{P}^n_F)$ satisfying $p, q \geq 0$ and $p + q \leq n$.

If Z is a closed subvariety of $\mathbb{P}^n_F$, we define the *degree* $\deg(Z)$ *of* Z as $\deg([Z])$.

Lemma 71.1. *Let $x \in \mathbb{P}^n_F$ be a closed point of degree $d > 1$ such that the field extension $F(x)/F$ is simple (i.e., generated by one element). Then there is a morphism $f : \mathbb{P}^1 \to \mathbb{P}^n$ with image a curve C satisfying $x \in C$ and $\deg(C) < d$.*

PROOF. Let u be a generator of the field extension $F(x)/F$. We can write the homogeneous coordinates s_i of x in the form $s_i = f_i(u)$, $i \in [0, n]$, where f_i are polynomials over F of degree less than d. Let k be the largest degree of the f_i and set $F_i(t_0, t_1) = t_1^k f_i(t_0/t_1)$. The polynomials F_i are all homogeneous of degree $k < d$. We may assume that all the F_i are relatively prime (by dividing out the gcd of the F_i). Consider the morphism $f : \mathbb{P}^1_F \to \mathbb{P}^n_F$ given by the polynomials F_i and let C be the image of f. Note that C contains x and $C(F) \neq \emptyset$. In particular, the map f is not constant. Therefore C is a closed curve in $\mathbb{P}^n_F$. We have $f_*([\mathbb{P}^1]) = r[C]$ for some $r \geq 1$.

Choose an index i such that F_i is a nonzero polynomial and consider the hyperplane H in $\mathbb{P}^n_F$ given by $s_i = 0$. The subscheme $f^{-1}(H) \subset \mathbb{P}^1_F$ is given by $F_i(t_0, t_1) = 0$, so $f^{-1}(H)$ is a 0-dimensional subscheme of degree $k = \deg F_i$. Hence H has a proper inverse image with respect to f. By Proposition 57.19, we have $f^*(h) = mp$, where p is the class of a point in $\mathbb{P}^1_F$ and $1 \leq m \leq k < d$. It follows from Proposition 56.9 that

$$h \cdot r[C] = h \cdot f_*\big([\mathbb{P}^1]\big) = f_*\big(f^*(h)\big) = f_*(mp) = mh^n.$$

Hence $\deg(C) = m/r \leq m < d$. □

We follow [**66**, Th. 3.2] in the proof of the next result.

Theorem 71.2. *Let X be an anisotropic (not necessarily smooth) quadric over F and let $x_0 \in X$ be a closed point of degree 2. Then for every closed point $x \in X$, we have $[x] = a[x_0]$ in $\mathrm{CH}_0(X)$ for some $a \in \mathbb{Z}$.*

PROOF. We induct on $d = \deg x$. Suppose first that there are no intermediate fields between F and $F(x)$. In particular, the field extension $F(x)/F$ is simple. The quadric X is a hypersurface in the projective space $\mathbb{P}^n_F$ for some n. By Lemma 71.1, there is an integral closed curve $C \subset \mathbb{P}^n_F$ of degree less than d with $C(F) \neq \emptyset$ and $x \in C$.

Let $g : X \to \mathbb{P}^n_F$ be the closed embedding. Since X is anisotropic and $C(F) \neq \emptyset$, C is not contained in X. Therefore, C has proper inverse image with respect to g. As $x \in C \cap X$, by Proposition 57.19,

$$g^\star([C]) = [x] + \alpha$$

in $\mathrm{CH}_0(X)$, where α is a nonnegative 0-dimensional cycle on X. By Proposition 56.11, we have

$$\deg\big(g^\star([C])\big) = \deg\big(g_* \circ g^\star([C])\big) = \deg([X] \cdot [C]) = 2\deg(C),$$

hence

$$\deg \alpha = 2 \deg C - \deg x = 2 \deg C - d < d.$$

Thus the cycle α is supported on closed points of degree less than d. By the induction hypothesis, $\alpha = b[x_0]$ in $\mathrm{CH}_0(X)$ for some $b \in \mathbb{Z}$. We also have $[C] = c[L]$ in $\mathrm{CH}_0(\mathbb{P}^n_F)$ where L is a line in $\mathbb{P}^n_F$ satisfying $x_0 \in L$ and $c = \deg C$. Since $L \cap X = \{x_0\}$, by Corollary 57.20, we have $g^\star([L]) = [x_0]$. Therefore,

$$[x] = g^\star([C]) - \alpha = cg^\star([L]) - b[x_0] = (c - b)[x_0].$$

Now suppose that there is a proper intermediate field L between F and $E = F(x)$. Let f denote the natural morphism $X_L \to X$. The morphism $\operatorname{Spec} E \to X$ induced by x and the inclusion of L into E define a closed point $x' \in X_L$ with $f(x') = x$ and $F(x') = E$. It follows that $f_*([x']) = [x]$.

Consider two cases:

Case 1: X_L is isotropic. Let $y \in X_L$ be a rational point. Since $\deg f_*\big([y]\big) = [L : F] < d$, by the induction hypothesis, $f_*\big([y]\big) \in \mathbb{Z} \cdot [x_0]$. By Corollary 70.4, we know that $\mathrm{CH}_0(X_L)$ is generated by the classes of rational points, hence $[x] = f_*([x']) \in \mathbb{Z} \cdot [x_0]$.

Case 2: X_L is anisotropic. Applying the induction hypothesis to the quadric X_L and the point x' of degree $[E : L] < d$, we have $[x'] = b[(x_0)_L]$ for some $b \in \mathbb{Z}$. Hence

$$[x] = f_*([x']) = bc[x_0],$$

where $c = [L : F]$. □

We therefore obtain another proof of Springer's Theorem 18.5.

Corollary 71.3 (Springer's Theorem). *If X is an anisotropic quadric, the image of the degree homomorphism* $\deg : \mathrm{CH}_0(X) \to \mathbb{Z}$ *is equal to* $2\mathbb{Z}$*, i.e., the degree of a finite field extension L/F with X_L isotropic, is even.*

In the case $\operatorname{char} F \neq 2$, the following important statement was proven in [**69**, Prop. 2.6] and by Swan in [**130**]. The case of characteristic 2 was considered by Totaro in [**132**].

Corollary 71.4. *For every anisotropic quadric X, the degree homomorphism* $\deg : \mathrm{CH}_0(X) \to \mathbb{Z}$ *is injective.*

72. The reduced Chow group

Let X be an arbitrary smooth projective quadric. We write $\mathrm{CH}(\bar{X}^r)$ for $\mathrm{CH}(X^r_E)$, where E is a field extension of F such that the quadric X_E is split. Note that for any field L containing E, the change of field homomorphism $\mathrm{CH}(X^r_E) \to \mathrm{CH}(X^r_L)$ of Example 49.14 is an isomorphism; therefore for any field extension E'/F with split $X_{E'}$, the groups $\mathrm{CH}(X^r_E)$ and $\mathrm{CH}(X^r_{E'})$ are canonically isomorphic, hence $\mathrm{CH}(\bar{X}^r)$ can be defined invariantly as the colimit of the groups $\mathrm{CH}(X^r_L)$, where L runs over *all* field extensions of F.

If D is even, an *orientation* of the quadric is the choice of one of two classes of d-dimensional linear subspaces in $\mathrm{CH}(\bar{X})$. An even-dimensional quadric with an orientation is called *oriented.*

The *reduced Chow group* $\overline{\mathrm{CH}}(X^r)$ is defined as the image of the change of field homomorphism $\mathrm{CH}(X^r) \to \mathrm{CH}(\bar{X}^r)$.

We say that an element of $\mathrm{CH}(\bar{X}^r)$ is *rational* if it lies in the subgroup $\overline{\mathrm{CH}}(X^r) \subset \mathrm{CH}(\bar{X}^r)$. More generally, for a field extension L/F, the elements of the subgroup $\overline{\mathrm{CH}}(X^r_L) \subset \mathrm{CH}(\bar{X}^r)$ are called *L-rational.*

Replacing the integral Chow group by the Chow group modulo 2 in the above definitions, we get the *modulo* 2 *reduced Chow group* $\overline{\mathrm{Ch}}(X^r) \subset \mathrm{Ch}(\bar{X}^r)$ and the corresponding notion of (L-)rational cycles modulo 2.

Abusing terminology, we shall often call elements of a Chow group, cycles. The basis described in Proposition 68.5 will be called the *basis for* $\mathrm{CH}(\bar{X}^r)$ and its elements *basis elements* or *basic cycles*. Similarly, this basis modulo 2 will be called the *basis for* $\mathrm{Ch}(\bar{X}^r)$ and its elements *basis elements* or *basic cycles.* We use the same notation for the basis elements of $\mathrm{CH}(\bar{X}^r)$ and for their reductions modulo 2. The *decomposition* of an element $\alpha \in \mathrm{Ch}(\bar{X}^r)$ will always mean its representation as a sum of basic cycles. We say that a basis cycle β is *contained* in the decomposition of α (or simply "is contained in α"), if β is a summand of the decomposition. More generally, for two cycles $\alpha', \alpha \in \mathrm{Ch}(\bar{X}^r)$, we say that α' *is contained* in α or that

α' is a *subcycle* of α (notation: $\alpha' \subset \alpha$), if every basis element contained in α' is also contained in α.

A basis element of $\mathrm{Ch}(\bar{X}^r)$ is called *nonessential*, if it is an external product of (internal) powers of h (including $h^0 = 1 = [\bar{X}]$); the other basis elements are called *essential*. An element of $\mathrm{Ch}(\bar{X}^r)$ that is a sum of nonessential basis elements, is called *nonessential* as well. Note that all nonessential elements are rational since h is rational. An element of $\mathrm{Ch}(\bar{X}^r)$ that is a sum of essential basis elements, is called *essential* as well. (The zero cycle is the only element which is essential and nonessential simultaneously). The group $\mathrm{Ch}(\bar{X}^r)$ is a direct sum of the subgroup of nonessential elements and the subgroup of essential elements. We call the essential component of an element $\alpha \in \mathrm{Ch}(\bar{X}^r)$ the *essence* of α. Clearly, the essence of a rational element is rational.

The group $\overline{\mathrm{Ch}}(X)$ is easy to compute. First of all, by Springer's theorem (Corollary 71.3), one has

Lemma 72.1. *If the quadric X is anisotropic (i.e., $X(F) = \emptyset$), then the element $l_0 \in \mathrm{Ch}(\bar{X})$ is not rational.*

Corollary 72.2. *If X is anisotropic, the group $\overline{\mathrm{Ch}}(X)$ is generated by the nonessential basis elements.*

PROOF. If the decomposition of an element $\alpha \in \overline{\mathrm{Ch}}(X)$ contains an essential basis element l_i for some $i \neq D/2$, then $l_i \in \overline{\mathrm{Ch}}(X)$ because l_i is the i-dimensional homogeneous component of α (and $\overline{\mathrm{Ch}}(X)$ is a graded subring of $\mathrm{Ch}(\bar{X})$). If the decomposition of an element $\alpha \in \overline{\mathrm{Ch}}(X)$ contains the essential basis element l_i for $i = D/2$, then $D/2 = d$, and the d-dimensional homogeneous component of α is either l_d or $l_d + h^d$ so we still have $l_i \in \overline{\mathrm{Ch}}(X)$. It follows that $l_0 = l_i \cdot h^i \in \overline{\mathrm{Ch}}(X)$, contradicting Lemma 72.1. □

Let V be the underlying vector space of φ and let $W \subset V$ be a totally isotropic subspace of dimension $a \leq d$. Let Y be the projective quadric of the quadratic form $\psi : W^{\perp}/W \to F$ induced by φ. Then ψ is nondegenerate, Witt-equivalent to φ, $\dim \psi = \dim \varphi - 2a$, and $\dim Y = \dim X - 2a$. A point of the product $Y \times X$ is a pair $(A/W, B)$, where $B \subset V$ is a totally isotropic subspace of dimension 1 and $A \subset W^{\perp}$ is a totally isotropic subspace of dimension $a+1$ containing W. Let $Z \subset Y \times X$ be the closed scheme of the pairs $(A/W, B)$ satisfying the condition $B \subset A$. Note that the composition $Z \hookrightarrow Y \times X \xrightarrow{pr_Y} Y$ is an a-dimensional projective bundle; in particular, Z is equidimensional (and Z is a variety if Y is) of dimension $\dim Z = \dim Y + a = \dim X - a$. Its class $\alpha = [Z] \in \mathrm{CH}(Y \times X)$ is called the *incidence correspondence*.

We first note that the inverse image $pr_X^{-1}\big(\mathbb{P}(W)\big)$ of the closed subvariety $\mathbb{P}(W) \subset X$ under the projection $pr_X : Y \times X \to X$ is contained in Z with complement a dense open subscheme of Z mapping under pr_X isomorphically onto $\big(\mathbb{P}(W^{\perp}) \cap X\big) \setminus \mathbb{P}(W)$.

We let $h^i = 0 = l_i$ for any negative integer i.

Lemma 72.3. *Let α be the incidence correspondence in* $\mathrm{CH}(Y \times X)$.

(1) *For any $i = 0, \dots, d-a$, the homomorphism $\alpha_* : \mathrm{CH}(\bar{Y}) \to \mathrm{CH}(\bar{X})$ takes h^i to h^{i+a} and l_i to l_{i+a}.*

(2) *For any* $i = 0, \dots, d$, *the homomorphism* $\alpha^* : \mathrm{CH}(\bar{X}) \to \mathrm{CH}(\bar{Y})$ *takes* h^i *to* h^{i-a} *and* l_i *to* l_{i-a}. *(In the case of even* D, *the two formulas involving* l_d *are true for an appropriate choice of orientations of* X *and of* Y.*)*

PROOF. For an arbitrary $i \in [0,\ d-a]$, let $L \subset W^\perp/W$ be a totally isotropic linear subspace of dimension $i+1$. Then $l_i = \big[\mathbb{P}(L)\big] \in \mathrm{CH}(\bar{Y})$. Let $p : W^\perp \to W^\perp/W$ be the projection. Since the dense open subscheme $\big(pr_Y^{-1}\big(\mathbb{P}(L)\big) \cap Z\big) \setminus pr_X^{-1}\big(\mathbb{P}(W)\big)$ of the intersection $pr_Y^{-1}\big(\mathbb{P}(L)\big) \cap Z$ maps under pr_X isomorphically onto $\mathbb{P}(p^{-1}(L)) \setminus \mathbb{P}(W)$, we have (using Proposition 57.21): $\alpha_*(l_i) = \big[\mathbb{P}(p^{-1}(L))\big] = l_{i+a} \in \mathrm{CH}(\bar{X})$. Similarly, for any linear subspace $H \subset W^\perp/W$ of codimension i, the element $h^i \in \mathrm{CH}(\bar{Y})$ is the class of the intersection $\mathbb{P}(H) \cap Y$ and maps under α_* to the class of $\big[\mathbb{P}(p^{-1}(H)) \cap X\big]$ which equals h^{i+a}.

To prove the statements on α^* for an arbitrary $i \in [a,\ d]$, let $L \subset V$ be an $(i+1)$-dimensional totally isotropic subspace satisfying $\dim(L \cap W^\perp) = \dim L - a$ and $L \cap W = 0$. (The second condition is, in fact, a consequence of the first one.) Then $l_i = \big[\mathbb{P}(L)\big] \in \mathrm{CH}(\bar{X})$ and the intersection $pr_X^{-1}\big(\mathbb{P}(L)\big) \cap Z$ maps under pr_Y isomorphically onto $\mathbb{P}\big(((L \cap W^\perp) + W)/W\big)$; consequently, $\alpha^*(l_i) = l_{i-a}$. Similarly, if $H \subset V$ is a linear subspace of codimension i satisfying $\dim(H \cap W^\perp) = \dim H - a$ and $H \cap W = 0$, then $h^i = \big[\mathbb{P}(H) \cap X\big] \in \mathrm{CH}(\bar{X})$ and the intersection $pr_X^{-1}\big(\mathbb{P}(H) \cap X\big) \cap Z$ maps under pr_Y isomorphically onto $\mathbb{P}\big(((H \cap W^\perp) + W)/W\big) \cap Y$; consequently, $\alpha^*(h^i) = h^{i-a}$. □

Corollary 72.4. *Assume that* X *is isotropic but not split and set* $a = \mathfrak{i}_0(X)$. *Let* X_0 *be the projective quadric given by an anisotropic quadratic form Witt-equivalent to* φ *(so* $\dim X_0 = D - 2a$*). Then*

(1) *The group* $\mathrm{Ch}_{D-a}(X \times X_0)$ *contains a correspondence "pr" such that the induced homomorphism* $pr_* : \mathrm{Ch}(\bar{X}) \to \mathrm{Ch}(\bar{X}_0)$ *takes* h^i *to* h^{i-a} *and* l_i *to* l_{i-a} *for* $i = 0, \dots, d$.

(2) *The group* $\mathrm{Ch}_{D-a}(X_0 \times X)$ *contains a correspondence "in" such that the induced homomorphism* $in_* : \mathrm{Ch}(\bar{X}_0) \to \mathrm{Ch}(\bar{X})$ *takes* h^i *to* h^{i+a} *and* l_i *to* l_{i+a} *for* $i = 0, \dots, d-a$.

Remark 72.5. Note that the homomorphisms in_* and pr_* of Corollary 72.4 map rational cycles to rational cycles. Since the composite $pr_* \circ in_*$ is the identity, it follows that $pr_*\big(\overline{\mathrm{Ch}}(X)\big) = \overline{\mathrm{Ch}}(X_0)$. More generally, for any $r \geq 1$ the homomorphisms

$$in_*^r : \mathrm{Ch}(\bar{X}_0^r) \to \mathrm{Ch}(\bar{X}^r) \quad \text{and} \quad pr_*^r : \mathrm{Ch}(\bar{X}^r) \to \mathrm{Ch}(\bar{X}_0^r),$$

induced by the rth tensor powers $in^r \in \mathrm{Ch}(X_0^r \times X^r)$ and $pr^r \in \mathrm{Ch}(X^r \times X_0^r)$ of the correspondences in and pr, map rational cycles to rational cycles and satisfy the relations $pr_*^r \circ in_*^r = \mathrm{id}$ and $pr_*^r\big(\overline{\mathrm{Ch}}(X^r)\big) = \overline{\mathrm{Ch}}(X_0^r)$.

We obtain now the following extension of Lemma 72.1.

Corollary 72.6. *Let* X *be an arbitrary quadric and let* i *be any integer. Then* $l_i \in \overline{\mathrm{Ch}}(X)$ *if and only if* $\mathfrak{i}_0(X) > i$.

PROOF. The "if" part of the statement is trivial. We prove the "only if" part by induction on i. The case $i = 0$ is Lemma 72.1.

We may assume that $i > 0$ and $l_i \in \overline{\mathrm{Ch}}(X)$. Since $l_i \cdot h = l_{i-1}$, the element l_{i-1} is also rational. Therefore $\mathfrak{i}_0(X) \geq i$ by the induction hypothesis. If $\mathfrak{i}_0(X) = i$ the image of $l_i \in \overline{\mathrm{Ch}}(X)$ under the map $pr_* : \mathrm{Ch}(\bar{X}) \to \mathrm{Ch}(\bar{X}_0)$ of Corollary 72.4

equals l_0 and is rational. Therefore, by Lemma 72.1, the quadric X_0 is isotropic, a contradiction. $\square$

The following observation is crucial:

Theorem 72.7. *The absolute and relative higher Witt indices of a nondegenerate quadratic form φ are determined by the group*

$$\overline{\mathrm{Ch}}(X^*) = \coprod_{r\geq 1} \overline{\mathrm{Ch}}(X^r).$$

PROOF. We first note that the group $\overline{\mathrm{Ch}}(X)$ determines $\mathfrak{i}_0(\varphi)$ by Corollary 72.6.

Let X_0 be as in Corollary 72.4. By Corollary 72.4 and Remark 72.5, the group $\overline{\mathrm{Ch}}(X_0^*)$ is recovered as the image of the group $\overline{\mathrm{Ch}}(X^*)$ under the homomorphism $\mathrm{Ch}(\bar{X}^*) \to \mathrm{Ch}(\bar{X}_0^*)$ induced by the tensor powers of the correspondence pr.

Let F_1 be the first field in the generic splitting tower of φ. The pull-back homomorphism $g_1^* : \mathrm{Ch}(X_0^r) \to \mathrm{Ch}\big((X_0)_{F_1}^{r-1}\big)$ with respect to the morphism of schemes $g_1 : (X_0)_{F_1}^{r-1} \to X_0^r$ given by the generic point of the first factor of X_0^r, is surjective (cf. Corollary 57.11). It induces an epimorphism $\overline{\mathrm{Ch}}(X_0^r) \twoheadrightarrow \overline{\mathrm{Ch}}\big((X_0)_{F_1}^{r-1}\big)$, which is the restriction of the epimorphism $\mathrm{Ch}(\bar{X}_0^r) \twoheadrightarrow \mathrm{Ch}(\bar{X}_0^{r-1})$ mapping each basis element of the form $h^0 \times \beta$, $\beta \in \mathrm{Ch}(\bar{X}_0^{r-1})$, to β and killing all other basis elements. Therefore, the group $\overline{\mathrm{Ch}}(X_0^*)$ determines the group $\overline{\mathrm{Ch}}\big((X_0)_{F_1}^*\big)$, and we finish by induction on the height $\mathfrak{h}(\varphi)$. $\square$

Remark 72.8. The proof of Theorem 72.7 shows that the statement of Theorem 72.7 can be made more precise in the following way: If for some $q \in [0, \mathfrak{h}(\varphi)]$ the absolute Witt indices $\mathfrak{j}_0, \dots, \mathfrak{j}_{q-1}$ are already known, then one determines $\mathfrak{j}_q$ by the formula

$$\mathfrak{j}_q = \max\big\{j \mid \text{the product } h^{\mathfrak{j}_0} \times h^{\mathfrak{j}_1} \times \cdots \times h^{\mathfrak{j}_{q-1}} \times l_{j-1} \text{ is contained in a rational cycle}\big\}.$$

73. Cycles on X^2

In this section, we study the groups $\overline{\mathrm{Ch}}_i(X^2)$ for $i \geq D$. After Lemma 73.2, we shall assume that X is anisotropic.

Most results of this section are simplified versions of original results on integral motives of quadrics due to Vishik in [**133**].

Lemma 73.1. *The sum*

$$\Delta = \sum_{i=0}^{d} (h^i \times l_i + l_i \times h^i) \in \mathrm{Ch}(\bar{X}^2)$$

is always rational.

PROOF. Either the composition with correspondence Δ or the composition with the correspondence $\Delta + h^d \times h^d$ (depending on whether l_d^2 is zero or not) induces the identity endomorphism of $\mathrm{Ch}(\bar{X}^2)$. Therefore, this correspondence is the class of the diagonal which is rational. $\square$

Lemma 73.2. *Suppose for some $i \in [1,\ d]$ at least one of the basis elements $l_d \times l_i$ or $l_i \times l_d$ of the group $\mathrm{Ch}(\bar{X}^2)$ appears in the decomposition of a rational cycle. Then X is hyperbolic.*

PROOF. Let α be a cycle in $\overline{\mathrm{Ch}}_{i+d}(X^2)$ containing $l_i \times l_d$ or $l_d \times l_i$. Replacing α by its transpose if necessary we may assume that $l_i \times l_d \in \alpha$. If $\beta \in \mathrm{CH}_{i+d}(\bar{X}^2)$ is a basic cycle, then

$$\beta_*(h^i) = \begin{cases} l_d & \text{if } \beta = l_i \times l_d, \\ h^d & \text{if } \beta = l_i \times h^d, \\ 0 & \text{otherwise,} \end{cases}$$

hence $\alpha_*(h^i)$ is equal to l_d or $h^d + l_d$. As $\alpha_*(h^i)$ is rational, the cycle l_d is rational and X is hyperbolic by Corollary 72.6. □

We assume now that X is anisotropic throughout the rest of this section.

Let $\alpha_1, \alpha_2 \in \overline{\mathrm{Ch}}_*(X^2)$. The *intersection* $\alpha_1 \cap \alpha_2$ denotes the sum of the basic cycles contained simultaneously in α_1 and α_2.

Lemma 73.3. *If* $\alpha_1, \alpha_2 \in \coprod_{i\geq 0} \overline{\mathrm{Ch}}_{D+i}(X^2)$, *then the cycle* $\alpha_1 \cap \alpha_2$ *is rational.*

PROOF. Clearly, we may assume that α_1 and α_2 are homogeneous of the same dimension $D+i$ and do not contain any nonessential basis element. Using Lemma 73.2 we see that the intersection $\alpha_1 \cap \alpha_2$ is the essence of the composite of rational correspondences $\alpha_2 \circ \big(\alpha_1 \cdot (h^0 \times h^i)\big)$ and hence is rational. □

Notation 73.4. We write $\overline{\mathrm{Che}}(\bar{X}^2)$ for the group of essential elements in

$$\coprod_{i\geq D} \mathrm{Ch}_i(\bar{X}^2)$$

and $\overline{\mathrm{Che}}(X^2)$ for the group of rational elements in $\overline{\mathrm{Che}}(\bar{X}^2)$.

Definition 73.5. A nonzero element of $\overline{\mathrm{Che}}(X^2)$ is called *minimal* if it does not contain any proper rational subcycle.

Note that a minimal cycle is always homogeneous.

Proposition 73.6. *Let X be a smooth anisotropic quadric. Then*

(1) *The minimal cycles form a basis of the group* $\overline{\mathrm{Che}}(X^2)$.
(2) *Two different minimal cycles intersect trivially.*
(3) *The sum of the minimal cycles of dimension D is equal to the sum*

$$\sum_{i=0}^{d} h^i \times l_i + l_i \times h^i$$

of all D-dimensional essential basis elements (excluding $l_d \times l_d$ in the case of even D).

PROOF. The first two statements of Proposition 73.6 follow from Lemma 73.3. The last statement follows from the previous ones together with Lemma 73.1. □

Let α be an element of $\mathrm{Ch}_{D+r}(\bar{X}^2)$ for some $r \geq 0$. For every $i \in [0,\ r]$, the products $\alpha \cdot (h^0 \times h^i),\ \alpha \cdot (h^1 \times h^{i-1}), \dots, \alpha \cdot (h^i \times h^0)$ will be called the (*ith order*) *derivatives* of α.

Note that all the derivatives of a rational cycle are also rational.

Lemma 73.7. (1) *A derivative of an essential basis element* $\beta \in \overline{\mathrm{Che}}_{D+r}(\bar{X}^2)$ *is an essential basis element.*

(2) *For any $r \geq 0$, nonnegative integers i_1, j_1, i_2, j_2 satisfying $i_1 + j_1 \leq r$, $i_2 + j_2 \leq r$, and nonzero essential cycle* $\beta \in \overline{\mathrm{Che}}_{D+r}(\bar{X}^2)$, *the two derivatives* $\beta \cdot (h^{i_1} \times h^{j_1})$ *and* $\beta \cdot (h^{i_2} \times h^{j_2})$ *of* β *coincide only if* $i_1 = i_2$ *and* $j_1 = j_2$.

(3) *For any $r \geq 0$, nonnegative integers i, j with $i+j \leq r$, and nonzero essential cycles $\beta_1, \beta_2 \in \overline{\mathrm{Ch}}\mathrm{e}_{D+r}(\bar{X}^2)$, the derivatives $\beta_1 \cdot (h^i \times h^j)$ and $\beta_2 \cdot (h^i \times h^j)$ of β_1 and β_2 coincide only if $\beta_1 = \beta_2$.*

PROOF. (1): If β is an essential basis element of $\overline{\mathrm{Ch}}\mathrm{e}_{D+r}(\bar{X}^2)$ for some $r > 0$, then up to transposition, $\beta = h^i \times l_{i+r}$ for some $i \in [0,\ d-r]$. An arbitrary derivative of β is equal to $\beta \cdot (h^{j_1} \times h^{j_2}) = h^{i+j_1} \times l_{i+r-j_2}$ for some $j_1, j_2 \geq 0$ such that $j_1 + j_2 \leq r$. It follows that the integers $i + j_1$ and $i + r - j_2$ are in the interval $[0,\ d]$; therefore, $h^{i+j_1} \times l_{i+r-j_2}$ is an essential basis element.

Statement (2) and (3) are left to the reader. □

Remark 73.8. To visualize the above, it is convenient to think of the essential basic cycles in $\coprod_{i \geq D} \mathrm{Ch}_i(\bar{X}^2)$ (with $l_{D/2} \times l_{D/2}$ excluded by Lemma 73.2) as points of two "pyramids". For example, if $D = 8$ or $D = 9$, we write

If we count the rows of the pyramids from the bottom starting with 0, the top row has number d, and for every $r = 0, \ldots, d$, the rth row of the left pyramid represents the essential basis elements $h^i \times l_{r+i}$, $i = 0, 1, \ldots, d-r$ of $\mathrm{Ch}_{D+r}(\bar{X}^2)$, while the rth row of the right pyramid represents the essential basis elements $l_{r+i} \times h^i$, $i = d-r, d-r-1, \ldots, 0$ (so that the basis elements of each row are ordered by the codimension of the first factor).

For any $\alpha \in \mathrm{Ch}(\bar{X}^2)$, we fill in the pyramids by putting a *mark* in the points representing basis elements contained in the decomposition of α; the picture thus obtained is the *diagram* of α. If α is homogeneous, the *marked points* (if any) lie in the same row. It is now easy to interpret the derivatives of α if α is homogeneous of dimension $\geq D$: the diagram of an ith order derivative is a translation of the marked points of the diagram of α moving them i rows lower. In particular, the diagram of every derivative of such an α has the same number of marked points as the diagram of α (cf. Lemma 73.7). The diagrams of any two different derivatives of the same order are shifts (to the right or to the left) of each other.

Example 73.9. Let $D = 8$ or $D = 9$. Let $\alpha \in \mathrm{Ch}_{D+1}(\bar{X}^2)$ be the essential cycle $\alpha = h^0 \times l_1 + h^2 \times l_3 + l_3 \times h^2$. Then the diagram of α is

There are precisely two first-order derivatives of α. They are given by $\alpha \cdot (h^0 \times h^1) = h^0 \times l_0 + h^2 \times l_2 + l_3 \times h^3$ and $\alpha \cdot (h^1 \times h^0) = h^1 \times l_1 + h^3 \times l_3 + l_2 \times h^2$. Their diagrams are as follows:

Lemma 73.10. *Let $\alpha \in \overline{\mathrm{Che}}(X^2)$. Then the following conditions are equivalent:*

(1) *The cycle α is minimal.*
(2) *All derivatives of α are minimal.*
(3) *At least one derivative of α is minimal.*

PROOF. Derivatives of a proper subcycle of α are proper subcycles of the derivatives of α; therefore, (3) $\Rightarrow$ (1).

In order to show that (1) $\Rightarrow$ (2), it suffices to show that the two first order derivatives $\alpha\cdot(h^0\times h^1)$ and $\alpha\cdot(h^1\times h^0)$ of a minimal cycle α are also minimal. If not, possibly replacing α by its transposition, we reduce to the case where the derivative $\alpha\cdot(h^0\times h^1)$ of a minimal α is not minimal. It follows that the cycle $\alpha\cdot(h^0\times h^i)$ with $i=\dim\alpha - D$ is also not minimal. Let α' be its proper subcycle. Taking the essence of the composite $\alpha\circ\alpha'$, we get a proper subcycle of α, a contradiction. □

Corollary 73.11. *The derivatives of a minimal cycle are disjoint.*

PROOF. The derivatives of a minimal cycle are minimal by Lemma 73.10 and pairwise different by Lemma 73.7. As two different minimal cycles are disjoint by Lemma 73.3, the result follows. □

We recall the notation of §25. If $F_0 = F, F_1, \dots, F_{\mathfrak{h}}$ is the generic splitting tower of φ with $\mathfrak{h}=\mathfrak{h}(\varphi)$ the height of φ, and $\varphi_i = (\varphi_{F_i})_{an}$ for $i\ge 0$, we let $X_i = X_{\varphi_i}$ be the projective quadric over F_i given by φ_i. Then $\mathfrak{i}_r=\mathfrak{i}_r(\varphi)=\mathfrak{i}_r(X)$ is the rth relative and $\mathfrak{j}_r=\mathfrak{j}_r(\varphi)=\mathfrak{j}_r(X)$ the rth absolute higher Witt index of φ, $r\in[0,\mathfrak{h}]$. We will also call these numbers the *relative* and the *absolute Witt indices of X* respectively.

Lemma 73.12. *If i,j are integers in the interval $[0,\ d]$ satisfying $i<\mathfrak{j}_q\le j$ for some $q\in[1,\ \mathfrak{h})$, then no element in $\overline{\mathrm{Ch}}(X^2)$ contains either $h^i\times l_j$ or $l_j\times h^i$.*

PROOF. Let i,j be integers of the interval $[0,\ d]$ such that $h^i\times l_j$ or $l_j\times h^i$ appears in the decomposition of some $\alpha\in\overline{\mathrm{Ch}}(X^2)$. Replacing α by its transpose if necessary, we may assume that $h^i\times l_j\in\alpha$. Replacing α by its homogeneous component containing $h^i\times l_j$, we reduce to the case that α is homogeneous.

Suppose q is an integer in $[1,\ \mathfrak{h})$ satisfying $i<\mathfrak{j}_q$. It suffices to show that $j<\mathfrak{j}_q$ as well.

Let L be a field extension of F with $\mathfrak{i}_0(X_L)=\mathfrak{j}_q$ (e.g., $L=F_q$). The cycles α and l_i are both L-rational. Therefore, so is the cycle $\alpha_*(l_i)=l_j$. It follows by Corollary 72.6 that $j<\mathfrak{j}_q$. □

Remark 73.13. In order to "see" the statement of Lemma 73.12, it is helpful to mark by a $*$ only those essential basis elements that are not "forbidden" by this lemma in the pyramids of basic cycles drawn in Remark 73.8 and to mark by a $\circ$ the remaining points of the pyramids. We will get isosceles triangles based on the lower row of these pyramids. For example, if X is a 34-dimensional quadric with relative higher Witt indices $4,2,4,8$, the picture looks as follows:

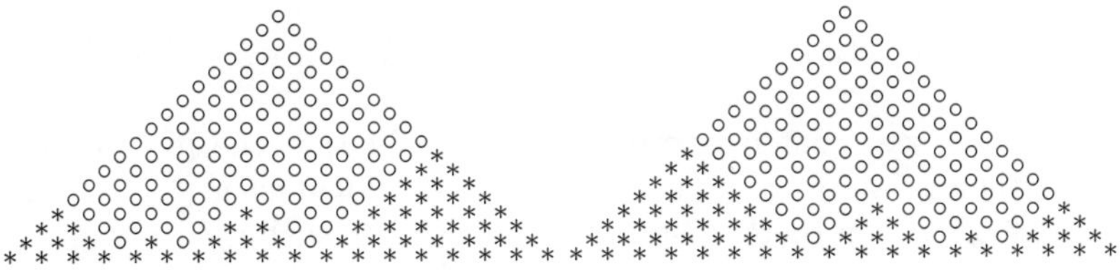

The triangles in the pyramids of Remark 73.13 will be called *shell triangles.* The shell triangles in the left pyramid are numbered from the left starting with 1. The shell triangles in the right pyramid are numbered from the right starting with 1 as well (so that the symmetric triangles have the same number; for any $q \in [1, \mathfrak{h}]$, the bases of the qth triangles have (each) $\mathfrak{i}_q$ points). The rows of the shell triangles are numbered from below starting with 0. The points of rows of the shell triangles (of the left ones as well as of the right ones) are numbered from the left starting with 1.

Lemma 73.14. *For every rational cycle $\alpha \in \coprod_{i \geq D} \overline{\mathrm{Ch}}_i(X^2)$, the number of essential basic cycles contained in α is even (i.e., the number of the marked points in the diagram of α is even).*

PROOF. We may assume that α is homogeneous, say, $\alpha \in \overline{\mathrm{Ch}}_{D+k}(X^2)$, $k \geq 0$. We may also assume that $k \leq d$ as there are no essential basic cycles of dimension $> D + d$. Let n be the number of essential basic cycles contained in α. The pull-back $\delta^*(\alpha)$ of α with respect to the diagonal $\delta : X \to X^2$ produces $n \cdot l_k \in \overline{\mathrm{Ch}}(X)$. It follows by Corollary 72.2 that n is even. □

Lemma 73.15. *Let $\alpha \in \overline{\mathrm{Ch}}(X^2)$ be a cycle containing the top of a qth shell triangle for some $q \in [1, \mathfrak{h}]$. Then α also contains the top of the other qth shell triangle.*

PROOF. We may assume that α contains the top of the *left* qth shell triangle. Replacing F by the field F_{q-1}, X by X_{q-1}, and α by $pr^2_*(\alpha)$, where $pr \in \mathrm{Ch}(X_{F_{q-1}} \times X_{q-1})$ is the correspondence of Corollary 72.4, we may assume that $q = 1$.

Replacing α by its homogeneous component containing the top of the left first shell triangle $\beta = h^0 \times l_{\mathfrak{j}_1 - 1}$, we may assume that α is homogeneous.

Suppose that the transpose of β is not contained in α. By Lemma 73.12, the element α does not contain any essential basic cycles having h^i with $0 < i < \mathfrak{i}_1$ as a factor. Since $\alpha \neq \beta$ by Lemma 73.14, we have $\mathfrak{h} > 1$. Moreover, the number of essential basis elements contained in α and the number of essential basis elements contained in $pr^2_*(\alpha) \in \overline{\mathrm{Ch}}(X_1^2)$ differ by 1. In particular, these two numbers have different parity. However, the number of the essential basis elements contained in α is even by Lemma 73.14. By the same lemma, the number of essential basis elements contained in $pr^2_*(\alpha)$ is even, too. □

Definition 73.16. A minimal cycle $\alpha \in \overline{\mathrm{Che}}(X^2)$ is called *primordial* if it is not a derivative of positive order of another rational cycle.

Lemma 73.17. *Let $\alpha \in \overline{\mathrm{Che}}(X^2)$ be a minimal cycle (as defined in* 73.5*) which also contains the top of a qth shell triangle for some $q \in [1, \mathfrak{h}]$. Then α is symmetric and primordial.*

PROOF. The cycle $\alpha \cap t(\alpha)$, where $t(\alpha)$ is the transpose of α (where the intersection of cycles is defined in Lemma 73.3) is symmetric, rational by Lemma 73.3, contained in α, and, by Lemma 73.15, still contains the tops $h^{\mathfrak{j}_{q-1}} \times l_{\mathfrak{j}_q - 1}$ and $l_{\mathfrak{j}_q - 1} \times h^{\mathfrak{j}_{q-1}}$ of both qth shell triangles. Therefore, it coincides with α by the minimality of α.

It is easy to "see" that α is primordial by looking at the picture in Remark 73.13. Nevertheless, we prove it. If there exists a rational cycle $\beta \neq \alpha$ such that α is a derivative of β, then there exists a rational cycle β' such that α is an order one derivative of β', i.e., $\alpha = \beta' \cdot (h^0 \times h^1)$ or $\alpha = \beta' \cdot (h^1 \times h^0)$. In the first case, β' would contain the basic cycle $h^{\mathfrak{j}_{q-1}} \times l_{\mathfrak{j}_q}$ while in the second case β' would contain $h^{\mathfrak{j}_{q-1}-1} \times l_{\mathfrak{j}_q-1}$. However, neither of these two cases is possible by Lemma 73.12. □

It is easy to see that a cycle α satisfying the hypothesis of Lemma 73.17 with $q = 1$ exists:

Lemma 73.18. *There exists a cycle in $\overline{\mathrm{Ch}}_{D+\mathfrak{i}_1-1}(X^2)$ containing the top $h^0 \times l_{\mathfrak{i}_1-1}$ of the first left shell triangle.*

PROOF. If $D = 0$, this follows by Lemma 73.1. So assume $D > 0$. Consider the pull-back homomorphism $\overline{\mathrm{Ch}}(X^2) \twoheadrightarrow \overline{\mathrm{Ch}}(X_{F(X)})$ with respect to the morphism $X_{F(X)} \to X^2$ produced by the generic point of the first factor of X^2. By Corollary 57.11, this is an epimorphism. It is also a restriction of the homomorphism $\mathrm{Ch}(\bar{X}^2) \to \mathrm{Ch}(\bar{X})$ mapping each basis element of the type $h^0 \times l_i$ to l_i and vanishing on all other basis elements. Therefore an arbitrary preimage of $l_{\mathfrak{i}_1-1} \in \overline{\mathrm{Ch}}(X_{F(X)})$ under the surjection $\overline{\mathrm{Ch}}(X^2) \twoheadrightarrow \overline{\mathrm{Ch}}(X_{F(X)})$ contains $h^0 \times l_{\mathfrak{i}_1-1}$. □

Lemma 73.19. *Let $\rho \in \overline{\mathrm{Ch}}_D(X^2)$, $q \in [1,\ \mathfrak{h}]$, and $i \in [1,\ \mathfrak{i}_q]$. Then the element $h^{\mathfrak{j}_{q-1}+i-1} \times l_{\mathfrak{j}_{q-1}+i-1}$ is contained in ρ if and only if the element $l_{\mathfrak{j}_q-i} \times h^{\mathfrak{j}_q-i}$ is contained in ρ.*

PROOF. Clearly, it suffices to prove Lemma 73.19 for $q = 1$. By Lemma 73.18, the basis element $h^0 \times l_{\mathfrak{i}_1-1}$ is contained in a rational cycle. Let α be the minimal cycle containing $h^0 \times l_{\mathfrak{i}_1-1}$. By Lemma 73.15, the cycle α also contains $l_{\mathfrak{i}_1-1} \times h^0$. Therefore, the derivative $\alpha \cdot (h^{i-1} \times h^{\mathfrak{i}_1-i})$ contains both $h^{i-1} \times l_{i-1}$ and $l_{\mathfrak{i}_1-i} \times h^{\mathfrak{i}_1-i}$. Since the derivative of a minimal cycle is minimal by Lemma 73.10, the lemma follows by Lemma 73.3. □

In the language of diagrams, the statement of Lemma 73.19 means that the ith point of the base of the qth *left* shell triangle in the diagram of ρ is marked if and only if the ith point of the base of the qth *right* shell triangle is marked.

Definition 73.20. The symmetric shell triangles (i.e., both qth shell triangles for some q) are called *dual*. Two points are called *dual*, if one of them is in a left shell triangle, while the other one is in the same row of the dual right shell triangle and has the same number as the first point.

Corollary 73.21. *In the diagram of an element of $\overline{\mathrm{Ch}}(X^2)$, any two dual points are simultaneously marked or not marked.*

PROOF. Let k be the number of the row containing two given dual points. The case of $k = 0$ is treated in Lemma 73.19 (while Lemma 73.15 treats the case of "locally maximal" k). The case of an arbitrary k is reduced to the case of $k = 0$ by taking a kth order derivative of α. □

Remark 73.22. By Corollary 73.21, it follows that the diagram of a cycle in $\overline{\mathrm{Ch}}(X^2)$ is determined by one (left or right) half of itself. By "shell triangles" we shall mean the left shell triangles. Note also that the transposition of a cycle acts symmetrically about the vertical axis on each shell triangle.

The following proposition generalizes Lemma 73.18.

Proposition 73.23. *Let $f : \overline{\mathrm{Ch}}(X^2) \to [1,\ \mathfrak{h}]$ be the map that assigns to each $\gamma \in \overline{\mathrm{Ch}}(X^2)$ the integer $q \in [1,\ \mathfrak{h}]$ such that the diagram of γ has a point in the qth shell triangle and has no points in the ith shell triangles for any $i < q$. For any $q \in f\big(\overline{\mathrm{Ch}}(X^2)\big)$, there exists an element $\alpha \in \overline{\mathrm{Ch}}(X^2)$ with $f(\alpha) = q$ so that α contains the top of the qth shell triangle.*

PROOF. We induct on q. If $q = 1$, the condition of Proposition 73.23 is automatically satisfied by Lemma 73.1 and the result follows by Lemma 73.18. So we may assume that $q > 1$.

Let γ be an element of $\overline{\mathrm{Ch}}(X^2)$ with $f(\gamma) = q$. Replacing γ by its appropriate homogeneous component, we may assume that γ is homogeneous. Replacing this homogeneous γ by any one of its maximal order derivatives, we may further assume that $\gamma \in \overline{\mathrm{Ch}}_D(X^2)$.

Let i be the smallest integer such that $h^{\mathfrak{j}_{q-1}+i} \times l_{\mathfrak{j}_{q-1}+i} \in \gamma$. We first prove that the group $\overline{\mathrm{Ch}}(X^2)$ contains a cycle γ' satisfying $f(\gamma') = q$ with γ' containing $h^{\mathfrak{j}_{q-1}+i} \times l_{\mathfrak{j}_q-1}$. (This is the point on the right side of the qth shell triangle such that the line connecting it with $h^{\mathfrak{j}_{q-1}+i} \times l_{\mathfrak{j}_q-1+i}$ is parallel to the left side of the shell triangle. If $i = 0$, then we can take $\alpha = \gamma'$ and finish the proof.)

Let

$$pr^2_* : \overline{\mathrm{Ch}}(X^2_{F(X)}) \to \overline{\mathrm{Ch}}(X^2_1) \quad \text{and} \quad in^2_* : \overline{\mathrm{Ch}}(X^2_1) \to \overline{\mathrm{Ch}}(X^2_{F(X)})$$

be the homomorphisms of Remark 72.5. Applying the induction hypothesis to the quadric X_1 and the cycle $pr^2_*(\gamma) \in \overline{\mathrm{Ch}}(X^2_1)$, we get a homogeneous cycle in $\overline{\mathrm{Ch}}_{D+\mathfrak{i}_q-1}(X^2_{F(X)})$ containing $h^{\mathfrak{j}_{q-1}} \times l_{\mathfrak{j}_q-1}$. Multiplying it by $h^i \times h^0$, we get a homogeneous cycle in $\overline{\mathrm{Ch}}(X^2_{F(X)})$ containing $h^{\mathfrak{j}_{q-1}+i} \times l_{\mathfrak{j}_q-1}$. Note that the quadric $X_{F(X)}$ is not hyperbolic (since $\mathfrak{h} \geq q > 1$) and therefore, by Lemma 73.2, the basis element $l_d \times l_d$ is not contained in this cycle. Therefore, the group $\overline{\mathrm{Ch}}(X^3)$ contains a homogeneous cycle μ containing $h^0 \times h^{\mathfrak{j}_{q-1}+i} \times l_{\mathfrak{j}_q-1}$ (and not containing $h^0 \times l_d \times l_d$). View μ as a correspondence of the middle factor of X^3 into the product of the two outer factors. Composing it with γ and taking the pull-back with respect to the partial diagonal map $\delta : X^2 \to X^3$, $(x_1, x_2) \mapsto (x_1, x_1, x_2)$, we get the required cycle γ' (more accurately, $\gamma' = \delta^*\big(t_{12}(\mu) \circ \gamma\big)$, where t_{12} is the automorphism of $\overline{\mathrm{Ch}}(X^3)$ given by the transposition of the first two factors of X^3).

The highest order derivative $\gamma' \cdot (h^{\mathfrak{i}_q-1-i} \times h^0)$ of γ' contains $h^{\mathfrak{j}_q-1} \times l_{\mathfrak{j}_q-1}$, the last point of the base of the qth shell triangle. Therefore, the transpose $t(\gamma')$ contains the first point $h^{\mathfrak{j}_{q-1}} \times l_{\mathfrak{j}_{q-1}}$ of the base of the qth shell triangle by Remark 73.22. Replacing γ by $t(\gamma')$, we are in the case that $i = 0$ (see the third paragraph of the proof), finishing the proof. □

Illustration 73.24. The following picture shows the displacements of the special marked point of the qth shell triangle in the proof of Proposition 73.23:

6
* *
* * *
* * * *
* * * * 3
* * * * * *
* * * * * * *
* * * * * * * *
* * 1 * * * * * *
* * * * * * * * * *
5 * * * 2 * * * * * 4

We start with a cycle $\gamma \in \overline{\mathrm{Ch}}(X^2)$ with $f(\gamma) = q$. It contains a point somewhere in the qth shell triangle, the point in Position 1. Then we modify γ in such a way that $f(\gamma)$ is always q and look at what happens with this point. Replacing γ by a maximal order derivative, we move the special point to the base of the shell triangle; for example, we can move it to Position 2. The heart of the proof is the movement from Position 2 to Position 3 (here we make use of the induction hypothesis). Again taking an appropriate derivative, we come to Position 4. Transposing this cycle, we come to Position 5. Finally, repeating the procedure used in the passage from Positions 2 to 3, we move from Position 5 to Position 6, arriving to the top.

Illustration 73.25. Let

$$\overline{\mathrm{Ch}}(X^2) \hookrightarrow \overline{\mathrm{Ch}}(X^2_{F(X)}) \xrightarrow{pr^2_*} \overline{\mathrm{Ch}}(X^2_1)$$

be the homomorphism used in the proof of Proposition 73.23. If $\alpha \in \overline{\mathrm{Ch}}(X^2)$, the diagram of $pr^2_*(\alpha)$ is obtained from the diagram of α by erasing the first shell triangle. An example is shown by the illustration with X as in Remark 73.13:

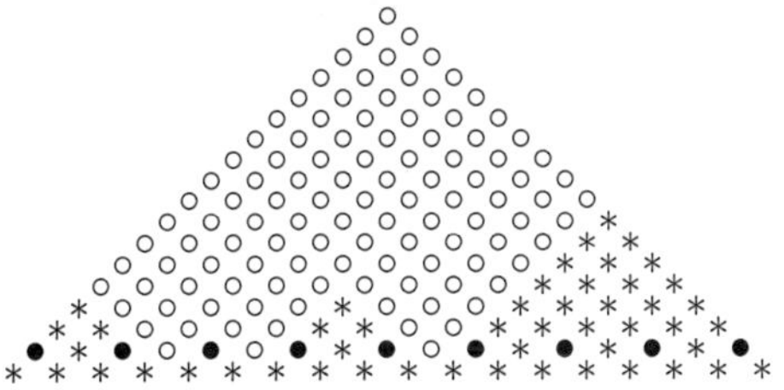

diagram of α

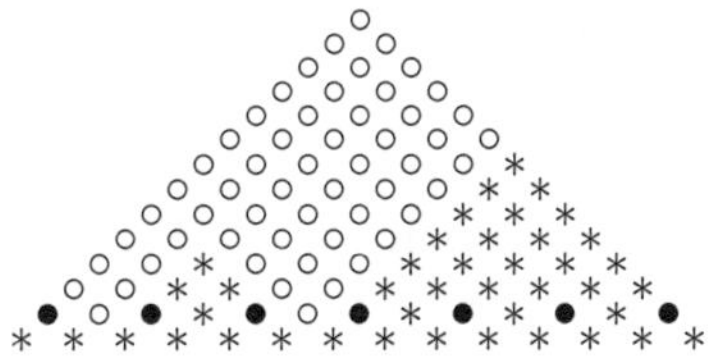

diagram of $pr^2_*(\alpha)$

Summarizing, we have the following structure result on $\overline{\mathrm{Che}}(X^2)$:

Theorem 73.26. *Let X be a smooth anisotropic quadric. The set of primordial cycles Π lying in $\overline{\mathrm{Che}}(X^2)$ has the following properties:*

(1) *All derivatives of all cycles in Π are minimal and pairwise disjoint, and the set of these forms a basis of $\overline{\mathrm{Che}}(X^2)$. In particular, the sum of all maximal order derivatives of the elements in Π is equal to the cycle*

$$\Delta = \sum_{i=0}^{d} (h^i \times l_i + l_i \times h^i) \in \mathrm{Ch}(\bar{X}^2).$$

(2) *Every cycle in Π is symmetric and has no points outside of the shell triangles.*

(3) *The map f in Proposition 73.23 is injective on Π. Every cycle $\pi \in \Pi$ contains the top of the $f(\pi)$th shell triangle and has no points in any shell triangle with number in $f(\Pi) \setminus \{f(\pi)\}$.*

(4) $1 \in f\big(\overline{\mathrm{Ch}}(X^2)\big) = f(\Pi)$.

Let f be as in Proposition 73.23. If $f(\alpha) = q$ for an element $\alpha \in \mathrm{Ch}(X^2)$, we say that α *starts in the qth shell triangle.* More specifically, if $f(\pi) = q$ for a *primordial* cycle π, we say that π is *q-primordial.*

The following statement is an additional property of 1-primordial cycles:

Proposition 73.27. *Let $\pi \in \overline{\mathrm{Ch}}(X^2)$ be a 1-primordial cycle. Suppose π contains $h^i \times l_{i+\mathfrak{i}_1-1}$ for some positive $i \leq d$. Then the smallest integer i with this property coincides with an absolute Witt index of φ, i.e., $i = \mathfrak{j}_{q-1}$ for some $q \in [2,\ \mathfrak{h}]$.*

PROOF. The cycle π contains $h^0 \times l_{\mathfrak{i}_1-1}$ as this is the top of the first shell triangle. By Lemma 73.12, π contains none of the cycles $h^1 \times l_{\mathfrak{i}_1}, \dots, h^{\mathfrak{i}_1-1} \times l_{2\mathfrak{i}_1-2}$. It follows that if $i \in [1,\ d]$ is the smallest integer satisfying $h^i \times l_{i+\mathfrak{i}_1-1} \in \pi$, then $i \geq \mathfrak{j}_1 = \mathfrak{i}_1$. Let $q \in [2,\ \mathfrak{h}]$ be the largest integer with $\mathfrak{j}_{q-1} \leq i$. We show that $\mathfrak{j}_{q-1} = i$. Suppose to the contrary that $\mathfrak{j}_{q-1} < i$.

Let X_1 be the quadric over $F(X)$ given by the anisotropic part of $\varphi_{F(X)}$ and

$$pr^2_* : \overline{\mathrm{Ch}}(X^2_{F(X)}) \to \overline{\mathrm{Ch}}(X_1^2),$$

the homomorphism of Remark 72.5. Then the element $pr^2_*(\pi)$ starts in the shell triangle number $q-1$ of X_1. Therefore, by Proposition 73.23, the quadric X_1 possesses a $(q-1)$-primordial cycle τ.

Let

$$in^2_* : \overline{\mathrm{Ch}}(X_1^2) \to \overline{\mathrm{Ch}}(X^2_{F(X)})$$

be the homomorphism of Remark 72.5. Then the cycle $\beta = in^2_*(\tau)$ in $\overline{\mathrm{Ch}}(X^2_{F(X)})$ contains $h^{\mathfrak{j}_{q-1}} \times l_{\mathfrak{j}_q-1}$ and does not contain any $h^j \times l_?$ with $j < \mathfrak{j}_{q-1}$.

Let $\eta \in \overline{\mathrm{Ch}}(X^3)$ be a preimage of β under the pull-back epimorphism

$$g^* : \overline{\mathrm{Ch}}(X^3) \twoheadrightarrow \overline{\mathrm{Ch}}(X^2_{F(X)})\ ,$$

where the morphism $g : X^2_{F(X)} \to X^3$ is induced by the generic point of the first factor of X^3 (cf. Corollary 57.11). The cycle η contains $h^0 \times h^{\mathfrak{j}_{q-1}} \times l_{\mathfrak{j}_q-1}$ and does not contain any $h^0 \times h^j \times l_?$ with $j < \mathfrak{j}_{q-1}$.

We consider η as a correspondence $X \rightsquigarrow X^2$. Define μ as the composition $\mu = \eta \circ \alpha$ with $\alpha = \pi \cdot (h^0 \times h^{\mathfrak{i}_1-1})$. The cycle α contains $h^0 \times l_0$ and does not contain any $h^j \times l_j$ with $j \in [1,\ i)$. In particular, since $\mathfrak{j}_{q-1} < i$, it does not contain any $h^j \times l_j$ with $j \in [1,\ \mathfrak{j}_{q-1}]$. Consequently, the cycle μ contains the basis element

$$h^0 \times h^{\mathfrak{j}_{q-1}} \times l_{\mathfrak{j}_q-1} = (h^0 \times h^{\mathfrak{j}_{q-1}} \times l_{\mathfrak{j}_q-1}) \circ (h^0 \times l_0)$$

and does not contain any $h^j \times h^? \times l_?$ with $j \in [1,\ \mathfrak{j}_{q-1}]$.

Let

$$\delta^* : \overline{\mathrm{Ch}}(X^3) \to \overline{\mathrm{Ch}}(X^2)$$

be the pull-back homomorphism with respect to the partial diagonal morphism

$$\delta : X^2 \to X^3,\quad (x_1 \times x_2) \mapsto (x_1 \times x_1 \times x_2)\ .$$

The cycle $\delta^*(\mu) \in \overline{\mathrm{Ch}}(X^2)$, contains the basis element

$$h^{\mathfrak{j}_{q-1}} \times l_{\mathfrak{j}_q-1} = \delta^*(h^0 \times h^{\mathfrak{j}_{q-1}} \times l_{\mathfrak{j}_q-1})$$

and does not contain any $h^j \times l_?$ with $j < \mathfrak{j}_{q-1}$. It follows that an appropriate derivative of the cycle $\delta^*(\mu)$ contains $h^i \times l_{i+\mathfrak{i}_1-1} \in \pi$ and does not contain $h^0 \times l_{\mathfrak{i}_1-1} \in \pi$. This contradicts the minimality of π. □

Remark 73.28. In the language of diagrams, Proposition 73.27 asserts that the point $h^i \times l_{i+\mathfrak{i}_1-1}$ lies on the left side of the qth shell triangle.

Definition 73.29. We say that the integer $q \in [2, \mathfrak{h}]$ occurring in Proposition 73.27 *is produced* by the 1-primordial cycle π.

CHAPTER XIV

The Izhboldin Dimension

Let X be an anisotropic smooth projective quadric over a field F (of arbitrary characteristic). The Izhboldin dimension $\dim_{\mathrm{Izh}} X$ of X is defined as

$$\dim_{\mathrm{Izh}} X := \dim X - \mathfrak{i}_1(X) + 1 ,$$

where $\mathfrak{i}_1(X)$ is the first Witt index of X.

Let Y be a complete (possibly singular) algebraic variety over F with all of its closed points of even degree and such that Y has a closed point of odd degree over $F(X)$. The main theorem of this chapter is Theorem 76.1 below. It states that $\dim_{\mathrm{Izh}} X \leq \dim Y$ and if $\dim_{\mathrm{Izh}} X = \dim Y$ the quadric X is isotropic over $F(Y)$.

An application of Theorem 76.1 is the positive solution of a conjecture of Izhboldin that states: if an anisotropic quadric Y becomes isotropic over $F(X)$, then $\dim_{\mathrm{Izh}} X \leq \dim_{\mathrm{Izh}} Y$ with equality if and only if X is isotropic over $F(Y)$.

The results of this chapter for the case that characteristic $\neq 2$ were obtained in [**76**].

74. The first Witt index of subforms

For the reader's convenience, we list some easy properties of the first Witt index:

Lemma 74.1. *Let φ be an anisotropic nondegenerate quadratic form over F of dimension at least two.*

(1) *The first Witt index $\mathfrak{i}_1(\varphi)$ coincides with the minimal Witt index of φ_E, where E runs over all field extensions of F such that the form φ_E is isotropic.*
(2) *Let ψ be a nondegenerate subform of φ of codimension r and let E/F be a field extension. Then $\mathfrak{i}_0(\psi_E) \geq \mathfrak{i}_0(\varphi_E) - r$. In particular, $\mathfrak{i}_1(\psi) \geq \mathfrak{i}_1(\varphi) - r$ (if $\mathfrak{i}_1(\psi)$ is defined, i.e., $\dim \psi \geq 2$).*

PROOF. The first statement is proven in Corollary 25.3. For the second statement, note that the intersection of a maximal isotropic subspace U (of dimension $\mathfrak{i}_0(\varphi_E)$) of the form φ_E with the space of the subform ψ_E is of codimension at most r in U. □

The following two statements are due to Vishik (in the case that characteristic $\neq 2$; cf. [**133**, Cor. 4.9]).

Proposition 74.2. *Let φ be an anisotropic nondegenerate quadratic form over F with $\dim \varphi \geq 2$. Let ψ be a nondegenerate subform of φ. If $\operatorname{codim}_\varphi \psi \geq \mathfrak{i}_1(\varphi)$, then the form $\psi_{F(\varphi)}$ is anisotropic.*

PROOF. Let $n = \operatorname{codim}_\varphi \psi$ and assume that $n \geq \mathfrak{i}_1(\varphi)$. If the form $\psi_{F(\varphi)}$ is isotropic, there exists a rational morphism $X \dashrightarrow Y$, where X and Y are the projective quadrics of φ and ψ respectively. We use the same notation as in §72. Let $\alpha \in \overline{\mathrm{Ch}}(X^2)$ be the class of the closure of the graph of the composition $X \dashrightarrow Y \hookrightarrow X$. Since the push-forward of α with respect to the first projection $X^2 \to X$ is nonzero, we have $h^0 \times l_0 \in \alpha$. On the other hand, as α is in the image of the push-forward homomorphism $\mathrm{Ch}(\bar{X} \times \bar{Y}) \to \mathrm{Ch}(\bar{X}^2)$ that maps any external product $\beta \times \gamma$ to $\beta \times in_*(\gamma)$, where the push-forward $in_* : \mathrm{Ch}(\bar{Y}) \to \mathrm{Ch}(\bar{X})$ maps h^i to h^{i+n}, and $n \geq \mathfrak{i}_1(\varphi)$, one has $l_{\mathfrak{i}_1(\varphi)-1} \times h^{\mathfrak{i}_1(\varphi)-1} \notin \alpha$, contradicting Lemma 73.19 (cf. also Corollary 73.21). □

Corollary 74.3. *Let φ be an anisotropic nondegenerate quadratic form and let φ' be a nondegenerate subform of φ of codimension n with $\dim \varphi' \geq 2$. If $n < \mathfrak{i}_1(\varphi)$, then $\mathfrak{i}_1(\varphi') = \mathfrak{i}_1(\varphi) - n$.*

PROOF. Let $\mathfrak{i}_1 = \mathfrak{i}_1(\varphi)$. By Lemma 74.1, we know that $\mathfrak{i}_1(\varphi') \geq \mathfrak{i}_1 - n$. Let ψ be a nondegenerate subform of φ' of dimension $\dim \varphi - \mathfrak{i}_1$. If $\mathfrak{i}_1(\varphi') > \mathfrak{i}_1 - n$, then the form $\psi_{F(\varphi)}$ is isotropic by Lemma 74.1 contradicting Proposition 74.2. □

Lemma 74.4. *Let φ be an anisotropic nondegenerate quadratic F-form satisfying $\dim \varphi \geq 3$ and let $\mathfrak{i}_1(\varphi) = 1$. Let $F(t)/F$ be a purely transcendental field extension of degree 1. Then there exists a nondegenerate subform ψ of $\varphi_{F(t)}$ of codimension 1 satisfying $\mathfrak{i}_1(\psi) = 1$.*

PROOF. First consider the case that $\mathrm{char}(F) \neq 2$. We can write $\varphi \simeq \varphi' \perp \langle a, b \rangle$ for some $a, b \in F^\times$ and some quadratic form φ'. Set

$$\psi = \varphi'_{F(t)} \perp \langle a + bt^2 \rangle .$$

This is clearly a subform of $\varphi_{F(t)}$ of codimension 1. Moreover, the fields $F(t)(\psi)$ and $F(\varphi)$ are isomorphic over F. In particular,

$$\mathfrak{i}_1(\psi) = \mathfrak{i}_0(\psi_{F(t)(\psi)}) \leq \mathfrak{i}_0(\varphi_{F(t)(\psi)}) = \mathfrak{i}_0(\varphi_{F(\varphi)}) = \mathfrak{i}_1(\varphi) = 1,$$

hence $\mathfrak{i}_1(\psi) = 1$.

Suppose $\mathrm{char}(F) = 2$. If $\dim \varphi$ is even then $\varphi \simeq \varphi' \perp [a, b]$ for some $a, b \in F$ and some even-dimensional nondegenerate quadratic form φ' by Proposition 7.31. In this case set

$$\psi = \varphi'_{F(t)} \perp \langle a + t + bt^2 \rangle .$$

If $\dim \varphi$ is odd, then $\varphi \simeq \varphi' \perp [a, b] \perp \langle c \rangle$ for some $c \in F^\times$, some $a, b \in F$, and some even-dimensional nondegenerate quadratic form φ'. In this case set

$$\psi = \varphi'_{F(t)} \perp [a, b + ct^2].$$

In either case, ψ is a nondegenerate subform of $\varphi_{F(t)}$ of codimension 1 such that the fields $F(t)(\psi)$ and $F(\varphi)$ are F-isomorphic. Therefore the argument above shows that $\mathfrak{i}_1(\psi) = 1$. □

75. Correspondences

Let X and Y be schemes over a field F. Suppose that X is equidimensional and let $d = \dim X$. Recall that a *correspondence* (*of degree zero*) $\alpha : X \rightsquigarrow Y$ *from* X *to* Y is an element $\alpha \in \mathrm{CH}_d(X \times Y)$ (cf. §62). A correspondence is called *prime* if it is represented by a prime cycle. Every correspondence is a linear combination of prime correspondences with integer coefficients.

Let $\alpha : X \rightsquigarrow Y$ be a correspondence. Assume that X is a variety and Y is complete. The projection morphism $p : X \times Y \to X$ is proper, hence the push-forward homomorphism

$$p_* : \mathrm{CH}_d(X \times Y) \to \mathrm{CH}_d(X) = \mathbb{Z} \cdot [X]$$

is defined (cf. Proposition 49.9). The integer $\mathrm{mult}(\alpha) \in \mathbb{Z}$ satisfying

$$p_*(\alpha) = \mathrm{mult}(\alpha) \cdot [X]$$

is called the *multiplicity* of α. Clearly, $\mathrm{mult}(\alpha + \beta) = \mathrm{mult}(\alpha) + \mathrm{mult}(\beta)$ for any two correspondences $\alpha, \beta : X \rightsquigarrow Y$.

A correspondence $\alpha : \operatorname{Spec} F \rightsquigarrow Y$ is represented by a 0-cycle z on Y. Clearly, $\mathrm{mult}(\alpha) = \deg(z)$, where $\deg : \mathrm{CH}_0(Y) \to \mathbb{Z}$ is the degree homomorphism defined in Example 57.7. More generally, we have the following statement.

Lemma 75.1. *The composition*

$$\mathrm{CH}_d(X \times Y) \to \mathrm{CH}_0\big(Y_{F(X)}\big) \xrightarrow{\deg} \mathbb{Z},$$

where the first map is the pull-back homomorphism with respect to the natural flat morphism $Y_{F(X)} \to X \times Y$ takes a correspondence α to $\mathrm{mult}(\alpha)$.

PROOF. The statement follows by Proposition 49.20 applied to the fiber product diagram

$$\begin{array}{ccc} Y_{F(X)} & \longrightarrow & X \times Y \\ \downarrow & & \downarrow \\ \operatorname{Spec} F(X) & \longrightarrow & X \end{array}$$

□

Lemma 75.2. *Let Y be a complete scheme and let $\tilde{F}/F$ be a purely transcendental field extension. Then*

$$\deg \mathrm{CH}_0(Y) = \deg \mathrm{CH}_0(Y_{\tilde{F}}).$$

PROOF. We may assume that $\tilde{F}$ is the function field of the affine line $\mathbb{A}^1$. By Proposition 49.20 applied to the fiber product diagram

$$\begin{array}{ccc} Y_{F(\mathbb{A}^1)} & \xrightarrow{f} & Y \\ \downarrow & & \downarrow \\ \operatorname{Spec} F(\mathbb{A}^1) & \longrightarrow & \operatorname{Spec} F \end{array}$$

it suffices to show that the pull-back homomorphism $f^* : \mathrm{CH}_0(Y) \to \mathrm{CH}_0\big(Y_{F(\mathbb{A}^1)}\big)$ is surjective. We have $f = p \circ q$ where $p : Y \times \mathbb{A}^1 \to Y$ is the projection and $q : Y_{F(\mathbb{A}^1)} \to Y \times \mathbb{A}^1$ is a natural morphism. The pull-back homomorphism p^* is an isomorphism by Theorem 57.13 and q^* is surjective by Corollary 57.11. As $f^* = q^* \circ p^*$ by Proposition 49.18, the map f^* is surjective. □

Corollary 75.3. *Let Y be a complete variety, X a smooth projective quadric, and $X' \subset X$ an arbitrary closed subvariety of X. Then*

$$\deg \mathrm{CH}_0\big(Y_{F(X)}\big) \subset \deg \mathrm{CH}_0\big(Y_{F(X')}\big).$$

PROOF. Since $F(X)$ is a subfield of $F(X \times X')$, we have

$$\deg \mathrm{CH}_0\big(Y_{F(X)}\big) \subset \deg \mathrm{CH}_0\big(Y_{F(X\times X')}\big).$$

As the quadric $X_{F(X')}$ is isotropic, the field extension $F(X \times X')/F(X')$ is purely transcendental. Hence

$$\deg \mathrm{CH}_0\big(Y_{F(X\times X')}\big) = \deg \mathrm{CH}_0\big(Y_{F(X')}\big)$$

by Lemma 75.2. □

Let X and Y be varieties over F with $\dim X = d$ and let $Z \subset X \times Y$ be a prime d-dimensional cycle of multiplicity $r > 0$. The generic point of Z defines a degree r closed point of the generic fiber $Y_{F(X)}$ of the projection $X \times Y \to X$ and vice versa. Hence there is a natural bijection of the following two sets for every $r > 0$:

(1) prime d-dimensional cycles on $X \times Y$ of multiplicity r,
(2) closed points of $Y_{F(X)}$ of degree r.

A rational morphism $X \dashrightarrow Y$ defines a prime correspondence $X \rightsquigarrow Y$ of multiplicity 1 by taking the closure of its graph. Conversely, a prime cycle $Z \subset X \times Y$ of multiplicity 1 is birationally isomorphic to X. Therefore, the projection to Y defines a rational map $X \dashrightarrow Z \to Y$. Hence there are natural bijections between the sets of:

(0) rational morphisms $X \dashrightarrow Y$,
(1) prime d-dimensional cycles on $X \times Y$ of multiplicity 1,
(2) rational points of $Y_{F(X)}$.

A prime correspondence $X \rightsquigarrow Y$ of multiplicity r can be viewed as a "generically r-valued map" between X and Y.

Let $\alpha : X \rightsquigarrow Y$ be a correspondence between varieties of dimension d. We write $\alpha^t : Y \rightsquigarrow X$ for the transpose of α (cf. §62).

Theorem 75.4. *Let X be an anisotropic smooth projective quadric with $\mathfrak{i}_1(X) = 1$. Let $\delta : X \rightsquigarrow X$ be a correspondence. Then* $\mathrm{mult}(\delta) \equiv \mathrm{mult}(\delta^t) \pmod 2$.

PROOF. The coefficient of $h^0 \times l_0$ in the decomposition of the class of δ in the modulo 2 reduced Chow group $\overline{\mathrm{Ch}}(X^2)$ is $\mathrm{mult}(\delta) \pmod 2$ (taking into account Lemma 73.2 in the case when $\dim X = 0$). Therefore, the theorem is a particular case of Corollary 73.21. (It is also a particular case of Lemma 73.19 and also of Lemma 73.15.)

We give another proof of Theorem 75.4. By Example 66.6, we have

$$\mathrm{CH}_d(X^2) \simeq \mathrm{CH}_d(X) \oplus \mathrm{CH}_{d-1}(\mathrm{Fl}) \oplus \mathrm{CH}_0(X)\ ,$$

where Fl is the flag variety of pairs (L, P), where L and P are a totally isotropic line and plane, respectively, satisfying $L \subset P$. It suffices to check the formula of Theorem 75.4 for δ lying in the image of any of these three summands.

Since the embedding $\mathrm{CH}_d(X) \hookrightarrow \mathrm{CH}_d(X^2)$ is given by the push-forward with respect to the diagonal map, its image is generated by the diagonal class for which the congruence clearly holds.

Since X is anisotropic, every element of $\mathrm{CH}_0(X)$ becomes divisible by 2 over an extension of F by Theorem 71.2 and Proposition 68.1. As multiplicity is not changed under a field extension homomorphism, we have $\mathrm{mult}(\delta) \equiv 0 \equiv \mathrm{mult}(\delta^t) \pmod 2$ for any δ in the image of $\mathrm{CH}_0(X)$.

Since the embedding $\mathrm{CH}_{d-1}(\mathrm{Fl}) \hookrightarrow \mathrm{CH}_d(X^2)$ is produced by a correspondence $\mathrm{Fl} \rightsquigarrow X^2$ of degree one, the image of $\mathrm{CH}_{d-1}(\mathrm{Fl})$ is contained in the image of the push-forward $\mathrm{CH}_d(\mathrm{Fl} \times X^2) \to \mathrm{CH}_d(X^2)$ with respect to the projection. Let $\delta \in \mathrm{CH}_d(X^2)$. By Lemma 75.1, the multiplicity of δ and of δ^t is the degree of the image of δ under the pull-back homomorphism $\mathrm{CH}_d(X^2) \to \mathrm{CH}_0\big(X_{F(X)}\big)$, given by the generic point of the appropriately chosen factor of X^2. As $\mathfrak{i}_1(X) = 1$, the degree of any closed point on $(\mathrm{Fl} \times X)_{F(X)}$ is even by Corollary 71.3. Consequently $\mathrm{mult}(\delta) \equiv 0 \equiv \mathrm{mult}(\delta^t) \pmod 2$ for any δ in the image of $\mathrm{CH}_{d-1}(\mathrm{Fl})$. □

Remark 75.5. The statement of Theorem 75.4 with X replaced by an anisotropic *nonsmooth* quadric (over a field of characteristic 2) was proved by Totaro in [**132**].

Corollary 75.6. *Let X be as in Theorem 75.4. Then any rational endomorphism $f : X \dashrightarrow X$ is dominant. In particular, the only point $x \in X$ admitting an F-embedding $F(x) \hookrightarrow F(X)$ is the generic point of X.*

PROOF. Let $\delta : X \rightsquigarrow X$ be the class of the closure of the graph of f. Then $\mathrm{mult}(\delta) = 1$. Therefore, the integer $\mathrm{mult}(\delta^t)$ is odd by Theorem 75.4. In particular, $\mathrm{mult}(\delta^t) \neq 0$; i.e., f is dominant. □

76. The main theorem

The main theorem of the chapter is

Theorem 76.1. *Let X be an anisotropic smooth projective quadric over F and let Y be a complete variety over F such that every closed point of Y is of even degree. If there is a closed point in $Y_{F(X)}$ of odd degree, then*

(1) $\dim_{\mathrm{Izh}} X \leq \dim Y$.
(2) *If* $\dim_{\mathrm{Izh}} X = \dim Y$, *then X is isotropic over $F(Y)$.*

PROOF. A closed point of Y over $F(X)$ of odd degree gives rise to a prime correspondence $\alpha : X \rightsquigarrow Y$ of odd multiplicity. By Springer's Theorem 71.3, to prove statement (2), it suffices to find a closed point of $X_{F(Y)}$ of odd degree, and equivalently, to find a correspondence $Y \rightsquigarrow X$ of odd multiplicity.

First assume that $\mathfrak{i}_1(X) = 1$, so $\dim_{\mathrm{Izh}} X = \dim X$. In this special case, we simultaneously prove both statements of Theorem 76.1 by induction on $n = \dim X + \dim Y$.

If $n = 0$, i.e., X and Y are both of dimension zero, then $X = \operatorname{Spec} K$ and $Y = \operatorname{Spec} L$ for some field extensions K and L of F with $[K : F] = 2$ and $[L : F]$ even. Taking the push-forward to $\operatorname{Spec}(F)$ of the correspondence α, we have

$$[K : F] \cdot \mathrm{mult}(\alpha) = [L : F] \cdot \mathrm{mult}(\alpha^t).$$

Since $\mathrm{mult}(\alpha)$ is odd, the correspondence $\alpha^t : Y \rightsquigarrow X$ is of odd multiplicity as needed.

So we may assume that $n > 0$. Let d be the dimension of X. We first prove (2), so we have $\dim Y = d > 0$. It suffices to show that $\mathrm{mult}(\alpha^t)$ is odd. Assume that the multiplicity of α^t is even. Let $x \in X$ be a closed point of degree 2. Since the multiplicity of the correspondence $[Y \times x] : Y \rightsquigarrow X$ is 2 and the multiplicity of $[x \times Y] : X \rightsquigarrow Y$ is zero, modifying α by adding an appropriate multiple of $[x \times Y]$, we can assume that $\mathrm{mult}(\alpha)$ is odd and $\mathrm{mult}(\alpha^t) = 0$.

The degree of the pull-back of α^t on $X_{F(Y)}$ is now zero by Lemma 75.1. By Corollary 71.4, the degree homomorphism

$$\deg : \mathrm{CH}_0\big(X_{F(Y)}\big) \to \mathbb{Z}$$

is injective. Therefore, by Proposition 52.9, there is a nonempty open subset $U \subset Y$ such that the restriction of α on $X \times U$ is trivial. Write Y' for the reduced scheme $Y \setminus U$, and let $i : X \times Y' \to X \times Y$ and $j : X \times U \to X \times Y$ denote the closed and open embeddings respectively. The sequence

$$\mathrm{CH}_d(X \times Y') \xrightarrow{i_*} \mathrm{CH}_d(X \times Y) \xrightarrow{j^*} \mathrm{CH}_d(X \times U)$$

is exact by Proposition 57.9. Hence there exists an $\alpha' \in \mathrm{CH}_d(X \times Y')$ such that $i_*(\alpha') = \alpha$. We can view α' as a correspondence $X \rightsquigarrow Y'$. Clearly, $\mathrm{mult}(\alpha') = \mathrm{mult}(\alpha)$, hence $\mathrm{mult}(\alpha')$ is odd. Since α' is a linear combination of prime correspondences, there exists a prime correspondence $\beta : X \rightsquigarrow Y'$ of odd multiplicity. The class β is represented by a prime cycle, hence we may assume that Y' is irreducible. Since $\dim Y' < \dim Y = \dim X = \dim_{\mathrm{Izh}} X$, we contradict statement (1) for the varieties X and Y' that holds by the induction hypothesis.

We now prove (1) when $\mathfrak{i}_1(X) = 1$. Assume that $\dim Y < \dim X$. Let $Z \subset X \times Y$ be a prime cycle representing α. Since $\mathrm{mult}(\alpha)$ is odd, the projection $Z \to X$ is surjective and the field extension $F(X) \hookrightarrow F(Z)$ is of odd degree. The restriction of the projection $X \times Y \to Y$ defines a proper morphism $Z \to Y$. Replacing Y by the image of this morphism, we assume that $Z \to Y$ is a surjection.

In view of Lemma 74.4, extending scalars to a purely transcendental extension of F, we can find a smooth subquadric X' of X of the same dimension as Y having $\mathfrak{i}_1(X') = 1$. By Lemma 75.2, all closed points on Y are still of even degree. Since purely transcendental extensions do not change Witt indices by Lemma 7.15, we still have $\mathfrak{i}_1(X) = 1$.

By Corollary 75.3, there exists a correspondence $X' \rightsquigarrow Y$ of odd multiplicity. Since $\dim X' < \dim X$, by the induction hypothesis, statement (2) holds for X' and Y, that is, X' has a point over Y, i.e., there exists a rational morphism $Y \dashrightarrow X'$. Composing this morphism with the embedding of X' into X, we get a rational morphism $f : Y \dashrightarrow X$.

Consider the rational morphism

$$h := \mathrm{id}_X \times f : X \times Y \dashrightarrow X \times X.$$

As the projection of Z to Y is surjective, Z intersects the domain of the definition of h. Let Z' be the closure of the image of Z under h. The composition of $Z \dashrightarrow Z'$ with the first projection to X yields a tower of field extensions $F(X) \subset F(Z') \subset F(Z)$. As $\big[F(Z) : F(X)\big]$ is odd, so is $\big[F(Z') : F(X)\big]$, i.e., the correspondence $\beta : X \rightsquigarrow X$ given by Z' is of odd multiplicity. The image of the second projection $Z' \to X$ is contained in X', hence $\mathrm{mult}(\beta^t) = 0$ as $\dim X' < \dim X$. This contradicts Theorem 75.4 and establishes Theorem 76.1 in the case $\mathfrak{i}_1(X) = 1$.

We now consider the general case. Let X' be a smooth subquadric of X with $\dim X' = \dim_{\mathrm{Izh}} X$. Then $\mathfrak{i}_1(X') = 1$ by Corollary 74.3, i.e., $\dim_{\mathrm{Izh}} X' = \dim_{\mathrm{Izh}} X$. By Corollary 75.3, the scheme $Y_{F(X')}$ has a closed point of odd degree since $Y_{F(X)}$ does. As $\mathfrak{i}_1(X') = 1$, we have shown in the first part of the proof that the statements (1) and (2) hold for X' and Y. In particular, $\dim_{\mathrm{Izh}} X = \dim_{\mathrm{Izh}} X' \leq \dim Y$ by (1) for X' and Y proving (1) for X and Y. If $\dim X' = \dim Y$, it follows from

(2) applied to X' and Y that X' is isotropic over $F(Y)$. Hence X is isotropic over $F(Y)$ proving (2) for X and Y. □

Remark 76.2. The statement of Theorem 76.1 with X replaced by an anisotropic *nonsmooth* quadric (over a field of characteristic 2) was proved by Totaro in [**132**].

A consequence of Theorem 76.1 is that an anisotropic smooth quadric X cannot be compressed to a variety Y of dimension smaller than $\dim_{\mathrm{Izh}} X$ with all closed points of even degree:

Corollary 76.3. *Let X be an anisotropic smooth projective F-quadric and let Y be a complete F-variety with all closed points of even degree. If $\dim_{\mathrm{Izh}} X > \dim Y$, then there are no rational morphisms $X \dashrightarrow Y$.*

Remark 76.4. Let X and Y be as in part (2) of Theorem 76.1. Suppose in addition that $\dim X = \dim_{\mathrm{Izh}} X$, i.e., $\mathfrak{i}_1(X) = 1$. Let $\alpha : X \rightsquigarrow Y$ be a correspondence of odd multiplicity. The proof of Theorem 76.1 shows $\operatorname{mult}(\alpha^t)$ is also odd.

Applying Theorem 76.1 to the special (but perhaps the most interesting) case where the variety Y is also a projective quadric, we prove the conjectures of Izhboldin:

Theorem 76.5. *Let X and Y be anisotropic smooth projective quadrics over F. Suppose that Y is isotropic over $F(X)$. Then*

(1) $\dim_{\mathrm{Izh}} X \le \dim_{\mathrm{Izh}} Y$.
(2) $\dim_{\mathrm{Izh}} X = \dim_{\mathrm{Izh}} Y$ *if and only if X is isotropic over $F(Y)$.*

PROOF. Choose a subquadric $Y' \subset Y$ with $\dim Y' = \dim_{\mathrm{Izh}} Y$. Since Y' becomes isotropic over $F(Y)$ by Lemma 74.1(2) and Y becomes isotropic over $F(X)$, the quadric Y' becomes isotropic over $F(X)$. By Theorem 76.1, we have $\dim_{\mathrm{Izh}} X \le \dim Y'$. Moreover, in the case of equality, X becomes isotropic over $F(Y')$ and hence over $F(Y)$. Conversely, if X is isotropic over $F(Y)$, interchanging the roles of X and Y, the argument above also yields $\dim_{\mathrm{Izh}} Y \le \dim_{\mathrm{Izh}} X$, hence equality holds. □

We have the following upper bound for the Witt index of Y over $F(X)$.

Corollary 76.6. *Let X and Y be anisotropic smooth projective quadrics over F. Suppose that Y is isotropic over $F(X)$. Then*

$$\mathfrak{i}_0(Y_{F(X)}) - \mathfrak{i}_1(Y) \le \dim_{\mathrm{Izh}} Y - \dim_{\mathrm{Izh}} X.$$

PROOF. If $\dim_{\mathrm{Izh}} X = 0$, the statement is trivial. Otherwise, let Y' be a smooth subquadric of Y of dimension $\dim_{\mathrm{Izh}} X - 1$. Since $\dim_{\mathrm{Izh}} Y' \le \dim Y' < \dim_{\mathrm{Izh}} X$, the quadric Y' remains anisotropic over $F(X)$ by Theorem 76.5(1). Therefore, $\mathfrak{i}_0(Y_{F(X)}) \le \operatorname{codim}_Y Y' = \dim Y - \dim_{\mathrm{Izh}} X + 1$ by Lemma 74.1, hence the inequality holds. □

We also have the following more precise version of Theorem 76.1:

Corollary 76.7. *Let X be an anisotropic smooth projective F-quadric and let Y be a complete variety over F such that every closed point of Y is of even degree. If there is a closed point in $Y_{F(X)}$ of odd degree, then there exists a closed subvariety $Y' \subset Y$ such that*

(1) $\dim Y' = \dim_{\mathrm{Izh}} X$.

(2) $Y'_{F(X)}$ *possesses a closed point of odd degree.*
(3) $X_{F(Y')}$ *is isotropic.*

PROOF. Let $X' \subset X$ be a smooth subquadric with $\dim X' = \dim_{\mathrm{Izh}} X$. Then $\dim_{\mathrm{Izh}} X' = \dim X'$ by Corollary 74.3. An odd degree closed point on $Y_{F(X)}$ determines a correspondence $X \rightsquigarrow Y$ of odd multiplicity which in turn gives a correspondence $X' \rightsquigarrow Y$ of odd multiplicity. We may assume that the latter correspondence is prime and take a prime cycle $Z \subset X' \times Y$ representing it. Let Y' be the image of the proper morphism $Z \to Y$. Clearly, $\dim Y' \leq \dim Z = \dim X' = \dim_{\mathrm{Izh}} X$. On the other hand, Z determines a correspondence $X' \rightsquigarrow Y'$ of odd multiplicity. Therefore $\dim Y' \geq \dim X'$ by Theorem 76.1, and condition (1) of Corollary 76.7 is satisfied. Moreover, $Y'_{F(X')}$ has a closed point of odd degree. Since the field $F(X \times X')$ is a purely transcendental extension of $F(X)$, Lemma 75.2 shows that there is a closed point on $Y'_{F(X)}$ of odd degree, i.e., condition (2) of Corollary 76.7 is satisfied. Finally, the quadric $X'_{F(Y')}$ is isotropic by Theorem 76.1; therefore $X_{F(Y')}$ is isotropic. □

77. Addendum: The Pythagoras number

Given a field F, its *pythagoras number* is defined to be

$$p(F) := \min\{n \mid D(n\langle 1\rangle) = D(\infty\langle 1\rangle)\}$$

or infinity if no such integer exists. If $\operatorname{char} F = 2$, then $p(F) = 1$ and if $\operatorname{char} F \neq 2$, then $p(F) = 1$ if and only if F is pythagorean. Let F be a field that is not formally real. Then the quadratic form $(s(F)+1)\langle 1\rangle$ is isotropic. In particular, $p(F) = s(F)$ or $s(F) + 1$ and each value is possible. So this invariant is only interesting when the field is formally real. For a given formally real field determining its pythagoras number is not easy. If F is an extension of a real closed field of transcendence degree n, then $p(F) \leq 2^n$ by Corollary 35.15. In particular, if $n = 1$ and F is not pythagorean then $p(F) = 2$. It is known that $p\big(\mathbb{R}(t_1, t_2)\big) = 4$ (cf. [**23**]), but in general, the value of $p\big(\mathbb{R}(t_1, \dots, t_n)\big)$ is not known. In this section, given any nonnegative integer n, we construct a formally real field having pythagoras number n. This was first done by Hoffmann in [**56**].

Lemma 77.1. *Let F be a formally real field and let φ be a quadratic form over F. If $P \in \mathfrak{X}(F)$, then P extends to an ordering on $F(\varphi)$ if and only if φ is indefinite at P, i.e., $|\operatorname{sgn}_P(\varphi)| < \dim \varphi$.*

PROOF. Suppose that φ is indefinite at P. Let F_P be the real closure of F with respect to P. Let $K = F_P(\varphi)$. As φ_{F_P} is isotropic, K/F_P is purely transcendental. Therefore the unique ordering on F_P extends to K. The restriction of this extension to $F(\varphi)$ extends P. The converse is clear. □

The following proposition is a consequence of the lemma and Theorem 76.5.

Proposition 77.2. *Let F be formally real and let $x, y \in D(\infty\langle 1\rangle)$. Let $\varphi \simeq m\langle 1\rangle \perp \langle -x\rangle$ and $\psi \simeq n\langle 1\rangle \perp \langle -y\rangle$ with $n > m \geq 0$. Then $F(\psi)$ is formally real. If, in addition, φ is anisotropic, then so is $\varphi_{F(\psi)}$.*

PROOF. As ψ is indefinite at every ordering, every ordering of F extends to $F(\psi)$. In particular, $F(\psi)$ is formally real. Suppose that φ is anisotropic. Since over each real closure of F both φ and ψ have Witt index 1, the first Witt index of φ and ψ must also be 1. As $\dim \varphi > \dim \psi$, the form $\varphi_{F(\psi)}$ is anisotropic by Theorem 76.5. □

Let F_0 be a formally real field. Let $F_1 = F_0(t_1, \dots, t_{n-1})$ and let $x = 1 + t_1^2 + \cdots + t_{n-1}^2 \in D(\infty\langle 1\rangle)$. By Corollary 17.13, the element x is a sum of n squares in F_1 but no fewer. In particular, $\varphi \simeq (n-1)\langle 1\rangle \perp \langle -x\rangle$ is anisotropic over F_1. For $i \geq 1$, inductively define F_{i+1} as follows:

Let

$$\mathfrak{A}_i := \{n\langle 1\rangle \perp \langle -y\rangle \mid y \in D(\infty\langle 1\rangle_{F_i})\}.$$

For any finite subset $S \subset \mathfrak{A}_i$, let X_S be the product of quadrics X_φ for all $\varphi \in S$. If $S \subset T$ are two subsets of $\mathfrak{A}_i$, we have the dominant projection $X_T \to X_S$ and therefore the inclusion of function fields $F(X_S) \to F(X_T)$. Set $F_{i+1} = \operatorname{colim} F_S$ over all finite subsets $S \subset \mathfrak{A}_i$. By construction, all quadratic forms $\varphi \in \mathfrak{A}_i$ are isotropic over the field extension F_i of F. Let $F = \bigcup F_i$. Then F has the following properties:

(1) F is formally real.
(2) $n\langle 1\rangle \perp \langle -y\rangle$ is isotropic for all $0 \neq y \in \sum (F^\times)^2$.

Consequently, $D(\infty\langle 1\rangle_F) \subset D(n\langle 1\rangle_F)$, so the pythagoras number $p(F) \leq n$. As $\varphi \simeq (n-1)\langle 1\rangle \perp \langle -x\rangle$ remains anisotropic over F, we have $p(F) \geq n$. So we have proved:

Theorem 77.3. *For every $n \geq 1$ there exists a formally real field F with $p(F) = n$.*

CHAPTER XV

Application of Steenrod Operations

Since Steenrod operations are not available in characteristic 2, throughout this chapter, the characteristic of the base field is assumed to be different from 2.

We write $v_2(n)$ for the 2-adic exponent of an integer n.

We shall use the notation of Chapter XIII. In particular, X is a smooth D-dimensional projective quadric over a field F given by a (nondegenerate) quadratic form φ, and $d = [D/2]$.

78. Computation of Steenrod operations

Recall that $h \in \mathrm{Ch}^1(X)$ is the class of a hyperplane section.

Lemma 78.1. *The modulo 2 total Chern class $c(T_X) : \mathrm{Ch}(X) \to \mathrm{Ch}(X)$ of the tangent vector bundle T_X of the quadric X is multiplication by $(1+h)^{D+2}$.*

PROOF. By Proposition 58.15, it suffices to show that $c(T_X)\big([X]\big) = (1+h)^{D+2}$. Let $i : X \hookrightarrow \mathbb{P}$ be the closed embedding of X into the $(D+1)$-dimensional projective space $\mathbb{P} = \mathbb{P}(V)$, where V is the underlying vector space of φ. We write $H \in \mathrm{Ch}^1(\mathbb{P})$ for the class of a hyperplane, so $h = i^*(H)$. Since X is a hypersurface in $\mathbb{P}$ of degree 2, the normal bundle N of the embedding i is isomorphic to $i^*(L^{\otimes 2})$, where L is the canonical line bundle over $\mathbb{P}$ (cf. §104.B). By Propositions 104.16 and 54.7, we have $c(T_X) \circ c\big(i^*(L^{\otimes 2})\big) = c(i^* T_{\mathbb{P}})$. By Example 61.16, we know that $c(T_{\mathbb{P}})$ is the multiplication by $(1+H)^{D+2}$ and by Propositions 54.3 and 57.26, that $c(L^{\otimes 2}) = \mathrm{id}$ modulo 2. It follows that

$$\begin{aligned} c(T_X)([X]) &= \big(c(i^*T_{\mathbb{P}}) \circ c\big(i^*(L^{\otimes 2})\big)^{-1}\big)\big(i^*([\mathbb{P}])\big) \\ &= \big(i^* \circ c(T_{\mathbb{P}}) \circ c(L^{\otimes 2})^{-1}\big)\big([\mathbb{P}]\big) = i^*(1+H)^{D+2} = (1+h)^{D+2} \end{aligned}$$

by Proposition 55.22. □

Corollary 78.2. *Suppose that $\mathfrak{i}_0(X) > n$ for some $n \geq 0$. Let $W \subset V$ be a totally isotropic $(n+1)$-dimensional subspace of V and let $\mathbb{P}$ be the n-dimensional projective space $\mathbb{P}(W)$. Let $i : \mathbb{P} \hookrightarrow X$ be the closed embedding. Then the modulo 2 total Chern class $c(N) : \mathrm{Ch}(\mathbb{P}) \to \mathrm{Ch}(\mathbb{P})$ of the normal bundle N of the embedding i is multiplication by $(1+H)^{D+1-n}$, where $H \in \mathrm{Ch}^1(\mathbb{P})$ is the class of a hyperplane.*

PROOF. By Propositions 104.16 and 54.7, we have $c(N) = c(T_{\mathbb{P}})^{-1} \circ c(i^* T_X)$; and by Proposition 55.22 and Lemma 78.1, we have

$$c\big(i^*T_X[\mathbb{P}]\big) = c(i^*T_X)(i^*[X]) = \big(i^* \circ c(T_X)\big)([X]) = i^*(1+h)^{D+2} = (1+H)^{D+2}.$$

Hence by Example 61.16, $c(T_{\mathbb{P}}) = (1+H)^{n+1}$. □

Corollary 78.3. *Under the hypothesis of Corollary 78.2, we have*

$$\mathrm{Sq}_X\big([\mathbb{P}]\big) = [\mathbb{P}] \cdot (1+h)^{D+1-n}.$$

PROOF. By the Wu Formula (Proposition 61.8), $\mathrm{Sq}_X([\mathbb{P}]) = i_*(c(N)[\mathbb{P}])$. By Corollary 78.2, we get

$$i_*(c(N)[\mathbb{P}]) = i_*((1+H)^{D+1-n}\cdot[\mathbb{P}]) = i_*(i^*(1+h)^{D+1-n}\cdot[\mathbb{P}]) = (1+h)^{D+1-n}\cdot i_*[\mathbb{P}]$$

using the Projection Formula (Proposition 56.9). □

We also have (cf. Example 61.16):

Lemma 78.4. $\mathrm{Sq}_X(h^i) = h^i \cdot (1+h)^i$ *for any* $i \geq 0$.

Corollary 78.5. *If the quadric X is split, then the ring endomorphism* $\mathrm{Sq}_X : \mathrm{Ch}(X) \to \mathrm{Ch}(X)$ *acts on the basis* $\{h^i, l_i\}_{i\in[0,\ d]}$ *of* $\mathrm{Ch}(X)$ *by the formulas*

$$\mathrm{Sq}_X(h^i) = h^i \cdot (1+h)^i \quad \text{and} \quad \mathrm{Sq}_X(l_i) = l_i \cdot (1+h)^{D+1-i}.$$

In particular, for any $j \geq 0$,

$$\mathrm{Sq}_X^j(h^i) = \binom{i}{j} h^{i+j} \quad \text{and} \quad \mathrm{Sq}_X^j(l_i) = \binom{D+1-i}{j} l_{i-j}.$$

Binomial coefficients modulo 2 are computed as follows (we leave the proof to the reader): Let $\mathbb{N}$ be the set of nonnegative integers, let $2^{\mathbb{N}}$ be the set of all subsets of $\mathbb{N}$, and let $\pi : \mathbb{N} \to 2^{\mathbb{N}}$ be the injection given by base 2 expansions. For any $n \in \mathbb{N}$, the set $\pi(n)$ consists of all those $m \in \mathbb{N}$ such that the base 2 expansion of n has 1 in the mth position. For two arbitrary nonnegative integers i and n, write $i \subset n$ if $\pi(i) \subset \pi(n)$.

Lemma 78.6. *For any* $i, n \in \mathbb{N}$, *the binomial coefficient* $\binom{n}{i}$ *is odd if and only if* $i \subset n$.

79. Values of the first Witt index

The main result of this section is Theorem 79.9 (conjectured by Hoffmann and originally proved in [**72**]); its main ingredient is given by Proposition 79.4. We begin with some observations.

Remark 79.1. By Theorem 61.9,

$$\begin{array}{ccc} \mathrm{Ch}(X^*) & \xrightarrow{\mathrm{Sq}_{X^*}} & \mathrm{Ch}(X^*) \\ \downarrow & & \downarrow \\ \mathrm{Ch}(\bar{X}^*) & \xrightarrow{\mathrm{Sq}_{\bar{X}^*}} & \mathrm{Ch}(\bar{X}^*) \end{array}$$

is commutative, hence we get an endomorphism $\overline{\mathrm{Ch}}(X^*) \to \overline{\mathrm{Ch}}(X^*)$ that we shall also call a Steenrod operation and denote it by Sq_{X^*}, even though it is a restriction of $\mathrm{Sq}_{\bar{X}^*}$ and not of Sq_{X^*}.

Remark 79.2. Let $l_n \times h^m \in \mathrm{Ch}(\bar{X}^2)$ be an essential basis element with $n \geq m$. Since $\mathrm{Sq}(l_n \times h^m) = \mathrm{Sq}(l_n) \times \mathrm{Sq}(h^m)$ by Theorem 61.14, we see by Corollary 78.5, that the value of $\mathrm{Sq}(l_n \times h^m)$ is a linear combination of the elements $l_i \times h^j$ with $i \leq n$ and $j \geq m$. If $m = 0$, one can say more: $\mathrm{Sq}(l_n \times h^0)$ is a linear combination of the elements $l_i \times h^0$ with $i \leq n$.

Of course, we have similar facts for the essential basis elements of type $h^m \times l_n$.

Representing essential basis elements of type $l_n \times h^m$ with $n \geq m$ as points of the right pyramid of Remark 73.8, we may interpret the above statements graphically as follows: the diagram of the value of the Steenrod operation on a point $l_n \times h^m$ is contained in the isosceles triangle based on the lower row of the pyramid whose top is the point $l_n \times h^m$ (an example of this is the picture on the left below). If $l_n \times h^m$ is on the right side of the pyramid, then the diagram of the value of the Steenrod operation is contained in the part of the right side of the pyramid, which is below the point (an example of this is the picture on the right below).

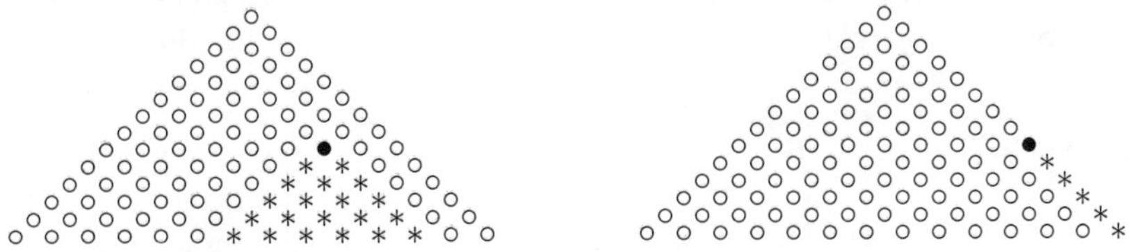

The next statement follows simply from Remark 79.2.

Lemma 79.3. *Assume that X is anisotropic. Let $\pi \in \mathrm{Ch}_{D+\mathfrak{i}_1-1}(X^2)$ be the 1-primordial cycle. For any $j \geq 1$, the element $S^j_{X^2}(\pi)$ has no points in the first shell triangle.*

PROOF. By the definition of π, the only point the cycle π has in the first (left as well as right) shell triangle is the top of the triangle. By Remark 79.2, the only point in the first left shell triangle that may be contained in $S^j(\pi)$ is the point on the left side of the triangle and the only point in the first right shell triangle that may be contained in $S^j(\pi)$ is the point on the right side of the triangle. Since these two points are not dual (points on the left side of the first left shell triangle are dual to points on the *left* side of the first right shell triangle), the lemma follows by Corollary 73.21. □

We shall obtain further information in Lemma 83.1 below. We write $\exp_2(a)$ for 2^a.

Proposition 79.4. *For any anisotropic quadratic form φ of $\dim \varphi \geq 2$,*

$$\mathfrak{i}_1(\varphi) \leq \exp_2 v_2\big(\dim \varphi - \mathfrak{i}_1(\varphi)\big).$$

PROOF. Let $r = v_2(\dim \varphi - \mathfrak{i}_1(\varphi))$. Apply the Steenrod operation $\mathrm{Sq}^{2^r}_{X^2} : \overline{\mathrm{Ch}}(X^2) \to \overline{\mathrm{Ch}}(X^2)$ to the 1-primordial cycle π. Since

$$\mathrm{Sq}^{2^r}_{X^2}(h^0 \times l_{\mathfrak{i}_1-1}) = h^0 \times \mathrm{Sq}^{2^r}_{X}(l_{\mathfrak{i}_1-1}) = \binom{\dim \varphi - \mathfrak{i}_1}{2^r} \cdot (h^0 \times l_{\mathfrak{i}_1-1-2^r})$$

by Theorem 61.14 and Corollary 78.5 and the binomial coefficient is odd by Lemma 78.6, we have $h^0 \times l_{\mathfrak{i}_1-1-2^r} \in \mathrm{Sq}^{2^r}_{X^2}(\alpha)$. It follows by Lemma 79.3 that $2^r \notin [1,\ \mathfrak{i}_1)$, i.e., that $2^r \geq \mathfrak{i}_1$. □

Remark 79.5. Let a be a positive integer written in base 2. A *suffix of a* is an integer written in base 2 that is obtained from a by deleting several (and at least one) consecutive digits starting from the left one. For example, all suffixes of 1011010 are 11010, 1010, 10 and 0.

Let $i < n$ be two nonnegative integers. Then the following are equivalent.

(1) $i \leq \exp_2 v_2(n-i)$.

(2) There exists an $r \geq 0$ satisfying $2^r < n$, $\quad i \equiv n \pmod{2^r}$, and $i \in [1,\ 2^r]$.

(3) $i-1$ is the remainder upon division of $n-1$ by an appropriate 2-power.

(4) The 2-adic expansion of $i-1$ is a suffix of the 2-adic expansion of $n-1$.

(5) The 2-adic expansion of i is a suffix of the 2-adic expansion of n or i is a 2-power divisor of n.

In particular, the integers $i = \mathfrak{i}_1(\varphi)$ and $n = \dim\varphi$ in Proposition 79.4 satisfy these conditions.

Corollary 79.6. *All relative higher Witt indices of an odd-dimensional quadratic form are odd. Any relative higher Witt index of an even-dimensional quadratic form is either even or 1.*

Example 79.7. Assume that φ is anisotropic and let $s \geq 0$ be the largest integer satisfying $\dim\varphi > 2^s$. Then it follows by Proposition 79.4 that $\mathfrak{i}_1(\varphi) \leq \dim\varphi - 2^s$ (e.g., by Condition (4) of Remark 79.5). In particular, if $\dim\varphi = 2^s + 1$, then $\mathfrak{i}_1(\varphi) = 1$.

The first statement of the following corollary is the Separation Theorem 26.5 (over a field of characteristic not 2); the second statement was originally proved by O. Izhboldin (using a different method) in [**63**, Th. 02]. A characteristic 2 version has been proven by Hoffmann and Laghribi in [**58**, Th. 1.3].

Corollary 79.8. *Let φ and ψ be two anisotropic quadratic forms over F.*

(1) *If $\dim\psi \leq 2^s < \dim\varphi$ for some $s \geq 0$, then the form $\psi_{F(\varphi)}$ is anisotropic.*
(2) *Suppose that $\dim\psi = 2^s + 1 \leq \dim\varphi$ for some $s \geq 0$. If the form $\psi_{F(\varphi)}$ is isotropic, then the form $\varphi_{F(\psi)}$ is also isotropic.*

PROOF. Let X and Y be the quadrics of φ and of ψ respectively. Then $\dim_{\mathrm{Izh}} X \geq 2^s - 1$ by Example 79.7. If $\dim\psi \leq 2^s$, then $\dim Y \leq 2^s - 2$. Therefore,

$$\dim_{\mathrm{Izh}} Y \leq \dim Y < 2^s - 1 \leq \dim_{\mathrm{Izh}} X$$

and $Y_{F(X)}$ is anisotropic by Theorem 76.5(1).

Suppose that $\dim\psi = 2^s + 1$. Then $\dim_{\mathrm{Izh}} Y = 2^s - 1 \leq \dim_{\mathrm{Izh}} X$. If $Y_{F(X)}$ is isotropic, then $\dim_{\mathrm{Izh}} Y = \dim_{\mathrm{Izh}} X$ by Theorem 76.5(1) and therefore $X_{F(Y)}$ is isotropic by Theorem 76.5(2). □

We show next that all values of the first Witt index not forbidden by Proposition 79.4 are possible by establishing the main result of this section:

Theorem 79.9. *Two nonnegative integers i and n satisfy $i \leq \exp_2 v_2(n-i)$ if and only if there exists an anisotropic quadratic form φ over a field of characteristic not 2 with*

$$\dim\varphi = n \quad \textit{and} \quad \mathfrak{i}_1(\varphi) = i.$$

PROOF. By Proposition 79.4 and Remark 79.5, we need only prove the necessity. Let i and n be two nonnegative integers satisfying $i \leq \exp_2 v_2(n-i)$. Let r be as in condition (2) of Remark 79.5. Write $n - i = 2^r \cdot m$ for some integer m.

Let F be any field of characteristic not 2 and consider the field

$$K = F(t_1, \dots, t_r)$$

of rational functions in r variables. By Corollary 19.6, the Pfister form $\pi = \langle\langle t_1, \dots, t_r \rangle\rangle$ is anisotropic over K. Let $L = K(s_1, \dots, s_m)$, where $s_1, \dots, s_m$ are variables. By Lemma 19.5, the quadratic L-form $\psi = \pi_L \otimes \langle 1, s_1, \dots, s_m \rangle$ is anisotropic.

We claim that $\mathfrak{i}_1(\psi) = 2^r$. Indeed, by Proposition 6.22, we have $\mathfrak{i}_1(\psi) \geq 2^r$. On the other hand, the field $E = L(\sqrt{-s_1})$ is purely transcendental over $K(s_2, \ldots, s_m)$ and therefore $\mathfrak{i}_0(\psi_E) = 2^r$. Consequently, $\mathfrak{i}_1(\psi) = 2^r$.

Let φ be an arbitrary subform of ψ of codimension $2^r - i$. Since $\dim \psi = 2^r \cdot (m+1) = n + (2^r - i)$, the dimension of φ is equal to n. As $2^r - i < 2^r = \mathfrak{i}_1(\psi)$, we have $\mathfrak{i}_1(\varphi) = i$ by Corollary 74.3. □

80. Rost correspondences

Recall that by abuse of notation we also denote the image of the element $h \in \mathrm{CH}^1(X)$ in the groups $\mathrm{CH}^1(\bar{X})$, $\mathrm{Ch}^1(X)$, and $\mathrm{Ch}^1(\bar{X})$ by the same symbol h. In the following lemma, h stands for the element of $\mathrm{Ch}(X)$.

Lemma 80.1. *Let n be the integer satisfying*

$$2^n - 1 \leq D \leq 2^{n+1} - 2.$$

Set $s = D - 2^n + 1$ and $r = 2^{n+1} - 2 - D$ (observe that $r + s = 2^n - 1$). If $\alpha \in \mathrm{Ch}_{r+s}(X)$, then

$$\mathrm{Sq}^X_{r+s}(\alpha) = h^r \cdot \alpha^2 \in \mathrm{Ch}_0(X).$$

Proof. By the definition of the cohomological Steenrod operation Sq_X, we have $\mathrm{Sq}_X = c(T_X) \circ \mathrm{Sq}^X$, where Sq^X is the homological Steenrod operation. Therefore, $\mathrm{Sq}^X = c(-T_X) \circ \mathrm{Sq}_X$. In particular,

$$\mathrm{Sq}^X_{r+s}(\alpha) = \sum_{i=0}^{r+s} c_i(-T_X) \circ \mathrm{Sq}_X^{r+s-i}(\alpha)$$

in $\mathrm{Ch}_0(X)$. By Lemma 78.1, we have $c_i(-T_X) = \binom{-D-2}{i} \cdot h^i$. As $\binom{-D-2}{i} = \pm\binom{D+i+1}{i}$, it follows from Lemma 78.6, that the latter binomial coefficient is even for any $i \in [r+1,\ r+s]$ and is odd for $i = r$. Since $\mathrm{Sq}_X^{r+s-i}(\alpha)$ is equal to 0 for $i \in [0,\ r-1]$ and is equal to α^2 for $i = r$ by Theorem 61.13, the required relation is established. □

Theorem 80.2. *Let n be the integer satisfying*

$$2^n - 1 \leq D \leq 2^{n+1} - 2.$$

Set $s = D - 2^n + 1$ and $r = 2^{n+1} - 2 - D$. Let X and Y be two anisotropic projective quadrics of dimension D over a field of characteristic not 2. Let $\bar{\rho}$ be a cycle in $\overline{\mathrm{Ch}}_{r+s}(X \times Y)$. Then $(pr_X)_*(\bar{\rho}) = 0$ if and only if $(pr_Y)_*(\bar{\rho}) = 0$, where $pr_X : X \times Y \to X$ and $pr_Y : X \times Y \to Y$ are the projections.

Proof. Let ρ be an element of the *nonreduced* Chow group $\mathrm{Ch}_{r+s}(X \times Y)$. Write $\bar{\rho}$ for the image of ρ in $\overline{\mathrm{Ch}}_{r+s}(X \times Y)$. The group $\overline{\mathrm{Ch}}_{r+s}(X)$ is generated by $h^s = h^s_X$ (if $s = d$, this is true as X is anisotropic, hence not split). Therefore, we have $(pr_X)_*(\bar{\rho}) = a_X h^s_X$ for some $a_X \in \mathbb{Z}/2\mathbb{Z}$. Similarly, $(pr_Y)_*(\bar{\rho}) = a_Y h^s_Y$ for some $a_Y \in \mathbb{Z}/2\mathbb{Z}$. To prove Theorem 80.2 we must show $a_X = a_Y$.

As X and Y are anisotropic, any closed point z in X or Y has even degree. In particular, the map

$$\frac{1}{2}\deg_Z : \mathrm{Ch}_0(Z) \to \mathbb{Z}/2\mathbb{Z}$$

given by $z \mapsto \frac{1}{2}\deg(z) \pmod 2$ is well-defined for $Z = X$ or Y. Consider the following diagram:

(80.3)

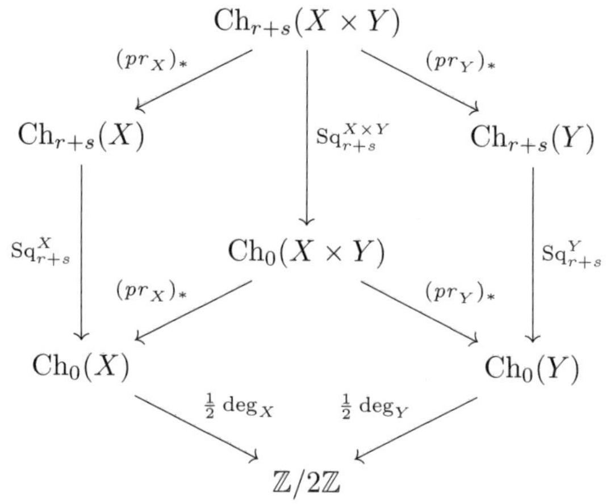

We show that the diagram (80.3) is commutative. The bottom diamond is commutative by the functorial property of the push-forward homomorphism (cf. Proposition 49.9 and Example 57.7). The left and the right parallelograms are commutative by Theorem 60.5. Therefore,

$$(\tfrac{1}{2}\deg_X) \circ \mathrm{Sq}^X_{r+s} \circ (pr_X)_*(\rho) = (\tfrac{1}{2}\deg_Y) \circ \mathrm{Sq}^Y_{r+s} \circ (pr_Y)_*(\rho).$$

Applying Lemma 80.1 to the element $\alpha = (pr_X)_*(\rho)$, we have

$$(\tfrac{1}{2}\deg_X) \circ \mathrm{Sq}^X_{r+s} \circ (pr_X)_*(\rho) = (\tfrac{1}{2}\deg_X)(h^r_X \cdot \alpha^2) = a_X.$$

Similarly $(\frac{1}{2}\deg_Y) \circ \mathrm{Sq}^Y_{r+s} \circ (pr_Y)_*(\rho) = a_Y$, proving the theorem. □

Exercise 80.4. Use Theorem 80.2 to prove the following generalization of Corollary 79.8(2): Let X and Y be two anisotropic projective quadrics satisfying $\dim X = \dim Y = D$. Let s be as in Theorem 80.2. If there exists a rational morphism $X \dashrightarrow Y$, then there exists a rational morphism $G_s(Y) \dashrightarrow G_s(X)$ where for an integer i, $G_i(X)$ is the scheme (variety, if $i \neq D/2$) of i-dimensional linear subspaces lying on X. (We shall study the scheme $G_d(X)$ in Chapter XVI.)

Remark 80.5. One can generalize Theorem 80.2 as follows. We replace Y by an arbitrary projective variety of an arbitrary dimension (and, in fact, Y need not be smooth nor of dimension $D = \dim X$). Suppose that every closed point of Y has even degree. Let $\rho \in \mathrm{Ch}_{r+s}(X \times Y)$ satisfy $(pr_X)_*(\bar\rho) \neq 0 \in \overline{\mathrm{Ch}}(X)$. Then $(pr_Y)_*(\rho) \neq 0 \in \mathrm{Ch}(Y)$ (note that this is in $\mathrm{Ch}(Y)$, not $\overline{\mathrm{Ch}}(Y)$). To prove this generalization, we use the commutative diagram 80.3. As before, we have $\deg_X \circ \mathrm{Sq}^X_{r+s} \circ (pr_X)_*(\rho) \neq 0$ provided that $pr_X(\bar\rho) \neq 0$. Therefore,

$$\deg_Y \circ \mathrm{Sq}^Y_{r+s} \circ (pr_Y)_*(\rho) \neq 0.$$

In particular, $(pr_Y)_*(\rho) \neq 0$.

Exercise 80.6. Show that one cannot replace the conclusion $(pr_Y)_*(\rho) \neq 0 \in \mathrm{Ch}(Y)$ by $(pr_Y)_*(\bar\rho) \neq 0 \in \overline{\mathrm{Ch}}(Y)$ in Remark 80.5. (Hint: Let Y be an anisotropic quadric with X a subquadric of Y satisfying $2\dim X < \dim Y$ and $\rho \in \mathrm{Ch}(X \times Y)$ the class of the diagonal of X.)

Taking $Y = X$ in Theorem 80.2, we have

Corollary 80.7. *Let X be an anisotropic quadric of dimension D and let s be as in Theorem 80.2. If a rational cycle in $\mathrm{Ch}(\bar{X}^2)$ contains $h^s \times l_0$, then it also contains $l_0 \times h^s$.*

Corollary 80.8. *Assume that X is an anisotropic quadric of dimension D and for some integer $i \in [0, d]$ the cycle $h^0 \times l_i + l_i \times h^0 \in \mathrm{Ch}(\bar{X}^2)$ is rational. Then the integer $\dim X - i + 1$ is a power of 2.*

PROOF. If the cycle $h^0 \times l_i + l_i \times h^0$ is rational, then, multiplying by $h^s \times h^i$, we see that the cycle $h^s \times l_0 + l_{i-s} \times h^i$ is also rational. By Corollary 80.7, it follows that $i = s$. Therefore, $\dim X - i + 1 = 2^n$ with n as in Theorem 80.2. □

Remark 80.9. By Lemma 73.12 and Corollary 73.21, the integer i in Corollary 80.8 is necessarily equal to $\mathfrak{i}_1(X) - 1$.

Recalling Definition 73.29, we have

Corollary 80.10. *If $\dim \varphi - \mathfrak{i}_1(\varphi)$ is not a 2-power, then the 1-primordial cycle on X^2 produces an integer.*

PROOF. If the 1-primordial cycle π does not produce any integer, then $\pi = h^0 \times l_{\mathfrak{i}_1 - 1} + l_{\mathfrak{i}_1 - 1} \times h^0$. Therefore, by Corollary 80.8, the integer $D - \big(\mathfrak{i}_1(\varphi) - 1\big) + 1 = \dim \varphi - \mathfrak{i}_1(\varphi)$ is a 2-power. □

The element $h^0 \times l_0 + l_0 \times h^0 \in \mathrm{Ch}(\bar{X}^2)$ is called the *Rost correspondence* of the quadric X.

Of course, the Rost correspondence of an isotropic quadric X is rational. A special case of Corollary 80.8 is given by:

Corollary 80.11. *If X is anisotropic and the Rost correspondence of X is rational, then $D + 1$ is a power of 2.*

By multiplying by $h^1 \times h^0$, we see that rationality of the Rost correspondence implies rationality of the element $h^1 \times l_0$. In fact, rationality of $h^1 \times l_0$ alone implies that $D + 1$ is a power of 2:

Corollary 80.12. *If X is anisotropic and the element $h^1 \times l_0 \in \mathrm{Ch}(\bar{X}^2)$ is rational, then $D + 1$ is a power of 2.*

PROOF. If $h^1 \times l_0$ is rational, then for any $i \geq 1$, the element $h^i \times l_0$ is also rational. Let s be as in Theorem 80.2. By Corollary 80.7 it follows that $s = 0$, i.e., $D = 2^n - 1$. □

Let A be a point of a shell triangle of a quadric. We write $A^\sharp$ for the dual point in the sense of Definition 73.20. The following statement was originally proved (in characteristic 0) by Vishik.

Corollary 80.13. *Let Y be another anisotropic projective quadric of dimension D. Basis elements of $\mathrm{Ch}(\bar{Y}^2)$ are in natural 1-1 correspondence with basis elements of $\mathrm{Ch}(\bar{X}^2)$. Assume that $\mathfrak{i}_1(Y) = s + 1$ with s as in Corollary 80.7. Let A be a point of a first shell triangle of Y and let $A^\sharp$ be its dual point. Then in the diagram of any element of $\overline{\mathrm{Ch}}(X^2)$, the point corresponding to A is marked if and only if the point corresponding to $A^\sharp$ is marked.*

PROOF. We may assume that A lies in the *left* first shell triangle of Y. Let $h^i \times l_j$ (with $0 \le i \le j \le s$) be the basis element represented by A. Then the basis element represented by $A^\sharp$ is $l_{s-i} \times h^{s-j}$. Let $\alpha \in \overline{\mathrm{Ch}}(X^2)$ and assume that α contains $h^i \times l_j$. Then the rational cycle $(h^{s-i} \times h^j) \cdot \alpha$ contains $h^s \times l_0$. Therefore, by Corollary 80.7, this rational cycle also contains $l_0 \times h^s$. It follows that α contains $l_{s-i} \times h^{s-j}$. □

Remark 80.14. The equality $\mathfrak{i}_1(Y) = s + 1$ holds if Y is excellent. By Theorem 79.9 this value of the first Witt index is maximal for all D-dimensional anisotropic quadrics.

81. On the 2-adic order of higher Witt indices, I

The main result of this section is Theorem 81.2 on a relationship between higher Witt indices and the integer produced by a 1-primordial cycle originally proved in [**75**]. This is used to establish a relationship between higher Witt indices of an anisotropic quadratic form (cf. Corollary 81.19).

Let φ be a nondegenerate (possibly isotropic) quadratic form of dimension D over a field F of characteristic not 2 and $X = X_\varphi$. Let $\mathfrak{h} = \mathfrak{h}(\varphi)$ be the height of φ (or X) and let

$$F = F_0 \subset F_1 \subset \cdots \subset F_{\mathfrak{h}}$$

be the generic splitting tower (cf. §25). For $q \in [0, \mathfrak{h}]$, let $\mathfrak{i}_q = \mathfrak{i}_q(\varphi)$, $\mathfrak{j}_q = \mathfrak{j}_q(\varphi)$, $\varphi_q = (\varphi_{F_q})_{an}$, and $X_q = X_{\varphi_q}$.

We shall use the following simple observation in the proof of Theorem 81.2:

Proposition 81.1. *Let α be a homogeneous element of* $\mathrm{Ch}(\bar{X}^2)$ *with* $\operatorname{codim} \alpha > d$. *Assume X is not split, i.e., $\mathfrak{h} > 0$ and that for some $q \in [0,\ \mathfrak{h} - 1]$ the cycle α is F_q-rational and contains neither $h^i \times l_?$ nor $l_? \times h^i$ for any $i < \mathfrak{j}_q$. Then $\delta_X^*(\alpha) = 0$ in* $\mathrm{Ch}(\bar{X})$, *where $\delta_X : X \to X^2$ is the diagonal morphism of X.*

PROOF. We may assume that $\dim \alpha = D + i$ with $i \ge 0$ (otherwise $\dim \delta_X^*(\alpha) < 0$). As X is not hyperbolic, $l_d \times l_d \notin \alpha$ by Lemma 73.2. Therefore, by Proposition 68.1, $\delta_X^*(\alpha) = n l_i$, where n is the number of essential basis elements contained in α. Since α contains neither $h^i \times l_?$ nor $l_? \times h^i$ with $i < \mathfrak{j}_q$, the number of essential basis elements contained in α coincides with the number of essential basis elements contained in $pr_*^2(\alpha)$, where

$$pr_*^2 : \overline{\mathrm{Ch}}(X^2_{F_q}) \to \overline{\mathrm{Ch}}(X_q^2)$$

is the homomorphism of Remark 72.5. In view of Corollary 72.4, the latter number is even by Lemma 73.14. □

We have defined minimal and primordial elements in $\overline{\mathrm{Ch}}(X^2)$ for an anisotropic quadric X (cf. Definitions 73.5 and 73.16). We extend these definitions to the case of an arbitrary quadric.

Let X be an arbitrary (smooth) quadric given by a quadratic form φ (not necessarily anisotropic) and let X_0 be the quadric given by the anisotropic part of φ. The images of minimal (respectively primordial) elements via the embedding in_*^2 : $\overline{\mathrm{Ch}}(X_0^2) \to \overline{\mathrm{Ch}}(X^2)$ of Remark 72.5 are called *minimal* (respectively *primordial*) elements of $\overline{\mathrm{Ch}}(X^2)$.

Theorem 81.2 (cf. [**75**, Th. 5.1]). *Let X be an anisotropic quadric of even dimension over a field of characteristic not 2. Let $\pi \in \overline{\mathrm{Ch}}(X^2)$ be the 1-primordial cycle. Suppose that π produces an integer $q \in [2, \mathfrak{h}]$ satisfying $v_2(\mathfrak{i}_2+\cdots+\mathfrak{i}_{q-1}) \geq v_2(\mathfrak{i}_1)+2$. Then $v_2(\mathfrak{i}_q) \leq v_2(\mathfrak{i}_1)+1$.*

PROOF. We fix the following notation:

$$\begin{aligned} a &= \mathfrak{i}_1, \\ b &= \mathfrak{i}_2 + \cdots + \mathfrak{i}_{q-1} = \mathfrak{j}_{q-1} - a, \\ c &= \mathfrak{i}_q. \end{aligned}$$

Set $n = v_2(\mathfrak{i}_1)$. Then $v_2(b) \geq n+2$.

Consider the $(a-1)$th derivative $\alpha = \pi \cdot (h^0 \times h^{a-1})$ of π. By Lemma 73.10 and Proposition 73.27, the cycle α is minimal since π is minimal and α contains the basis elements $h^0 \times l_0$ and $h^{a+b} \times l_{a+b}$.

Suppose the result is false, i.e., $v_2(c) \geq n+2$. Proposition 81.3 below contradicts the minimality of α, hence proving Theorem 81.2. To state Proposition 81.3, we need the following morphisms:

$$g_1 : X^2_{F(X)} \to X^3,$$

the morphism given by the generic point of the first factor of X^3;

$$t_{12} : \overline{\mathrm{Ch}}(X^3) \to \overline{\mathrm{Ch}}(X^3),$$

the automorphism given by the transposition of the first two factors of X^3;

$$\delta_{X^2} : X^2 \to X^4, \quad (x_1, x_2) \mapsto (x_1, x_2, x_1, x_2),$$

the diagonal morphism of X^2. We also use the pairing

$$\circ : \mathrm{Ch}(\bar{X}^r) \times \mathrm{Ch}(\bar{X}^s) \to \mathrm{Ch}(\bar{X}^{r+s-2})$$

(for various $r, s \geq 1$) given by a composition of correspondences, where the elements of $\mathrm{Ch}(\bar{X}^s)$ are considered as correspondences $\bar{X}^{s-1} \rightsquigarrow \bar{X}$ and the elements of $\mathrm{Ch}(\bar{X}^r)$ are considered as correspondences $\bar{X} \rightsquigarrow \bar{X}^{r-1}$.

Note that applying Proposition 73.23 to the quadric X_1 with cycle $pr^2_*(\pi) \in \overline{\mathrm{Ch}}(X_1^2)$, there exists a homogeneous essential symmetric cycle $\beta \in \overline{\mathrm{Ch}}(X^2_{F(X)})$ containing the basis element $h^{a+b} \times l_{a+b+c-1}$ and none of the basis elements having h^i with $i < a+b$ as a factor.

Proposition 81.3. *In the notation above let $\eta \in \overline{\mathrm{Ch}}(X^3)$ be a preimage of β under the pull-back epimorphism g_1^* and let μ be the essence of the composition $\eta \circ \alpha$. Then the cycle*

$$(h^0 \times h^{c-a-1}) \cdot \delta^*_{X^2}\Big(t_{12}(\mu) \circ \big(\mathrm{Sq}^{2a}_{X^3}(\mu) \cdot (h^0 \times h^0 \times h^{c-a-1})\big)\Big) \in \overline{\mathrm{Ch}}(X^2)$$

contains $h^{a+b} \times l_{a+b}$ and does not contain $h^0 \times l_0$.

PROOF. Recall that $b \geq 0$ and 2^{n+2} divides b and c, where $n = v_2(a)$. By Proposition 79.4, we also have that 2^{n+2} divides $\dim \varphi_{q-1}$, so 2^{n+2} divides $\dim \varphi_1$. We then see, again using Proposition 79.4, that $a = 2^n$. In addition, $\dim \varphi \equiv 2a \pmod{2^{n+2}}$ so

$$\mathrm{Sq}^{2a}_X(l_{a+b+c-1}) = 0 \tag{81.4}$$

by Corollary 78.5 and Lemma 78.6.

Let $\mathrm{Sym}(\rho) = \rho + \rho^t$ for a cycle ρ on $\bar{X}^2$ be the *symmetrization* operation. The cycle β is homogeneous, essential, symmetric, and does not contain any basis element having h^i with $i < a+b$ as a factor. Consequently, we can write $\beta = \beta_0 + \beta_1$ with

(81.5) $$\beta_0 = \mathrm{Sym}\left(h^{a+b} \times l_{a+b+c-1}\right),$$

(81.6) $$\beta_1 = \mathrm{Sym}\left(\sum_{i\in I} h^{i+a+b} \times l_{i+a+b+c-1}\right)$$

for some set of positive integers I. Furthermore, since α does not contain any of $h^i \times l_i$ with $i \in (0,\ a+b)$, we have

(81.7) $$\mu = h^0 \times \beta + h^{a+b} \times \gamma + \nu$$

for some essential cycle $\gamma \in \mathrm{Ch}_{D+a+b+c-1}(\bar{X}^2)$ and some cycle $\nu \in \mathrm{Ch}(\bar{X}^3)$ such that the first factor of every basis element included in ν is of codimension $> a+b$. We can decompose $\gamma = \gamma_0 + \gamma_1$ with

(81.8) $$\gamma_0 = x \cdot \left(h^0 \times l_{a+b+c-1}\right) + y \cdot \left(l_{a+b+c-1} \times h^0\right),$$

(81.9) $$\gamma_1 = \sum_{j\in J} h^j \times l_{j+a+b+c-1} + \sum_{j\in J'} l_{j+a+b+c-1} \times h^j$$

for some modulo 2 integers $x, y \in \mathbb{Z}/2\mathbb{Z}$ and some sets of integers $J, J' \subset (0,\ +\infty)$.

We need the following:

Lemma 81.10. *In the above, we have $x = y = 1$, $I \subset [c,\ +\infty)$, and $J, J' \subset [a+b+c,\ +\infty)$.*

PROOF. To determine y, consider the cycle $\delta^*(\mu)\cdot(h^0 \times h^{c-1}) \in \overline{\mathrm{Ch}}(X^2)$ where $\delta : X^2 \to X^3$ is the morphism $(x_1, x_2) \mapsto (x_1, x_2, x_1)$. This rational cycle does not contain $h^0 \times l_0$, while the coefficient of $h^{a+b} \times l_{a+b}$ equals $1 + y$. Consequently, $y = 1$ by the minimality of α.

Similarly, using the morphism $X^2 \to X^3$, $(x_1, x_2) \mapsto (x_1, x_1, x_2)$ instead of δ, one sees that $x = 1$ (although the value of x is not important for our future purposes).

To show that $I \subset [c,\ +\infty)$, assume to the contrary that $0 < i < c$ for some $i \in I$. Then $l_{i+a+b} \in \overline{\mathrm{Ch}}(X_{F_q})$ for this i and therefore the cycle

$$l_{i+a+b+c-1} = (pr_3)_*\left((l_0 \times l_{i+a+b} \times h^0)\cdot \mu\right)$$

is F_q-rational, where $pr_3 : X^3 \to X$ is the projection onto the third factor. This contradicts Corollary 72.6 because $i + a + b + c - 1 \geq a + b + c = \mathfrak{j}_q(X) = \mathfrak{i}_0(X_{F_q})$.

To prove the statement for J, assume to the contrary that there exists a $j \in J$ with $0 < j < a+b+c$. Then $l_j \in \overline{\mathrm{Ch}}(X_{F_q})$, hence

$$l_{j+a+b+c-1} = (pr_3)_*\left((l_{a+b} \times l_j \times h^0)\cdot \mu\right) \in \overline{\mathrm{Ch}}(X_{F_q}),$$

a contradiction. The statement for J' is proved similarly. □

Lemma 81.11. *The cycle β is F_1-rational. The cycles γ and γ_1 are F_q-rational.*

PROOF. Since $F_1 = F(X)$, the cycle β is F_1-rational by definition.

Let $pr_{23} : X^3 \to X^2$, $(x_1, x_2, x_3) \mapsto (x_2, x_3)$ be the projection onto the product of the second and the third factors of X^3. The cycle l_{a+b} is F_q-rational; therefore $\gamma = (pr_{23})_*\left((l_{a+b} \times h^0 \times h^0)\cdot \mu\right)$ is also F_q-rational. The cycle γ_0 is F_q-rational as $l_{a+b+c-1}$ is F_q-rational. It follows that γ_1 is F_q-rational as well. □

Define

$$\xi(\chi) := \delta_{X^2}^*\Big(t_{12}^*(\chi) \circ \big(\mathrm{Sq}^{2a}_{X^3}(\chi) \cdot (h^0 \times h^0 \times h^{c-a-1})\big)\Big) \quad \text{for any } \chi \in \mathrm{Ch}(\bar{X}^3).$$

We must prove that the cycle $\xi(\mu) \cdot (h^0 \times h^{c-a-1}) \in \overline{\mathrm{Ch}}(X^2)$ contains $h^{a+b} \times l_{a+b}$ and does not contain $h^0 \times l_0$, i.e., we have to show that $h^{a+b} \times l_{b+c-1} \in \xi(\mu)$ and $h^0 \times l_{c-a-1} \notin \xi(\mu)$.

If $h^0 \times l_{c-a-1} \in \xi(\mu)$, then, passing from F to $F_1 = F(X)$, we have

$$l_{c-a-1} = (pr_2)_*\big((l_0 \times h^0) \cdot \xi(\mu)\big) \in \overline{\mathrm{Ch}}(X_{F(X)}),$$

where $pr_2 : X^2 \to X$ is the projection onto the second factor of X^2, contradicting Corollary 72.6 as $c - a - 1 \geq a = \mathfrak{i}_1(X) = \mathfrak{i}_0(X_{F(X)})$.

It remains to show that $h^{a+b} \times l_{b+c-1} \in \xi(\mu)$. For any $\chi \in \mathrm{Ch}(\bar{X}^2)$, write $\mathrm{coeff}(\chi) \in \mathbb{Z}/2\mathbb{Z}$ for the coefficient of $h^{a+b} \times l_{b+c-1}$ in χ. Since $\mathrm{coeff}(\nu) = 0$, it follows from (81.7) that

$$\mathrm{coeff}\,\big(\xi(\mu)\big) = \mathrm{coeff}\,\big(\xi(h^0 \times \beta + h^{a+b} \times \gamma)\big).$$

We claim that

$$\text{(81.12)} \qquad \mathrm{coeff}\,\big(\xi(h^0 \times \beta)\big) = 0 = \mathrm{coeff}\,\big(\xi(h^{a+b} \times \gamma)\big).$$

Indeed, since $\mathrm{Sq}^{2a}_{X^3}(h^0 \times \beta) = h^0 \times \mathrm{Sq}^{2a}_{X^2}(\beta)$ by Theorem 61.14, we have

$$\xi(h^0 \times \beta) = h^0 \times \delta_X^*\Big(\beta \circ \big(\mathrm{Sq}^{2a}_{X^2}(\beta) \cdot (h^0 \times h^{c-a-1})\big)\Big)$$

where $\delta_X : X \to X^2$ is the diagonal morphism of X. Hence $\mathrm{coeff}\,\big(\xi(h^0 \times \beta)\big) = 0$.

Since $\mathrm{Sq}^{2a}_{X^3}(h^{a+b} \times \gamma)$ is $h^{a+b} \times \mathrm{Sq}^{2a}_{X^2}(\gamma)$ plus terms having h^j with $j > a+b$ as the first factor by Remark 79.2, we have

$$\mathrm{coeff}\big(\xi(h^{a+b} \times \gamma)\big) = \mathrm{coeff}\,\Big(h^{2a+2b} \times \delta_X^*\Big(\gamma \circ \big(\mathrm{Sq}^{2a}_{X^2}(\gamma) \cdot (h^0 \times h^{c-a-1})\big)\Big)\Big) = 0.$$

This proves the claim.

It follows by claim (81.12) that

$$\text{(81.13)} \quad \mathrm{coeff}\,\big(\xi(\mu)\big) = \mathrm{coeff}\,\big(\xi(h^0 \times \beta + h^{a+b} \times \gamma) - \xi(h^0 \times \beta) - \xi(h^{a+b} \times \gamma)\big).$$

To compute the right-hand side in (81.13), we need only the terms $h^{a+b} \times \mathrm{Sq}^{2a}_{X^2}(\gamma)$ in the formula for $\mathrm{Sq}^{2a}_{X^3}(h^{a+b} \times \gamma)$ since the other terms do not effect coeff. Therefore, we see that the right-hand side coefficient in (81.13) is equal to

$$\mathrm{coeff}\,\Big(h^{a+b} \times \delta_X^*\Big(\gamma \circ \big(\mathrm{Sq}^{2a}_{X^2}(\beta) \cdot (h^0 \times h^{c-a-1})\big) + \beta \circ \big(\mathrm{Sq}^{2a}_{X^2}(\gamma) \cdot (h^0 \times h^{c-a-1})\big)\Big)\Big).$$

Consequently, to prove Proposition 81.3, it remains to prove

Lemma 81.14.

$$\delta_X^*\Big(\gamma \circ \big(\mathrm{Sq}^{2a}_{X^2}(\beta) \cdot (h^0 \times h^{c-a-1})\big) + \beta \circ \big(\mathrm{Sq}^{2a}_{X^2}(\gamma) \cdot (h^0 \times h^{c-a-1})\big)\Big) = l_{b+c-1}.$$

PROOF. We start by showing that

$$\text{(81.15)} \qquad \delta_X^*\Big(\beta \circ \big(\mathrm{Sq}^{2a}_{X^2}(\gamma) \cdot (h^0 \times h^{c-a-1})\big)\Big) = 0.$$

Note that Sq^{2a} vanishes on $h^0 \times l_{a+b+c-1}$ by (81.4). Therefore $\mathrm{Sq}^{2a}(\gamma) = \mathrm{Sq}^{2a}(\gamma_1)$ by (81.8). By Lemma 81.10 we may assume that $\dim X \geq 4(a+b+c)-2$ (we shall need this assumption in order to apply Proposition 81.1), otherwise $\gamma_1 = 0$.

Looking at the exponent of the first factor of the basis elements contained in $\mathrm{Sq}^{2a}(\gamma_1)$ and using Lemma 81.10, we see that none of the basis elements $h^j \times l_{j+b+c-1}$ and $l_{j+b+c-1} \times h^j$ with $j < a+b+c$ are present in $\beta \circ \big(\mathrm{Sq}^{2a}(\gamma_1) \cdot (h^0 \times h^{c-a-1})\big)$. As γ_1 is F_q-rational by Lemma 81.11, Equation (81.15) holds by Proposition 81.1.

We compute $\mathrm{Sq}^{2a}(\beta_0)$ where β_0 is as in (81.5). By Corollary 78.5 and Lemma 78.6, we have $\mathrm{Sq}^0(h^{a+b}) = h^{a+b}$, $\mathrm{Sq}^a(h^{a+b}) = h^{2a+b}$, and $\mathrm{Sq}^j(h^{a+b}) = 0$ for all others, $j \leq 2a$. Moreover, we have shown in (81.4) that $\mathrm{Sq}^{2a}(l_{a+b+c-1}) = 0$. Therefore, $\mathrm{Sq}^{2a}(\beta_0) = \mathrm{Sym}\,(h^{2a+b} \times l_{b+c-1})$ by Theorem 61.14.

Using Lemma 81.10, we have

$$\gamma_0 \circ \big(\mathrm{Sq}^{2a}(\beta_0) \cdot (h^0 \times h^{c-a-1})\big) = l_{b+c-1} \times h^0$$

and

$$\delta_X^*\Big(\gamma_0 \circ \big(\mathrm{Sq}^{2a}(\beta_0) \cdot (h^0 \times h^{c-a-1})\big)\Big) = l_{b+c-1}. \tag{81.16}$$

The composition $\gamma_0 \circ \big(\mathrm{Sq}^{2a}(\beta_1) \cdot (h^0 \times h^{c-a-1})\big)$ is trivial. Indeed, by Lemma 81.10, every basis element of the cycle $\mathrm{Sq}^{2a}(\beta_1) \cdot (h^0 \times h^{c-a-1})$ has (as the second factor) either l_j with $j \geq 2a+b+c > 0$ or h^j with $j \geq b+2c-1 > a+b+c-1$, while the two basis elements of γ_0 have h^0 and $l_{a+b+c-1}$ as the first factor. Consequently

$$\delta_X^*\Big(\gamma_0 \circ \big(\mathrm{Sq}^{2a}(\beta_1) \cdot (h^0 \times h^{c-a-1})\big)\Big) = 0. \tag{81.17}$$

Looking at the exponent of the first factor of the basis elements contained in γ_1 and using Lemma 81.10, we see that none of the basis elements $h^j \times l_{j+b+c-1}$ and $l_{j+b+c-1} \times h^j$ with $j < a+b+c$ is present in $\gamma_1 \circ \big(\mathrm{Sq}^{2a}(\beta) \cdot (h^0 \times h^{c-a-1})\big)$. Therefore, the relation

$$\delta_X^*\Big(\gamma_1 \circ \big(\mathrm{Sq}^{2a}(\beta) \cdot (h^0 \times h^{c-a-1})\big)\Big) = 0 \tag{81.18}$$

holds by Proposition 81.1 in view of Lemma 81.11.

Taking the sum of the relations in (81.15)–(81.18), we have established the proof of Lemma 81.14. □

This completes the proof of Proposition 81.3. □

Theorem 81.2 is proved. □

Corollary 81.19 (cf. [**75**, Th. 1.1]). *Let φ be an anisotropic quadratic form over a field of characteristic not* 2. *If $\mathfrak{h} = \mathfrak{h}(\varphi) > 1$, then*

$$v_2(\mathfrak{i}_1) \geq \min\big(v_2(\mathfrak{i}_2), \dots, v_2(\mathfrak{i}_{\mathfrak{h}})\big) - 1.$$

PROOF. For any odd-dimensional φ, the statement is trivial, as all $\mathfrak{i}_q$ are odd by Corollary 79.6. Assume that the inequality fails for an even-dimensional anisotropic φ. Note that in this case the difference

$$\dim\varphi - \mathfrak{i}_1 = \mathfrak{i}_1 + 2(\mathfrak{i}_2 + \cdots + \mathfrak{i}_{\mathfrak{h}})$$

cannot be a power of 2 because it is larger than 2^n and not divisible by 2^{n+1} for $n = v_2(\mathfrak{i}_1)$. Therefore, by Corollary 80.10, the 1-primordial cycle on X^2 does produce an integer. Therefore, the assumptions of Theorem 81.2 are satisfied, leading to a contradiction. □

Example 81.20. For an anisotropic quadratic form of dimension 6 and of trivial discriminant, we have $\mathfrak{h} = 2$, $\mathfrak{i}_1 = 1$, and $\mathfrak{i}_2 = 2$. Therefore, the lower bound on $v_2(\mathfrak{i}_1)$ in Corollary 81.19 is exact.

82. Holes in I^n

Recall that F is a field of characteristic not 2. For every integer $n \geq 1$, we set

$$\dim I^n(F) := \{\dim \varphi \mid \varphi \in I^n(F) \text{ and anisotropic}\}.$$

and

$$\dim I^n := \bigcup \dim I^n(F),$$

where the union is taken over *all* fields F (of characteristic not 2).

In this section, we determine the set $\dim I^n$. Theorem 82.8 states that $\dim I^n$ is the set of even nonnegative integers without the following disjoint open intervals (which we call *holes in* I^n):

$$U_{n-i} = (2^{n+1} - 2^{i+1},\ 2^{n+1} - 2^i)\ ,\quad i = n, n-1, \dots, 1.$$

The statement that $U_0 \cap \dim I^n = \emptyset$ is the Hauptsatz (Theorem 23.7). This was originally proved by Arason and Pfister [**10**, Hauptsatz]. The statement for $U_1 \cap \dim I^n$ with $n = 3$ was originally proved in 1966 by Pfister [**110**, Satz 14], with $n = 4$ in 1998 by Hoffmann [**55**, Main Theorem], and for arbitrary n in 2000 by Vishik [**133**, Th. 6.4]. The statement that $U_{n-i} \cap \dim I^n = \emptyset$ for any n and i was conjectured by Vishik [**133**, Conj. 6.5]. A positive solution of the conjecture was announced by Vishik in 2002 but the proof is not available; a proof was given in [**73**].

Proposition 82.1. *Let φ be a nonzero anisotropic form of even dimension with $\deg \varphi = n \geq 1$. If $\dim \varphi < 2^{n+1}$, then $\dim \varphi = 2^{n+1} - 2^{i+1}$ for some $i \in [0,\ n-1]$.*

PROOF. We use the notation of §81. We prove the statement by induction on $\mathfrak{h} = \mathfrak{h}(\varphi)$. The case of $\mathfrak{h} = 1$ is trivial.

So assume that $\mathfrak{h} > 1$. As $\dim \varphi_1 < \dim \varphi < 2^{n+1}$ and $\deg \varphi_1 = \deg \varphi$, where φ_1 is the first (anisotropic) kernel form of φ, the induction hypothesis implies

$$\dim \varphi_1 = 2^{n+1} - 2^{i+1} \quad \text{for some } i \in [1,\ n-1].$$

Therefore, $\dim \varphi = 2^{n+1} - 2^{i+1} + 2\mathfrak{i}_1$. Since $\dim \varphi < 2^{n+1}$, we have $\mathfrak{i}_1 < 2^i$. In particular, $v_2(\dim \varphi - \mathfrak{i}_1) = v_2(\mathfrak{i}_1)$. As $\mathfrak{i}_1 \leq \exp_2 v_2(\dim \varphi - \mathfrak{i}_1)$ by Proposition 79.4, it follows that $\mathfrak{i}_1$ is a 2-power, say $\mathfrak{i}_1 = 2^j$ for some $j \in [0,\ i-1]$.

It follows by the induction hypothesis that each of the integers

$$\dim \varphi_1, \dots, \dim \varphi_{\mathfrak{h}}$$

is divisible by 2^{i+1}. Therefore, $v_2(\mathfrak{i}_q) \geq i$ for all $q \in [2, \mathfrak{h}]$ and hence by Corollary 81.19, we have $j \geq i-1$. Consequently, $j = i-1$, so $\dim \varphi = 2^{n+1} - 2^i$. □

Corollary 82.2. *Let φ be an anisotropic quadratic form such that $\varphi \in I^n(F)$ for some $n \geq 1$. If $\dim \varphi < 2^{n+1}$, then $\dim \varphi = 2^{n+1} - 2^{i+1}$ for some $i \in [0,\ n]$.*

PROOF. We may assume that $\varphi \neq 0$. We have $\deg \varphi \geq n$ by Corollary 25.12. Since $2^{\deg \varphi} \leq \dim \varphi < 2^{n+1}$, we must have $\deg \varphi = n$. The result follows from Proposition 82.1. □

Corollary 82.3. *Let $\varphi \neq 0$ be an anisotropic quadratic form in $I^n(F)$ with* $\dim \varphi < 2^{n+1}$. *Then the higher Witt indices of φ are the successive 2-powers:*

$$\mathfrak{i}_1 = 2^i,\ \mathfrak{i}_2 = 2^{i+1}, \dots,\ \mathfrak{i}_{\mathfrak{h}} = 2^{n-1},$$

where $i = \log_2(2^{n+1} - \dim \varphi) - 1$ is an integer.

PROOF. By Corollary 82.2, we have $\dim \varphi = 2^{n+1} - 2^{i+1}$ for i as in the statement of Corollary 82.3, and $\dim \varphi_1 = 2^{n+1} - 2^{j+1}$ for some $j > i$. It follows by Proposition 79.4 that $\mathfrak{i}_1 = 2^i$. We proceed by induction on $\dim \varphi$. □

We now show that every even value of $\dim \varphi$ for $\varphi \in I^n(F)$ not forbidden by Corollary 82.2 is possible over some F. We start with some preliminary work.

Lemma 82.4. *Let φ be a nonzero anisotropic quadratic form in $I^n(F)$ and let* $\dim \varphi < 2^{n+1}$ *for some $n \geq 1$. Then the 1-primordial cycle is the only primordial cycle in $\overline{\mathrm{Ch}}(X^2)$.*

PROOF. We induct on $\mathfrak{h} = \mathfrak{h}(\varphi)$. The case $\mathfrak{h} = 1$ is trivial, so we assume that $\mathfrak{h} > 1$. Let $pr_*^2 : \overline{\mathrm{Ch}}(X^2) \to \overline{\mathrm{Ch}}(X_1^2)$ be the homomorphism of Remark 72.5. Since the integer $\dim \varphi - \mathfrak{i}_1$ lies inside the open interval $(2^n,\ 2^{n+1})$, it is not a 2-power. Hence by Corollary 80.10, we have $pr_*^2(\pi) \neq 0$, where $\pi \in \overline{\mathrm{Ch}}(X^2)$ is the 1-primordial cycle. Therefore, by the induction hypothesis, the diagram of $pr_*^2(\pi)$ has points in every shell triangle. Thus, the diagram of π itself has points in every shell triangle. By Theorem 73.26, this means that π is the unique primordial cycle in $\overline{\mathrm{Ch}}(X^2)$. □

Corollary 82.5. *Let φ be a nonzero anisotropic quadratic form in $I^n(F)$ and let* $\dim \varphi = 2^{n+1} - 2$ *for some $n \geq 1$. Then for any $i > 0$, the group $\overline{\mathrm{Ch}}_{D+i}(X^2)$ contains no essential element.*

PROOF. By Lemma 82.4, the 1-primordial cycle is the only primordial cycle in $\overline{\mathrm{Ch}}(X^2)$. Since $\mathfrak{i}_1 = 1$ by Corollary 82.3, we have $\dim \pi = D$. To finish we apply Theorem 73.26. □

Lemma 82.6. *Let F_0 be a field (of* $\operatorname{char} F_0 \neq 2$*) and let*

$$F = F_0(t_{1j}, t_{2j})_{1 \leq j \leq n}$$

be the field of rational functions in $2n$ variables. Then the quadratic form

$$\langle\langle t_{11}, \dots, t_{1n}\rangle\rangle' \perp - \langle\langle t_{21}, \dots, t_{2n}\rangle\rangle'$$

over F is anisotropic (where the prime denotes the pure subform of the Pfister form).

PROOF. For any $i = 0, 1, \dots, n$, we set

$$\varphi_i = \langle\langle t_{11}, \dots, t_{1i}\rangle\rangle \text{ and } \psi_i = \langle\langle t_{21}, \dots, t_{2i}\rangle\rangle.$$

We prove that the form $\varphi_i' \perp - \psi_i'$ is anisotropic by induction on i. For $i = 0$ the statement is trivial. For $i \geq 1$, we have:

$$\varphi_i' \perp - \psi_i' \simeq (\varphi_{i-1}' \perp - \psi_{i-1}') \perp t_{1i}\varphi_{i-1} \perp - t_{2i}\psi_{i-1}.$$

The summand $\varphi_{i-1}' \perp - \psi_{i-1}'$ is anisotropic by the induction hypothesis, while the forms φ_{i-1} and ψ_{i-1} are so by Corollary 19.6. Applying Lemma 19.5 repeatedly we conclude that the whole form is anisotropic. □

Define the *anisotropic pattern* of a quadratic form φ over F to be the set of integers $\dim(\varphi_K)_{an}$ for all field extensions K/F. By Proposition 25.1, the anisotropic pattern of a form φ coincides with the set

$$\{\dim \varphi - 2\mathfrak{j}_q(\varphi) \mid q \in [0,\ \mathfrak{h}(\varphi)]\}.$$

The following result is due to Vishik.

Proposition 82.7. *Let F_0 be a field (of* char $F_0 \neq 2$*) and let $n \geq 1$ and $m \geq 2$ be integers. Let*

$$F = F_0(t_i, t_{ij})_{1 \leq i \leq m,\ 1 \leq j \leq n}$$

be the field of rational functions in variables t_i and t_{ij}. Then the anisotropic pattern of the quadratic form

$$\varphi = t_1 \cdot \langle\langle t_{11}, \dots, t_{1n}\rangle\rangle \perp \dots \perp t_m \cdot \langle\langle t_{m1}, \dots, t_{mn}\rangle\rangle$$

over F is the set

$$\{2^{n+1} - 2^i \mid i \in [1,\ n+1]\} \cup \big(2\mathbb{Z} \cap [2^{n+1},\ m \cdot 2^n]\big).$$

PROOF. We first show that all the integers $2^{n+1} - 2^i$ are in the anisotropic pattern of φ. Indeed, the anisotropic part of φ over the field E obtained from F by adjoining the square roots of $t_{31}, t_{41}, \dots, t_{m1}$, of t_1 and of $-t_2$, is isomorphic to the form

$$\langle\langle t_{11}, \dots, t_{1n}\rangle\rangle' \perp - \langle\langle t_{21}, \dots, t_{2n}\rangle\rangle'$$

of dimension $2^{n+1} - 2$. This form is anisotropic by Lemma 82.6. The anisotropic pattern of this form is $\{2^{n+1} - 2^i \mid i \in [1, n+1]\}$ by Corollary 82.3.

Now assume that there is an even integer in the interval $[2^{n+1},\ m \cdot 2^n]$ *not* in the anisotropic pattern of φ. Among all such integers take the smallest one and call it a. Let $b = a - 2$ and let c be the smallest integer greater than a and lying in the anisotropic pattern of φ. Let E be the field in the generic splitting tower of φ such that $\dim \psi = c$ where $\psi = (\varphi_E)_{an}$ and let Y be the projective quadric given by the quadratic form ψ. Let $\pi \in \overline{\mathrm{Ch}}(Y^2)$ be the 1-primordial cycle. We claim that

$$\pi = h^0 \times l_{\mathfrak{i}_1 - 1} + l_{\mathfrak{i}_1 - 1} \times h^0$$

where $\mathfrak{i}_1 = \mathfrak{i}_1(Y)$. Indeed, since $\mathfrak{i}_1 = (c - b)/2 > 1$ and $\mathfrak{i}_q(Y) = 1$ for all q such that $\dim \psi_q \in [2^{n+1} - 2,\ b - 2]$, the diagram of the cycle π does not have any point in the qth shell triangle for such q. For the integer q satisfying $\dim \psi_q = 2^{n+1} - 2$, the cycle $pr^2_*(\pi) \in \overline{\mathrm{Ch}}(Y_q^2)$ has dimension $> \dim Y_q$, hence it is 0 by Corollary 82.5. The relation $pr^2_*(\pi) = 0$ means that π has no point in any shell triangle with number $> q$.

It follows that $\pi = h^0 \times l_{\mathfrak{i}_1 - 1} + l_{\mathfrak{i}_1 - 1} \times h^0$. By Corollary 80.8, the integer $\dim Y - \mathfrak{i}_1 + 2$ is a power of 2, say 2^p. Since

$$\dim Y - \mathfrak{i}_1 + 2 = (c - 2) - (c - b)/2 + 2 = (b + c)/2,$$

the integer 2^p lies inside the open interval $(b,\ c)$. It follows that the integer 2^p satisfies $2^{n+1} \leq 2^p < m \cdot 2^n$ and is *not* in the splitting pattern of the quadratic form φ. But every integer $\leq m \cdot 2^n$ divisible by 2^n is evidently *in* the anisotropic pattern of φ. This contradiction establishes Proposition 82.7. □

Summarizing, we have

Theorem 82.8. *For any integer $n \geq 1$,*

$$\dim I^n = \{2^{n+1} - 2^i \mid i \in [1,\ n+1]\} \cup \big(2\mathbb{Z} \cap [2^{n+1},\ +\infty)\big).$$

PROOF. The inclusion $\subset$ is given by Corollary 82.2, while the inclusion $\supset$ follows by Proposition 82.7. □

Remark 82.9. The case of dimension $2^{n+1} - 2^i$ can be realized directly by the difference of two $(i-1)$-linked n-fold Pfister forms (cf. Corollary 24.3).

83. On the 2-adic order of higher Witt indices, II

Throughout this section, X is an anisotropic quadric of dimension D over a field of characteristic not 2. We write $\mathfrak{i}_1, \dots, \mathfrak{i}_{\mathfrak{h}}$ and $\mathfrak{j}_1, \dots, \mathfrak{j}_{\mathfrak{h}}$ for the relative and absolute higher Witt indices of X, respectively, where $\mathfrak{h}$ is the height of X (cf. §81).

The main result of this section is Theorem 83.3 originally proved in [**75**]. It is used to establish further relations between higher Witt indices in Corollary 83.4.

We first establish further special properties of the 1-primordial cycle in addition to those in Proposition 73.27 and Theorem 81.2.

Lemma 83.1. *Let $\pi \in \overline{\mathrm{Ch}}(X^2)$ be the 1-primordial cycle. Then $\mathrm{Sq}^j_{X^2}(\pi) = 0$ for all $j \in (0,\ \mathfrak{i}_1)$.*

PROOF. Let $\mathrm{Sq} = \mathrm{Sq}_{X^2}$. Assume that $\mathrm{Sq}^j(\pi) \neq 0$ for some $j \in (0,\ \mathfrak{i}_1)$. By Remark 79.2, one sees that $\mathrm{Sq}^j(\pi)$ has a nontrivial *intersection* with an appropriate jth order derivative of π. As the derivative of π is minimal by Theorem 73.26, the cycle $\mathrm{Sq}^j(\pi)$ contains this derivative. It follows that $\mathrm{Sq}^j(\pi)$ has a point in the first left shell triangle, contradicting Lemma 79.3. □

Proposition 83.2. *Let i be an integer such that $h^i \times l_?$ is contained in the 1-primordial cycle. Then i is divisible by 2^{n+1} for any $n \geq 0$ satisfying $\mathfrak{i}_1 > 2^n$.*

PROOF. Assume that the statement is false. Let i be the minimal integer not divisible by 2^{n+1} and such that $h^i \times l_?$ is contained in the 1-primordial cycle $\pi \in \overline{\mathrm{Ch}}(X^2)$.

Note that π contains only essential basis elements and is symmetric. As $\dim \pi = D + \mathfrak{i}_1 - 1$, we have $h^i \times l_{i+\mathfrak{i}_1-1} \in \pi$.

For any nonnegative integer k divisible by 2^{n+1}, the binomial coefficient $\binom{k}{l}$ with a nonnegative integer l is odd only if l is divisible by 2^{n+1} by Lemma 78.6. Therefore, $\mathrm{Sq}_X(h^k) = h^k(1+h)^k$ is a sum of powers of h with exponents divisible by 2^{n+1}. It follows that the value $\mathrm{Sq}^j_X(\pi)$ contains the element $\mathrm{Sq}^0_X(h^i) \times \mathrm{Sq}^j_X(l_{i+\mathfrak{i}_1-1}) = h^i \times \mathrm{Sq}^j_X(l_{i+\mathfrak{i}_1-1})$ for any integer j. Since $\mathrm{Sq}^j_X(\pi) = 0$ for $j \in (0,\ \mathfrak{i}_1)$ by Lemma 83.1, we have

$$\mathrm{Sq}^j_X(l_{i+\mathfrak{i}_1-1}) = 0 \quad \text{for } j \in (0,\ \mathfrak{i}_1).$$

Now look at the specific value $\mathrm{Sq}^{2^{v_2(i)}}_X(l_{i+\mathfrak{i}_1-1})$. Since i is not divisible by 2^{n+1} and $\mathfrak{i}_1 > 2^n$, the degree $2^{v_2(i)}$ of the Steenrod operation lies in the interval $(0,\ \mathfrak{i}_1)$. By Corollary 78.5, the value $\mathrm{Sq}^{2^{v_2(i)}}_X(l_{i+\mathfrak{i}_1-1})$ is equal to $l_{i+\mathfrak{i}_1-1-2^{v_2(i)}}$ multiplied by the binomial coefficient

$$\binom{D - i - \mathfrak{i}_1 + 2}{2^{v_2(i)}}.$$

The integer $D - \mathfrak{i}_1 + 2 = \dim\varphi - \mathfrak{i}_1$ is divisible by 2^{n+1} by Proposition 79.4 as $\mathfrak{i}_1 > 2^n$. Therefore the binomial coefficient is odd by Lemma 78.6. This is a contradiction, so it establishes the result. □

Theorem 83.3 (cf. [**75**, Th. 3.3]). *Let X be an anisotropic quadric over a field of characteristic not 2. Suppose that the 1-primordial cycle $\pi \in \overline{\mathrm{Ch}}(X^2)$ produces the integer q. Then $v_2(\mathfrak{i}_q) \geq v_2(\mathfrak{i}_1)$.*

PROOF. Let $n = v_2(\mathfrak{i}_1)$. Then the integer 2^n divides $\dim\varphi - \mathfrak{i}_1$ by Proposition 79.4. Therefore, 2^n divides $\dim\varphi$ as well.

We have $h^{\mathfrak{j}_{q-1}} \times l_{\mathfrak{j}_{q-1}+\mathfrak{i}_1-1} \in \pi$ by definition of q. Consequently, by Proposition 83.2, the integer $\mathfrak{j}_{q-1}$ is divisible by 2^n. It follows that 2^n divides $\dim\varphi_{q-1} = \dim\varphi - 2\mathfrak{j}_{q-1}$, where φ_{q-1} is the $(q-1)$th (anisotropic) kernel of φ. If $m < n$ for $m = v_2(\mathfrak{i}_q)$, then applying Proposition 79.4, we have $\mathfrak{i}_q = \mathfrak{i}_1(\varphi_{q-1})$ is equal to 2^m and, in particular, smaller than $\mathfrak{i}_1$. Therefore the 1-primordial cycle π has no points in the qth shell triangle. But the point $h^{\mathfrak{j}_{q-1}} \times l_{\mathfrak{j}_{q-1}+\mathfrak{i}_1-1} \in \pi$ lies in the qth shell triangle. This contradiction establishes the theorem. □

Corollary 83.4 (cf. [**75**, Th. 1.1]). *We have $v_2(\mathfrak{i}_1) \leq \max\big(v_2(\mathfrak{i}_2), \dots, v_2(\mathfrak{i}_{\mathfrak{h}})\big)$ if the integer*

$$\dim\varphi - \mathfrak{i}_1 = \mathfrak{i}_1 + 2(\mathfrak{i}_2 + \cdots + \mathfrak{i}_{\mathfrak{h}})$$

is not a power of 2.

PROOF. If the integer $\dim\varphi - \mathfrak{i}_1$ is not a 2-power, then the 1-primordial cycle does produce an integer by Corollary 80.10. The result follows by Theorem 83.3. □

84. Minimal height

Every nonnegative integer n is uniquely representable in the form of an alternating sum of 2-powers:

$$n = 2^{p_0} - 2^{p_1} + 2^{p_2} - \cdots + (-1)^{r-1}2^{p_{r-1}} + (-1)^r 2^{p_r}$$

for some integers $p_0, p_1, \dots, p_r$ satisfying $p_0 > p_1 > \cdots > p_{r-1} > p_r + 1 > 0$. We shall write $P(n)$ for the set $\{p_0, p_1, \dots, p_r\}$. Note that p_r coincides with the 2-adic order $v_2(n)$ of n. For $n = 0$ our representation is the empty sum, so $P(0) = \emptyset$.

Define the *height* $\mathfrak{h}(n)$ of the integer n as the number of positive elements in $P(n)$. Therefore, $\mathfrak{h}(n)$ is the number $|P(n)|$, the cardinality of the set $P(n)$ if n is even, while $\mathfrak{h}(n) = |P(n)| - 1$ if n is odd.

In this section we prove the following theorem conjectured by Rehmann and originally proved in [**59**]:

Theorem 84.1. *Let φ be an anisotropic quadratic form over a field of characteristic not 2. Then*

$$\mathfrak{h}(\varphi) \geq \mathfrak{h}(\dim\varphi).$$

Remark 84.2. Let $n \geq 0$ and let φ be an anisotropic excellent quadratic form of dimension n. It follows from Proposition 28.5 that $\mathfrak{h}(\varphi) = \mathfrak{h}(n)$. Therefore, the bound in Theorem 84.1 is sharp.

We shall see (cf. Corollary 84.5) that Theorem 84.1 in odd dimensions is a consequence of Proposition 79.4. In even dimensions we shall also need Theorems 81.2 and 83.3. We shall also need several combinatorial arguments to prove Theorem 84.1.

Suppose φ is anisotropic. Let φ_i be the ith (anisotropic) kernel form of φ, and let $n_i = \dim \varphi_i$, $0 \le i \le \mathfrak{h}(\varphi)$.

Lemma 84.3. *For any $i \in [1,\ \mathfrak{h}]$, the difference $\mathfrak{d}(i) := \mathfrak{h}(n_{i-1}) - \mathfrak{h}(n_i)$ satisfies the following:*

(I) *If the dimension of φ is odd, then $|\mathfrak{d}(i)| = 1$.*

(II) *If the dimension of φ is even, then $|\mathfrak{d}(i)| \le 2$. Moreover,*

(+2) *If $\mathfrak{d}(i) = 2$, then $P(n_i) \subset P(n_{i-1})$ and $v_2(n_i) \ge v_2(n_{i-1}) + 2$.*

(+1) *If $\mathfrak{d}(i) = 1$, the set difference $P(n_i) \setminus P(n_{i-1})$ is either empty or consists of a single element p, in which case both integers $p-1$ and $p+1$ lie in $P(n_{i-1})$.*

(0) *If $\mathfrak{d}(i) = 0$, the set difference $P(n_i) \setminus P(n_{i-1})$ consists of one element p and either $p-1$ or $p+1$ lies in $P(n_{i-1})$.*

(−1) *If $\mathfrak{d}(i) = -1$, the set difference $P(n_i) \setminus P(n_{i-1})$ consists either of two elements $p-1$ and $p+1$ for some $p \in P(n_{i-1})$ or the set difference consists of one element.*

(−2) *If $\mathfrak{d}(i) = -2$, the set difference $P(n_i) \setminus P(n_{i-1})$ consists of two elements, i.e., $P(n_i) \supset P(n_{i-1})$. Moreover, in this case one of these two elements is equal to $p+1$ for some $p \in P(n_{i-1})$.*

PROOF. Write $p_0, p_1, \dots, p_r$ for the elements of $P(n_{i-1})$ in descending order. We have $n_i = n_{i-1} - 2\mathfrak{i}_i$. We also know by Proposition 79.4 that there exists a nonnegative integer m such that $2^m < n_{i-1}$, $\mathfrak{i}_i \equiv n_{i-1} \pmod{2^m}$, and $1 \le \mathfrak{i}_i \le 2^m$. The condition $2^m < n_{i-1}$ implies $m < p_0$. Let p_s be the element with maximal even s satisfying $m < p_s$.

If $m = p_s - 1$, then $\mathfrak{i}_i = 2^{p_s - 1} - 2^{p_{s+1}} + 2^{p_{s+2}} - \dots$ and, therefore,

$$n_i = 2^{p_0} - 2^{p_1} + \dots - 2^{p_{s-1}} + 2^{p_{s+1}} - 2^{p_{s+2}} + \dots + (-1)^{r-1} 2^{p_r}.$$

If $s = r$ and $p_{r-1} + 1 = p_{r-2}$, then $P(n_i)$ equals $P(n_{i-1})$ without p_{r-2} and p_r. Otherwise, $P(n_i)$ equals $P(n_{i-1})$ without p_s.

Therefore, we may assume that $m < p_s - 1$.

If $s = r$, then $\mathfrak{i}_i = 2^m$ and $n_i = n_{i-1} - 2^{m+1}$. If $m = p_r - 2$, we have $P(n_i)$ is obtained from $P(n_{i-1})$ by replacing p_r with $p_r - 1$. If $m < p_r - 2$, we have $P(n_i)$ equals $P(n_{i-1})$ with $m+1$ added.

Therefore, we may assume in addition that $s < r$.

If $p_s - 1 > m > p_{s+1}$, then $\mathfrak{i}_i = 2^m - 2^{p_{s+1}} + 2^{p_{s+2}} - \dots$ and, therefore,

$$n_i = 2^{p_0} - 2^{p_1} + \dots - 2^{p_{s-1}} + 2^{p_s} - 2^{m+1} + 2^{p_{s+1}} - 2^{p_{s+2}} + \dots + (-1)^{r+1} 2^{p_r}.$$

This is the correct representation of n_i and, therefore, $P(n_i)$ equals $P(n_{i-1})$ with $m+1$ added.

It remains to consider the case with $m \le p_{s+1}$ while $s < r$. In this case, first assume that $s = r - 1$. Then $\mathfrak{i}_i = 2^m$ and $n_i = n_{i-1} - 2^{m+1}$.

If $m < p_r - 2$, then $P(n_i)$ equals $P(n_{i-1})$ with $p_r + 1$ and $m+1$ added.

If $m = p_r - 2$, then $P(n_i)$ equals $P(n_{i-1})$ with p_r removed and $p_r + 1$ and $p_r - 1$ added.

If $m = p_r - 1$, one has two possibilities. If $p_{r-1} > p_r + 2$, then $P(n_i)$ equals $P(n_{i-1})$ with p_r removed and $p_r + 1$ added. If $p_{r-1} = p_r + 2$, then $P(n_i)$ equals $P(n_{i-1})$ with p_r and p_{r-1} removed while $p_r + 1$ added.

Finally, if $m = p_r$, then either $p_{r-1} = p_r + 2$ and $P(n_i)$ equals $P(n_{i-1})$ without p_{r-1}, or $P(n_i)$ equals $P(n_{i-1})$ with $p_r + 2$ added.

To finish the proof we may assume that $m \leq p_{s+1}$ and $s < r-1$. We have: $\mathfrak{i}_i = 2^{p_{s+2}} - 2^{p_{s+3}} + \cdots + (-1)^r 2^{p_r}$ and

$$n_i = 2^{p_0} - 2^{p_1} + \cdots + 2^{p_s} - 2^{p_{s+1}+1} + 2^{p_{s+1}} - 2^{p_{s+2}} + \cdots + (-1)^{r+1} 2^{p_r}.$$

Thus, if $p_s > p_{s+1} + 1$, then $P(n_i)$ equals $P(n_{i-1})$ with $p_{s+1}+1$ added; otherwise $P(n_i)$ is $P(n_{i-1})$ with p_s removed. □

Corollary 84.4. *Let φ be an anisotropic odd-dimensional quadratic form and let $i \in [1,\ \mathfrak{h}(\varphi)]$. Then*

$$\mathfrak{h}(n_{i-1}) - \mathfrak{h}(n_i) \leq 1.$$

Corollary 84.5. *Let φ be an anisotropic quadratic form of odd dimension n. Then $\mathfrak{h}(\varphi) \geq \mathfrak{h}(n)$.*

PROOF. Let $\mathfrak{h} := \mathfrak{h}(\varphi)$. As $\dim \varphi$ is odd, $n_{\mathfrak{h}} = 1$. Then $\mathfrak{h}(n_{\mathfrak{h}}) = 0$ and by Corollary 84.4, we have $\mathfrak{h}(n_{i-1}) - \mathfrak{h}(n_i) \leq 1$ for every $i \in [1,\ \mathfrak{h}]$. Therefore, $\mathfrak{h}(n_0) \leq \mathfrak{h}$. Since φ is anisotropic, $n = n_0$, and the result follows. □

Remark 84.6. By Lemma 84.3, for any quadratic form φ of odd dimension n, we have $\mathfrak{h}(n_i) = \mathfrak{h}(n_{i-1}) \pm 1$. Therefore $\mathfrak{h}(\varphi) \equiv \mathfrak{h}(n) \pmod 2$.

Proposition 84.7. *Let φ be an anisotropic quadratic form of even dimension n and $\mathfrak{h} := \mathfrak{h}(\varphi)$. Suppose that $v_2(n_i) \geq v_2(n_{i-1}) + 2$ for some $i \in [1,\ \mathfrak{h})$. Then the open interval $(i,\ \mathfrak{h})$ contains an integer i' satisfying $|v_2(n_{i'}) - v_2(n_{i-1})| \leq 1$.*

PROOF. It suffices to consider the case $i = 1$. Note that $\mathfrak{h} \geq 2$. Set $p = v_2(n_0)$. By assumption, we have $v_2(n_1) \geq p+2$. Therefore, $v_2(\mathfrak{i}_1) = p-1$. Clearly, the integer $n_0 - \mathfrak{i}_1 = \mathfrak{i}_1 + n_1$ is not a power of 2. Therefore, by Corollary 80.10, the 1-primordial cycle of $\overline{\mathrm{Ch}}(X^2)$ produces an integer $j \in [2,\ \mathfrak{h}]$. We shall show that either $v_2(n_{j-1})$ or $v_2(n_j)$ lies in $[p-1,\ p+1]$ for this j. We then take $i' = j-1$ in the first case and $i' = j$ in the second case. Note that $i' \neq 1$ and $i' \neq \mathfrak{h}$ as $v_2(n_1) \geq p+2$, while $v_2(n_{\mathfrak{h}}) = \infty$.

By Theorem 83.3, we have $v_2(\mathfrak{i}_j) \geq p-1$. Consequently, $v_2(n_{j-1}) \geq p-1$ by Proposition 79.4 as well. Since $n_1 = 2(\mathfrak{i}_2 + \cdots + \mathfrak{i}_{j-1}) + n_{j-1}$, it follows that $v_2(\mathfrak{i}_2 + \cdots + \mathfrak{i}_{j-1}) + 1 \geq p-1$. If $v_2(\mathfrak{i}_2 + \cdots + \mathfrak{i}_{j-1}) < p+1$, then $v_2(n_{j-1}) = v_2(\mathfrak{i}_2 + \cdots + \mathfrak{i}_{j-1}) + 1 \in [p-1,\ p+1]$. So, we may assume that $v_2(\mathfrak{i}_2 + \cdots + \mathfrak{i}_{j-1}) \geq p+1$ and apply Theorem 81.2 to conclude that $v_2(\mathfrak{i}_j) \leq p$. We have $v_2(\mathfrak{i}_j) \in \{p-1, p\}$. If $v_2(n_{j-1}) > v_2(\mathfrak{i}_j) + 1$, then $v_2(n_j) = v_2(\mathfrak{i}_j) + 1 \in \{p, p+1\}$. If $v_2(n_{j-1}) = v_2(\mathfrak{i}_j) + 1$, then $v_2(n_{j-1}) \in \{p, p+1\}$. Finally, if $v_2(n_{j-1}) < v_2(\mathfrak{i}_j) + 1$, then $v_2(n_{j-1}) = v_2(\mathfrak{i}_j)$, hence $v_2(n_{j-1}) \in \{p-1, p\}$. □

Corollary 84.8. *Let φ be an anisotropic quadratic form of even dimension n. Suppose that $v_2(n_i) \geq v_2(n_{i-1}) + 2$ for some $i \in [1,\ \mathfrak{h})$. Set $p = v_2(n_{i-1})$. Then there exists $i' \in (i,\ \mathfrak{h})$ such that the set $P(n_{i'})$ contains an element p' with $|p' - p| \leq 1$.*

PROOF. Let i' be the integer in the conclusion of Proposition 84.7. Then $p' = v_2(n_{i'})$ works. □

We now prove Theorem 84.1.

PROOF OF THEOREM 84.1. By Corollary 84.5, we need only prove Theorem 84.1 for even-dimensional forms. Let $\mathfrak{h} := \mathfrak{h}(\varphi) \geq 1$ and let $\{n_0 > n_1 > \cdots > n_{\mathfrak{h}}\}$ with $n_i = \dim \varphi_i$ be the anisotropic pattern of φ with $n = n_0$ even.

Let H be the set $\{1, 2, \dots, \mathfrak{h}\}$. For any $i \in H$, let $\mathfrak{d}(i) := \mathfrak{h}(n_{i-1}) - \mathfrak{h}(n_i)$. Recall that $\mathfrak{d}(i) \leq 2$ for any $i \in H$ by Lemma 84.3. Let C be the subset of H consisting of all those $i \in H$ such that $\mathfrak{d}(i) = 2$. We shall construct a map $f : C \to H$ satisfying $\mathfrak{d}(j) \leq 1 - |f^{-1}(j)|$ for any $j \in f(C)$. In particular, we shall have $f(C) \subset H \setminus C$. Once such a map is constructed, we establish Theorem 84.1 as follows: The subsets $f^{-1}(j) \cup \{j\} \subset H$, where j runs over $H \setminus C$, are disjoint and cover H. In addition, the average value of $\mathfrak{d}$ on each such subset is ≤ 1, so the average value $\big(\sum_{i \in H} \mathfrak{d}(i)\big)/\mathfrak{h} = \mathfrak{h}(n)/\mathfrak{h}$ of $\mathfrak{d}$ on H is ≤ 1, i.e., $\mathfrak{h}(n) \leq \mathfrak{h}$.

So it remains to define the map f with the desired properties. Let $i \in C$. By Lemma 84.3, we have $v_2(n_i) \geq v_2(n_{i-1}) + 2$. Therefore, by Corollary 84.8, there exists $i' \in (i,\ \mathfrak{h})$ such that the set $P(n_{i'})$ contains an element p' satisfying $|p' - p| \leq 1$ for $p = v_2(n_{i-1})$. Taking the minimal i' with this property, set $f(i) = i'$. We also define $g(i)$ to be the minimal element of $P(n_{f(i)})$ satisfying $|g(i) - p| \leq 1$.

This defines the map f. To finish, we must show that f has the desired property.

First observe that by the definition of f, for any $i \in C$ and any $j \in [i,\ f(i) - 1]$ the set $P(n_j)$ does not contain any element p with $|p - v_2(n_{i-1})| \leq 1$. It follows that if $f(i_1) = f(i_2)$ for some $i_1 \neq i_2$, then for $p_1 = v_2(n_{i_1-1})$ with $p_2 = v_2(n_{i_2-1})$ one has $|p_2 - p_1| \geq 2$. Moreover, if $g(i_1) = g(i_2)$, then $|p_1 - p| \leq 1$ and $|p_2 - p| \leq 1$ for $p = g(i_1) = g(i_2)$ by definition of g. Therefore, we have

$$\text{(84.9)} \qquad \begin{aligned} &\text{if } f(i_1) = f(i_2) \text{ and } g(i_1) = g(i_2) \text{ for some } i_1 \neq i_2, \\ &\text{then } |p_2 - p_1| = 2 \text{ for } p_1 = v_2(n_{i_1-1}) \text{ and } p_2 = v_2(n_{i_2-1}). \end{aligned}$$

Let $j \in f(C)$. By the definition of f, the set difference $P(n_j) \setminus P(n_{j-1})$ is nonempty. Then $\mathfrak{d}(j) \neq 2$ by Lemma 84.3(II)(+2). Moreover, the above set difference contains an element p satisfying $\{p-1,\ p+1\} \not\subset P(n_{j-1})$. Consequently, $\mathfrak{d}(j) \neq 1$ by Lemma 84.3(II)(+1). Therefore, $\mathfrak{d}(j) \leq 0$ by Lemma 84.3.

Now let j be an element of $f(C)$ with $|f^{-1}(j)| \geq 2$. Let $i_1 < i_2$ be two different elements of $f^{-1}(j)$. Note that $i_1 < i_2 < j$. Moreover, if $p_1 = v_2(n_{i_1-1})$ and $p_2 = v_2(n_{i_2-1})$, then, by the definition of $f(i_1)$, we have $|p_2 - p_1| > 1$. We shall show that $\mathfrak{d}(j) \leq -1$. We already know that $\mathfrak{d}(j) \leq 0$. If $\mathfrak{d}(j) = 0$, then by Lemma 84.3(II)(−0), the set difference $P(n_j) \setminus P(n_{j-1})$ consists of one element p' and either $p' - 1$ or $p' + 1$ lies in $P(n_{j-1})$. Since the difference $P(n_j) \setminus P(n_{j-1})$ consists of one element p', we have $p' = g(i_1) = g(i_2)$. It follows that $\{p_1, p_2\} = \{p' - 1, p' + 1\}$. Consequently, the set $P(n_{j-1})$ contains neither $p' - 1$ nor $p' + 1$, a contradiction. Thus we have proved that $\mathfrak{d}(j) \leq -1$ if $|f^{-1}(j)| \geq 2$.

Now let j be an element of $f(C)$ with $|f^{-1}(j)| \geq 3$. Let i_1, i_2, i_3 be three different elements of $f^{-1}(j)$. The equalities $g(i_1) = g(i_2) = g(i_3)$ cannot take place simultaneously, as otherwise, by (84.9), we would have $|p_2 - p_1| = 2$, $|p_3 - p_2| = 2$, and $|p_1 - p_3| = 2$, a contradiction. However, the set difference $P(n_j) \setminus P(n_{j-1})$ can have at most two elements. Therefore, we may assume that $g(i_1) = g(i_2)$ and that $g(i_3)$ is different from $g(i_1) = g(i_2)$. Set $p' = g(i_1) = g(i_2)$. We shall show that $\mathfrak{d}(j) = -2$. We already know that $\mathfrak{d}(j) \leq -1$. If $\mathfrak{d}(j) = -1$, then by Lemma 84.3(II)(−1), the set difference $P(n_j) \setminus P(n_{j-1})$ consists of $\tilde{p} - 1$ and $\tilde{p} + 1$ for some $\tilde{p} \in P(n_{j-1})$. However, p' is neither $\tilde{p} - 1$ nor $\tilde{p} + 1$, a contradiction.

We finish the proof by showing that $|f^{-1}(j)|$ is never ≥ 4. Indeed, if $|f^{-1}(j)| \geq 4$, then the set difference $P(n_j) \setminus P(n_{j-1})$ contains two elements p' and p'' with none of $p' \pm 1$ or $p'' \pm 1$ lying in $P(n_{j-1})$, contradicting Lemma 84.3. □

CHAPTER XVI

The Variety of Maximal Totally Isotropic Subspaces

So far in this book, the projective quadric has been the only variety associated with a quadratic form. In this chapter, we introduce another variety, the variety of maximal isotropic subspaces.

85. The variety $\mathrm{Gr}(\varphi)$

Let φ be a nondegenerate quadratic form on V over F. In this chapter, we study the scheme $\mathrm{Gr}(\varphi)$ of maximal totally isotropic subspaces of V. We view $\mathrm{Gr}(\varphi)$ as a closed subscheme of the Grassmannian variety of V. Let n be the integer part of $(\dim\varphi - 1)/2$, so $\dim\varphi = 2n+1$ or $2n+2$.

Example 85.1. If $\dim\varphi = 1$, we have $\mathrm{Gr}(\varphi) = \operatorname{Spec} F$. If $\dim\varphi = 2$ or 3, then $\mathrm{Gr}(\varphi)$ coincides with the quadric of φ; that is $\mathrm{Gr}(\varphi) = \operatorname{Spec} C_0(\varphi)$ if $\dim\varphi = 2$ and $\mathrm{Gr}(\varphi)$ is the conic curve associated to the quaternion algebra $C_0(\varphi)$ if $\dim\varphi = 3$.

The orthogonal algebraic group $\mathbf{O}(V,\varphi)$ acts transitively on $\mathrm{Gr}(\varphi)$. We write $\mathbf{O}^+(V,\varphi)$ for the (connected) special orthogonal group (cf. [**86**, §23]). If $\dim\varphi$ is odd, $\mathbf{O}^+(V,\varphi)$ acts transitively on $\mathrm{Gr}(\varphi)$ and therefore, $\mathrm{Gr}(\varphi)$ is a smooth projective variety over F.

Suppose that $\dim\varphi = 2n+2$ is even. Then the group $\mathbf{O}(V,\varphi)$ has two connected components, one is $\mathbf{O}^+(V,\varphi)$. The factor group $\mathbf{O}(V,\varphi)/\,\mathbf{O}^+(V,\varphi)$ is identified with the Galois group over F of the center Z of the even Clifford algebra $C_0(V,\varphi)$. Recall that Z is an étale quadratic F-algebra (cf. Proposition 11.6). The class of $Z(\varphi)$ in $\mathrm{\acute{E}t}_2(F)$ is the discriminant of φ (cf. §13).

A point of $\mathrm{Gr}(\varphi)$ over a commutative ring R is a totally isotropic direct summand P of rank $n+1$ of the R-module $V_R = V\otimes_F R$. Since $p^2 = 0$ in the Clifford algebra $C(V,\varphi)_R$ for every $p\in P$, the inclusion of P into V_R gives rise to an injective R-module homomorphism $h : \Lambda^{n+1}(P) \to C(V,\varphi)_R$. Let W be the image of h. Since $ZW = W$, left multiplication by elements of the center Z of $C_0(V,\varphi)$ defines an F-algebra homomorphism $Z \to \operatorname{End}_R(W) = R$. Therefore, we have a morphism $\mathrm{Gr}(\varphi) \to \operatorname{Spec} Z$, so $\mathrm{Gr}(\varphi)$ is a scheme over Z.

If the discriminant of φ is trivial, i.e., $Z = F\times F$, the scheme $\mathrm{Gr}(\varphi)$ has two smooth (irreducible) connected components Gr_1 and Gr_2 permuted by the quotient $\mathbf{O}(V,\varphi)/\,\mathbf{O}^+(V,\varphi)$. More precisely, they are isomorphic under any reflection of V. If Z is a field, the discriminant of φ_Z is trivial and therefore $\mathrm{Gr}(\varphi)$ is isomorphic to a connected component of $\mathrm{Gr}(\varphi_Z)$.

The varieties of even- and odd-dimensional forms are related by the following statement.

Proposition 85.2. *Let φ be a nondegenerate quadratic form on V over F of dimension $2n+2$ and trivial discriminant and let φ' be a nondegenerate subform of φ on a subspace $V' \subset V$ of codimension 1. Let Gr_1 be a connected component of $\mathrm{Gr}(\varphi)$. Then the assignment $U \mapsto U \cap V'$ gives rise to an isomorphism $\mathrm{Gr}_1 \xrightarrow{\sim} \mathrm{Gr}(\varphi')$.*

PROOF. Since both varieties Gr_1 and $\mathrm{Gr}(\varphi')$ are smooth, it suffices to show that the assignment induces a bijection on points over any field extension L/F. Moreover, we may assume that $L = F$. Let $U' \subset V'$ be a totally isotropic subspace of dimension n. Then the orthogonal complement $U'^{\perp}$ of U' in V is $(n+2)$-dimensional and the induced quadratic form on $H = U'^{\perp}/U'$ has trivial discriminant (i.e., H is a hyperbolic plane). The space H has exactly two isotropic lines permuted by a reflection. Therefore, the pre-images of these lines in V are two totally isotropic subspaces of dimension $n+1$ living in different components of $\mathrm{Gr}(\varphi)$. Thus exactly one of them represents a point of Gr_1 over F. □

Let φ' be a nondegenerate subform of codimension 1 of a nondegenerate quadratic form φ of even dimension. Let Z be the discriminant of φ. By Proposition 85.2, we have that $\mathrm{Gr}(\varphi')_Z$ is isomorphic to a connected component Gr_1 of $\mathrm{Gr}(\varphi_Z)$ and therefore, $\mathrm{Gr}(\varphi) \simeq \mathrm{Gr}_1 \simeq \mathrm{Gr}(\varphi')_Z$.

Example 85.3. If $\dim \varphi = 4$, then $\mathrm{Gr}(\varphi)$ is the conic curve (over Z) associated to the quaternion algebra $C_0(\varphi)$.

Exercise 85.4. Show that if $3 \le \dim \varphi \le 6$, then $\mathrm{Gr}(\varphi)$ is isomorphic to the Severi-Brauer variety associated to the even Clifford algebra $C_0(\varphi)$.

86. The Chow ring of $\mathrm{Gr}(\varphi)$ in the split case

In this section, we present the calculation of the Chow groups of $\mathrm{Gr}(\varphi)$ given by Vishik in [**134**]. Set $\mathrm{Gr} := \mathrm{Gr}(\varphi)$ and $r := \dim \varphi - n - 1$, where n is the integer part of $(\dim \varphi - 1)/2$. We define the *tautological* vector bundle E over Gr of rank r to be the restriction of the tautological vector bundle over the Grassmannian variety of V, i.e., variety E is the closed subvariety of the trivial bundle $V\mathbb{1} := V \times \mathrm{Gr}$ consisting of pairs (u, U) such that $u \in U$. The projective bundle $\mathbb{P}(E)$ is a closed subvariety of $X \times \mathrm{Gr}$, where X is the (smooth) projective quadric of φ.

Let $E^{\perp}$ be the kernel of the natural morphism $V\mathbb{1} \to E^{\vee}$ given by the polar bilinear form $\mathfrak{b}_{\varphi}$. If $\dim \varphi = 2n+2$, we have $U^{\perp} = U$ for any totally isotropic subspace $U \subset V$ of dimension $n+1$, hence $E^{\perp} = E$.

In the case when $\dim \varphi = 2n+1$, the situation is as follows. For any totally isotropic subspace $U \subset V$ of dimension n, the orthogonal complement $U^{\perp}$ contains U as a subspace of codimension 1. Therefore, $E^{\perp}$ is a vector bundle over Gr of rank $n+1$ containing E. The fiber of $E^{\perp}$ over U is the orthogonal complement $U^{\perp}$.

Now suppose that φ is isotropic. Choose an isotropic line $L \subset V$. Set $\widetilde{V} = L^{\perp}/L$ and let $\tilde{\varphi}$ be the quadratic form on $\widetilde{V}$ induced by φ. Denote the projective quadric of $\tilde{\varphi}$ by $\widetilde{X}$.

A totally isotropic subspace of $\widetilde{V}$ of dimension $r-1$ is of the form U/L, where U is a totally isotropic subspace of V of dimension r containing L. Therefore, we can view the variety $\widetilde{\mathrm{Gr}} := \mathrm{Gr}(\tilde{\varphi})$ of maximal totally isotropic subspaces of $\widetilde{V}$ as a closed subvariety of Gr. Denote by $i : \widetilde{\mathrm{Gr}} \to \mathrm{Gr}$ the closed embedding.

Let U be a totally isotropic subspace of V of dimension r that does not contain L. Then $\dim(U \cap L^\perp) = r - 1$ and $\big((U \cap L^\perp) + L\big)/L$ is a totally isotropic subspace of $\widetilde{V}$ of dimension $r - 1$.

Lemma 86.1. *The morphism* $f : \mathrm{Gr} \setminus \widetilde{\mathrm{Gr}} \to \widetilde{\mathrm{Gr}}$ *that takes* U *to* $\big((U \cap L^\perp) + L\big)/L$ *is an affine bundle.*

PROOF. We use the criterion of Lemma 52.12. Let R be a local commutative F-algebra. An F-morphism $\operatorname{Spec} R \to \widetilde{\mathrm{Gr}}$, or equivalently an R-point of $\widetilde{\mathrm{Gr}}$, is given by a direct summand $\widetilde{U}$ of the R-module $V_R = V \otimes_F R$ of rank r with $L_R \subset \widetilde{U}$.

An R-point of $\mathrm{Gr} \setminus \widetilde{\mathrm{Gr}}$ is a totally isotropic direct summand U of V_R of rank r with $U + L_R$ a direct summand U of V_R of rank $r+1$. If $f(U) = \widetilde{U}/L_R$, then $\widetilde{U} \subset U + L_R$. The assignment $U \mapsto (U + L_R)/\widetilde{U}$ gives rise to an isomorphism between the fiber $\operatorname{Spec} R \times_{\widetilde{\mathrm{Gr}}} (\mathrm{Gr} \setminus \widetilde{\mathrm{Gr}})$ of f over $\widetilde{U}/L_R$ and $\mathbb{P}_R(V_R/\widetilde{U}) \setminus \mathbb{P}_R(L_R^\perp/\widetilde{U}) \simeq \mathbb{A}^n_R$. By Lemma 52.12, we have f is an affine bundle. $\square$

Note that $\dim \mathrm{Gr} = \dim \widetilde{\mathrm{Gr}} + n$, so $\dim \mathrm{Gr} = n(n+1)/2$.

By Lemma 86.1, Gr is a cellular variety with the short filtration $\widetilde{\mathrm{Gr}} \subset \mathrm{Gr}$, hence by Corollary 66.4, we have a decomposition of Chow motives:

$$M(\mathrm{Gr}) = M(\widetilde{\mathrm{Gr}}) \oplus M(\widetilde{\mathrm{Gr}})(n). \tag{86.2}$$

The morphism $M(\widetilde{\mathrm{Gr}}) \to M(\mathrm{Gr})$ is induced by the embedding $i : \widetilde{\mathrm{Gr}} \to \mathrm{Gr}$ and the morphism $M(\widetilde{\mathrm{Gr}})(n) \to M(\mathrm{Gr})$ is given by the transpose of the closure of the graph of f, the class of which we shall denote by $\beta \in \mathrm{CH}(\widetilde{\mathrm{Gr}} \times \mathrm{Gr})$. The correspondence β is given by the scheme of pairs $(W/L, U)$, where U is a totally isotropic r-dimensional subspace of V, W is a totally isotropic r-dimensional subspace of $L^\perp$ containing L, and $\dim(U + W) \le r + 1$.

For the rest of this section, we shall assume that φ is split. It follows by induction, using (86.2) and Example 85.1, that $\mathrm{CH}(\mathrm{Gr})$ is a free abelian group of rank 2^r. We shall determine the multiplicative structure of $\mathrm{CH}(\mathrm{Gr})$.

Since the motive of X (and also Gr) is split, we have $\mathrm{CH}(X \times \mathrm{Gr}) = \mathrm{CH}(X) \otimes \mathrm{CH}(\mathrm{Gr})$ by Proposition 64.3. In other words, $\mathrm{CH}(X \times \mathrm{Gr})$ is a free module over $\mathrm{CH}(\mathrm{Gr})$ with basis

$$\big\{h^k \times [\mathrm{Gr}],\ l_k \times [\mathrm{Gr}] \mid k \in [0,\ n-1]\big\} \quad \text{if} \dim\varphi = 2n+1,$$
$$\big\{h^k \times [\mathrm{Gr}],\ l_k \times [\mathrm{Gr}],\ l_n \times [\mathrm{Gr}],\ l'_n \times [\mathrm{Gr}] \mid k \in [0,\ n-1]\big\} \quad \text{if} \dim\varphi = 2n+2.$$

Note that in the even-dimensional case, we have assumed that X is oriented.

In both cases, $\mathbb{P}(E)$ is a closed subvariety of $X \times \mathrm{Gr}$ of codimension n. Therefore, in the odd-dimensional case there are unique elements $e_k \in \mathrm{CH}^k(\mathrm{Gr})$, $k \in [0,\ n]$, satisfying

$$[\mathbb{P}(E)] = l_{n-1} \times e_0 + \sum_{k=1}^{n} h^{n-k} \times e_k \tag{86.3}$$

in $\mathrm{CH}(X \times \mathrm{Gr})$. Pulling this back with respect to the canonical morphism $X_{F(\mathrm{Gr})} \to X \times \mathrm{Gr}$, we see that $e_0 = 1$.

In the even-dimensional case, there are unique elements $e_k \in \mathrm{CH}^k(\mathrm{Gr})$, $k \in [0,\ n]$ and $e'_0 \in \mathrm{CH}^0(\mathrm{Gr})$, satisfying

$$(86.4) \qquad [\mathbb{P}(E)] = l_n \times e_0 + l'_n \times e'_0 + \sum_{k=1}^{n} h^{n-k} \times e_k$$

in $\mathrm{CH}(X \times \mathrm{Gr})$. Choose a totally isotropic subspace $U \subset V$ of dimension $n+1$ so that $[\mathbb{P}(U)] = l_n$ in $\mathrm{CH}(X)$ and let U' be a reflection of U. It follows from Exercise 68.4 that $[\mathbb{P}(U')] = l'_n$. Let g denote the generic point of Gr whose closure contains $[U]$. Let g' be another generic point of Gr whose closure contains $[U']$. Note that $\mathrm{CH}^0(\mathrm{Gr}) = \mathbb{Z}[g] \oplus \mathbb{Z}[g']$. Pulling back equation (86.4) with respect to the two morphisms $X \to X \times \mathrm{Gr}$ given by the points $[U]$ and $[U']$ respectively, we see that $e_0 = [g]$ and $e'_0 = [g']$. In particular, e_0 and e'_0 are orthogonal idempotents of $\mathrm{CH}^0(\mathrm{Gr})$, hence $e_0 + e'_0 = 1$.

It follows that for every totally isotropic subspace $W \subset V$ of dimension $n+1$ with $[W]$ in the closure of g (resp. g'), we have $[W] = l_n$ (respectively $[W] = l'_n$). In particular, to give an orientation of X is to choose one of the two connected components of Gr.

The multiplication rule in Proposition 68.1 for $\mathrm{CH}(X)$ implies that in both cases

$$e_k = p_*\big((l_{n-k} \times 1) \cdot [\mathbb{P}(E)]\big)$$

for $k \in [1,\ n]$, where $p : X \times \mathrm{Gr} \to \mathrm{Gr}$ is the projection.

We view the cycle $\gamma = [\mathbb{P}(E)]$ in $\mathrm{CH}(X \times \mathrm{Gr})$ as the *incidence correspondence* $X \rightsquigarrow \mathrm{Gr}$. It follows from Proposition 63.2 that the induced homomorphism $\gamma_* : \mathrm{CH}(X) \to \mathrm{CH}(\mathrm{Gr})$ takes l_{n-k} to e_k.

Let $s : \mathbb{P}(E) \to \mathrm{Gr}$ and $t : \mathbb{P}(E) \to X$ be the two projections. Proposition 62.7 provides the following simple formula for e_k:

$$(86.5) \qquad e_k = s_* \circ t^*(l_{n-k}).$$

Lemma 86.6. *We have* $e_n = [\widetilde{\mathrm{Gr}}]$ *in* $\mathrm{CH}^n(\mathrm{Gr})$.

PROOF. The element $t^*(l_0)$ coincides with the cycle of the intersection $(\{L\} \times \mathrm{Gr}) \cap \mathbb{P}(E) = \{L\} \times \widetilde{\mathrm{Gr}}$. It follows from (86.5) that $[\widetilde{\mathrm{Gr}}] = s_* \circ t^*(l_0) = e_n$. □

We write $\tilde{h}$ and $\tilde{l}_i$ for the standard generators of $\mathrm{CH}(\widetilde{X})$. Recall that the incidence correspondence $\alpha : \widetilde{X} \rightsquigarrow X$ is given by the scheme of pairs $(A/L, B)$ of one-dimensional isotropic subspaces of $\widetilde{V}$ and V respectively with $B \subset A$. By Lemma 72.3, we can orient $\widetilde{X}$ (in the case $\dim \varphi$ is even) so that $\alpha_*(\tilde{l}_{k-1}) = l_k$ and $\alpha^*(l_k) = \tilde{l}_{k-1}$ for all k.

Denote by $\tilde{e}_k \in \mathrm{CH}^k(\widetilde{\mathrm{Gr}})$ the elements given by (86.3) or (86.4) for $\widetilde{\mathrm{Gr}}$. Similarly, we have the incidence correspondence $\tilde{\gamma} : \widetilde{X} \rightsquigarrow \widetilde{\mathrm{Gr}}$ with $\tilde{\gamma}_*(\tilde{l}_{n-1-k}) = \tilde{e}_k$.

Lemma 86.7. *The diagram of correspondences*

$$\begin{array}{ccccc} \widetilde{X} & \overset{\alpha}{\rightsquigarrow} & X & \overset{\alpha^t}{\rightsquigarrow} & \widetilde{X} \\ {\scriptstyle\tilde{\gamma}}\wr\downarrow & & {\scriptstyle\gamma}\wr\downarrow & & {\scriptstyle\tilde{\gamma}}\wr\downarrow \\ \widetilde{\mathrm{Gr}} & \overset{\beta}{\rightsquigarrow} & \mathrm{Gr} & \overset{i^t}{\rightsquigarrow} & \widetilde{\mathrm{Gr}} \end{array}$$

is commutative.

PROOF. By Corollary 57.22, all calculations can be done on the level of cycles representing the correspondences. By definition of the composition of correspondences, the compositions $\gamma \circ \alpha$ and $\beta \circ \tilde{\gamma}$ coincide with the cycle of the subscheme of $\widetilde{X} \times \mathrm{Gr}$ consisting of all pairs $(A/L, U)$ with $\dim(A+U) \leq r+1$. Similarly, the compositions $\tilde{\gamma} \circ \alpha^t$ and $i^t \circ \gamma$ coincide with the cycle of the subscheme of $X \times \widetilde{\mathrm{Gr}}$ consisting of all pairs $(B, \widetilde{U}/L)$ with $B \subset \widetilde{U}$. □

Corollary 86.8. *We have $\beta_*(\tilde{e}_k) = e_k$ and $i^*(e_k) = \tilde{e}_k$ for all $k \in [0,\ n-1]$.*

PROOF. The equalities $\beta_*(\tilde{e}_0) = e_0$ and $i^*(e_0) = \tilde{e}_0$ follow from the fact that X and $\widetilde{X}$ have compatible orientations. If $k \geq 1$, we have by Lemma 86.7,

$$\beta_*(\tilde{e}_k) = \beta_* \circ \tilde{\gamma}_*(\tilde{l}_{n-1-k}) = \gamma_* \circ \alpha_*(\tilde{l}_{n-1-k}) = \gamma_*(l_{n-k}) = e_k,$$

and

$$\begin{aligned} i^*(e_k) &= i^t_*(e_k) = i^t_* \circ \gamma_*(l_{n-k}) \\ &= \tilde{\gamma}_* \circ \alpha^t_*(l_{n-k}) = \tilde{\gamma}_* \circ \alpha^*(l_{n-k}) = \tilde{\gamma}_*(\tilde{l}_{n-1-k}) = \tilde{e}_k. \end{aligned}$$ □

For a subset I of $[0,\ n]$ let e_I be the product of e_k for all $k \in I$. Similarly, we define $\tilde{e}_J$ for any subset $J \subset [0,\ n-1]$.

Corollary 86.9. *We have $i_*(\tilde{e}_J) = e_J \cdot e_n = e_{J \cup \{n\}}$ for every $J \subset [0,\ n-1]$.*

PROOF. By Corollary 86.8, we have $i^*(e_J) = \tilde{e}_J$. It follows from Lemma 86.6 and the Projection Formula (Proposition 56.9) that

$$i_*(\tilde{e}_J) = i_*(i^*(e_J) \cdot 1) = e_J \cdot i_*(1) = e_J \cdot e_n = e_{J \cup \{n\}}.$$ □

Corollary 86.10. *The monomial $e_{[0,\ n]} = e_0 e_1 \cdots e_n$ is the class of a rational point in $\mathrm{CH}_0(\mathrm{Gr})$.*

PROOF. The statement follows from the formula $e_{[0,\ n]} = i_*(\tilde{e}_{[0,\ n-1]})$ and by induction on n. □

Let $j : \mathrm{Gr} \setminus \widetilde{\mathrm{Gr}} \to \mathrm{Gr}$ be the open embedding. Let $f : \mathrm{Gr} \setminus \widetilde{\mathrm{Gr}} \to \widetilde{\mathrm{Gr}}$ be the morphism in Lemma 86.1.

Lemma 86.11. *We have $f^*(\tilde{e}_J) = j^*(e_J)$ for any $J \subset [0,\ n-1]$.*

PROOF. It suffices to prove that $f^*(\tilde{e}_k) = j^*(e_k)$ for all $k \in [0,\ n-1]$. By the construction of β (cf. §66), we have $\beta^t \circ j = f$. It follows from Corollary 86.8 that $f^*(\tilde{e}_k) = j^* \circ (\beta^t)^*(\tilde{e}_k) = j^* \circ \beta_*(\tilde{e}_k) = j^*(e_k)$. □

Theorem 86.12. *Let φ be a nondegenerate quadratic form on V over F of dimension $2n+1$ or $2n+2$ and let* Gr *be the scheme of maximal totally isotropic subspaces of V. Then the set of monomials e_I for all 2^n subsets $I \subset [1,\ n]$ is a basis of $\mathrm{CH}(\mathrm{Gr})$ over $\mathrm{CH}^0(\mathrm{Gr})$.* φ split?

PROOF. We induct on n. The localization property gives the exact sequence (cf. the proof of Theorem 66.2)

$$0 \to \mathrm{CH}(\widetilde{\mathrm{Gr}}) \xrightarrow{i_*} \mathrm{CH}(\mathrm{Gr}) \xrightarrow{j^*} \mathrm{CH}(\mathrm{Gr} \setminus \widetilde{\mathrm{Gr}}) \to 0.$$

By the induction hypothesis and Corollary 86.9, the set of monomials e_I for all I containing n is a basis of the image of i_*. Since $f^* : \mathrm{CH}(\widetilde{\mathrm{Gr}}) \to \mathrm{CH}(\mathrm{Gr} \setminus \widetilde{\mathrm{Gr}})$ is an isomorphism by Theorem 52.13, again by the induction hypothesis and Lemma

86.11, the set of all the elements $j^*(e_I)$ with $n \notin I$ is a basis of $\mathrm{CH}(\mathrm{Gr} \setminus \widetilde{\mathrm{Gr}})$. The statement follows. □

We now can compute the Chern classes of the tautological vector bundle E over Gr.

Proposition 86.13. *We have* $c_k(V\mathbb{1}/E) = c_k(E^\vee) = 2e_k$ *and* $c_k(E) = (-1)^k 2e_k$ *for all* $k \in [1,\ n]$.

PROOF. Let $s : X \to \mathbb{P}(V)$ be the closed embedding. Let H denote the class of a hyperplane in $\mathbb{P}(V)$. We have $s_*(h^k) = 2H^{k+1}$ for all $k \geq 0$.

First suppose that $\dim \varphi = 2n + 1$. It follows from (86.3) that

$$[\mathbb{P}(E)] = s_*(l_{n-1}) \times 1 + \sum_{k=1}^{n} 2H^{n+1-k} \times e_k$$

in $\mathrm{CH}\big(\mathbb{P}(V) \times \mathrm{Gr}\big)$.

On the other hand, by Proposition 58.10 applied to the subbundle E of $V\mathbb{1}$, we have

$$[\mathbb{P}(E)] = \sum_{k=0}^{n+1} H^{n+1-k} \times c_k(V\mathbb{1}/E).$$

It follows from the Projective Bundle Theorem 53.10 that $c_k(V\mathbb{1}/E) = 2e_k$ for $k \in [1,\ n]$.

By duality, $V\mathbb{1}/E^\perp \simeq E^\vee$. Note that the line bundle $E^\perp/E$ carries a nondegenerate quadratic form, hence is isomorphic to its dual. Since $\mathrm{Pic}(\mathrm{Gr}) = \mathrm{CH}^1(\mathrm{Gr})$ is torsion-free, we conclude that $E^\perp/E \simeq \mathbb{1}$. Therefore,

$$c(E^\vee) = c(V\mathbb{1}/E^\perp) = c(V\mathbb{1}/E).$$

The proof in the case $\dim \varphi = 2n+2$ proceeds along similar lines: one uses the equality (86.4) and the duality isomorphism $V\mathbb{1}/E \simeq E^\vee$. □

Remark 86.14. Let φ be a nondegenerate quadratic form that is not necessarily split. By Proposition 86.13, the classes $2e_k$, $k \geq 1$, that are a priori defined over a splitting field of φ, are in fact defined over F.

In order to determine the multiplicative structure of CH(Gr), we present the set of defining relations between the e_k. For convenience, we set $e_k = 0$ if $k > n$.

Since $c(V\mathbb{1}/E) \circ c(E) = c(V\mathbb{1}) = 1$ and CH(Gr) is torsion-free, it follows from Proposition 86.13 that

$$e_k^2 - 2e_{k-1}e_{k+1} + 2e_{k-2}e_{k+2} - \cdots + (-1)^{k-1}2e_1e_{2k-1} + (-1)^k e_{2k} = 0 \tag{86.15}$$

for all $k \geq 1$.

Proposition 86.16. *The equalities* (86.15) *form a set of defining relations between the generators* e_k *of the ring* CH(Gr) *over* $\mathrm{CH}^0(\mathrm{Gr})$.

PROOF. Let A be the factor ring of the polynomial ring $\mathbb{Z}[t_1, t_2, \dots, t_n]$ modulo the ideal generated by polynomials giving the relations (86.15). We claim that the ring homomorphism $A \to \mathrm{CH}(\mathrm{Gr})$ taking t_k to e_k is an isomorphism.

Call a monomial $t_1^{r_1} t_2^{r_2} \cdots t_n^{r_n}$ with $r_i \geq 0$ *basic* if $r_k = 0$ or 1 for every k. By Theorem 86.12, it is sufficient to prove that the ring A is generated by classes of basic monomials.

We define the *weight* $w(m)$ of a monomial $m = t_1^{r_1} t_2^{r_2} \cdots t_n^{r_n}$ by the formula

$$w(m) = \sum_{k=1}^{n} k^2 \cdot r_k$$

and the weight of a polynomial $f(t_1, \dots, t_n)$ over $\mathbb{Z}$ as the minimum of weights of its nonzero monomials. Clearly, $w(m \cdot m') = w(m) + w(m')$. For example, in the formula (86.15), we have $w(t_k^2) = 2k^2$, $w(t_{k-i}t_{k+i}) = 2k^2 + 2i^2$, and $w(t_{2k}) = 4k^2$. Thus, t_k^2 is the monomial of the lowest weight in the formula (86.15).

Let f be a polynomial representing an element of the ring A. Applying formula (86.15) to the square of a variable t_k in a nonbasic monomial of f of the lowest weight increases the weight but not the degree of f. Since the weight of a polynomial of degree d is at most $n^2 d$, we eventually reduce to a polynomial having only basic monomials. □

The relations (86.15) look particularly simple modulo 2:

$$e_k^2 \equiv e_{2k} \quad \mod \ 2\,\mathrm{CH}(\mathrm{Gr}) \quad \text{for all} \quad k \geq 1.$$

Proposition 86.17. *Let φ be a split nondegenerate quadratic form on V over F of dimension $2n+2$ and let φ' be a nondegenerate subform of φ on a subspace $V' \subset V$ of codimension 1. Let f denote the morphism $\mathrm{Gr}(\varphi) \to \mathrm{Gr}(\varphi')$ taking U to $U \cap V'$, and let e'_k, $k \geq 1$, denote the standard generators of $\mathrm{CH}\big(\mathrm{Gr}(\varphi')\big)$. Then $f^*(e'_k) = e_k$ for all $k \in [1, n]$.*

PROOF. Denote by $E \to \mathrm{Gr}(\varphi)$ and $E' \to \mathrm{Gr}(\varphi')$ the tautological vector bundles of ranks $n+1$ and n respectively. The line bundle

$$E/f^*(E') = E/(V'\mathbb{1} \cap E) \simeq (E + V'\mathbb{1})/V'\mathbb{1} = V\mathbb{1}/V'\mathbb{1}$$

is trivial. In particular, $c(E) = c(f^*E') = f^*c(E')$. It follows from Proposition 86.13 that

$$2f^*(e'_k) = (-1)^k f^*\big(c_k(E')\big) = (-1)^k c_k\big(f^*(E')\big) = (-1)^k c_k(E) = 2e_k.$$

The result follows since $\mathrm{CH}\big(\mathrm{Gr}(\varphi)\big)$ is torsion-free. □

87. The Chow ring of Gr(φ) in the general case

Let φ be a nondegenerate quadratic form of dimension $2n+1$ on V over an arbitrary field F and let $X = X_\varphi$. Let Y be a smooth proper scheme over F and let $h : Y \to \mathrm{Gr} = \mathrm{Gr}(\varphi)$ be a morphism. We set $E' = h^*(E)$, where E is the tautological vector bundle over Gr, and view $\mathbb{P}(E')$ as a closed subscheme of $X \times Y$.

Proposition 87.1. *The $\mathrm{CH}(Y)$-module $\mathrm{CH}(X \times Y)$ is free with basis h^k, $h^k \cdot [\mathbb{P}(E')]$ with $k \in [1, n-1]$.*

PROOF. Let $V\mathbb{1}$ denote the trivial vector bundle $V \times Y$ over Y. We claim that the restriction $f : T = (X \times Y) \setminus \mathbb{P}(E') \to \mathbb{P}(V\mathbb{1}/E'^{\perp})$ of the natural morphism $f : \mathbb{P}(V\mathbb{1}) \setminus \mathbb{P}(E'^{\perp}) \to \mathbb{P}(V\mathbb{1}/E'^{\perp})$ is an affine bundle. To do so we use the criterion of Lemma 52.12.

Let R be a local commutative F-algebra. An F-morphism

$$\operatorname{Spec} R \to \mathbb{P}(V\mathbb{1}/E'^{\perp}),$$

equivalently, an R-point of $\mathbb{P}(V\mathbb{1}/E'^{\perp})$, determines a pair (U, W) where U is a totally isotropic direct summand of V_R of rank n and W is a direct summand of

V_R of rank $n+2$ containing $U^\perp$. Since $\operatorname{rank} W^\perp = n-1$, one can choose an R-basis of W so that the restriction of the quadratic form φ on $W/W^\perp$ is equal to $xy + az^2$ for some $a \in R^\times$ and U is given by $x = z = 0$ in W. Therefore, the fiber $\operatorname{Spec} R \times_{\mathbb{P}(V\mathbb{1}/E'^\perp)} T$ is given by the equation $y/x = a(z/x)^2$ over R and hence is isomorphic to an affine space over R. By Lemma 52.12, f is an affine bundle.

Thus $X \times Y$ is equipped with the structure of a cellular scheme. In particular, we have a (split) exact sequence

$$0 \to \mathrm{CH}\big(\mathbb{P}(E')\big) \xrightarrow{i_*} \mathrm{CH}(X \times Y) \to \mathrm{CH}(T) \to 0$$

and the isomorphism

$$f^* : \mathrm{CH}\big(\mathbb{P}(V\mathbb{1}/E'^\perp)\big) \xrightarrow{\sim} \mathrm{CH}(T).$$

The restriction of the canonical line bundle over $\mathbb{P}(V)$ to $X \times Y$ and $\mathrm{CH}\big(\mathbb{P}(E')\big)$ are also canonical bundles. It follows from the Projective Bundle Theorem 53.10 and the Projection Formula (Proposition 56.9) that the image of i_* is a free $\mathrm{CH}(Y)$-module with basis $h^k \cdot [\mathbb{P}(E')]$, $k \in [0,\ n-1]$.

The geometric description of the canonical line bundle given in §104.C shows that the pull-back with respect to f of the canonical line bundle is the restriction to T of the canonical bundle on $X \times Y$. Again, it follows from the Projective Bundle Theorem 53.10 that $\mathrm{CH}(T)$ is a free $\mathrm{CH}(Y)$-module with basis the restrictions of h^k, $k \in [0,\ n-1]$, on T. The statement readily follows. □

Remark 87.2. The proof of Proposition 87.1 gives the motivic decomposition

$$M(X \times Y) = M\big(\mathbb{P}(E')\big) \oplus M\big(\mathbb{P}(V\mathbb{1}/E'^\perp)\big)(n).$$

As in the case of quadrics, we write $\mathrm{CH}(\overline{\mathrm{Gr}})$ for the colimit of $\mathrm{CH}(\mathrm{Gr}_L)$ over all field extensions L/F and $\overline{\mathrm{CH}}(\mathrm{Gr})$ for the image of $\mathrm{CH}(\mathrm{Gr})$ in $\mathrm{CH}(\overline{\mathrm{Gr}})$. We say that a cycle α in $\mathrm{CH}(\overline{\mathrm{Gr}})$ is *rational* if it belongs to $\overline{\mathrm{CH}}(\mathrm{Gr})$. We use similar notations and definitions for the cycles on Gr^2, the classes of cycles modulo 2, etc.

Corollary 87.3. *The elements* $(e_k \times 1) + (1 \times e_k)$ *in* $\mathrm{CH}(\overline{\mathrm{Gr}}^{\,2})$ *are rational for all* $k \in [1,\ n]$.

Proof. Let E_1 and E_2 be the two pull-backs of E on $\mathrm{Gr}^2 := \mathrm{Gr} \times \mathrm{Gr}$. Pulling the formula (86.3) back to $\overline{X} \times \overline{\mathrm{Gr}}^{\,2}$, we get in $\mathrm{CH}(\overline{X} \times \overline{\mathrm{Gr}}^{\,2})$:

$$[\mathbb{P}(E_1)] = l_{n-1} \times 1 \times 1 + \sum_{k=1}^{n} h^{n-k} \times e_k \times 1,$$

$$[\mathbb{P}(E_2)] = l_{n-1} \times 1 \times 1 + \sum_{k=1}^{n} h^{n-k} \times 1 \times e_k.$$

Therefore the cycle

$$[\mathbb{P}(E_1)] - [\mathbb{P}(E_2)] = \sum_{k=1}^{n} h^{n-k} \times (e_k \times 1 - 1 \times e_k)$$

is rational. Applying Proposition 87.1 to the variety Gr^2, we have the cycles $(e_k \times 1) - (1 \times e_k)$ are also rational. Note that by Proposition 86.13, the cycles $2e_k$ are rational. □

Now consider the Chow groups Ch(Gr), Ch($\overline{\mathrm{Gr}}$) modulo 2 and write $\overline{\mathrm{Ch}}(\mathrm{Gr})$ for the image of Ch(Gr) in Ch($\overline{\mathrm{Gr}}$). We still write e_k for the class of the generator in $\mathrm{Ch}^k(\overline{\mathrm{Gr}})$.

For every subset $I \subset [1,\ n]$, the rational correspondence

$$x_I = \prod_{k\in I} \big[(e_k \times 1) + (1 \times e_k)\big] \in \overline{\mathrm{Ch}}(\mathrm{Gr}^2) \tag{87.4}$$

defines an endomorphism $(x_I)_*$ of $\mathrm{Ch}(\overline{\mathrm{Gr}})$ taking $\overline{\mathrm{Ch}}(\mathrm{Gr})$ into $\overline{\mathrm{Ch}}(\mathrm{Gr})$.

Lemma 87.5. *For any subsets $I, J \subset [1,\ n]$, we have*

$$(x_J)_*(e_I) = \begin{cases} e_{I\cap J} & \text{if } I \cup J = [1,\ n], \\ 0 & \text{otherwise,} \end{cases}$$

in $\mathrm{Ch}(\overline{\mathrm{Gr}})$.

PROOF. We have $x_J = \sum e_{J_1} \times e_{J_2}$, where the sum is taken over all subsets J_1 and J_2 of $[1,\ n]$ such that J is the disjoint union of J_1 and J_2. Hence

$$(x_J)_*(e_I) = \sum \deg(e_I \cdot e_{J_1})\, e_{J_2}$$

and the statement is implied by the following lemma.

Lemma 87.6. *For any subsets $I, J \subset [1,\ n]$,*

$$\deg(e_I \cdot e_J) \equiv \begin{cases} 1 \mod 2 & \text{if } J = [1,\ n] \setminus I, \\ 0 \mod 2 & \text{otherwise.} \end{cases}$$

PROOF. If $J = [1,\ n] \setminus I$, the product $e_I \cdot e_J = e_{[1,\ n]}$ is the class of a rational point of $\overline{\mathrm{Gr}}$ by Corollary 86.10, hence $\deg(e_I \cdot e_J) = 1$. Otherwise modulo 2, $e_I \cdot e_J$ is either zero or the monomial e_K for some K different from $[1,\ n]$ (one uses the relations between the generators modulo 2). Hence $\deg(e_I \cdot e_J) \equiv 0 \mod 2$. □

The following statement is due to Vishik. We give a proof different from the one in [**134**].

Theorem 87.7. *Let* Gr *be the variety of maximal isotropic subspaces of a nondegenerate quadratic form of dimension $2n + 1$ or $2n + 2$. Then the ring $\overline{\mathrm{Ch}}(\mathrm{Gr})$ is generated by all e_k, $k \in [0,\ n]$, such that $e_k \in \overline{\mathrm{Ch}}(\mathrm{Gr})$.*

PROOF. By Propositions 85.2 and 86.17, it suffices to consider the case of dimension $2n + 1$. It follows from Theorem 86.12 that every element $\alpha \in \overline{\mathrm{Ch}}(\mathrm{Gr})$ can be written in the form $\alpha = \sum a_I e_I$ with $a_I \in \mathbb{Z}/2\mathbb{Z}$. It suffices to prove the following:

Claim: For every I satisfying $a_I = 1$, we have $e_k \in \overline{\mathrm{Ch}}(\mathrm{Gr})$ for any $k \in I$.

To prove the claim, we may assume that α is homogeneous. We establish the claim by induction on the number of nonzero coefficients of α. Choose I with largest $|I|$ such that $a_I = 1$ and set $J = ([1,\ n] \setminus I) \cup \{k\}$. By Lemma 87.5, $(x_J)_*(\alpha) = e_k$ or $1 + e_k$. Indeed, if $a_{I'} = 1$ for some $I' \subset [1,\ n]$ with $I' \cup J = [1,\ n]$, then either $I' = [1,\ n] \setminus J$ and hence $(x_J)_*(e_{I'}) = e_\emptyset = 1$, or $I' = ([1,\ n] \setminus J) \cup \{l\}$ for some l. But since α is homogeneous, we must have $l = k$. Therefore $I' = I$ and $(x_J)_*(e_{I'}) = e_k$.

We have shown that $e_k \in \overline{\mathrm{Ch}}(\mathrm{Gr})$ for all $k \in I$. Therefore, $e_I \in \overline{\mathrm{Ch}}(\mathrm{Gr})$ and $\alpha - e_I \in \overline{\mathrm{Ch}}(\mathrm{Gr})$. By the induction hypothesis, the claim holds for $\alpha - e_I$ and therefore for α. □

Exercise 87.8. Prove that the tangent bundle of Gr is canonically isomorphic to $\Lambda^2(V\mathbb{1}/E)$.

88. The invariant $J(\varphi)$

In this section, we define a new invariant of nondegenerate quadratic forms. It differs slightly from the one defined in [**134**].

Let φ be a nondegenerate quadratic form of dimension $2n+1$ or $2n+2$ and set $\mathrm{Gr} = \mathrm{Gr}(\varphi)$. As before, let e_k be viewed as the class of the generator for $\mathrm{Ch}(\overline{\mathrm{Gr}})$. We define a new discrete *J-invariant* $J(\varphi)$ as follows:

$$J(\varphi) = \{k \in [0,\ n] \quad \text{with} \quad e_k \notin \overline{\mathrm{Ch}}(\mathrm{Gr})\}.$$

Recall that $e_0 = 1$ if $\dim\varphi = 2n+1$, hence in this case $J(\varphi) \subset [1,\ n]$. When $\dim\varphi = 2n+2$, we have $0 \in J(\varphi)$ if and only if the discriminant of φ is not trivial.

If $\dim\varphi = 2n+2$ and φ' is a nondegenerate subform of φ of codimension 1, then

$$J(\varphi) = \begin{cases} J(\varphi') & \text{if } \operatorname{disc}\varphi \text{ is trivial,} \\ \{0\} \cup J(\varphi') & \text{otherwise.} \end{cases}$$

For a subset $I \subset [0,\ n]$ let $||I||$ denote the sum of all $k \in I$.

Proposition 88.1. *The smallest dimension i such that $\overline{\mathrm{Ch}}_i(\mathrm{Gr}) \neq 0$ is equal to $||J(\varphi)||$.*

PROOF. By Theorem 87.7, the product of all e_k satisfying $k \notin J(\varphi)$ is a nontrivial element of $\overline{\mathrm{Ch}}(\mathrm{Gr})$ of the smallest dimension equal to $||J(\varphi)||$. □

Proposition 88.2. *A nondegenerate quadratic form φ is split if and only if $J(\varphi) = \emptyset$.*

PROOF. The "only if" part follows from the definition. Suppose the set $J(\varphi)$ is empty. Since all the e_k are rational, the class of a rational point of $\overline{\mathrm{Gr}}$ belongs to $\overline{\mathrm{Ch}}_0(\mathrm{Gr})$ by Corollary 86.10. It follows that Gr has a closed point of odd degree, i.e., φ is split over an odd degree finite field extension. By Springer's Theorem (Corollary 18.5), the form φ is split. □

Lemma 88.3. *Let $\varphi = \tilde{\varphi} \perp \mathbb{H}$. Then $J(\varphi) = J(\tilde{\varphi})$.*

PROOF. Suppose that $\dim\varphi = 2n+1$. Note first that the cycle $e_n = [\mathrm{Gr}(\tilde{\varphi})]$ is rational so that $n \notin J(\varphi)$. Let $k \leq n-1$. It follows from the decomposition (86.2) that $\mathrm{CH}^k(\mathrm{Gr}) \simeq \mathrm{CH}^k\,\mathrm{Gr}(\tilde{\varphi})$ and e_k corresponds to $\tilde{e}_k$ by Lemma 86.11. Hence $e_k \in J(\varphi)$ if and only if $\tilde{e}_k \in J(\tilde{\varphi})$. The case of the even dimension is similar. □

Corollary 88.4. *Let φ and φ' be Witt-equivalent quadratic forms. Then $J(\varphi) = J(\varphi')$.*

Lemma 88.5. *Let X be a variety, Y a scheme, and n an integer such that the natural homomorphism $\mathrm{CH}_i(X) \to \mathrm{CH}_i\big(X_{F(y)}\big)$ is surjective for every point $y \in Y$ and $i \geq \dim X - n$. Then $\mathrm{CH}_j(Y) \to \mathrm{CH}_j\big(Y_{F(X)}\big)$ is surjective for every $j \geq \dim Y - n$.*

PROOF. Using a localization argument similar to that used in the proof of Proposition 52.10, one checks that the top homomorphism in the commutative diagram

$$\begin{array}{ccc} \mathrm{CH}(X)\otimes \mathrm{CH}(Y) & \longrightarrow & \mathrm{CH}(X\times Y) \\ \downarrow & & \downarrow \\ \mathrm{CH}(Y) & \longrightarrow & \mathrm{CH}\big(Y_{F(X)}\big) \end{array}$$

is surjective in dimensions $\geq \dim X + \dim Y - n$ by induction on $\dim Y$. Since the right vertical homomorphism is surjective, so is the bottom homomorphism in dimensions $\geq \dim Y - n$. □

Let φ be a quadratic form of dimension $2n+1$ or $2n+2$.

Corollary 88.6. *The canonical homomorphism* $\mathrm{CH}^i(\mathrm{Gr}) \to \mathrm{CH}^i\big(\mathrm{Gr}_{F(X)}\big)$ *is surjective for all* $i \leq n-1$.

PROOF. Note that X is split over $F(y)$ for every $y \in \mathrm{Gr}$. Therefore, $\mathrm{CH}^k(X_{F(y)})$ is generated by h^k for all $k \leq n-1$ and, hence, the homomorphism $\mathrm{CH}^k(X) \to \mathrm{CH}^k\big(X_{F(y)}\big)$ is surjective. □

Corollary 88.7. $J(\varphi) \cap [0,\ n-1] \subset J(\varphi_{F(X)}) \subset J(\varphi)$.

The following proposition relates the set $J(\varphi)$ and the absolute Witt indices of φ. It follows from Corollaries 88.4 and 88.7.

Proposition 88.8. *Let φ be a nondegenerate quadratic form of dimension $2n+1$ or $2n+2$. Then*

$$J(\varphi) \subset \big\{n - \mathfrak{j}_0(\varphi), n - \mathfrak{j}_1(\varphi), \dots, n - \mathfrak{j}_{\mathfrak{h}(\varphi)-1}(\varphi)\big\}.$$

In particular, $|J(\varphi)| \leq \mathfrak{h}(\varphi)$.

Remark 88.9. One can impose further restrictions on $J(\varphi)$. Choose a nondegenerate form ψ such that one of the forms φ and ψ is a subform of the other of codimension 1 and the dimension of the largest form is even. Then the sets $J(\varphi)$ and $J(\psi)$ differ by at most one element 0. Therefore, the inclusion in Proposition 88.8 applied to the form ψ gives

$$J(\varphi) \subset \big\{0, n - \mathfrak{j}_0(\psi), n - \mathfrak{j}_1(\psi), \dots, n - \mathfrak{j}_{\mathfrak{h}(\psi)-1}(\psi)\big\}.$$

Example 88.10. Suppose that φ is an anisotropic m-fold Pfister form, $m \geq 1$. Then $J(\varphi) = \{2^{m-1} - 1\}$. Indeed, $\mathfrak{h}(\varphi) = 1$, hence $J(\varphi) \subset \{2^{m-1} - 1\}$ by Proposition 88.8. But $J(\varphi)$ is not empty by Proposition 88.2.

We write n_{Gr} for the gcd of $\deg(g)$ taken over all closed points $g \in \mathrm{Gr}$. The ideal $n_{\mathrm{Gr}} \cdot \mathbb{Z}$ is the image of the degree homomorphism $\mathrm{CH}(\mathrm{Gr}) \to \mathbb{Z}$. Since φ splits over a field extension of F of degree a power of 2, the number n_{Gr} is a 2-power.

Proposition 88.11. *Let φ be a nondegenerate quadratic form of odd dimension. Then*

$$2^{|J(\varphi)|} \cdot \mathbb{Z} \subset n_{\mathrm{Gr}} \cdot \mathbb{Z} \subset \mathrm{ind}\big(C_0(\varphi)\big) \cdot \mathbb{Z}.$$

PROOF. For every $k \notin J(\varphi)$, let f_k be a cycle in $\overline{\mathrm{CH}}^k(\mathrm{Gr})$ satisfying $f_k \equiv e_k$ modulo $2\,\mathrm{CH}^k(\overline{\mathrm{Gr}})$. By Remark 86.14, we have $2e_k \in \overline{\mathrm{CH}}^k(\mathrm{Gr})$ for all k. Let α be the product of all the f_k satisfying $k \notin J(\varphi)$ and all the $2e_k$ satisfying $k \in J(\varphi)$.

Clearly, α is a cycle in $\overline{\mathrm{CH}}(\mathrm{Gr})$ of degree $2^{|J(\varphi)|}m$, where m is an odd integer. The first inclusion now follows from the fact that n_{Gr} is a 2-power.

Let L be the residue field $F(g)$ of a closed point $g \in \mathrm{Gr}$. Since φ splits over L, so does the even Clifford algebra $C_0(\varphi)$. It follows that $\operatorname{ind} C_0(\varphi)$ divides $[L:F] = \deg g$ for all g and therefore divides n_{Gr}. □

Propositions 88.8 and 88.11 yield

Corollary 88.12. *Let φ be a nondegenerate quadratic form of dimension $2n+1$. Consider the following statements:*

(1) *$C_0(\varphi)$ is a division algebra.*
(2) *$n_{\mathrm{Gr}} = 2^n$.*
(3) *$J(\varphi) = [1,\ n]$.*
(4) *$\mathfrak{j}_k = k$ for all $k = 0, 1, \dots, n$.*

Then (1) ⇒ (2) ⇒ (3) ⇒ (4).

The following statement is a refinement of the implication (1) ⇒ (3).

Corollary 88.13. *Let φ be a nondegenerate quadratic form of odd dimension and $\operatorname{ind} C_0(\varphi) = 2^k$. Then $[1,\ k] \subset J(\varphi)$.*

PROOF. We proceed by induction on $\dim \varphi = 2n+1$. If $k = n$, i.e., $C_0(\varphi)$ is a division algebra, the statement follows from Corollary 88.12. So we may assume that $k < n$. Let φ' be a form over $F(\varphi)$ Witt-equivalent to $\varphi_{F(\varphi)}$ of dimension less than $\dim \varphi$. The even Clifford algebra $C_0(\varphi')$ is Brauer-equivalent to $C_0(\varphi)_{F(\varphi)}$. Since $C_0(\varphi)$ is not a division algebra, it follows from Corollary 30.10 that $\operatorname{ind}\big(C_0(\varphi')\big) = \operatorname{ind}\big(C_0(\varphi)\big) = 2^k$. By the induction hypothesis, $[1,\ k] \subset J(\varphi')$. By Corollaries 88.4 and 88.7, we have $J(\varphi') = J(\varphi_{F(\varphi)}) \subset J(\varphi)$. □

Exercise 88.14. Let φ be a quadratic form of odd dimension.

(1) Prove that $1 \in J(\varphi)$ if and only if the even Clifford algebra $C_0(\varphi)$ is not split.

(2) Prove that $2 \in J(\varphi)$ if and only if $\operatorname{ind} C_0(\varphi) > 2$.

89. Steenrod operations on $\mathrm{Ch}\big(\mathrm{Gr}(\varphi)\big)$

We calculate the Steenrod operations on $\mathrm{Ch}\big(\mathrm{Gr}(\varphi)\big)$ as given by Vishik in [**134**]. Let φ be a nondegenerate quadratic form on V over F of dimension $2n+1$ or $2n+2$, let X be the projective quadric of φ, let Gr be the variety of maximal totally isotropic subspaces of V, and let E be the tautological vector bundle over Gr. Let $s : \mathbb{P}(E) \to \mathrm{Gr}$ and $t : \mathbb{P}(E) \to X$ be the projections. There is an exact sequence of vector bundles over $\mathbb{P}(E)$:

$$0 \to \mathbb{1} \to L_c \otimes s^*(E) \to T_s \to 0,$$

where L_c is the canonical line bundle over $\mathbb{P}(E)$ and T_s is the relative tangent bundle of s (cf. Example 104.20). Note that L_c is the pull-back with respect to t of the canonical line bundle over X, hence $c(L_c) = 1 + t^*(h)$, where $h \in \mathrm{CH}^1(X)$ is the class of a hyperplane section of X. By Proposition 86.13, $c_i(E)$ is divisible by 2 for all $i > 0$. It follows that

$$c(T_s) = c\big(L_c \otimes s^*(E)\big) \equiv c(L_c \otimes \mathbb{1}^n) = c(L_c)^n = \big(1 + t^*(h)\big)^n \bmod 2.$$

Theorem 89.1. *Let* $\operatorname{char} F \neq 2$ *and* $\mathrm{Gr} = \mathrm{Gr}(\varphi)$ *with* φ *a nondegenerate quadratic form of dimension* $2n+1$ *or* $2n+2$. *Then the Steenrod operation* $\mathrm{Sq}_{\mathrm{Gr}} : \mathrm{Ch}(\mathrm{Gr}) \to \mathrm{Ch}(\mathrm{Gr})$ *of cohomological type satisfies*

$$\mathrm{Sq}^i_{\mathrm{Gr}}(e_k) = \binom{k}{i} e_{k+i}$$

for all i *and* $k \in [1,\ n]$.

PROOF. We have $\mathrm{Sq}_X(l_{n-k}) = (1+h)^{n+k} \cdot l_{n-k}$ by Corollary 78.5. It follows from (86.5), Theorem 61.9, and Proposition 61.10 that

$$\begin{aligned}
\mathrm{Sq}_{\mathrm{Gr}}(e_k) &= \mathrm{Sq}_{\mathrm{Gr}} \circ s_* \circ t^*(l_{n-k}) \\
&= s_* \circ c(-T_s) \circ \mathrm{Sq}_{\mathbb{P}(E)} \circ t^*(l_{n-k}) \\
&= s_*\big((1+t^*h)^{-n} \cdot t^* \circ \mathrm{Sq}_X(l_{n-k})\big) \\
&= s_* \circ t^*\big((1+h)^{-n} \cdot (1+h)^{n+k} \cdot l_{n-k}\big) \\
&= s_* \circ t^*\big((1+h)^k \cdot l_{n-k}\big) \\
&= \sum_{i \geq 0} \binom{k}{i} s_* \circ t^*(l_{n-k-i}) \\
&= \sum_{i \geq 0} \binom{k}{i} e_{k+i}. \qquad \square
\end{aligned}$$

Exercise 89.2. Let φ be an anisotropic quadratic form of even dimension and height 1. Using Steenrod operations, give another proof of the fact that $\dim(\varphi)$ is a 2-power. (Hint: Use Propositions 88.2 and 88.8.)

90. Canonical dimension

Let F be a field and let $\mathcal{C}$ be a class of field extensions of F. Call a field $E \in \mathcal{C}$ *generic* if for any $L \in \mathcal{C}$ there is an F-place $E \rightharpoonup L$ (cf. §103).

Example 90.1. Let X be a scheme over F. A field extension L of F is called an *isotropy field of* X if $X(L) \neq \emptyset$. If X is a smooth variety, it follows from §103 that the field $F(X)$ is generic in the class of all isotropy fields of X.

The *canonical dimension* $\operatorname{cdim}(\mathcal{C})$ of the class $\mathcal{C}$ is defined to be the minimum of the $\operatorname{tr.deg}_F E$ over all generic fields $E \in \mathcal{C}$. If X is a scheme over F, we write $\operatorname{cdim}(X)$ for $\operatorname{cdim}(\mathcal{C})$, where $\mathcal{C}$ is the class of fields as defined in Example 90.1. If X is smooth, then $\operatorname{cdim}(X) \leq \dim X$.

Let p be a prime integer and let $\mathcal{C}$ be a class of field extensions of F. A field $E \in \mathcal{C}$ is called *p-generic* if for any $L \in \mathcal{C}$ there is an F-place $E \rightharpoonup L'$ for some finite extension L' of L of degree prime to p. The *canonical p-dimension* $\operatorname{cdim}_p(\mathcal{C})$ of $\mathcal{C}$ and $\operatorname{cdim}_p(X)$ of a scheme X over F are defined similarly. Clearly, $\operatorname{cdim}_p(\mathcal{C}) \leq \operatorname{cdim}(\mathcal{C})$ and $\operatorname{cdim}_p(X) \leq \operatorname{cdim}(X)$.

The following theorem answers an old question of Knebusch (cf. [**83**, §4]):

Theorem 90.2. *For an arbitrary anisotropic smooth projective quadric* X,

$$\operatorname{cdim}_2(X) = \operatorname{cdim}(X) = \dim_{\mathrm{Izh}} X.$$

PROOF. Let Y be a smooth subquadric of X of dimension $\dim Y = \dim_{\mathrm{Izh}} X$. Note that $\mathfrak{i}_1(Y) = 1$ by Corollary 74.3. Clearly, the function field $F(Y)$ is an isotropy field of X. Moreover, if L is an isotropy field of X, then by Lemma 74.1, we have $Y(L) \neq \emptyset$. Since the variety Y is smooth, there is an F-place $F(Y) \rightharpoonup L$ (cf. §103). Therefore, $F(Y)$ is a generic isotropy field of X.

Suppose that E is an arbitrary 2-generic isotropy field of X. We show that $\operatorname{tr.deg}_F E \geq \dim Y$ which will finish the proof.

Since E and $F(Y)$ are both generic isotropy fields of the same X, we have F-places $\pi : F(Y) \rightharpoonup E$ and $\varepsilon : E \rightharpoonup E'$, where E' is an odd degree field extension of $F(Y)$. Let y and y' be the centers of π and $\varepsilon \circ \pi$ respectively. Clearly, y' is a specialization of y and therefore,

$$\dim y' \leq \dim y \leq \operatorname{tr.deg}_F E.$$

The morphism $\operatorname{Spec} E' \to Y$ induced by $\varepsilon \circ \pi$ gives rise to a prime correspondence $\delta \colon Y \rightsquigarrow Y$ with odd $\operatorname{mult}(\delta)$ so that $p_{2*}(\delta) = [y']$, where $p_2 : Y \times Y \to Y$ is the second projection. By Theorem 75.4, $\operatorname{mult}(\delta^t)$ is odd, hence y' is the generic point of Y and $\dim y' = \dim Y$. □

The rest of this section is dedicated to determining the canonical 2-dimension of the class $\mathcal{C}$ of all splitting fields of a nondegenerate quadratic form φ. Note for this class $\mathcal{C}$, we have $\operatorname{cdim}(\mathcal{C}) = \operatorname{cdim}(\mathrm{Gr})$ and $\operatorname{cdim}_2(\mathcal{C}) = \operatorname{cdim}_2(\mathrm{Gr})$, where $\mathrm{Gr} = \mathrm{Gr}(\varphi)$, since $L \in \mathcal{C}$ if and only if $\mathrm{Gr}(L) \neq \emptyset$. Hence

$$\operatorname{cdim}_2(\mathrm{Gr}) \leq \operatorname{cdim}(\mathrm{Gr}) \leq \dim \mathrm{Gr}.$$

Theorem 90.3. *Let φ be a nondegenerate quadratic form over F. Then*

$$\operatorname{cdim}_2\bigl(\mathrm{Gr}(\varphi)\bigr) = ||J(\varphi)||.$$

PROOF. Let E be a 2-generic isotropy field of Gr such that $\operatorname{tr.deg}_F E = \operatorname{cdim}_2(\mathrm{Gr})$. As E is an isotropy field, there is a morphism $\operatorname{Spec} E \to \mathrm{Gr}$ over F. Let Y be the closure of the image of this morphism. We view $F(Y)$ as a subfield of E. Clearly, $\operatorname{tr.deg}_F E \geq \dim Y$.

Since E is 2-generic, there is a field extension $L/F(\mathrm{Gr})$ of odd degree and an F-place $E \rightharpoonup L$. Restricting this place to the subfield $F(Y)$, we get a morphism $f : \operatorname{Spec} L \to Y$ as Y is complete. Let $g : \operatorname{Spec} L \to \mathrm{Gr}$ be the morphism induced by the field extension $L/F(\mathrm{Gr})$. Then the closure Z of the image of the morphism $(f, g) : \operatorname{Spec} L \to Y \times \mathrm{Gr}$ is of odd degree $[L : F(\mathrm{Gr})]$ when projecting to Gr. Therefore, the image of $[Z]$ under the composition

$$\mathrm{Ch}(Y \times \mathrm{Gr}) \xrightarrow{(i \times 1_{\mathrm{Gr}})_*} \mathrm{Ch}(\mathrm{Gr} \times \mathrm{Gr}) \xrightarrow{q_*} \mathrm{Ch}(\mathrm{Gr}),$$

where $i : Y \to \mathrm{Gr}$ is the closed embedding and q is the second projection, is equal to $[\mathrm{Gr}]$. In particular, $(i \times 1_{\mathrm{Gr}})_*([Z]) \neq 0$, hence $(i \times 1_{\mathrm{Gr}})_* \neq 0$.

We claim that the push-forward homomorphism $i_* : \mathrm{Ch}(Y) \to \mathrm{Ch}(\mathrm{Gr})$ is also nontrivial. Let L be the residue field of a point of Y. Consider the induced morphism $j : \operatorname{Spec} L \to \mathrm{Gr}$. The pull-back of the element x_I in $\overline{\mathrm{Ch}}(\mathrm{Gr}^2)$ (defined in (87.4)) with respect to the morphism $j \times 1_{\mathrm{Gr}} : \mathrm{Gr}_L \to \mathrm{Gr}^2$ is equal to $e_I \in \overline{\mathrm{Ch}}(\mathrm{Gr}_L) = \mathrm{Ch}(\mathrm{Gr}_L)$. Since the elements e_I generate $\mathrm{Ch}(\mathrm{Gr}_L)$ by Theorem 86.12, the pull-back homomorphism $\mathrm{Ch}(\mathrm{Gr}^2) \to \mathrm{Ch}(\mathrm{Gr}_L)$ is surjective. Applying Proposition 58.18 to the projection $p : Y \times \mathrm{Gr} \to Y$ and the embedding

$i \times 1_{\mathrm{Gr}} : Y \times \mathrm{Gr} \to \mathrm{Gr}^2$, we conclude that the product

$$h_Y : \mathrm{Ch}(Y) \otimes \mathrm{Ch}(\mathrm{Gr}^2) \to \mathrm{Ch}(Y \times \mathrm{Gr}), \quad \alpha \otimes \beta \mapsto p^*(\alpha) \cdot \beta$$

is surjective.

By Propositions 49.20 and 58.17, the diagram

$$\begin{array}{ccc} \mathrm{Ch}(Y) \otimes \mathrm{Ch}(\mathrm{Gr}^2) & \xrightarrow{h_Y} & \mathrm{Ch}(Y \times \mathrm{Gr}) \\ {\scriptstyle i_* \otimes 1}\downarrow & & \downarrow{\scriptstyle (i \times 1_{\mathrm{Gr}})_*} \\ \mathrm{Ch}(\mathrm{Gr}) \otimes \mathrm{Ch}(\mathrm{Gr}^2) & \xrightarrow{h_{\mathrm{Gr}}} & \mathrm{Ch}(\mathrm{Gr} \times \mathrm{Gr}) \end{array}$$

is commutative. As $(i \times 1_{\mathrm{Gr}})_*$ is nontrivial, we conclude that i_* is also nontrivial. This proves the claim.

By Proposition 88.1, we have $\dim Y \geq ||J(\varphi)||$, hence

$$\mathrm{cdim}_2(\mathrm{Gr}) = \mathrm{tr.deg}_F E \geq \dim Y \geq ||J(\varphi)||.$$

It follows from Proposition 88.1 that there is a closed subvariety $Y \subset \mathrm{Gr}$ of dimension $||J(\varphi)||$ such that $[Y] \neq 0$ in $\overline{\mathrm{Ch}}(\mathrm{Gr}) \subset \mathrm{Ch}(\mathrm{Gr}_{F(\mathrm{Gr})})$. By Lemma 87.6, there is a $\beta \in \mathrm{Ch}(\mathrm{Gr}_{F(\mathrm{Gr})})$ satisfying $[Y] \cdot \beta \neq 0$ in $\mathrm{Ch}_0(\mathrm{Gr}_{F(\mathrm{Gr})})$. It follows from Proposition 56.11 that the product $[Y] \cdot \beta$ belongs to the image of the push-forward homomorphism

$$\mathrm{Ch}_0\big(Y_{F(\mathrm{Gr})}\big) \to \mathrm{Ch}_0\big(\mathrm{Gr}_{F(\mathrm{Gr})}\big),$$

therefore $\mathrm{Ch}_0\big(Y_{F(\mathrm{Gr})}\big) \neq 0$. In other words, there is a closed point $y \in Y_{F(\mathrm{Gr})}$ of odd degree. Let Z be the closure of the image of y under the canonical morphism $Y_{F(\mathrm{Gr})} \to Y \times \mathrm{Gr}$. Note that that the projection $Z \to \mathrm{Gr}$ is of odd degree $\deg(y)$, hence $F(Z)$ is an extension of $F(\mathrm{Gr})$ of odd degree. Let Y' denote the image of the projection $Z \to Y$, so $F(Y')$ is isomorphic to a subfield of $F(Z)$.

We claim that $F(Y')$ is a 2-generic splitting field of Gr. Indeed, since Y' is a subvariety of Gr, the field $F(Y')$ is a splitting field of Gr. Let L be another splitting field of Gr. By Lemma 103.2, a geometric F-place $F(\mathrm{Gr}) \rightharpoonup L$ can be extended to an F-place $F(Z) \rightharpoonup L'$ where L' is an extension of L of odd degree. Restricting to $F(Y')$, we get an F-place $F(Y') \rightharpoonup L'$. This proves the claim. Therefore, we have

$$\mathrm{cdim}_2(\mathrm{Gr}) \leq \dim Y' \leq \dim Y = ||J(\varphi)||. \qquad \square$$

Theorem 90.3 and Corollary 88.12 yield

Corollary 90.4 (cf. [**74**, Th. 1.1]). *Let φ be a nondegenerate quadratic form of dimension $2n+1$ such that $J(\varphi) = [1,\, n]$ (e.g., if $C_0(\varphi)$ is a division algebra or if $n_{\mathrm{Gr}} = 2^n$). Then*

$$\mathrm{cdim}_2(\mathrm{Gr}) = \mathrm{cdim}(\mathrm{Gr}) = \dim(\mathrm{Gr}) = \frac{n(n+1)}{2}.$$

Example 90.5. Let φ be an anisotropic m-fold Pfister form with $m \geq 1$. Since the class of splitting fields of φ coincides with the class of isotropy fields, we have $\mathrm{cdim}(\mathrm{Gr}) = \dim_{\mathrm{Izh}}(X) = 2^{m-1} - 1$. By Theorem 90.3 and Example 88.10, we have $\mathrm{cdim}_2(\mathrm{Gr}) = ||J(\varphi)|| = 2^{m-1} - 1$.

We next compute the canonical dimensions $\mathrm{cdim}(\mathrm{Gr})$, $\mathrm{cdim}_2(\mathrm{Gr})$ and determine the set $J(\varphi)$ for an excellent quadratic form φ. Write the dimension of φ in the form

$$\dim \varphi = 2^{p_0} - 2^{p_1} + 2^{p_2} - \cdots + (-1)^{s-1} 2^{p_{s-1}} + (-1)^s 2^{p_s} \tag{90.6}$$

with some integers $p_0, p_1, \dots, p_s$ satisfying $p_0 > p_1 > \cdots > p_{s-1} > p_s + 1 > 0$. Note that the height $\mathfrak{h}$ of φ equals $s+1$ for even $\dim \varphi$, while $\mathfrak{h} = s$ if $\dim \varphi$ is odd.

Let ψ be the $p_{\mathfrak{h}-1}$-fold Pfister form $\rho_{\mathfrak{h}-1}$ over F defined in Theorem 28.3 for the form φ. Since φ and ψ have the same classes of splitting fields, we have $\operatorname{cdim} \operatorname{Gr}(\varphi) = \operatorname{cdim} \operatorname{Gr}(\psi)$ and $\operatorname{cdim}_2 \operatorname{Gr}(\varphi) = \operatorname{cdim}_2 \operatorname{Gr}(\psi)$. By Example 90.5,

$$\operatorname{cdim}\big(\operatorname{Gr}(\varphi)\big) = \operatorname{cdim}_2\big(\operatorname{Gr}(\varphi)\big) = 2^{p_{\mathfrak{h}-1}-1} - 1. \tag{90.7}$$

Proposition 90.8. *Let φ be an anisotropic excellent form of height $\mathfrak{h}$. Then $J(\varphi) = \{2^{p_{\mathfrak{h}-1}-1} - 1\}$, where the integer $p_{\mathfrak{h}-1}$ is determined in* (90.6).

PROOF. Note that $j_{\mathfrak{h}-1} = (\dim \varphi - \dim \psi)/2$ if $\dim \varphi$ is even and $j_{\mathfrak{h}-1} = (\dim \varphi - \dim \psi + 1)/2$ if $\dim \varphi$ is odd. Hence by Proposition 88.8, every element of $J(\varphi)$ is at least $2^{p_{\mathfrak{h}-1}-1} - 1$. By Theorem 90.3, we have $\operatorname{cdim}_2\big(\operatorname{Gr}(\varphi)\big) = ||J(\varphi)||$. It follows from (90.7) that $J(\varphi) = \{2^{p_{\mathfrak{h}-1}-1} - 1\}$. $\square$

The notion of canonical dimension of algebraic groups was introduced by Berhuy and Reichstein in [**17**]. Here we have presented the more general notion of canonical dimension and p-canonical dimension of a class of field extensions of a given field (cf. [**77**]). The computation of $\operatorname{cdim}_2\big(\operatorname{Gr}(\varphi)\big)$ given in Theorem 90.3 is new. It is conjectured in [**134**, Conj. 6.6] that $\operatorname{cdim}\big(\operatorname{Gr}(\varphi)\big) = \operatorname{cdim}_2\big(\operatorname{Gr}(\varphi)\big)$.

CHAPTER XVII

Motives of Quadrics

91. Comparison of some discrete invariants of quadratic forms

In this section, F is an arbitrary field, n a positive integer, V a vector space over F of dimension $2n$ or $2n+1$, φ a nondegenerate quadratic form on V, X the projective quadric of φ. For any positive integer i, we write G_i for the scheme of i-dimensional totally isotropic subspaces of V. In particular, $G_1 = X$ and $G_i = \emptyset$ for $i > n$.

We write $\mathrm{Ch}(Y)$ for the Chow group modulo 2 of an F-scheme Y; $\mathrm{Ch}(\bar{Y})$ is the colimit of $\mathrm{Ch}(Y_L)$ over all field extensions L/F, $\overline{\mathrm{Ch}}(Y)$ is the reduced Chow group, i.e., the image of the homomorphism $\mathrm{Ch}(Y) \to \mathrm{Ch}(\bar{Y})$.

We write $\overline{\mathrm{Ch}}(G_*)$ for the direct sum $\coprod_{i\geq 1} \overline{\mathrm{Ch}}(G_i)$. We recall that $\overline{\mathrm{Ch}}(X^*)$ stands for $\coprod_{i\geq 1} \overline{\mathrm{Ch}}(X^i)$, where X^i is the direct product of i copies of X. We consider $\overline{\mathrm{Ch}}(G_*)$ and $\overline{\mathrm{Ch}}(X^*)$ as invariants of the quadratic form φ. Note that their components $\overline{\mathrm{Ch}}(G_i)$ and $\overline{\mathrm{Ch}}(X^i)$ are subsets of the finite sets $\mathrm{Ch}(\bar{G}_i)$ and $\mathrm{Ch}(\bar{X}^i)$ respectively, depending only on $\dim \varphi$.

These invariants are not independent. A relation between them is described in the following theorem:

Theorem 91.1. *The following three invariants of a nondegenerate quadratic form φ of a fixed dimension are equivalent in the sense that if φ' is another nondegenerate quadratic form with $\dim \varphi = \dim \varphi'$ and the values of one of the invariants for φ and φ' are equal, then the values of any other of the invariants for φ and φ' are also equal.*

(i) $\overline{\mathrm{Ch}}(X^*)$,
(ii) $\overline{\mathrm{Ch}}(X^n)$,
(iii) $\overline{\mathrm{Ch}}(G_*)$.

Remark 91.2. Although the equivalence of the above invariants means that any of them can be expressed in terms of any other, it does not seem to be possible to get manageable formulas relating (iii) with (ii) or (i).

We need some preparation to prove Theorem 91.1. For $i \geq 1$, write Fl_i for the scheme of flags $V_1 \subset \cdots \subset V_i$ of totally isotropic subspaces $V_1, \dots, V_i$ of V, where $\dim V_j = j$. In particular, $\mathrm{Fl}_1 = X$ and $\mathrm{Fl}_i = \emptyset$ for $i > n$. The following lemma generalizes Example 66.6:

Lemma 91.3. *For any $i \geq 1$, the product $\mathrm{Fl}_i \times X$ has the canonical structure of a relative cellular scheme where the bases of the cells are as follows: a rank $i-1$ projective bundle over Fl_i, the scheme Fl_{i+1}, and the scheme Fl_i taken i times.*

PROOF. We construct a descending cellular filtration

$$Y = Y^{(0)} \supset Y^{(1)} \supset \cdots \supset Y^{(i+1)} \supset Y^{(i+2)} = \emptyset$$

on the scheme $Y = \mathrm{Fl}_i \times X$ as follows: for $j \in [1,\ i]$ the scheme $Y^{(j)}$ is the subscheme of pairs

$$(V_1 \subset \cdots \subset V_i, W) \tag{91.4}$$

such that the subspace $W + V_j$ is totally isotropic and $Y^{(i+1)}$ is the subscheme of the pairs (91.4) with $W \subset V_i$.

The natural projection $Y^{(j-1)} \setminus Y^{(j)} \to \mathrm{Fl}_i$ is a vector bundle for $j \in [1,\ i]$. The map $Y^{(i)} \setminus Y^{(i+1)} \to \mathrm{Fl}_{i+1}$ taking a pair (91.4) to the flag $V_1 \subset \cdots \subset V_i \subset V_i + W$ is also a vector bundle. The projection of the scheme $Y^{(i+1)}$ onto Fl_i is a (rank $i-1$) projective bundle. Of course, if $i \geq n$, then $Y^{(i+1)} = Y^{(i)}$ (and the base of the empty "cell" $Y^{(i)} \setminus Y^{(i+1)}$ is the empty scheme Fl_{i+1}). □

Corollary 91.5. *The motive of the product* $\mathrm{Fl}_i \times X$ *for* $i \leq n$ *canonically decomposes as a direct sum, where each summand is some shift of the motive of the scheme* Fl_i *or of the scheme* Fl_{i+1}*. Moreover, a shift of the motive of* Fl_i *occurs in the decomposition and a shift of the motive of* Fl_{i+1} *also occurs (if* $i+1 \leq n$*).*

PROOF. By Corollary 66.4 and Lemma 91.3, the motive of $\mathrm{Fl}_i \times X$ decomposes into a direct sum of summands that are shifts of the motives of $Y^{(i+1)}$, Fl_{i+1}, and Fl_i, where $Y^{(i+1)}$ is a projective bundle over Fl_i. By Theorem 63.10, the motive of $Y^{(i+1)}$ is also a direct sum of shifts of the motive of Fl_i. □

Corollary 91.6. *For any* $r \geq 1$*, the motive of* X^r *canonically decomposes into a direct sum, where each summand is a shift of the motive of some* Fl_i *with* $i \in [1,\ r]$*. Moreover, for any* $i \in [1,\ r]$ *with* $i \leq n$*, a shift of the motive of* Fl_i *occurs.*

PROOF. We induct on r. Since $X^1 = X = \mathrm{Fl}_1$, the case $r = 1$ is immediate. If the statement is proved for some $r \geq 1$, then the statement for X^{r+1} follows by Corollary 91.5. □

Lemma 91.7. *For any* $i \geq 1$*, the motive of* Fl_i *canonically decomposes into a direct sum, where each summand is a shift of the motive of the scheme* G_i*.*

PROOF. For each $j \in [1,\ i]$, write Φ_j for the scheme of flags $V_1 \subset \cdots \subset V_{i-j} \subset V_i$ of totally isotropic subspaces V_k of V satisfying $\dim V_k = k$ for any k. In particular, $\Phi_1 = \mathrm{Fl}_i$ and $\Phi_i = G_i$. The projections

$$\mathrm{Fl}_i = \Phi_1 \to \Phi_2 \to \cdots \to \Phi_i = G_i$$

are projective bundles. Therefore, the lemma follows from Theorem 63.10. □

Combining Corollary 91.6 with Lemma 91.7, we get

Corollary 91.8. *For any* $r \geq 1$*, the motive of* X^r *canonically decomposes into a direct sum, where each summand is a shift of the motive of some* G_i *with* $i \in [1,\ r]$*. Moreover, for any* $i \in [1,\ r]$ *with* $i \leq n$*, a shift of the motive of* G_i *occurs.*

PROOF OF THEOREM 91.1. The equivalences *(i)* ⇔ *(iii)* and *(ii)* ⇔ *(iii)* are given by Corollary 91.8. □

Remark 91.9. One may say that the invariant $\overline{\mathrm{Ch}}(X^n)$ is a "compact form" of the invariant $\overline{\mathrm{Ch}}(X^*)$ and also that the invariant $\overline{\mathrm{Ch}}(G_*)$ is a "compact form" of $\overline{\mathrm{Ch}}(X^n)$. However, some properties of these invariants are easier formulated and proven on the level of $\overline{\mathrm{Ch}}(X^*)$. Among such properties (used many times above), we have the stability of $\overline{\mathrm{Ch}}(X^*) \subset \mathrm{Ch}(\bar{X}^*)$ under partial operations on $\mathrm{Ch}(\bar{X}^*)$ given by permutations of factors of any X^r as well as pull-backs and push-forwards with respect to partial projections and partial diagonals between X^r and X^{r+1}. It is also easier to describe a basis of $\mathrm{Ch}(\bar{X}^*)$ and compute multiplication and Steenrod operations (giving further restrictions on $\overline{\mathrm{Ch}}(X^*)$) in terms of this basis, than to do a similar analysis for $\mathrm{Ch}(\bar{G}_*)$.

92. The Nilpotence Theorem for quadrics

Let Λ be a commutative ring. We shall work in the categories $\mathrm{CR}_*(F,\Lambda)$ and $\mathrm{CR}(F,\Lambda)$, introduced in §63.

Let $\mathcal{C}$ be a class of smooth complete schemes over field extensions of F closed under taking finite disjoint unions (of schemes over the same field), taking connected components, and scalar extensions. We say that $\mathcal{C}$ is *tractable* if for any variety X in $\mathcal{C}$ having a rational point and of positive dimension, there is a scheme X' in $\mathcal{C}$ satisfying $\dim X' < \dim X$ and $M(X') \simeq M(X)$ in $\mathrm{CR}_*(F,\Lambda)$. A scheme is called *tractable*, if it is member of a tractable class.

Our primary example of a tractable scheme will be any smooth projective quadric over F, the tractable class being the class of (all finite disjoint unions of) all smooth projective quadrics over field extensions of F (cf. Example 66.7).

A smooth projective scheme is called *split* if its motive in $\mathrm{CR}_*(F,\Lambda)$ is isomorphic to a finite direct sum of several copies of the motive Λ. Any tractable scheme X splits over an extension of the base field. Moreover, the number of copies of Λ in the corresponding decomposition is an invariant of X. We call this the *rank* of X and denote it by $\operatorname{rk} X$. The number of components of any tractable scheme does not exceed its rank.

Exercise 92.1. Let X/F be a smooth complete variety such that for any field extension E/F satisfying $X(E) \neq \emptyset$, the scheme X_E is split (for example, the variety of the maximal totally isotropic subspace of a nondegenerate odd-dimensional quadratic form considered in Chapter XVI). Show that X is tractable.

Exercise 92.2. Show that the product of two tractable schemes is tractable.

Remark 92.3. As shown in [**25**], the class of all projective homogeneous varieties (under the action of an algebraic group) is tractable.

The following theorem was initially proved by Rost in the case of quadrics (cf. Theorem 67.1). The more general case of a projective homogeneous variety was done in [**25**].

Theorem 92.4 (Nilpotence Theorem for tractable schemes). *Let X be a tractable scheme over F with $M(X)$ its motive in $\mathrm{CR}_*(F,\Lambda)$ or in $\mathrm{CR}(F,\Lambda)$ and $\alpha \in \operatorname{End} M(X)$ a correspondence. If $\alpha_E \in \operatorname{End} M(X_E)$ vanishes for some field extension E/F, then α is nilpotent.*

PROOF. It suffices to consider the case of the category $\mathrm{CR}_*(F,\Lambda)$ as the functor (63.3) is faithful. We fix a tractable class of schemes containing X. We shall construct a map

$$N : [0,\ +\infty) \times [1,\ \operatorname{rk} X] \to [1,\ +\infty)$$

with the following properties. If Y is a scheme in the tractable class with $\operatorname{rk} Y \leq \operatorname{rk} X$ and $\alpha \in \mathrm{CH}(Y^2;\Lambda)$ a correspondence vanishing over some field extension of F, then $\alpha^{N(i,j)} = 0$ if $\dim Y \leq i$ and the number of i-dimensional connected components of Y is at most j.

If $\dim Y = 0$, then any extension of scalars induces an injection of $\mathrm{CH}(Y^2; \Lambda)$. In this case, we set $N(0, j) = 1$ for any $j \geq 1$.

Now order the set $[0, +\infty) \times [1, \operatorname{rk} X]$ lexicographically. Let (i, j) be a pair with $i \geq 1$. Assume that N has been defined on all pairs smaller than (i, j).

Let Y be an arbitrary scheme in the class such that $\dim Y = i$ and the number of the i-dimensional components of Y is j. To simplify the notation, we shall assume that the field of definition of Y is F. Let Y_1 be a fixed i-dimensional component of Y and set Y_0 to be the union of the remaining components of Y. We take an arbitrary correspondence $\alpha \in \mathrm{CH}(Y^2; \Lambda)$ vanishing over a scalar extension and replace it by $\alpha^{N(i',j')}$, where (i', j') is the pair preceding (i, j) in the lexicographical order. Then for any point $y \in Y_1$, we have $\alpha_{F(y)} = 0$, since the motive of the scheme $Y_{F(y)}$ is isomorphic to the motive of another scheme having $j - 1$ components of dimension i. Applying Theorem 67.1, we see that

$$\alpha^{i+1} \circ \mathrm{CH}(Y_1 \times Y; \Lambda) = 0.$$

In particular, the composite of the inclusion morphism $M(Y_1) \to M(Y)$ with α^{i+1} is trivial. Replace α by α^{i+1}. We can view α as a 2×2 matrix according to the decomposition $M(Y) \simeq M(Y_0) \oplus M(Y_1)$. Its entries corresponding to $\operatorname{Hom}\big(M(Y_1), M(Y_0)\big)$ and to $\operatorname{End} M(Y_1)$ are 0. Moreover, the matrix entry corresponding to $\operatorname{End} M(Y_0)$ is nilpotent with $N(i', j')$ its nilpotence exponent, as the number of the i-dimensional components of Y_0 is at most $j - 1$. Replacing α by $\alpha^{N(i',j')}$, we may assume that α has only one possibly nonzero entry, namely, the (nondiagonal) entry corresponding to $\operatorname{Hom}\big(M(Y_0), M(Y_1)\big)$. Therefore, $\alpha^2 = 0$. Set $N(i, j) = 2(i+1)N(i', j')^2$. We have shown that for any scheme Y in the tractable class with $\operatorname{rk} Y \leq \operatorname{rk} X$ and any correspondence $\alpha \in \mathrm{CH}(Y^2; \Lambda)$ vanishing over some field extension of F we have $\alpha^{N(i,j)} = 0$ if $\dim Y = i$ and the number of i-dimensional connected components of Y is j. Since $N(i, j) \geq N(i', j')$, one also has $\alpha^{N(i,j)} = 0$ if $\dim Y \leq i$ and the number of i-dimensional connected components of Y is smaller than j. □

Corollary 92.5. *Let X be a tractable scheme over F and let E/F be a field extension. If $q \in \operatorname{End} M(X_E)$ is an idempotent lying in the image of the restriction $\operatorname{End} M(X) \to \operatorname{End} M(X_E)$ in the motivic category $\mathrm{CR}_*(F, \Lambda)$ or $\mathrm{CR}(F, \Lambda)$, then there exists an idempotent $p \in \operatorname{End} M(X)$ satisfying $p_E = q$.*

PROOF. Choose a correspondence $p' \in \operatorname{End} M(X)$ satisfying $p'_E = q$. Let A (respectively, B) be the (commutative) subring of

$$\operatorname{End} M(X) \quad (\text{respectively, } \operatorname{End}\big(M(X_E)\big))$$

generated by p' (respectively, q). By Theorem 92.4, the kernel of the ring epimorphism $A \twoheadrightarrow B$ consists of nilpotent elements. It follows that the map $\operatorname{Spec} B \to \operatorname{Spec} A$ is a homeomorphism and, in particular, induces a bijection of the sets of the connected components of these topological spaces. Therefore, the homomorphism $A \to B$ induces a bijection of the sets of idempotents of these rings (cf. [**18**, Ch. II, §4.3, Prop. 15, Cor. 1]), and we can find a required p inside of A. □

Exercise 92.6. Show that one can take for p some power of p'. (Hint: Prove and use the fact that the kernel of $\operatorname{End} X \to \operatorname{End} X_E$ is annihilated by a positive integer.)

Corollary 92.7. *Let X and Y be tractable schemes and let $p \in \operatorname{End} M(X)$ and $q \in \operatorname{End} M(Y)$ be idempotents in the motivic category $\mathrm{CR}_*(F,\Lambda)$ or $\mathrm{CR}(F,\Lambda)$. Let f be a morphism $(X,p) \to (Y,q)$ in the category $\mathrm{CM}_*(F,\Lambda)$ or $\mathrm{CM}(F,\Lambda)$ respectively. If f_E is an isomorphism for some field extension E/F, then f is also an isomorphism.*

PROOF. By Proposition 63.4, it suffices to prove the result for the category $\mathrm{CR}_*(F,\Lambda)$.

First suppose that $Y = X$ and $q = p$. We may assume that the scheme X_E is split and we have fixed an isomorphism of the motive (X_E, p_E) with the direct sum of n copies of Λ for some n. Then $\operatorname{Aut}(X_E, p_E) = \mathrm{GL}_n(\Lambda)$. Let $P(t) \in \Lambda[t]$ be the characteristic polynomial of the matrix f_E, hence $P(f_E) = 0$. If $Q(t) \in \Lambda[t]$ satisfies $P(t) = P(0) + tQ(t)$, the endomorphism

$$f_E \circ Q(f_E) = Q(f_E) \circ f_E = P(f_E) - P(0) = -P(0) = \pm \det f_E$$

is multiplication by an invertible element $\varepsilon = \pm \det f_E$ in the coefficient ring Λ. By Theorem 92.4, the endomorphism $\alpha \in \operatorname{End}(X,p)$ satisfying $f \circ Q(f) = Q(f) \circ f = \varepsilon + \alpha$ is nilpotent. Thus the composites $f \circ Q(f)$ and $Q(f) \circ f$ are automorphisms. Consequently f is an automorphism.

In the general case, consider the transpose $f^t : (Y,q) \to (X,p)$ of f. Since f_E is an isomorphism, f^t_E is also an isomorphism. It follows by the previous case that the composites $f \circ f^t$ and $f^t \circ f$ are automorphisms. Thus f is an isomorphism. □

Corollary 92.8. *Let X be a tractable scheme with $p, p' \in \operatorname{End} M(X)$ idempotents satisfying $p_E = p'_E$ for some field $E \supset F$. Then the motives (X,p) and (X,p') are canonically isomorphic.*

PROOF. The morphism $p' \circ p : (X,p) \to (X,p')$ is an isomorphism because it becomes an isomorphism over E. □

93. Criterion of isomorphism

In this section, we let $\Lambda = \mathbb{Z}/2\mathbb{Z}$.

Theorem 93.1. *Let X and Y be smooth projective quadrics over F. Then the motives of X and Y in the category $\mathrm{CR}(F,\mathbb{Z}/2\mathbb{Z})$ are isomorphic if and only if $\dim X = \dim Y$ and $\mathfrak{i}_0(X_L) = \mathfrak{i}_0(Y_L)$ for any field extension L/F.*

PROOF (cf. [**71**]). The "only if" part of the statement is easy: the motive $M(X)$ of X in $\mathrm{CR}(F,\mathbb{Z}/2\mathbb{Z})$ determines the graded group $\mathrm{Ch}^*(X)$ which in turn determines $\dim X$ and $\mathfrak{i}_0(X)$ by Corollary 72.6. To prove the "if" part assume that $\dim X = \dim Y$ and $\mathfrak{i}_0(X_L) = \mathfrak{i}_0(Y_L)$ for any field extension L/F. As before we write D for $\dim X$ and set $d = [D/2]$.

The case of split X and Y is trivial. Note that in the split case an isomorphism $M(X) \to M(Y)$ is given by the cycle $c_{XY} + \deg(l_d^2)(h^d \times h^d)$, where

$$c_{XY} = \sum_{i=0}^{d} (h^i \times l_i + l_i \times h^i) \in \mathrm{Ch}(X \times Y)$$

(cf. Lemma 73.1). By Corollary 92.7, it follows in the nonsplit case that the motives of X and Y are isomorphic if the cycle $c_{XY} \in \mathrm{Ch}(\bar{X} \times \bar{Y})$ is rational.

To prove Theorem 93.1 in the general case, we show by induction on D that the cycle c_{XY} is rational.

If X (and therefore Y) is isotropic, then the cycle $c_{X_0Y_0}$ is rational by the induction hypothesis, where X_0 and Y_0 are the anisotropic parts of X and Y respectively. It follows that the cycle c_{XY} is rational in the isotropic case. We may therefore assume that X and Y are anisotropic.

To finish the proof we need two results. For their proofs we introduce some special notation and terminology. Write N for the set of the symbols

$$\{h^i \times l_i,\ l_i \times h^i\}_{i \in [0,\ d]}.$$

For any subset $I \subset N$, write $c_{XY}(I)$ for the sum of the basis elements of $\mathrm{Ch}^D(\bar{X} \times \bar{Y})$ corresponding to the symbols of I. Similarly, define the cycles $c_{YX}(I) \in \mathrm{Ch}^D(\bar{Y} \times \bar{X})$, $c_{XX}(I) \in \mathrm{Ch}^D(\bar{X}^2)$, and $c_{YY}(I) \in \mathrm{Ch}^D(\bar{Y}^2)$.

We call a subset $I \subset N$ *admissible*, if the cycles $c_{XY}(I)$ and $c_{YX}(I)$ are rational and *weakly admissible* if $c_{XX}(I)$ and $c_{YY}(I)$ are rational.

Since the set N is weakly admissible by Lemma 73.1, the complement $N \setminus I$ of any weakly admissible set I is also weakly admissible.

Call a subset $I \subset N$ *symmetric*, if it is stable under transposition, i.e., $I^t = I$. For any $I \subset N$, the set $I \cup I^t$ is the smallest symmetric set containing I; we call it the *symmetrization* of I.

Lemma 93.2. (1) *Any admissible set is weakly admissible.*

(2) *The symmetrization of an admissible set is admissible.*

(3) *A union of admissible sets is admissible.*

PROOF. (1): This follows from the formulas which hold up to the addition of $h^d \times h^d$:

$$c_{XX}(I) = c_{YX}(I) \circ c_{XY}(I) \quad \text{and} \quad c_{YY}(I) = c_{XY}(I) \circ c_{YX}(I).$$

(3): Let I and J be admissible sets. The cycle $c_{XY}(I \cup J)$ is rational as

$$c_{XY}(I \cup J) = c_{XY}(I) + c_{XY}(J) + c_{XY}(I \cap J)$$

and up to the addition of $h^d \times h^d$, we have $c_{XY}(I \cap J) = c_{XY}(J) \circ c_{XX}(I)$. The rationality of $c_{YX}(I \cup J)$ is proved analogously.

(2): The transpose I^t of an admissible set $I \subset N$ is admissible. Therefore, by (3), the union $I \cup I^t$ is admissible. □

The key observation is:

Proposition 93.3. *Let I be a weakly admissible set and let $h^r \times l_r \in I$ be the element with smallest r. Then $h^r \times l_r$ is contained in an admissible set.*

Assuming Proposition 93.3, we finish the proof of Theorem 93.1 by showing that the set N is admissible.

Note that $\emptyset$ is a symmetric admissible set. Let I_0 be a symmetric admissible set. It suffices to show that if $I_0 \neq N$, then I_0 is contained in a strictly bigger symmetric admissible set I_1.

By Lemma 93.2(1), the set I_0 is weakly admissible. Therefore, the set $I := N \setminus I_0$ is weakly admissible as well. Since the set I is nonempty and symmetric, $h^i \times l_i \in I$ for some i. Take the smallest r with $h^r \times l_r \in I$. By Proposition 93.3, there is an admissible set J containing $h^r \times l_r$. By Lemma 93.2(3), the union $I_0 \cup J$ is also an admissible set. Let I_1 be its symmetrization. Then the set I_1 is admissible by Lemma 93.2(2). It is symmetric and contains I_0 properly because $h^r \times l_r \in I_1 \setminus I_0$.

So to finish, we must only prove Proposition 93.3.

PROOF OF PROPOSITION 93.3. Multiplying the generic point morphism

$$\operatorname{Spec} F(X) \to X$$

by $X \times Y$ (on the left), we get a flat morphism

$$(X \times Y)_{F(X)} \to X \times Y \times X.$$

This induces a surjective pull-back homomorphism

$$f : \operatorname{Ch}^D(\bar{X} \times \bar{Y} \times \bar{X}) \to \operatorname{Ch}^D(\bar{X} \times \bar{Y})$$

mapping each basis element of the form $\beta_1 \times \beta_2 \times h^0$ to $\beta_1 \times \beta_2$ and vanishing on the remaining basis elements. Note that this homomorphism maps the subgroup of F-rational cycles *onto* the subgroup of $F(X)$-rational cycles (cf. Corollary 57.11).

Since the quadrics $X_{F(X)}$ and $Y_{F(X)}$ are isotropic, the cycle $c_{XY}(N)$ is $F(X)$-rational. Therefore, the set $f^{-1}(c_{XY}(N))$ contains a rational cycle. Any cycle in this set has the form

$$c := c_{XY}(N) \times h^0 + \sum \alpha \times \beta \times \gamma, \tag{93.4}$$

where the sum is taken over some homogeneous cycles α, β, γ with $\operatorname{codim} \gamma$ positive. In the following, we assume that (93.4) is a rational cycle.

Let I and r be as in the statement of Proposition 93.3. Viewing the cycle (93.4) as a correspondence from $\bar{X}$ to $\bar{Y} \times \bar{X}$, we may take the composition $c \circ c_{XX}(I)$. The result is a rational cycle on $\bar{X} \times \bar{Y} \times \bar{X}$ that (up to the addition of $h^d \times h^d$) is equal to

$$c' := c_{XY}(I) \times h^0 + \sum \alpha \times \beta \times \gamma, \tag{93.5}$$

where the sum is taken over some (other) homogeneous cycles α, β, γ with $\operatorname{codim} \gamma > 0$ and $\operatorname{codim} \alpha \geq r$. Take the pull-back of the cycle c' with respect to the morphism $\bar{X} \times \bar{Y} \to \bar{X} \times \bar{Y} \times \bar{X}$, given by $(x, y) \mapsto (x, y, x)$, that is induced by the diagonal of $\bar{X}$. The result is a rational cycle on $\bar{X} \times \bar{Y}$ that is equal to

$$c'' := c_{XY}(I) + \sum (\alpha \cdot \gamma) \times \beta, \tag{93.6}$$

where $\operatorname{codim}(\alpha \cdot \gamma) > r$. It follows that $c'' = c_{XY}(J')$ for some set J' containing $h^r \times l_r$.

Repeating the procedure with X and Y interchanged, we can find a set J'' containing $h^r \times l_r$ with the cycle $c_{YX}(J'')$ rational. Then the set $J := J' \cap J''$ contains $h^r \times l_r$ and is admissible as $c_{XY}(J)$ coincides (up to addition of $h^d \times h^d$) with the composition $c_{XY}(J') \circ c_{YX}(J'') \circ c_{XY}(J')$ and a similar equality holds for $c_{YX}(J)$. □

The proof of Theorem 93.1 is now complete. □

Remark 93.7. By Theorem 27.3, an isomorphism of motives of odd-dimensional quadrics gives rise to an isomorphism of the corresponding varieties. The question whether for a given *even* n the conditions

$$n = \dim \varphi = \dim \psi \quad \text{and} \quad \mathfrak{i}_0(\varphi_L) = \mathfrak{i}_0(\psi_L) \text{ for any } L$$

imply that φ and ψ are similar is answered positively if characteristic is not 2 and $n \leq 6$ in [**62**], and negatively for all $n \geq 8$ but 12 in [**63**]. It remains open for $n = 12$.

94. Indecomposable summands

In this section, $\Lambda = \mathbb{Z}/2\mathbb{Z}$ and we work in the category $\mathrm{CM}(F, \mathbb{Z}/2\mathbb{Z})$ of graded motives. Let X be a smooth anisotropic projective quadric of dimension D. We write P for the set of idempotents in $\mathrm{Ch}_D(X^2) = \mathrm{End}\, M(X)$. We shall provide some information about objects (X, p) (with $p \in P$) in the category $\mathrm{CM}(F, \mathbb{Z}/2\mathbb{Z})$. For such p (or, more generally, for any element $p \in \mathrm{Ch}_D(X^2)$), let $\bar{p}$ stand for the *essence* (as defined in §72) of the image of p in the reduced Chow group $\overline{\mathrm{Ch}}(X^2)$. We write $[(X, p)]$ for the isomorphism class of the motive (X, p).

Theorem 94.1. (1) *The map*

$$\{[(X,p)]\}_{p\in P} \to \overline{\mathrm{Ch}}_D(X^2) \quad \textit{given by} \quad [(X,p)] \mapsto \bar{p}$$

is well-defined and injective; its image is the group $\overline{\mathrm{Che}}_D(X^2)$ *of all D-dimensional essential cycles.*

(2) *Let* $p, p_1, p_2 \in P$. *Then* $(X, p) \simeq (X, p_1) \oplus (X, p_2)$ *if and only if* $\bar{p}$ *is a disjoint union of* $\bar{p}_1$ *and* $\bar{p}_2$, *i.e.,* $\bar{p}_1$ *and* $\bar{p}_2$ *do not intersect and* $\bar{p} = \bar{p}_1 + \bar{p}_2$. *In particular, the motive* (X, p) *is indecomposable if and only if the cycle* $\bar{p}$ *is minimal.*

(3) *For any* $p, p' \in P$, *the motives* (X, p) *and* (X, p') *are isomorphic to twists of each other if and only if* $\bar{p}$ *and* $\bar{p}'$ *are* derivatives *of the same rational cycle. More precisely, if* $i \geq 0$, *then* $(X, p) \simeq (X, p')(i)$ *if and only if* $\bar{p} = (h^0 \times h^i) \cdot \alpha$ *and* $\bar{p}' = (h^i \times h^0) \cdot \alpha$ *for some* $\alpha \in \overline{\mathrm{Ch}}_{D+i}(X^2)$.

PROOF. Let E/F be a field extension such that the quadric X_E is split.

(1): By Corollary 92.7, we have $[(X, p)] = [(X, p')]$ if and only if $[(X, p)_E] = [(X, p')_E]$. Assume that D is odd. Then the basis of $\mathrm{Ch}(X_E^2)$ is a system of orthogonal idempotents. Therefore any element $\alpha \in \mathrm{Ch}(X_E^2)$ is an idempotent and the motive (X_E, α) is isomorphic to the direct sum of the motives (X_E, β) over all basis elements β appearing in the decomposition of α. Besides, $(X_E, h^i \times l_i) \simeq \mathbb{Z}/2\mathbb{Z}(i)$ and $(X_E, l_i \times h^i) \simeq \mathbb{Z}/2\mathbb{Z}(D-i)$ for any $i \in [0,\ d]$. It follows that the map in (1) is well-defined and injective. The statement on the image of the map is a consequence of Corollary 92.5.

Now assume that D is even. In this case the basis elements

$$\{h^i \times l_i, l_i \times h^i\}_{i\in[0,\ d-1]}$$

of $\mathrm{Ch}(X_E^2)$ are orthogonal idempotents, $(X_E, h^i \times l_i) \simeq \mathbb{Z}/2\mathbb{Z}(i)$ and $(X_E, l_i \times h^i) \simeq \mathbb{Z}/2\mathbb{Z}(D-i)$ for any $i \in [0,\ d-1]$. Moreover, each of these idempotents is orthogonal to each of the remaining basis elements $h^d \times l_d$, $l_d \times h^d$, $h^d \times h^d$, and $l_d \times l_d$. The elements $h^d \times l_d$, $h^d \times l_d + h^d \times h^d$, $l_d \times h^d$ and $l_d \times h^d + h^d \times h^d$ are also idempotents, each of the four motives produced by these idempotents is isomorphic to $\mathbb{Z}/2\mathbb{Z}(d)$.

If the even integer D is not divisible by 4, then $l_d^2 = 0$ (cf. Exercise 68.3) and it follows that the idempotents $h^d \times l_d$ and $l_d \times h^d$ are orthogonal and their sum is the fifth (the last) nonzero idempotent in the span of $h^d \times l_d$, $l_d \times h^d$ and $h^d \times h^d$. If D is divisible by 4, then $l_d^2 = l_0$ (cf. Exercise 68.3) and it follows that the idempotents $h^d \times l_d$ and $l_d \times h^d + h^d \times h^d$ are orthogonal and their sum is the fifth (the last) nonzero idempotent in the span of $h^d \times l_d$, $l_d \times h^d$ and $h^d \times h^d$. The statement (1) follows now from Lemmas 73.2 and 73.19.

(2): $(X,p) \simeq (X,p_1) \oplus (X,p_2)$ if and only if $(X,p)_E \simeq (X,p_1)_E \oplus (X,p_2)_E$ if and only if $\bar{p}$ is a disjoint union of $\bar{p}_1$ and $\bar{p}_2$.

(3): A correspondence $\alpha \in \mathrm{Ch}_{D+i}(X^2)$ determines an isomorphism $(X,p)_E \to (X,p')(i)_E$ if and only if $\bar{p} = (h^0 \times h^i) \cdot \bar{\alpha}$ and $\bar{p}' = (h^i \times h^0) \cdot \bar{\alpha}$. □

Corollary 94.2. *The motive of any anisotropic smooth projective quadric X decomposes into a direct sum of indecomposable summands. Moreover, such a decomposition is unique and the number of summands coincides with the number of the minimal cycles in $\overline{\mathrm{Ch}}_D(X^2)$, where $D = \dim X$.*

Exercise 94.3 (Rost motives). Let π be an anisotropic n-fold Pfister form. Show the following:

(1) The decomposition of the motive of the projective quadric of π into a sum of indecomposable summands has the form $\coprod_{i=0}^{2^{n-1}-1} R_\pi(i)$ for some motive R_π uniquely determined by π. The motive R_π is called the *Rost motive associated to* π.

(2) For any splitting field extension E/F of π, we have

$$(R_\pi)_E \simeq \mathbb{Z}/2\mathbb{Z} \oplus \mathbb{Z}/2\mathbb{Z}(2^{n-1}-1).$$

(3) The motive of the quadric given by any 1-codimensional subform of π decomposes as $\coprod_{i=0}^{2^{n-1}-2} R_\pi(i)$.

(4) Let φ be a $(2^{n-1}+1)$-dimensional nondegenerate subform of π. Find a smooth projective quadric X such that the motive of the quadric of φ decomposes as $M(X)(1) \oplus R_\pi$. Finally, reprove all this for motives with integral coefficients.

Theorems 93.1 and 94.1 hold in the more general case of motives with integral coefficients. This was done by Vishik in [**133**]. We note that the integral versions follow from the $\mathbb{Z}/2\mathbb{Z}$-versions by [**51**] (cf. also [**66**, Th. 11.2(c)] or [**67**, Ch. E]).

Appendices

95. Formally real fields

In this section, we review the Artin-Schreier theory of formally real fields. These results and their proofs, may be found in the books by Lam [**89**] and Scharlau [**121**].

Let F be a field, $P \subset F$ a subset. We say that P is a *preordering* of F if P satisfies all of the following:

$$P + P \subset P, \qquad P \cdot P \subset P, \qquad -1 \notin P, \qquad \text{and} \qquad \sum F^2 \subset P.$$

A preordering P of F is called an *ordering* if, in addition,

$$F = P \cup -P.$$

A field F is called *formally real* if

$$\widetilde{D}(\infty\langle 1\rangle) := \{x \in F \mid x \text{ is a sum of squares in } F\}$$

is a preordering of F, equivalently if -1 is not a sum of squares in F, i.e., the polynomial $t_1^2 + \cdots + t_n^2$ has no nontrivial zero over F for any (positive) integer n. Clearly, if F is formally real, then the characteristic of F must be zero. (If $\operatorname{char} F \neq 2$, then F is not formally real if and only if $F = \widetilde{D}(\infty\langle 1\rangle)$.) One checks that a preordering is an ordering if and only if it is maximal with respect to set inclusion in the set of preorderings of F. By Zorn's lemma, maximal preorderings and therefore orderings exist for F if it is formally real. In particular, a field F is formally real if and only if the *space of orderings* on F,

$$\mathfrak{X}(F) := \{P \mid P \text{ is an ordering of } F\}$$

is not empty. Every $P \in \mathfrak{X}(F)$ (if any) contains the preordering $\widetilde{D}(\infty\langle 1\rangle)$. Let $P \in \mathfrak{X}(F)$ and $0 \neq x \in F$. If $x \in P$, then x is called *positive* (respectively, *negative*) *with respect to* P and we write $x >_P 0$ (respectively, $x <_P 0$). Elements that are positive (respectively, negative) with respect to all orderings of F (if any) are called *totally positive* (respectively, *totally negative*). In fact, we have

Proposition 95.1 (cf. [**89**, Th. VIII.1.12] or [**121**, Cor. 3.1.7]). *Suppose that F is formally real. Then $\widetilde{D}(\infty\langle 1\rangle) = \bigcap_{P\in\mathfrak{X}(F)} P$, i.e., a nonzero element of F is totally positive if and only if it is a sum of squares.*

It follows that a formally real field has precisely one ordering if and only if $\widetilde{D}(\infty\langle 1\rangle)$ is an ordering in F, e.g., $\mathbb{Q}$ or $\mathbb{R}$. The field of real numbers even has $\mathbb{R}^2$ as an ordering. A formally real field F having F^2 as an ordering is called *euclidean*. For such a field every element is either a square or the negative of a square. For example, the field of real constructible numbers is euclidean.

A formally real field is called *real closed* if it has no proper algebraic extension that is formally real. If F is such a field, then it must be euclidean. Let K/F be an algebraic field extension with K real closed. Then $K^2 \cap F$ is an ordering on F.

Let $Q \in \mathfrak{X}(K)$. The pair (K, Q) is called an *ordered field*. Let K/F be a field extension with K formally real. If $P \in \mathfrak{X}(F)$ satisfies $P = Q \cap F$, then $(K, Q)/(F, P)$ is called an *extension* of ordered fields and Q is called an *extension* of P. If, in addition, K/F is algebraic and there exist no extension $(L, R)/(K, Q)$ with L/K nontrivial algebraic, we call (K, Q) a *real closure* of (F, P).

Proposition 95.2 (cf. [**121**, Th. 3.1.14]). *If (K, Q) is a real closure of (F, P), then K is real closed and $Q = K^2$.*

The key to proving this is

Theorem 95.3 (cf. [**121**, Th. 3.1.9]). *Let (F, P) be an ordered field.*

(1) *Let $d \in F$ and $K = F(\sqrt{d})$. Then there exists an extension of P to K if and only if $d \in P$.*
(2) *If K/F is finite of odd degree, then there exists an extension of P to K.*

The main theorem of Artin-Schreier Theory is

Theorem 95.4 (cf. [**89**, Th. VIII.2.8] or [**121**, Th. 3.1.13 and Th. 3.2.8]). *Every ordered field (F, P) has a real closure $(\overline{F}, \overline{F}^2)$ and this real closure is unique up to a canonical F-isomorphism and this isomorphism is order-preserving.*

Because of the last results, if we fix an algebraic closure $\widetilde{F}$ of a formally real field F and $P \in \mathfrak{X}(F)$, then there exists a unique real closure $(\overline{F}, \overline{F}^2)$ of (F, P) with $\overline{F} \subset \widetilde{F}$. We denote $\overline{F}$ by F_P.

96. The space of orderings

We view the space of orderings $\mathfrak{X}(F)$ on a field F as a subset of the space of functions $\{\pm 1\}^{F^\times}$ by the embedding

$$\mathfrak{X}(F) \to \{\pm 1\}^{F^\times} \quad \text{via} \quad P \mapsto (\operatorname{sign}_P : x \mapsto \operatorname{sign}_P x)$$

(the sign of x in F rel P). Giving $\{\pm 1\}$ the discrete topology, we have $\{\pm 1\}^{F^\times}$ is Hausdorff and by Tychonoff's Theorem, compact. The collection of *clopen* (i.e., open and closed) sets given by

$$H_\varepsilon(a) := \{g \in \{\pm 1\}^{F^\times} \mid g(a) = -\varepsilon\} \tag{96.1}$$

for $a \in F^\times$ and $\varepsilon \in \{\pm 1\}$ forms a subbase for the topology of $\{\pm 1\}^{F^\times}$, hence $\{\pm 1\}^{F^\times}$ is also totally disconnected. Consequently, $\{\pm 1\}^{F^\times}$ is a *boolean space* (i.e., a compact totally disconnected Hausdorff space). Let $\mathfrak{X}(F)$ have the induced topology arising from the embedding $f : \mathfrak{X}(F) \to \{\pm 1\}^{F^\times}$.

Theorem 96.2. *$\mathfrak{X}(F)$ is a boolean space.*

Proof. It suffices to show that $\mathfrak{X}(F)$ is closed in $\{\pm 1\}^{F^\times}$. Take any element $s \in \{\pm 1\}^{F^\times} \setminus f(\mathfrak{X}(F))$. First suppose that s is the constant function ε. Then the clopen set $H_\varepsilon(\varepsilon)$ is disjoint from $f(\mathfrak{X}(F))$ and contains s, so it separates s from $f(\mathfrak{X}(F))$. Assume that s is not a constant function, hence, is surjective. Since $s^{-1}(1)$ is not an ordering on F, there exist $a, b \in F^\times$ such that $s(a) = 1 = s(b)$ (i.e., a, b are "positive") but either $s(a + b) = -1$ or $s(ab) = -1$. Let $c = ab$ if $s(ab) = -1$, otherwise let $c = a + b$. As there cannot be an ordering in which a and b are positive but c negative, $H_1(-a) \cap H_1(-b) \cap H_{-1}(-c)$ is disjoint from $f(\mathfrak{X}(F))$ and contains s, so it separates s from $f(\mathfrak{X}(F))$. □

As we are identifying $\mathfrak{X}(F)$ with its image in $\{\pm 1\}^{F^\times}$, we see that the collection of sets

$$H(a) = H_F(a) := H_1(a) \subset \mathfrak{X}(F), \quad a \in F^\times,$$

forms a subbasis of clopen sets for the topology of $\mathfrak{X}(F)$ called the *Harrison subbasis*. So $H(a)$ is the set of orderings on which a is negative. It follows that the collection of sets

$$H(a_1,\dots,a_n)=H_F(a_1,\dots,a_n):=\bigcap_{i=1}^{n}H(a_i),\quad a_1,\dots,a_n\in F^\times$$

forms a basis for the topology of $\mathfrak{X}(F)$.

97. C_n-fields

We call a homogeneous polynomial of (total) degree d a *d-form*. A field F is called a *C_n-field* if every d-form over F in at least d^n+1 variables has a nontrivial zero over F.

For example, a field is algebraically closed if and only if it is a C_0-field. Every finite field is a C_1-field by the Chevalley-Warning Theorem (cf. [**123**], I.2, Th. 3).

An n-form in n-variables over F is called a *normic form* if it has no nontrivial zero. For example, let E/F be a finite field extension of degree n. Let $\{x_1,\dots,x_n\}$ be an F-basis for E. Then the norm form of the extension E/F is the polynomial

$$N_{E(t_1,\dots,t_n)/F(t_1,\dots,t_n)}(t_1x_1+\dots+t_nx_n)$$

over F in the variables $t_1,\dots,t_n$ is of degree n and has no nontrivial zero, hence it is normic (the reason for the name).

Lemma 97.1. *Let F be a nonalgebraically closed field. Then there exist normic forms of arbitrarily large degree.*

PROOF. There exists a normic form φ of degree n for some $n>1$. Having defined a normic form φ_s of degree n^s, let

$$\varphi_{s+1}:=\varphi(\varphi_s|\varphi_s|\dots|\varphi_s).$$

This notation means that new variables are to be used after each occurrence of $|$. The form φ_{s+1} of degree n^{s+1} has no nontrivial zero. □

Theorem 97.2. *Let F be a C_n-field and let $f_1,\dots,f_r$ be d-forms in N common variables. If $N>rd^n$, then the forms have a common nontrivial zero in F.*

PROOF. Suppose first that $n=0$ (i.e., F is algebraically closed) or $d=1$. As $N>r$, it follows from [**125**, Ch. I, §6.2, Prop.] that the forms have a common nontrivial zero over F.

So we may assume that $n>0$ and $d>1$. By Lemma 97.1, there exists a normic form φ of degree at least r. We define a sequence of forms φ_i, $i\ge 1$, of degree d_i in N_i variables as follows. Let $\varphi_1=\varphi$. Assuming that φ_i is defined let

$$\varphi_{i+1}=\varphi(f_1,\dots,f_r\mid f_1,\dots,f_r\mid\dots\mid f_1,\dots,f_r\mid 0,\dots,0),$$

where zeros occur in $<r$ places. The forms f_i between two consecutive signs $|$ have the same sets of variables.

If $x\in\mathbb{R}$, let $[x]$ denote the largest integer $\le x$. We have

$$d_{i+1}=dd_i\quad\text{and}\quad N_{i+1}=N\Big[\frac{N_i}{r}\Big]. \tag{97.3}$$

Note that since $N>rd^n\ge 2r$, we have $N_i\to\infty$ as $i\to\infty$.

Set

$$\alpha_i=\frac{r}{N_i}\Big[\frac{N_i}{r}\Big]. \tag{97.4}$$

We have $\alpha_i \to 1$ as $i \to \infty$. It follows from (97.3) and (97.4) that

$$\frac{N_{i+1}}{d_{i+1}^n} = \frac{N_i}{d_i^n} \cdot \frac{\alpha_i N}{rd^n}.$$

Since $N > rd^n$ and $\alpha_i \to 1$, there is a $\beta > 1$ and an integer s such that $\dfrac{\alpha_i N}{rd^n} > \beta$ if $i \geq s$. Therefore, we have

$$\frac{N_{i+1}}{d_{i+1}^n} > \frac{N_i}{d_i^n} \cdot \beta,$$

if $i \geq s$. It follows that $N_k > d_k^n$ for some k. As F is a C_n-field, the form φ_k has a nontrivial zero. Choose the smallest k with this property. By definition of φ_k, a nontrivial zero of φ_k gives rise to a nontrivial common zero of the forms $f_1, \ldots, f_r$. □

Corollary 97.5. *Let F be a C_n-field and let K/F be an algebraic field extension. Then K is a C_n-field.*

PROOF. Let f be a d-form over K in N variables with $N > d^n$. The coefficients of f belong to a finite field extension of F, so we may assume that K/F is a finite extension. Let $\{x_1, \ldots, x_r\}$ be an F-basis for K. Choose variables t_{ij}, $i = 1, \ldots, N$, $j = 1, \ldots, r$ over F and set

$$t_i = t_{i1}x_1 + \cdots + t_{ir}x_r$$

for every i. Then

$$f(t_1, \ldots, t_N) = f_1(t_{ij})x_1 + \cdots + f_r(t_{ij})x_r$$

for some d-forms f_j in rN variables. Since $rN > rd^n$, it follows from Theorem 97.2 that the forms f_j have a nontrivial common zero over F which produces a nontrivial zero of f over K. □

Corollary 97.6. *Let F be a C_n-field. Then $F(t)$ is a C_{n+1}-field.*

PROOF. Let f be a d-form in N variables over $F(t)$ with $N > d^{n+1}$. Clearing denominators of the coefficients of f we may assume that all the coefficients are polynomials in t. Choose variables t_{ij}, $i = 1, \ldots, N$, $j = 0, \ldots, m$ for some m and set

$$t_i = t_{i0} + t_{i1}t + \cdots + t_{im}t^m$$

for every i. Then

$$f(t_1, \ldots, t_N) = f_0(t_{ij})t^0 + \cdots + f_{dm+r}(t_{ij})t^{dm+r}$$

for some d-forms f_j in $N(m+1)$ variables over F and $r = \deg_t(f)$. Since $N > d^{n+1}$, one can choose m such that $N(m+1) > (dm + r + 1)d^n$. By Theorem 97.2, the forms f_j have a nontrivial common zero over F which produces a nontrivial zero of f over $F(t)$. □

Corollaries 97.5 and 97.6 yield

Theorem 97.7. *Let F be a C_n-field and let K/F be a field extension of transcendence degree m. Then K is a C_{n+m}-field.*

As algebraically closed fields are C_0-fields, the theorem shows that a field of transcendence degree n over an algebraically closed field is a C_n-field. In particular, we have the classical Tsen Theorem:

Theorem 97.8. *If F is algebraically closed and K/F is a field extension of transcendence degree* 1*, then the Brauer group* $\mathrm{Br}(K)$ *is trivial.*

PROOF. Let A be a central division algebra over K of degree $d > 1$. The reduced norm form Nrd of D is a form of degree d in d^2 variables. By Theorem 97.7, K is a C_1-field, hence Nrd has a nontrivial zero, a contradiction. □

98. Algebras

For more details see [**86**] and [**52**].

98.A. Semisimple, separable and étale algebras. Let F be a field. A finite-dimensional (associative, unital) F-algebra A is called *simple* if A has no nontrivial (two-sided) ideals. By Wedderburn's theorem, every simple F-algebra is isomorphic to the matrix algebra $\mathbb{M}_n(D)$ for some n and a division F-algebra D uniquely determined by A up to isomorphism over F.

An F-algebra A is called *semisimple* if A is isomorphic to a (finite) product of simple algebras.

An F-algebra A is called *separable* if the L-algebra $A_L := A \otimes_F L$ is semisimple for every field extension L/F. This is equivalent to A being a finite product of the matrix algebras $\mathbb{M}_n(D)$, where D is a division F-algebra with center a finite separable field extension of F. Separable algebras satisfy the following descent condition: If A is an F-algebra and E/F is a field extension, then A is separable if and only if A_E is separable as an E-algebra.

Let A be a finite-dimensional commutative F-algebra. If A is separable, it is called *étale*. Consequently, A is étale if and only if A is a finite product of finite separable field extensions of F. An étale F-algebra A is called *split* if A is isomorphic to a product of several copies of F. Let A be a commutative (associative, unital) F-algebra. The determinant (respectively, the trace) of the linear endomorphism of A given by left multiplication by an element $a \in A$ is called the *norm* $\mathrm{N}_A(a)$ (respectively, the *trace* $\mathrm{Tr}_A(a)$). We have $\mathrm{Tr}_A(a + a') = \mathrm{Tr}_A(a) + \mathrm{Tr}_A(a')$ and $\mathrm{N}_A(aa') = \mathrm{N}_A(a)\,\mathrm{N}_A(a')$ for all $a, a' \in A$. Every $a \in A$ satisfies the *characteristic polynomial* equation

$$a^n - \mathrm{Tr}_A(a)a^{n-1} + \cdots + (-1)^n\,\mathrm{N}_A(a) = 0$$

where $n = \dim A$.

98.B. Quadratic algebras. A *quadratic* algebra A over F is an F-algebra of dimension 2. A quadratic algebra is necessarily commutative. Every element $a \in A$ satisfies the quadratic equation

$$a^2 - \mathrm{Tr}_A(a)a + \mathrm{N}_A(a) = 0. \tag{98.1}$$

For every $a \in A$, set $\bar{a} := \mathrm{Tr}_A(a) - a$. We have $\overline{aa'} = \bar{a}\bar{a}'$ for all $a, a' \in A$. Indeed, since $\dim A = 2$, it suffices to check the equality when $a \in F$ or $a' \in F$ (this is obvious) and when $a' = a$ (it follows from the quadratic equation). Thus the map $a \mapsto \bar{a}$ is an algebra automorphism of A of exponent 2. We have

$$\mathrm{Tr}_A(a) = a + \bar{a} \quad \text{and} \quad \mathrm{N}_A(a) = a\bar{a}.$$

A quadratic F-algebra A is étale if A is either a quadratic separable field extension of F or A is split, i.e., is isomorphic to $F \times F$.

Let A and B be two quadratic étale F-algebras. The subalgebra $A \star B$ of the tensor product $A \otimes_F B$ consisting of all elements stable under the automorphism of $A \otimes_F B$ defined by $x \otimes y \mapsto \bar{x} \otimes \bar{y}$ is also a quadratic étale F-algebra. The operation $\star$ on quadratic étale F-algebras yields a (multiplicative) group structure on the set $\acute{\mathrm{E}}\mathrm{t}_2(F)$ of isomorphism classes $[A]$ of quadratic étale F-algebras A. Thus $[A] \cdot [B] = [A \star B]$. Note that $\acute{\mathrm{E}}\mathrm{t}_2(F)$ is an abelian group of exponent 2.

Example 98.2. If $\operatorname{char} F \neq 2$, every quadratic étale F-algebra is isomorphic to

$$F_a := F[t]/(t^2 - a)$$

for some $a \in F^\times$. Let j be the class of t in F_a. For every $u = x + yj$, we have

$$\bar{u} = x - yj, \qquad \mathrm{Tr}(u) = 2x, \quad \text{and} \quad \mathrm{N}(u) = x^2 - ay^2.$$

The assignment $a \mapsto [F_a]$ gives rise to an isomorphism $F^\times/F^{\times 2} \cong \acute{\mathrm{E}}\mathrm{t}_2(F)$.

Example 98.3. If $\operatorname{char} F = 2$, every quadratic étale F-algebra is isomorphic to

$$F_a := F[t]/(t^2 + t + a)$$

for some $a \in F$. Let j be the class of t in F_a. For every $u = x + yj$, we have

$$\bar{u} = x + y + yj, \quad \mathrm{Tr}(u) = y, \quad \text{and} \quad \mathrm{N}(u) = x^2 + xy + ay^2.$$

The assignment $a \mapsto [F_a]$ induces an isomorphism $F/\operatorname{Im} \wp \cong \acute{\mathrm{E}}\mathrm{t}_2(F)$, where the *Artin-Schreier map* $\wp : F \to F$ is defined by $\wp(x) = x^2 + x$.

98.C. Brauer group. An F-algebra A is called *central* if $F \cdot 1$ coincides with the center of A. A central simple F-algebra A is called *split* if $A \cong \mathbb{M}_n(F)$ for some n.

Two central simple F-algebras A and B are called *Brauer equivalent* if $\mathbb{M}_n(A) \cong \mathbb{M}_m(B)$ for some n and m. For example, all split F-algebras are Brauer equivalent.

The set $\mathrm{Br}(F)$ of all Brauer equivalence classes of central simple F-algebras is a torsion abelian group with respect to the tensor product operation $A \otimes_F B$, called the *Brauer group of F*. The identity element of $\mathrm{Br}(F)$ is the class of split F-algebras.

The class of a central simple F-algebra A will be denoted by $[A]$ and the product of $[A]$ and $[B]$ in the Brauer group, represented by the tensor product $A \otimes_F B$, will be denoted by $[A] \cdot [B]$.

The inverse class of A in $\mathrm{Br}(F)$ is given by the class of the *opposite* algebra A^{op}. The order of $[A]$ in $\mathrm{Br}(F)$ is called the *exponent* of A and will be denoted by $\exp(A)$. In particular, $\exp(A)$ divides 2 if and only if $A^{op} \cong A$, i.e., A has an anti-automorphism.

For an integer m, we write $\mathrm{Br}_m(F)$ for the subgroup of all classes $[A] \in \mathrm{Br}(F)$ such that $[A]^m = 1$.

Let A be a central simple algebra over F and let L/F be a field extension. Then A_L is a central simple algebra over L. (In particular, every central simple F-algebra is separable.) The correspondence $[A] \mapsto [A_L]$ gives rise to a group homomorphism

$$r_{L/F} : \mathrm{Br}(F) \to \mathrm{Br}(L).$$

We set $\mathrm{Br}(L/F) := \operatorname{Ker} r_{L/F}$. The class A is said to be *split over* L (and L/F is called a *splitting field extension of* A) if the algebra A_L is split, equivalently $[A] \in \mathrm{Br}(L/F)$.

A central simple F-algebra A is isomorphic to $\mathbb{M}_k(D)$ for a central division F-algebra D, unique up to isomorphism. The integers $\sqrt{\dim D}$ and $\sqrt{\dim A}$ are called the *index* and the *degree* of A respectively and denoted by $\operatorname{ind}(A)$ and $\deg(A)$.

Fact 98.4. *Let A be a central simple algebra over F and L/F a finite field extension. Then*

$$\operatorname{ind}(A_L) \mid \operatorname{ind}(A) \mid \operatorname{ind}(A_L) \cdot [L : F].$$

Corollary 98.5. *Let A be a central simple algebra over F and L/F a finite field extension. Then*

(1) *If L is a splitting field of A, then* $\operatorname{ind}(A)$ *divides* $[L : F]$.
(2) *If $[L : F]$ is relatively prime to* $\operatorname{ind}(A)$, *then* $\operatorname{ind}(A_L) = \operatorname{ind}(A)$.

Fact 98.6. *Let A be a central division algebra over F.*

(1) *A subfield $K \subset A$ is maximal if and only if $[K : F] = \operatorname{ind}(A)$. In this case K is a splitting field of A.*
(2) *Every splitting field of A of degree* $\operatorname{ind}(A)$ *over F can be embedded into A over F as a maximal subfield.*

98.D. Severi-Brauer varieties. Let A be a central simple F-algebra of degree n. Let r be an integer dividing n. The *generalized Severi-Brauer variety* $\mathrm{SB}_r(A)$ *of A* is the variety of right ideals of dimension rn in A [**86**, 1.16]. We simply write $\mathrm{SB}(A)$ for $\mathrm{SB}_1(A)$.

If A is split, i.e., $A = \operatorname{End}(V)$ for a vector space V of dimension n, every right ideal I in A of dimension rn has the form $I = \operatorname{Hom}(V, U)$ for a uniquely determined subspace $U \subset V$ of dimension r. Thus the correspondence $I \mapsto U$ yields an isomorphism $\mathrm{SB}_r(A) \cong \mathrm{Gr}_r(V)$, where $\mathrm{Gr}_r(V)$ is the Grassmannian variety of r-dimensional subspaces in V. In particular, $\mathrm{SB}(A) \cong \mathbb{P}(V)$.

Proposition 98.7 ([**86**, Prop. 1.17]). *Let A be a central simple F-algebra and let r be an integer dividing* $\deg(A)$. *Then the Severi-Brauer variety $X = \mathrm{SB}_r(A)$ has a rational point over an extension L/F if and only if* $\operatorname{ind}(A_L)$ *divides r. In particular,* $\mathrm{SB}(A)$ *has a rational point over L if and only if A is split over L.*

Let V_1 and V_2 be vector spaces over F of finite dimension. The *Segre closed embedding* is the morphism

$$\mathbb{P}(V_1) \times \mathbb{P}(V_2) \to \mathbb{P}(V_1 \otimes_F V_2)$$

taking a pair of lines U_1 and U_2 in V_1 and V_2 respectively to the line $U_1 \otimes_F U_2$ in $V_1 \otimes_F V_2$.

Example 98.8. The Segre embedding identifies $\mathbb{P}^1_F \times \mathbb{P}^1_F$ with a projective quadric in $\mathbb{P}^3_F$.

The Segre embedding can be generalized as follows. Let A_1 and A_2 be two central simple algebras over F. Then the correspondence $(I_1, I_2) \mapsto I_1 \otimes I_2$ yields a closed embedding

$$\mathrm{SB}(A_1) \times \mathrm{SB}(A_2) \to \mathrm{SB}(A_1 \otimes_F A_2).$$

98.E. Quaternion algebras. Let L/F be a Galois quadratic field extension with Galois group $\{e, g\}$ and $b \in F^\times$. The F-algebra $Q = (L/F, b) := L \oplus Lj$, where the symbol j satisfies $j^2 = b$ and $jl = g(l)j$ for all $l \in L$, is central simple of dimension 4 and is called a *quaternion algebra*. We have Q is either split, i.e., isomorphic to the matrix algebra $\mathbb{M}_2(F)$, or a division algebra. The algebra Q carries a *canonical involution* $\bar{\ } : Q \to Q$ satisfying $\bar{j} = -j$ and $\bar{l} = g(l)$ for all $l \in L$.

Using the canonical involution, we define the linear *reduced trace map* as

$$\mathrm{Trd} : Q \to F \text{ by } \mathrm{Trd}(q) = q + \bar{q},$$

and the quadratic *reduced norm map* as

$$\mathrm{Nrd} : Q \to F \text{ by } \mathrm{Nrd}(q) = q \cdot \bar{q}.$$

An element $q \in Q$ is called a *pure quaternion* if $\mathrm{Trd}(q) = 0$, or equivalently, $\bar{q} = -q$. Denote by Q' the 3-dimensional subspace of all pure quaternions. We have $\mathrm{Nrd}(q) = -q^2$ for any $q \in Q'$.

Proposition 98.9. *Every central division algebra of dimension* 4 *is isomorphic to a quaternion algebra.*

PROOF. Let $L \subset Q$ be a separable quadratic subfield. By the Skolem-Noether Theorem, the only nontrivial automorphism g of L over F extends to an inner automorphism of Q, i.e., there is $j \in Q^\times$ such that $jlj^{-1} = g(l)$ for all $l \in L$. Clearly, $Q = L \oplus Lj$ and j^2 commutes with j and L. Hence j^2 belongs to the center of Q, i.e., $b := j^2 \in F^\times$. Therefore, Q is isomorphic to the quaternion algebra $Q = (L/F, b)$. □

Example 98.10. If $\operatorname{char} F \neq 2$, a separable quadratic subfield L of a quaternion algebra $Q = (L/F, b)$ is of the form $L = F(i)$ with $i^2 = a \in F^\times$. Hence Q has a basis $\{1, i, j, k := ij\}$ with multiplication table

$$i^2 = a, \quad j^2 = b, \quad ji + ij = 0,$$

for some $b \in F^\times$. We shall denote the algebra generated by i and j with these relations by $\left(\frac{a,b}{F}\right)$.

The space of pure quaternions has $\{i, j, k\}$ as a basis. For every $q = x + yi + zj + wk$ with $x, y, z, w \in F$, we have

$$\bar{q} = x - yi - zj - wk, \qquad \mathrm{Trd}(q) = 2x, \quad \text{and} \quad \mathrm{Nrd}(q) = x^2 - ay^2 - bz^2 + abw^2.$$

Example 98.11. If $\operatorname{char} F = 2$, a separable quadratic subfield L of a quaternion algebra $Q = (L/F, b)$ is of the form $L = F(s)$ with $s^2 + s + c = 0$ for some $c \in F$. Set $i = sj/b$. We have $i^2 = a := c/b$. Hence Q has a basis $\{1, i, j, k := ij\}$ with the multiplication table

$$i^2 = a, \quad j^2 = b, \quad ji + ij = 1. \tag{98.12}$$

We shall denote by $\left[\frac{a,b}{F}\right]$ the algebra given by the generators i and j and the relations (98.12). Note that this is also a quaternion algebra (in fact split) when $b = 0$.

The space of pure quaternions has $\{1, i, j\}$ as a basis. For every $q = x + yi + zj + wk$ with $x, y, z, w \in F$, we have

$$\bar{q} = (x + w) + yi + zj + wk, \quad \mathrm{Trd}(q) = w,$$
$$\text{and} \quad \mathrm{Nrd}(q) = x^2 + ay^2 + bz^2 + abw^2 + xw + yz.$$

The classes of quaternion F-algebras satisfy the following relations in $\mathrm{Br}(F)$:

Fact 98.13. *Suppose that* $\operatorname{char} F \neq 2$. *Then*

(1) $\left(\frac{aa',b}{F}\right) = \left(\frac{a,b}{F}\right) \cdot \left(\frac{a',b}{F}\right)$ *and* $\left(\frac{a,bb'}{F}\right) = \left(\frac{a,b}{F}\right) \cdot \left(\frac{a,b'}{F}\right)$.

(2) $\left(\frac{a,b}{F}\right) = \left(\frac{b,a}{F}\right)$.

(3) $\left(\frac{a,b}{F}\right)^2 = 1$.

(4) $\left(\frac{a,b}{F}\right) = 1$ *if and only if* a *is a norm of the quadratic étale extension* F_b/F.

Fact 98.14. *Suppose that* $\operatorname{char} F = 2$. *Then*

(1) $\left[\frac{a+a',b}{F}\right] = \left[\frac{a,b}{F}\right] \cdot \left[\frac{a',b}{F}\right]$ *and* $\left[\frac{a,b+b'}{F}\right] = \left[\frac{a,b}{F}\right] \cdot \left[\frac{a,b'}{F}\right]$.

(2) $\left[\frac{ab,c}{F}\right] \cdot \left[\frac{bc,a}{F}\right] \cdot \left[\frac{ca,b}{F}\right] = 1$.

(3) $\left[\frac{a,b}{F}\right] = \left[\frac{b,a}{F}\right]$.

(4) $\left[\frac{a,b}{F}\right]^2 = 1$.

(5) $\left[\frac{a,b}{F}\right] = 1$ *if and only if* a *is a norm of the quadratic étale extension* F_{ab}/F.

We shall need the following properties of quaternion algebras.

Lemma 98.15 (Chain Lemma). *Let* $\left(\frac{a,b}{F}\right)$ *and* $\left(\frac{c,d}{F}\right)$ *be isomorphic quaternion algebras over a field* F *of characteristic not* 2. *Then there is an* $e \in F^\times$ *satisfying* $\left(\frac{a,b}{F}\right) \simeq \left(\frac{a,e}{F}\right) \simeq \left(\frac{c,e}{F}\right) \simeq \left(\frac{c,d}{F}\right)$.

PROOF. Note that if x and y are pure quaternions in a quaternion algebra Q that are orthogonal with respect to the reduced trace bilinear form, i.e., $\operatorname{Trd}(xy) = 0$, then $Q \simeq \left(\frac{x^2,y^2}{F}\right)$. Let $Q = \left(\frac{a,b}{F}\right)$. By assumption, there are pure quaternions x, y satisfying $x^2 = a$ and $y^2 = c$. Choose a pure quaternion z orthogonal to x and y. Setting $e = z^2$, we have $Q \simeq \left(\frac{a,e}{F}\right) \simeq \left(\frac{c,e}{F}\right)$. □

Lemma 98.16. *Let* Q *be a quaternion algebra over a field* F *of characteristic* 2. *Suppose that* Q *is split by a purely inseparable field extension* K/F *such that* $K^2 \subset F$. *Then* $Q \cong \left[\frac{a,b}{F}\right]$ *with* $a \in K^2$.

PROOF. First suppose that $K = F(\sqrt{a})$ is a quadratic extension of F. By Fact 98.6, we know that K can be embedded into Q. Therefore there exists an $i \in Q \setminus F$ such that $i^2 = a \in K^2$. Note that i is a pure quaternion in $Q' \setminus F$. The linear map $Q' \to F$ taking an x to $ix + xi$ is nonzero, hence there is a $j \in Q'$ such that $ij + ji = 1$. Hence, $Q \cong \left[\frac{a,b}{F}\right]$ where $b = j^2$.

In the general case, write $Q = \left[\frac{c,d}{F}\right]$. By Property (5) of Fact 98.14, we have $c = x^2 + xy + cdy^2$ for some $x, y \in K$. Since x^2 and y^2 belong to F, we have $xy \in F$. Hence the extension $E = F(x, y)$ splits Q and $[E : F] \le 2$. The statement now follows from the first part of the proof. □

Let σ be an automorphism of a ring R. Denote by $R[t, \sigma]$ the ring of σ-*twisted polynomials* in the variable t with multiplication defined by $tr = \sigma(r)t$ for all $r \in R$. For example, if σ is the identity, then $R[t, \sigma]$ is the ordinary polynomial ring $R[t]$ over R. Observe that if R has no zero divisors, then neither does $R[t, \sigma]$.

Example 98.17. Let A be a central division algebra over a field F. Consider an automorphism σ of the polynomial ring $A[x]$ defined by $\sigma(a) = a$ for all $a \in A$ and

$$\sigma(x) = \begin{cases} -x & \text{if char } F \neq 2, \\ x+1 & \text{if char } F = 2. \end{cases}$$

Let B be the quotient ring of $A[x][t, \sigma]$. The ring B is a division algebra over its center E where

$$E = \begin{cases} F(x^2, t^2) & \text{if char } F \neq 2, \\ F(x^2 + x, t^2) & \text{if char } F = 2. \end{cases}$$

Moreover, $B = A \otimes_F Q$, where Q is a quaternion algebra over E satisfying

$$Q = \begin{cases} \left(\frac{x^2, t^2}{E}\right) & \text{if char } F \neq 2, \\ \left[\frac{(x^2+x)/t^2, t^2}{E}\right] & \text{if char } F = 2. \end{cases}$$

Iterating the construction in Example 98.17 yields the following

Proposition 98.18. *For any field F and integer $n \geq 1$, there is a field extension L/F and a central division L-algebra that is a tensor product of n quaternion algebras.*

We now study interactions between two quaternion algebras.

Theorem 98.19. *Let Q_1 and Q_2 be division quaternion algebras over F. Then the following conditions are equivalent:*

(1) *The tensor product $Q_1 \otimes_F Q_2$ is not a division algebra.*
(2) *Q_1 and Q_2 have isomorphic separable quadratic subfields.*
(3) *Q_1 and Q_2 have isomorphic quadratic subfields.*

PROOF. (1) $\Rightarrow$ (2): Write X_1, X_2, and X for Severi-Brauer varieties of Q_1, Q_2, and $A := Q_1 \otimes_F Q_2$ respectively. The morphism $X_1 \times X_2 \to X$ taking a pair of ideals I_1 and I_2 to the ideal $I_1 \otimes I_2$ identifies $X_1 \times X_2$ with a twisted form of a 2-dimensional quadric in X (cf. §98.D).

Let Y be the generalized Severi-Brauer variety of rank 8 ideals in A. A rational point of Y, i.e., a right ideal $J \subset A$ of dimension 8, defines the closed curve C_J in X comprising all ideals of rank 4 contained in J. In the split case, Y is the Grassmannian variety of planes and C_J is the projective line (the projective space of the plane corresponding to J) properly intersecting the quadric $X_1 \times X_2$ in two points. Thus there is a nonempty open subset $U \subset Y$ with the following property: for any rational point $J \in U$, we have $C_J \cap (X_1 \times X_2) = \{x\}$, where x is a point of degree 2 with residue field L a separable quadratic field extension of F. By assumption, there is a right ideal $I \subset A$ of dimension 8, i.e., $Y(F) \neq \emptyset$. The algebraic group G of invertible elements of A acts transitively on Y, i.e., the morphism $G \to Y$ taking an a to the ideal aI is surjective. As rational points of G are dense in G, we have rational points of Y are dense in Y. Hence U possesses a rational point J.

As $X_1(L) \times X_2(L) = (X_1 \times X_2)(L) \neq \emptyset$, it follows that the field L splits both Q_1 and Q_2 and therefore L is isomorphic to quadratic subfields in Q_1 and Q_2.

(2) $\Rightarrow$ (3) is trivial.

(3) $\Rightarrow$ (1): Let L/F be a common quadratic subfield of both Q_1 and Q_2. It follows that Q_1 and Q_2 and, hence, A are split by L. It follows from Corollary 98.5 that $\operatorname{ind}(A) \leq 2$, i.e., A is not a division algebra. $\square$

99. Galois cohomology

For more details see [**124**].

99.A. Galois modules and Galois cohomology groups. Let Γ be a profinite group and let M be a (left) discrete Γ-module. For any $n \in \mathbb{Z}$, let $H^n(\Gamma, M)$ denote the *nth cohomology group of* Γ *with coefficients in* M. In particular, the group $H^n(\Gamma, M)$ is trivial, if $n < 0$ and

$$H^0(\Gamma, M) = M^\Gamma := \{m \in M \mid \gamma m = m \quad \text{for all} \quad \gamma \in \Gamma\},$$

the subgroup of Γ-invariant elements of M.

An exact sequence $0 \to M' \to M \to M'' \to 0$ gives rise to an infinite long exact sequence of cohomology groups

$$0 \to H^0(\Gamma, M') \to H^0(\Gamma, M) \to H^0(\Gamma, M'') \to H^1(\Gamma, M') \to H^1(\Gamma, M) \to \dots.$$

Let F be a field. Denote by Γ_F the *absolute Galois group of* F, i.e., the Galois group of a separable closure F_{sep} of the field F over F. A discrete Γ_F-module is called a *Galois module over* F. For a Galois module M over F, we write $H^n(F, M)$ for $H^n(\Gamma_F, M)$.

Example 99.1. (1) Every abelian group A can be viewed as a Galois module over F with trivial action. We have $H^0(F, A) = A$ and $H^1(F, A) = \operatorname{Hom}_c(\Gamma_F, A)$, the group of continuous homomorphisms (where A is viewed with discrete topology). In particular, $H^1(F, A)$ is trivial if A is torsion-free, e.g., $H^1(F, \mathbb{Z}) = 0$.

The group $H^1(F, \mathbb{Q}/\mathbb{Z}) = \operatorname{Hom}_c(\Gamma_F, \mathbb{Q}/\mathbb{Z})$ is called the *character group of* Γ_F and will be denoted by $\chi(\Gamma_F)$.

The cohomology group $H^n(F, M)$ is torsion for every Galois module M and any $n \geq 1$.

Since the group $\mathbb{Q}$ is uniquely divisible, we have $H^n(F, \mathbb{Q}) = 0$ for all $n \geq 1$. The cohomology exact sequence for the short exact sequence of Galois modules with trivial action

$$0 \to \mathbb{Z} \to \mathbb{Q} \to \mathbb{Q}/\mathbb{Z} \to 0$$

then gives an isomorphism $H^n(F, \mathbb{Q}/\mathbb{Z}) \xrightarrow{\sim} H^{n+1}(F, \mathbb{Z})$ for any $n \geq 1$. In particular, $H^2(F, \mathbb{Z}) \cong \chi(\Gamma_F)$.

Let m be a natural integer. The cohomology exact sequence for the short exact sequence

$$0 \to \mathbb{Z} \xrightarrow{m} \mathbb{Z} \to \mathbb{Z}/m\mathbb{Z} \to 0$$

gives an isomorphism of $H^1(F, \mathbb{Z}/m\mathbb{Z})$ with the subgroup $\chi_m(\Gamma_F)$ of characters of exponent m.

(2) The cohomology groups $H^n(F, F_{sep})$ with coefficients in the additive group F_{sep} are trivial if $n > 0$. If $\operatorname{char} F = p > 0$, the cohomology exact sequence for the short exact sequence

$$0 \to \mathbb{Z}/p\mathbb{Z} \to F_{sep} \xrightarrow{\wp} F_{sep} \to 0,$$

where $\wp$ is the Artin-Schreier map defined by $\wp(x) = x^p - x$, yields canonical isomorphisms

$$H^n(F, \mathbb{Z}/p\mathbb{Z}) \cong \begin{cases} \mathbb{Z}/p\mathbb{Z} & \text{if } n = 0, \\ F/\wp(F) & \text{if } n = 1, \\ 0 & \text{otherwise.} \end{cases}$$

In fact, $H^n(F, M) = 0$ for all $n \geq 2$ and every Galois module M over F of characteristic p satisfying $pM = 0$.

(3) We have the following canonical isomorphisms for the cohomology groups with coefficients in the multiplicative group $F_{sep}^{\times}$:

$$H^n(F, F_{sep}^{\times}) \cong \begin{cases} F^{\times} & \text{if } n = 0, \\ 1 & \text{if } n = 1 \text{ (Hilbert Theorem 90)}, \\ \mathrm{Br}(F) & \text{if } n = 2. \end{cases}$$

(4) The group $\mu_m = \mu_m(F_{sep})$ of mth roots of unity in F_{sep} is a Galois submodule of $F_{sep}^{\times}$. We have the following exact sequence of Galois modules:

$$1 \to \mu_m \to F_{sep}^{\times} \to F_{sep}^{\times} \to F_{sep}^{\times}/F_{sep}^{\times m} \to 1, \tag{99.2}$$

where the middle homomorphism takes x to x^m.

If m is not divisible by $\operatorname{char} F$, we have $F_{sep}^{\times}/F_{sep}^{\times m} = 1$. Therefore, the cohomology exact sequence (99.2) yields isomorphisms

$$H^n(F, \mu_m) \cong \begin{cases} \mu_m(F), & \text{if } n = 0, \\ F^{\times}/F^{\times m}, & \text{if } n = 1, \\ \mathrm{Br}_m(F), & \text{if } n = 2. \end{cases}$$

For any $a \in F^{\times}$, we shall write $(a)_m$ or simply (a) for the element of $H^1(F, \mu_m)$ corresponding to the coset $aF^{\times m}$ in $F^{\times}/F^{\times m}$.

If $p = \operatorname{char} F > 0$, we have $\mu_p(F_{sep}) = 1$ and the cohomology exact sequence (99.2) gives an isomorphism

$$H^1(F, F_{sep}^{\times}/F_{sep}^{\times p}) \cong \mathrm{Br}_p(F).$$

Example 99.3. Let $\xi \in \chi_2(\Gamma_F)$ be a nontrivial character. Then $\mathrm{Ker}(\xi)$ is a subgroup of Γ_F of index 2. By Galois theory, it corresponds to a Galois quadratic field extension F_ξ/F. The correspondence $\xi \mapsto F_\xi$ gives rise to an isomorphism $\chi_2(\Gamma_F) \xrightarrow{\sim} \acute{\mathrm{E}}\mathrm{t}_2(F)$.

99.B. Cup-products. Let M and N be Galois modules over F. Then the tensor product $M \otimes_{\mathbb{Z}} N$ is also a Galois module via the diagonal Γ_F-action. There is a pairing

$$H^m(F, M) \otimes H^n(F, N) \to H^{m+n}(F, M \otimes_{\mathbb{Z}} N), \quad \alpha \otimes \beta \mapsto \alpha \cup \beta$$

called the *cup-product*. When $m = n = 0$ the cup-product coincides with the natural homomorphism $M^{\Gamma_F} \otimes N^{\Gamma_F} \to (M \otimes_{\mathbb{Z}} N)^{\Gamma_F}$.

Fact 99.4. (Cf. [**24**, Ch. IV, §7].) *Let* $0 \to M' \to M \to M'' \to 0$ *be an exact sequence of Galois modules over* F. *Suppose that for a Galois module* N *the sequence*

$$0 \to M' \otimes_{\mathbb{Z}} N \to M \otimes_{\mathbb{Z}} N \to M'' \otimes_{\mathbb{Z}} N \to 0$$

is exact. Then the diagram

$$\begin{array}{ccc} H^n(F, M'') \otimes H^m(F, N) & \xrightarrow{\ \cup\ } & H^{n+m}(F, M'' \otimes_{\mathbb{Z}} N) \\ \Big\downarrow{\scriptstyle \partial \otimes \mathrm{id}} & & \Big\downarrow{\scriptstyle \partial} \\ H^{n+1}(F, M') \otimes H^m(F, N) & \xrightarrow{\ \cup\ } & H^{n+m+1}(F, M' \otimes_{\mathbb{Z}} N) \end{array}$$

is commutative.

Example 99.5. The cup-product

$$H^0(F, F_{sep}^\times) \otimes H^2(F, \mathbb{Z}) \to H^2(F, F_{sep}^\times)$$

yields a pairing

$$F^\times \otimes \acute{\mathrm{E}}\mathrm{t}(F) \to \mathrm{Br}_2(F).$$

If char $F \neq 2$, we have $a \cup [F_b] = \left(\frac{a,b}{F}\right)$ for all $a, b \in F^\times$. In the case that char $F = 2$, we have $a \cup [F_{ab}] = \left[\frac{a,b}{F}\right]$ for all $a \in F^\times$ and $b \in F$.

Suppose that char $F \neq 2$. Then $\mu_2 \simeq \mathbb{Z}/2\mathbb{Z}$. The cup-product

$$H^1(F, \mathbb{Z}/2\mathbb{Z}) \otimes H^1(F, \mathbb{Z}/2\mathbb{Z}) \to H^2(F, \mathbb{Z}/2\mathbb{Z})$$

gives rise to a pairing

$$F^\times/F^{\times 2} \otimes F^\times/F^{\times 2} \to \mathrm{Br}_2(F).$$

We have $(a) \cup (b) = \left(\frac{a,b}{F}\right)$ for all $a, b \in F^\times$. In particular, $(a) \cup (1-a) = 0$ for every $a \neq 0, 1$ by Fact 98.13(4).

99.C. Restriction and corestriction homomorphisms. Let M be a Galois module over F and K/F an arbitrary field extension. Separable closures of F and K can be chosen so that $F_{sep} \subset K_{sep}$. The restriction then yields a continuous group homomorphism $\Gamma_K \to \Gamma_F$. In particular, M has the structure of a discrete Γ_K-module and we have the *restriction homomorphism*

$$r_{K/F} : H^n(F, M) \to H^n(K, M).$$

If K/F is a finite separable field extension, then Γ_K is an open subgroup of finite index in Γ_F. For every $n \geq 0$ there is a natural *corestriction homomorphism*

$$c_{K/F} : H^n(K, M) \to H^n(F, M).$$

In the case $n = 0$, the map $c_{K/F} : M^{\Gamma_K} \to M^{\Gamma_F}$ is given by $x \to \sum \gamma(x)$ where the sum is over a left transversal of Γ_K in Γ_F. The composition $c_{K/F} \circ r_{K/F}$ is multiplication by $[K : F]$.

Let K/F be an arbitrary finite field extension and let M be a Galois module over F. Let E/F be the maximal separable sub-extension in K/F. As the restriction map $\Gamma_K \to \Gamma_E$ is an isomorphism, we have a canonical isomorphism $s : H^n(K, M) \xrightarrow{\sim} H^n(E, M)$. We define the corestriction homomorphism $c_{K/F} : H^n(K, M) \to H^n(F, M)$ as $[K : E]$ times the composition $c_{E/F} \circ s$.

Example 99.6. The corestriction homomorphism $c_{K/F} : H^1(K, \mu_m) \to H^1(F, \mu_m)$ takes a class $(x)_m$ to $\big(\mathrm{N}_{K/F}(x)\big)_m$.

Example 99.7. The restriction map in Galois cohomology agrees with the restriction map for Brauer groups defined in §98.C. The corestriction in Galois cohomology yields a map $c_{K/F} : \mathrm{Br}(K) \to \mathrm{Br}(F)$ for a finite field extension K/F. Since the composition $c_{K/F} \circ r_{K/F}$ is the multiplication by $m = [K : F]$ we have $\mathrm{Br}(K/F) \subset \mathrm{Br}_m(F)$.

Let K/F be a finite separable field extension and let M be a Galois module over K. We view Γ_K as a subgroup of Γ_F. Denote by $\mathrm{Ind}_{K/F}(M)$ the group $\mathrm{Map}_{\Gamma_K}(\Gamma_F, M)$ of Γ_K-equivariant maps $\Gamma_F \to M$, i.e., maps $f : \Gamma_F \to M$ satisfying $f(\rho\delta) = \rho f(\delta)$ for all $\rho \in \Gamma_K$ and $\delta \in \Gamma_F$. The group $\mathrm{Ind}_{K/F}(M)$ has a structure of Galois module over F defined by $(\gamma f)(\delta) = f(\delta\gamma)$ for all $f \in \operatorname{Ind}_{K/F}(M)$ and $\gamma, \delta \in \Gamma_F$. Consider the Γ_K-module homomorphisms

$$M \xrightarrow{u} \operatorname{Ind}_{K/F}(M) \xrightarrow{v} M$$

defined by $v(f) = f(1)$ and

$$u(m)(\gamma) = \begin{cases} m & \text{if } \gamma \in \Gamma_K, \\ 0 & \text{otherwise.} \end{cases}$$

Fact 99.8. *Let M be a Galois module over F and let K/F be a finite separable field extension. Then the compositions*

$$H^n\big(F, \operatorname{Ind}_{K/F}(M)\big) \xrightarrow{r_{K/F}} H^n\big(K, \operatorname{Ind}_{K/F}(M)\big) \xrightarrow{H^n(K,v)} H^n(K, M),$$

$$H^n(K, M) \xrightarrow{H^n(K,u)} H^n\big(K, \operatorname{Ind}_{K/F}(M)\big) \xrightarrow{c_{K/F}} H^n\big(F, \operatorname{Ind}_{K/F}(M)\big)$$

are isomorphisms inverse to each other.

Suppose, in addition, that M is a Galois module over F. Consider the Γ_F-module homomorphisms

$$0 \to M \xrightarrow{w} \operatorname{Ind}_{K/F}(M) \xrightarrow{t} M \to 0 \tag{99.9}$$

defined by $w(m)(\gamma) = \gamma m$ and

$$t(f) = \sum \gamma\big(f(\gamma^{-1})\big),$$

where the sum is taken over a left transversal of Γ_K in Γ_F.

Corollary 99.10. (1) *The composition*

$$H^n(F, M) \xrightarrow{H^n(F,w)} H^n\big(F, \operatorname{Ind}_{K/F}(M)\big) \xrightarrow{\sim} H^n(K, M)$$

coincides with $r_{K/F}$.

(2) *The composition*

$$H^n(K, M) \xrightarrow{\sim} H^n\big(F, \operatorname{Ind}_{K/F}(M)\big) \xrightarrow{H^n(F,t)} H^n(K, M)$$

coincides with $c_{K/F}$.

99.D. Residue homomorphism. Let m be an integer. A Galois module M over F is said to be *m-periodic* if $mM = 0$. If m is not divisible by $\operatorname{char} F$, we write $M(-1)$ for the Galois module $\operatorname{Hom}(\mu_m, M)$ with the action of Γ_F given by $(\gamma f)(\xi) = \gamma f(\gamma^{-1}\xi)$ for every $f \in M(-1)$ (the construction is independent of the choice of m). For example, $\mu_m(-1) = \mathbb{Z}/m\mathbb{Z}$.

Let L be a field with a discrete valuation v and residue field F. Suppose that the inertia group of an extension of v to L_{sep} acts trivially on M. Then M has a natural structure of a Galois module over F.

Fact 99.11 (Cf. [**46**, §7]). *Let L be a field with a discrete valuation v and residue field F. Let M be an m-periodic Galois module L with m not divisible by* $\operatorname{char} F$ *such that the inertia group of an extension of v to L_{sep} acts trivially on M. Then there exists a residue homomorphism*

$$\partial_v : H^{n+1}(L, M) \to H^n\big(F, M(-1)\big)$$

satisfying

(1) *If $M = \mu_m$ and $n = 0$, then $\partial_v\big((x)_m\big) = v(x) + m\mathbb{Z}$ for every $x \in L^\times$.*

(2) *For every* $x \in L^\times$ *with* $v(x) = 0$, *we have* $\partial_v\big(\alpha \cup (x)_m\big) = \partial_v(\alpha) \cup (\bar{x})_m$, *where* $\alpha \in H^{n+1}(L, M)$ *and* $\bar{x} \in F^\times$ *is the residue of* x.

Let X be a variety (integral scheme) over F and let $x \in X$ be a regular point of codimension 1. The local ring $\mathcal{O}_{X,x}$ is a discrete valuation ring with quotient field $F(X)$ and residue field $F(x)$. For any m-periodic Galois module M over F let

$$\partial_x : H^{n+1}\big(F(X), M\big) \to H^n\big(F(x), M(-1)\big)$$

denote the residue homomorphism ∂_v of the associated discrete valuation v on $F(X)$.

Fact 99.12 (Cf. [**46**, Th. 9.2]). *For every field* F, *the sequence*

$$0 \to H^{n+1}(F, M) \xrightarrow{r} H^{n+1}\big(F(t), M\big) \xrightarrow{(\partial_x)} \coprod_{x \in \mathbb{P}^1} H^n\big(F(x), M(-1)\big) \xrightarrow{c} H^n\big(F, M(-1)\big) \to 0,$$

where c *is the direct sum of the corestriction homomorphisms* $c_{F(x)/F}$, *is exact.*

99.E. A long exact sequence. Let $K = F(\sqrt{a})$ be a quadratic field extension of a field F of characteristic not 2. Let M be a 2-periodic Galois module over F.

We have the exact sequence (99.9) of Galois modules over F. By Corollary 99.10, the induced exact sequence of Galois cohomology groups reads as follows:

$$\dots \xrightarrow{\partial} H^n(F, M) \xrightarrow{r_{K/F}} H^n(K, M) \xrightarrow{c_{K/F}} H^n(F, M) \xrightarrow{\partial} H^{n+1}(F, M) \to \dots.$$

We now compute the connecting homomorphisms ∂. If $n = 0$ and $M = \mathbb{Z}/2\mathbb{Z}$, we have the exact sequence

$$\mathbb{Z}/2\mathbb{Z} \xrightarrow{0} \mathbb{Z}/2\mathbb{Z} \xrightarrow{\partial} F^\times/F^{\times 2} \to K^\times/K^{\times 2}.$$

The kernel of the last homomorphism is the cyclic group $\{1, (a)\}$. It follows that $\partial(1 + 2\mathbb{Z}) = (a)$. By Fact 99.4, the homomorphisms $\partial : H^n(F, M) \to H^{n+1}(F, M)$ coincide with the cup-product by (a).

We have proven

Theorem 99.13. *Let* $K = F(\sqrt{a})$ *be a quadratic field extension of a field* F *of characteristic not* 2 *and let* M *be a* 2*-periodic Galois module over* F. *Then the following sequence*

$$\dots \xrightarrow{\cup(a)} H^n(F, M) \xrightarrow{r_{K/F}} H^n(K, M) \xrightarrow{c_{K/F}} H^n(F, M) \xrightarrow{\cup(a)} H^{n+1}(F, M) \xrightarrow{r_{K/F}} \dots$$

is exact.

100. Milnor K-theory of fields

A more detailed exposition on the Milnor K-theory of fields is available in [**43**, Ch. IX] and [**47**, Ch. 7].

100.A. Definition. Let F be a field. Let T_* denote the tensor ring of the multiplicative group $F^\times$. That is a graded ring with T_n the nth tensor power of $F^\times$ over $\mathbb{Z}$. For instance, $T_0 = \mathbb{Z}$, $T_1 = F^\times$, $T_2 = F^\times \otimes_{\mathbb{Z}} F^\times$, etc. The *graded Milnor ring* $K_*(F)$ *of* F is the factor ring of T_* by the ideal generated by tensors of the form $a \otimes b$ with $a + b = 1$.

The class of a tensor $a_1 \otimes a_2 \otimes \ldots \otimes a_n$ in $K_*(F)$ is denoted by $\{a_1, a_2, \ldots, a_n\}_F$ or simply by $\{a_1, a_2, \ldots, a_n\}$ and is called a *symbol*. We have $K_n(F) = 0$ if $n < 0$, $K_0(F) = \mathbb{Z}$, $K_1(F) = F^\times$. For $n \geq 2$, $K_n(F)$ is generated (as an abelian group) by the symbols $\{a_1, a_2, \ldots, a_n\}$ with $a_i \in F^\times$ that are subject to the following defining relations:

(M1) (*Multilinearity*)

$$\{a_1, \ldots, a_i a_i', \ldots, a_n\} = \{a_1, \ldots, a_i, \ldots, a_n\} + \{a_1, \ldots, a_i', \ldots, a_n\};$$

(M2) (*Steinberg Relation*) $\{a_1, a_2, \ldots, a_n\} = 0$ if $a_i + a_{i+1} = 1$ for some $i \in [1, n-1]$.

Note that the operation in the group $K_n(F)$ is written additively. In particular, $\{ab\} = \{a\} + \{b\}$ in $K_1(F)$ where $a, b \in F^\times$.

The product in the ring $K_*(F)$ is given by the rule

$$\{a_1, a_2, \ldots, a_n\} \cdot \{b_1, b_2, \ldots, b_m\} = \{a_1, a_2, \ldots, a_n, b_1, b_2, \ldots, b_m\}.$$

Fact 100.1. *Let* $a_1, \ldots, a_n \in F^\times$.

(1) *For a permutation* $\sigma \in S_n$, *we have*

$$\{a_{\sigma(1)}, a_{\sigma(2)}, \ldots, a_{\sigma(n)}\} = \operatorname{sgn}(\sigma)\{a_1, a_2, \ldots, a_n\}.$$

(2) $\{a_1, a_2, \ldots, a_n\} = 0$ *if* $a_i + a_j = 0$ *or* 1 *for some* $i \neq j$.

A field homomorphism $F \to L$ induces the *restriction graded ring homomorphism* $r_{L/F} : K_*(F) \to K_*(L)$ taking a symbol $\{a_1, a_2, \ldots, a_n\}_F$ to the symbol $\{a_1, a_2, \ldots, a_n\}_L$. In particular, $K_*(L)$ has a natural structure of a left and right graded $K_*(F)$-module. The image $r_{L/F}(\alpha)$ of an element $\alpha \in K_*F$ is also denoted by α_L.

If E/L is another field extension, then $r_{E/F} = r_{E/L} \circ r_{L/F}$. Thus, K_* is a functor from the category of fields to the category of graded rings.

Proposition 100.2. *Let* L/F *be a quadratic field extension. Then*

$$K_n(L) = r_{L/F}\big(K_{n-1}(F)\big) \cdot K_1(L)$$

for every $n \geq 1$, *i.e.,* $K_*(L)$ *is generated by* $K_1(L)$ *as a left* $K_*(F)$*-module.*

PROOF. It is sufficient to treat the case $n = 2$. Let $x, y \in L \setminus F$. If $x = cy$ for some $c \in F^\times$, then $\{x, y\} = \{-c, y\} \in r_{L/F}\big(K_1(F)\big) \cdot K_1(L)$. Otherwise, as x, y, and 1 are linearly dependent over F, there are $a, b \in F^\times$ such that $ax + by = 1$. We have

$$0 = \{ax, by\} = \{x, y\} + \{x, b\} + \{a, by\},$$

hence $\{x, y\} = \{b\}_L \cdot \{x\} - \{a\}_L \cdot \{by\} \in r_{L/F}\big(K_1(F)\big) \cdot K_1(L)$. □

We write $k_*(F)$ for the graded ring $K_*(F)/2K_*(F)$. Abusing notation, if $\{a_1, \ldots, a_n\}$ is a symbol in $K_n(F)$, we shall also write it for its coset $\{a_1, \ldots, a_n\} + 2K_n(F)$ in $k_n(F)$.

We need some relations among symbols in $k_2(F)$.

Lemma 100.3. *We have the following relations in $k_2(F)$:*

(1) $\{a, x^2 - ay^2\} = 0$ *for all* $a \in F^\times$, $x, y \in F$ *satisfying* $x^2 - ay^2 \neq 0$.
(2) $\{a, b\} = \{a + b, ab(a + b)\}$ *for all* $a, b \in F^\times$ *satisfying* $a + b \neq 0$.

PROOF. (1): The statement follows from Fact 100.1 if $x = 0$. Suppose $x \neq 0$. By the Steinberg relation, we have

$$0 = \{a(yx^{-1})^2, 1 - a(yx^{-1})^2\} = \{a, x^2 - ay^2\}.$$

(2): Since $a(a+b) + b(a+b)$ is a square, by (1) we have

$$0 = \{a(a+b), b(a+b)\} = \{a, b\} + \{a + b, ab(a+b)\}. \qquad \square$$

100.B. Residue homomorphism. Let L be a field with a discrete valuation v and residue field F. The homomorphism $L^\times \to \mathbb{Z}$ given by the valuation can be viewed as a homomorphism $K_1(L) \to K_0(F)$. More generally, for every $n \geq 0$, there is the *residue homomorphism*

$$\partial_v : K_{n+1}(L) \to K_n(F)$$

uniquely determined by the following condition:

If $a_0, a_1, \dots, a_n \in L^\times$ satisfying $v(a_i) = 0$ for all $i = 1, 2, \dots, n$, then

$$\partial_v\big(\{a_0, a_1, \dots, a_n\}\big) = v(a_0) \cdot \{\bar{a}_1, \dots, \bar{a}_n\},$$

where $\bar{a} \in F$ denotes the residue of a.

Fact 100.4. *Let L be a field with a discrete valuation v and residue field F.*

(1) *If $\alpha \in K_*(L)$ and $a \in L^\times$ satisfies $v(a) = 0$, then*

$$\partial_v(\alpha \cdot \{a\}) = \partial_v(\alpha) \cdot \{\bar{a}\} \quad \textit{and} \quad \partial_v(\{a\} \cdot \alpha) = -\{\bar{a}\} \cdot \partial_v(\alpha).$$

(2) *Let K/L be a field extension and let u be a discrete valuation of K extending v with residue field E. Let e denote the ramification index. Then for every $\alpha \in K_*(L)$,*

$$\partial_u\big(r_{K/L}(\alpha)\big) = e \cdot r_{E/F}\big(\partial_v(\alpha)\big).$$

100.C. Milnor's theorem. Let X be a variety (integral scheme) over F and let $x \in X$ be a regular point of codimension 1. The local ring $\mathcal{O}_{X,x}$ is a discrete valuation ring with quotient field $F(X)$ and residue field $F(x)$. Denote by

$$\partial_x : K_{*+1}\big(F(X)\big) \to K_*\big(F(x)\big)$$

the residue homomorphism of the associated discrete valuation on $F(X)$.

The following description of the K-groups of the function field $F(t) = F(\mathbb{A}^1_F)$ of the affine line is known as Milnor's Theorem.

Fact 100.5. (Milnor's Theorem) *For every field F, the sequence*

$$0 \to K_{n+1}(F) \xrightarrow{r_{F(t)/F}} K_{n+1}\big(F(t)\big) \xrightarrow{(\partial_x)} \coprod_{x \in \mathbb{A}^1_{(0)}} K_n\big(F(x)\big) \to 0$$

is split exact.

100.D. Specialization. Let L be a field and let v be a discrete valuation on L with residue field F. If $\pi \in L^\times$ is a prime element, i.e., $v(\pi) = 1$, we define the *specialization homomorphism*

$$s_\pi : K_*(L) \to K_*(F)$$

by the formula $s_\pi(u) = \partial(\{-\pi\} \cdot u)$. We have

$$s_\pi\big(\{a_1, a_2, \dots, a_n\}\big) = \{\bar{b}_1, \bar{b}_2, \dots, \bar{b}_n\},$$

where $b_i = a_i/\pi^{v(a_i)}$.

Example 100.6. Consider the discrete valuation v of the field of rational functions $F(t)$ given by the irreducible polynomial t. For every $u \in K_*(F)$, we have $s_t(u_{F(t)}) = u$. In particular, the homomorphism $K_*(F) \to K_*\big(F(t)\big)$ is split injective as stated in Fact 100.5.

100.E. Norm homomorphism. Let L/F be a finite field extension. The standard norm homomorphism $L^\times \to F^\times$ can be viewed as a homomorphism $K_1(L) \to K_1(F)$. In fact, there exists the *norm homomorphism*

$$c_{L/F} : K_n(L) \to K_n(F)$$

for every $n \geq 0$ defined as follows.

Suppose first that the field extension L/F is simple, i.e., L is generated by one element over F. We identify L with the residue field $F(y)$ of a closed point $y \in \mathbb{A}^1_F$. Let $\alpha \in K_n(L) = K_n\big(F(y)\big)$. By Milnor's Theorem 100.5, there is $\beta \in K_{n+1}\big(F(\mathbb{A}^1_F)\big)$ satisfying

$$\partial_x(\beta) = \begin{cases} \alpha & \text{if } x = y, \\ 0 & \text{otherwise.} \end{cases}$$

Let v be the discrete valuation of the field $F(\mathbb{P}^1_F) = F(\mathbb{A}^1_F)$ associated with the infinite point of the projective line $\mathbb{P}^1_F$. We set $c_{L/F}(\alpha) = \partial_v(\beta)$.

In the general case, we choose a sequence of simple field extensions

$$F = F_0 \subset F_1 \subset \dots \subset F_n = L$$

and set

$$c_{L/F} = c_{F_1/F_0} \circ c_{F_2/F_1} \circ \dots \circ c_{F_n/F_{n-1}}.$$

It turns out that the norm map $c_{L/F}$ is well-defined, i.e., it does not depend on the choice of the sequence of simple field extensions and the identifications with residue fields of closed points of the affine line (cf. [**43**, Ch. IX, §3], [**47**, Ch. 7, §3]).

The following theorem is the direct consequence of the definition of the norm map and Milnor's Theorem 100.5.

Theorem 100.7. *For every field F, the sequence*

$$0 \to K_{n+1}(F) \xrightarrow{r_{F(t)/F}} K_{n+1}\big(F(t)\big) \xrightarrow{(\partial_x)} \coprod_{x \in \mathbb{P}^1_{(0)}} K_n\big(F(x)\big) \xrightarrow{c} K_n(F) \to 0$$

is exact where c is the direct sum of the corestriction homomorphisms $c_{F(x)/F}$.

Fact 100.8. (1) (Transitivity) *Let L/F and E/L be finite field extensions. Then* $c_{E/F} = c_{L/F} \circ c_{E/L}$.

(2) *The norm map $c_{L/F} : K_0(L) \to K_0(F)$ is multiplication by $[L:F]$ on $\mathbb{Z}$. The norm map $c_{L/F} : K_1(L) \to K_1(F)$ is the classical norm $L^\times \to F^\times$.*

(3) (Projection Formula) *Let L/F be a finite field extension. Then for every $\alpha \in K_*F$ and $\beta \in K_*(L)$ we have*

$$c_{L/F}(r_{L/F}(\alpha)\cdot\beta) = \alpha \cdot c_{L/F}(\beta),$$

i.e., if we view $K_*(L)$ as a $K_*(F)$-module via $r_{L/F}$, then $c_{L/F}$ is a homomorphism of $K_*(F)$-modules. In particular, the composition $c_{L/F} \circ r_{L/F}$ is multiplication by $[L:F]$.

(4) *Let L/F be a finite field extension and let v be a discrete valuation on F. Let $v_1, v_2, \dots, v_s$ be all the extensions of v to L. Then the following diagram is commutative:*

$$\begin{array}{ccc} K_{n+1}(L) & \xrightarrow{(\partial_{v_i})} & \coprod_{i=1}^s K_n(L(v_i)) \\ {\scriptstyle c_{L/F}}\downarrow & & \downarrow{\scriptstyle \sum c_{L(v_i)/F(v)}} \\ K_{n+1}(F) & \xrightarrow{\partial_v} & K_n(F(v)). \end{array}$$

(5) *Let L/F be a finite and E/F an arbitrary field extension. Let $P_1, P_2, \dots, P_k$ be all the prime (maximal) ideals of the ring $R = L \otimes_F E$. For every $i \in [1,k]$, let R_i denote the residue field R/P_i and let l_i be the length of the localization ring R_{P_i}. Then the following diagram is commutative:*

$$\begin{array}{ccc} K_n(L) & \xrightarrow{(r_{R_i/L})} & \coprod_{i=1}^k K_n(R_i) \\ {\scriptstyle c_{L/F}}\downarrow & & \downarrow{\scriptstyle \sum l_i \cdot c_{R_i/E}} \\ K_n(F) & \xrightarrow{r_{E/F}} & K_n(E). \end{array}$$

We now turn to fields of positive characteristic.

Fact 100.9 (Cf. [**60**, Th. A]). *Let F be a field of characteristic $p > 0$. Then the p-torsion part of $K_*(F)$ is trivial.*

Fact 100.10 (Cf. [**60**, Cor. 6.5]). *Let F be a field of characteristic $p > 0$. Then the natural homomorphism*

$$K_n(F)/pK_n(F) \to H^0\big(F, K_n(F_{sep})/pK_n(F_{sep})\big)$$

is an isomorphism.

Now consider the case of purely inseparable quadratic extensions.

Lemma 100.11. *Let L/F be a purely inseparable quadratic field extension. Then the composition $r_{L/F} \circ c_{L/F}$ on $K_n(L)$ is the multiplication by 2.*

Proof. The statement is obvious if $n = 1$. The general case follows from Proposition 100.2 and Fact 100.8(3). □

Write $k_n(E) := K_n(E)/2K_n(E)$ for a field E.

Proposition 100.12. *Let L/F be a purely inseparable quadratic field extension. Then the sequence*

$$k_n(F) \xrightarrow{r_{L/F}} k_n(L) \xrightarrow{c_{L/F}} k_n(F) \xrightarrow{r_{L/F}} k_n(L)$$

is exact.

PROOF. Let $\alpha \in K_n(F)$ satisfy $\alpha_L = 2\beta$ for some $\beta \in K_n(L)$. By Proposition 100.8,

$$2\alpha = c_{L/F}(\alpha_L) = c_{L/F}(2\beta) = 2c_{L/F}(\beta),$$

hence $\alpha = c_{L/F}(\beta)$ in view of Fact 100.9.

Let $\beta \in K_n(L)$ satisfy $c_{L/F}(\beta) = 2\alpha$ for some $\alpha \in K_n(F)$. It follows from Lemma 100.11 that

$$2\beta = c_{L/F}(\beta)_L = 2\alpha_L,$$

hence $\beta = \alpha_L$, again by Fact 100.9. □

101. The cohomology groups $H^{n,i}(F, \mathbb{Z}/m\mathbb{Z})$

Let F be a field. For all $n, m, i \in \mathbb{Z}$ with $m > 0$, we define the group $H^{n,i}(F, \mathbb{Z}/m\mathbb{Z})$ as follows (cf. [**46**]): If m is not divisible by char F, we set

$$H^{n,i}(F, \mathbb{Z}/m\mathbb{Z}) = H^n\big(F, \mu_m^{\otimes i}\big),$$

where $\mu_m^{\otimes i}$ is the ith tensor power of μ_m if $i \geq 0$ and $\mu_m^{\otimes i} = \operatorname{Hom}(\mu_m^{\otimes -i}, \mathbb{Z}/m\mathbb{Z})$ if $i < 0$. If char $F = p > 0$ and m is a power of p, we set

$$H^{n,i}(F, \mathbb{Z}/m\mathbb{Z}) = \begin{cases} K_i(F)/mK_i(F) & \text{if } n = i, \\ H^1\big(F, K_i(F_{sep})/mK_i(F_{sep})\big) & \text{if } n = i+1, \\ 0 & \text{otherwise.} \end{cases}$$

In the general case, write $m = m_1 m_2$, where m_1 is not divisible by char F and m_2 is a power of char F if char $F > 0$, and set

$$H^{n,i}(F, \mathbb{Z}/m\mathbb{Z}) = H^{n,i}(F, \mathbb{Z}/m_1\mathbb{Z}) \oplus H^{n,i}(F, \mathbb{Z}/m_2\mathbb{Z}).$$

Note that if char F does not divide m and $\mu_m \subset F^\times$, we have a natural isomorphism

$$H^{n,i}(F, \mathbb{Z}/m\mathbb{Z}) \simeq H^{n,0}(F, \mathbb{Z}/m\mathbb{Z}) \otimes \mu_m^{\otimes i}.$$

In particular, the groups $H^{n,i}(F, \mathbb{Z}/m\mathbb{Z})$ and $H^{n,0}(F, \mathbb{Z}/m\mathbb{Z})$ are (noncanonically) isomorphic.

Example 101.1. For an arbitrary field F, we have canonical isomorphisms

(1) $H^{0,0}(F, \mathbb{Z}/m\mathbb{Z}) \simeq \mathbb{Z}/m\mathbb{Z}$,
(2) $H^{1,1}(F, \mathbb{Z}/m\mathbb{Z}) \simeq F^\times/F^{\times m}$,
(3) $H^{1,0}(F, \mathbb{Z}/m\mathbb{Z}) \simeq \operatorname{Hom}_c(\Gamma_F, \mathbb{Z}/m\mathbb{Z})$, $\quad H^{1,0}(F, \mathbb{Z}/2\mathbb{Z}) \cong \operatorname{\acute{E}t}_2(F)$,
(4) $H^{2,1}(F, \mathbb{Z}/m\mathbb{Z}) \simeq \operatorname{Br}_m(F)$.

If L/F is a field extension, there is the *restriction homomorphism*

$$r_{L/F} : H^{n,i}(F, \mathbb{Z}/m\mathbb{Z}) \to H^{n,i}(L, \mathbb{Z}/m\mathbb{Z}).$$

If L is a finite over F, we define the *corestriction homomorphism*

$$c_{L/F} : H^{n,i}(L, \mathbb{Z}/m\mathbb{Z}) \to H^{n,i}(F, \mathbb{Z}/m\mathbb{Z})$$

as follows: It is sufficient to consider the following two cases.

(1) If L/F is separable, then $c_{L/F}$ is the corestriction homomorphism in Galois cohomology.

(2) If L/F is purely inseparable, then $\Gamma_L = \Gamma_F$, we have $[L_{sep} : F_{sep}] = [L : F]$, and $c_{L/F}$ is induced by the corestriction homomorphism $K_*(L_{sep}) \to K_*(F_{sep})$.

Example 101.2. Let L/F be a finite field extension. By Example 99.6, the map

$$c_{L/F} : L^\times/L^{\times m} = H^{1,1}(L, \mathbb{Z}/m\mathbb{Z}) \to H^{1,1}(F, \mathbb{Z}/m\mathbb{Z}) = F^\times/F^{\times m}$$

is induced by the norm map $\mathrm{N}_{L/F} : L^\times \to F^\times$. If $\operatorname{char} F = p > 0$, it follows from Example 99.1(2) that the map

$$c_{L/F} : L/\wp(L) = H^{1,0}(L, \mathbb{Z}/p\mathbb{Z}) \to H^{1,0}(F, \mathbb{Z}/p\mathbb{Z}) = F/\wp(F)$$

is induced by the trace map $\mathrm{Tr}_{L/F} : L \to F$.

Let $l, m \in Z$. If $\operatorname{char} F$ does not divide l and m, we have a natural exact sequence of Galois modules

$$1 \to \mu_l^{\otimes i} \to \mu_{lm}^{\otimes i} \to \mu_m^{\otimes i} \to 1$$

for every i. If l and m are powers of $\operatorname{char} F > 0$, then by Fact 100.9, the sequence of Galois modules

$$0 \to K_n(F_{sep})/lK_n(F_{sep}) \to K_n(F_{sep})/lmK_n(F_{sep}) \to K_n(F_{sep})/mK_n(F_{sep}) \to 0$$

is exact. Taking the long exact sequences of Galois cohomology groups yields the following proposition.

Proposition 101.3. *For any $l, m, n, i \in \mathbb{Z}$ with $l, m > 0$, there is a natural long exact sequence*

$$\cdots \to H^{n,i}(F, \mathbb{Z}/l\mathbb{Z}) \to H^{n,i}(F, \mathbb{Z}/lm\mathbb{Z}) \to H^{n,i}(F, \mathbb{Z}/m\mathbb{Z}) \to H^{n+1,i}(F, \mathbb{Z}/l\mathbb{Z}) \to \cdots .$$

The cup-product in Galois cohomology and the product in the Milnor ring induce a structure of the graded ring on the graded abelian group

$$H^{*,*}(F, \mathbb{Z}/m\mathbb{Z}) = \coprod_{i,j \in \mathbb{Z}} H^{i,j}(F, \mathbb{Z}/m\mathbb{Z})$$

for every $m \in \mathbb{Z}$. The product in this ring will be denoted by $\cup$.

101.A. Norm residue homomorphism. Let symbol $(a)_m$ denote the element in $H^{1,1}(F, \mathbb{Z}/m\mathbb{Z})$ corresponding to $a \in F^\times$ under the isomorphism in Example 101.1(2).

Lemma 101.4 (Steinberg Relation). *Let $a, b \in F^\times$ satisfy $a + b = 1$. Then $(a)_m \cup (b)_m = 0$ in $H^{2,2}(F, \mathbb{Z}/m\mathbb{Z})$.*

PROOF. We may assume that $\operatorname{char} F$ does not divide m. Let $K = F[t]/(t^m - a)$ and let $\alpha \in K^\times$ be the class of t. We have $a = \alpha^m$ and $\mathrm{N}_{K/F}(1 - \alpha) = b$, hence $1 - \alpha \in K^\times$. It follows from the projection formula and Example 99.6 that

$$(a)_m \cup (b)_m = c_{K/F}\big(r_{K/F}(a)_m \cup (1 - \alpha)_m\big) = 0$$

since $r_{K/F}(a)_m = m(\alpha)_m = 0$ in $H^{1,1}(K, \mathbb{Z}/m\mathbb{Z})$. □

It follows from Lemma 101.4 that for every $n, m \in \mathbb{Z}$ there is a unique *norm residue homomorphism*

$$h_F^{n,m} : K_n(F)/mK_n(F) \to H^{n,n}(F, \mathbb{Z}/m\mathbb{Z}) \tag{101.5}$$

taking the class of a symbol $\{a_1, a_2, \dots, a_n\}$ to the cup-product $(a_1)_m \cup (a_2)_m \cup \dots \cup (a_n)_m$.

The norm residue homomorphism allows us to view $H^{*,*}(F, \mathbb{Z}/m\mathbb{Z})$ as a module over the Milnor ring $K_*(F)$.

By Example 99.1, the map $h_F^{n,m}$ is an isomorphism for $n = 0$ and 1. Bloch and Kato conjectured that $h_F^{n,m}$ is always an isomorphism.

For every $l, m \in \mathbb{Z}$, we have a commutative diagram

$$\begin{array}{ccc} K_n(F)/lmK_n(F) & \longrightarrow & K_n(F)/mK_n(F) \\ {\scriptstyle h_F^{n,lm}}\downarrow & & \downarrow{\scriptstyle h_F^{n,m}} \\ H^{n,n}(F, \mathbb{Z}/lm\mathbb{Z}) & \longrightarrow & H^{n,n}(F, \mathbb{Z}/m\mathbb{Z}) \end{array}$$

with top map the natural surjective homomorphism.

The following important result was originally conjectured by Milnor in [**106**] and was proven in [**136**] by Voevodsky.

Fact 101.6. *Let F be a field of characteristic not 2. If m is a power of 2, then the norm residue homomorphism $h_F^{n,m}$ is an isomorphism.*

Proposition 101.3 and commutativity of the diagram above yield

Corollary 101.7. *Let l and m be powers of 2. Then the natural homomorphism $H^{n,n}(F, \mathbb{Z}/lm\mathbb{Z}) \to H^{n,n}(F, \mathbb{Z}/m\mathbb{Z})$ is surjective and the sequence*

$$0 \to H^{n+1,n}(F, \mathbb{Z}/l\mathbb{Z}) \to H^{n+1,n}(F, \mathbb{Z}/lm\mathbb{Z}) \to H^{n+1,n}(F, \mathbb{Z}/m\mathbb{Z})$$

is exact for any n.

Now consider the case $m = 2$. We shall write h_F^n for $h_F^{n,2}$ and $H^n(F)$ for $H^{n,n}(F, \mathbb{Z}/2\mathbb{Z})$.

The norm residue homomorphisms commute with field extension homomorphisms. They also commute with residue and corestriction homomorphisms as the following two propositions show.

Proposition 101.8. *Let L be a field with a discrete valuation v and residue field F of characteristic different from 2. Then the diagram*

$$\begin{array}{ccc} k_{n+1}(L) & \xrightarrow{\partial_v} & k_n(F) \\ {\scriptstyle h_L^{n+1}}\downarrow & & \downarrow{\scriptstyle h_F^{n}} \\ H^{n+1}(L) & \xrightarrow{\partial_v} & H^n(F) \end{array}$$

is commutative.

Proof. Fact 99.11(1) shows that the diagram is commutative when $n = 0$. The general case follows from Fact 99.11(2) as the group $k_{n+1}(L)$ is generated by symbols $\{a_0, a_1, \dots, a_n\}$ with $v(a_1) = \dots = v(a_n) = 0$. □

Proposition 101.9. *Let L/F be a finite field extension. Then the diagram*

$$\begin{array}{ccc} k_n(L) & \xrightarrow{c_{L/F}} & k_n(F) \\ {\scriptstyle h^n_L}\downarrow & & \downarrow{\scriptstyle h^n_F} \\ H^n(L) & \xrightarrow{c_{L/F}} & H^n(F) \end{array}$$

is commutative.

PROOF. We may assume that L/F is a simple field extension. The statement follows from the definition of the norm map for the Milnor K-groups, Fact 99.12, and Proposition 101.8. □

Proposition 101.10. *Let F be a field of characteristic different from 2 and let $L = F(\sqrt{a})/F$ be a quadratic extension with $a \in F^\times$. Then the following infinite sequence*

$$\dots \to k_{n-1}(F) \xrightarrow{\{a\}} k_n(F) \xrightarrow{r_{L/F}} k_n(L) \xrightarrow{c_{L/F}} k_n(F) \xrightarrow{\{a\}} k_{n+1}(F) \to \dots$$

is exact.

PROOF. It follows from Proposition 101.9 that the diagram

$$\begin{array}{ccccccccc} k_{n-1}(F) & \xrightarrow{\{a\}} & k_n(F) & \xrightarrow{r_{L/F}} & k_*(L) & \xrightarrow{c_{L/F}} & k_n(F) & \xrightarrow{\{a\}} & k_{n+1}(F) \\ {\scriptstyle h^{n-1}_F}\downarrow & & {\scriptstyle h^n_F}\downarrow & & {\scriptstyle h^n_L}\downarrow & & {\scriptstyle h^n_F}\downarrow & & \downarrow{\scriptstyle h^{n+1}_F} \\ H^{n-1}(F) & \xrightarrow{\{a\}} & H^n(F) & \xrightarrow{r_{L/F}} & H^*(L) & \xrightarrow{c_{L/F}} & H^n(F) & \xrightarrow{\{a\}} & H^{n+1}(F) \end{array}$$

is commutative. By Fact 101.6, the vertical homomorphisms are isomorphisms. By Theorem 99.13, the bottom sequence is exact. The result follows. □

Now consider the case $\operatorname{char} F = 2$. The product in the Milnor ring and the cup-product in Galois cohomology yield a pairing

$$K_*(F) \otimes H^*(F) \to H^*(F)$$

making $H^*(F)$ a module over $K_*(F)$.

Example 101.11. By Example 99.5, we have $\{a\} \cdot [F_{ab}] = \left[\frac{a,b}{F}\right]$ in $\operatorname{Br}_2(F)$ for all $a \in F^\times$ and $b \in F$.

Proposition 101.12. *Let F be a field of characteristic 2 and let L/F be a separable quadratic field extension. Then the following sequence*

$$0 \to k_n(F) \xrightarrow{r_{L/F}} k_n(L) \xrightarrow{c_{L/F}} k_n(F)$$
$$\xrightarrow{\cdot [L]} H^{n+1}(F) \xrightarrow{r_{L/F}} H^{n+1}(L) \xrightarrow{c_{L/F}} H^{n+1}(F) \to 0$$

is exact where the middle map is multiplication by the class of L in $H^1(F)$.

PROOF. We shall show that the sequence in question coincides with the exact sequence in Theorem 99.13 for the quadratic field extension L/F and the Galois module $k_n(F_{sep})$ over F. Indeed, by Fact 100.10, we have $H^0(E, k_n(E_{sep})) \simeq k_n(E)$ and by definition, $H^1(E, k_n(E_{sep})) \simeq H^{n+1}(E)$ for every field E. Note that $H^2(F, k_n(F_{sep})) = 0$ by Example 99.1(2). The connecting homomorphism in the sequence in Theorem 99.13 is multiplication by the class of L in $H^1(F)$. □

Now let F be a field of characteristic different from 2. The connecting homomorphism

$$\mathfrak{d}_n : H^n(F) \to H^{n+1}(F)$$

with respect to the short exact sequence

$$(101.13) \qquad 0 \to \mathbb{Z}/2\mathbb{Z} \to \mathbb{Z}/4\mathbb{Z} \to \mathbb{Z}/2\mathbb{Z} \to 0$$

is called the *Bockstein map*.

Proposition 101.14. *The Bockstein map is trivial if n is even and coincides with multiplication by (-1) if n is odd.*

PROOF. If n is even or $-1 \in F^{\times 2}$, then $\mu_4^{\otimes n} \simeq \mathbb{Z}/4\mathbb{Z}$ and the statement follows from Corollary 101.7.

Suppose that n is odd and $-1 \notin F^{\times 2}$. In this case $\mu_4^{\otimes n} \simeq \mu_4$. Consider the field $K = F(\sqrt{-1})$. By Theorem 99.13, the connecting homomorphism $H^n(F) \to H^{n+1}(F)$ with respect to the exact sequence (99.9) is the cup-product with (-1). The classes of the sequences (101.13) and (99.9) differ in $\operatorname{Ext}^1_\Gamma(\mathbb{Z}/2\mathbb{Z}, \mathbb{Z}/2\mathbb{Z})$ by the class of the sequence

$$0 \to \mathbb{Z}/2\mathbb{Z} \to \mu_4^{\otimes n} \to \mathbb{Z}/2\mathbb{Z} \to 0.$$

By Corollary 101.7, the connecting homomorphism $H^n(F) \to H^{n+1}(F)$ with respect to this exact sequence is trivial. It follows that $\mathfrak{d}_n$ is the cup-product with (-1). □

101.B. Cohomological dimension and p-special fields. Let p be a prime integer. A field F is called *p-special* if the degree of every finite field extension of F is a power of p.

The following property of p-special fields is very useful.

Proposition 101.15. *Let F be a p-special field and let L/F be a finite field extension. Then there is a tower of field extensions*

$$F = F_0 \subset F_1 \subset \cdots \subset F_{n-1} \subset F_n = L$$

satisfying $[F_{i+1} : F_i] = p$ for all $i \in [0, n-1]$.

PROOF. The result is clear if L/F is purely inseparable. So we may assume that L/F is a separable extension. Let E/F be a normal closure of L/F. Set $G = \operatorname{Gal}(E/F)$ and $H = \operatorname{Gal}(E/L)$. As G is a p-group, there is a sequence of subgroups

$$G = H_0 \supset H_1 \supset \cdots \supset H_{n-1} \supset H_n = H$$

with the property $[H_i : H_{i+1}] = p$ for all $i \in [0, n-1]$. Then the fields $F_i = L^{H_i}$ satisfy the required property. □

Proposition 101.16. *For every prime integer p and field F, there is a field extension L/F satisfying*

(1) *L is p-special.*
(2) *The degree of every finite sub-extension of L/F is not divisible by p.*

PROOF. If $\operatorname{char} F = q > 0$ and different from p, we set $F' := \bigcup F^{q^{-n}}$ (in a fixed algebraic closure of F), otherwise $F' := F$. Let Γ be the Galois group of F'_{sep}/F' and let $\Delta \subset \Gamma$ be a Sylow p-subgroup. The field $L = (F'_{sep})^\Delta$ satisfies the required conditions. □

We call the field L in Proposition 101.16 a *p-special closure of F*.

Let F be a field and let p be a prime integer. The *cohomological p-dimension of F*, denoted $\operatorname{cd}_p(F)$, is the smallest integer such that for every $n > \operatorname{cd}_p(F)$ and every finite field extension L/F we have $H^{n,n-1}(L, \mathbb{Z}/p\mathbb{Z}) = 0$.

Example 101.17. (1) $\operatorname{cd}_p(F) = 0$ if and only if F has no separable finite field extensions of degree a power of p.

(2) $\operatorname{cd}_p(F) \leq 1$ if and only if $\operatorname{Br}_p(L) = 0$ for all finite field extensions L/F.

(3) If F is p-special, then $\operatorname{cd}_p(F) < n$ if and only if $H^{n,n-1}(F, \mathbb{Z}/p\mathbb{Z}) = 0$.

102. Length and Herbrand index

102.A. Length. Let A be a commutative ring and let M be an A-module of finite length. The length of M is denoted by $l_A(M)$. The ring A is *artinian* if the A-module $M = A$ is of finite length. We write $l(A)$ for $l_A(A)$.

Lemma 102.1. *Let C be a flat B-algebra where B and C are commutative local artinian rings. Then for every finitely generated B-module M, we have*

$$l_C(M \otimes_B C) = l(C/QC) \cdot l_B(M),$$

where Q is the maximal ideal of B.

PROOF. Since C is flat over B, both sides of the equality are additive in M. Thus, we may assume that M is a simple B-module, i.e., $M = B/Q$. We have $M \otimes_B C \simeq C/QC$ and the equality follows. □

Setting $M = B$ we obtain

Corollary 102.2. *In the conditions of Lemma* 102.1, *one has* $l(C) = l(C/QC) \cdot l(B)$.

Lemma 102.3. *Let B be a commutative A-algebra and let M be a B-module of finite length over A. Then*

$$l_A(M) = \sum l_{B_Q}(M_Q) \cdot l_A(B/Q),$$

where the sum is taken over all maximal ideals $Q \subset B$.

PROOF. Both sides are additive in M, so we may assume that $M = B/Q$, where Q is a maximal ideal of B. The result follows. □

102.B. Herbrand index. Let M be a module over a commutative ring A and let $a \in A$. Suppose that the modules M/aM and let ${}_aM := \operatorname{Ker}(M \xrightarrow{a} M)$ have finite length. The integer

$$h(a, M) = l_A(M/aM) - l_A({}_aM)$$

is called the *Herbrand index of M relative to a*.

We collect simple properties of the Herbrand index in the following lemma.

Lemma 102.4. (1) *Let* $0 \to M' \to M \to M'' \to 0$ *be an exact sequence of A-modules. Then* $h(a, M) = h(a, M') + h(a, M'')$.

(2) *If M has finite length, then* $h(a, M) = 0$.

Lemma 102.5. *Let S be a 1-dimensional Noetherian local ring and let $P_1, \dots, P_m$ be all the minimal prime ideals of S. Let M be a finitely generated S-module and $s \in S$ not belonging to any of P_i. Then*

$$h(s, M) = \sum_{i=1}^{m} l_{S_{P_i}}(M_{P_i}) \cdot l\big(S/(P_i + sS)\big).$$

PROOF. Since $s \notin P_i$, the coset of s in S/P_i is not a zero divisor. Hence

$$l_S\big(S/(P_i + sS)\big) = h(s, S/P_i).$$

Both sides of the equality are additive in M. Since M has a filtration with factors S/P, where P is a prime ideal of S, we may assume that $M = S/P$. If P is maximal, then $M_{P_i} = 0$ and $h(s, M) = 0$ since M is simple. If $P = P_i$ for some i, then $l_{S_{P_j}}(M) = 1$ if $i = j$ and zero otherwise. The equality holds in this case too. □

103. Places

Let K be a field. A *valuation ring* R of K is a subring $R \subset K$ such that for any $x \in K \setminus R$, we have $x^{-1} \in R$. A valuation ring is a local domain. A trivial example of a valuation ring is the field K itself.

Given two fields K and L, a *place* $\pi : K \rightharpoonup L$ is a local ring homomorphism $f : R \to L$ of a valuation ring $R \subset K$. We say that the place π *is defined on* R. An embedding of fields is a trivial example of a place defined everywhere. A place $K \rightharpoonup L$ is called *surjective* if f is surjective.

If K and L are extensions of a field F, we say that a place $K \rightharpoonup L$ is an F-*place* if π is defined and the identity on F.

Let $K \rightharpoonup L$ and $L \rightharpoonup E$ be two places, given by ring homomorphisms $f : R \to L$ and $g : S \to E$ respectively, where $R \subset K$ and $S \subset L$ are valuation rings. Then the ring $T = f^{-1}(S)$ is a valuation ring of K and the composition $T \xrightarrow{f|_T} S \xrightarrow{g} E$ defines the *composition place* $K \rightharpoonup E$. In particular, any place $L \rightharpoonup E$ can be restricted to any subfield $K \subset L$.

A composition of two F-places is an F-place. Every place is a composition of a surjective place and a field embedding.

A place $K \rightharpoonup L$ is said to be *geometric* if it is a composition of (finitely many) places each defined on a discrete valuation ring. An embedding of fields will also be viewed as a geometric place.

Let Y be a complete variety over F and let $\pi : F(Y) \rightharpoonup L$ be an F-place. The valuation ring R of the place dominates a unique point $y \in Y$, i.e., $\mathcal{O}_{Y,y} \subset R$ and the maximal ideal of $\mathcal{O}_{Y,y}$ is contained in the maximal ideal M of R. The induced homomorphism of fields $F(y) \to R/M \to L$ over F gives rise to an L-point of Y, i.e., to a morphism $f : \operatorname{Spec}(L) \to Y$ with image $\{y\}$. We say that y is the *center of* π and f *is induced by* π.

Let X be a regular variety over F and let $f : \operatorname{Spec}(L) \to X$ be a morphism over F. Choose a regular system of parameters $a_1, a_2, \dots, a_n$ in the local ring $R = \mathcal{O}_{X,x}$, where $\{x\}$ is the image of f. Let M_i be the ideal of R generated by $a_1, \dots, a_i$ and set $R_i = R/M_i$, $P_i = M_{i+1}/M_i$. Denote by F_i the quotient field of R_i, in particular, $F_0 = F(X)$ and $F_n = F(x)$. The localization ring $(R_i)_{P_i}$ is a discrete valuation ring with quotient field F_i and residue field F_{i+1} therefore, it determines a place

$F_i \rightharpoonup F_{i+1}$. The composition of places

$$F(X) = F_0 \rightharpoonup F_1 \rightharpoonup \dots \rightharpoonup F_n = F(x) \hookrightarrow L$$

is a geometric place constructed (noncanonically) out of the point $f \in X(L)$.

Lemma 103.1. *Let K be an arbitrary field, let K'/K be an odd degree field extension, and let L/K be an arbitrary field extension. Then there exists a field L', containing K' and L, such that the extension L'/L is of odd degree.*

PROOF. We may assume that K'/K is a simple extension, i.e., K' is generated over K by one element. Let $f(t) \in F[t]$ be its minimal polynomial. Since the degree of f is odd, there exists an irreducible divisor $g \in L[t]$ of f over L with odd $\deg(g)$. We set $L' = L[t]/(g)$. □

Lemma 103.2. *Let K be a field extension of F of finite transcendence degree over F, let $K \rightharpoonup L$ be a geometric F-place and let K' be a finite field extension of K of odd degree. Then there exists an odd degree field extension L'/L such that the place $K \rightharpoonup L$ extends to a place $K' \rightharpoonup L'$.*

PROOF. By Lemma 103.1, it suffices to consider the case of a surjective place $K \rightharpoonup L$ given by a discrete valuation ring R. It also suffices to consider only two cases: (1) K'/K is purely inseparable and (2) K'/K is separable.

In the first case, the degree $[K' : K]$ is a power of an odd prime p. Let R' be an arbitrary valuation ring of K' *lying over* R, i.e., such that $R' \cap K = R$ and with the embedding $R \to R'$ local (such an R' exists in the case of an arbitrary field extension K'/K by [**140**, Ch. VI, Th. 5′]). We have a surjective place $K' \rightharpoonup L'$, where L' is the residue field of R'. We claim that L' is purely inseparable over L (and therefore $[L' : L]$, being a power of p, is odd). Indeed if $l \in L'$, choose a preimage $k \in R'$ of l. Then $k^{p^n} \in K$ for some n, hence $l^{p^n} \in L$, i.e., L'/L is a purely inseparable extension.

In the second case, consider all the valuation rings $R_1, \dots, R_r$ of K' lying over R (the number of such valuation rings is finite by [**140**, Ch. VI, Th. 12, Cor. 4]). The residue field of each R_i is a finite extension of L. Moreover, $\sum_{i=1}^r e_i n_i = [K' : K]$ by [**140**, Ch. VI, Th. 20 and p. 63], where n_i is the degree over L of the residue field of R_i, and e_i is the ramification index of R_i over R (cf. [**140**, Def. on pp. 52–53]). It follows that at least one of the n_i is odd. □

104. Cones and vector bundles

The word "scheme" in the next two sections means a separated scheme of finite type over a field.

104.A. Definition of a cone. Let X be a scheme over a field F and let $S^\bullet = S^0 \oplus S^1 \oplus S^2 \oplus \dots$ be a sheaf of graded $\mathcal{O}_X$-algebras. We assume that

(1) The natural morphism $\mathcal{O}_X \to S^0$ is an isomorphism.

(2) The $\mathcal{O}_X$-module S^1 is coherent.

(3) The sheaf of $\mathcal{O}_X$-algebras $S^\bullet$ is generated by S^1.

The *cone of* $S^\bullet$ is the scheme $C = \operatorname{Spec}(S^\bullet)$ over X and $\mathbb{P}(C) = \operatorname{Proj}(S^\bullet)$ is called the *projective cone of* C. Recall that $\operatorname{Proj}(S^\bullet)$ has a covering by the principal open subschemes $D(s) := \operatorname{Spec}(S_{(s)})$ over all $s \in S^1$, where $S_{(s)}$ is the subring of degree 0 elements in the localization S_s.

We have natural morphisms $C \to X$ and $\mathbb{P}(C) \to X$. The canonical homomorphism $S^\bullet \to S^0$ of $\mathcal{O}_X$-algebras induces the *zero section* $X \to C$.

If C and C' are cones over X, then $C \times_X C'$ has a natural structure of a cone over X. We denote it by $C \oplus C'$.

Example 104.1. A coherent $\mathcal{O}_X$-module P defines the cone $C(P) = \operatorname{Spec}\big(S^\bullet(P)\big)$ over X, where $S^\bullet$ stands for the symmetric algebra. If the sheaf P is locally free, the cone $E := C(P)$ is called the *vector bundle over* X *with the dual sheaf of sections* $P^\vee := \operatorname{Hom}_{\mathcal{O}_X}(P, \mathcal{O}_X)$. The projective cone $\mathbb{P}(E)$ is called the *projective bundle of* E. The assignment $P \mapsto C(P^\vee)$ gives rise to an equivalence between the category of locally free coherent $\mathcal{O}_X$-modules and the category of vector bundles over X. In particular, such operations over the locally free $\mathcal{O}_X$-modules as the tensor product, symmetric power, dual sheaf etc., and the notion of an exact sequence translate to the category of vector bundles. We write $K_0(X)$ for the Grothendieck group of the category of vector bundles over X. The group $K_0(X)$ is the abelian group given by generators, the isomorphism classes $[E]$ of vector bundles E over X and relations $[E] = [E'] + [E'']$ for every exact sequence $0 \to E' \to E \to E'' \to 0$ of vector bundles over X.

Example 104.2. The trivial line bundle $X \times \mathbb{A}^1 \to X$ will be denoted by $\mathbb{1}$.

Example 104.3. Let $f : Y \to X$ be a closed embedding and let $I \subset \mathcal{O}_X$ be the sheaf of ideals of the image of f in X. The cone

$$C_f = \operatorname{Spec}\big(\mathcal{O}_X/I \oplus I/I^2 \oplus I^2/I^3 \oplus \dots\big)$$

over Y is called the *normal cone of* Y *in* X. If X is a scheme of pure dimension d, then C_f is also a scheme of pure dimension d [**45**, B.6.6].

Example 104.4. If $f : X \to C$ is the zero section of a cone C, then $C_f = C$.

Example 104.5. The cone $T_X := C_f$ of the diagonal embedding $f : X \to X \times X$ is called the *tangent cone of* X. If X is a scheme of pure dimension d, then the tangent cone T_X is a scheme of pure dimension $2d$ (cf. Example 104.3).

Let U and V be vector spaces over a field F and let

$$U = U_0 \supset U_1 \supset U_2 \supset \dots \quad \text{and} \quad V = V_0 \supset V_1 \supset V_2 \supset \dots$$

be two filtrations by subspaces. Consider the filtration on $U \otimes V$ defined by

$$(U \otimes V)_k = \sum_{i+j=k} U_i \otimes V_j.$$

The following lemma can be proven by a suitable choice of bases of U and V.

Lemma 104.6. *The canonical linear map*

$$\coprod_{i+j=k} (U_i/U_{i+1}) \otimes (V_j/V_{j+1}) \to (U \otimes V)_k/(U \otimes V)_{k+1}$$

is an isomorphism for every $k \geq 0$.

Proposition 104.7. *Let* $f : Y \to X$ *and* $g : S \to T$ *be closed embeddings. Then there is a canonical isomorphism* $C_f \times C_g \simeq C_{f \times g}$.

PROOF. We may assume that $X = \operatorname{Spec}(A)$, $Y = \operatorname{Spec}(A/I)$, and $T = \operatorname{Spec}(B)$, $S = \operatorname{Spec}(B/J)$, where $I \subset A$ and $J \subset B$ are ideals. Then $X \times T = \operatorname{Spec}(A \otimes B)$ and $Y \times S = \operatorname{Spec}(A \otimes B)/K$, where $K = I \otimes B + A \otimes J$.

Consider the vector spaces $U_i = I^i$ and $V_j = J^j$. We have $(U \otimes V)_k = K^k$. By Lemma 104.6,

$$C_f \times C_g = \operatorname{Spec}\Big(\coprod_{i \geq 0} I^i/I^{i+1} \otimes \coprod_{j \geq 0} J^j/J^{j+1}\Big) \simeq \operatorname{Spec}\Big(\coprod_{k \geq 0} K^k/K^{k+1}\Big) = C_{f \times g}. \quad \square$$

Corollary 104.8. *If X and Y are two schemes, then $T_{X \times Y} = T_X \times T_Y$.*

104.B. Regular closed embeddings. Let A be a commutative ring. A sequence $\mathfrak{a} = (a_1, a_2, \dots, a_d)$ of elements of A is called *regular* if the ideal generated by $a_1, \dots, a_d$ is different from A and the coset of a_i is not a zero divisor in the factor ring $A/(a_1 A + \dots + a_{i-1} A)$ for all $i \in [1, d]$. We write $l(\mathfrak{a}) = d$.

Let Y be a scheme and let $d : Y \to \mathbb{Z}$ be a locally constant function. A closed embedding $f : Y \to X$ is called *regular of codimension* d if for every point $y \in Y$ there is an affine neighborhood $U \subset X$ of $f(y)$ such that the ideal of $f(Y) \cap U$ in $F[U]$ is generated by a regular sequence of length $d(y)$.

Let $f : Y \to X$ be a closed embedding and let I be the sheaf of ideals of Y in $\mathcal{O}_X$. The embedding of I/I^2 into $\coprod_{k \geq 0} I^k/I^{k+1}$ induces a surjective $\mathcal{O}_Y$-algebra homomorphism $S^\bullet(I/I^2) \to \coprod_{k \geq 0} I^k/I^{k+1}$ and therefore a closed embedding of cones $\varphi_f : C_f \to C(I/I^2)$ over Y.

Proposition 104.9. (Cf. [**48**, Cor. 16.9.4, Cor. 16.9.11].) *A closed embedding $f : Y \to X$ is regular of codimension d if and only if the $\mathcal{O}_Y$-module I/I^2 is locally free of rank d and the natural morphism $\varphi_f : C_f \to C(I/I^2)$ is an isomorphism.*

Corollary 104.10. *Let $f : Y \to X$ be a regular closed embedding of codimension d and let I be the sheaf of ideals of Y in $\mathcal{O}_X$. Then the normal cone C_f is a vector bundle over Y of rank d with the sheaf of sections naturally isomorphic to $(I/I^2)^\vee$.*

We shall write N_f for the normal cone C_f of a regular closed embedding f and call N_f the *normal bundle of* f.

Proposition 104.11. *Let $f : Y \to X$ be a closed embedding and let $g : X' \to X$ be a faithfully flat morphism. Then f is a regular closed embedding if and only if the closed embedding $f' : Y' = Y \times_X X' \to X'$ is regular.*

PROOF. Let I be the sheaf of ideals of Y in $\mathcal{O}_X$. Then $I' = g^*(I)$ is the sheaf of ideals of Y' in $\mathcal{O}_{X'}$. Moreover

$$g^*(I^k/I^{k+1}) = I'^k/I'^{k+1}, \quad C_f \times_Y Y' = C_{f'}, \quad C(I/I^2) \times_Y Y' = C(I'/I'^2),$$

and $\varphi_f \times_Y 1_{Y'} = \varphi_{f'}$. By faithfully flat descent, I/I^2 is locally free and φ_f is an isomorphism if and only if I'/I'^2 is locally free and $\varphi_{f'}$ is an isomorphism. The statement follows by Proposition 104.9. $\square$

Proposition 104.12 (Cf. [**48**, Cor. 17.12.3]). *Let $g : X \to Y$ be a smooth morphism of relative dimension d and let $f : Y \to X$ be a section of g, i.e., $g \circ f = 1_Y$. Then f is a regular closed embedding of codimension d and $N_f = f^* T_g$, where $T_g := \operatorname{Ker}(T_X \to g^* T_Y)$ is the relative tangent bundle of g over X.*

Corollary 104.13. *Let X be a smooth scheme. Then the diagonal embedding $X \to X \times X$ is regular. In particular, the tangent cone T_X is a vector bundle over X.*

PROOF. The diagonal embedding is a section of any of the two projections $X \times X \to X$. □

If X is a smooth scheme, the vector bundle T_X is called the *tangent bundle over* X.

Corollary 104.14. *Let $f : X \to Y$ be a morphism where Y is a smooth scheme of pure dimension d. Then the morphism $h = (1_X, f) : X \to X \times Y$ is a regular closed embedding of codimension d with $N_h = f^*T_Y$.*

PROOF. Applying Proposition 104.12 to the smooth projection $p : X \times Y \to X$ of relative dimension d, we have the closed embedding h is regular of codimension d. The tangent bundle T_p is equal to q^*T_Y, where $q : X \times Y \to Y$ is the other projection. Since $q \circ h = f$, we have

$$N_h = h^*T_p = h^* \circ q^*T_Y = f^*T_Y. \qquad \square$$

Proposition 104.15 (Cf. [**48**, Prop. 19.1.5]). *Let $g : Z \to Y$ and $f : Y \to X$ be regular closed embeddings of codimension s and r respectively. Then $f \circ g$ is a regular closed embedding of codimension $r + s$ and the natural sequence of normal bundles over Z,*

$$0 \to N_g \to N_{f \circ g} \to g^*N_f \to 0,$$

is exact.

Proposition 104.16 (Cf. [**48**, Th. 17.12.1, Prop. 17.13.2]). *A closed embedding $f : Y \to X$ of smooth schemes is regular and the natural sequence of vector bundles over Y,*

$$0 \to T_Y \to f^*T_X \to N_f \to 0,$$

is exact.

104.C. Canonical line bundle. Let $C = \operatorname{Spec}(S^\bullet)$ be a cone over X. The cone $\operatorname{Spec}(S^\bullet[t]) = C \times \mathbb{A}^1$ coincides with $C \oplus \mathbb{1}$. Let $I \subset S^\bullet[t]$ be the ideal generated by S^1. The closed subscheme of $\mathbb{P}(C \oplus \mathbb{1})$ defined by I is isomorphic to $\operatorname{Proj}(S^0[t]) = \operatorname{Spec}(S^0) = X$. Thus we get a canonical closed embedding (canonical section) of X into $\mathbb{P}(C \oplus \mathbb{1})$.

Set $L_c := \mathbb{P}(C \oplus \mathbb{1}) \setminus X$. The inclusion of $S^\bullet_{(s)}$ into $S^\bullet[t]_{(s)}$ for every $s \in S^1$ induces a morphism $L_c \to \mathbb{P}(C)$.

Proposition 104.17. *The morphism $L_c \to \mathbb{P}(C)$ has a canonical structure of a line bundle.*

PROOF. We have $S^\bullet[t]_{(s)} = S^\bullet_{(s)}[\frac{t}{s}]$, hence the preimage of $D(s)$ is isomorphic to $D(s) \times \mathbb{A}^1$. For any other element $s' \in S^1$ we have $\frac{t}{s'} = \frac{s}{s'}\frac{t}{s}$, i.e., the change of coordinate function is linear. □

The line bundle $L_c \to \mathbb{P}(C)$ is called the *canonical line bundle over* $\mathbb{P}(C)$.

A section of L_c over the open set $D(s)$ is given by an $S^\bullet_{(s)}$-algebra homomorphism $S^\bullet_{(s)}[\frac{t}{s}] \to S^\bullet_{(s)}$ that is uniquely determined by the image a_s of $\frac{t}{s}$. The element $sa_s \in S_s$ of degree 1 agrees with $s'a_{s'}$ on the intersection $D(s) \cap D(s')$. Therefore the sheaf of sections of L_c coincides with $\mathcal{O}(1) := \widetilde{S^\bullet}(1)$ where $\widetilde{S^\bullet}$ denotes the sheaf associated to $S^\bullet$ (cf. [**50**, Ch. II, §5]).

The scheme $\mathbb{P}(C)$ can be viewed as a locally principal divisor of $\mathbb{P}(C \oplus \mathbb{1})$ given by t. The open complement $\mathbb{P}(C \oplus \mathbb{1}) \setminus \mathbb{P}(C)$ is canonically isomorphic to C. The image of the canonical section $X \to \mathbb{P}(C \oplus \mathbb{1})$ is contained in C (and in fact is equal to the image of the zero section of C), hence it does not intersect $\mathbb{P}(C)$. Moreover, $\mathbb{P}(C \oplus \mathbb{1}) \setminus (\mathbb{P}(C) \cup X)$ is canonically isomorphic to $C \setminus X$.

If C is a cone over X, we write C° for $C \setminus X$ where X is viewed as a closed subscheme of C via the zero section. We have shown that C° is canonically isomorphic to L_c°. Note that C is a cone over X and L_c is a cone (in fact, a line bundle) over $\mathbb{P}(C)$.

For every $s \in S^1$, the localization S_s is the Laurent polynomial ring $S_{(s)}[s, s^{-1}]$ over $S_{(s)}$. Hence the inclusion of $S_{(s)}$ into S_s induces a flat morphism $C^\circ \to \mathbb{P}(C)$ of relative dimension 1.

104.D. Tautological line bundle. Let $C = \operatorname{Spec}(S^\bullet)$ be a cone over X. Consider the tensor product $T^\bullet = S^\bullet \otimes_{S^0} S^\bullet$ as a graded ring with respect to the second factor. We have

$$\operatorname{Proj}(T^\bullet) = C \times_X \mathbb{P}(C).$$

Let J be the ideal of $T^\bullet$ generated by $x \otimes y - y \otimes x$ for all $x, y \in S^1$ and set

$$L_t := \operatorname{Proj}(T^\bullet / J).$$

Thus L_t is a closed subscheme of $C \times_X \mathbb{P}(C)$ and we have natural projections $L_t \to C$ and $L_t \to \mathbb{P}(C)$.

Proposition 104.18. *The morphism $L_t \to \mathbb{P}(C)$ has a canonical structure of a line bundle.*

PROOF. Let $s \in S^1$. The preimage of $D(s)$ in L_t coincides with

$$\operatorname{Spec}(T^\bullet_{(1\otimes s)} / J_{(1\otimes s)}),$$

where $J_{(1\otimes s)} = J_{1\otimes s} \cap T^\bullet_{(1\otimes s)}$. The homomorphism $T^\bullet \to S^\bullet_s[t]$, where t is a variable, defined by $x \otimes y \mapsto \frac{xy}{s^n} \cdot t^n$ for any $x \in S^n$ and $y \in S^\bullet$, gives rise to an isomorphism between $T^\bullet_{(1\otimes s)} / J_{(1\otimes s)}$ and $S^\bullet_{(s)}[t]$. Hence the preimage of $D(s)$ is isomorphic to $D(s) \times \mathbb{A}^1$. □

The line bundle $L_t \to \mathbb{P}(C)$ is called the *tautological line bundle over* $\mathbb{P}(C)$.

Example 104.19. If L is a line bundle over X, then $\mathbb{P}(L) = X$ and $L_t = L \times_X \mathbb{P}(L) = L$.

Similar to the case of the canonical line bundle, a section of L_t over the open set $D(s)$ is given by an element $a_s \in S^\bullet_{(s)}$ and the element $a_s/s \in S^\bullet_s$ of degree -1 agrees with $a_{s'}/s'$ on the intersection $D(s) \cap D(s')$. Therefore the sheaf of sections of L_t coincides with $\widetilde{S^\bullet}(-1) = \mathcal{O}(-1)$. In particular, the tautological line bundle is dual to the canonical line bundle, $L_t = L_c{}^\vee$.

The ideal $I = S^{>0}$ in $S^\bullet$ defines the image of the zero section of C. The graded ring $T^\bullet/J$ is isomorphic to $S^\bullet \oplus I \oplus I^2 \oplus \cdots$. Therefore, the canonical morphism $L_t \to C$ is the blowup of C along the image of the zero section of C. The exceptional divisor in L_t is the image of the zero section of L_t. Hence the induced morphism $L_t^\circ \to C^\circ$ is an isomorphism.

Example 104.20. Let $F[\varepsilon]$ be the F-algebra of dual number over F. The tangent space $T_{\mathbb{P}(V),L}$ of the point of the projective space $\mathbb{P}(V)$ given by a line $L \subset V$ coincides with the fiber over L of the map $\mathbb{P}(V)(F[\varepsilon]) \to \mathbb{P}(V)(F)$ induced by the ring homomorphism $F[\varepsilon] \to F$, $\varepsilon \mapsto 0$. For example, the $F[\varepsilon]$-submodule $L \oplus L\varepsilon$ of $V[\varepsilon] := V \otimes F[\varepsilon]$ represents the zero vector of the tangent space $T_{\mathbb{P}(V),L}$.

For a linear map $h : L \to V$ let W_h be the $F[\varepsilon]$-submodule of $V[\varepsilon]$ generated by the elements $v + h(v)\varepsilon$, $v \in L$. Since $W_h/\varepsilon W_h \simeq L$, we can view W_h as a point of $T_{\mathbb{P}(V),L}$. The map $\mathrm{Hom}_F(L, V) \to T_{\mathbb{P}(V),L}$ given by $h \mapsto W_h$ yields an exact sequence of vector spaces

$$0 \to \mathrm{Hom}_F(L, L) \to \mathrm{Hom}_F(L, V) \to T_{\mathbb{P}(V),L} \to 0.$$

In other words,

$$T_{\mathbb{P}(V),L} = \mathrm{Hom}_F(L, V/L).$$

Since the fiber of the tautological line bundle L_t over the point given by L coincides with L, we get an exact sequence of vector bundles over $\mathbb{P}(V)$:

$$0 \to \mathrm{Hom}(L_t, L_t) \to \mathrm{Hom}(L_t, \mathbb{1} \otimes_F V) \to T_{\mathbb{P}(V)} \to 0.$$

The first term of the sequence is isomorphic to $\mathbb{1}$ and the second term is isomorphic to $L_c \otimes_F V \simeq (L_c)^{\oplus n}$, where $n = \dim V$. It follows that

$$[T_{\mathbb{P}(V)}] = n[L_c] - 1 \in K_0\big(\mathbb{P}(V)\big).$$

More generally, if $E \to X$ is a vector bundle, then there is an exact sequence of vector bundles over $\mathbb{P}(E)$:

$$0 \to \mathbb{1} \to L_c \otimes q^*E \to T_q \to 0,$$

where $q : \mathbb{P}(E) \to X$ is the natural morphism and T_q is the relative tangent bundle of q.

104.E. Deformation to the normal cone. Let $f : Y \to X$ be a closed embedding of schemes. First suppose that X is an affine scheme, $X = \mathrm{Spec}(A)$, and Y is given by an ideal $I \subset A$. Set $Y = \mathrm{Spec}(A/I)$. Consider the subring

$$\widetilde{A} = \coprod_{n \in \mathbb{Z}} I^{-n} t^n$$

of the Laurent polynomial ring $A[t, t^{-1}]$, where the negative powers of the ideal I are understood as equal to A. The scheme $D_f := \mathrm{Spec}(\widetilde{A})$ is called the *deformation scheme of the closed embedding* f. In the general case, in order to define D_f, we cover X by open affine subschemes and glue together the deformation schemes of the restrictions of f to the open sets of the covering.

The inclusion of $A[t]$ in $\widetilde{A}$ induces a morphism $g : D_f \to \mathbb{A}^1 \times X$. Denote by C_f the inverse image $g^{-1}(\{0\} \times X)$. In the affine case,

$$C_f = \mathrm{Spec}(A/I \oplus I/I^2 \cdot t^{-1} \oplus I^2/I^3 \cdot t^{-2} \oplus \cdots).$$

Thus, C_f is the normal cone of f (cf. Example 104.3). If f is a regular closed embedding of codimension d, then C_f is a vector bundle over Y of rank d. We write N_f for C_f in this case.

The open complement $D_f \setminus C_f$ is the inverse image $g^{-1}(\mathbb{G}_m \times X)$. In the affine case, it is the spectrum of the ring $\widetilde{A}[t^{-1}] = A[t, t^{-1}]$. Hence the inverse image is canonically isomorphic to $\mathbb{G}_m \times X$ via g, i.e.,

$$D_f \setminus C_f \simeq \mathbb{G}_m \times X.$$

In the affine case, the natural ring homomorphism $A[t] \to (A/I)[t]$ extends canonically to a ring homomorphism $\widetilde{A} \to (A/I)[t]$. Hence the morphism $f \times \mathrm{id} :$

$\mathbb{A}^1 \times Y \to \mathbb{A}^1 \times X$ factors through the canonical morphism $h : \mathbb{A}^1 \times Y \to D_f$ over $\mathbb{A}^1$. The fiber of h over a point $t \neq 0$ of $\mathbb{A}^1$ is naturally isomorphic to the morphism f. The fiber of h over $t = 0$ is isomorphic to the zero section $Y \to C_f$ of the normal cone C_f of f. Thus we can view h as a family of closed embeddings parameterized by $\mathbb{A}^1$ deforming the closed embedding f into the zero section $Y \to C_f$ as the parameter t "approaches 0". We have the following diagram of natural morphisms:

$$\begin{array}{ccccc} Y & \longrightarrow & \mathbb{A}^1 \times Y & \longleftarrow & \mathbb{G}_m \times Y \\ \downarrow & & \downarrow & & \downarrow \\ C_f & \longrightarrow & D_f & \longleftarrow & \mathbb{G}_m \times X \\ \downarrow & & \downarrow & & \| \\ Y & \longrightarrow & \mathbb{A}^1 \times X & \longleftarrow & \mathbb{G}_m \times X. \end{array}$$

Note that the normal cone C_f is the principal divisor in D_f of the function t.

Consider a fiber product diagram

$$\begin{array}{ccc} Y' & \xrightarrow{f'} & X' \\ {\scriptstyle g}\downarrow & & \downarrow{\scriptstyle h} \\ Y & \xrightarrow{f} & X \end{array} \tag{104.21}$$

where f and f' are closed embeddings. It induces the fiber product diagram of open and closed embeddings:

$$\begin{array}{ccccc} C_{f'} & \longrightarrow & D_{f'} & \longleftarrow & \mathbb{G}_m \times X' \\ {\scriptstyle k}\downarrow & & {\scriptstyle l}\downarrow & & \downarrow{\scriptstyle \mathrm{id}\times h} \\ C_f & \longrightarrow & D_f & \longleftarrow & \mathbb{G}_m \times X. \end{array} \tag{104.22}$$

Proposition 104.23. *In the notation of* (104.21), *there are natural closed embeddings $D_{f'} \to D_f \times_X X'$ and $C_{f'} \to C_f \times_X X'$. These embeddings are isomorphisms if h is flat.*

PROOF. We may assume that all schemes are affine and h is given by a ring homomorphism $A \to A'$. The scheme Y is defined by an ideal $I \subset A$ and Y' is given by $I' = IA' \subset A'$. The natural homomorphism $I^n \otimes_A A' \to (I')^n$ is surjective, hence $\widetilde{A} \otimes_A A' \to \widetilde{A'}$ is surjective. Consequently, $D_{f'} \to D_f \times_X X'$ and $C_{f'} \to C_f \times_X X'$ are closed embeddings. If A' is flat over A, the homomorphism $I^n \otimes_A A' \to (I')^n$ is an isomorphism. □

104.F. Double deformation space. Let A be a commutative ring.

Lemma 104.24. *Let I be the ideal of A generated by a regular sequence $\mathfrak{a} = (a_1, a_2, \dots, a_d)$ and let $a \in A$ be such that $a + I$ is not a zero divisor in A/I. If $ax \in I^m$ for some $x \in A$ and m, then $x \in I^m$.*

PROOF. By Proposition 104.9, multiplication by $a + I$ on I^n/I^{n+1} is injective for any n. The statement of the lemma follows by induction on m. □

Let $\mathfrak{a} = (a_1, a_2, \dots, a_d)$ and $\mathfrak{b} = (b_1, b_2, \dots, b_e)$ be two sequences of elements of A. We write $\mathfrak{a} \subset \mathfrak{b}$ if $d \leq e$ and $a_i = b_i$ for all $i \in [1, d]$. Clearly, if $\mathfrak{a} \subset \mathfrak{b}$ and $\mathfrak{b}$ is regular, so is $\mathfrak{a}$.

Let $I \subset J$ be ideals of A. We define the ideals $I^n J^m$ for $n < 0$ or $m < 0$ by

$$I^n J^m = \begin{cases} J^{n+m} & \text{if } n < 0, \\ I^n & \text{if } m < 0. \end{cases}$$

Proposition 104.25. *Let $\mathfrak{a} \subset \mathfrak{b}$ be two regular sequences in a ring A and let $I \subset J$ be the ideals of A generated by $\mathfrak{a}$ and $\mathfrak{b}$ respectively. Then*

$$I^n J^m \cap I^{n+1} = I^{n+1} J^{m-1},$$
$$I^n J^m \cap J^{n+m+1} = I^n J^{m+1},$$

for all n and m.

PROOF. We prove the first equality. The proof of the second one is similar.

We proceed by induction on m. The case $m \leq 1$ is clear. Suppose $m \geq 2$. As the inclusion "$\supset$" is easy, we need to prove that

$$I^n J^m \cap I^{n+1} \subset I^{n+1} J^{m-1}.$$

Let $\mathfrak{d}$ be a sequence such that $\mathfrak{a} \subset \mathfrak{d} \subset \mathfrak{b}$ and let L be the ideal generated by $\mathfrak{d}$, so $I \subset L \subset J$. By descending induction on the length $l(\mathfrak{d})$ of the sequence $\mathfrak{d}$, we prove that

$$I^n J^m \cap I^{n+1} \subset L^{n+1} J^{m-1}. \tag{104.26}$$

When $l(\mathfrak{d}) = l(\mathfrak{a})$, i.e., $\mathfrak{d} = \mathfrak{a}$ and $L = I$, we get the desired inclusion.

The case $l(\mathfrak{d}) = l(\mathfrak{b})$, i.e., $\mathfrak{d} = \mathfrak{b}$ and $L = J$ is obvious. Let $\mathfrak{c}$ be the sequence satisfying $\mathfrak{a} \subset \mathfrak{c} \subset \mathfrak{d}$ and $l(\mathfrak{c}) = l(\mathfrak{d}) - 1$. Let K be the ideal generated by $\mathfrak{c}$. We have $L = K + aA$ where a is the last element of the sequence $\mathfrak{d}$. Assuming (104.26), we shall prove that

$$I^n J^m \cap I^{n+1} \subset K^{n+1} J^{m-1}.$$

Let $x \in I^n J^m \cap I^{n+1}$. By assumption,

$$x \in L^{n+1} J^{m-1} = \sum_{k=0}^{n+1} a^{n+1-k} K^k J^{m-1},$$

hence

$$x = \sum_{k=0}^{n+1} a^{n+1-k} x_k$$

for some $x_k \in K^k J^{m-1}$. For any $s \in [0, n+1]$, set

$$y_s = \sum_{k=0}^{s} a^{s-k} x_k.$$

We claim that $y_s \in K^s J^{m-1}$ for $s \in [0, n+1]$. We prove the claim by induction on s. The case $s = 0$ is obvious since $y_0 = x_0 \in J^{m-1}$. Suppose $y_s \in K^s J^{m-1}$ for some $s < n + 1$. We have

$$x = a^{n+1-s} y_s + \sum_{k=s+1}^{n+1} a^{n+1-k} x_k,$$

where $x_k \in K^k J^{m-1} \subset K^{s+1}$ if $k \geq s+1$ and $x \in I^{n+1} \subset K^{s+1}$. Hence $a^{n+1-s} y_s \in K^{s+1}$ and therefore $y_s \in K^{s+1}$ by Lemma 104.24. Thus $y_s \in K^s J^{m-1} \cap K^{s+1}$. By the first induction hypothesis, the latter ideal is equal to $K^{s+1} J^{m-2}$ and $y_s \in K^{s+1} J^{m-2}$. Since $x_{s+1} \in K^{s+1} J^{m-1}$, we have $y_{s+1} = a y_s + x_{s+1} \in K^{s+1} J^{m-1}$. This proves the claim. By the claim, $x = y_{n+1} \in K^{n+1} J^{m-1}$. □

Let $Z \xrightarrow{g} Y \xrightarrow{f} X$ be regular closed embeddings. We have closed embeddings

$$i : (N_f)|_Z \to N_f \quad \text{and} \quad j : N_g \to N_{fg}.$$

We shall construct the *double deformation scheme* $D = D_{f,g}$ and a morphism $D \to \mathbb{A}^2$ satisfying all of the following:

(1) $D|_{\mathbb{A}^1 \times \mathbb{G}_m} = D_f \times \mathbb{G}_m$.

(2) $D|_{\mathbb{G}_m \times \mathbb{A}^1} = \mathbb{G}_m \times D_{fg}$.

(3) $D|_{\mathbb{A}^1 \times \{0\}} = D_j$.

(4) $D|_{\{0\} \times \mathbb{A}^1} = D_i$.

(5) $D|_{\{0\} \times \{0\}} = N_i \simeq N_j$.

As in the case of an ordinary deformation space, it suffices to consider the affine case: $X = \operatorname{Spec}(A)$, $Y = \operatorname{Spec}(A/I)$, and $Z = \operatorname{Spec}(A/J)$, where $I \subset J$ are the ideals of A generated by regular sequences. Consider the subring

$$\widehat{A} = \coprod_{n,m \in \mathbb{Z}} I^n J^{m-n} \cdot t^{-n} s^{-m}$$

of the Laurent polynomial ring $A[t, s, t^{-1}, s^{-1}]$ and set $D = \operatorname{Spec}(\widehat{A})$. Since $\widehat{A}$ contains the polynomial ring $A[t, s]$, there are natural morphisms $D \to X \times \mathbb{A}^2 \to \mathbb{A}^2$.

We have

$$\widehat{A}[s^{-1}] = \coprod_{n,m \in \mathbb{Z}} I^n \cdot t^{-n} s^{-m} = \Big(\coprod_{n,m \in \mathbb{Z}} I^n \cdot t^{-n} \Big)[s, s^{-1}],$$

$$\widehat{A}[t^{-1}] = \coprod_{n,m \in \mathbb{Z}} J^m \cdot t^{-n} s^{-m} = \Big(\coprod_{n,m \in \mathbb{Z}} J^m \cdot s^{-m} \Big)[t, t^{-1}].$$

This proves (1) and (2).

To prove (3) consider the rings

$$\widehat{A}/s\widehat{A} = \coprod_{n,m \in \mathbb{Z}} \big[I^n J^{m-n} / I^n J^{m-n+1} \big] \cdot t^{-n},$$

$$R = \coprod_{m \in \mathbb{Z}} \big[J^m / J^{m+1} \big] \cdot s^{-m},$$

$$S = \coprod_{m \in \mathbb{Z}} \big[(J^m + I)/(J^{m+1} + I) \big] \cdot s^{-m}.$$

We have $\operatorname{Spec}(R) = N_{fg}$ and $\operatorname{Spec}(S) = N_g$. The natural surjection $R \to S$ corresponds to the embedding $j : N_g \to N_{fg}$.

Let $\widetilde{I} = \operatorname{Ker}(R \to S)$. By Proposition 104.25, $J^m \cap I = I J^{m-1}$, hence

$$\widetilde{I} = \coprod_{m \in \mathbb{Z}} \big[I J^{m-1} + J^{m+1} / J^{m+1} \big] \cdot s^{-m}$$

and

$$\widetilde{I}^n = \coprod_{m \in \mathbb{Z}} \big[I^n J^{m-n} + J^{m+1} / J^{m+1} \big] \cdot s^{-m}.$$

Therefore, D_j is the spectrum of the ring

$$\coprod_{m \in \mathbb{Z}} \big[I^n J^{m-n} + J^{m+1} / J^{m+1} \big] \cdot t^{-n} s^{-m}.$$

It follows from Proposition 104.25 that this ring coincides with $\widehat{A}/s\widehat{A}$, hence (3).

To prove (4) consider the ring

$$\widehat{A}/t\widehat{A} = \coprod_{n,m\in\mathbb{Z}} \left[I^n J^{m-n}/I^{n+1}J^{m-n-1}\right] \cdot s^{-m}.$$

The normal bundle N_f is the spectrum of the ring

$$T = \coprod_{n\in\mathbb{Z}} \left[I^n/I^{n+1}\right] \cdot u^{-m}.$$

Let $\widetilde{J}$ be the ideal of T of the closed subscheme $(N_f)|_Z$. We have

$$\widetilde{J}^m = \coprod_{n\in\mathbb{Z}} \left[I^n J^m + I^{n+1}/I^{n+1}\right] \cdot u^{-m}.$$

The deformation scheme D_i is the spectrum of the ring

$$U = \coprod_{n,m\in\mathbb{Z}} \left[I^n J^m + I^{n+1}/I^{n+1}\right] \cdot u^{-n}s^{-m}.$$

We define the surjective ring homomorphism $\varphi : \widehat{A}/t\widehat{A} \to U$ taking

$$(x + I^{n+1}J^{m-n-1}) \cdot t^{-n}s^{-m} \qquad \text{to} \qquad (x + I^{n+1}) \cdot u^{-n}s^{-m+n}.$$

By Proposition 104.25, the map φ is also injective. Hence φ gives the identification (4). Property (5) follows from (3) and (4).

105. Group actions on algebraic schemes

In this section all schemes are quasi-projective over a field F. We write $G = \{1, \sigma\}$ for a cyclic group of order 2.

105.A. G-schemes. A *G-scheme* is a scheme Y together with a G-action on Y. As Y is a quasi-projective scheme, every pair of points of Y belong to an open affine subscheme. It follows that there is an open G-invariant affine covering of such Y. Therefore, in most of the constructions and proofs, we may restrict to the class of affine G-schemes.

In particular, to define a subscheme $Y^G \subset Y$ for a G-scheme Y, we may assume that Y is affine, i.e., $Y = \operatorname{Spec} R$, where R is a *G-algebra*, i.e., a commutative F-algebra with G acting on R by F-algebra automorphisms. Consider the ideal $I \subset R$ generated by $\sigma(r) - r$ for all $r \in R$ and set

$$Y^G := \operatorname{Spec}(R/I).$$

Example 105.1. For any scheme X over F, the group G acts on $X \times X$ by permutation of the factors. Then $(X\times X)^G$ coincides with the image of the diagonal closed embedding $\Delta : X \to X \times X$. Indeed, if $X = \operatorname{Spec}(A)$, then $X \times X = \operatorname{Spec}(A \otimes_F A)$ and the ideal $I \subset A \otimes_F A$ is generated by $\sigma(a \otimes a') - a \otimes a' = a' \otimes a - a \otimes a'$ for all $a, a' \in A$. Hence it coincides with the kernel of the product map $A \otimes A \to A$.

Exercise 105.2. Prove that $(X \times Y)^G = X^G \times Y^G$ for G-schemes X and Y.

Proposition 105.3. *Let Y be a G-scheme and let B be the blowup of Y along Y^G. Then G acts naturally on B and the subscheme B^G of B coincides with the exceptional divisor $\mathbb{P}(C)$, where C is the normal cone of the closed embedding $Y^G \to Y$.*

PROOF. We may assume that $Y = \operatorname{Spec} R$ for a G-algebra R. Then $Y^G = \operatorname{Spec}(R/I)$, where I is the ideal of R generated by $\sigma(r) - r$ for all $r \in R$, and $B = \operatorname{Proj} S$, where $S = R \oplus I \oplus I^2 \oplus \cdots$. The scheme B is covered by open G-invariant subschemes $\operatorname{Spec}(S_{(s)})$ with $s = \sigma(r) - r$, $r \in R$. The intersection of $\operatorname{Spec}(S_{(s)})$ with the exceptional divisor $\mathbb{P}(C)$ is given by the ideal $sS_{(s)}$ in $S_{(s)}$. Therefore, it suffices to show that the ideal J in $S_{(s)}$ generated by elements of the form $\sigma(x) - x$ with $x \in S_{(s)}$ coincides with $sS_{(s)}$. Clearly $s = \sigma(r) - r \in J$. Conversely, let $x = \frac{r}{s^n}$ with $r \in I^n$ be an element in $S_{(s)}$. As σ acts on I^n/I^{n+1} by multiplication by $(-1)^n$, we have $t := (-1)^n\sigma(r) - r \in I^{n+1}$ and hence

$$\sigma(x) - x = \frac{t}{s^n} = s\,\frac{t}{s^{n+1}} \in sS_{(s)}. \qquad \square$$

Let Y be a G-scheme. To define a scheme Y/G we may assume that $Y = \operatorname{Spec} R$. Set

$$Y/G := \operatorname{Spec}(R^G).$$

The natural morphism $p : Y \to Y/G$ is called a *G-torsor* if $Y^G = \emptyset$.

Example 105.4. Let Y be a G-scheme and let $Y' = Y \setminus Y^G$. Then the natural morphism $Y' \to Y'/G$ is a G-torsor.

Proposition 105.5. *Let $p : Y \to Y/G$ be a G-torsor. Then*

(1) *$p_*(\mathcal{O}_Y)$ is a locally free $\mathcal{O}_{Y/G}$-module of rank 2.*
(2) *The morphism p is étale.*

PROOF. We may assume that $Y = \operatorname{Spec}(R)$ for a G-algebra R. As $Y^G = \emptyset$, the elements $s = \sigma(r) - r$ with $r \in R$ generate the unit ideal of R, i.e., the G-invariant open sets $\operatorname{Spec}(R_s)$ cover Y. Replacing R by R_s we may assume that s is invertible in R. Then $\{1, r\}$ is a basis of R over R^G since for every $a \in R$ we have $a = f + gr$ with $f = \big(a\sigma(r) - \sigma(a)r\big)s^{-1} \in R^G$ and $g = \big(\sigma(a) - a\big)s^{-1} \in R^G$.

Moreover, r is the root of the quadratic polynomial $h(t) = t^2 - bt + c$ over R^G with $b = \sigma(r) + r$ and $c = \sigma(r)r$. As the element $h'(r) = 2r - b = -s$ is invertible in R, the morphism p is étale (cf. [**105**, Ch. I, Ex. 3.4]). $\square$

Example 105.6. Let $p : Y \to Y/G$ be a G-torsor with $Y/G = \operatorname{Spec} F$. Then $Y = \operatorname{Spec} K$, where K is an étale quadratic F-algebra.

Proposition 105.7. *Suppose* $\operatorname{char} F \neq 2$. *Let Y be a G-scheme and let $U = Y \setminus Y^G$. Then*

(1) *The composition $Y^G \to Y \xrightarrow{p} Y/G$ is a closed embedding with the complement U/G.*
(2) *If $I \subset \mathcal{O}_{Y/G}$ is the sheaf of ideals of Y^G in Y/G, then $p^*(I) = J^2$, where $J \subset \mathcal{O}_Y$ is the sheaf of ideals of Y^G in Y.*
(3) *If Y is regular and Y^G is a regular divisor in Y, then Y/G is also regular.*

PROOF. We may assume that $Y = \operatorname{Spec} R$ for a G-algebra R.

(1): The composition $Y^G \to Y \to Y/G$ is given by the composition of algebra homomorphisms $R^G \to R \to R/I$, where I is the ideal of R generated by $\sigma(r) - r$ for all $r \in R$. The latter composition is surjective as any $r \in R$ is the image of $\big(\sigma(r) + r\big)/2 \in R^G$.

The scheme U (respectively U/G) is covered by the open sets $\operatorname{Spec}(R_s)$ (respectively $\operatorname{Spec}(R^G_{s^2})$) with $s = \sigma(r) - r$, $r \in R$. Therefore, the second statement in (1) follows from the equality $(R_s)^G = (R^G)_{s^2}$.

(2): We need to prove that $(I \cap R^G)R = I^2$. The generators $s = \sigma(r) - r$ of I satisfy $\sigma(s) = -s$. Therefore, the ideal I^2 is generated by G-invariant elements. Hence $I^2 \subset (I \cap R^G)R$. Conversely, let $a \in I \cap R^G$. We can write $a = \sum r_i s_i$ with $r_i \in R$ and $s_i \in I$ satisfying $\sigma(s_i) = -s_i$. Hence

$$a = \frac{1}{2}\big(a + \sigma(a)\big) = \frac{1}{2}\sum \big(r_i - \sigma(r_i)\big)s_i \in I^2.$$

(3): Let $z \in Y/G$ and let $y \in Y$ be a point above z. Suppose $y \in Y \setminus Y^G$. The natural morphism $Y \setminus Y^G \to \big(Y \setminus Y^G\big)/G$ is a G-torsor by Example 105.4, hence it is flat by Proposition 105.5. As the local ring $\mathcal{O}_{Y,y}$ is regular, so is $\mathcal{O}_{Y/G,z}$ by [**48**, Prop. 17.3.3(i)].

Suppose now that $y \in Y^G$. We may assume that F is algebraically closed and y is a rational point. The tangent space $T_{Y^G,y}$ has codimension 1 in $T_{Y,y}$ and coincides with the subspace of all G-invariant vectors in $T_{Y,y}$. Hence there is a basis $\bar{a}_1, \bar{a}_2, \dots, \bar{a}_n$ of $\mathfrak{m}_{Y,y}/\mathfrak{m}^2_{Y,y}$ satisfying $\sigma(\bar{a}_1) = -\bar{a}_1$ and $\sigma(\bar{a}_i) = \bar{a}_i$ for $i \geq 2$. Lift the $\bar{a}_i$ to a local system of parameters $a_i \in \mathfrak{m}_{Y,y}$ with $\sigma(a_1) = -a_1$ and $\sigma(a_i) = a_i$ for $i \geq 2$. The completion of the local ring $\mathcal{O}_{Y/G,z}$ coincides with

$$\widehat{\mathcal{O}}_{Y/G,z} = \big(\widehat{\mathcal{O}}_{Y,y}\big)^G = F\,[[a_1, a_2, \dots, a_n]]^G = F\,[[a_1^2, a_2, \dots, a_n]],$$

and hence is regular. Therefore, z is a regular point on Y/G. □

105.B. The scheme B_X. Let X be a scheme. Write B_X for the blowup of $X^2 \times \mathbb{A}^1 := X \times X \times \mathbb{A}^1$ along $\Delta(X) \times \{0\}$. Since the normal cone of $\Delta(X) \times \{0\}$ in $X^2 \times \mathbb{A}^1$ is $T_X \oplus \mathbb{1}$ (cf. Proposition 104.7), the projective cone $\mathbb{P}\big(T_X \oplus \mathbb{1}\big)$ is the exceptional divisor in B_X (cf. [**45**, B.6.6]).

Let G act on $X^2 \times \mathbb{A}^1$ by $\sigma(x, x', t) = (x', x, -t)$. If $\operatorname{char} F \neq 2$, we have

$$(X^2 \times \mathbb{A}^1)^G = (X^2)^G \times (\mathbb{A}^1)^G = \Delta(X) \times \{0\}.$$

Set $U_X = (X^2 \times \mathbb{A}^1) \setminus \big(\Delta(X) \times \{0\}\big)$. By Proposition 105.3, the group G acts naturally on B_X so that $(B_X)^G = \mathbb{P}\big(T_X \oplus \mathbb{1}\big)$ and $B_X \setminus \mathbb{P}\big(T_X \oplus \mathbb{1}\big)$ is canonically isomorphic to U_X.

Proposition 105.8. *Suppose* $\operatorname{char} F \neq 2$. *If X is smooth, then so are B_X and B_X/G.*

PROOF. The scheme B_X is a blowup of a smooth scheme along smooth center, hence it is smooth. As the scheme $\mathbb{P}\big(T_X \oplus \mathbb{1}\big) = (B_X)^G$ is a smooth divisor in B_X, the second statement follows from Proposition 105.7(3). □

Bibliography

[1] A. A. Albert, *Structure of algebras*, Revised printing. American Mathematical Society Colloquium Publications, Vol. XXIV, American Mathematical Society, Providence, RI, 1961.

[2] M. Amer, *Quadratische Formen über Funktionenkörpern*, Ph.D. thesis, Johannes Gutenberg-Universität Mainz, Germany, 1976.

[3] J. K. Arason, *On Krüskemper's theorem*, Unpublished.

[4] ———, *Cohomologische Invarianten quadratischer Formen*, J. Algebra **36** (1975), no. 3, 448–491.

[5] ———, *A proof of Merkurjev's theorem*, Quadratic and Hermitian forms (Hamilton, Ont., 1983), CMS Conf. Proc., vol. 4, Amer. Math. Soc., Providence, RI, 1984, pp. 121–130.

[6] J. K. Arason and R. Baeza, *Relations in* I^n *and* I^nW_q *in characteristic* 2, J. Algebra **314** (2007), no. 2, 895–911.

[7] J. K. Arason and R. Elman, *Nilpotence in the Witt ring*, Amer. J. Math. **113** (1991), no. 5, 861–875.

[8] ———, *Powers of the fundamental ideal in the Witt ring*, J. Algebra **239** (2001), no. 1, 150–160.

[9] J. K. Arason and M. Knebusch, *Über die Grade quadratischer Formen*, Math. Ann. **234** (1978), no. 2, 167–192.

[10] J. K. Arason and A. Pfister, *Beweis des Krullschen Durchschnittsatzes für den Wittring*, Invent. Math. **12** (1971), 173–176.

[11] ———, *Zur Theorie der quadratischen Formen über formalreellen Körpern*, Math. Z. **153** (1977), no. 3, 289–296.

[12] R. Aravire and R. Baeza, *Milnor's k-theory and quadratic forms over fields of characteristic two*, Comm. Algebra **20** (1992), no. 4, 1087–1107.

[13] ———, *The behavior of quadratic and differential forms under function field extensions in characteristic two*, J. Algebra **259** (2003), no. 2, 361–414.

[14] ———, *Annihilators of quadratic and bilinear forms over fields of characteristic two*, J. Algebra **299** (2006), no. 1, 294–308.

[15] R. Baeza, *Quadratic forms over semilocal rings*, Springer-Verlag, Berlin, 1978, Lecture Notes in Mathematics, Vol. 655.

[16] ———, *Comparing u-invariants of fields of characteristic* 2, Bol. Soc. Brasil. Mat. **13** (1982), no. 1, 105–114.

[17] G. Berhuy and Z. Reichstein, *On the notion of canonical dimension for algebraic groups*, Adv. Math. **198** (2005), no. 1, 128–171.

[18] N. Bourbaki, *Commutative algebra. Chapters 1–7*, Elements of Mathematics (Berlin), Springer-Verlag, Berlin, Germany, 1998, Translated from the French, Reprint of the 1989 English translation.

[19] L. Bröcker, *Zur Theorie der quadratischen Formen über formal reellen Körpern*, Math. Ann. **210** (1974), 233–256.

[20] P. Brosnan, *Steenrod operations in Chow theory*, Trans. Amer. Math. Soc. **355** (2003), no. 5, 1869–1903 (electronic).

[21] A. Brumer, *Remarques sur les couples de formes quadratiques*, C. R. Acad. Sci. Paris Sér. A-B **286** (1978), no. 16, A679–A681.

[22] J. W. S. Cassels, *On the representation of rational functions as sums of squares*, Acta Arith. **9** (1964), 79–82.

[23] J. W. S. Cassels, W. J. Ellison, and A. Pfister, *On sums of squares and on elliptic curves over function fields*, J. Number Theory **3** (1971), 125–149.

[24] J. W. S. Cassels and A. Fröhlich (eds.), *Algebraic number theory*, London, Academic Press Inc. [Harcourt Brace Jovanovich Publishers], 1986, Reprint of the 1967 original.

[25] V. Chernousov, S. Gille, and A. S. Merkurjev, *Motivic decomposition of isotropic projective homogeneous varieties*, Duke Math. J. **126** (2005), no. 1, 137–159.

[26] C. M. Cordes, *The Witt group and the equivalence of fields with respect to quadratic forms*, J. Algebra **26** (1973), 400–421.

[27] A. Delzant, *Définition des classes de Stiefel-Whitney d'un module quadratique sur un corps de caractéristique différente de* 2, C. R. Acad. Sci. Paris **255** (1962), 1366–1368.

[28] M. A. Dickmann and F. Miraglia, *On quadratic forms whose total signature is zero mod* 2^n, Invent. Math. **133** (1998), no. 2, 243–278.

[29] ———, *Lam's conjecture*, Algebra Colloq. **10** (2003), no. 2, 149–176.

[30] R. Elman, *Pfister forms and K-theory of fields*, Ph.D. thesis, University of California, Berkeley, California, 1972.

[31] ———, *Quadratic forms and the u-invariant.* III, Conference on Quadratic Forms–1976 (Proc. Conf., Queen's Univ., Kingston, Ont., 1976), Queen's Univ., Kingston, Ont., 1977, pp. 422–444. Queen's Papers in Pure and Appl. Math., No. 46.

[32] R. Elman and T. Y. Lam, *Pfister forms and K-theory of fields*, J. Algebra **23** (1972), 181–213.

[33] ———, *Quadratic forms over formally real fields and Pythagorean fields*, Amer. J. Math. **94** (1972), 1155–1194.

[34] ———, *Quadratic forms and the u-invariant.* I, Math. Z. **131** (1973), 283–304.

[35] ———, *Quadratic forms and the u-invariant.* II, Invent. Math. **21** (1973), 125–137.

[36] ———, *Classification theorems for quadratic forms over fields*, Comment. Math. Helv. **49** (1974), 373–381.

[37] ———, *Quadratic forms under algebraic extensions*, Math. Ann. **219** (1976), no. 1, 21–42.

[38] R. Elman, T. Y. Lam, and A. Prestel, *On some Hasse principles over formally real fields*, Math. Z. **134** (1973), 291–301.

[39] R. Elman, T. Y. Lam, and A. R. Wadsworth, *Amenable fields and Pfister extensions*, Conference on Quadratic Forms–1976 (Proc. Conf., Queen's Univ., Kingston, Ont., 1976), Queen's Univ., Kingston, Ont., 1977, pp. 445–492. With an appendix "Excellence of $F(\varphi)/F$ for 2–fold Pfister forms" by J. K. Arason. Queen's Papers in Pure and Appl. Math., No. 46.

[40] R. Elman and H.-K. Lee, *On a relative u-invariant*, Preprint.

[41] R. Elman and C. Lum, *On the cohomological* 2*-dimension of fields*, Comm. Algebra **27** (1999), no. 2, 615–620.

[42] R. Elman and A. Prestel, *Reduced stability of the Witt ring of a field and its Pythagorean closure*, Amer. J. Math. **106** (1984), no. 5, 1237–1260.

[43] I. B. Fesenko and S. V. Vostokov, *Local fields and their extensions*, second ed., Translations of Mathematical Monographs, vol. 121, American Mathematical Society, Providence, RI, 2002, With a foreword by I. R. Shafarevich.

[44] R. W. Fitzgerald, *Function fields of quadratic forms*, Math. Z. **178** (1981), no. 1, 63–76.

[45] W. Fulton, *Intersection theory*, second ed., Ergebnisse der Mathematik und ihrer Grenzgebiete. 3. Folge. A Series of Modern Surveys in Mathematics [Results in Mathematics and Related Areas. 3rd Series. A Series of Modern Surveys in Mathematics], vol. 2, Springer-Verlag, Berlin, Germany, 1998.

[46] S. Garibaldi, A. S. Merkurjev, and J.-P. Serre, *Cohomological invariants in Galois cohomology*, University Lecture Series, vol. 28, American Mathematical Society, Providence, RI, 2003.

[47] P. Gille and T. Szamuely, *Central simple algebras and Galois cohomology*, Cambridge Studies in Advanced Mathematics, vol. 101, Cambridge University Press, Cambridge, UK, 2006.

[48] A. Grothendieck, *Éléments de géométrie algébrique. IV. Étude locale des schémas et des morphismes de schémas.* IV, Inst. Hautes Études Sci. Publ. Math. (1967), no. 32, 361.

[49] D. K. Harrison, *Witt rings*, Lecture Notes, Department of Mathematics, University of Kentucky, Lexington, KY, 1970.

[50] R. Hartshorne, *Algebraic geometry*, Springer-Verlag, New York, 1977, Graduate Texts in Mathematics, No. 52.

[51] O. Haution, *Lifting of coefficients for Chow motives of quadrics*, Preprint, Linear Algebraic Groups and Related Structures (preprint server), 2007.

[52] I. N. Herstein, *Noncommutative rings*, Mathematical Association of America, Washington, DC, 1994, Reprint of the 1968 original, with an afterword by Lance W. Small.

[53] D. W. Hoffmann, *Isotropy of quadratic forms over the function field of a quadric*, Math. Z. **220** (1995), no. 3, 461–476.

[54] ———, *On Elman and Lam's filtration of the u-invariant*, J. Reine Angew. Math. **495** (1998), 175–186.

[55] ———, *On the dimensions of anisotropic quadratic forms in I^4*, Invent. Math. **131** (1998), no. 1, 185–198.

[56] ———, *Pythagoras numbers of fields*, J. Amer. Math. Soc. **12** (1999), no. 3, 839–848.

[57] D. W. Hoffmann and A. Laghribi, *Quadratic forms and Pfister neighbors in characteristic* 2, Trans. Amer. Math. Soc. **356** (2004), no. 10, 4019–4053 (electronic).

[58] ———, *Isotropy of quadratic forms over the function field of a quadric in characteristic* 2, J. Algebra **295** (2006), no. 2, 362–386.

[59] J. Hurrelbrink, N. A. Karpenko, and U. Rehmann, *The minimal height of quadratic forms of given dimension*, Arch. Math. (Basel) **87** (2006), no. 6, 522–529.

[60] O. T. Izhboldin, *On p-torsion in K_*^M for fields of characteristic p*, Algebraic K-theory, Amer. Math. Soc., Providence, RI, 1991, pp. 129–144.

[61] ———, *On the cohomology groups of the field of rational functions*, Mathematics in St. Petersburg, Amer. Math. Soc., Providence, RI, 1996, pp. 21–44.

[62] ———, *Motivic equivalence of quadratic forms*, Doc. Math. **3** (1998), 341–351 (electronic).

[63] ———, *Motivic equivalence of quadratic forms.* II, Manuscripta Math. **102** (2000), no. 1, 41–52.

[64] ———, *Fields of u-invariant* 9, Ann. of Math. (2) **154** (2001), no. 3, 529–587.

[65] B. Jacob and M. Rost, *Degree four cohomological invariants for quadratic forms*, Invent. Math. **96** (1989), no. 3, 551–570.

[66] B. Kahn, *Formes quadratiques et cycles algébriques (d'après Rost, Voevodsky, Vishik, Karpenko et al.)*, Astérisque (2006), no. 307, Exp. No. 941, vii, 113–163, Séminaire Bourbaki. Vol. 2004/2005.

[67] ———, *Formes quadratiques sur un corps*, Cours Spécialisés [Specialized Courses], Société Mathématique de France, Paris, to appear.

[68] I. Kaplansky, *Quadratic forms*, J. Math. Soc. Japan **5** (1953), 200–207.

[69] N. A. Karpenko, *Algebro-geometric invariants of quadratic forms*, Algebra i Analiz **2** (1990), no. 1, 141–162.

[70] ———, *Cohomology of relative cellular spaces and of isotropic flag varieties*, Algebra i Analiz **12** (2000), no. 1, 3–69.

[71] ———, *Criteria of motivic equivalence for quadratic forms and central simple algebras*, Math. Ann. **317** (2000), no. 3, 585–611.

[72] ———, *On the first Witt index of quadratic forms*, Invent. Math. **153** (2003), no. 2, 455–462.

[73] ———, *Holes in I^n*, Ann. Sci. École Norm. Sup. (4) **37** (2004), no. 6, 973–1002.

[74] ———, *Canonical dimension of orthogonal groups*, Transform. Groups **10** (2005), no. 2, 211–215.

[75] ———, *A relation between higher Witt indices*, Proceedings of the St. Petersburg Mathematical Society. Vol. XI (Providence, RI), Amer. Math. Soc. Transl. Ser. 2, vol. 218, Amer. Math. Soc., 2006, pp. 77–86.

[76] N. A. Karpenko and A. S. Merkurjev, *Essential dimension of quadrics*, Invent. Math. **153** (2003), no. 2, 361–372.

[77] ———, *Canonical p-dimension of algebraic groups*, Adv. Math. **205** (2006), no. 2, 410–433.

[78] K. Kato, *Symmetric bilinear forms, quadratic forms and Milnor K-theory in characteristic two*, Invent. Math. **66** (1982), no. 3, 493–510.

[79] ———, *Milnor K-theory and the Chow group of zero cycles*, Applications of algebraic K-theory to algebraic geometry and number theory, Parts I, II (Boulder, Colo., 1983), Contemp. Math., vol. 55, Amer. Math. Soc., Providence, RI, 1986, pp. 241–253.

[80] Y. Kitaoka, *Arithmetic of quadratic forms*, Cambridge Tracts in Mathematics, vol. 106, Cambridge University Press, Cambridge, UK, 1993.

[81] M. Knebusch, *Ein Satz über die Werte von quadratischen Formen über Körpern*, Invent. Math. **12** (1971), 300–303.

[82] ———, *Specialization of quadratic and symmetric bilinear forms, and a norm theorem*, Acta Arith. **24** (1973), 279–299, Collection of articles dedicated to Carl Ludwig Siegel on the occasion of his seventy-fifth birthday, III.

[83] ———, *Generic splitting of quadratic forms.* I, Proc. London Math. Soc. (3) **33** (1976), no. 1, 65–93.

[84] ———, *Generic splitting of quadratic forms.* II, Proc. London Math. Soc. (3) **34** (1977), no. 1, 1–31.

[85] M. Knebusch and W. Scharlau, *Algebraic theory of quadratic forms*, DMV Seminar, vol. 1, Birkhäuser Boston, Mass., 1980, Generic methods and Pfister forms, Notes taken by Heisook Lee.

[86] M.-A. Knus, A. S. Merkurjev, M. Rost, and J.-P. Tignol, *The book of involutions*, American Mathematical Society Colloquium Publications, vol. 44, Amer. Math. Soc., Providence, RI, 1998, with a preface in French by J. Tits.

[87] M. Krüskemper, *On real local-global principles*, Math. Z. **204** (1990), no. 1, 145–151.

[88] T. Y. Lam, *Ten lectures on quadratic forms over fields*, Conference on Quadratic Forms–1976 (Proc. Conf., Queen's Univ., Kingston, Ont., 1976), Queen's Univ., Kingston, Ont., 1977, pp. 1–102. Queen's Papers in Pure and Appl. Math., No. 46.

[89] ———, *Introduction to quadratic forms over fields*, Graduate Studies in Mathematics, vol. 67, Amer. Math. Soc., Providence, RI, 2005.

[90] S. Lang, *Algebra*, third ed., Graduate Texts in Mathematics, vol. 211, Springer-Verlag, New York, 2002.

[91] H.-K. Lee, *Relative u-invariants of algebraic quadratic forms*, Ph.D. thesis, University of California, Los Angeles, CA, 1995.

[92] D. B. Leep, *The Amer-Brumer theorem over arbitrary fields*, Preprint, 1971.

[93] ———, *Systems of quadratic forms*, J. Reine Angew. Math. **350** (1984), 109–116.

[94] D. W. Lewis, *New proofs of the structure theorems for Witt rings*, Exposition. Math. **7** (1989), no. 1, 83–88.

[95] F. Lorenz and J. Leicht, *Die Primideale des Wittschen Ringes*, Invent. Math. **10** (1970), 82–88.

[96] P. Mammone, R. Moresi, and A. R. Wadsworth, *u-invariants of fields of characteristic* 2, Math. Z. **208** (1991), no. 3, 335–347.

[97] M. Marshall, *A reduced theory of quadratic forms*, Conference on Quadratic Forms–1976 (Proc. Conf., Queen's Univ., Kingston, Ont., 1976), Queen's Univ., Kingston, Ont., 1977, pp. 569–579. Queen's Papers in Pure and Appl. Math., No. 46.

[98] H. Matsumoto, *Sur les sous-groupes arithmétiques des groupes semi-simples déployés*, Ann. Sci. École Norm. Sup. (4) **2** (1969), 1–62.

[99] H. Matsumura, *Commutative algebra*, second ed., Mathematics Lecture Note Series, vol. 56, Benjamin/Cummings Publishing Co., Inc., Reading, Mass., 1980.

[100] A. S. Merkurjev, *On the norm residue symbol of degree* 2, Dokl. Akad. Nauk SSSR **261** (1981), no. 3, 542–547.

[101] ———, *Kaplansky's conjecture in the theory of quadratic forms*, Zap. Nauchn. Sem. Leningrad. Otdel. Mat. Inst. Steklov. (LOMI) **175** (1989), no. Koltsa i Moduli. 3, 75–89, 163–164.

[102] ———, *Simple algebras and quadratic forms*, Izv. Akad. Nauk SSSR Ser. Mat. **55** (1991), no. 1, 218–224.

[103] A. S. Merkurjev and A. A. Suslin, *The group K_3 for a field*, Izv. Akad. Nauk SSSR Ser. Mat. **54** (1990), no. 3, 522–545.

[104] A. Meyer, *Über die Auflösung der Gleichung $ax^2 + by^2 + cz^2 + du^2 + ev^2 = 0$ in ganzen Zahlen*, Vierteljahrs. Naturfors. Gesells **29** (1884), 220–222.

[105] J. S. Milne, *Étale cohomology*, Princeton University Press, Princeton, NJ, 1980.

[106] J. Milnor, *Algebraic K-theory and quadratic forms*, Invent. Math. **9** (1969/1970), 318–344.

[107] D. Orlov, A. Vishik, and V. Voevodsky, *An exact sequence for $K^M_*/2$ with applications to quadratic forms*, Ann. of Math. (2) **165** (2007), no. 1, 1–13.

[108] A. Pfister, *Multiplikative quadratische Formen*, Arch. Math. **16** (1965), 363–370.

[109] ———, *Zur Darstellung von −1 als Summe von Quadraten in einem Körper*, J. London Math. Soc. **40** (1965), 159–165.

[110] ———, *Quadratische Formen in beliebigen Körpern*, Invent. Math. **1** (1966), 116–132.

[111] ———, *Quadratic forms with applications to algebraic geometry and topology*, London Mathematical Society Lecture Note Series, vol. 217, Cambridge University Press, Cambridge, UK, 1995.

[112] A. Prestel, *Remarks on the Pythagoras and Hasse number of real fields*, J. Reine Angew. Math. **303/304** (1978), 284–294.

[113] D. Quillen, *Higher algebraic K-theory*. I, Algebraic K-theory, I: Higher K-theories (Proc. Conf., Battelle Memorial Inst., Seattle, Wash., 1972), Springer, Berlin, 1973, pp. 85–147. Lecture Notes in Math., Vol. 341.

[114] U. Rehmann and U. Stuhler, *On K_2 of finite-dimensional division algebras over arithmetical fields*, Invent. Math. **50** (1978/79), no. 1, 75–90.

[115] M. Rost, *Hilbert* 90 *for K_3 for degree-two extensions*, Preprint, 1986.

[116] ———, *On quadratic forms isotropic over the function field of a conic*, Math. Ann. **288** (1990), no. 3, 511–513.

[117] ———, *Chow groups with coefficients*, Doc. Math. **1** (1996), No. 16, 319–393 (electronic).

[118] C. H. Sah, *Symmetric bilinear forms and quadratic forms*, J. Algebra **20** (1972), 144–160.

[119] W. Scharlau, *Zur Pfisterschen Theorie der quadratischen Formen*, Invent. Math. **6** (1969), 327–328.

[120] ———, *Quadratic reciprocity laws*, J. Number Theory **4** (1972), 78–97.

[121] ———, *Quadratic and Hermitian forms*, Grundlehren der Mathematischen Wissenschaften [Fundamental Principles of Mathematical Sciences], vol. 270, Springer-Verlag, Berlin, 1985.

[122] C. Schubert, *Going up of the u-invariant over formally real fields*, Math. Ann. **338** (2007), no. 3, 757–765.

[123] J.-P. Serre, *A course in arithmetic*, Springer-Verlag, New York, 1973, Translated from the French, Graduate Texts in Mathematics, No. 7.

[124] ———, *Galois cohomology*, English ed., Springer-Verlag, Berlin, 2002, Translated from the French by Patrick Ion and revised by the author.

[125] I. R. Shafarevich, *Basic algebraic geometry*. 1, second ed., Springer-Verlag, Berlin, 1994, Varieties in projective space, Translated from the 1988 Russian edition and with notes by Miles Reid.

[126] T. A. Springer, *Sur les formes quadratiques d'indice zéro*, C. R. Acad. Sci. Paris **234** (1952), 1517–1519.

[127] ———, *Quadratic forms over fields with a discrete valuation. I. Equivalence classes of definite forms*, Nederl. Akad. Wetensch. Proc. Ser. A. **58** = Indag. Math. **17** (1955), 352–362.

[128] A. A. Suslin, *Quaternion homomorphism for the field of functions on a conic*, Dokl. Akad. Nauk SSSR **265** (1982), no. 2, 292–296.

[129] ———, *Torsion in K_2 of fields*, K-Theory **1** (1987), no. 1, 5–29.

[130] R. G. Swan, *Zero cycles on quadric hypersurfaces*, Proc. Amer. Math. Soc. **107** (1989), no. 1, 43–46.

[131] J.-P. Tignol, *Réduction de l'indice d'une algèbre simple centrale sur le corps des fonctions d'une quadrique*, Bull. Soc. Math. Belg. Sér. A **42** (1990), no. 3, 735–745, Algebra, groups and geometry.

[132] B. Totaro, *Birational geometry of quadrics in characteristic* 2, J. Algebraic Geom. **17** (2008), no. 3, 577–597.

[133] A. Vishik, *Motives of quadrics with applications to the theory of quadratic forms*, Geometric methods in the algebraic theory of quadratic forms, Lecture Notes in Math., vol. 1835, Springer, Berlin, 2004, pp. 25–101.

[134] ———, *On the Chow groups of quadratic Grassmannians*, Doc. Math. **10** (2005), 111–130 (electronic).

[135] ———, *Fields of u-invariant* 2^r+1, Algebra, Arithmetic and Geometry - Manin Festschrift, Progr. Math., Birkhäuser Boston, Boston, MA, in press.

[136] V. Voevodsky, *Motivic cohomology with* $\mathbf{Z}/2$*-coefficients*, Publ. Math. Inst. Hautes Études Sci. (2003), no. 98, 59–104.

[137] ———, *Reduced power operations in motivic cohomology*, Publ. Math. Inst. Hautes Études Sci. (2003), no. 98, 1–57.

[138] A. R. Wadsworth, *Merkurjev's elementary proof of Merkurjev's theorem*, Applications of algebraic K-theory to algebraic geometry and number theory, Part I, II (Boulder, Colo., 1983), Contemp. Math., vol. 55, Amer. Math. Soc., Providence, RI, 1986, pp. 741–776.

[139] E. Witt, *Theorie der quadratischen Formen in beliebigen Körpern*, J. Reine Angew. Math. **176** (1937), 31–44.

[140] O. Zariski and P. Samuel, *Commutative algebra. Vol. II*, Springer-Verlag, New York, 1975, Reprint of the 1960 edition, Graduate Texts in Mathematics, Vol. 29.

Notation

Terminology

Titles in This Series

56 **Richard Elman, Nikita Karpenko, and Alexander Merkurjev,** The algebraic and geometric theory of quadratic forms, 2008

55 **Alain Connes and Matilde Marcolli,** Noncommutative geometry, quantum fields and motives, 2007

54 **Barry Simon,** Orthogonal polynomials on the unit circle, 2005

53 **Henryk Iwaniec and Emmanuel Kowalski,** Analytic number theory, 2004

52 **Dusa McDuff and Dietmar Salamon,** J-holomorphic curves and symplectic topology, 2004

51 **Alexander Beilinson and Vladimir Drinfeld,** Chiral algebras, 2004

50 **E. B. Dynkin,** Diffusions, superdiffusions and partial differential equations, 2002

49 **Vladimir V. Chepyzhov and Mark I. Vishik,** Attractors for equations of mathematical physics, 2002

48 **Yoav Benyamini and Joram Lindenstrauss,** Geometric nonlinear functional analysis, Volume 1, 2000

47 **Yuri I. Manin,** Frobenius manifolds, quantum cohomology, and moduli spaces, 1999

46 **J. Bourgain,** Global solutions of nonlinear Schrödinger equations, 1999

45 **Nicholas M. Katz and Peter Sarnak,** Random matrices, Frobenius eigenvalues, and monodromy, 1999

44 **Max-Albert Knus, Alexander Merkurjev, and Markus Rost,** The book of involutions, 1998

43 **Luis A. Caffarelli and Xavier Cabré,** Fully nonlinear elliptic equations, 1995

42 **Victor Guillemin and Shlomo Sternberg,** Variations on a theme by Kepler, 1990

41 **Alfred Tarski and Steven Givant,** A formalization of set theory without variables, 1987

40 **R. H. Bing,** The geometric topology of 3-manifolds, 1983

39 **N. Jacobson,** Structure and representations of Jordan algebras, 1968

38 **O. Ore,** Theory of graphs, 1962

37 **N. Jacobson,** Structure of rings, 1956

36 **W. H. Gottschalk and G. A. Hedlund,** Topological dynamics, 1955

35 **A. C. Schaeffer and D. C. Spencer,** Coefficient regions for Schlicht functions, 1950

34 **J. L. Walsh,** The location of critical points of analytic and harmonic functions, 1950

33 **J. F. Ritt,** Differential algebra, 1950

32 **R. L. Wilder,** Topology of manifolds, 1949

31 **E. Hille and R. S. Phillips,** Functional analysis and semigroups, 1957

30 **T. Radó,** Length and area, 1948

29 **A. Weil,** Foundations of algebraic geometry, 1946

28 **G. T. Whyburn,** Analytic topology, 1942

27 **S. Lefschetz,** Algebraic topology, 1942

26 **N. Levinson,** Gap and density theorems, 1940

25 **Garrett Birkhoff,** Lattice theory, 1940

24 **A. A. Albert,** Structure of algebras, 1939

23 **G. Szegö,** Orthogonal polynomials, 1939

22 **C. N. Moore,** Summable series and convergence factors, 1938

21 **J. M. Thomas,** Differential systems, 1937

20 **J. L. Walsh,** Interpolation and approximation by rational functions in the complex domain, 1935

19 **R. E. A. C. Paley and N. Wiener,** Fourier transforms in the complex domain, 1934

18 **M. Morse,** The calculus of variations in the large, 1934

TITLES IN THIS SERIES

17 **J. M. Wedderburn,** Lectures on matrices, 1934
16 **G. A. Bliss,** Algebraic functions, 1933
15 **M. H. Stone,** Linear transformations in Hilbert space and their applications to analysis, 1932
14 **J. F. Ritt,** Differential equations from the algebraic standpoint, 1932
13 **R. L. Moore,** Foundations of point set theory, 1932
12 **S. Lefschetz,** Topology, 1930
11 **D. Jackson,** The theory of approximation, 1930
10 **A. B. Coble,** Algebraic geometry and theta functions, 1929
9 **G. D. Birkhoff,** Dynamical systems, 1927
8 **L. P. Eisenhart,** Non-Riemannian geometry, 1927
7 **E. T. Bell,** Algebraic arithmetic, 1927
6 **G. C. Evans,** The logarithmic potential, discontinuous Dirichlet and Neumann problems, 1927
5.1 **G. C. Evans,** Functionals and their applications; selected topics, including integral equations, 1918
5.2 **O. Veblen,** Analysis situs, 1922
4 **L. E. Dickson,** On invariants and the theory of numbers
W. F. Osgood, Topics in the theory of functions of several complex variables, 1914
3.1 **G. A. Bliss,** Fundamental existence theorems, 1913
3.2 **E. Kasner,** Differential-geometric aspects of dynamics, 1913
2 **E. H. Moore,** Introduction to a form of general analysis
M. Mason, Selected topics in the theory of boundary value problems of differential equations
E. J. Wilczyński, Projective differential geometry, 1910
1 **H. S. White,** Linear systems of curves on algebraic surfaces
F. S. Woods, Forms on noneuclidean space
E. B. Van Vleck, Selected topics in the theory of divergent series and of continued fractions, 1905